Additive Manufacturing Handbook

Systems Innovation Series

Series Editor
Adedeji B. Badiru
Air Force Institute of Technology (AFIT) – Dayton, Ohio

PUBLISHED TITLES

Additive Manufacturing Handbook: Product Development for the Defense Industry,
Adedeji B. Badiru, Vhance V. Valencia, & David Liu

Carbon Footprint Analysis: Concepts, Methods, Implementation, and Case Studies,
Matthew John Franchetti & Defne Apul

Cellular Manufacturing: Mitigating Risk and Uncertainty, *John X. Wang*

Communication for Continuous Improvement Projects, *Tina Agustiady*

Computational Economic Analysis for Engineering and Industry, *Adedeji B. Badiru &
Olufemi A. Omitaomu*

Conveyors: Applications, Selection, and Integration, *Patrick M. McGuire*

Culture and Trust in Technology-Driven Organizations, *Frances Alston*

Design for Profitability: Guidelines to Cost Effectively Management the Development Process
of Complex Products, *Salah Ahmed Mohamed Elmoselhy*

Global Engineering: Design, Decision Making, and Communication, *Carlos Acosta, V. Jorge Leon,
Charles Conrad, & Cesar O. Malave*

Global Manufacturing Technology Transfer: Africa–USA Strategies, Adaptations, and Management,
Adedeji B. Badiru

Guide to Environment Safety and Health Management: Developing, Implementing, and
Maintaining a Continuous Improvement Program, *Frances Alston & Emily J. Millikin*

Handbook of Construction Management: Scope, Schedule, and Cost Control,
Abdul Razzak Rumane

Handbook of Emergency Response: A Human Factors and Systems Engineering Approach,
Adedeji B. Badiru & LeeAnn Racz

Handbook of Industrial Engineering Equations, Formulas, and Calculations, *Adedeji B. Badiru &
Olufemi A. Omitaomu*

Handbook of Industrial and Systems Engineering, Second Edition, *Adedeji B. Badiru*

Handbook of Military Industrial Engineering, *Adedeji B. Badiru & Marlin U. Thomas*

Industrial Control Systems: Mathematical and Statistical Models and Techniques,
Adedeji B. Badiru, Oye Ibidapo-Obe, & Babatunde J. Ayeni

Industrial Project Management: Concepts, Tools, and Techniques, *Adedeji B. Badiru,
Abidemi Badiru, & Adetokunboh Badiru*

Inventory Management: Non-Classical Views, *Mohamad Y. Jaber*

Kansei Engineering — 2-volume set
- Innovations of Kansei Engineering, *Mitsuo Nagamachi & Anitawati Mohd Lokman*
- Kansei/Affective Engineering, *Mitsuo Nagamachi*

Kansei Innovation: Practical Design Applications for Product and Service Development,
Mitsuo Nagamachi & Anitawati Mohd Lokman

Knowledge Discovery from Sensor Data, *Auroop R. Ganguly, João Gama, Olufemi A. Omitaomu,
Mohamed Medhat Gaber, & Ranga Raju Vatsavai*

Learning Curves: Theory, Models, and Applications, *Mohamad Y. Jaber*

Managing Projects as Investments: Earned Value to Business Value, *Stephen A. Devaux*

Additive Manufacturing Handbook
Product Development for the Defense Industry

Editors

Adedeji B. Badiru

Vhance V. Valencia

David Liu

CRC Press
Taylor & Francis Group
Boca Raton London New York

CRC Press is an imprint of the
Taylor & Francis Group, an **informa** business

MATLAB® is a trademark of The MathWorks, Inc. and is used with permission. The MathWorks does not warrant the accuracy of the text or exercises in this book. This book's use or discussion of MATLAB® software or related products does not constitute endorsement or sponsorship by The MathWorks of a particular pedagogical approach or particular use of the MATLAB® software.

CRC Press
Taylor & Francis Group
6000 Broken Sound Parkway NW, Suite 300
Boca Raton, FL 33487-2742

First issued in paperback 2019

© 2017 by Taylor & Francis Group, LLC
CRC Press is an imprint of Taylor & Francis Group, an Informa business

No claim to original U.S. Government works

ISBN-13: 978-1-4822-6408-1 (hbk)
ISBN-13: 978-0-367-87121-5 (pbk)

**Visit the Taylor & Francis Web site at
http://www.taylorandfrancis.com**

**and the CRC Press Web site at
http://www.crcpress.com**

Dedication

To our families, who continue to support and embrace us in all we do.

Contents

Preface

The *Additive Manufacturing Handbook* is a comprehensive collection of chapters written by selected experts from their respective fields. This handbook is unique and unrivaled in its focus and orientation. Popularly known as 3D Printing, additive manufacturing involves direct digital manufacturing of products. This handbook focuses primarily on defense applications, but it is widely applicable to general applications in business, industry, education, research, government, and policy making. Even home-based readers will benefit from this handbook for its rich collection of knowledge-enhancement topics. It is a reference material that anyone interested in this topic will need to have. This handbook is organized into sections that cover both theory and practice of additive manufacturing. It is suitable for students, instructors, researchers, practitioners, and policy makers. All areas of engineering, business, and industry that interface with defense applications can benefit from this handbook. It covers all the essential topics in one volume. Topics covered in the 48 chapters in this handbook include elements of direct digital manufacturing, properties of additive manufacturing, research and development planning for 3D printing, systems engineering framework for 3D printing, product re-configurability, modularity, reliability, adaptability, and reusability using 3D printing, 3D printing using the design, evaluation, justification, and integration (DEJI) system model, hybrid systems of new product development, 3D printing technical workforce development, organizational adaptation for 3D printing, managing 3D printing projects and programs, digital manufacturing sustainability, outsourcing, insourcing, open-sourcing of 3D manufacturing, advanced educational foundation for 3D printing, energy strategies for sustaining 3D technology development, 3D technology transfer model and modes, leveraging 3D printing and big data, global partnership strategies for direct digital manufacturing, manufacturing technology transfer using additive manufacturing techniques, 3D printing consortia and development centers for defense applications, quality inspection in 3D printing, learning and forgetting in an additive manufacturing enterprise, and 3D printing applications in the defense industry.

Adedeji B. Badiru

Vhance V. Valencia

David Liu

Acknowledgments

A comprehensive and technical handbook of this magnitude and complexity requires plenty of support and coordination. We could not have done this without the exceptional, dedicated, and consistent support of several people. Our great appreciation goes to our capable Air Force Institute of Technology (AFIT) support personnel, students, and colleagues, including Seth Poulsen, Patrick Deering, Afolabi Akingbe, Annabelle Sharp, Anna Maloney, Samantha Bozada, Matthew Loh, John McCrea, Walid Basraoui, Karson Roberts, Preston Green, David Buehler, Songmi Berarducci, and Nancy Parks. Also, we thank Ms. Cindy Carelli, Executive Editor, and all her exceptional colleagues at Taylor & Francis Group, for the innumerable support and guidance they provided throughout the production of this monumental handbook. They are the best!

Editors

Dr. Adedeji B. Badiru is a professor of systems engineering at the Air Force Institute of Technology (AFIT), Wright-Patterson Air Force Base, Ohio. He is a registered professional engineer (PE). He is a fellow of the Institute of Industrial & Systems Engineers and a fellow of the Nigerian Academy of Engineering. He is also a certified project management professional (PMP). He is PhD in industrial engineering from the University of Central Florida, Orlando, Florida. Dr. Badiru is the author of several books and technical journal articles. His areas of interest include manufacturing systems, technology transfer, project management, mathematical modeling and simulation, economic analysis, learning curve analysis, quality engineering, and productivity improvement.

Dr. Vhance V. Valencia is an assistant professor in the Systems Engineering and Management Department, Air Force Institute of Technology (AFIT), Wright-Patterson Air Force Base, Ohio. He earned his PhD in systems engineering (2013) from AFIT; an MS in engineering management (2007) from AFIT; and a BS in mechanical engineering (2001) from San Diego State University, San Diego, California. Dr. Valencia is also a military officer and has held various engineering positions within the United States Air Force including facility construction and infrastructure program management, project management, and various staff and other leadership positions. A prolific writer, he has coauthored one book and written numerous journal articles and conference papers. His research interests include engineering applications for additive manufacturing, management of infrastructure assets and systems, systems engineering, and systems modeling and analysis. Dr. Valencia is a registered professional engineer and a member of the Society of American Military Engineers.

Dr. David Liu is an aerospace engineer at the Weapons Directorate, Air Force Lifecycle Management Center (AFLCMC) on Eglin Air Force Base (AFB), Florida. He also serves as an adjunct assistant professor of aerospace engineering at the Air Force Institute of Technology (AFIT), Wright-Patterson AFB, Ohio. He is also a member of the America Institute for Aeronautics and Astronautics (AIAA) and is currently on the Survivability Technical Committee. Dr. Liu is the author of several technical journal articles on the subject of aircraft survivability, ballistic effects, propulsion, and additive manufacturing.

Contributors

Rachel Abrahams
US Air Force Research Laboratory
Dayton, Ohio

Adeola Adediran
The Bredesen Center, University of
    Tennessee
Knoxville, Tennessee

Abdulrahman Sulaiman Alwabel
Department of Systems Engineering &
    Management
Air Force Institute of Technology
Dayton, Ohio

Andrew T. Anderson
Lawrence Livermore National Laboratory
Livermore, California

Omotunji Badiru
Department of Mechanical Engineering
Wright State University
Fairborn, Ohio

Valmik Bhavar
Bharat Forge Limited
Pune, India

C. D. Boley
Lawrence Livermore National Laboratory
Livermore, California

Kim Brand
3D Parts Manufacturing, LLC
Indianapolis, Indiana

Chad Cooper
i3D MFG
The Dalles, Oregon

Ronald A. Coutu, Jr.
Marquette University
Milwaukee, Michigan

Marcelo J. Dapino
The Ohio State University
Columbus, Ohio

Allison Dempsey
US Air Force

Philip J. Depond
Lawrence Livermore National Laboratory
Livermore, California

Jason Deters
General Dynamics Land Systems
Sterling Heights, Michigan

Alkan Donmez
National Institute of Standards and
    Technology
Washington, DC

Larry Dosser
Universal Technology Corporation
Dayton, Ohio

B. Dutta
DM3D Technology
Auburn Hills, Michigan

Abiodun A. Fasoro
Department of Manufacturing Engineering
Central State University
Wilberforce, Ohio

Shaw Feng
National Institute of Standards and
    Technology
Washington, DC

R. M. Ferencz
Lawrence Livermore National Laboratory
Livermore, California

Ronnie Fesperman
National Institute of Standards and
    Technology
Washington, DC

Francis H. Froes
Consultant to Titanium Industry
Tacoma, Washington

Julien Gardan
University of Technology of Troyes
France/Engineering School
Troyes, France (UTT/EPF)

Hugh E. Gardenier
US Air Force

Ali P. Gordon
University of Central Florida
Orlando, Florida

William T. Graves, Jr.
United States Naval Academy
Annapolis, Maryland

Nathan Greiner
US Air Force

Kiran Gujar
Bharat Forge Limited
Pune, India

Gabe Guss
Lawrence Livermore National Laboratory
Livermore, California

Evan Hanks
US Air Force Academy
Colorado Springs, Colorado

Ian D. Harris
EWI Organization
Columbus, Ohio

Kevin Hartke
Universal Technology Corporation
Dayton, Ohio

J. Y. Hascoet
Ecole Centrale de Nantes
Nantes, France

N. E. Hodge
Lawrence Livermore National
    Laboratory
Livermore, California

Eric S. Holm
US Air Force

John L. Irwin
Department of Mechanical Engineering
Michigan Technological University
Houghton, Michigan

Ron Jacobson
Universal Technology Corporation
Dayton, Ohio

Alan Jennings
Raytheon Missile Systems
Arlington, Virginia

C. Kamath
Lawrence Livermore National
    Laboratory
Livermore, California

Michael D. Kass
Oak Ridge National Lab
Oak Ridge, Tennessee

Prakash Kattire
Bharat Forge Limited
Pune, India

O. Kerbrat
Ecole Normale Supérieure de Rennes
Rennes, France

Saad A. Khairallah
Lawrence Livermore National Laboratory
Livermore, California

Shreyans Khot
Bharat Forge Limited
Pune, India

Wayne E. King
Lawrence Livermore National Laboratory
Livermore, California

Martin Kumke
Volkswagen Aktiengesellschaft
Wolfsburg, Germany

Brandon Lane
National Institute of Standards and
    Technology
Washington, DC

R. L. Lanser
US Air Force

Reid A. Larson
US Air Force

Tod V. Laurvick
Air Force Institute of Technology
Dayton, Ohio

Alex Li
Air Force Institute of Technology
Dayton, Ohio

Junghsen Lieh
Department of Mechanical Engineering
Wright State University
Fairborn, Ohio

Mahesh Mani
National Institute of Standards and
    Technology
Washington, DC

Manyalibo J. Matthews
Lawrence Livermore National
    Laboratory
Livermore, California

John Mark Mattox
US Army

Maria T. Meeks
US Air Force

P. Mognol
Ecole Normale Supérieure de Rennes
Rennes, France

Shawn Moylan
National Institute of Standards and
    Technology
Washington, DC

Sean Murphy
US Air Force

Mark W. Noakes
Oak Ridge National Laboratory
Oak Ridge, Tennessee

Akinola Oyedele
The Bredesen Center, University of
    Tennessee
Knoxville, Tennessee

William Page
US Air Force

Anthony N. Palazotto
Department of Aeronautics and
    Astronautics
Air Force Institute of Technology
Dayton, Ohio

Vinaykumar Patil
Bharat Forge Limited
Pune, India

Sarah Payne
Universal Technology Corporation
Dayton, Ohio

Glen P. Perram
Department of Engineering Physics
Air Force Institute of Technology
Dayton, Ohio

Grady T. Phillips
Air Force Institute of Technology
Dayton, Ohio

R. Ponche
Ecole Centrale de Nantes
Nantes, France

Seth N. Poulsen
US Air Force

Kevin D. Rekedal
US Air Force

Hayden K. Richards
US Air Force

Alexander M. Rubenchik
Lawrence Livermore National
    Laboratory
Livermore, California

Marina B. Ruggles-Wrenn
Air Force Institute of Technology
Dayton, Ohio

Vipul Sharma
US Air Force

Bradford L. Shields
Air Force Institute of Technology
Dayton, Ohio

Sanna F. Siddiqui
University of Central Florida
Orlando, Florida

Rajkumar Singh
Bharat Forge Limited
Pune, India

Shesh Srivatsa
Srivatsa Consulting, LLC
Cincinnati, Ohio

Erin Stone
i3D MFG
The Dalles, Oregon

Jonathan Torres
University of Central Florida
Orlando, Florida

Shane Veitenheimer
US Air Force

Thomas Vietor
Technische Universität Braunschweig
Braunschweig, Germany

David Walker
US Air Force

Bin Wang
Department of Mechanical Engineering
Wright State University
Fairborn, Ohio

Hagen Watschke
Technische Universität Braunschweig
Braunschweig, Germany

Leo Williams
Oak Ridge National Laboratory
Oak Ridge, Tennessee

Paul J. Wolcott
General Motors
Detroit, Michigan

section one

Introductory section

chapter one

From traditional manufacturing to additive manufacturing

Adedeji B. Badiru

Contents

Manufacturing has undergone a major advancement and technology shift in recent years. Traditional manufacturing relies on tools and techniques developed and honed over several decades of making things. From a technological standpoint (Raman and Wadke, 2014), manufacturing involves the making of products from raw materials through the use of human labor and resources that include machines, tools, and facilities. It could be more generally regarded as the conversion of an unusable state into a usable state by adding value along the way. For instance, a log of wood serves as the raw material for making lumber, which, in turn, is the raw material to produce chairs. The value added is usually represented in terms of cost and/or time. The term *manufacturing* originates from the Latin word *manufactus,* which means *made by hand.* Manufacturing has seen several advances over the past three centuries: mechanization, automation, and, most recently, computerization leading to the emergence of direct digital manufacturing, which is popularly known as 3D printing.

Processes that used to be predominantly done by hand and hand tools have evolved into sophisticated processes making use of cutting edge technology and machinery. A steady improvement in quality has resulted with today's specifications even on simple toys exceeding those that were achievable just a few years ago. Mass production, a concept developed by Henry Ford, has advanced so much that it is now a complex, highly agile, and highly automated manufacturing enterprise. There are several managerial and technical aspects of embracing new technologies to advance manufacturing. Sieger and Badiru (1993), Badiru (1989, 1990, 2005), and others address the emergence and leveraging of past manufacturing-centric techniques of expert systems, flexible manufacturing systems (FMS), nanomanufacturing, and artificial neural networks. In each case, strategic implementation, beyond hype and fad, is essential for securing the much touted long-term benefits. It is envisioned that the contents of this handbook will expand the knowledge of readers to facilitate strategic embrace of 3D printing.

1.1   What is 3D printing?

3D printing, also known as additive manufacturing or direct digital manufacturing, is a process for making a physical object from a three-dimensional digital model, typically by laying down many successive thin layers of a material. The successive layering of materials constitutes the technique of additive manufacturing. Thus, the term *direct digital manufacturing* stems from the process of going from a digital blueprint of a product to a finished physical product. Manufacturers can use 3D printing to make prototypes of products before going for full production. In educational settings, faculty and students use this process to make project-related prototypes. Open-source and consumer-level 3D printers allow for creating products at home, thus advancing the concept of distributed additive manufacturing. Defense-oriented and aerospace products are particularly feasible for the application of 3D printing. The military often operates in remote regions of the world, where quick replacement of parts may be difficult to accomplish. With 3D printing, rapid production of routine replacement parts can be achieved at a low cost onsite to meet urgent needs. The military civil engineering community is particularly fertile for the application of 3D printing for military asset management purposes.

By all accounts, 3D printing is now energizing the world of manufacturing. The concept of 3D printing was initially developed by Charles W. Hull in the 1980s (U.S. Patent, 1986) as a stereolithography (SL) tool for making basic polymer objects. Today, the process is used to make intricate aircraft and automobile components. We are now seeing more and more applications in making prostheses. The first commercial 3D-printing product came out in 1988 and was proved a hit among auto manufacturers and aerospace companies. Design of medical equipment has also enjoyed a boost due to 3D-printing capabilities. The possibilities appear endless from home-printed implements to the printing of complex parts in an outer space. The patent abstract (U.S. Patent, 1986) for 3D printing states the following:

> A system for generating three-dimensional objects by creating a cross-sectional pattern of the object to be formed at a selected surface of a fluid medium capable of altering its physical state in response to appropriate synergistic stimulation by impinging radiation, particle bombardment, or chemical reaction, successive adjacent laminae,

> representing corresponding successive adjacent cross-sections of the object, being automatically formed and integrated together to provide a step-wise laminar buildup of the desired object, whereby a three-dimensional object is formed and drawn from a substantially planar surface of the fluid medium during the forming process.

In order to understand and appreciate the full impact and implications of direct digital manufacturing (3D printing), we need to understand the traditional manufacturing processes that 3D printing is rapidly replacing. The following sections are reprinted (with permission) from Sirinterlikci (2014).

1.2   *Definition of manufacturing and its impact on nations*

In his book, *Manufacturing Systems Engineering*, Hitomi (1996) differentiated between the terms *production* and *manufacturing*. According to him, production encompasses both making tangible products and providing intangible services, while manufacturing is the transformation of raw materials into tangible products. Manufacturing is driven by a series of energy applications, each of which causes well-defined changes in the physical and chemical characteristics of the materials (Dano, 1966).

Manufacturing has a history of several thousand years and may impact humans and their nations in the following ways (Hitomi, 1994):

- *Providing basic means for human existence*: Without manufacturing products and goods, humans are unable to live, and this is becoming more and more critical in our modern society.
- *Creating wealth of nations*: The wealth of a nation is impacted greatly by manufacturing. A country with a diminished manufacturing sector becomes poor and cannot provide a desired high standard of living to its people.
- *Moving toward human happiness and stronger world's peace*: Prosperous countries can provide better welfare and happiness to their people in addition to stronger security while posing less of a threat to their neighbors and each other.

In 1991, the National Academy of Engineering/Sciences in Washington, DC rated manufacturing as one of the three critical areas necessary for America's economic growth and national security, the others being science and technology (Hitomi, 1996). In recent history, nations that became active in lower level manufacturing activities have grown into higher level advanced manufacturing and have stronger research standing in the world (Gallager, 2012).

As the raw materials are converted into tangible products by manufacturing activities, the original value (monetary worth) of the raw materials is increased (Kalpakjian and Schmid, 2006). Thus, a wire coat hanger has a greater value than its raw material—the wire. Manufacturing activities may produce *discrete products* such as engine components, fasteners, gears, or *continuous products* like sheet metal, plastic tubing, and conductors that are later used in making discrete products. Manufacturing occurs in a complex environment that connects multiple other activities: product design, process planning and tool engineering, materials engineering, purchasing and receiving, production control, marketing and sales, shipping, customer and support services (Kalpakjian and Schmid, 2006).

1.3   Manufacturing processes and process planning

1.3.1   Manufacturing processes

Today's manufacturing processes are extensive and continuously expanding while presenting multiple choices for manufacturing a single part of a given material (Kalpakjian and Schmid, 2006). The processes can be classified as traditional and nontraditional before they can be divided into their mostly physics-based categories. Although most of the traditional processes have been around for a long time, some of the nontraditional processes may have been in existence for some time as well, such as in the case of electrical discharge machining (EDM), but not utilized as a controlled manufacturing method until a few decades ago.

Traditional manufacturing processes can be categorized as follows:

1. Casting and molding processes
2. Bulk and sheet-forming processes
3. Polymer processing
4. Machining processes
5. Joining processes
6. Finishing processes

Nontraditional processes include the following:

1. Electrically based machining
2. Laser machining
3. Ultrasonic welding
4. Water-jet cutting
5. Powder metallurgy
6. Small-scale manufacturing
7. Additive manufacturing
8. Biomanufacturing

1.3.2   Process planning and design

Selection of a manufacturing process or a sequence of processes depends on a variety of factors including the desired shape of a part and its material properties for performance expectations (Kalpakjian and Schmid, 2006). Mechanical properties such as strength, toughness, ductility, hardness, elasticity, fatigue, and creep; physical properties such as density, specific heat, thermal expansion and conductivity, melting point, and magnetic and electrical properties; and chemical properties such as oxidation, corrosion, general degradation, toxicity, and flammability may play a major role in the duration of the service life of a part and recyclability. The manufacturing properties of materials are also critical, since they determine whether the material can be cast, deformed, machined, or heat treated into the desired shape. For example, brittle and hard materials cannot be deformed without failure or high-energy requirements, whereas they cannot be machined unless a nontraditional method, such as EDM, is employed. Table 1.1 depicts general manufacturing characteristics of various alloys and can be utilized in selection of processes based on the material requirements of parts.

Table 1.1 Amenability of alloys for manufacturing processes

Type of alloy	Amenability for		
	Casting	Welding	Machining
Aluminum	Very high	Medium	High to very high
Copper	Medium to high	Medium	Medium to high
Gray cast iron	Very high	Low	High
White cast iron	High	Very low	Very low
Nickel	Medium	Medium	Medium
Steels	Medium	Very high	Medium
Zinc	Very high	Low	Very high

Each manufacturing process has its characteristics, advantages, and constraints including production rates and costs. For example, conventional blanking and piercing process used in making sheet metal parts can be replaced by its laser-based counterparts if the production rates and costs can justify such a switch. Eliminating the need for tooling will also be a plus as long as the surfaces delivered by the laser-cutting process is comparable or better than that of the conventional method (Kalpakjian and Schmid, 2006). Quality is a subjective metric in general (Raman and Wadke, 2006). However, in manufacturing, it often implies *surface finish and tolerances, both dimensional and geometric.* The economics of any process is again very important and can be conveniently decomposed with the analysis of manufacturing operations and their tasks. A manufactured part can be broken into its features; the features can be meshed with certain operations; and operations can be separated into their tasks. Since several possible operations may be available and multiple sequences of operations coexist, several viable process plans can be made (Raman and Wadke, 2006).

Process routes are a sequence of operations through which raw materials are converted into parts and products. They must be determined after completion of production planning and product design according to the conventional wisdom (Hitomi, 1996). However, newer concepts, such as concurrent or simultaneous engineering or design for manufacture and assembly (DFMA), are encouraging simultaneous execution of part and process design and planning processes and additional manufacture-related activities. Process planning includes the following two basic steps (Timms and Pohlen, 1970):

1. *Process design* is a macroscopic decision making for an overall process route for the manufacturing activity.
2. *Operation/task design* is a microscopic decision making for individual operations and their detail tasks within the process route.

The main problems in process and operation design are as follows: analyzing the workflow (flow-line analysis) for the manufacturing activity and selecting the workstations for each operation within the workflow (Hitomi, 1996). These two problems are interrelated and must be resolved at the same time. If the problem to be solved is for an existing plant, the decision is made within the capabilities of that plant. On the contrary, an optimum workflow is determined, and then the individual workstations are developed for a new plant within the financial and physical constraints of the manufacturing enterprise (Hitomi, 1996).

Workflow is a sequence of operations for the manufacturing activity. It is determined by manufacturing technologies and forms the basis for operation design and layout planning. Before an analysis of workflow is completed, certain factors have to be defined including precedence relationships and workflow patterns. There are two possible relationships between any two operations of the workflow (Hitomi, 1996):

1. A partial order, *precedence*, exists between two operations such as in the case of counterboring. Counterboring must be conducted after drilling.
2. No precedence exists between two operations if they can be performed in parallel or concurrently. Two sets of holes with different sizes in a part can be made sequentially or concurrently.

Harrington (1973) identifies three different workflow patterns: *sequential* (tandem) process pattern of gear manufacturing, *disjunctive* (decomposing) pattern of coal or oil refinery processes, and *combinative* (synthesizing) process pattern in assembly processes.

According to Hitomi (1996), there are several alternatives for workflow analysis depending on production quantity (demand volume, economic lot size), existing production capacity (available technologies, degree of automation), product quality (surface finish, dimensional accuracy, and tolerances), and raw materials (material properties, manufacturability). The best workflow is selected by evaluating each alternative based on a criterion that minimizes *the total production (throughput) time* or *total production cost* defined in Equations 1.1 and 1.2. *Operation process* or *flow process charts* can be used to define and present information for the workflow of the manufacturing activity. Once an optimum workflow is determined, the detail design process of each operation and its tasks are conducted. A *break-even analysis* may be needed to select the right equipment for the workstation. Additional tools such as *man–machine analysis* as well as *human factors analysis* are also used to define the details of each operation. *Operation sheets* are another type of tool used to communicate about the requirements of each task making up individual operations.

$$\text{Total production time} = \Sigma \left[\begin{array}{l} \text{Transfer time between stages} + \text{waiting time} \\ + \text{setup time} + \text{operation time} + \text{inspection time} \end{array} \right] \quad (1.1)$$

$$\text{Total production cost} = \text{Material cost} + \Sigma \left[\begin{array}{l} \text{cost of transfer between stages} \\ + \text{setup cost} + \text{operation cost} \\ + \text{tooling cost} + \text{inspection cost} \\ + \text{work-in-process inventory cost} \end{array} \right] \quad (1.2)$$

where Σ represents all stages of the manufacturing activity.

Industrial engineering and operations management tools have been used to determine optimum paths for the workflow. Considering the amount of effort involved in the complex structure of today's manufacturing activities, computer-aided process planning (CAPP) systems have become very attractive in order to generate feasible sequences and to minimize the lead time and nonvalue-added costs (Raman and Wadke, 2006).

1.4 Traditional manufacturing processes

1.4.1 Casting and molding processes

Casting and molding processes can be classified into the following four categories (Kalpakjian and Schmid, 2006):

1. *Permanent mold based*: Permanent molding, (high and low pressure) die casting, centrifugal casting, and squeeze casting
2. *Expandable mold and permanent pattern based*: Sand casting, shell mold casting, and ceramic mold casting
3. *Expandable mold and expandable pattern*: Investment casting, lost foam casting, and single-crystal investment casting
4. *Other processes*: Melt spinning

These processes can be further classified based on their molds: permanent and expandable mold-type processes (Raman and Wadke, 2006). The basic concept behind these processes is to superheat a metal or metal alloy beyond its melting point or range, then pour or inject it into a die or mold, and finally allow it to solidify and cool within the tooling. Upon solidification and subsequent cooling, the part is removed from the tooling and is finished accordingly. The expandable mold processes destroy the mold during the removal of the part or parts such as in sand casting (Figure 1.1) and investment casting (Figure 1.2). Investment casting results in better surface finishes and tighter tolerances than sand casting. Die casting (Figure 1.3) and centrifugal-casting processes also result in good finishes but are permanent mold processes. In these processes, the preservation of tooling is a major concern since they are reused over and over, sometimes for hundreds of thousands of parts (Raman and Wadke, 2006). Thermal management of tooling through spraying and cooling channels is also imperative since thermal fatigue is a major failure mode for this type of tooling (Sirinterlikci, 2000).

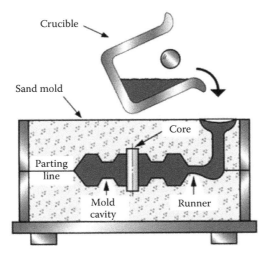

Figure 1.1 Sand casting.

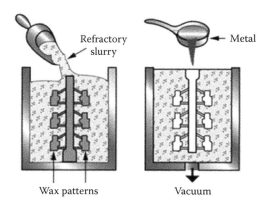

Figure 1.2 Investment casting.

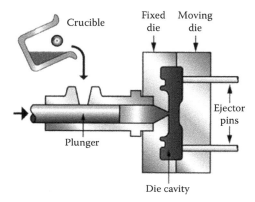

Figure 1.3 High-pressure die casting.

Common materials that are cast include metals such as aluminum, magnesium, copper, and their low-melting point alloys including zinc alloys, cast iron, and steel (Raman and Wadke, 2006). The tooling used has simple ways of introducing liquid metal and feeding it into the cavity, and it has mechanisms to exhaust air or gas entrapped within, to prevent defects such as shrinkage porosity, and to promote solid castings and easy removal of the part or parts. Cores and cooling channels are also included for making voids in the parts and controlled cooling of them to reduce cycle times for the process. The die or mold design, metal fluidity, and solidification patterns are all critical to obtain high-quality castings. Suitable provisions are made through allowances to compensate for shrinkage and finishing (Raman and Wadke, 2006).

1.4.2   *Bulk forming processes*

Forming processes include bulk-metal forming as well as sheet-metal operations. No matter what the type or the nature of the process used, forming is mainly applicable to metals that are workable by plastic deformation. This constraint makes brittle materials not eligible for forming. Bulk forming is the combined application of temperature and pressure to modify the shape of a solid object (Raman and Wadke, 2006). While cold-forming

processes conducted near room temperature require higher pressures, hot-working processes take advantage of the decrease in material strength. Consequent pressure and
energy requirements are also much lower in hot working, especially when the material
is heated above its recrystallization temperature—60% of the melting point (Raman and
Wadke, 2006). Net shape and near net shape processes accomplish part dimensions that
are exact or close to specification requiring little or no secondary finishing operations.

A group of operations can be included in the classification of bulk-forming processes
(Kalpakjian and Schmid, 2006):

1. *Rolling processes*: Flat rolling, shape rolling, ring rolling, and roll forging
2. *Forging processes*: Open-die forging, closed-die forging, heading, and piercing
3. *Extrusion and drawing processes*: Direct extrusion, cold extrusion, drawing, and tube
   drawing

In the flat-rolling process, two rolls rotating in opposite directions are utilized in reducing the thickness of a plate or sheet metal. This thickness reduction is compensated for
by an increase in the length, and when the thickness and width are close in dimension,
an increase in both width and length occurs based on the preservation of the volume of
the parts (Raman and Wadke, 2006). A similar process, shape rolling (Figure 1.4), is used
for obtaining different shapes or cross sections. Forging is used for shaping objects in
a press and additional tooling and may involve more than one preforming operation,
including blocking, edging, and fullering (Raman and Wadke, 2006). Open-die forging is done on a flat anvil, and closed-die forging process (Figure 1.5) uses a die with a
distinct cavity for shaping. Open-die forging is less accurate but can be used in making
extremely large parts due to its ease on pressure and consequent power requirements.
Although mechanical hammers deliver sudden loads, hydraulic presses apply gradually
increasing loads. Swaging is a rotary variation of the forging process, where a diameter
of a wire is reduced by reciprocating movement of one or two opposing dies (Figure 1.6).
Extrusion is the forcing of a billet out of a die opening similar to squeezing toothpaste
out of its tube (Figure 1.7), either directly or indirectly. This process enables fabrication
of different cross sections on long pieces (Raman and Wadke, 2006). In coextrusion, two
different materials are extruded at the same time and bond with each other. On the contrary, drawing process is based on pulling of a material through an orifice to reduce the
diameter of the material.

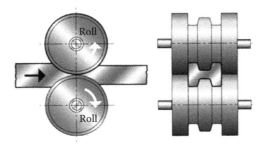

Figure 1.4 Shape-rolling process.

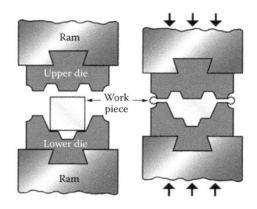

Figure 1.5 Open-die forging.

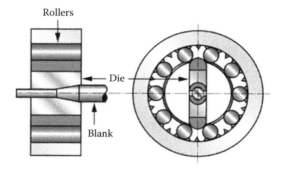

Figure 1.6 Swaging.

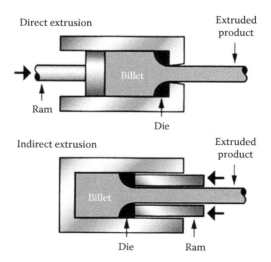

Figure 1.7 Extrusion.

1.4.3   Sheet-forming processes

Stamping is a generic term used for sheet-metal processes. They include processes such as blanking, punching (piercing), bending, stretching, deep drawing, bending, and coining (Figure 1.8). Stamping processes are executed singularly or consecutively to obtain complex sheet-metal parts with a uniform sheet metal thickness. Progressive dies allow multiple operations to be performed at the same station. Since these tooling elements are dedicated, their costs are high and expected to perform without failure for the span of the production.

Sheet metal pieces are profiled by a number of processes based on shear fracture of the sheet metal. In punching, a circular or a shaped hole is obtained by pushing the hardened die (punch) through the sheet metal (Figure 1.9). In a similar process called perforating, a group of punches are employed in making a hole pattern. In the blanking process, the aim

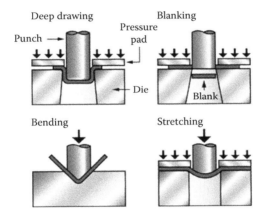

Figure 1.8 Stamping processes.

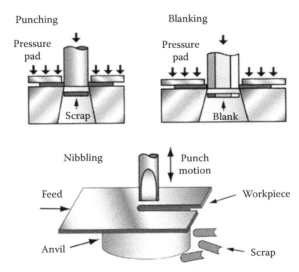

Figure 1.9 Punching, blanking, and nibbling processes.

Pass 1   Pass 2

Pass 3   Pass 4

Pass 5

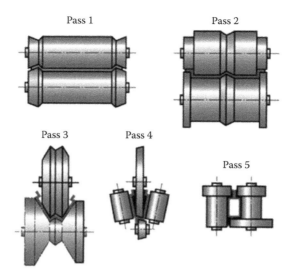

Figure 1.10 Roll forming of sheet metal.

is to keep the part that is punched out by the punch, not the rest of the sheet metal with the punched hole in the punching process (Figure 1.9). In the nibbling process, a sheet supported by an anvil is cut to a shape by successive bites of a punch similar to the motion of the sewing machine head (Figure 1.9).

After the perforation of a sheet metal part, multiple different stamping operations may be applied to it. In simple bending, the punch bends the blank on a die (Figure 1.8). Stretching may be accomplished while strictly holding the sheet metal piece by the pressure pads and forming it in a die, whereas the sheet metal piece is allowed to be deeply drawn into the die while being held by the pressure pads in deep drawing (Figure 1.8). More sophisticated geometries can be obtained by roll forming (Figure 1.10) in a progressive setting.

1.4.4   Polymer processes

Polymer extrusion is a process for making semifinished polymer products such as rods, tubes, sheets, and film in mass quantities. Raw materials such as pellets or beads are fed into the barrel to be heated and extruded at temperatures as high as 370°C (698°F) (Figure 1.11). The extrusion is then air or water cooled and may later be drawn into smaller

Extruded   Heating   Hopper
product   jacket

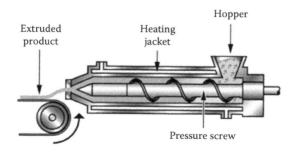

Pressure screw

Figure 1.11 Polymer extrusion.

cross sections. Variations of this process are film blowing, extrusion blow molding, and filament forming. This extrusion process is used in making blended polymer pellets and becomes a postprocess for other processes such as injection molding. It is also utilized in coating metal wires in high speeds.

Injection molding (Figure 1.12) is a process similar to polymer extrusion with one main difference, the extrusion being forced into a metal mold for solidification under pressure and cooling. Feeding systems into the mold include the sprue area, runners, and gate. Thermoplastics, thermosets, elastomers, and even metal materials are being injection molded. Coinjection process allows molding of parts with different material properties including colors and features. While injection foam molding with inert gas or chemical blowing agents results in making large parts with solid skin and cellular internal structure, reaction injection molding (RIM) mixes low-viscosity chemicals under low pressures (0.30–0.70 MPa) (43.51–101.53 psi) to be polymerized via a chemical reaction inside a mold (Figure 1.13). The RIM process can produce complex geometries and works with thermosets, such as polyurethane or other polymers such as nylons, and epoxy resins. The RIM process is also adapted to fabricate fiber-reinforced composites.

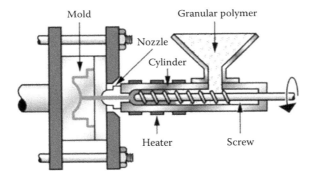

Figure 1.12 Injection molding.

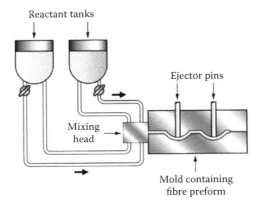

Figure 1.13 Reaction injection molding.

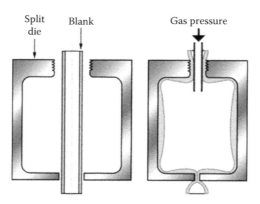

Figure 1.14 Blow molding.

Adapted from the glass-blowing technology, the blow-molding process utilizes hot air to push the polymer against the mold walls to be frozen (Figure 1.14). The process has multiple variations including extrusion and stretch blow molding. The generic blow-molding process allows inclusion of solid handles and has better control over the wall thickness compared with its extrusion variant.

Thermoforming processes are used in making large sheet-based moldings (Figure 1.15). Vacuum thermoforming applies vacuum to draw the heated and softened sheet into the mold surface to form the part. Drape thermoforming takes advantages of the natural sagging of the heated sheet in addition to the vacuum, whereas plug-assisted variant of thermoforming supplements the vacuum with a plug by pressing on the sheet. In addition, pressure thermoforming applies a few atmospheres (atm) to push the heated sheet into the mold. Various molding materials employed in the thermoforming processes included wood, metal, and polymer foam.

A wide variety of other polymer-processing methods are available including, but not limited to, rotational molding, compression molding, and resin transfer molding (RTM).

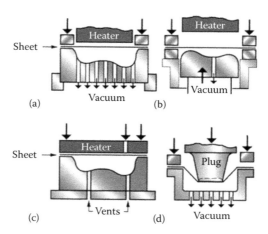

Figure 1.15 Thermoforming: (a) vacuum forming, (b) drape forming, (c) pressure forming, and (d) plug-assisted.

1.4.5 Machining processes

Machining processes use a cutting tool to remove material from the workpiece in the form of chips (Raman and Wadke, 2006). The cutting process requires plastic deformation and consequent fracture of the workpiece material. The type of chip impacts both the removal of the material and the quality of surface generated. The size and the type of the chip are dependents of the type of machining operation and cutting parameters. The chip types are continuous, discontinuous, continuous with a built-up edge, and serrated (Raman and Wadke, 2006). The critical cutting parameters include the cutting speed (in revolutions per minute—rpm's, surface feet per minute—sfpm, or millimeters per minute—mm/min), the feed rate (inches per minute—ipm or millimeters per minute—mm/min), and the depth of cut (inches—in or millimeters—mm). These parameters affect the workpiece, the tool, and the process itself (Raman and Wadke, 2006). The conditions of the forces, stresses, and temperatures of the cutting tool are determined by these parameters. Typically, the workpiece or tool are rotated or translated such that there is relative motion between the two. A primary zone of deformation causes shear of material separating a chip from the workpiece. A secondary zone is also developed based on the friction between the chip and cutting tool (Raman and Wadke, 2006). While rough machining is an initial process to obtain the desired geometry without accurate dimensions and surface finish, finish machining is a precision process capable of great dimensional accuracy and surface finish. Besides metals, stones, bricks, wood, and plastics can be machined.

There are usually three types of chip-removal operations: single-point, multipoint (fixed geometry), and multipoint (random geometry) (Raman and Wadke, 2006). Random geometry multipoint operation is also referred to as abrasive machining process and includes operations such as grinding (Figure 1.16), honing, and lapping. The cutting tool in a single-point operation resembles a wedge with several angles and radii to aid cutting. The cutting-tool geometry is characterized by the rake angle, lead, or main cutting edge angle, nose radius, and edge radius. Common single-point operations include turning, boring, and facing (Raman and Wadke, 2006). Turning is performed to make round parts (Figure 1.17), facing makes flat features, and boring fabricates nonstandard diameters and internal cylindrical surfaces. Multipoint (fixed geometry) operations include milling and drilling. Milling operations (Figure 1.17) can be categorized into face milling, peripheral (slab) milling, and end milling. The face milling uses the face of the tool, whereas slab milling uses the periphery of the cutter to generate the cutting action (Figure 1.18). These are typically applied to make flat features at a rate of material removal significantly higher than single-point operations such as shaping and planning (Raman and Wadke, 2006). End milling cuts along

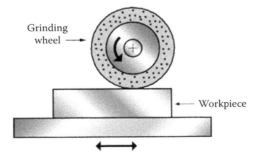

Figure 1.16 Grinding—used for finishing in general or machining hard materials.

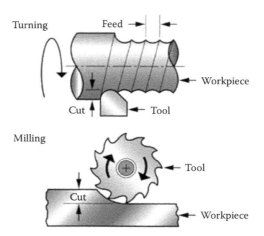

Figure 1.17 Machining processes.

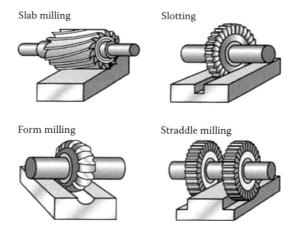

Figure 1.18 Milling operations.

with both the face and periphery and is used for making slots and extensive contours (Figure 1.18). Drilling is used to make standard-sized holes with a cutter with multiple active cutting edges (flutes). Rotary end of the cutter is used in the material removal process. Drilling has been the fastest and most economical method of making holes into a solid object. A multitude of drilling and relevant operations are available including core drilling, step (peck) drilling, counterboring and countersinking, and reaming and tapping (Figure 1.19).

1.4.6 Assembly and joining processes

Joining processes are employed in manufacturing multipiece parts and assemblies (Raman and Wadke, 2006). These processes encompass mechanical fastening through removable bolting (Figure 1.20), nonremoval riveting (Figure 1.21), adhesive bonding, and welding processes.

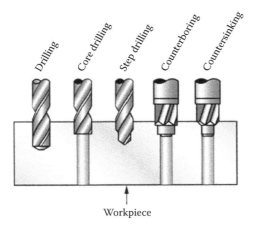

Figure 1.19 Drilling and other relevant operations.

Figure 1.20 Threaded fasteners.

Figure 1.21 Rivets and staples.

Welding processes use different heat sources to cause localized melting of the metal parts to be joined or the melting of a filler to develop a joint between mainly two metals— also being heated. Welding of plastics has also been established. Cleaned surfaces are joined together through a butt weld or a lap weld, although other configurations are also feasible (Raman and Wadke, 2006). Two other joining processes are brazing (Figure 1.22) and soldering, which differ from each other in the process temperatures and are not as strong as welding (Figure 1.22).

Arc welding utilizes an electric arc between two electrodes to generate the required heat for the process. One electrode is the plate to be joined, whereas the other electrode is the consumable one. Stick welding is most common and is also called shielded metal arc welding (SMAW) (Figure 1.23). Metal inert gas (MIG) or gas metal arc welding (Figure 1.24) uses a consumable electrode (Raman and Wadke, 2006) as well. The electrode provides the filler, and the inert gas provides an atmosphere such that contamination of the weld pool is prevented and the consequent weld quality is obtained. A steady flow of electrode is accomplished automatically to maintain the arc gap by sequentially controlling the temperature of the arc (Raman and Wadke, 2006). Gas tungsten

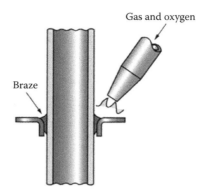

Figure 1.22 Brazing.

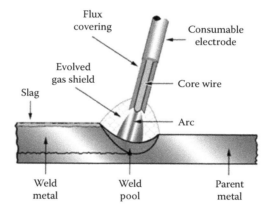

Figure 1.23 Stick (SMAW) welding.

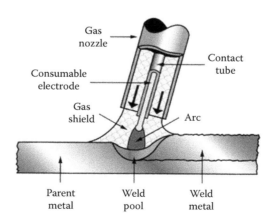

Figure 1.24 Metal inert gas welding.

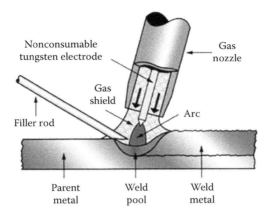

Figure 1.25 Tungsten inert gas welding.

arc welding or tungsten inert gas (TIG) welding uses a nonconsumable electrode, and a filler is required for the welding (Figure 1.25). In resistance welding, the resistance is generated by the air gap between the surfaces to obtain and maintain the flow of electric current between two fixed electrodes. The electrical current is then used to generate the heat required for welding (Raman and Wadke, 2006). This process results in spot or seam welds. Gas welding (Figure 1.26) typically employs acetylene (fuel) and oxygen (catalyzer) to develop different temperatures to heat workpieces or fillers for welding, brazing, and soldering. If the acetylene is in excess, a reducing (carbonizing) flame is obtained. The reducing flame is used in hard-facing or backhand pipe welding operations. On the contrary, if oxygen is in excess, then an oxidizing flame is generated. The oxidizing flame is used in braze-welding and welding of brasses or bronzes. Finally, if equal proportions of the two are used, a neutral flame results. The neutral flame is used in welding or cutting. Other solid-state processes include thermit welding, ultrasonic welding, and friction welding.

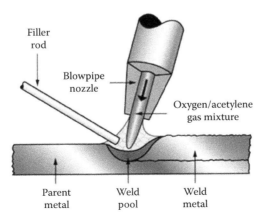

Figure 1.26 Oxy-fuel welding.

1.4.7   Finishing processes

Finishing processes include surface treatment processes and material removal processes such as polishing, shot peening, sand blasting, cladding and electroplating (Figure 1.27), and coating and painting (Raman and Wadke, 2006). Polishing involves very little material removal and is also classified under machining operations. Shot and sand blasting are used to improve surface properties and cleanliness of parts. Chemical vapor deposition (CVD) or physical vapor deposition (PVD) (Figure 1.28) methods are also applied to improve surface properties (Kalpakjian and Schmid, 2006). Hard coatings such as CVD or PVD are applied to softer substrates to improve wear resistance while retaining fracture resistance (Raman and Wadke, 2006) as in the case of metal-coated polymer injection molding inserts (Noorani, 2006). The coatings are less than 10 mm thick in many

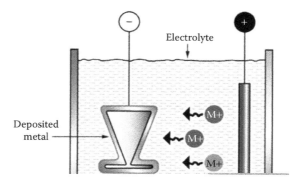

Figure 1.27 Electroplating.

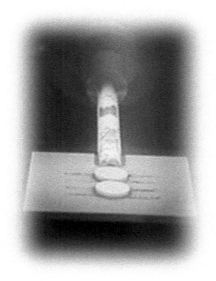

Figure 1.28 Physical vapor deposition.

cases. On the other hand, cladding is done, as in the case of aluminum cladding on stainless steel, to improve its heat conductivity for thermally critical applications (Raman and Wadke, 2006).

1.5   *Nontraditional manufacturing processes*

There are many nontraditional manufacturing processes. Nontraditional processes include the following:

1. *Electrically based machining*: These processes include the EDM (Figure 1.29) and electrochemical machining (ECM). In the plunge EDM process, the workpiece is held in a workholder submerged in a dielectric fluid. Rapid electric pulses are discharged between the graphite electrode and the workpiece, causing plasma to erode the workpiece. The dielectric fluid then carries the debris. The wire EDM uses mainly a brass wire in place of the graphite electrode but functions in a similar way (Figure 1.30).

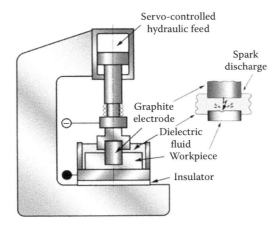

Figure 1.29 Electrodischarge machining.

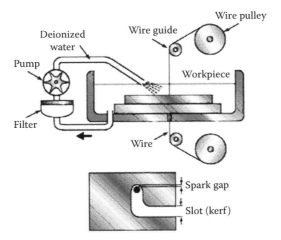

Figure 1.30 Wire electrodischarge machining.

The ECM (also called reverse electroplating) process is similar to the EDM processes, but it does not cause any tool wear, nor can any sparks be seen. Both processes can be used in machining very hard materials that are electrically conductive.

2. *Laser machining*: Lasers are used in a variety of applications ranging from cutting of complex 3D contours (i.e., today's coronary stents) to etching or engraving patterns on rolls for making texture on rolled parts. Lasers are also effectively used in hole making, precision micromachining, removal of coating, and ablation. Laser transformation and shock-hardening processes make workpiece surfaces very hard, whereas the laser surface-melting process produces refined and homogenized microstructures (Figure 1.31).

3. *Ultrasonic welding*: Ultrasonic-welding process requires an ultrasonic generator, a converter, a booster, and a welding tool (Figure 1.32). The generator converts 50 Hz into 20 KHz. These higher frequency signals are then transformed into mechanical oscillations through the reverse piezoelectric effect. The booster and the welding tool transmit these oscillations into the welding area causing vibrations of 10–30 μm in amplitude. Meanwhile, a static pressure of 2–15 MPa is applied to the workpieces as they slide, get heated, and bonded.

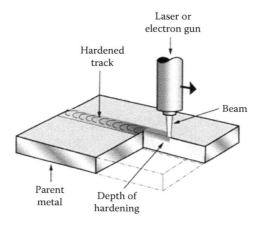

Figure 1.31 Laser surface hardening and melting.

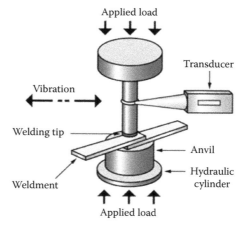

Figure 1.32 Ultrasonic welding.

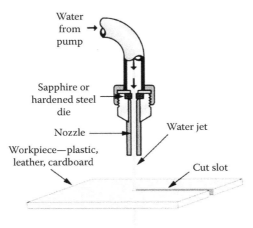

Figure 1.33 Water-jet cutting.

4. *Water-jet cutting*: Water-jet cutting is an abrasive machining process that employs an abrasive slurry in a jet of water to machine hard-to-machine materials (Raman and Wadke, 2006) (Figure 1.33). Water is pumped at high pressures such as 400 MPa and can reach to speeds of 850 m/s (3,060 km/h or 1901.4 mph). Abrasive slurries are not needed when cutting softer materials.

5. *Powder metallurgy*: Powder-based fabrication methods are critical in employing materials with higher melting points due to the hardship of casting them (Figure 1.34). Once compressed under pressures using different methods and temperatures, compacted (green) powder parts are sintered (fused) usually at 2/3 of their melting points. Common powder metallurgy materials are ceramics and refractory metals, stainless steel, and aluminum.

6. *Small-scale manufacturing*: The past two decades have seen marriage of microscale electronics device and their manufacturing with mechanical systems—leading to the design and manufacturing of microelectromechanical systems (MEMS).

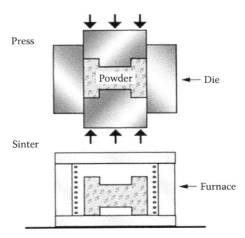

Figure 1.34 Powder metallurgy.

Even newer cutting-edge technologies have emerged in a smaller scale such as nanotechnology or molecular manufacturing (Raman and Wadke, 2006). The ability to modify and construct products at a molecular level makes nanotechnology very attractive and usable. Single-wall nanotubes are one of the biggest innovations for building future transistors and sensors. Variants of the nano area include nanomanufacturing, such as ultrahigh precision machining or adding ionic antimicrobials to biomedical devices (Raman and Wadke, 2006; Sirinterlikci et al., 2012).

7. *Additive manufacturing*: Additive manufacturing has been driven from additive rapid prototyping technology. Since the late 1980s, rapid prototyping technologies have intrigued scientists and engineers. In the early years, there were attempts of obtaining 3D geometries using various layered approaches as well as direct 3D geometry generation by robotic plasma spraying. Today, a few processes, such as fused deposition modeling (FDM), SLA, laser-sintering processes (selective or direct metal), and 3D printing, have become household names. There are also other very promising processes such as Objet's Polyjet (Inkjet) 3D-printing technology. In the past two decades, the rapid prototyping technology has seen an increase in the number of materials available for processing; layer thicknesses have become less, whereas control systems have improved to better the accuracy of the parts. The end result is shortened cycle times and better quality functional parts. In addition, there have been many successful applications of rapid tooling and manufacturing (Figures 1.35 and 1.36).

8. *Biomanufacturing*: Biomanufacturing may encompass biological and biomedical applications. Thus manufacturing of human vaccinations may use biomass from plants, whereas biofuels are extracted from corn or other crops. Hydraulic oils, printer ink-technology, paints, cleaners, and many other products are taking advantage of the developments in biomanufacturing (Sirinterlikci et al., 2010). On the other hand, biomanufacturing is working with nanotechnology, additive manufacturing, and other emerging technologies to improve the biomedical engineering field (Sirinterlikci et al., 2012).

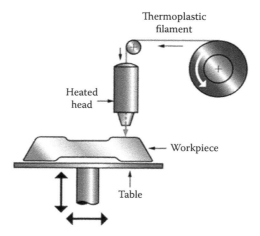

Figure 1.35 Fused deposition modeling.

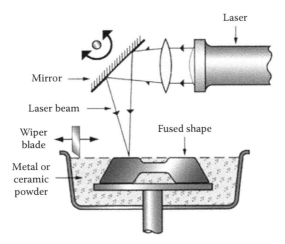

Figure 1.36 Selective laser sintering.

1.6   *Manufacturing systems*

Manufacturing systems are systems that are employed to manufacture products and the parts assembled into those parts (Groover, 2001). A manufacturing system is the collection of equipment, people, information, processes, and procedures to realize the manufacturing targets of an enterprise. Groover (2001) defines manufacturing systems in two parts:

1. Physical facilities include the factory, the equipment within the factory, and the way the equipment is arranged (the layout).
2. Manufacturing support systems are a set of procedures of the company to manage its manufacturing activities and to solve technical and logistics problems including product design and some business functions.

Production quantity is a critical classifier of a manufacturing system, whereas the way the system operates is another one including *production to order* (i.e., Just in Time—JIT manufacturing) and *production for stock* (Hitomi, 1996). According to Groover (2001) and Hitomi (1996), a manufacturing system can be categorized into three groups based on its production volume:

1. *Low production volume—jobbing systems*: In the range of 1–100 units per year, associated with a job shop, fixed position layout, or process layout.
2. *Medium production volume—intermittent or batch systems*: With the production range of 100–10,000 units annually, associated with a process layout or cell.
3. *High production volume—mass manufacturing systems*: 10,000 units to millions per year, associated with a process or product layout (flow line).

Each classification is associated inversely with a product variety level. Thus, when product variety is high, production quantity is low, and when product variety is low, production quantity is high (Groover, 2001). Hitomi (1996) states that only 15% of the manufacturing activities in the late 1990s were coming from mass manufacturing systems, whereas small-batch multiproduct systems had a share more than 80%, perhaps due to diversification of human demands.

1.7 Interchangeability and assembly operations

While each manufacturing process is important for fabrication of a single part, assembling these piece parts into subassemblies and assemblies also presents a major challenge. The concept that makes assembly a reality is interchangeability. Interchangeability relies on standardization of products and processes (Raman and Wadke, 2006). Besides facilitating easy assembly, interchangeability also enables easy and affordable replacement of parts within subassemblies and assemblies (Raman and Wadke, 2006).

The key factors for interchangeability are *tolerances* and *allowances* (Raman and Wadke, 2006). Tolerance is the permissible variation of geometric features and part dimensions on manufactured parts with the understanding that perfect parts are hard to be made, especially repeatedly. Even if the parts could be manufactured perfectly, current measurement tools and systems may not be able to verify their dimensions and features accurately (Raman and Wadke, 2006). Allowances determine the degree of looseness or tightness of a fit in an assembly of two parts (i.e., a shaft and its bearing) (Raman and Wadke, 2006). Depending on the allowances, the fits are classified into *clearance, interference,* and *transition* fits. Since most commercial products and systems are based on assemblies, tolerances (dimensional or geometric) and allowances must be suitably specified to promote interchangeable manufacture (Raman and Wadke, 2006).

1.8 Systems metrics and manufacturing competitiveness

According to Hitomi (1996), efficient and economical execution of manufacturing activities can be achieved by completely integrating the material flow (manufacturing processes and assembly), information flow (manufacturing management system), and cost flow (manufacturing economics). A manufacturing enterprise needs to serve for the welfare of the society by not harming the people and the environment (green manufacturing, environmentally conscious manufacturing) along with targeting its profit objectives (Hitomi, 1996). A manufacturing enterprise needs to remain competitive and thus has to evaluate its products' values and/or the effectiveness of its manufacturing system using the following three metrics (Hitomi, 1996):

1. Function and quality of products
2. Production costs and product prices
3. Production quantities (productivity) and on-time deliveries
4. Following any industry regulations

A variety of means are needed to support these metrics including process planning and control; quality control; costing; safety, health, and environment (SHE); production planning; scheduling; and control including assurance of desired cycle/takt and throughput times. Many other metrics are also used in the detailed design and execution of systems including machine utilization, floor space utilization, and inventory turnover rates.

1.9   Automation of systems

Some parts of a manufacturing system need to be automated, whereas other parts remain under manual or clerical control (Groover, 2001). Either the actual physical manufacturing system or its support system can be automated as long as the cost of automation is justified. If both systems are automated, a high level of integration can also be reached as in the case of computer-integrated manufacturing (CIM) systems. Groover (2001) lists the following examples of automation in a manufacturing system:

- Automated machines
- Transfer lines that perform multiple operations
- Automated assembly operations
- Robotic manufacturing and assembly operations
- Automated material handling and storage systems
- Automated inspection stations

Automated manufacturing systems are classified into the following three groups (Groover, 2001):

1. *Fixed automation systems*: Used for high production rates or volumes and very little product variety as in the case of a welding fixture used in making circular welds around pressure vessels.
2. *Programmable automation systems*: Used for batch production with small volumes and high variety as in the case of a robotic welding cell.
3. *Flexible automation systems*: Medium production rates and varieties can be covered as in the case of flexible manufacturing cells with almost no lost time for changeover from one part family to another one.

Groover lists some of the reasons for automating a manufacturing entity:

1. To increase labor productivity and costs by substituting human workers with automated means
2. To mitigate skilled labor shortages as in welding and machining
3. To reduce routine and boring manual and clerical tasks
4. To improve worker safety by removing them from the point of operation in dangerous tasks such nuclear, chemical, or high energy
5. To improve product quality and repeatability
6. To reduce manufacturing lead times
7. To accomplish processes that cannot be done manually
8. To avoid the high cost of not automating

1.10   Conclusions

Manufacturing is the livelihood and future of every nation and is mainly misunderstood. It also has a great role in driving engineering research and development, in addition to wealth generation. Utilization of efficient and effective methods is crucial for any manufacturing enterprise to remain competitive in this very global market, with intense international collaboration and rivalry. Especially, the importance of industrial engineering tools in optimizing processes and integrating those processes into systems needs to be grasped to better the manufacturing processes and their systems. Automation is still a valid medium for improving the manufacturing enterprise as long as the costs of doing it are justified. Tasks too difficult to automate, life cycles that are too short, products that are very customized, and cases where demands are very unpredictably varying cannot justify application of automation (Hitomi, 1996).

References

Badiru, A. B., Analysis of data requirements for FMS implementation is crucial to success, *Industrial Engineering*, 22(10), 29–32, 1990.

Badiru, A. B., How to Plan and Manage Manufacturing Automation Projects, in *Proceedings of 1989 IIE Fall Conference*, Atlanta, GA, November 13–15, 1989, pp. 540–545.

Badiru, A. B., Product Planning and Control for Nano-Manufacturing: Application of Project Management, in *Distinguished Lecture Series*, University of Central Florida, Orlando, FL, March 2005.

Dano, S., *Industrial Production Models: A Theoretical Study*, Springer, Vienna, Austria, 1966.

Gallager, P., *Presentation at the National Network of Manufacturing Innovation Meeting II*, Cuyahoga Community College, Cleveland, OH, July 9, 2012.

Groover, M., *Automation, Production Systems, and Computer-Integrated Manufacturing*, Prentice Hall, Upper Saddle River, NJ, 2001.

Harrington, J. Jr., *Computer Integrated Manufacturing*, Industrial Press, New York, 1973.

Hitomi, K., *Manufacturing Systems Engineering: A Unified Approach to Manufacturing Technology, Production Management, and Industrial Economics*, 2nd Edition, Taylor & Francis, Bristol, PA, 1996.

Hitomi, K., Moving toward manufacturing excellence for future production perspectives, *Industrial Engineering*, 26(6), 7–11, 1994.

Kalpakjian, S. and Schmid, S. R., *Manufacturing Engineering and Technology*, 5th Edition, Pearson Prentice Hall, Upper Saddle River, NJ, 2006.

Noorani, R., *Rapid Prototyping and Applications*, John Wiley and Sons, Hoboken, NJ, 2006.

Raman, S. and Wadke, A., Manufacturing Technology, in Badiru, A. B. (Ed.), *Handbook of Industrial and Systems Engineering*, 2nd Edition, pp. 337–349, CRC Press/Taylor and Francis, Boca Raton, FL, 2014.

Raman, S. and Wadke, A., Manufacturing Technology, *Handbook of Industrial and Systems Engineering*, A. Badiru, (Ed.), CRC Press, Boca Raton, FL, 2006.

Sieger, D. B. and Badiru, A. B., An artificial neural network case study: Prediction versus classification in a manufacturing application, *Computers and Industrial Engineering*, 25(1–4), 381–384, 1993.

Sirinterlikci, A., Manufacturing Processes and Systems, in Badiru, A. B. (Ed.), *Handbook of Industrial and Systems Engineering*, 2nd Edition, pp. 371–397, CRC Press/Taylor and Francis, Boca Raton, FL, 2014.

Sirinterlikci, A., *Thermal Management and Prediction of Heat Checking in H-13 Die-Casting Dies*, PhD Dissertation, The Ohio State University, 2000.

Sirinterlikci, A., Acors, C., Pogel, S., Wissinger, J., and Jimenez, M., Antimicrobial Technologies in Design and Manufacturing of Medical Devices, in *SME Nanomanufacturing Conference*, Boston, MA, 2012.

Sirinterlikci, A., Karaman, A., Imamoglu, O., Buxton, G., Badger, P., and Dress, B., Role of Biomaterials in Sustainability, in *Proceedings of the 2nd Annual Conference of the Sustainable Enterprises of the Future*, Pittsburgh, PA, 2010.

Timms, H. L. and Pohlen, M. F., *The Production Function in Business—Decision Systems for Production and Operations Management*, 3rd Edition, Irwin, Homewood, IL, 1970.

U.S. Patent, Apparatus for production of three-dimensional objects by stereolithography, U.S. 4575330 A, March 11, 1986, http://www.google.com/patents/US4575330, accessed December 22, 2014.

chapter two

A novice's guide to 3D printing

Making the process less magical and more understandable

Kim Brand

Contents

Few manufacturing news stories get picked up by the mainstream media, let alone make it to the cover of *The Economist* magazine, which called 3D printing the "Third Industrial Revolution." The media also fixated on a gun which could be made with an inexpensive 3D printer. No wonder I get asked if 3D printing will put machine shops out of business.

To paraphrase Mark Twain, the demise of traditional manufacturing, at the hands of 3D printing, has been greatly exaggerated.

First, let us get the name correct. There is very little about 3D printing that resembles printing. Those in the know, refer to it by its proper name: additive manufacturing (AM). That umbrella term collects many technologies under one metaphor and more accurately describes what is going on. Slowly, layer-by-layer, parts are *grown* on these machines by adding or solidifying material according to instructions derived from a three-dimensional model of the part designed on a computer. Described this way, 3D printing is less magical, but more understandable.

2.1   3D printing is growing

Terry Wohlers, a 28-year veteran 3D printing industry analyst, reports that the market for AM products and services worldwide grew at a compound annual growth rate of 35.2% to $4.1 billion in 2014, expanding by more than $1 billion over 2013. Nearly 50 manufacturers produce industrial-grade AM machines—those selling for more than $5,000.

It would be hard to estimate the comparable sales for the subtractive manufacturing market, but *Modern Machine Shop* (*mmsonline.com/articles/American-manufacturing-on-the-rise*) reports that "machine tool sales should rise to $7.442 billion in 2014, an increase of 19 percent over 2013." But wait—that only includes the sales of machine tools. The AM products

Figure 2.1 **(See color insert.)** 3D Systems' ProX 400 is capable of printing in more than a dozen alloys, including stainless steel, aluminum, cobalt chrome, titanium, and maraging steel. (Courtesy of 3D systems.)

and services sales figures include not only the cost of the machines but also the value of the products produced on the machines. With global manufacturing representing 15% of a $70 trillion economy, AM represents 0.04% of the manufacturing economy (Figure 2.1).

Despite its potential for enormous growth, it is a vanishingly small factor in the manufacturing business today. There is an alphabet soup of AM methods that have been devised over the past 30 plus years. The big players are 3D Systems and Stratasys, which have a combined capitalization of $4.15 billion and have been gobbling up competitors and service bureaus and accumulating patent portfolios at a tremendous rate. Unfortunately for them, their stock prices have declined nearly as fast lately. The industry is cautiously anticipating the entry of Hewlett. Packard in 2016 with a new technology that is said to be more capable. We will see.

2.2 *3D printing is cool*

For sure, 3D printing can make really cool things. As it produces parts directly from a design file, it eliminates front-end investment (dollars and time) in tooling and fixturing. Rapid pro-totyping, quicker design iterations, and *lot size one* manufacturing are key benefits. Designers can focus more on function and less on fabrication. Making parts in layers, rather than whit-tling away at the outside, allows for the creation of complex internal structures like cool-ing channels or weight-reducing honeycombs that yield savings in materials and reducing waste. Supply chain benefits include part consolidation; a multipart assembly can now be made in one piece. That reduces investment in quality control (QC), inventory, and labor.

Early adopters of AM include industries that make high-value/low-volume products: aerospace and medical devices are two. The former values weight savings, the latter is interested in biomimicry. Both leverage the ability of 3D printing to make complex or unique parts in short lots. (But both have long certification cycles, which have frustrated rapid adoption.) Marketing departments love 3D printing to create replica parts when production lines do not exist. 3D-printed parts, molds, and mold masters for short runs put R&D cycles on steroids, reducing time-to-market and enhancing competitive advantage.

2.3   New tools, new rules

These new tools impose new rules. For one, gravity is not your friend. Most of the additive technologies require the introduction of support structures, which keep cantilevers and cavities from collapsing under their own weight during manufacture. Designing these supports requires attention to their material cost, build time, orientation of the part in the build space, and removal methods. This not-so-minor detail requires experience and experimentation and adds value to an AM service bureau relationship.

Checking the size and shape of those internal cavities frustrates legacy QC methods; so a new generation of metrology is needed. Another QC conundrum is process control. Less is known about how variation in the 3D-printing process affects part quality. In performance or safety critical applications, there remain many unknowns (Figure 2.2).

Near-net printing is a strategy that combines AM and traditional machining to achieve a result better than either can produce on its own. Some 3D methods leave parts with unsuitable finishes that require postprocessing. We like to say that industrial 3D printing is a team sport.

Heating and cooling introduced in some AM processes produce stress, and those stresses can warp or deform the product. Steel parts require postprocess heat treating to relieve these stresses … not to mention that steel parts emerge from direct metal laser

Figure 2.2 **(See color insert.)** Six materials for 3D printing. New material choices are contributing to the growth of additive manufacturing. (Courtesy of 3D systems.)

The Alphabet Soup of Additive Manufacturing

FDM: Fused deposition modeling was patented by Stratasys in 1989. Stratasys bought Makerbot in 2013 for over $400 million. This technology has become wildly popular with hobbyists as patents have expired, and a *gold rush* of companies and individuals have begun to make low-cost printers that fabricate parts from ABS (acrylonitrile butadiene styrene, a common thermoplastic polymer—Legos are made from the same material), PLA (polylactide, a biodegradable thermoplastic a liphatic polyester derived from renewable resources), and a range of other thermoplastics. Think: A glue gun controlled by a robot financed on Kickstarter.

FFF: Fused filament fabrication is the equivalent of FDM, but the term is unrestrained by the trademark Stratasys owns on FDM.

SLA: Stereolithography was patented by 3D Systems in 1986. This method uses photopolymers exposed to UV lightor lasers to harden tiny elements of a liquid goo, which, when aggregated, create a solid objeact.

SLS/DMLS/SLM: Selective laser sintering/direct metal laser sintering/selective laser melting are processes that use focused lasers to melt powders (plastic or metal) into tiny pools of material, which then cool and aggregate into parts. The battle for patent rights may continue until they have all expired.

CJP: Colorjet printing was invented at MIT in 1993 and marketed by ZCorp until it was acquired by 3D Systems in 2012. In this process, a layer of powder is infused with a liquid binder and cured to create the part. This method is notable because it works like a color inkjet printer.

PolyJet: Invented by Objet Geometries in 1998, PolyJet was acquired by Stratasys in 2011. It is a 3D-printing system that uses two or more photopolymer resins deposited in tiny droplets, like an inkjet printer, that are mixed in real time and cured with UV light to create a solid object. The PolyJet technology can create over 100 types of durable plastic materials including hard, soft, clear, and full color.

Figure 2.3 Selected glossary of additive manufacturing.

sintering (DMLS) systems welded to the build plate and usually require electrical discharge machining (EDM) to remove the product (Figure 2.3).

Of note to the gas industry, 3D printing with metals is akin to welding, and for best results it is performed in inert environments, usually argon.

One shortcoming of AM is the fairly sparse selection of materials—several dozens of all types are available in total for the industry across technologies. The choices represent a broad cross section, and most users will be satisfied with the options; but when compared to the diversity of materials available for subtractive, it may be a deal killer for your project. Characterization of new metals, for example, can take more than a year and cost $1 million. Also, the materials are likely to be developed, certified, and marketed by the industrial AM equipment vendors. This situation creates lock-in and reduces competition for consumables, which are a significant portion of finished part cost (Figure 2.4).

Finally, the rate at which parts are produced today using AM methods is pitifully slow. Creation of solid volumes by 3D printers is very inefficient—like coloring in large areas of an outline with a crayon over and over. This fact eliminates the use of 3D printing in

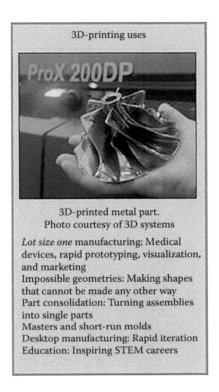

3D-printing uses

3D-printed metal part.
Photo courtesy of 3D systems

Lot size one manufacturing: Medical devices, rapid prototyping, visualization, and marketing
Impossible geometries: Making shapes that cannot be made any other way
Part consolidation: Turning assemblies into single parts
Masters and short-run molds
Desktop manufacturing: Rapid iteration
Education: Inspiring STEM careers

Figure 2.4 3D-printing uses.

many high-volume industries. 3D Parts Manufacturing's EOS DMLS system can build at a maximum rate of 8 h/inch at a 20 μ layer height. Production rates of hours per inch of "Z" height are typical.

2.4   Now the good news

Turning ideas into parts may be a bit of an exaggeration, but it is not far from the truth. With a 3D model and an inexpensive FDM printer, you can prototype a design and hold the results in your hand—today! (Well, in a few days for sure.) Companies are using 3D printing as a communication tool to share product design details with vendors to reduce errors and lead times. A 3D-printed sample can create credibility for a new product idea or convince a prospect that you mean business or can persuade an investor that your invention will work. Manufacturers are considering the use of 3D printing to produce parts near the point-of-use and on-demand to slash logistics costs. For example, NASA is operating a 3D printer in space to evaluate parts production where spare parts just cannot be inventoried.

3D printing is used by clinicians to create assistive devices that overcome disabilities or reproduce the function of missing limbs. Bioprinting technologies are being developed that work with living cells to replace organs and body tissues.

Schools are using 3D printers to inspire students to think. There is nothing so engaging or motivating as seeing a thing you have designed take shape right before your eyes. We have witnessed dramatic turnarounds when kids gain confidence through the experience

Figure 2.5 **(See color insert.)** With a 3D printer, you can prototype a design and hold the results in your hand in just days. (Courtesy of Stratasys.)

of making something, no matter how imperfect. Pride of ownership fuels a *fix it* attitude. The next generation of students will consider access to 3D printing as common as PCs—and we can hardly imagine what they will create.

New material choices, improved production readiness, simpler operation, better reliability, and lower prices will contribute to continued dramatic growth of 3D printing across all industry sectors (Figure 2.5).

If you want one of something, if it includes *crazy* geometries, or if you want to make it on your own desktop today, then 3D printing is the best new thing to happen to manufacturing since electricity. Only time will tell if it deserves being described as the *Third Industrial Revolution*. But for many companies, consumers, and students, it is inspiring new thinking about product design, unleashing creativity, democratizing *making*, and keeping publicists, patent attorneys, and pundits very busy!

Reference

"History of Additive Manufacturing" (*wohlersassociates.com/history2013.pdf*), Wohlers Associates (*wohlersassociates.com*), and Metal-AM (*metal-am.com*).

chapter three

Comprehensive project management of high-end additive manufacturing equipment

Adedeji B. Badiru

Contents

Project management for additive manufacturing (PM-4-AM) is the premise of this chapter. The chapter presents a comprehensive project management approach for installing high-end additive manufacturing (AM) equipment. The core of the comprehensive model is the Triple C model of project management, which presents a systematic structure for communication, cooperation, and coordination across high-tech assets. Like all advanced manufacturing endeavors, the funding, purchase, installation, maintenance, utilization, and decommissioning of high-end AM equipment requires strategic implementation of project management techniques. The level of communication, cooperation, and coordination required for effective acquisition adoption of AM can be facilitated and enhanced by comprehensive project management practices from a system's perspective.

3.1   Introduction

AM (also known as 3D printing or direct digital manufacturing) is quickly emerging as the new technology of choice in manufacturing. The new global business model necessitates that products have to be developed across geographically disparate regions. This creates new challenges for the technical and managerial aspects of developing new products. For a product to be competitive in the new market place, its design process must be agile and adaptable to the changing environment. AM makes this possible. However, this new technological tool must be managed just like any conventional project. A comprehensive project management approach (Badiru and Pulat, 1995) holds good promise for enhancing the adoption of AM.

Badiru (2012) defines project management as the process of managing, allocating, and timing resources in order to achieve a given objective in an expeditious manner. The objective may be in terms of time, monetary, or technical results. Project management is the process of achieving objectives by utilizing the combined capabilities of available resources. It represents a systematic execution of tasks needed to achieve project objectives. In a new technology environment, the basic functions of project management cover the following:

1. Planning
2. Organizing
3. Scheduling
4. Control

Because of the complexity often encountered when installing new high-tech equipment, the steps of the design process require thinking outside of the conventional project box. It has been shown again and again that the majority of technology failures can be traced to communication failures at the initial stages of a project. Thus, communication constitutes an important foundation for achieving success in AM technology projects. When embarking on the purchase, installation, and utilization of high-end AM equipment, some of the issues of crucial consideration include the following:

- Purchasing process and contracting requirement
- Delivery timeline
- Safety concerns
- Training requirements
- Maintenance
- Skilled operators
- Service contract
- Space requirements (equipment footprint and supporting infrastructure)
- Power supply
- Water needs
- Heating, ventilation and air conditioning (HVAC) needs
- Operational requirements
- Occupational safety and health administration (OSHA) requirements
- Sustained utilization
- Funding (initial and subsequent)
- Vibration control
- Facilities upkeep (housekeeping around equipment)
- Production level requirements
- Minimum acceptable quality

All of these, and some more not listed here, require a whole lot of coordinated project management. Essentially, a comprehensive project management is required.

3.2   Basics of the Triple C model for additive manufacturing project management

The Triple C model introduced by (Badiru, 2008) is an effective project-planning tool that has been successfully utilized for projects of all types. It can be particularly effective for a distributed product development environment, such as AM, where personnel coordination is very crucial. The model states that project management can be enhanced by implementing it within the integrated functions of

- Communication
- Cooperation
- Coordination

The Triple C model facilitates a systematic approach to planning, organizing, scheduling, and control. The model is shown graphically in Figure 3.1. It highlights what must be done and when. It can also help to identify the resources, such as personnel, peripheral equipment, facilities, power supply, and space requirements, associated with the AM equipment.

Typical questions to be addressed in PM-4-AM include the following:

Who: Who is the point of contact for the new equipment? Who made the selection? Who else is involved? Who has been informed? Who will run the equipment? Who are the users? Who will maintain the equipment? Who is proving the funding for all the needs affiliated with the equipment?

What: What is being purchased? What will the equipment be used for? What are the options? What will be equipment replace or supplement? What peripheral installation needs are involved? Safety concerns? Security concerns? Power supply needs?

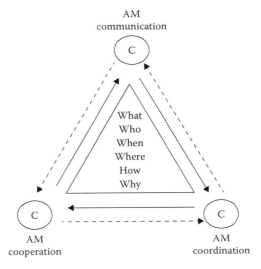

Figure 3.1 Triple C project management framework for additive manufacturing.

Fire suppressant? Water supply needs? Lighting needs? HVAC needs? Vibrant concerns? Emission concerns? Stability concerns?

Which: Which functional and/or administrative units are responsible for the equipment?

When: When will the equipment be purchased? When is the delivery timeline? When is the contracting timeline, if applicable?

Where: Where will the equipment be placed? Is colocation with other organization facilities possible?

How: How will the equipment be used? How will the equipment be maintained? How will the equipment utilization be sustained? How will the equipment be decommissioned, when applicable?

Why: Why is the equipment needed at all?

3.3   Communication

Communication facilitates team work. The communication function of project management involves making all those concerned become aware of project requirements and progress. Those who will be affected by the project directly or indirectly, as direct participants or as beneficiaries, should be informed regarding the following:

- Scope of the product
- Personnel contribution required
- Expected cost and merits of the project
- Project organization and implementation plan
- Potential adverse effects if the project should fail
- Alternatives, if any, for achieving the project goal
- Potential direct and indirect benefits of the product development project

The communication channel must be kept open throughout the project life cycle. In addition to internal communication, appropriate external sources should also be consulted. This is particularly essential for a distributed product design environment where design participants may be geographically dispersed over large distances. Figure 3.2 presents a specific application to intermodule communication in AM product development.

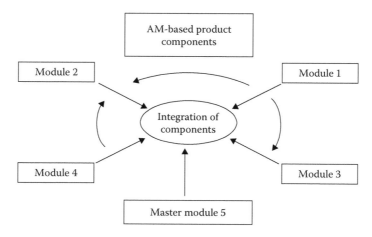

Figure 3.2 AM product intermodule communication channels.

Using Triple C helps to clarify the following questions, particularly when the modules are designed at geographically dispersed locations:

- Does each product development participant know what the objective is?
- Does each product development participant know his or her role in achieving the objective?
- What obstacles may prevent a participant from playing his or her role effectively?

Some of the sources of communication problems for high-tech technology project management are summarized below:

Social environment: Communication problems sometimes arise because people have been conditioned by their prevailing social environment to interpret certain issues in unique ways, particularly when new pieces of technological equipment are being contemplated. Vocabulary, idioms, organizational status, social stereotypes, and economic situation are among the social factors that can impede effective communication in advanced manufacturing organizations. AM is not immune to these adverse scenarios.

Cultural background: Cultural differences are among the most pervasive barriers to technological project communications, especially in today's multinational organizations. Language and cultural idiosyncrasies often determine how communication is approached, received, and interpreted.

Semantic and syntactic factors: Semantic and syntactic barriers to communications usually occur in written documents. Semantic factors are those that relate to the intrinsic knowledge of the subject of the communication. Syntactic factors are those that relate to the form in which the communication is presented. The problems created by these factors become acute in situations where response, feedback, or reaction to the communication cannot be observed directly or face-to-face. Explicit efforts must be made to bring everybody on board for AM equipment installation.

Organizational structure: Frequently, the organization structure within which a technical project is housed has a direct influence on the flow of information and, consequently, on the effectiveness of communication. Organization hierarchy may determine how different personnel levels perceive specific information. One key aspect to keep in mind is the proverbial guide of *the higher the level of management, the lower the level of details needed*. An overly technical presentation of an AM project can quickly lose the interest of management. This is particularly important where funding decisions are involved.

Communication medium: The method of transmitting a message may also affect the value ascribed to the message and, consequently, how it is interpreted or used. With the excessive prevalent of e-mail communications nowadays, it is essential to determine where and when direct face-to-face communication is better than email transmission of critical information about a proposed AM equipment.

Figure 3.3 shows a condensed sample of multidimensional communication matrix for AM environment. Actual users will include all the pertinent elements for their specific operating environment. Communication across various functional lines is important to bring everyone on board for a cohesive AM effort. Of particular importance is the need to keep end-user requirements in mind throughout the development process. The cells in the communication matrix indicate the source-to-target communication linkages as well as specific topic of communication. This helps to identify not only who is communicating with whom, but also what is expected to be communicated.

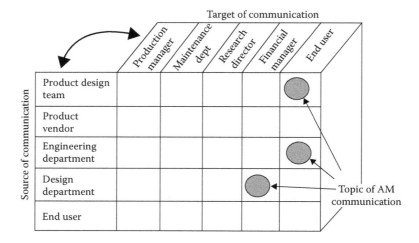

Figure 3.3 A template for communication matrix for additive manufacturing project management.

3.4 Cooperation

Cooperation of the personnel involved in AM must be elicited using explicit means. Merely, voicing consent for a project is not an enough assurance of full cooperation. Participants and beneficiaries of the project must be convinced of the merits of the project. The pros and cons should be addressed. Never shy away from the *cons* of a project. Rather than being a source of ire for team members, a specification of the *cons* may be vital for garnering support, as long as individuals know what to expect and what not to expect. Some of the factors that influence cooperation in a project environment include personnel requirements, resource requirements, budget limitations, past experiences, conflicting priorities, space limitation, resource-sharing constraints, and lack of uniform organizational support. A structured approach to seeking cooperation for AM should clarify the following:

- The level and type of cooperative efforts required
- Precedents for collaborative projects
- The possible implication of lack of cooperation
- The criticality of cooperation to project success
- The expected organizational impact of cooperation
- The time frame involved in the project
- The organizational benefits of cooperation
- The personal benefits or rewards of cooperation

The types of cooperation required for a successful product development include functional cooperation, social cooperation, legal cooperation, administrative cooperation, proximity cooperation, dependency cooperation, lateral cooperation, vertical cooperation, and imposed cooperation. Some of these are possible only in certain types of project scenarios. Below are some guidelines for securing cooperation for AM:

- Establish achievable goals for the project.
- Clearly outline individual commitments required.

- Integrate project priorities with existing priorities.
- Allay the fear of job loss due to AM products compared to traditional manufacturing.
- Anticipate and preempt potential sources of resource conflicts.
- Remove skepticism by referring to earlier communication of the merits of the project.

3.5   Coordination

After communication and cooperation functions have been initiated successfully, the efforts of the project personnel must be coordinated. Many projects fail because the project team anxiously jumps to the coordination stage. But where there has not been sufficient communication and there is a lack of cooperation, coordination cannot be accomplished effectively. Coordination facilitates congruent organization of efforts. The construction of a responsibility chart can be very helpful at this stage. A responsibility chart is a matrix consisting of columns of individual or functional departments and rows of required actions. Cells within the matrix are filled with relationship codes that indicate who is responsible for what. The matrix helps to avoid neglecting crucial communication requirements and obligations. It helps resolve questions such as:

- Who is to do what?
- How long will it take?
- Who is to inform whom of what?
- Whose approval is needed for what?
- Who is responsible for which results?
- What personnel interfaces are required?
- What support is needed from whom and when?

When implemented as an integrated process, the Triple C model can help avoid conflicts in new high-end equipment installation. When conflicts do develop, it can help in resolving the conflicts. Several sources of conflicts can exist in complex technical projects, including the following:

Schedule conflict: Conflicts can develop because of improper timing or sequencing of project tasks. This is particularly common in large multiple projects spread over multiple locations. Procrastination can lead to having too much to do at once, thereby creating a clash of project functions and discord among team members. Inaccurate estimates of time requirements may also lead to infeasible activity schedules.

Cost conflict: Product development cost may not be generally acceptable to the clients of a project. This will lead to project conflicts. Even if the initial cost of the product development is acceptable, a lack of cost control during implementation can lead to conflicts. Poor budget allocation approaches and the lack of a financial feasibility study will cause cost conflicts later in the product development process. One area of concern for AM is the cost of supplies to sustain the operation of the AM equipment. Adequately funding the purchase of AM equipment is one thing, but funding the recurring purchase of supplies is an entirely different thing.

Performance conflict: If clear performance requirements are not established, AM product performance conflicts will develop. Lack of clearly defined quality standards and expectations can lead each person to evaluate his or her own performance based on personal value judgments. In order to uniformly evaluate quality of AM outputs and monitor project progress, performance standards should be established based on the intended scope of the AM project.

Management conflict: There must be a two-way alliance between management and the AM team. The views of management should be understood by the team. The views of the team should be appreciated by management. If this does not happen, management conflicts will develop.

Technical conflict: If the technical basis of a project is not sound, technical conflicts will develop. New manufacturing projects are particularly prone to technical conflicts because of their significant dependence on technology. Lack of a comprehensive technical feasibility study will lead to technical conflicts. AM is relatively new in industrial practice. Consequently, many technical issues remain to be ironed out. Clear communication, solid cooperation, and tight coordination can help defuse the adverse impacts of technical conflicts.

Priority conflict: Priority conflicts can develop if project objectives are not defined properly and applied uniformly across a project. A lack of a direct project definition can lead each project member to define his or her own goals which may be in conflict with the intended goal of the project. A lack of consistency of the project mission is another potential source of priority conflicts. Over-assignment of responsibilities with no guidelines for relative significance levels can also lead to priority conflicts. One person taking on the task of what should be a team effort is a sure basis for priority conflict. Again, using the Triple C model can help preempt or resolve priority conflicts.

Resource conflict: Resource allocation problems are a major source of conflicts in any project management. Competition for resources, including personnel, tools, hardware, software, space, and so on, can lead to disruptive conflicts.

Power conflict: Project politics lead to a power play which can adversely affect the progress of a project. Project authority and project power should be clearly delineated. Project authority is the control that a person has by virtue of his or her functional position. Project power relates to the clout and influence, which a person can exercise due to connections within the administrative structure of an organization. People with popular personalities can often wield a lot of project power in spite of low or nonexistent project authority.

Personality conflict: Personality conflict is a common problem in projects involving a large group of people. The larger the project, the larger the size of the management team needed to keep things running. Unfortunately, the larger management team creates an opportunity for personality conflicts. Communication and cooperation can help defuse personality conflicts.

3.6   *Distributed additive manufacturing product development*

This section covers the fundamentals of distributed product development in AM. Figure 3.4 presents the product development process in a distributed environment across functional areas. The inputs are in terms of capital, raw material, and labor. At the

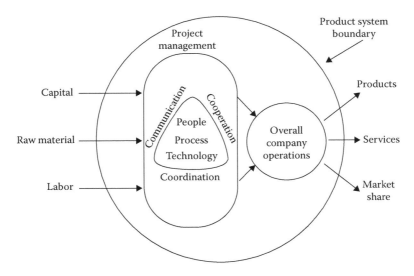

Figure 3.4 Input–output framework for distributed AM product development.

output end, the physical products are complemented by organizational services and a metric of market share. The project management approach embodies technology, people, and work process. In this environment, the Triple C model serves as the tool to integrate the various project management efforts.

3.7   Analysis of additive manufacturing project requirements

A typical project is undertaken to create a unique *product*, *service*, or *result*. In the case of AM, the project output is a certain *product*, hopefully of high quality, that meets the market needs of the organization. The key to getting everyone on board with the AM process is to ensure that product objectives are clear and comply with the principle of *SMART* as outlined below:

- *Specific*: Task objective must be specific. Project objectives must be specific, explicit, and unambiguous. Objectives that are not specific are subject to misinterpretations and misuse.
- *Measurable*: Task objective must be measurable. Project objectives should be designed to be measurable. Any factor that cannot be measured cannot be tracked, evaluated, or controlled.
- *Aligned*: Task objective must be achievable and aligned with overall project goal.
- *Realistic*: Task objective must be realistic and relevant to the organization. A project's goals and objectives must be aligned with the core strategy of an organization and relevant to prevailing needs. If not aligned, an objective will have misplaced impacts. A project and its essential elements must be realistic and achievable. It is good to *dream* and have lofty ideas of what can be achieved. But if those pursuits are not realistic, a project will just end up *spinning wheels* without any significant achievements.

- *Timed*: Task objective must have a time basis. Timing is the standardized basis for work accomplishment. If project expectations are not normalized against time, there will be no basis for an accurate assessment of performance.

If a task has the above intrinsic characteristics, then the function of communicating the task will more likely lead to personnel cooperation. A SMART approach to developing and communicating AM objectives can ensure the cooperation of everyone. Specific means that an observable action, behavior, or achievement is described. It also means that the work links to a rate of performance, frequency, percentage, or other quantifiable measure. For some jobs, being specific can, itself, be nebulous. However, to whatever extent possible and reasonable, we should try to achieve specificity. That is exactly what project management seeks to achieve. This ensures that the leadership team, operators, staff, and customers all share the same expectations.

The word *measurable* means observable or verifiable, which implies that a method or procedure must be in place to track and assess the behavior or action on which the objective focuses and the quality of the outcome. As not all work lends itself to measurability, objectives can be written in a way that focuses on observable or verifiable behavior or results, rather than on measurable results. If no measurement system exists, the project manager must be able to monitor performance to ensure that it complies with the specified objective.

An aligned objective provides a conceptual basis to draw a linkage line from the objective to other factors throughout the project. It means that the objectives throughout the organization pull in the same direction. In this way, the performance of the project team and whole organization is improved.

Project managers must have a clear understanding of their own objectives before they can work with project team members to establish their job objectives. This is one of the key building blocks of performance assessment in project management. If managers know the functions on which people actually are spending time, they can make meaningful improvements in organizational performance by ensuring effort is focused on work that is valued by the organization and by eliminating inefficient processes. Job objectives align work with organizational goals and the mission, drawing the line of sight between the employee's work, the work unit's goals, the project functions, and the organization's success. The letter "R" in SMART has two meanings that are both important: realistic and relevant.

Realistic has two meanings:

- The achievement of an objective is something an employee or a team can do that will support a work unit's goal. The objective should be sufficiently complex to challenge the individual or team but not so complex that it cannot be accomplished. At the same time, it should not be so easy that it does not bring value to the individual or the team.
- The objective should be achievable within the time and resources available to the project, which is usually expressed as triple constraints of time, cost, and quality.

Relevant implies that it is important for the advancement of the employee and the organization.

Figure 3.5 illustrates the application of the Triple C approach of project management in the context of using the SMART principle of project performance assessment.

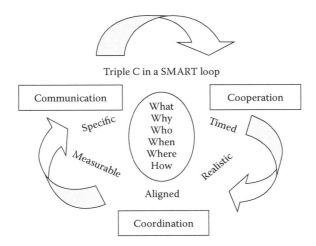

Figure 3.5 Application of Triple C in a SMART loop.

3.8   Conclusion

This chapter has presented general principles of project management and applicability to AM. The knowledge areas compiled by the Project Management Institute (PMI) are generally applicable to the theme of this chapter. Readers are encouraged to seek more in-depth techniques of project management within the specific knowledge areas listed below, based on PMI's project management body of knowledge (PMBOK®):

1. Project *integration* management
2. Project *scope* management
3. Project *time* management
4. Project *cost* management
5. Project *quality* management
6. Project *human* resource management
7. Project *communications* management
8. Project *risk* management
9. Project *procurement* management

The above segments of the body of knowledge of project management cover the range of functions associated with any project, particularly complex ones, such as AM. Multinational projects particularly pose unique challenges pertaining to reliable power supply, efficient communication systems, credible government support, dependable procurement processes, consistent availability of technology, progressive industrial climate, trustworthy risk mitigation infrastructure, regular supply of skilled labor, uniform focus on quality of work, global consciousness, hassle-free bureaucratic processes, coherent safety and security system, steady law and order, unflinching focus on customer satisfaction, and fair labor relations. Assessing and resolving concerns about these issues in a step-by-step fashion will create a foundation of success for a large project. Although no system can be perfect and satisfactory in all aspects, a tolerable trade-off on the factors is essential for project success. That is what this chapter advocates for new endeavors of AM.

References

Badiru, Adedeji B. (2008), *Triple C Model of Project Management: Communication, Cooperation, and Coordination*, CRC Press/Taylor & Francis, Boca Raton, FL.

Badiru, Adedeji B. (2012), *Project Management: Systems, Principles, and Applications*, CRC Press/Taylor & Francis, Boca Raton, FL.

Badiru, Adedeji B. and P. S. Pulat (1995), *Comprehensive Project Management: Integrating Optimization Models, Management Principles, and Computers*, Prentice-Hall, Englewood Cliffs, NJ.

chapter four

3D-printing impacts on systems engineering in defense industry

Jason Deters

Contents

Although the fundamental science has existed for decades, 3D printing has gained incredible momentum in recent years for applications like rapid prototyping—and is now being adopted for niche production applications. 3D printing holds tremendous potential, but considerable challenges exist before this process can truly revolutionize the future of manufacturing. This paper highlights examples of how 3D printing is being used in aerospace and defense industries today, how it may be applied in the future, and what obstacles must be overcome before widespread applications become mainstream.

From a business standpoint, the primary overarching benefits that 3D printing offers are related to economies of scale and scope, essentially increasing the variety of products a unit of capital can produce. For this reason, 3D printing is particularly well suited for the aerospace and defense industries; thanks to its ability to produce multiple design iterations on a single machine setup. Similar to other organizations providing equipment to the U.S. Armed Forces, general dynamics land systems (GDLS) constantly searches for innovative methods to improve performance, reduce lead time, and deliver meaningful value where it matters most. 3D printing can undoubtedly contribute to these objectives, but fully realizing this potential will take strategic vision, technical innovation, and willingness to move beyond traditional process limitations.

4.1   Near-term applications abound

Although many people picture 3D printing as a single technology, there are actually several unique processes that use distinctly different methods to build a one-layer component at a time from a computer-aided design (CAD) model. A common process used today for the templates, prototypes, and tooling mentioned above is called fused deposition modeling (FDM), which has gained popularity in recent years as 3D printing of plastic parts evolves beyond traditional stereolithography. A wide variety of organizations currently use high-temperature polymers for 3D printing; dramatically reducing lead time and cost while enabling design freedoms is not possible with traditional processes. From a systems engineering standpoint, the ability to quickly and easily 3D-print prototypes holds substantial promise in reducing design cycle time, comparing multiple design iterations, and optimizing overall packaging layout. This rapid prototyping capability enables design teams to identify issues in the process early and to optimize component geometry for seamless integration. A variety of advanced polymers and similar materials for this type of 3D printing have emerged in recent years, each with unique properties suited for particular applications.

GDLS has utilized FDM technology in a number of applications that illustrate the capability of 3D printing to provide measurable efficiency gains without actually being used for production components. One example is a fit-check model for a radio that requires weeks of lead time to obtain. Since the design team was only interested in how the radio integrated into the vehicle to identify any interference or access issues, the 3D-printed plastic model of the radio satisfied all the objectives (Figure 4.1).

This part went from CAD model to 3D-printed model in two days and prevented costly delay to the program schedule. Another recent example is a wiring harness connector that was on back order, posing a program delay of several weeks. Instead of accepting this negative schedule impact, GDLS was able to quickly print plastic versions of the component and use them in place of the back-ordered item in an engine test application at over 350°F. This is not only an illustration of the rapid turnaround time that 3D printing can deliver, but also shows that the engineered polymers being used in 3D printing are robust enough to handle demanding conditions, such as this high-temperature test.

Given the rapid manufacturing capability for FDM to create low-cost plastic components with optimized geometry, GDLS has investigated the possibility of incorporating a

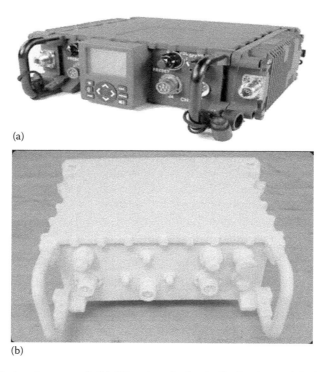

(a)

(b)

Figure 4.1 (a) Radio hardware and (b) 3D-printed plastic fit-check model. (Courtesy of General Dynamics Land Systems.)

thin, high-strength structural metal cladding to be applied to a 3D-printed plastic component. This combination could offer performance similar to a metal component at a fraction of the weight but would likely only apply to noncritical areas, such as brackets, and so on. Extensive testing is necessary to better understand the long-term properties and failure modes of both the 3D-printed polymer core and the structural cladding. If this hybrid solution is proven to be a viable option, it opens up a number of possibilities for applying 3D-printed *semimetallic* components for end item use in the near term.

To further leverage the systems engineering benefits that 3D printing can provide throughout the design process, GDLS has also developed sophisticated reverse-engineering capabilities that employ a 3D-scanning unit that can quickly capture the geometry of a physical part or assembly and can convert it to a CAD model and 3D print as needed. Current applications at GDLS are using the scan data to 3D-print plastic parts for instances where CAD data are not readily available, but in the future this same capability can be used for metallic components as well.

4.2   *Tremendous opportunities and substantial challenges*

Applying 3D printing for solid metallic components, on the other hand, presents a host of technical obstacles that are the focus of ongoing development efforts in industry, academia, and government organizations. The primary challenge to qualifying 3D-printed metallic components for end-item production use, especially in aerospace and defense, is difficulty in controlling process inputs to achieve consistent outputs. Industry standards are being developed to regulate the composition and quality of raw materials used for 3D printing, and similar efforts are underway to create a centralized database for material-specific process parameters as well. This level of industry standardization is crucial to the advancement of 3D printing in metal-based production applications and is the focus of substantial research and development. These development initiatives are encouraging, but until they are approved for broad use on aerospace and defense components, it is impossible to qualify a component without conducting a detailed analysis of each individual part. A total of 100% inspection would obviously be cost prohibitive and limits the use of 3D-printing metal components to niche applications.

GDLS has committed substantial effort in recent years to better characterize how and when 3D-printing technology will impact the ground combat vehicle industry. Partnering with industry and academia, GDLS is pursuing 3D-printing production solutions in a variety of areas with a prime example being part consolidation on complex-welded components. Part consolidation, such as converting a multipiece weldment to a one-piece 3D-printed part, reduces complexity, production cost, and component weight. In addition, 3D printing enables optimization of the design by eliminating constraints imposed by traditional fabrication methods.

In a recent demonstration, GDLS selected a nine-piece steel weldment and partnered with academic and industry partners to 3D print it as a single piece, which GDLS then machined to final configuration. Consider all the process costs associated with fabricating a multipiece weldment, such as the one shown in Figure 4.2. A part like this requires several time-consuming steps, including material transfer, plate cutting, forming, machining, welding, and multiple inspection processes. The component can, in theory, be built as a one-piece *preform* on a 3D printer, then it can have the critical interfaces, such as tapped holes and key datum surfaces, cleaned up on a machining center. For this specific demonstration, using 3D printing reduced the part's complexity, but a prohibitive amount of machining was required to meet the end-item requirements. Since the demonstration

Figure 4.2 **(See color insert.)** 3D-printed titanium component (on right) with nine-piece welded steel version. (Courtesy of General Dynamics Land Systems.)

was only intended to illustrate the work envelope and material properties possible with electron beam additive manufacturing (EBAM), the component design was not optimized for 3D printing.

The large amount of postprocessing was required because the part is designed for traditional processes, such as bending, welding, and machining. As a result, the project did not fully leverage the advantages that the 3D printing offers in geometry optimization and material efficiency. The next step in this effort aims to redesign and optimize this same component specifically for 3D printing, reducing the amount of postprocessing required, while still satisfying performance and integration requirements. By leveraging advanced design software capabilities, this component can be redesigned for 3D printing with a process called *topology optimization*. This capability captures the performance and integration design requirements for the component and optimizes material layout for the specified loads within the given design space. This capability represents an advanced design for manufacturing (DFM) exercise, except in this case a number of the traditional process constraints have been removed (Figure 4.3).

Another example application where GDLS has explored the application of 3D printing is for reduction of heat-induced failures in electronic housings. High-end electronics are used in virtually all aerospace and defense vehicle platforms, and excessive heat is a common cause of failure for these expensive components. Many electronic enclosures used in these industries are produced from cast aluminum and feature simple *fins* to increase surface area on the exterior of the housing, which helps to dissipate heat.

3D printing eliminates geometric constraints, enabling unique and optimized passive cooling features instead of standard cooling fins (see Figure 4.4). These innovative geometries can significantly improve thermal characteristics of the housing, thereby protecting its sensitive contents. In addition, the geometric freedom afforded by 3D printing allows designers to incorporate small conformal cooling channels into the wall thickness of the electronic housing, which further improves heat dissipation (Figure 4.5).

Although traditional machining can only create straight-line cooling channels, 3D printing enables the channels to conform to the specific shape of the housing for maximum

Figure 4.3 **(See color insert.)** 3D-printed titanium preform before final machining. (Courtesy of General Dynamics Land Systems.)

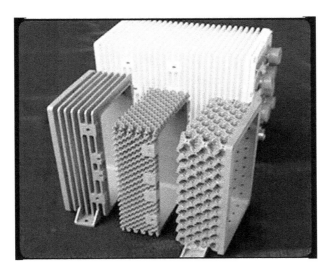

Figure 4.4 **(See color insert.)** 3D-printed concepts with *cooling fins* version. (Courtesy of General Dynamics Land Systems.)

efficiency. An effective solution in this area will apply to many aerospace and defense platforms, where heat generated by high-density electronics is an important design consideration. Low production volumes often associated with these industries also enable 3D printing to be cost-competitive with traditional castings.

As 3D printing continues to open up new opportunities for innovative applications, an interesting potential application is embedded electronics and sensors. This application consists of encapsulating a rugged sensor or electronic device within a printed component. Applicability in harsh environments is likely a long-term possibility, but the potential impacts of this capability would be tremendous. For example, the ability to embed

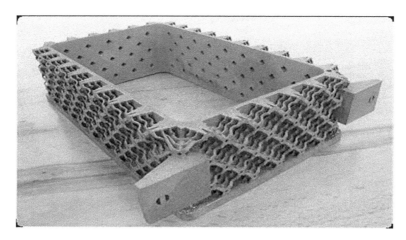

Figure 4.5 **(See color insert.)** 3D-printed titanium preform before final machining. (Courtesy of General Dynamics Land Systems.)

sensors in a structure to measure and communicate damage from a ballistic or mine-blast event would provide valuable real-time data to improve the mission effectiveness.

4.3   *3D printing offers far-reaching potential*

As these examples illustrate, 3D printing of metallic components has a game-changing potential in aerospace and defense applications, but this technology is relatively immature and several technical obstacles exist. Commercially available material options are limited, and the lack of industry-standard material and process parameter specifications makes part qualification very difficult. Although there are a number of organizations partnering to develop a common set of specifications to effectively standardize 3D printing, it is expected to take several years for 3D printing of metallic components to become a widely accepted production process.

An important business and security aspect of 3D printing that cannot be overlooked is the ownership and sharing of intellectual property related to chemical composition, process parameters, and communication of design information. At this point, most large manufacturing companies developing 3D-printing processes to build their products are protecting this information as proprietary, which provides short-term competitive advantage but at the same time slows the effort to develop industry-wide process standardization. This industry standardization is critical to successfully implement 3D printing for the end-use components in combat vehicles and many other business sectors.

To further leverage the systems engineering benefits that 3D printing can provide throughout the design process, GDLS has also acquired sophisticated reverse-engineering capabilities that employ a high-fidelity 3D-scanning unit that can quickly capture the geometry of a physical part or assembly and can convert it into a CAD model that can then be 3D printed with an impressive degree of accuracy (See Figure 4.6). Current applications at GDLS are using the scan data to 3D-print plastic parts for instances where CAD data are not readily available, but in the future this same capability can be used for metallic components as well. One example application envisioned for this technology is building replacements for obsolete components where models and drawings are not available.

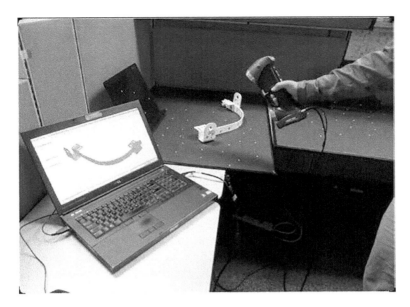

Figure 4.6 3D scanning capability for parts without models. (Courtesy of General Dynamics Land Systems.)

Figure 4.7 **(See color insert.)** On-demand spare parts will dramatically enhance the ability to quickly service combat vehicles. (Courtesy of General Dynamics Land Systems.)

The scanning and printing aspects of this capability have been proven, but this technology will have limited application until 3D printing of metallic end-use components becomes a viable option (Figure 4.7).

Although not directly related to systems engineering, another promising application for 3D printing in defense applications is cost-effective, on-demand spare parts.

Developments are underway to create an integrated system to 3D print, finish-machine, paint, and assemble components on an as-needed basis, with minimal human interaction. When this process capability matures, it has the potential to revolutionize the infrastructure and logistics involved with supplying spare parts to the front line. This *beyond the supply chain* capability will dramatically cut inventory cost, reduce obsolescence waste, and can be used for repair/refurbishment of worn parts as well. Instead of filling and maintaining a warehouse for spare parts, a 3D printer can conceivably print the desired part when and where it is needed.

This sounds too good to be true for good reason, as there are several technical limitations that must be overcome before this concept can revolutionize the way in which spare parts are supplied to the front line. For a simple metal bracket or enclosure, the solution may be relatively straightforward, but for more complex components that require multiple materials, surface treatments, precision machining, assembly processes, and so on, the solution becomes much more complex. Adding to the complexity is the important consideration for how technical data, such as 3D models, are owned, transferred, and securely maintained at a remote base or similar setting. This is sure an area of further development, as 3D-printing technology and communications' infrastructure continue to evolve and mature.

4.4   Conclusion

3D printing will not completely replace traditional processes in our lifetime, but instead represents a new tool in our toolbox—from both a design and a manufacturing standpoint. While the examples described in this paper show significant potential for 3D printing to transform systems engineering, true adoption of this technology must start in the earliest stages of product and process design. Adding 3D printing to our existing capability set requires a paradigm shift in how we develop a concept, prove it out, and ultimately manufacture it. This change in mindset is gradually taking place in a variety of industries and will gain broader acceptance as cost and technical barriers are overcome. Streamlining the design and manufacturing process to this extent can provide dramatic reduction in production cost and lead time, both of which are critical in supplying the best capability and value to the brave men and women who defend the freedom of the United States and its allies.

chapter five

3D-printing design using systems engineering

Bradford L. Shields and Vhance V. Valencia

Contents

5.1 Systems engineering background

5.1.1 Chapter overview

As product designs and large-scale projects become more and more complicated, the need for a comprehensive design process becomes even more evident. This process must bring together all of the stakeholders to identify risk, point out design flaws, and in the end, present the best product for the customer. Systems engineering is that integrated approach focusing on the necessary details of a project design through the cooperation of the stakeholders and a step-by-step approach through the entire design process (Blanchard & Fabrycky, 2011). The intent of this chapter is to provide a successful technique for navigating the entire design process by explaining the background of system's engineering, discussing one of the most common models used within systems engineering, and then providing simple examples of system's engineering at work.

5.1.2   Systems engineering processes

5.1.2.1   What is systems engineering?

The implementation of a systematic and iterative design procedure is critical to ensuring the final system being produced adequately meets all customer requirements. This methodical approach or systems' thinking is defined as systems engineering, which helps identify and reduce uncertain risk over the course of a project's life cycle. The most accepted definition of systems engineering comes from the *Systems Engineering Management Guide*:

> The application of scientific and engineering efforts to (a) transform an operation need into a description of system performance parameters and a system configuration through the use of an iterative process of definition, synthesis, analysis, design, test, and evaluation; (b) integrate related technical parameters and ensure compatibility of all physical, functional, and program interfaces in a manner that optimizes the total system definition and design; and (c) integrate reliability, maintainability, safety, survivability, human engineering, and other such factors into the total engineering effort to meet cost, schedule, supportability, and technical performance objectives. (DSMC, 1990)

Systems engineering uses multiple factors for the evaluation of a design. Those factors include, but are not limited to, user needs, requirements, functionality, design constraints, and the actual design itself. Being originally created for use in the development of software and weapons' acquisitions, systems engineering has stepped through various methods and approaches throughout its short lifetime (Blanchard & Fabrycky, 2011). The initial concept focused on the visualization and conceptualization of how to systematically step through a product's entire design process. The most applicable and most commonly used systems engineering methods included the waterfall process model, the *Vee* process model, and the spiral process model. All these three processes are discussed briefly in the previous chapter; however, this chapter focuses solely on the use of the spiral process model and its application within the design process.

5.1.2.2   Spiral process model overview

The most common model used by systems engineers today is the spiral process model, presented in Figure 5.1. Originally developed in 1968, the model was "intended to introduce a risk-driven approach for the development of products or systems" (Blanchard & Fabrycky, 2011). Using the constant feedback provided by the *Vee* process method, the spiral model makes the process of requirements, design, and conception cyclical while adding in a factor of risk. The need for risk analysis was the main component lacking from the *Vee* process model and the basis for what drove the design of the spiral model. The spiral has four separate phases: planning, risk analysis, engineering, and evaluation (Munassar & Govardhan, 2010). The phases allow the design team for any prototype development to continually walk through each process in the chain to ensure it meets all the desired specifications. Based on the spiral design, the angular component represents the progress of the design, whereas the radius of the spiral represents cost (Munassar & Govardhan, 2010). At the end of each cycle, prior to moving to the next cycle, the design team is mandated to evaluate their prototypes and alternatives, solicit suggestions and changes from stakeholders, evaluate the inputs, and decide what changes to make.

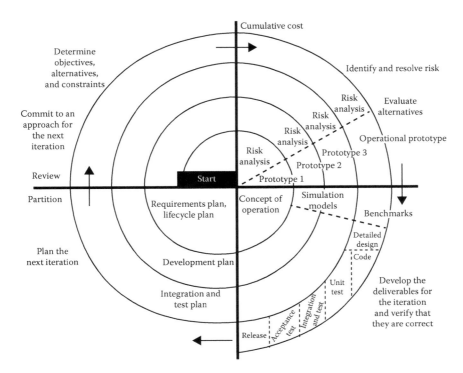

Figure 5.1 Spiral process model.

Software developers and design teams agree that the spiral model's focus on risk analysis is an advantage, it works well for large and mission-critical projects, and it is iterative and extremely flexible. Disadvantages of the model include that it is costly to use, requires highly specific expertise, is incompatible with smaller projects, and exquisitely dependent on the risk analysis phase (Munassar & Govardhan, 2010). Even with its liabilities, the process quickly became the model of choice for design teams walking through a cradle-to-grave design process (Blanchard & Fabrycky, 2011).

5.1.2.2.1   Spiral process model: Planning   The most important step at the beginning of the design process is to ensure it starts out on the right foot. This is done through the development of specific requirements from the user. These requirements identify exactly what the end product is supposed to do when it comes to solving the problem or meeting the objective. The basics of a requirement include the actual need of the user, what the product is intending to accomplish for the user, and what initial design constraints may factor into the design of the problem. Many users cannot differentiate between an actual need and a requirement, because they are very similar.

To begin, a need or want is simply a broad definition of the overall requirements: the high-level description, *30,000-foot view,* and so on, of the problem without identifying any specifics. Examples of a need or want include the following:

- The football team needs to get better.
- The expanding family wants a house.
- The environmental department at a school needs something to securely hold and shake their glass jars for experiments.

All of these identify an overall goal but do not address any specifics. This is because the need addresses only the high-level task.

The next level, which is the identification of the requirements, is more specific. Requirements delve into *what* the product must do to accomplish the need, as in the following examples:

- To get better, the football team must acquire a top 30 recruiting class, state-of-the-art facilities, and depth at key lineman and skill positions.
- The expanding family is in need of at least a three-bedroom, two-bathroom home with a two-car garage, and fenced backyard for their dog.
- The agitator has to carry four glass bottles, the necks of the bottles are 3 inches in diameter, and the unit will spin at 30 revolutions per minute instead of shaking.

Each of these specifics falls under the broad *need*, and each plays an important role in the overall design of the product by providing specificity. The sooner requirements are agreed upon by the user and designer, the more smoothly the project will go. Some projects are hindered by requirements that increase during the design's evolution, causing delays and frustrations to both the user and the designer.

Even with specific requirements, every project has design constraints that limit the design in some manner. Clear constraints can be identified quickly and up front in initial meetings; however, more often than not, constraints are found as the design progresses and prototypes are tested. Design constraints can range from material type, color, size, placement, or any other factor that affects the design in some manner. Although constraints do limit the design, a thorough identification and consideration of all constraints make for a better product for the end user.

The foundation for a smooth design and implementation process starts with a proper analysis of the user's need, all the requirements, and any constraints limiting the design. From there, the designer and builder move on to the project's risk analysis, and the design, construction, and testing phases of the process.

5.1.2.2.2 Spiral process model: Risk analysis   Risk analysis is done throughout the project, but it is even more important before the start of the initial design cycle. Risk is defined as the "potential that something will go wrong as a result of one event or a series of events" (Blanchard & Fabrycky, 2011). Through the use of certain analysis tools, like discrete event simulation (DES) and control theory, certain risks can be investigated and mitigated (de Weck, Roos, & Magee, 2011); however, not all risk can be mitigated within a project. There are four basic types of risks associated with a project (Blanchard & Fabrycky, 2011):

1. *Technical risk*: This risk is associated with engineering designs and specific performance requirements. When constructed, the owner is taking the risk that the contractor is technically capable of providing a usable product. If the designer contracts out certain designs, then the designer is taking the risk that the designs will be technically sound and meet all performance requirements. These risks are ameliorated through design checks, bonding requirements, and experience levels that ensure specific performance and technical requirements are met.
2. *Cost risk*: For any project, there is always a risk that the costs will exceed the original amount that was bid or estimated. Depending on the type of contract, this risk could

be more to the owner or to the contractor. Detailed plans and cost estimates are built prior to the project start to mitigate any possible cost overruns.

3. *Schedule risk*: Any deadline runs the risk of exceeding the projected completion date. Contractors can be pressed not to go over their completion date through incentives or delay penalties written into the contract. Detailed schedules with Gantt charts and task lists are created to help a project stay on schedule.

4. *Programmatic risk*: In a large organization, this type of risk is much more prevalent. This is the risk of certain events imposed on the project/program, which is a result of external influences. Either from leadership, external supply factors, or any other outside influences, this type of risk can be the most unforeseen and challenging to plan for (Blanchard & Fabrycky, 2011).

In the end, risk analysis is an identification-and-mitigation process for the risks assumed to be the most prevalent surrounding a project. Good project managers, designers, and builders construct risk management plans (RMPs) to document the ways they go about mitigating certain risks. Most RMPs also include a broad plan identifying what the project manager would do in the event when certain risk events took place on their project site.

5.1.2.2.3 Spiral process model: Engineering   The engineering process within the spiral process model involves the design, production, and testing of the product. No longer will stakeholders debate the requirements of the project, but now they focus on the mechanical properties, the materials, the shape, and the overall effectiveness of the product. The initial design constraints and desired shape should also already be fleshed out through the constant communication between the stakeholders and the design team. Most designs will contain multiple iterations through the engineering phase of the design. Typically, a small sample of the questions asked during this phase includes the following:

- What should the design look like?
- What are the highest priority material properties?
- What production limitations will inhibit the geometries or size of the part?

After hammering out the initial desired design criteria, the design team launches into developing the first design of the product. For the different iterations of the design that follow, the team should continue to discuss the overall design, how the material properties of the design add to or take away from the desired end goal, and how the current production method may limit the actual optimization of the product. During each of these sessions, alternative ideas and avenues should be discussed by the team in order to find the best route for the design to follow.

5.1.2.2.3.1 Design/Building—Iteration #1, 2, 3…   A project may include numerous design-and-building processes because the spiral model allows for prototypes to be passed across design and testing phases multiple times until a useful product is created, and all requirements are satisfied. Within the design-and-building process, requirements and constraints are taken into account, and a product is born. The first model may look nothing like the final product, but the iterative process stepping through the design and testing phases allows the designer to create a product and test it for the user and designer to visually inspect. Once the designer makes any changes from the previous testing cycle and finishes the iterative design, it is tested once again.

5.1.2.2.3.2 Testing—Iteration #1, 2, 3... Following any design-and-building process, the testing phase ensures the product meets the requirements identified within the first step. During this phase, the user and the designer are able to see the prototype of the product in use. Sometimes, visualization of the actual designed product can result in additional requirements and constraints on the project due to unforeseen visual or functional problems with the prototype. A part that does not pass the testing phase is sent back to the design-and-building phase for modifications.

Within the testing phase, the viability of the product must also be verified. For this research, the process to accomplish this step will be determining the usability of each product identified and designed. The overall method for verification will be discussed further in this chapter, whereas the verification results were outlined in Chapter 4. All of the user's technical and functional specifications must be met for the design to pass the testing phase and move on to the final handoff and integration.

5.1.2.2.4 Spiral process model: Evaluation The spiral model is just an iterative spiral that walks through each phase numerous times throughout the design. This allows the stakeholders to reevaluate changes and risks after iteration, a form of change management that ensures nothing is missed between different design iterations. The most important part of this phase hinges on the performance metrics used by the design team and/or stakeholders when evaluating the designs and prototypes following each iteration. Within this research, which will be discussed in the next section, a survey tested the overall usability of each designed part; however, the respective metric should be selectively chosen based on the part being designed. Following an acceptable test of the product, the product is then given to the unit for use.

5.1.2.2.4.1 Final handoff, integration, and maintenance Once through the testing phase, products reach the final handoff and integration phase. This is very different for all products, as some may just be handed to the customer and other products may go to a manufacturer for production. Either way, the customer receives a usable product that must be integrated into his processes. This may involve training and analysis of existing operations and requires ensuring that procedures are in place for its maintenance and care.

Each of these steps, from the identification of requirements to the implementation and maintenance of the product, is essential to the *life cycle* mindset of systems engineering design. Thinking holistically can cause designs to become more flexible and adaptable than before, allowing for improvement of processes and lowering life-cycle costs (de Weck et al., 2011). This systems engineering design process can be adapted and implemented into the analysis and thinking of all design practices to provide a more efficient model for product development.

5.1.2.3 Systems engineering "-ilities"

At the turn of the twentieth century, when automobiles, electricity, and the airplane were all at the forefront of the technological innovations, these inventions were *designed for first use*. This means that the primary aim of the designer was to create an invention that would fulfill its primary function at the time when it was first turned on, and little thought was given to the indirect consequences of the invention in the future (de Weck et al., 2011). This time period was referred to as the *epoch of great inventions and artifacts* (de Weck et al., 2011).

As the innovations and inventions became more common, the focus of the designers began to change, leading to the *epoch of engineering systems*. During this time period, designers placed greater emphasis on understanding the systems engineering properties that affected the long-term utility of their products (de Weck et al., 2011). The change in thinking was because the customer began to understand the concept of downstream life-cycle outcomes and therefore began placing more responsibility on the designers of the product (Blanchard & Fabrycky, 2011). This increase in responsibility led designers to consider the product's systems engineering properties, commonly referred to as -ilities, more carefully. A technical definition of -ilities has been stated:

> … The desired properties of systems, such as flexibility or maintainability, that often manifest themselves after a system has been put to its initial use. These properties are not the primary functional requirements of a system's performance, but typically concern wider system impacts with respect to time and stakeholders than are embodied in those primary functional requirements. The -ilities do not include factors that are always present, including size and weight. (de Weck et al., 2011)

The properties most commonly analyzed in products today include the following: quality, reliability, safety, flexibility, robustness, durability, scalability, adaptability, usability, interoperability, sustainability, maintainability, testability, modularity, resilience, extensibility, agility, manufacturability, repairability, and evolvability (de Weck et al., 2011). Each of these -ilities is specifically defined and can be analyzed through different techniques; however, each -ility's individual definition is highly dependent on other -ilities within the list. The main -ility being discussed and analyzed within this research is usability; however, the definition of usability requires the testing of product quality, flexibility, durability, adaptability, interoperability, maintainability, testability, manufacturability, repairability, and evolvability.

5.1.2.4   Usability

Usability, slightly different from operability, "deals with an individual's ability to accomplish specific tasks or achieve broader goals while 'using' whatever it is that is being investigated, improved, or designed—including services that don't even involve a 'thing' like a doorknob" (Reiss, 2012). The analysis of usability relies both on the performance of the product and on the human factors required to operate the product. Human factors are the "properties of human capability and the cognitive needs and limitations of humans" (de Weck et al., 2011). The usability of a computer program would be zero for a group that had no idea how to use a computer, even if the program is state of the art.

Usability is most commonly analyzed as an -ility for computer interfaces and programs, because there is a definite relationship between the program's purpose and the customer's ability to properly use the program. Within computer program design, usability analysis is normally broken into six different objectives or goals. The six measured objectives include the product's being (Preece, Rogers, & Sharp, 2007)

1. Effective to use (effectiveness).
2. Efficient to use (efficiency).
3. Safe to use (safety).
4. In good utility (utility).
5. Easy to learn (learnability).
6. Easy to remember how to use (memorability).

The goals related to analyzing the usability of a product are normally operationalized as questions to help provide an exact method of assessing the numerous aspects of the interactive product and the customer experience. The more detailed the questions, the more likely the designer is to find unforeseen problems within the design. Having a clear definition and understanding of the different usability objectives helps the designer develop the questions for the analysis.

5.1.2.4.1 Usability: Effectiveness   Based on the requirements identified at the beginning of the design process, the effectiveness of the product determines "how good the product is at doing what it was designed to do" (Preece et al., 2007). If the designer correctly identified all the requirements of the user in the beginning and was able to incorporate all those requirements into the design, the effectiveness of the product should be simple to determine.

5.1.2.4.2 Usability: Efficiency   The efficiency of the product is determined by the user's level of productivity once the product has been learned (Nielson, 1993). Within the definition of efficiency, other -ilities like quality, flexibility, maintainability, and durability are subsequently tested. This is because a product that is hard to maintain or a product that is of poor quality will result in a lack of efficiency over time for the user.

5.1.2.4.3 Usability: Safety   Safety involves multiple tiers of ensuring the product is safe for the customer. The first part of analyzing safety is determining whether the product will place the customer into a hazardous or dangerous environment (Preece et al., 2007). For computer systems that are near hazardous areas like X-rays or chemicals, the program should allow the user to access it remotely.

The second part of the safety analysis is determining whether the product causes the user to carry out unwanted actions accidentally or the ease with which a customer can make an error (Nielson, 1993). This normally occurs due to buttons being too close together, toggles being too sensitive, or a lack of understanding of all the abilities of the product. By ensuring proper safeguards are in place to minimize mistakes and quell any fears by the customer, the safety of the product is addressed (Preece et al., 2007).

Safety can involve analyzing other -ilities including durability, interoperability, repairability, and flexibility. Understanding the dependency on each of the -ilities will make for a better quality product overall.

5.1.2.4.4 Usability: Utility   Utility refers to the "extent to which the product provides the right kind of functionality so that users can do what they need or want to do" (Preece et al., 2007). The difference between low utility and high utility is based on the user's ability to complete everything needed within the task using the tool provided or needing to use other tools and devices to complement the product to solve the problem. By testing utility, one is also looking at flexibility, adaptability, and evolvability.

5.1.2.4.5 Usability: Learnability   Learnability is the ease with which the user is able to learn to use the product (Preece et al., 2007). When designers are looking at a program to design, they ensure the system is easy enough for the user to have the ability to begin doing productive work within a reasonable amount of time without extensive, in-depth training (Nielson, 1993). Learnability is ensured through quality assurance practices by the owners and the design companies, as well as iterative testing with those who will use the product.

5.1.2.4.6 Usability: Memorability   Finally, memorability deals with the user's ability to retain the training and skills necessary to still effectively use the product. If a user is able to return to the system after an extended period of time and immediately begin using the product efficiently, then the product is said to have high memorability (Nielson, 1993). A usable system has a higher memorability if continuous training is not required to stay proficient on the system or product. In a way, memorability tests quality, testability, interoperability, and agility. By testing a product's usability through the given objectives, a researcher is able to analyze how usable a product is for the crowd for whom the product is most intended.

5.1.3   Summary

Within this research, the development of useful products capable of solving the needs within today's Air Force becomes more streamlined and efficient due to the successful use of the systems engineering design methodology. Through the implementation of the spiral model, the research specifically targets unique applications within the engineering community to provide possible solutions through the use of additive manufacturing. By seeking out applications, identifying the unit's specific requirements, conducting a comprehensive risk analysis, and then working to design, print, and test a successful tool, the research looks to validate the ability of additive manufacturing technology to provide solutions to unique challenges in the current resource limited environment. With the systems engineering design methodology described above, the actual tools used within the research design process will now be examined further.

5.2   Additive manufacturing designs

5.2.1   Additive-manufactured applications using systems engineering

The adaptability of the systems engineering spiral model encompasses all sorts of possible designs, including software and prototype design. This research utilized the spiral model to walk through the model's different phases and additively manufacture usable products for civil engineers and computer engineers at Wright-Patterson AFB, OH. This chapter discusses the designing, printing, and testing of six parts, including a brief overview of the explosive ordnance disposal (EOD) bracket, and summarizing how the final parts were evaluated using a usability survey. Overall, this section provides a sample of how systems engineering, paired with AM technology, has the ability to completely change the way the Air Force solves some of its unique future challenges.

5.2.1.1   Computer engineering microchip bracket

Even though this research focuses on civil engineering applications for additive manufacturing, potential uses for the technology far outreach the boundaries of the Civil Engineering (CE) career field. Students in computer engineering, another graduate degree focus at Air Force Institute of Technology (AFIT), use a specially made jig to hold their microchips in place to make modifications and repairs. The jig, which is difficult to order and bulky in size, does not fit every model of microchip used in their research. The need is for a jig that securely holds various microchips and that fastens to the bed of the testing equipment. Microchips of multiple sizes are used by the engineers, and one jig did not have to fit all. The engineers required a jig that fits the chips currently being used; however, they also wanted a saved design that could easily and quickly be changed and printed when needed. As the engineers had were using a jig, which is where iteration 1 began.

As before, the original jig was large, bulky, and inefficient. The area required to hold a microchip accounted for only 10%–15% of the overall material area, so building that jig was

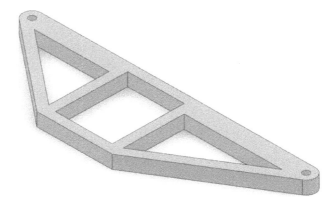

Figure 5.2 Computer engineering jig, iteration 2 design.

wasteful of material. Iteration 1 was designed around the required microchip area and the location of the bolts for the bed of the equipment. These were the only areas where material was required. Once the microchip area was outlined and the straightest path to the bolts was built, iteration 1 was complete. The part was printed and went into the testing phase. Within this phase, a new requirement surfaced that was incorporated into iteration 2.

During the modification and repair of the microchips, sensors are placed on the microchip to monitor certain functions. These sensors are delicately connected to the chip and rest on the jig during the repair process. The *excessive* area removed from the design of the old jig left nowhere for the sensors to rest, which resulted in their falling off from the jig and disconnection from the microchip. The second design, shown in Figure 5.2, provides an adequate area for the sensors to rest adjacent to the microchip. The second prototype passed all tests and is now in use within the computer engineering department.

The printed jig and the CAD design were handed over following final testing. According to the requirements identified at the beginning of the design, the user is now able to quickly change the needed dimensions of the jig and reprint the tool. The time table for the actual tool, shown in Figure 5.3, from the identification of requirements to handing off the part was approximately one month. The next section discusses the results of the usability survey evaluating the computer engineering jig.

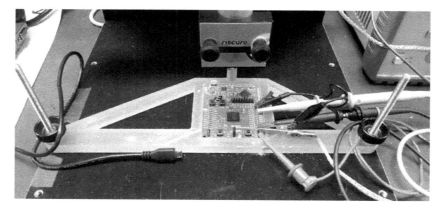

Figure 5.3 Computer engineering jig in operational use (photo courtesy by author).

5.2.1.2   Unmanned ground vehicle brackets

Among the numerous innovations being researched at AFIT, another researcher is looking into the possibility of being able to conduct underground utility infrastructure inspections through the use of completely autonomous vehicles which is intriguing for many aspects within civil engineering. Similar to the current condition of aboveground infrastructure, much of the infrastructure below the surface is just as degraded and well beyond its useful life. The ability to conduct accurate inspections and then pinpoint where the next failure will occur has the potential to mitigate millions of dollars in contingency spending for broken water mains and utility lines. Due to the nature of this research, the researcher designed a vehicle, shown in Figure 5.4, to be placed inside utility pipes and carry out inspections based on specific inspection parameters. Several pieces of equipment designed to go on the robot were more difficult to attach than originally expected. The lightweight and customizable benefits provided by additive manufacturing enabled the researcher to continue their research without compromising their vehicle due to the equipment limitations. The parts designed and printed for the autonomous vehicle project are discussed in the following sections. An overview and interpretation of the usability survey results follow the discussion of the design for the autonomous vehicle parts.

5.2.1.2.1   *Front camera bracket design*   The first piece of equipment designed for the autonomous utility inspection vehicle was a dual bracket intended to go on the front of the vehicle and to hold both light, detection, and radar (LIDAR) sensor and a digital video camcorder. The Hokuyo® URG-04X-UG01 LIDAR sensor, conducts a 270° scan of the pipe, and its programmed algorithm detects any anomalies. Prior to the additive manufacturing bracket design, attaching the front LIDAR sensor required running a plastic cord (cable tie) through holes in the base plate of the vehicle, shown in Figure 5.4; however, the geometry of the sensor caused it not to sit exactly level, which made it difficult to orient the sensor perfectly level with the vehicle and limited the reliability and range of the sensor, which is critical to the accuracy of data it generates. The sensor must also be far enough forward of the vehicle, so that its sensor can freely perform the perpendicular 270° scan around the diameter of the pipe. The digital camcorder, pictured in Figures 5.4 and 5.5, provides both light and a video feed to the inspector. The required tilt of the camera was exactly 39° based on the focus specifications and the inspector's need to see approximately 10 inches

Figure 5.4 Autonomous utility inspection vehicle (photo courtesy by author).

Figure 5.5 Prosilica GC1290C Camera. (AVT, *Prosilica Cameras*, n.d. Retrieved from https://www.alliedvision.com/en/digital-industrial-camera-solutions.html.)

in front of the vehicle. The camera has to sit up high for a good picture but must not block the scan from the LIDAR sensor below it. These requirements were taken into account as the additive manufacturing design began taking shape.

From the start of the design, due to the conditions of the inspection, the most important risk analysis factor was the difficulty of retrieving any item that fell off the vehicle if the printed bracket failed during an inspection. Based on this, the connection of the bracket to the vehicle was rated equally important as securing the equipment to the printed part. The main design constraint was the limited number of areas available for attaching the part to the robot.

From a distance, the LIDAR sensor looks like a cube with a lens on the front; however, the rear of the unit is larger than the front, so it points slightly down when set on a level surface. The original design, shown in Figure 5.6, developed a box slightly sloped from front to back to hold the LIDAR sensor completely leveled. The rest of the box surrounded the sensor and fit snug. The design placed the camera on top of the LIDAR sensor sloped at the required 39° and provided a hole for the camera lens to slide through. The approximate size of the camera lens was equal to the height and width of the other parts of the camera; therefore, the design intended the attachment of the lens to take place prior to placing it into the bracket. This actually held the camera in place and did not require any other constraints to fasten the camera to the bracket. Iteration 1 worked well during testing; however, the researcher needed the camera height to be increased and an area cut out for cabling to be connected to ports on the right side of the LIDAR sensor. Iteration #2 took into account those design changes.

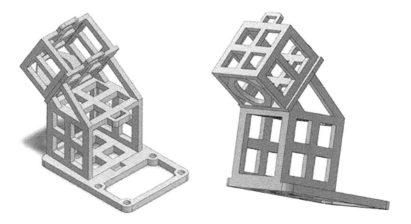

Figure 5.6 Autonomous vehicle front LIDAR and camera bracket iteration #1.

The second iteration of the design for the front LIDAR and camera bracket elevated the camera and the port connection area on the right side of the LIDAR sensor. The designed part, shown in Figure 5.7, has four connection points rearward the LIDAR area. These points will bolt to the frame of the autonomous vehicle and hold the entire bracket in place. Testing of the bracket proved successful, and this and a spare were handed over to advance the civil engineering autonomous vehicle research. Pictures showing the testing of the bracket can be seen in Figure 5.8. The total time for the identification of requirements, design of the part, printing, and testing took approximately two weeks. The survey results for this part are discussed in a later section.

5.2.1.2.2   Rear LIDAR bracket design   The autonomous vehicle required a separate LIDAR sensor, the pulsed light, Inc® LIDAR Lite™ unidirectional laser range finder, on the rear of the vehicle for the purpose of determining specific distance and location measurement. The sensor, shown in Figure 5.9, shows four separate connection points; however they are perpendicular to the base to the vehicle. Again, prior to an additive manufacturing solution, this LIDAR sensor was cable tied to the base near the rear of the vehicle. Due to the sensors having zero requirements for placement on the vehicle, the original location unnecessarily took up valuable space on the base plate. The design of the LIDAR bracket was aimed to free space for the robot by strategically removing the sensor from the footprint of the base plate and hanging it from the rear of the vehicle.

This component required single design iteration (Figure 5.10) and included four connection points for attaching the bracket to the vehicle and four connection points for

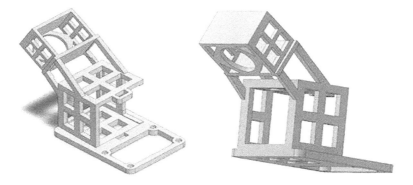

Figure 5.7 Autonomous vehicle front LIDAR and camera bracket iteration #2.

Figure 5.8 Front LIDAR and camera bracket testing (photos courtesy by author).

Figure 5.9 LIDAR Lite™ Range Finder. (RobotShop, n.d., Retrieved January 1, 2016, from http://www.robotshop.com/en/lidar-lite-2-laser-rangefinder-pulsedlight.html.)

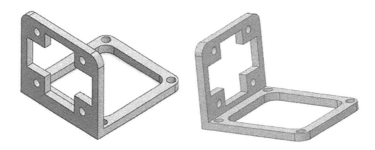

Figure 5.10 Autonomous vehicle rear LIDAR camera design.

attaching the sensor to the bracket. The testing of the rear LIDAR bracket proved extremely successful and provided more reliable results from the LIDAR sensor than those in previous tests. Since this design allowed the connection of the sensor to the vehicle without taking up critical space, the researcher was able to improve the location of certain other pieces of equipment on the vehicle. The design process for the rear LIDAR camera took approximately one week. The bracket, successfully attached to the robot, is seen on the far right hand side of Figure 5.11. Following the completion of the bracket, those taking part in the autonomous vehicle research took part in the usability survey. Their results solely described their feelings regarding the process surrounding the design and printing of the rear LIDAR bracket and are discussed in a later section.

5.2.1.2.3 Large and small battery receptacle design   The autonomous vehicle and all the equipment it carries are powered by numerous batteries of different shapes and sizes. The two batteries powering all the equipment and causing limitations for the vehicle have dimensions of 7 in × 3 in × 1.5 in and 5 in × 2.5 in × 1.25 in. Prior to an additively manufactured solution, no practical method of securing the batteries to the vehicle was available. During test runs with the vehicle, the batteries were simply placed on top without

Figure 5.11 Autonomous utility inspection vehicle rear LIDAR test.

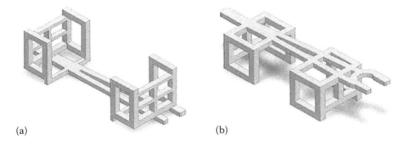

(a)   (b)

Figure 5.12 Autonomous vehicle large battery receptacle design (a) and small battery design (b).

any constraints; however, the batteries tended to fall off when the vehicle was subjected to rough terrain. The design of the battery imposed minimal requirements about the placement, except that they are spread out as widely as possible to distribute their weight. This was taken into account during the initial design process.

Two long connection pieces beneath the base plate of the autonomous vehicle snap into place to hold other vehicle pieces in place. The design from that connection piece was adapted to place two additional battery receptacles on top, the larger one on the left side of the vehicle and the smaller one on the right. The design, shown in Figure 5.12, created a box wherein the batteries are securely held and easily connected to the vehicle. The orientation of the two designs was due to how each side of vehicle connected to the long piece of the bracket.

Additional design iterations did not change the design, only strengthened the walls for more support. Testing of the printed pieces resulted in successful prints, and the two brackets were handed to the student for her research. The design and printing of each bracket, including the different iterations, took approximately two weeks. Following the design process, those close to the research took part in the usability survey for the two brackets. Although the two brackets were discussed concurrently due to their similar requirements, each had its own design process; therefore, two separate surveys were conducted to provide the most accurate results. The attached brackets are seen in Figures 5.13 and 5.14.

Figure 5.13 Autonomous vehicle battery receptacle during testing (photo courtesy by author).

Figure 5.14 Autonomous vehicle battery receptacle attached to robot (photo courtesy by author).

5.2.1.3   *Overall part evaluation*

The usability survey, given to those members for whom a part was designed and printed, resulted in identifying that each part is undoubtedly usable in the terms specified by the seven components of usability. Although each part may be usable for the need for which it was designed, the question of 3D printing's ability to provide a usable product for developing unique solutions for problems within the CE career field still stands. Each bracket's usability components were rolled up to calculate an aggregate usability confidence interval in Table 5.1, which provides a measure of usability for each bracket.

The only component to score an overall perfect score was safety, which was not surprising due to the nature of the survey question and the bracket's being designed. Memorability was a close second, with a mean score of 6.92 and a standard deviation of 0.39. Only one member, the EOD technician, believed that the bracket itself required retraining for members who had been away for a certain time period. As discussed in the autonomous vehicle section, the utility and learnability components were the lowest of

Table 5.1 Overall usability results ($n = 13$)

Objective	# of Questions	Evaluator scores			
		High	Low	Mean	St dev
Quality	3	7	5	6.56	0.79
Efficiency	–	7	0	6.62	1.39
Effectiveness	4	7	–	6.81	0.69
Utility	3	7	0	5.23	2.18
Learnability	–	7	–	5.83	1.16
Safety	1	7	7	7.00	0.00
Memorability	2	7	5	6.92	0.39
Aggregate	19			6.48	**1.33**
90% CI Range		7.00	4.78		

any of the usability objectives. This was due to the questions regarding additional uses for the tool. Since each tool was designed for a specific purpose, these objectives had the largest variance in responses. In the end, the mean overall response was 6.48, with a standard deviation of only 1.33. Based on the survey from all 13 users of the designed and printed brackets, the 90% Confidence Interval (CI) suggests that any bracket designed and printed using additive manufacturing technology would most likely fall somewhere between 4.78 and 7.0 on the usability scale. This score interval, along with the fact that every printed part is currently being used within the intended operation, provides evidence of the usability of additive manufacturing technology as a capable tool for solving problems within CE.

5.2.2   *Summary*

This small sample of products designed, printed, and tested by incorporating systems engineering design practices and AM technology only provides further proof that AM, coupled with systems engineering methodologies, can provide real-time solutions to some of the problems seen in the Department of Defense (DoD) today. The survey provided a tool for evaluating the prints; however, the real test was whether the units continued to use the printed products for their intended purpose. The five printed parts described in this chapter, as well as the EOD bracket described previously, are all continuing to be used by their respective units and have made each process more efficient.

5.3   *Conclusion*

Contract modifications, poor project estimates, and ambiguity of requirements plague the DoD projects year in and year out. The ability to adapt systems engineering practices for any design whether a prototype, software, or weapons' system provides a standardized, iterative process of identifying the system requirements, mitigating certain risks, and ensuring the end product meets the real intent. This chapter identified and discussed the most common systems engineering model, and then showed how it can be adapted to AM designs. Within the DoD, the applications for AM exist; however, it is going to take a concerted effort based solely on the education of all members to seek out and identify where those applications truly exist.

References

AVT. (n.d.). *Prosilica Cameras*. Retrieved from https://www.alliedvision.com/en/digital-industrial-camera-solutions.html (accessed on February 22nd, 2016).

Blanchard, B., & Fabrycky, W. (2011). *Systems Engineering and Analysis*. Englewood Cliffs, NJ: Prentice-Hall.

de Weck, O. L., Roos, D., & Magee, C. L. (2011). *Engineering Systems: Meeting Human Needs in a Complex Technological World*. Cambridge, MA: Massachusetts Institute of Technology.

DSMC. (1990). *Systems Engineering Management Guide*. Washington, DC: Defense Systems Management College, Superintendent of Documents.

Munassar, N. M., & Govardhan, A. (2010). A comparison between five models of software engineering. *International Journal of Computer Science, 5*, 94–101.

Nielson, J. (1993). *Usability Engineering*. Cambridge, MA: Academic Press.

Preece, J., Rogers, Y., & Sharp, H. (2007). *Interactive Design: Beyond Human-Computer Interaction*. West Sussex, England: John Wiley & Sons.

Reiss, E. (2012). *Usable Usability*. Indianapolis, IN: John Wiley & Sons.

RobotShop. (n.d.). Retrieved January 1, 2016, from http://www.robotshop.com/en/lidar-lite-2-laser-rangefinder-pulsedlight.html.

chapter six

Evaluation of existing modeling software*

Shesh Srivatsa

Contents

This chapter provides an evaluation of existing modeling software used in the additive manufacturing (AM) process. Here, we explore both commercial software packages and provide an overview of software and projects at selected universities. For commercial software, this chapter evaluates their viability for the AM process and outlines improvements that should be made. In selected universities and national laboratories, we highlight some interesting software and projects which represent the state of the art and future of modeling software.

* "Additive Manufacturing (AM) Design and Simulation Tools Study" April 2014, written by Shesh Srivatsa under contract to Air Force Research Laboratory Materials and Manufacturing Directorate, Wright-Patterson Air Force Base, OH 45433, Air Force Material Command, United States Air Force (Case Number: 88ABW-2014-3753). Approved for public release; distribution is unlimited.

  The field of additive manufacturing is rapidly evolving. The information contained in this article was current at the time of writing (2014), and it has to be supplemented with developments which have occurred since then. A number of universities and national laboratories (both within and outside the USA) are active in this area and the reader is referred to their latest work to keep abreast of recent developments. In addition, Government funded SBIR programs and other programs are a good source for recent developments.

Modeling software is important because before AM can be employed, the object desired to be created requires modeling in computer-aided design (CAD) software. To be effective, the software should be able to model different material properties such as fluid flow and heat transfer to help predict how the materials will react during the AM process. This chapter is divided into two parts. First, four commercial packages, ANSYS, COMSOL, DEFORM, and engineering systems international (ESI), are evaluated and compared with each other to assess the advantages of each one as well as the aspects of each program that need improvement. Second, this chapter explores the work of several universities and national laboratories that are interested in AM. This chapter documents their activities and research plans to better understand and improve AM technology.

6.1   Commercial software

To effectively model AM processes, the software should be able to model heat transfer, stress analysis, fluid flow, chemical reactions (for polymers), melting, solidification, and phase change. This limits the field of commercially available software packages which can be potentially used for modeling various features of AM. The development effort that is needed to model AM depends on which features the user needs to model, and which commercial software is used.

Four packages (ANSYS, COMSOL, DEFORM, ESI) were considered to evaluate the general state-of-the-art. Although DEFORM cannot model fluid flow, it was considered since it is widely used in the aerospace industry for process modeling.

Since it was not possible to evaluate a large number of commercial codes, a representative sample was chosen. ANSYS is a general-purpose program similar to ABAQUS and others. ESI was chosen because of PROCAST™ (melting and solidification) features which would be useful for AM modeling. COMSOL was chosen because of its strength in tightly coupled multiphysics which is needed for AM. Finally, DEFORM was chosen because of its widespread use in the aerospace supply chain and because of the extensive residual stress validation performed in several metals affordability initiative (MAI) programs.

A brief overview of selected commercial software packages is provided in the following sections:

- ANSYS
- COMSOL
- DEFORM
- ESI

The comparison of the various commercial codes is based on information obtained from meeting or talking with the technical representatives, from the open literature, and from the respective web sites. It is a best-effort attempt at comparing the codes. The best way to further evaluate the relative performance of the various codes is to model a demonstration case which exercises all the major AM features.

6.1.1   ANSYS

Website: http://ansys.com/

Analysis system (ANSYS) offers general-purpose capabilities from both the computational fluid dynamics (CFD) and finite element analysis (FEA) disciplines. The ANSYS multifield solver solves a wide variety of coupled physics problems, that is, thermal–structural, thermal–electric–magnetic, electromagnetic–structural, and fluid–structure interactions. With sequential coupling, each physics discipline is solved sequentially, and results are passed from one physics discipline to another. Results can be passed across a dissimilar mesh interface between the physics disciplines. A dissimilar mesh interface allows the user to optimize the mesh for each individual physics discipline. The ANSYS workbench™ platform supports a collaborative environment for developing multiphysics solutions. The ANSYS engineering knowledge manager™ (EKM) supports the seamless sharing of product specifications, performance metrics, and other critical engineering insights—so that the entire team is equipped with the same real-time information. ANSYS has developed a parametric, computer-aided engineering (CAE) platform to support robust design initiatives. This feature allows users to vary a range of parameters—including geometry, material properties, model controls, and operating conditions—to identify the most critical parameters. ANSYS is widely used in the aerospace industry for component design. The user needs to license various modules to fully model AM. ANSYS has been used to model certain features of some AM processes.

ANSYS capabilities for metal processing are available at: http://www.ansys.com/ Industries/Materials+&+Chemical+Processing/Metal.

6.1.2   COMSOL

Website: http://www.comsol.com/

The COMSOL Group provides software solutions for multiphysics modeling. The product line includes a suite of discipline-specific modules for structural mechanics, high- and low-frequency electromagnetics, fluid flow, heat transfer, chemical reactions, microelectromechanical systems (MEMS), acoustics, and more. COMSOL multiphysics simulations can be integrated with CAD models. Multipurpose tools are available to boost the functionality of the software and to verify and optimize the solutions. Since COMSOL is basically a partial differential equation (PDE) solver, the user has the freedom to add extra equations and boundary conditions for any new application. COMSOL is used widely around the world by researchers and engineers working for technical enterprises, research labs, and universities. The user needs to license various modules to fully model AM. COMSOL has been used to model certain features of some AM processes.

6.1.3   DEFORM

Website: http://www.deform.com/

Design environment for forming (DEFORM) is an engineering software that enables designers to analyze metal forming, heat treatment, machining, and mechanical joining processes. Process simulation using DEFORM has been instrumental in cost, quality, and delivery improvements at leading companies for two decades. DEFORM has been used in a wide range of research and industrial applications in many countries. DEFORM is widely used in the forging industry for modeling both forming and heat treatment processes. DEFORM does not have fluid flow features, and it has not been used to model AM.

6.1.4 ESI

Website: https://www.esi-group.com/

ESI offers a broad range of software and services tailored to meet the industry's needs of product and process development. ESI provides know-how in virtual product engineering based on an integrated suite of industry-oriented applications, which include castings, composites, sheet metal forming, welding, and assembly. The ACE+ Suite is a CFD and multiphysics software package serving the semiconductor, microfluidics, biotech, energy, automotive, and aerospace applications. The user needs to license various modules to fully model AM. ESI software has been used to model certain features of some AM processes (e.g., powder blown processes).

6.2 Summary of commercial software features

Common features of the four commercial software packages are as follows:

- Solve the basic equations of mass, momentum, and energy conservation.
- Provide an interface with CAD systems to import 2D and 3D geometries via standard formats, e.g., initial graphics exchange specification (IGES), stereolithography (ST), and so on.
- Use an element activation (or birth) scheme to introduce new material and a deactivation (or death) scheme to remove material.
- An extensive industrial use for a wide range of applications, each having particular market segments.
- Provide customer support to licensed users.
- Produce standard postprocessing plots of thermomechanical variables or new user variables.
- Need modeling postprocessing subroutines to correlate the formation of defects (balling, cracks, porosity, etc.) to thermomechanical processing conditions.
- Need modeling postprocessing subroutines to correlate the prediction of microstructure and mechanical properties to thermomechanical processing conditions. These can be simple regression equations, neural net models, fast-acting models with simplified semiempirical formulations, or full-blown fundamental physics-based models (impractical due to long computer run times).
- Need verification, validation, and uncertainty quantification (UQ) for AM processes.
- Need several hours to days of compute time (on a GPU machine) to model a full AM build process; if the code has been used for AM, only simple shapes and a few scans have been modeled.
- Licensing costs given in the table below are estimates provided by the different companies.

The advantages of each of the commercial software packages and the enhancements needed to effectively model the entire AM process are summarized in Table 6.1. The complete table is shown in Chapter 7.

Table 6.1 Comparison of commercial software features

Software	Advantages	Improvements needed
ANSYS	• Standard material constitutive equations • Can set up sensitivity analysis and parametric study runs • Widely used by the aircraft engine design community—can facilitate interaction between Original Equipment Manufacturer (OEM) designers and AM companies	• Efficient mesh generation for extremely complex 3D geometries and ~1000 deposited layers • Adaptive fine mesh in heat-affected zone • Library of templates for lattice structures • Has most of the physics in a suite of codes • Weakly coupled physics: data passed between different codes, e.g., stress—fluid code • Cannot model postprocessing hot isostatic pressing (HIP) or cold isostatic pressing (CIP) • Integration of different steps in AM difficult, since data have to be passed between different codes each using its own mesh and solution procedure • Adding new equations and boundary conditions can involve some effort • Incorporating different material constitutive models can involve some effort • Incorporation of special-purpose user features can involve some effort • Licensing cost high (~$40K/year) • Estimate ~5 years of effort to get integrated AM package
COMSOL	• Easy integration of all AM steps—all calculations done in one tightly coupled code • Open architecture enables user to add new equations and boundary conditions • Different material types: already present or can be easily added, elastic–plastic at room temperature to pure viscous behavior above melting point • Has most of the physics in a single code • Tightly coupled physics in one single code • Can build models with different levels of fidelity to address run-time issues • Can set up sensitivity analysis and parametric study runs • Incorporation of special-purpose user features straight forward • Licensing cost low (~$10K/year)	• Efficient mesh generation for extremely complex 3D geometries and ~1000 deposited layers • Adaptive fine mesh in heat-affected zone • Library of templates for lattice structures • Not as widely used as DEFORM or ANSYS for process modeling or design work in the aerospace industry

6.3   University programs

This section describes the AM activities at selected universities and national laboratories:

- Carnegie Mellon University
- Georgia Institute of Technology
- Loughborough University
- University of Louisville
- Missouri University of Science and Technology
- Pennsylvania State University
- University of Texas
- America Makes Institute
- Oak Ridge National Laboratory (ORNL)
- Lawrence Livermore National Laboratory (LLNL)
- Granta Data Base
- Other

6.3.1   Carnegie Mellon University

Website: http://www.cit.cmu.edu/media/feature/2013/04_15_namii.html

Carnegie Mellon University and Case Western Reserve University are leading a team of five universities, five companies, and two national laboratories in two projects to better control and understand 3D-printing technologies for fabricating metal components. The projects are part of the America Makes Institute to help develop AM technology.

The objective of the project is to control and understand microstructure and mechanical properties of parts made with two powder bed AM processes for the aerospace and medical industries: EOS laser sintering and the Arcam electron beam melting.

A second project led by Penn State will use thermal imaging to help determine how heat impacts metal AM processes. Process mapping technologies developed at Carnegie Mellon University will allow control of the material microstructure based on thermal images of the melt pool.

6.3.2   Georgia Institute of Technology

Website: ddm.me.gatech.edu/

A Georgia Tech startup has developed 3D-printing technology to transform the way in which costly metal parts, such as aircraft engine turbine blades, are designed and made. Using AM technology, direct digital manufacturing (DDM) systems can reduce the time it takes to make first castings of prototype turbine engine components. The core technology is licensed from Georgia Tech and based on $4.65 million in funding from the defense advanced research projects agency (DARPA) through its disruptive manufacturing technologies program.

By directly printing ceramic molds for casting, not only cost and lead times are reduced, but it also opens the prospect of creating advanced complex designs which have previously been considered impossible to manufacture. DDM systems' current focus is on turbine-engine airfoils used in jet engines currently made by investment casting. Their approach is to build ceramic molds using AM. A technique called large area maskless photopolymerization (LAMP) uses ultraviolet light to bind ceramic particles, layer-by-layer, into a desired shape. The result is a ceramic structure into which molten metals, such as nickel-based superalloys or titanium-based alloys, are poured, producing a highly accurate casting. A first generation LAMP machine is building six high-pressure turbine-blade molds in 6 hours. A second-generation machine will produce more than 100 molds and cores at a time in about 24 hours.

In addition, modeling and simulation projects are being conducted at Georgia Tech.

6.3.3   Loughborough University

Website: www.add3d.co.uk

The additive manufacturing research group (AMRG) activity spans both fundamental and applied research focused on AM process and materials development, software, teaching, and business management. The AMRG engages in projects ranging from large-scale collaborative research funded by bodies such as the Engineering and Physical Sciences Research Council (EPSRC) and European Framework programs, through individual short-term applications research, supported directly by AM systems' vendors and technology users.

The AMRG laboratory houses AM facilities for polymeric, metallic, and ceramic hardware, in addition to 3D laser-scanning equipment, and a suite of 3D solid modeling and design software tools.

6.3.4   University of Louisville

Website: http://louisville.edu/speed/rpc

The University of Louisville's Rapid Prototyping Center (RPC) is a joint academic-industrial consortium providing access to capabilities in AM via laser and electron beam powder bed processes for metals, plastics, and ceramics. The RPC is equipped with software for solid modeling and part design of new components, in addition to reverse engineering of existing parts. The facility is capable of producing prototype parts and low-volume components using laser sintering, direct metal laser sintering, electron beam melting, ultrasonic consolidation, fused deposition modeling, 3D printing, and stereolithography.

6.3.5   Missouri University of Science and Technology

Website: http://www.3ders.org/articles/20130918-nasa-funds-research-to-create-stronger-materials-using-additive-manufacturing.html

Researchers at Missouri University of Science and Technology are running computer simulations of processes that could lead to stronger, more durable materials. The process involves the use of high-powered lasers to melt small particles of powdered materials as they exit a nozzle to create three-dimensional shapes, layer-by-layer. The additive approach applies to a broad range of manufacturing from the fabrication of large aircraft components to minuscule biomaterials used in surgical procedures. With hybrid manufacturing, researchers could apply an AM technique to create aircraft components from two different metals and then could smooth the parts' rough edges using automated computer-numerical control machining. The models will lead to a greater understanding of how layered materials bond to the surface on which they are deposited.

6.3.6 Pennsylvania State University

Website: http://www.cimp-3d.org/

The Pennsylvania State University along with its partners Battelle Memorial Institute and Sciaky Corporation operate the Center for Innovative Materials Processing through Direct Digital Deposition (CIMP-3D) for advancing and deploying AM technology for both metallic and advanced material systems. CIMP-3D seeks to provide technical assistance to industry and promote AM technology through training, education, and dissemination of information.

The CIMP-3D also includes government, industry, and academic partners having the common goal of advancing and implementing this technology within the industrial base. The additive manufacturing demonstration facility (MDF) includes several AM systems capable of full consolidation of polymeric, metallic, and ceramic material systems, as well as a state-of-the-art design studio and prototyping laboratory which includes a host of characterization techniques.

Current research programs include: the development of advanced materials and material architectures applicable to AM; development of new engineering design methodologies enabled by AM; the establishment of manufacturing models that allow rapid determination of process and economic metrics; the advancement of process simulation techniques that enable accurate representation of thermal history, stress state, and microstructural evolution required for establishing a foundational understanding of the process, supporting the realization of advanced sensing and control strategies; progression of cyber-enabled methods for distributed manufacturing and analysis; and development of machine and system designs for the next generation of AM.

CIMP-3D possesses extensive modeling, simulation, and analysis capabilities involving all aspects of the AM process (Figure 6.1). These capabilities are being utilized to develop an integrated, comprehensive, physics-based approach to describe and link important relationships that govern AM. Practical ramifications of this effort include the ability to optimize designs based on process, performance, and economic considerations; to advance process control techniques for improved reliability; and to influence resultant properties and characteristics for producing components and structures used for demanding applications.

6.3.7 University of Texas

Website: http://utwired.engr.utexas.edu/lff/

The Laboratory for Freeform Fabrication (LFF) was founded in 1988 at the University of Texas at Austin. The research group is active in diverse areas related to AM. They have several commercial selective laser sintering (SLS) stations as well as a number of research machines constructed on campus. The LFF is host to the Solid Freeform Fabrication Symposium, first held in 1990 and the longest continuously running annual meeting dealing with research in freeform fabrication.

Researchers in the LFF represent considerable depth and breadth, including process development, materials, applications, and modeling. Research includes major funding from national funding agencies as well as industrial projects of varying sizes and durations. An Industrial Affiliates Program provides special opportunities for industry to interact with the Lab.

The LFF is part of the Advanced Manufacturing Center, which was established in 2004 to initiate, support, and coordinate research and education in advanced manufacturing and materials processing, to disseminate the results of this research to potential users, and to promote and provide resources for education in this field.

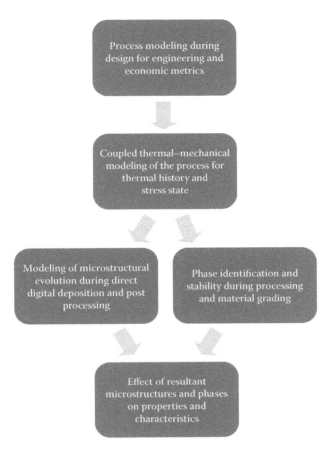

Figure 6.1 CIMP-3D modeling strategy for additive manufacturing. (From http://www.cimp-3d. org/)

6.3.8   America Makes Institute

Website: https://americamakes.us/

In March 2012, President Obama announced plans to revitalize the U.S. manufacturing base with the creation of the National Network for Manufacturing Innovation (NNMI). The NNMI will have as many as 15 institutes throughout the country. America Makes was founded in August 2012 as the pilot institute to serve as a prototype for subsequent NNMI institutes. Based in Youngstown, Ohio, America Makes is the National Additive Manufacturing Innovation Institute. Being structured as a public–private partnership with members from industry, academia, government, nongovernment agencies, the Institute is working to innovate and accelerate AM to increase the U.S. global manufacturing competitiveness.

The mission of the Institute is to help bridge the gap between basic research and mature product development, to provide shared assets to help companies to access latest technologies, and to educate and train students and workers in AM skills.

The Institute has developed a National Additive Roadmap which defines a path to advance 3D-printing technologies and help drive demand for 3D products. Projects related to the roadmap are proposed by member teams and funded by America Makes with matching funds from the team members. Full project results will be available to members of America Makes. Nonmembers will have access to report abstracts.

The following projects were awarded in March 2013 and are underway:

1. Maturation of fused deposition modelion (FDM) component manufacturing
2. Thermal imaging for process monitoring and control of AM
3. Sparse-build rapid tooling by FDM for composite manufacturing and hydroforming
4. Rapid qualification methods for powder bed direct metal AM processes
5. FDM for complex composites tooling
6. Qualification of AM processes for repurposing and rejuvenation of tooling
7. Maturation of high-temperature SLS

In January 2014, additional 15 projects were awarded in Project Call #2, with America Makes funding totaling about $4.5 million. With cost share from participating companies, total funding for these projects totaled more than $9 million.

6.3.9   *Oak Ridge National Laboratory*

Website: http://www.ornl.gov/science-discovery/advanced-materials/research-areas/ materials-synthesis-from-atoms-to-systems/additive-manufacturing

ORNL works with AM equipment manufacturers and end users. The research and development in this field are enabling a wealth of opportunities for product customization, improved performance, multifunctionality, and lower overall manufacturing costs. Through collaboration with every aspect of the manufacturing supply chain, ORNL identifies critical equipment and materials advancements required to establish AM as a mainstream manufacturing process. Research and development projects focus on advanced material development for both metal and polymer-based systems. ORNL is exploring the manufacture and integration of carbon nanofibers into thermoplastic materials for fused deposition modeling. They are also developing process conditions for lightweight and refractive metal alloys for electron beam powder bed deposition systems and are also using neutron characterization techniques to analyze geometric tolerances and mapping residual stress in complex components. These efforts are leading to novel, lightweight, and high-strength materials that have the potential to significantly impact a variety of application areas, including aerospace, automotive, biomedical, and nuclear.

Work is being conducted on AM technologies at the ORNL MDF sponsored by the U.S. Department of Energy's (DOE) Advanced Manufacturing Office. The MDF is focusing on R&D of metal and polymer AM pertaining to *in situ* process monitoring and closed-loop controls and implementation of advanced materials in AM technologies. ORNL recently completed the first step toward optimizing the final design and manufacture of a component part using CAD tools, FEA and simulations, and internally developed optimization software.

6.3.10   *Lawrence Livermore National Laboratory (LLNL)*

Website: https://acamm.llnl.gov/

LLNL is developing physics-based models that relate microstructure, properties, and process (including postprocessing) parameters to performance. The goal of the project is to provide accelerated certification for additively manufactured metals through computational tools and closed-loop process controls.

This is a well-integrated plan which addresses most of the features of AM modeling and simulation. However, the plan relies on the use of very fundamental physics-based models which take days or weeks to run in a highly parallelized multiprocessor environment. Such powerful computers and the necessary expertise to run these codes will not be available to a typical AM company. Surrogate models developed from the fundamental models are more likely to find immediate use at the AM companies. The fundamental models will likely find more widespread use many years later.

LLNL is developing an effective medium model that simulates the process at the scale of the part and a mesoscale model that simulates the process at the scale of the powder and feeds information to the effective medium model. This approach builds on LLNL models that have been developed for forming, rolling, and casting. The overall strategy is shown in Figure 6.2.

The effective medium model (Figure 6.3) predicts temperature, residual stress, and distortions during and after the build. It models melting, solidification, solid state phase transformations, and thermomechanical material behavior (ignoring fluid flow). Localized mesh smoothing and other techniques are used for computational efficiency. With LLNL's parallel codes and multiprocessor computer resources, it takes a few days for modeling a build of 20 layers. A limited amount of thermal and residual stress validation has been performed.

Microstructure will be predicted using phase field models, and properties will be predicted using crystal plasticity and dislocation dynamics. The incorporation of process optimization, data mining, and UQ will guide the AM process to yield optimized properties and performance. Results of simulations will be validated against measured material properties and data acquired from real-time *in situ* process monitors using design of experiments. Integrated in-process sensing, monitoring, and control technologies will be developed to ensure the end-processed material properties and component performance. The material being studied is 316L steel.

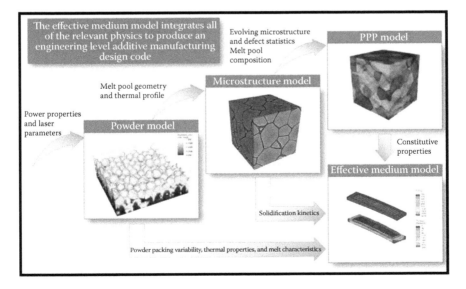

Figure 6.2 **(See color insert.)** LLNL modeling strategy for additive manufacturing. (From https://acamm.llnl.gov/)

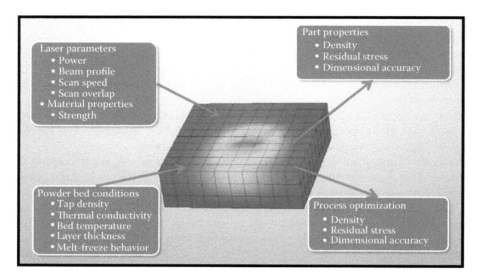

Figure 6.3 **(See color insert.)** LLNL-effective medium model. (From https://acamm.llnl.gov/)

The powder model will predict the melting of powder and its resulting densification and provide data to the effective medium model. Laser–material interaction is treated via ray tracing and a physics-based absorption model. It models melting of the powder, flow of the liquid, and behavior of trapped gas. This model treats the powder as a discrete system of particles and uses a Lattice Boltzmann approach to investigate melting and resolidification. It is computationally very intensive.

The process–property–performance model will be used to compute the constitutive properties for input to the effective medium model and to develop a connection among process, properties, and observed performance of the material.

Data mining and UQ will be used to gain insights into the data produced by experiments and simulations and to understand how uncertainties in inputs influence the output. Surrogate models will be constructed for simpler analyses. Sensitivity analyses will be used to determine which parameters contribute most to the output uncertainties. Calibration will use experimental data to find the best parameter values.

After enough simulations and experimental data are generated, probabilistic techniques can be used to incorporate uncertainties into the analysis.

6.3.11   Granta Data Base

Granta Design has announced the details of the company's latest advances in simulation and materials information management to help additive manufacturers improve research, design, testing, and simulation. These developments are the result of Granta's involvement in two European Framework Seven projects (NANOMICRO and AMAZE). This is an area which requires effective management of large quantities of materials information. Granta technology can help to capture complex processing histories and other data and has been validated through these projects.

Granta is also a partner in the ongoing project additive manufacturing aiming towards zero waste & efficient production of high-tech metal products (AMAZE). Led by the European Space Agency, this 30 partner project seeks to rapidly produce large defect-free additively manufactured metallic components up to 2 meters in size for use in aeronautics, space, nuclear fusion, automotive, and tooling. Granta is helping project partners manage the materials processing and test information for analysis and simulation tailored uniquely for AM, as AMAZE seeks to achieve 50% reduction in power, consumables, raw materials, machining, and cost for finished parts, compared to traditional processing.

6.3.12   Others

A lot of AM research personnel in Europe are actively working on developing new technologies, on improving performance, and on modeling and design tools. Joint multiyear and multimillion dollar efforts are being conducted between the Industry, Government, and Universities and between countries to transit the technology to manufacturing parts. Rolls-Royce Aero Engines is collaborating with the UK universities in AM research. Some of the major activities are at

- Fraunhofer Institute, Aachen, Germany.
- University of Freiburg, Germany.
- University of Loughborough, United Kingdom.
- Manchester Institute of Science and Technology, United Kingdom.
- University of Liverpool, United Kingdom.
- University of Nottingham, United Kingdom.
- University of Leeds, United Kingdom.

In China, Tsinghua University and Huazhong University of Science and Technology have large research groups actively involved in various aspects of AM research.

6.4   Summary

AM equipment use 3D models constructed in CAD software to understand how to create an object. In this chapter, a description of existing commercial software tools was first provided. The various commercial tools were compared for the features relevant to the modeling of AM. There are areas where commercial software needs improvements to make them excellent tools for AM. In the second portion of this chapter, the projects of different universities and national laboratories in the area of AM modeling software are reported. Together, this chapter represents the state of the art in AM modeling software and future directions needed to advance the field.

chapter seven

Additive manufacturing research and development needs*

Shesh Srivatsa

Contents

* "Additive Manufacturing (AM) Design and Simulation Tools Study" April 2014, written by Shesh Srivatsa under contract to Air Force Research Laboratory Materials and Manufacturing Directorate, Wright-Patterson Air Force Base, OH 45433, Air Force Material Command, United States Air Force (Case Number: 88ABW-2014-3753). Approved for public release; distribution is unlimited.

In this section, additive manufacturing (AM) research and development needs are described. We also provide recommendations for filling the gaps in AM modeling and simulation.

AM uses modeling and simulation to understand the process and detect any problems that may arise. Control of these processes helps to achieve desired mechanical properties and other important variables. Properties such as temperature and deposition layer thickness are needed to help guarantee a robust process. However, there are some models that need to be developed or improved upon before being beneficial to the AM industry.

Details of the required modeling and simulation research and development needs are provided in the following sections:

- Process design
- Sensors and controls
- Model implementation
- Postprocessing
- Input material
- Additional model features
- Other needs (cannot be directly addressed by modeling and simulation)

7.1   Summary of AM modeling and simulation literature survey

A summary of the state-of-the-art in AM modeling and simulation is presented in this section. The summary of the literature survey is limited to the modeling of the powder bed fusion process for metals which is the most widely used process for aerospace components.

The summary is divided into the following sections:

- Thermomechanical process models
- Microstructure models
- Structure-property models
- Other models
- Material property data for modeling
- Model validation

7.1.1   Thermomechanical process models

Control of AM processes is important to achieve the desired microstructure, mechanical properties, residual stresses, distortions, and other important output variables. Modeling of these processes is an important activity for achieving any degree of control and optimization. Such models can help achieve a better understanding of the process and the problems which may arise.

Analytical and finite element models for the prediction of temperature, powder flow, residual stress, and deposition geometry have been developed. These models have aided the understanding of AM processes in terms of the effect of various process parameters on outputs of interest (thermal and stress responses). These models have used either in-house codes or commercial codes like ABAQUS, COMSOL, and ANSYS (FLUENT). Some are university codes which have good educational value in training students but are of limited industrial value. Most of these models involve a number of simplifying assumptions and investigate only thermal or some other features of the process. Many of them lack user-friendly pre and postprocessors which make the model difficult to use. An overall integrated model is not available. Existing models are claimed by some to be extendable in principle to a real part geometry and process, but that is a big leap of faith.

There is an abundance of literature describing models for various aspects of welding including weld pool and weld bead geometries. There is a commonality in the melt flow between AM and welding. Modeling techniques for modeling weld melt pools can be adapted to AM melt pools.

7.1.2   Microstructure models

A 3D dislocation density based-thermomechanical finite element framework has been developed. Crystal plasticity methods have been used to provide details of the crystal structure. Microscopic cellular automata (CA) methods have been coupled with finite element and macroscopic thermodynamic models and used to predict the dendritic grain size, structure, and morphological evolution during the solidification phase of the deposition process. Finite element heat transfer calculations have been coupled with transformation kinetic theory to predict the microstructure and properties.

Such calculations, while interesting, are computationally intractable for real-life problems. They also have a number of material data inputs, which are difficult to measure and have to be calibrated. Microstructural and property models are still in the research and development stage even for more mature conventional manufacturing processes such as casting and forging, which have a simpler thermomechanical processing history than AM processes. These models have to mature first for the conventional processes before their widespread use for AM can be considered. These models have long-term potential. Research should continue in the development of these models, speeding up the computations, and bringing them to a more production ready stage.

7.1.3   Structure-property models

Statistically based design of experiments (DOE) and multiple regression analysis methods have been applied to quantitatively establish relationships between common process parameters and resulting outputs such as build-up height, thickness, and surface roughness.

Temperature gradient, G, and solidification rate, R, are the two most important parameters controlling the solidification microstructure. Microstructure and defect generation models are similar to those in casting modeling—lack of fusion, shrinkage, porosity, and cracking. Numerous semianalytical and theoretical relationships have been proposed to describe the dependence of mechanical properties on porosity.

The process map approach defines the region of feasible operating parameters—via modeling and experimentation. Models were used to obtain processing maps relating the scanning speed, idle time between the deposition of consecutive layers and substrate temperature to the microstructure, hardness, and Young's modulus in parts produced by laser powder deposition.

7.1.4   Other models

Models have been developed for the spreading and compaction of powder by a rotating roller in the powder bed fusion process. The pressure from the roller during the leveling process results in some densification of the powder. Friction from the roller can result in several line traces instead of a flat surface. If these forces are large enough, they can cause excessive part movement after which the build process has to be stopped. The effects of friction, layer thickness, roller diameter, and initial density on the powder bed's relative density were investigated using the model.

Surface finish cannot be directly predicted by finite element process models.

Effective cost models are needed which include all factors contributing to cost of AM and exploit the unique features of AM.

7.1.5   Material property data for modeling

Various methods are available for the measurement of high-temperature material properties needed for modeling. Since the measurement of these properties is expensive and time consuming, a modeling sensitivity analysis is recommended to establish the importance of the various properties and the level of accuracy to which they need to be measured for any particular application. Many commercial software packages (COMSOL, ANSYS, etc.) have the ability to automatically set up sensitivity runs by varying the selected input parameters over specified uncertainty ranges and evaluating the effect on outputs of interest. Estimation methods such as Thermocalc and JMATPRO can be used to obtain reasonable estimates of various properties. Inverse methods can also be used to calibrate the required properties and boundary conditions.

Table 7.1 Expanded comparison of commercial software packages (COMSOL, ANSYS, etc.)

Software	Advantages	Improvements needed
ANSYS	• Standard material constitutive equations • Can setup sensitivity analysis, parametric study runs • Widely used by the aircraft engine design community—can facilitate interaction between OEM designers and AM companies	• Efficient mesh generation for extremely complex • 3D geometries and ~1000 deposited layers • Adaptive fine mesh in heat affected zone • Library of templates for lattice structures • Has most of the physics in a suite of codes • Weakly coupled physics: data passed between different codes, e.g., stress—fluid code • Cannot model post-processing HIP or CIP • Integration of different steps in AM difficult since data has to be passed between different codes each using its own mesh and solution procedure • Adding new equations and boundary conditions can involve some effort • Incorporating different material constitutive models can involve some effort • Incorporation of special-purpose user features can involve some effort • Licensing cost high (~$40K/year) • Estimate ~5 years of effort to get integrated AM package
COMSOL	• Easy integration of all AM steps—all calculations done in one tightly coupled code • Open architecture enables user to add new equations and boundary conditions • Different material types: already present or can be easily added: elastic-plastic at room temperature to pure viscous behavior above melting point • Has most of the physics in a single code • Tightly coupled physics in one single code • Can build models with different levels of fidelity to address run time issues • Can setup sensitivity analysis, parametric study runs	• Efficient mesh generation for extremely complex 3D geometries and ~1000 deposited layers • Adaptive fine mesh in heat affected zone • Library of templates for lattice structures • Not as widely used as DEFORM or ANSYS for process modeling or design work in the aerospace industry

(*Continued*)

Table 7.1 (Continued) Expanded comparison of commercial software packages (COMSOL, ANSYS, etc.)

Software	Advantages	Improvements needed
	• Incorporation of special-purpose user features straight forward • Licensing cost low (~$10K/year) • An easy-to-use customized code for AM can be constructed in ~2 years with low licensing cost; this has been done for CMCs	
DEFORM	• Easy integration of all AM steps—all calculations done in one tightly coupled code (except no fluid flow) • New boundary conditions and material data can be added through user subroutines • Preliminary framework available for sensitivity analysis and parametric study; will need additional work for AM • Special-purpose user features can be input through user subroutine • Licensing cost medium (~$20K.year) ~3 years effort for a commercially ready package (without fluid flow)	• Efficient mesh generation for extremely complex 3D geometries and ~1000 deposited layers • Adaptive fine mesh in heat affected zone • Activate/deactivate elements as needed • Library of templates for lattice structures • Fluid flow related features not available; • Otherwise tightly coupled physics in one code
ESI	• Whole computational domain is pre-meshed and elements activated as time progresses • Licensing Cost High (>$40K/year) • 3–5 years for a package like PROCAST	• Efficient mesh generation for extremely complex 3D geometries and ~1000 deposited layers • Adaptive fine mesh in heat affected zone • Library of templates for lattice structures • Has most of the physics in a suite of codes • Weakly coupled physics: data passed between different codes, e.g., stress—fluid code • Cannot model post-processing HIP or CIP • Integration of different steps in AM difficult since data has to be passed between different codes each using its own mesh and solution procedure • Limited flexibility with boundary conditions and material data; Node searching with moving heat source is computationally costly • Sensitivity Analysis, Parametric Study—not automated, through text files • User features, code architecture—limited user control

7.1.6   Model validation

Microstructure and mechanical properties (strength and fatigue) have been measured with various AM processes, various processing parameters, and various materials (steels, Ti, and Ni alloys). Experimental DOEs have been conducted to evaluate the influence of processing parameters on microstructure and mechanical properties. There are many publications with a lot of experimental data which can be used in the future for validation of models. However, little data exist on fatigue.

7.2   Required AM research and development

During the course of this program, the team visited or talked with several AM companies, universities, national laboratories, and commercial software developers to get their inputs on how to develop a model which would be beneficial to the AM industry. The team looked at what are the most important things for a model to predict and to what accuracy. How should the model development activity be planned in order to ensure that it benefits the industry and gets implemented in the industry? What is the industry looking for from modeling and simulation?

Based on interactions with various AM practitioners, the needs which can be addressed through modeling and simulation are listed below (roughly in order of importance) and described in the following sections:

Process design

- Reduce residual stresses and subsequent distortions
- Reduce support structures
- Reduce material envelopes
- Reduce wall thickness and feature size which can be manufactured

Sensors and controls

- Devise closed-loop feedback control.
- Ensure a robust process (repeatable parts).
- Efficient models are needed for real time monitoring and control.

Model development

- Select right software for meeting needs
- Perform model verification, calibration, and validation
- Perform sensitivity analysis, parametric studies, and DOE
- Understand effects of variability in process parameters and inputs
- Identify, optimize, and control the critical factors that influence process output
- Generate material property database—for model input and for design data

Postprocessing

- Reduce postprocessing (HIP, stress relief, machining)
- Improve surface finish (especially with electron beam [EB])

Input material

- Understand effects of input feedstock powder characteristics on the process parameters and the final product

Additional model features

- Predict defect formation: porosity, lack of fusion, balling, and so on
- Predict microstructure and mechanical properties in the product
- Improve energy efficiency

Other needs (which cannot be directly addressed by modeling and simulation)

- Form consortia for precompetitive collaborative efforts
- Improve computer-aided design (CAD) systems and geometry representation and interface with process models
- Develop standards and design guides and tools: exploitation of unique features of AM
- Develop process and material standards to control part-to-part consistency
- Develop sensors to monitor the process and provide input to control system
- Institute statistical quality control procedures
- Develop nondestructive evaluation (NDE) and inspection techniques for AM
- Standardize paths to rapid certification and qualification
- Develop effective cost models
- Build bigger, faster, and more capable AM machines
- Reduce input powder material cost
- Develop process for newer materials
- Develop education programs: acceptance of AM by management and corporations

7.2.1   Process design

The needs for process design are as follows:

- Reduce residual stresses and subsequent distortions
- Reduce support structures
- Reduce material envelopes
- Reduce wall thickness and feature size which can be manufactured

Complete information on the thermal behavior at each location of the part becomes essential in order to understand the mechanisms of the microstructure formation and the mechanical properties in the finished part. Despite considerable progress to date, several key aspects of the process are still unclear, such as the temperature profile in the molten pool, the molten pool geometry for each layer as a function of the process parameters, and the overall spatial and temporal variations of the temperature field in the part. A careful calibration and validation of the thermal model are required in order to obtain reliable numerical predictions. The calibrated thermal model can then be coupled to a metallurgical model to predict the microstructure or to a mechanical model to predict the residual stress in the fabricated part.

The melt pool shape affects the local temperature and residual stresses. However, further away, the thermal and mechanical behaviors are governed by the overall shape and extent of the deposition process. Thus, the result obtained using a simplification for deposit cross section is generally adequate.

Support structures can be reduced with a good thermomechanical model which can predict distortions during and after the build. Modeling trial and error is needed to see the influence of any given set of supports and then selectively eliminate or modify the supports to arrive at an acceptable level of distortion.

Material envelopes can be reduced again with a good thermomechanical model which can predict distortions after final machining. The extra envelope is there to ensure that the built part will clean up and machine to the final required dimensions and also to account for part to part variability. To be on the safer side, more material envelope than is necessary is added on. With a good predictive model, the envelope can be reduced while still ensuring final parts at the right dimensions.

Wall thickness and feature size can again be reduced with a good thermal and stress model. The temperature history in thin wall sections will strongly influence the resulting microstructure and properties. The stress history will determine the resulting distortions or the potential for cracking of thin sections.

7.2.2   *Sensors and controls*

The needs for sensors and controls are as follows:

- Devise closed-loop feedback control.
- Ensure a robust process (repeatable parts).
- Efficient models are needed for real-time monitoring and control.

Accurate process variable data offer detailed information to understand the process physics, to monitor the process characteristics and performance, and the part quality and consistency. Process variable measurements also serve the need to validate the process modeling/simulations. Measurements in metal-based AM is very challenging because of high temperatures, temperature gradients, complex material states (solid/liquid), and limited access to the process chamber (especially vacuum chambers for EB processes).

New sensors need to be developed or enhanced to provide real-time feedback on the build environment to ensure consistent part quality. Current sensor technologies including infrared imagers and high-speed cameras need to be enhanced in terms of speed and spatial resolution of measurement for AM applications.

A possible control strategy includes using optical sensors which record layer-by-layer information, such as temperatures, geometry, distortion, and porosity, and feeding this information to a control system which makes the needed adjustments to beam parameters for the next layer. Process maps coupled with real-time thermal imaging of the melt pool can be the basis for an automated feedback control system for the deposition process. Until control systems can act rapidly (within the timescale for one layer buildup), the measured data can be used only to monitor the process and not to provide on-the-fly feedback.

Some publications mention using process models for process control. Since detailed finite element process models for any realistic shaped part take hours to run, using such a model for process control is not feasible. Only simple fast-acting process models can be used for this purpose. As long as the simple model can predict the right trends in the output variables, the closed-loop feedback control system can take care of the approximations in the model.

In order to ensure a robust process with repeatable parts, the needs for measurements and control are

- Temperature
- Melt pool size and shape
- Deposition layer thickness
- Surface roughness with micron level accuracy
- Composition of the deposited layer
- Detection of defects such as cracks, porosity, underfill, residual stress, and distortion
- Detecting and predicting final geometry relative to CAD intent
- Nondestructive detection and measurement of anisotropic properties
- Monitoring and controlling grain size and direction

7.2.3   Model development

Model development steps are described in the following sections:

- Select right software for meeting needs
- Perform model verification, validation (V & V), and calibration
- Perform sensitivity analysis, parametric studies, and DOE
- Understand effects of variability in process parameters and inputs
- Identify, optimize, and control the critical factors that influence process output
- Generate material property database—for model input and for design data

7.2.4   Software selection

Table 7.2 shows a modeling and simulation gap analysis for powder bed fusion processes. It shows the current state and a desired future state for various features:

- Geometry-related features
  - Dimensionality
  - Geometry
  - Number of scans
  - Mesh
- AM physics
- Postprocessing steps
- Integrated model of full process
- Boundary conditions
- Material data
- Defect formation
- Model verification, validation, and calibration
- Sensitivity analysis and parametric study
- Software
- Computer run times
- Neural network models

Table 7.2 Modeling and simulation gap analysis (powder bed fusion processes)

Feature	Current state	Desired state
Dimensionality	1D, 2D, and 3D (with highly simplified geometries)	3D (with realistic geometries)
Geometry	Simple block type geometries	Integration with CAD packages for realistic geometries
Number of scans	• Mostly single scans • Maximum 20 scans due to computational and model size limitations	~1000 scans for actual components
Mesh	Limited use of mesh optimization to reduce overall model size and computer run times	• Efficient mesh generation for extremely complex 3D geometries and ~1000 deposited layers • Adaptive fine mesh in heat affected zone • Activate/deactivate elements as needed • Library of templates for lattice structures
Additive manufacturing physics	• Partial combination of various physical features • Full model with all physics does not exist • For example, some models ignore convection in the melt pool; others ignore solidification or re-melting effects, powder porosity, etc.	Coupled multi-physics solutions incorporating: • Melting of powder • Powder bed absorptivity for laser and EB = f(powder characteristics) • Distribution of energy absorbed over the layer thickness • Moving heat source with Gaussian (or other) distribution • Solidification • Solidification shrinkage and latent heat • Re-melting of solid material • Thermal and compositional Marangoni flow • Thermal and compositional buoyancy • Capillary forces • Melt pool free surface tracking • Flow in mushy regime • Turbulent flow in melt pool • Transient behavior at melt pool free edge (related to balling) • Radiation heat transfer • Light element (e.g., Al, Cr) evaporation • Phase transformations (transformational plasticity) • Heat transfer through porous powder • Material addition in layers • Etc.
AM post-processing steps	• Limited work in cold and hot isostatic pressing • Limited work in stress relief	Thermal-stress-transformation coupled models for cold and hot isostatic pressing Thermal-stress-transformation-creep coupled models for stress relief

(Continued)

Table 7.2 (Continued) Modeling and simulation gap analysis (powder bed fusion processes)

Feature	Current state	Desired state
Integrated model of full process	• Individual steps have been modeled • Integrated model linking all processing steps does not exist	• Material deposition • Fluid Flow • Heat transfer • Residual Stress • Distortion • Microstructure • Properties • Etc.
Boundary conditions	• Several simplifications • Not tied to a CAD system	• Realistic heat transfer boundary conditions at deposition face • Moving heat source linked to CAD geometry • Realistic stress boundary conditions including base plate and supports
Material data	• Variation of properties with temperature, porosity and/or composition often ignored • Little description of high temperature data used	• Temperature, porosity and composition dependent properties • Elasto-plastic at room temperature to pure viscous behavior above melting point • Isotropic or kinematic hardening effects • Non-isotropic properties • Metallurgical phase changes (transformational plasticity) • Creep data for post-processing steps (CIP, HIP, stress relief)
Defect formation	Attempts made to correlate process conditions to defect formation	• Melt pool instability—ball formation • Effect of surface tension forces at the melt pool surface on adjacent unmolten powder particles and on surface roughness • Residual stress and distortion predictions
Model verification, validation, and calibration	• Lack of validated physics-based predictive models • Validation data: Temperature measurement; limited residual stress measurement • Model input calibration with limited experimental data	• Standardized procedures for verification, validation, and calibration • Well defined verification and validation plan • Benchmark cases • Calibration procedure for input data difficult to measure
Sensitivity analysis, parametric study, uncertainty quantification	Limited work	• Automated procedure for setting up and running sensitivity analyses and parametric studies • Automated procedure for extracting outputs of interest • Statistical analysis of modeling results to identify critical variables • Established procedure for uncertainty quantification (many inputs will be known only approximately)

(Continued)

Table 7.2 (Continued) Modeling and simulation gap analysis (powder bed fusion processes)

Feature	Current state	Desired state
Software	• Commercial codes (ANSYS, COMSOL, ABAQUS, SYSWELD, etc.) • University in-house codes	Full integrated package based on supported commercial code: • Software (solver) • Model pre-processor: user-friendly problem setup • Input data • Model post-processor: user-friendly output interpretation • Validation of outputs • Measurement of inputs and outputs
Computer run times	Present software would take days or weeks to model an actual part, so this has not been attempted	• One-day turnaround • Judicious simplifications appropriate to the specific objectives of a particular study • GPU-based computer platforms (~$5000)
Software architecture	Some features available in some commercial codes	• Easy incorporation of special-purpose user features: new equations, boundary conditions, and material behavior • Different fidelity models by activating—deactivating select features appropriate to specific objectives of a particular study
Neural network models	Limited work	• Simple models for quick results • Relate microstructure and properties to process variables • Models get better as more data is obtained

7.2.5   Sensitivity analysis, parametric studies, DOE

No part always performs as intended since there are inherent material and process variations that affect the part properties. Conventional simulations assume all inputs are known and maximize the desired performance characteristics at a single design point. This ignores the fact that parts are never completely produced at nominal geometry or process conditions. To ensure product robustness, one must progress from a single design point to exploring a multitude of design points. Simulations must be performed in a parametric way to identify the best possible overall product design by considering sources of uncertainty and variation. Robust design enables the prediction and control of part performance even with many variations and uncertainties. Quantifying and controlling variation, uncertainty, and risk are critical to improving performance, part yield, and quality.

A sensitivity analysis is needed to account for various sources of uncertainty inherent in materials' behavior, manufacturing processes, models, and so on, to arrive at a robust control strategy to ensure minimal variability in the component characteristics. Using simulation to understand how products will actually perform over a broad range of product geometries, boundary conditions, and materials' types, design uncertainty, and the risk of failure can be reduced. Sensitivity analysis identifies, optimizes, and controls the parameters which contribute most to the output uncertainties. One can then focus on the key factors that affect product performance and reliability and ignore others that are unimportant and thus manage the key sources of variation. Off-design conditions can also be assessed effectively.

This is called reliability-driven product development, robust design, design for variation, or design for six sigma.

As a follow-on step, the input parameters can be set to *worst-case* values to determine product and process performance under the worst possible conditions.

Many commercial software packages (COMSOL, ANSYS, etc.) provide the facility to automatically set up and run sensitivity analyses and parametric studies. DOE techniques should be used to statistically define the modeling runs. Techniques like the Latin Hypercube method should be used to minimize the number of modeling runs needed to complete a sensitivity analysis. Key output parameters are then extracted, and a statistical analysis identifies the critical variables.

7.2.6   Material property data base

There are two parts: One is material property data needed as modeling inputs, and the other is material property data needed for product design.

The properties of many aerospace alloys are very sensitive to the production methods used; so the use of alternative methods of manufacture has to be carefully considered, and the property–process relationship is understood.

For manufacturing, it is essential to determine the full set of mechanical, thermal, and other properties. It will also be important to measure property shifts. The data will be useful in finite element analysis. Generally, the filled materials are stronger in the x–y direction than they are in the z direction. This needs to be addressed in the design of the parts.

Generating design allowables—A database of the mechanical properties of materials produced by AM must be established. The AM parts do not necessarily have to match or exceed the mechanical properties of parts made by traditional manufacturing because as long as the mechanical properties are known and are reproducible, the part may still be engineered to specification. The digitally manufactured component requires statistically robust property levels that meet design requirements for a broad array of components. Quantify the base product mechanical properties, including differences due to orientation, section thickness, and surface finish. The effects of normal parameter variation will need to be characterized.

Laser forming is stable and repeatable enough to generate a design allowable with the parameters allowed by the fixed process. Generation of general allowables for the full range of direct metal deposition processes is still needed. The potential that different processes will yield consistent properties is by no means certain. It is recommended that an allowables' program can be embarked upon. The idea is that by having a large number of heats, processes, and heat treatments, there is sufficient testing to generate an allowable that is representative of the majority of the processes. This matrix can be expanded for additional processes. The use of common parts provides the ability to compare and contrast different processes.

Once the part is produced with a known set of parameters, the challenge is to define the mechanical and metallurgical properties to be measured and ensure part-to-part and machine-to-machine consistency. As parts produced by AM are anisotropic and vary by fabrication method, careful consideration must be made in measuring and reporting mechanical property values. Documented databases of mechanical properties and the establishment of industry specifications and standards for parts produced by AM are crucial.

7.2.7   Postprocessing steps

Parts produced by AM often require post processing to improve surface finish and mechanical properties. In metals, the dimensional precision of additive processes is not yet sufficient to produce a part that can meet tight tolerances without further processing. Finishing operations include cleaning, stress relief and aging, hot isostatic pressing (HIP), and final machining.

Cleaning and finishing parts can be a challenge. To make AM work for manufacturing, streamlined methods must be developed for removing excess material and finishing the parts. Removing support material from holes, slots, and trapped areas normally requires expensive labor, so these material removal problems must be resolved. Hand sanding is also time consuming and expensive and could become a *show stopper* if the task cannot be automated. In some cases, the parts may require strengthening using one method or another, such as infiltration. Parts may also require paint or one or more clear coats.

Stress relief and aging are needed to relieve build residual stresses and to generate the required microstructures and mechanical properties. Control of these process parameters is needed to ensure parts with consistent mechanical properties and finish dimensions. Due to the highly nonequilibrium deposition processes typical of AM systems, there is a great deal of work to be done with regards to understanding microstructural evolution and what constitutes optimal thermal postprocessing for the majority of material systems typically used.

Metal-based AM systems typically melt the metal particles with the goal of achieving 100% dense parts. Some systems achieve this goal, whereas others reach only 99+% density due to the presence of small random voids. Some of these voids are inherent in the powdered raw material. Because AM is similar to a fusion welding process, it also suffers from some of the defects of fusion welding, namely lack of fusion which can result in porosity. One process for eliminating internal discontinuities, HIP, uses high temperatures and pressures to heal discontinuities in titanium and other alloys. HIP can heal porosity, but entrapped Ar gas in a lack of fusion region can prevent healing. This step adds cost and time in the manufacture of AM parts.

The near-net geometry and fixturing of AM parts typically make the part more flexible than conventionally made parts. Therefore, when machining these parts, care must be taken to avoid distortion and chatter of the parts, which can result in poor surface finish and dimensional variation and also reduced tool life. The complex geometry and distortion of the parts also complicate fixturing and alignment. All these factors can increase machining cost and time and also impact part reproducibility.

7.2.8   Input material

Incoming feedstock powder properties must be tightly controlled. Particle size distribution has a significant impact on the success of AM processes. These controls on powder properties also limit the amount of unconsolidated powder which can be recycled. This adds to material cost and makes AM less attractive. The effects of input feedstock powder characteristics on the process parameters and the final product need to be understood and quantified.

7.2.9   Additional model features

- Predict defect formation: porosity, lack of fusion, balling, and so on
- Predict microstructure and mechanical properties in the product
- Improve energy efficiency

7.2.10   Other needs (besides modeling and simulation)

The following is a list of AM research and development needs which cannot be directly addressed by modeling and simulation:

- Form consortia for precompetitive collaborative efforts
- Improve CAD systems and geometry representation and interface with process models
- Develop standards and design guides and tools: exploitation of unique features of AM
- Develop process and material standards to control part-to-part consistency
- Develop sensors to monitor the process and provide input to control system
- *In situ* measurement of composition, temperature, and distortion
- Institute statistical quality control procedures
- Develop NDE and inspection techniques for AM
- Standardize paths to certification and qualification
- Develop effective cost models
- Build bigger, faster, and more capable AM machines
- Reduce input powder material cost
- Develop process for newer materials
- Develop education programs: acceptance of AM by management and corporations

7.2.11   Consortia and collaboration

Noncompetitive collaboration has played an important role in the development of many industries. Types of collaboration include user groups, online forums, industry roadmaps, industrial consortia, conferences, and workshops. Collaborations also occur among educational entities and working groups dedicated to establishing industry standards and educational curricula. These have provided a forum for the dissemination of best practices and new applications.

National Consortia are expected to bring together industry, universities, community colleges, federal agencies, and states to accelerate innovation. This model has been successfully implemented in other countries. Such consortia can

- Bridge the gap between basic research and product development.
- Provide shared assets to help small manufacturers to access cutting-edge capabilities and equipment.
- Create an environment to educate and train students and the workforce in advanced manufacturing skills.

An industry-led, government-enabled, university, and national lab-supported initiative will help AM flourish in the United States. For example, in the semiconductor industry, semiconductor manufacturing technology (SEMATECH) is a consortium that performs research and development to advance chip manufacturing. It is funded by member dues and members including chipmakers, equipment and material suppliers, universities, research institutes, and government partners. The group solves common manufacturing problems and enables competitiveness for the U.S. semiconductor industry. SEMATECH was funded over five years by defense advanced research projects agency (DARPA) for a total of $500 million. It became a self-sustaining system driven principally by private resources in less than a decade.

AM can benefit in the same way by adopting a SEMATECH-like model. It would use government funding to get companies to work together with universities and national laboratories on helping produce and implement AM technology. The America Makes Institute in Youngstown, Ohio, received a $30 million federal grant and will connect manufacturers with universities and government departments to accelerate innovation in key areas of high-tech manufacturing. Additional funding from National Science Foundation (NSF), National Institute of Standards and Technology (NIST), DARPA, and United States Air Force (USAF) can accelerate this process.

Boeing, Electro Optical Systems (EOS), Evonik Industries, MCP HEK Tooling, and the University of Paderborn in Germany formed the Direct Manufacturing Research Center (DMRC) in 2008, for the development of AM processes and systems. EOS is a leading manufacturer of laser-sintering systems. Evonik Industries produces polymer-based materials as well as material solutions tailored for AM. MCP HEK Tooling is a supplier to the aircraft and automotive industries. The DMRC, located at Paderborn, builds on the expertise of the industrial partners ranging from aerospace, material production, and equipment manufacturing and on the research capabilities of the University of Paderborn. Initial research focused on improvement of the processes for laser sintering/melting technology for metal and plastic powder, and industry requirements for materials, training, and standards development. The DMRC was funded by the member companies and by the local governments. America Makes can form a good starting point for such collaborative efforts. Most of their current programs do not involve modeling and simulation. However, given enough member interest, a valid business case, and a viable implementation plan, modeling, and simulation programs can be proposed and funded.

A robust supply chain is also needed to support this industry. Such a supply chain would comprise a diverse group of technologists including AM machine original equipment manufacturers (OEMs), metallurgists, manufacturing engineers, nondestructive testing experts, and specialized software developers.

7.2.12 Standards and design guides

AM being a new technology does not yet have adequate design practices and standards. Industry standards are becoming increasingly important as AM is applied to the production of final products. Design guides must be developed for AM. The standards need to include areas such as material input, preparation, processing, postprocessing, and machine qualification to ensure part-to-part consistency. Standards for modeling and measurement procedures, material data, and boundary condition inputs need to be prepared.

Industry standards published by American Society for Testing and Materials (ASTM) as of 2013 are as follows:

- ASTM F2792-12a "Standard Terminology for Additive Manufacturing Technologies" standardizes terminology, including process definitions and other terms associated with the industry.
- ASTM F2921-11e2 "Standard Terminology for Additive Manufacturing—Coordinate Systems and Test Methodologies" defines terminology for coordinate systems and test methods.

- ASTM F2915-12 "Specification for Additive Manufacturing File Format (AMF)" defines the AMF, which serves as an alternative to the stereolithography (ST) file format. The AMF supports units, color, textures, curved triangles, lattice structures, and functionally graded materials.
- ASTM F2924-12a "Standard Specification for Additive Manufacturing Ti-6Al-4V with Powder Bed Fusion" helps producers and purchasers of Ti-6Al-4V parts made using the powder bed fusion process to define requirements and ensure consistent part properties.
- ASTM F3001-13 "Standard Specification for Additive Manufacturing Ti-6Al-4V extra low interstitial (ELI) with Powder Bed Fusion" helps producers and purchasers of Ti-6Al-4V ELI parts made using the powder bed fusion process to define requirements and ensure consistent part properties.

Other standards (issued or in preparation):

- AMS 4999-DED EBM, Ti-6Al-4V
- MIL-STD-3049-DED Repair
- WK Working Standards in process at ASTM:
  - WK 30107 Reporting Test Data
  - WK 40606 Powder testing
  - WK 27752 PBF Polymers
  - WK 33833 PBF CoCr
  - WK 37658 PBF IN625
  - WK 33776 PBF IN718

In late 2011, the International Organization for Standardization (ISO) and ASTM International signed an agreement to increase cooperation in developing international standards for AM. The Partner Standards Development Organization (PSDO) cooperation agreement provides opportunities for the two organizations to adopt and jointly develop international standards that serve the global marketplace. This will benefit both AM committees: ISO Technical Committee (TC) 261, and ASTM committee F42.

7.2.13   Summary

So far we presented a summary of the state-of-the-art in AM modeling and simulation. The comparison between the current state of powder bed fusion processes and the desired future state identifies the gaps and the required research and development needed to mature models that would help the AM industry. Details of the modeling and simulation needs were provided for process design, sensors and controls, model implementation, postprocessing steps, input material, and additional model features. Based on these needs, future plans and recommendations will be presented.

7.3   Future plan and recommendations

7.3.1   Technical need for modeling and simulation

The technology gaps and barriers to widespread implementation of AM were discussed in the preceding sections. The top business gaps are funding, education, limited supplier base, cost modeling, and a change of designer and management mindset. The top technical gaps are process control and robustness, NDE practices, material property database, surface finish, control of distortions and defects, and machine size, speed, cost, and efficiency. In this section, the technical need for modeling and simulation of AM processes is presented.

The development and implementation of new materials and manufacturing processes for aerospace application are often hindered by the high cost and long time span associated with current qualification procedures. The data requirements necessary for material and process qualification are extensive and often require millions of dollars and multiple years to complete. This burden is a serious impediment to the pursuit of revolutionary new materials and more affordable processing methods for aerospace components. The application of integrated computational materials engineering (ICME) methods to this problem can help to reduce the barriers to rapid insertion of new materials and processes. By establishing predictive capability for the development of process parameters, microstructural features and mechanical properties, a streamlined approach to qualification is possible.

Recent advances in ICME and manufacturing process models, use of probabilistic methods and uncertainty quantification (UQ) offer methods to improve quality and affordability and enable accelerated maturation of new manufacturing processes and technology. Rapid maturation and qualification would enable more effective transition of technology into production use for new systems, would enable faster, more capable response to design changes, and would reduce risk of insertion of new materials and manufacturing processes.

Significant benefits in manufacturing processes have been realized by modeling tools (e.g., DEFORM™ for metal forming and PROCAST™ for casting) which are in routine industrial use today. It is expected that similar benefits will derive from the AM models.

Currently, limited AM modeling work has been performed with several simplifying assumptions. These models illustrate trends but do not provide sufficient predictive capability for process design or the prediction of major defects, microstructures, and mechanical properties. Also the methodologies vary, lack standardization, and are not sufficiently validated to support their quantitative application. The program described here fills these gaps and addresses model development and validation, defect prediction, parametric studies to identify critical variables and the generation of material data which are vital for modeling accuracy.

7.3.2 Modeling challenges

The development of an accurate predictive model for AM is very complicated due to the coupling between various physical processes, numerous process parameters, and hard to measure materials' properties and boundary conditions required for modeling.

Physical phenomena associated with AM processes are complex, including absorption of energy from laser or EBs, melting/solidification, mushy zone formation, solidification shrinkage, vaporization phase changes, phase transformations, phase separation, microstructural evolution, wetting and dewetting, sintering, capillary forces, surface tension-driven free-surface Marangoni flow which determines the part shape and smoothness, heat transfer (convection, conduction, and radiation with a moving heat source), and mass transfer. Transport phenomena occur across a wide range of time and length scales and involve the solution of the governing equations for mass, momentum, energy, and species conservation.

The thermal and composition gradients and fluid flow (heating, melting, and cooling cycles) are key factors controlling the microstructure, porosity, surface finish, residual stresses, and the final properties of the part.

The computational domain includes the substrate, melt pool, remelted zone, deposited layer, and part of the gas region, depending on the subprocess being modeled.

The process also involves geometrical complexities of adjacent areas of thick and thin cross section, overhanging unsupported features, and high-aspect ratios of geometrical features. Improved computational design tools for AM are needed to efficiently represent the geometry and physics at multiple time and length scales.

Models reported in the literature have made a number of simplifying assumptions to make the problem solvable, thereby restricting their benefit to the AM industry which has had to rely on the conventional empirical trial and error approach.

The full-bed problem appears intractable unless computations can be speeded up by orders of magnitude. The computational effort involved in modeling a real-life component with typical values is given in the following:

Build height	100 mm
Computational time step	10 microseconds
Layer thickness	20 microns
Number of layers	5000
Build time	20 hours
Cool down time	5–20 hours (metals, varies for laser or e-beam)
	~40 hours for non-metals
Number of computational time steps	7.2×10^9 (build time)
Estimated computer time	several days (with graphics processing unit [GPU] machines)

7.3.3 Modeling objectives

The overall objective of AM modeling and simulation is to establish standard material characterization, measurement, and modeling methods to ensure accurate and repeatable simulations. The goal of modeling is to develop a fundamental understanding of the relationships between process variables (energy input, traverse velocity, preheating temperatures, and part geometry) on the key process features (melt pool size, residual stress, microstructure, and properties). The deliverable should be a suite of tools to model the sequence of unit operations and materials' response during AM. This will allow the manufacturers to design robust processing sequences, trouble-shoot process problems, and refine processes to reduce costs without sacrificing material quality.

The material behavior and process models should be verified and validated by working with a supplier to model a typical production-scale process. Project activities should include development of an integrated, comprehensive, physics-based model, verification and validation on subscale and full-scale components, streamlining, and integration of commercial codes for user-friendly and easy industrial implementation, and developing industry guidelines for model usage. Finally, the models should be useful for several alloy systems.

The goal will be to develop a design practices document and standard in the form of an AMS specification in support of accelerated certification, rapid maturation, and risk reduction for introducing AM for aerospace components. The document should be aligned with current technology readiness level (TRL) and manufacturing readiness level (MRL) gated processes. Developed *Best Practices* should be aimed at producing consistent results, independent of the user, with acceptable accuracy.

7.3.4   Desired model features

Some of the desired model features are listed below. These features will depend on the intended application of the model:

- Integrated multiscale, multiphysics 3D models with options for simplification of physics, geometry, dimensionality (2D and 1D) by activating/deactivating select features as appropriate to the specific objectives of a particular study.
- Meshing techniques to keep computational effort manageable:
  - Robust mesh generation for several hundred or thousand layers of deposit.
  - Mesh consolidation well below the current deposit layer.
  - Large aspect ratio elements both in the deposit plane and in the build direction.
  - Tight robust coupling with CAD system geometry.
- Right level of fidelity for the intended application. The model need not be based on a *first principles* completely physics-based approach since high precision may not always be necessary, especially for preliminary design estimates. A correlation matrix approach with a high regression coefficient between input and output may be adequate for certain applications, provided it predicts the right output trends with changes in critical input parameters and can be calibrated and refined by incorporating empirical data. Neural network or data mining tool that models complex relationships between multiple inputs/outputs can identify patterns in complex data sets and may fulfill the needs for some applications. Fundamental physics-based models which take days or weeks to run in a highly parallelized multiprocessor environment are unlikely to be used by a typical AM company. The lower fidelity models developed from the fundamental models are more likely to find immediate use at the AM companies. The fundamental models will likely find more widespread use many years later.
- Reduced order models for quick ball-park evaluations and for use in process control. A computer simulation is a translation of real-world physical laws into a virtual form. How much simplification takes place in the translation process helps to determine the accuracy of the resulting model. Models in increasing order of complexity:
  - Algebraic equation: stress = force/area
  - Handbooks
  - Spread sheet
  - Slip-line method
  - Slab method
  - Upper/lower bound method
  - Finite element method
  - Physics based multiscale models
- Balance of computational time, accuracy, and user-friendliness. For routine industrial use, the goal should be no slower than a one-day turnaround on GPU-based computer platforms (~$5000).
- Easy incorporation of special-purpose user features: new equations, boundary conditions, and material behavior. The software architecture should be open enough for an advanced user to add in special features through user subroutines or other means.
- Inclusion of statistical process data and probabilistic methods to assess confidence intervals of the predictions.
- Exploration of comparative *what if* scenarios.
- Reduction of experimental effort by identifying key parameters.

- The software package should contain a preprocessor customized for AM for user-friendly setup, a core solver which gives rapid converged solutions, and a postprocessor for customized graphical representations to aid visualizing output for rapid decision making. The software must be maintained, continually enhanced with the latest developments, and supported. All these ingredients are needed for successful industrial implementation. Many university and research and development codes do not offer the complete package needed by the industry.

7.3.5 Proposed modeling approach

The input variables to a general AM model and the output variables required from the model are shown in Figure 7.1. The formulation of a model to obtain the outputs for the given inputs is described below.

A 2D section of the overall model is shown schematically in Figure 7.2 for the powder bed fusion process. The various modeling inputs needed are listed below:

Geometry: The model consists of the full powder bed and the component being manufactured.

Finite element mesh: Details are described in Phase I.

Processing inputs: Details are shown in Figure 7.1

Thermal boundary conditions: At the top is the current deposited layer. A heat source is applied to a portion of the top surface which will end up in the component being manufactured. This heat source needs to be distributed through a few layers of material depending on whether the material below it is powder or previously

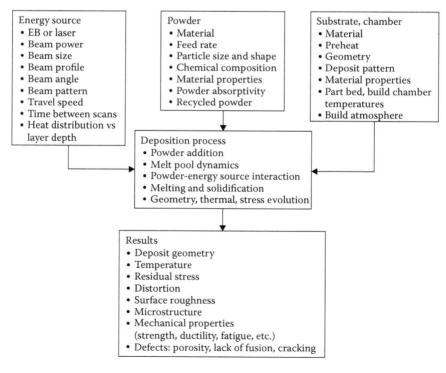

Figure 7.1 Model inputs and outputs.

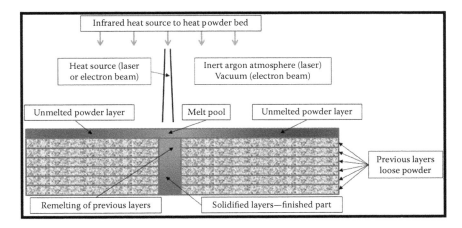

Figure 7.2 **(See color insert.)** Model formulation.

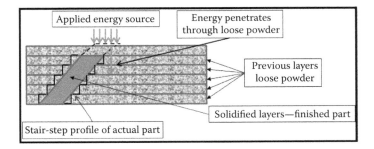

Figure 7.3 **(See color insert.)** The energy has to be distributed through layers of powder beneath the area where it is applied.

solidified material (Figure 7.3). All the needed energy source parameters are shown in Figure 7.1. In addition, in some machines there is an infrared heat source which preheats the whole or selective parts of the powder bed, and this appears as a heat flux boundary condition in the model. Radiation heat losses from the exposed faces need to be included.

Deformation boundary conditions: Appropriate displacement constraints need to be applied at the supports and at the substrate material.

Material properties: Depending on the problem being modeled (thermal, coupled thermal–stress or coupled thermal–stress–fluid), the necessary material properties have to be specified. These properties should cover the full temperature, composition, and porosity range of the material and are needed for both the solid and liquid states. The needed material properties are listed below:

Thermal:

- Thermal conductivity
- Specific heat
- Density
- Radiation emissivity

- Powder bed absorptivity
- Latent heat of solidification and melting
- Solidus and liquidus temperatures

Stress:

- Thermal expansion coefficient
- Solidification/melting volume change
- Elastic modulus
- Stress–strain constitutive behavior (elastic–plastic at room temperature to pure viscous behavior above the melting point)
- Poisson's ratio

Fluid:

- Viscosity
- Volumetric expansion coefficient
- Surface tension
- Capillary effects

Other:

- Evaporation and microstructure related if being modeled

Thermomechanical and fluids solution: Depending on the outputs required, the thermal model, or the coupled thermal–stress, or the coupled thermal–stress–fluids equations are solved. The overall steps for solution are shown in the modeling flowchart in Figure 7.4. The coupling between the thermal, stress, and microstructural features is shown in Figure 7.5.

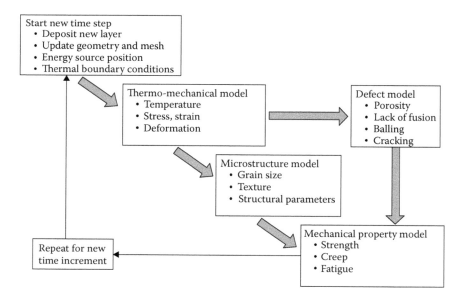

Figure 7.4 Model flowchart.

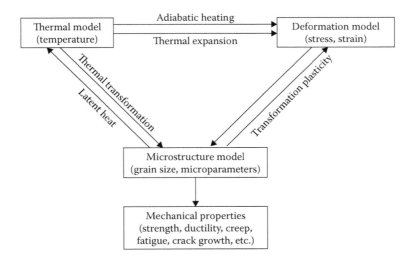

Figure 7.5 Thermal–stress–microstructure–property interaction.

7.4 Overall modeling strategy

In view of the challenges and the current state-of-the-art as described in the previous sections, a phased modeling approach is recommended. This approach will ensure a number of intermediate deliverables all through the development effort. Therefore, the developed tools can be implemented in the industry at an early stage without having to wait for many years for a fully developed model. Some of the Phases (e.g., VI, VII, VIII) can be performed in parallel with Phases I, II, III, or IV:

Phase I	CAD and finite element method (FEM) integration
Phase II	Thermal model
Phase III	Coupled thermomechanical model
Phase IV	Coupled thermomechanical, fluid mechanics model
Phase V	Particle bed models
Phase VI	Defect prediction models (semiempirical correlations to thermomechanical output)
Phase VII	Microstructural models (semiempirical correlations to thermomechanical output)
Phase VIII	Mechanical property models (semiempirical correlations to thermomechanical output)

In addition to the above sequential phases, some activities should occur in parallel during all the phases:

- Documentation, commercialization, and implementation
- Material data generation
- Verification, validation, and calibration
- UQ
- Process monitoring and control

Finally, a few very long term items (>3 years) to consider are

- Powder model
- Probabilistic methods
- Fundamental microstructural models (phase field, CA, crystal plasticity, etc.)
- Fundamental mechanical property models (dislocation dynamics, etc.)

These activities are described below and summarized in Table 7.3. The computational effort goal is what would be required for the model to be implemented in the industry for a typical problem.

A notional time line for model development is shown in Table 7.4. It is estimated that the overall program will take about five years to complete with a team of investigators working the various tasks, some in parallel.

Table 7.3 Model development strategy

Phase	Analysis type	Outputs (all functions to time and space)	Computational effort goal (CPU machine)	Commercial codes
I	CAD-FEM Integration; Mesh Generation	Complex geometry in digital layers imported from CAD into FEM model and meshed	~1 hour	All
II	Thermal	Temperature	~2–4 hours	COMSOL, DEFORM, ESI, ANSYS
III	Coupled thermomechanical	Temperature, stress, strain, distortion, and propensity for cracking	~12–24 hours	COMSOL, DEFORM, ESI, ANSYS
IV	Coupled thermomechanical, Fluid mechanics	Melt flow characteristics in addition to above	~24–36 hours	COMSOL, ANSYS
V	Particle bed behavior	Particle bed interaction with energy source	~24–36 hours	COMSOL
VI	Defects	Porosity, lack of fusion, surface finish, balling (from semi-empirical correlations)	Additional 1–2 hours of post-processing of thermo-mechanical output	Add-on to any of above
VII	Microstructure	Grain size, microstructural features (from semi-empirical correlations)	Additional 1–2 hours of post-processing of thermo-mechanical output	Add-on to any of above
VIII	Mechanical properties	Strength, creep, fatigue (from semi-empirical correlations)	Additional 1–2 hours of post-processing of thermo-mechanical output	Add-on to any of above

Table 7.4 Notional time line for model development

Phase	Task description	Year 1	Year 2	Year 3	Year 4	Year 5
I	CAD-FEM Integration; Mesh generation	x				
II	Thermal	x				
III	Coupled thermos-mechanical		x			
IV	Coupled thermos-mechanical, fluids		x	x		
V	Particle bed behavior			x		
VI	Defects				x	
VII	Microstructure					x
VIII	Mechanical properties					x
In parallel tasks						
	Documentation, commercialization, and implementation	x	x	x	x	x
	Material data generation	x	x	x	x	x
	Verification, validation, and calibration	x	x	x	x	x
	Uncertainty quantification	x	x	x	x	x
	Process monitoring and control	x	x	x	x	x
Longer range tasks						
	Powder model					
	Probabilistic methods					
	Fundamental microstructural models					
	Fundamental mechanical property models					

7.4.1   Phase I: CAD and FEM integration

Almost all the models reported in the literature have dealt with very simple geometries and a very few layers of deposition (typically 2 or 3, at the most 20) and need several hours of computational effort in spite of these simplifications. In order to model an actual component with any level of geometrical complexity, the first step is that the CAD system and the finite element model should be tightly integrated. The geometry of the component, sliced into several layers, has to be transmitted seamlessly to the finite element model. The geometry has to be linked to the energy input boundary conditions and to an automatic 3D finite element mesh generator.

A proposed meshing scheme is described below. It is intended to concentrate small-size elements in the vicinity of the build region where the thermomechanical effects are most active and gradually increase the element size away from the build region. This strategy ensures that the total number of elements (which is directly related to the computational effort) is kept manageable without sacrificing numerical accuracy.

The proposed meshing scheme is described for

- The build direction.
- In the build plane.
- Adjustment due to density changes (melting).

2D cross sections are shown to keep the figures simple and understandable. The meshing strategies are illustrated with typical numbers. There are no hard and fast rules, and the mesh for each case has to be decided based on the tradeoff between computational effort and accuracy. The vertical build direction is referred to as the z-direction, and the horizontal build plane is referred to as the x–y plane.

7.4.1.1   Meshing strategy in the building direction

In the z-direction, at least 2–4 elements are recommended in the current layer and for a few layers below it where the thermal effects are still going through heating and cooling cycles. As one gets further down away from these layers, the mesh can be consolidated in the z-direction. Several elements are gradually combined to form a larger element. Typically, the layer thickness is around 20 microns. With four elements in a layer, each element will be 5 microns thick. The part size is typically in millimeters. In order to avoid a sudden change in element size, the consolidation of elements should progress in geometric progression until the size reaches about 1 mm (or about 10% the part dimension in the x–y build plane) (Figure 7.6).

Suppose five layers, each 20 microns thick, are going through thermal cycling and need four elements (5 microns thick) through the thickness each. The sixth layer below, with two elements consolidated into one, will have elements 10 microns thick. Following this sequence, the element z-thicknesses in the various layers are:

Layers 1–5	5 microns (going through thermal cycling)
Consolidated layer 6	10 microns
Consolidated layer 8	20 microns
Consolidated layer 9	40 microns
Consolidated layer 10	80 microns
Consolidated layer 11	160 microns
Consolidated layer 12	320 microns
Consolidated layer 13	640 microns
Consolidated layer 14	1280 microns, or 1.28 mm; the mesh coarsening can stop here.

Thus, about 10 transition layers are needed to transition from elements 5 microns thick to about 1 mm thick in the z-direction.

When the elements are consolidated, care should be taken in averaging the field variables over the element volume to ensure that energy conservation is preserved.

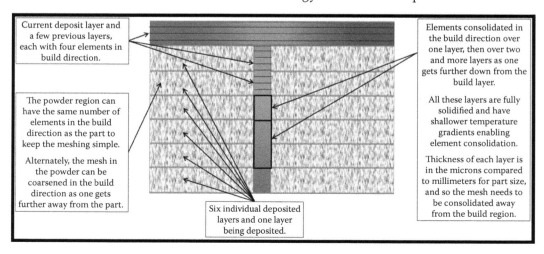

Figure 7.6 **(See color insert.)** Meshing strategy in the build direction.

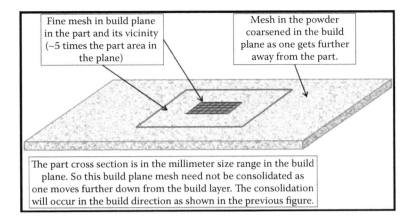

Figure 7.7 (**See color insert.**) Meshing strategy in the build plane.

7.4.1.2 Meshing strategy in the build plane

If the part cross section area in the *x–y* build plane is about 5,000 mm² and each element in the build plane is 1 × 1 mm square, the number of elements in the build plane is 5,000. This is a rather fine mesh and can possibly be reduced depending on the individual part geometry. The mesh should be kept to a similar size surrounding an area about five times the part cross section area and gradually coarsened out to about 5 × 5 mm square size elements (Figure 7.7).

Note that in the build layer, the typical element thickness in the *z*-direction is 5 microns, and so the aspect ratio of the elements is about 200:1. If an attempt is made to get aspect ratios of close to unity, it will require an inordinately large number of elements which may not improve the accuracy since the gradients are much steeper in the build direction and that is the direction where the element size is the smallest.

7.4.1.3 Mesh adjustment due to density changes (melting of powder layer)

After the current layer melts, the powder at 60%–70% relative density is transformed into 100% dense liquid with an associated volume reduction. Therefore, the level will drop slightly in the region where melting has occurred. This should be compensated for in the mesh for the next layer. A correspondingly larger amount of powder per unit area will be deposited over the molten area to keep the top of the layer horizontal (Figure 7.8).

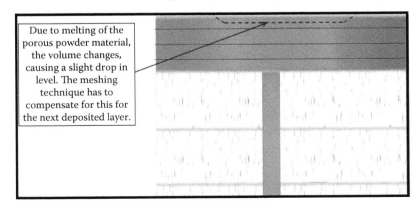

Figure 7.8 (**See color insert.**) Mesh adjustment due to melting of powder layer.

For a typical part, the build sizes were assumed to be:

Build height	100 mm
Powder bed area (300 × 300 mm)	90,000 mm²
Part cross-section area in any build plane	5,000 mm²
Area occupied by powder	85,000 mm²
Layer thickness	20 microns

With the meshing scheme described, the total number of elements for this typical size is part is estimated as follows:

1.	Active layers with a fine z-mesh	5 (100 microns deep)
2.	Consolidated layers with a transition z-mesh	10 (~2 mm deep)
3.	Consolidated layers with a coarse z-mesh	100 (100 mm deep)
4.	Build-plane fine elements in part cross section	5,000
5.	Build-plane transition elements near part	5,000
6.	Build-plane coarse elements outside part	3,500 (85,000/25 mm²)
7.	Fine elements (a·d)	25,000
8.	Transition elements (b·e)	50,000
9.	Coarse elements (c·f)	350,000
10.	Total 3D elements (g+h+i)	425,000

This is a large but still manageable number. The largest contributor to the total is the number of coarse elements in the relatively inactive powder volume. If the element size is further increased to 10 × 10 mm square, the total number of elements can be reduced to about 150,000. Computations with a turn-around time goal of one day can be performed with this meshing strategy.

The EOS M 400, which is one of the larger machines currently made, has a build volume of 400 × 400 × 400 mm. If the part x–y cross section is assumed to occupy 10% of the total area (16,000 mm²), the total number of 3D finite elements, using the above scheme, is estimated at 2.64 million which is beyond current computational limits for GPU-based machines. For such large parts, larger element sizes have to be used to bring the number of elements down to a range of about 1,00,000–2,00,000. Note that the number of coarse elements in the relatively inactive powder volume can be easily reduced without sacrificing accuracy.

1.	Active layers with a fine z-mesh	5 (100 microns deep)
2.	Consolidated layers with a transition z-mesh	10 (~2 mm deep)
3.	Consolidated layers with a coarse z-mesh	400 (100 mm deep)
4.	Build-plane fine elements in part cross section	16,000
5.	Build-plane transition elements near part	16,000
6.	Build-plane coarse elements outside part	6,000 (144,000/25 mm²)
7.	Fine elements (a·d)	80,000
8.	Transition elements (b·e)	160,000
9.	Coarse elements (c·f)	2,400,000
10.	Total 3D elements (g+h+i)	2,640,000

7.4.2   Phase II: Thermal model

This phase involves a purely thermal analysis to predict the temperature at all locations in the powder bed (and part) through the build process and subsequent cool down to room temperature.

Some or all of the following features need to be added on to a commercial thermal analysis code in order to model AM processes:

- Coupling of geometry description with a CAD system
- Meshing features unique to AM
- Moving heat input distribution for each layer from CAD data
- Heat flux distribution within the energy beam
- Distribution of heat input through the thickness of the current layer and below
- Porous media heat transfer
- Temperature, porosity, and composition-dependent properties

Several standard commercial codes, besides the ones listed in Table 7.3, can be used as a starting point for developing the thermal model.

7.4.3   Phase III: Coupled thermomechanical model

This phase involves a coupled thermomechanical analysis to predict the temperature, stress, strain, and distortion at all locations in the powder bed (and part) through the build process and subsequent cool down to room temperature. The transient stress field relative to the strength of the material at any location and its temperature can be used to evaluate the propensity for cracking. Final residual stresses and distortions in the part are obtained after cool down to room temperature. Postprocessing operations (stress relief and hipping) are also included in this phase.

Some or all of the following features need to be added on to a commercial thermome-chanical analysis code in order to model AM processes:

- Features listed for Phase I
- Boundary conditions at supports and base plate
- Porous media constitutive behavior
- Material behavior ranging from elastic–plastic at room temperature to pure viscous behavior above melting point
- Isotropic or kinematic hardening effects
- Nonisotropic properties
- Metallurgical phase changes (transformational plasticity)
- Creep model for postprocessing steps (stress relief)
- Compaction model for HIP or CIP (hot and cold isostatic pressing)
- Robust numerical solvers to ensure convergence of the tightly coupled nonlinear governing partial differential equations

Few standard commercial codes can be used as a starting point for developing the coupled thermomechanical model. Many of the commercial codes lack the ability to handle the appropriate material behavior all the way from room temperature to above the melting point, and/or the ability to model the postprocessing steps of hot or cold isostatic pressing.

7.4.4 Phase IV: Coupled thermomechanical, fluid mechanics model

This phase involves a coupled thermomechanical and fluids analysis to predict the melt pool characteristics (temperature, fluid flow, melting, solidification, and mushy zone) in addition to Phase II outputs. The fluid flow is generally restricted to a very small fraction of the build volume. Besides the current layer being deposited, there is remelting for a few layers below. If fluid flow is ignored, the predicted maximum temperature is generally too high since fluid mixing tends to even out the temperature field. Fluid flow is also important for predicting defects like balling and for predicting microstructure development.

The addition of fluid flow adds considerable complexity to the model and increases the computational effort significantly. Each case should be evaluated for the potential benefits of considering fluid flow versus the added effort.

Some or all of the following features need to be added on to a commercial thermomechanical/fluids analysis code in order to model AM processes:

- Features listed for Phase II
- Solidification shrinkage and latent heat
- Thermal and compositional Marangoni flow
- Thermal and compositional buoyancy
- Capillary forces
- Melt pool-free surface tracking
- Flow in mushy regime
- Turbulent flow in melt pool
- Transient behavior at melt pool-free edge (related to balling)
- Light element (e.g., Al, Cr) evaporation
- Effect of surface tension forces at the melt pool surface on adjacent unmolten powder particles and on surface roughness
- Robust numerical solvers to ensure convergence of the tightly coupled nonlinear governing partial differential equations

Few standard commercial codes can be used as a starting point for developing the coupled thermomechanical, fluid mechanics model. Besides lacking the features mentioned in Phase III, many of the commercial codes (e.g., ANSYS) do not provide a tight coupling between fluid and solid mechanics. These analyses are performed in separate codes with data exchange between them. COMSOL is one of the few software packages where all the physics is tightly coupled in one code.

7.4.5 Phase V: Particle bed model

This phase involves the development of a model to predict the absorptivity of the powder bed for both laser and EB energy inputs. The absorptivity determines the fraction of input energy absorbed by the bed. In addition, the distribution of the energy absorbed over the layer thickness has to be determined.

Some publications have modeled the laser–material interaction via ray tracing and a physics-based absorption model. It models melting of the powder, flow of the liquid, and behavior of trapped gas. This model treats the powder as a discrete system of particles and uses a Lattice Boltzmann approach to investigate melting and resolidification.

Fundamental models are computationally very intensive. A combination of experimental data and lower fidelity models is recommended to determine the particle bed thermal properties. These properties do not need to be determined for each part; they have to be determined once for each type of material, powder particle size and distribution, and other parameters.

7.4.6 *Phase VI: Defect prediction model*

Major AM process defects are

- Balling
- Delamination
- Porosity
- High residual stresses

7.4.6.1 *Balling*

In balling, instabilities occur in forming an even and smooth layer of solidified material. Small spheres of material form with the approximate diameter of the laser beam. Surface tension affects the metal flow at the edges of the molten pool causing molten metal to resolidify as a series of balls. Surface tension forces at the melt pool surface can drag adjacent unmolten powder particles into the pull causing rough surfaces. The melt pool size and its aspect ratio are critical surface tension variables which affect balling.

The prediction of balling using modeling is difficult. At best, some process regimes can be identified via both modeling and experiments which need to be avoided to prevent balling. Balling can be avoided by selecting laser powers and scan speeds that do not promote instabilities in the melt pool, for example, Rayleigh instabilities.

7.4.6.2 *Delamination*

In delamination, two consecutive layers tend to separate due to the development of tensile stresses during cooling. A thermomechanical model can be used to predict the stress field and the onset of delamination.

7.4.6.3 *Porosity*

Pores or voids can form due to lack of complete fusion or during solidification of the molten material and require processes like HIP to reduce porosity within the part. Porosity can vary significantly across the build due to the variations in the heat dissipation at different locations of the build. The laser power, scan speed, and powder flow rate can be adjusted during deposition to rectify this problem.

Porosity models are similar to those in casting modeling. Numerous semianalytical and theoretical relationships have been proposed to describe the dependence of mechanical properties on porosity.

7.4.6.4 *Residual stresses*

Residual stresses develop in the component during additive layering and cooling and can cause excessive distortion or reduce the fatigue life of parts.

7.4.7 *Phase VII: Microstructural model*

Microstructure models similar to those in casting modeling can be used for AM also. Temperature gradient, G, and solidification rate, R, are the two most important parameters controlling the solidification microstructure. These models take the temperature history at any given point and relate that to microstructural features using empirical correlations. Statistically based DOE and multiple regression analysis methods have been applied to quantitatively establish relationships between common process parameters and resulting microstructural outputs. The use of neural net-based models is recommended since these models get better as more data becomes available. Neural net or regression models can be embedded into the thermomechanical model through user subroutines. These models are computationally fast, and the additional computational effort is generally insignificant.

The use of more fundamental models is a long-range goal. For the present, surrogate models should be constructed and used for simpler analyses.

7.4.8   Phase VIII: Mechanical property model

Mechanical property models take the temperature history at any given point and relate that to a property (strength, creep, fatigue, etc.) using empirical correlations. Statistically based DOE and multiple regression analysis methods have been applied to quantitatively establish relationships between common process parameters and resulting mechanical property outputs. The use of neural net-based models is recommended since these models get better as more data become available. Neural net or regression models can be embedded into the thermomechanical model through user subroutines. These models are computationally fast, and the additional computational effort is generally insignificant.

The use of more fundamental models is a long-range goal. For the present, surrogate models should be constructed and used for simpler analyses.

7.5   Parallel development activities

In addition to the above sequential phases, some activities should occur in parallel during all the phases:

- Documentation, commercialization, and implementation
- Material data generation
- Verification, validation, and calibration
- UQ
- Process monitoring and control
- Design tools

7.5.1   Documentation, commercialization, and implementation

All software need to be documented and prepared for commercialization to minimize maintenance and enhancement risks. Conditions affecting the accuracy and the variability of the modeling predictions should be identified. Use cases, which codify the methodology and describe the problem-solving steps, have been used successfully in prior programs to demonstrate the modeling framework. Use cases should be defined to design the process and identify optimum parameters. Regular training sessions should be held for industrial users and their feedback should be utilized in planning subsequent model development.

7.5.2   Material data generation

Material data need to be acquired from literature or measured or calculated from thermodynamic principles. The material properties (some of which can be nonisotropic) are dependent on temperature, concentration, particle size distribution, powder particle morphology, and density. Difficult to measure properties and boundary conditions have to be obtained by calibrating the model with experimental data.

7.5.3   Verification, validation, and calibration

One of the key areas for successful model development is verification and validation, including the need to develop guidelines and standards for this activity. The rigorous, systematic verification and validation (V & V) of models is critical to their successful implementation and acceptance.

The key gaps in AM modeling programs are: A number of uncertain model parameters need to be quantified using calibration methods, no formal verification plan, and only limited validation. More formalism and generic V & V guidelines are needed to insure a robust assessment of the models. Engineering disciplines, such as Fluid and Solid Mechanics, have mature models and have developed guidelines (e.g., ASME V & V Guide 10-2006) over many years of effort.

Verification is the process of determining if a computational model accurately represents the underlying mathematical model and its solution, that is, *solving the equations right*. This should be accomplished by comparing the model solutions with known analytical or closed-form exact solutions. Verification of a complex modeling system should be done individually for each of the submodels.

Validation is the process of determining the degree to which a model is an accurate representation of the real world from the perspective of the intended uses of the model; it is a physics issue: *Solving the right equations for the current problem*. Validation of a complex modeling system should be pursued hierarchically from parts to the whole model. Thus with AM, the thermal model should be validated first, followed by the stress and distortion model, and then the defect, microstructure, and property models. Benchmark test cases should be defined for validation, taking into account the fact that multiple validation experiments may be necessary to validate various elements of complex or sequential models. Validation experiments should be conducted under well-controlled conditions with all the inputs accurately defined to minimize uncertainties and variability. The use of DOE techniques is recommended to generate validation data over a range of operating conditions with a minimum number of experiments. Appropriate validation metrics should be defined.

Calibration is the process of determining modeling inputs which are difficult to measure or not known precisely by comparing key modeling outputs with experimental data. Material properties and boundary condition data fall in this category. Inverse methods can also be used to calibrate the required properties. In these methods, the model outputs with assumed input variables are compared with experimental measurements. The difference between the model predictions and the experimental measurements is minimized by adjusting the values of the input variables.

When using this method, it is vital that the effect of measurement errors be quantified and that the variable being calibrated has a significant effect on the modeling output being used for calibration. Depending on the quality of the experimental data, the variables can be obtained only to some level of accuracy.

7.5.4   Uncertainty quantification

UQ involves the quantitative assessment of the contribution of uncertainty in model inputs and internal parameters to overall uncertainty of model output. UQ is needed since many inputs will be known only approximately especially with a model of a complex process like AM. UQ affects how one assesses the level of agreement of a model relative to both input and output data, as well as the variation in model results.

UQ studies should be performed to identify important process variables and input data and performance metrics. Guidelines should be developed for use of UQ for parameter sensitivity analysis and for selected example validation or benchmark cases. In some models, Kennedy O'Hagan Bayesian hybrid modeling methods have been used to support calibration and UQ analysis. The calibration includes uncertainties in the model, input variables, and test data. It also yields global sensitivity information, which can be used to identify the key parameters at each step and at the system level and enable targeted testing.

7.5.5   Process monitoring and control

Integrated in-process sensing, monitoring, and control technologies should be developed to ensure that the material properties and component performance meet the requirements.

7.5.6   Design tools

There is a need for the development of additional software tools for the AM industry. Software is needed for part consolidation. A tool of this type would analyze and recommend which assemblies, or subassemblies, are candidates for being consolidated into a single, complex part manufactured by 3D printing.

Important AM functions that could be performed by software include optimization of part orientation, especially as it relates to optimizing the packing density of a build. Another important task is the optimal generation of anchors for the metal powder bed processes. Anchors need to be strong enough to prevent part distortion from thermal stresses, but not overly thick and heavy, which increases powder consumption, power use, and build time. Today, this optimization function is mainly a manual process performed by an experienced user.

With the multimaterial jetting processes from Stratasys and 3D systems and all directed energy deposition (blown powder) processes, parts could be defined, and built, on a voxel-by-voxel basis. Yet software tools to design multimaterial parts are not available. The development of software tools to fully support these multimaterial processes is required.

7.6   Longer term development

Finally, a few long-term items (>3 years) to consider are:

7.6.1   Powder model

The powder material consists of particles of different sizes and geometries. Modeling of single particles is impractical. Stochastic modeling of the material as a continuum or as a collection of discrete particles is a possible solution. Lagrangian (moving grid) methods or

Eulerian (fixed grid) methods or a combination of the two are needed to find the shape of the free surface. Lattice Boltzmann approaches can be used to determine the interaction of the powder bed with the energy source. The powder model will predict the melting of powder and its resulting densification. This model will be computationally very intensive. However, it needs to be exercised only a few times to generate the inputs to the thermomechanical models.

7.6.2 Probabilistic methods

After enough simulations and experimental data are generated, probabilistic techniques can be used to incorporate uncertainties into the analysis. A properly verified, validated, and calibrated deterministic model is needed before probabilistic analyses can be attempted. As such, this is a much longer term activity.

7.6.3 Fundamental microstructural and property models

Various fundamental microstructure and mechanical property models have been reported in the literature: dislocation density, crystal plasticity, CA, phase field, and so on. These methods have been coupled with finite element and macroscopic thermodynamic models, and used to predict the dendritic grain size, microstructure, morphological evolution, and mechanical properties (strength, creep, fatigue) during the deposition process.

Such calculations, while interesting, are today computationally intractable for real-life problems. They also have a number of material data inputs which are difficult to measure and have to be calibrated. Microstructural and property models are still in the research and development stage even for more mature conventional manufacturing processes such as casting and forging which have a simpler thermomechanical processing history than AM processes. These models have to mature first for the conventional processes before their widespread use for AM can be considered. These models have long-term potential. Research should continue in the development of these models, speeding up the computations, and bringing them to a more production ready stage. Particular attention should be given to quantifying variability in microstructure and constitutive response.

7.6.4 Summary of recommendations

Commercial software is recommended to ensure that issues of support, updates, maintenance, and so on, are all addressed. Commercially available and maintained software will enable rapid and efficient deployment throughout the aerospace supply chain. University codes can be used to test physics and solution methods but eventually need to be transferred into commercial codes for implementation in a production environment. All software need to be prepared for commercialization to minimize maintenance and enhancement risks. Models need to be validated, made user-friendly, robust, reliable, accurate, and integrated to seamlessly work together. The model of a complex process like AM will involve multiple simulation components, each dealing with a particular aspect of the overall process physics. A combination of modeling components should be selected to provide a balance between accuracy and computational effort depending on the application. Conditions affecting the accuracy and the variability of the modeling predictions should be identified. Use cases, which codify the methodology and describe the problem-solving steps, have been used successfully in prior programs to demonstrate the modeling framework. Use cases should be defined to design the process and identify optimum parameters.

The following recommendations are made for commercial software:

- COMSOL is recommended for the modeling of all aspects of AM for the following reasons:
  - Tightly coupled physics in one code and no problem of passing data between codes over dissimilar meshes and geometries and time steps. This enables easy integration of all AM steps into one code which already contains most of the needed physics.
  - Coupled multiphysics analysis is one of the strengths of COMSOL. By default, COMSOL will solve all active multiphysics equations simultaneously. But if some of the physics is one-way coupled, COMSOL can be set up to solve in a stepwise fashion saving memory and computational time. For example, typically the temperature field affects the stress field through the thermal expansion coefficient. The stress field has little or no influence on the temperature field. Therefore, this multiphysics problem can be solved sequentially and COMSOL can be set up to do just.
  - Interfaces with standard CAD systems exist.
  - Flexible architecture permits easy incorporation of user sensitive information into a generic common framework. New equations, boundary conditions, material data, and special purpose user features can be easily added.
  - Lower licensing costs than most comparable commercial software.
  - Program structure enables the creation of a customized AM modeling pre and postprocessor to facilitate user-friendly implementation.
  - Has a chemical reaction module which can be used to model curing of polymer and plastics in AM.
  - Can build models with different levels of fidelity to address run-time issues.
  - Can set up sensitivity analysis, and parametric study runs automatically.
- If fluid flow can be ignored, DEFORM is recommended for the following reasons:
  - Widely used in the aerospace industry and supply chain which makes implementation easier.
  - Fast efficient solver and good meshing techniques.
  - Has been extensively validated for heat treatment residual stresses and machining distortions in various USAF MAI (metals affordability initiative) programs.

Funding estimate: For an accurate funding estimate, the team needs to be formed and all the individual tasks need to be costed in detail. For planning purposes, a rough order of magnitude cost estimate is $5–10 million to execute Phases I through VIII over five years. This includes model development and the generation of experimental data for model calibration and validation.

There are a number of special-purpose commercial codes which have a lot of features related to a specific process, for example, PROCAST™ for casting. Some of these special-purpose features may not exist in a general-purpose code like COMSOL. It should be noted that special-purpose versions of COMSOL have also been created, for example, for composites. Special-purpose codes might offer an advantage when dealing with certain process-specific features for which they were designed. However, the recommendations made here are based on looking at the AM process as a whole and determining which code can get to the desired goal the fastest and most economical way.

7.7   Summary

Modeling and simulation are used to understand how the materials utilized in AM would react during the processes. This allows the processes to be more predictable and repeatable. In the first part of this chapter, the current and future desired states of AM models were discussed. Although there have been some advancement in data modeling, verification, and validation of AM processes, there are still gaps that need to be researched to satisfy industry standards such as boundary conditions and defect formation.

A plan was presented for AM model development, verification, and validation. The plan was limited to the powder bed fusion process for metals which is the most widely used process for aerospace components. Technical needs for modeling and simulation, the challenges involved, the overall modeling objectives for a phased development effort were presented. Desired model features and a proposed model formulation to get from the current state to a future state of an integrated AM model were described. Recommendations were made on the overall modeling approach. The idea is to figure out the areas where AM needs improvement and determine which commercial code and model features would benefit the AM industry.

chapter eight

Operational aspects and regulatory gaps in additive manufacturing

Adeola Adediran and Akinola Oyedele

Contents

8.1 Introduction

Manufacturing processes have earlier been predominantly subtractive, that is, three-dimensional objects were created by successively cutting material away from a solid block of material, either by scraping, machining, turning, or dissolving. Additive manufacturing (AM) or three-dimensional (3D) printing, in contrast, is controlled material addition, implemented by successively depositing layers of material until a predesigned shape is formed. AM represents an innovative technology in manufacturing and is certainly set to transform production processes from the design to manufacture and to eventual distribution to end users. The unique capability of this technology to produce intricate geometries with customizable material properties has made it a widely interesting and welcome development among scientists, industry, and the general public. However, until now, most attention has been focused solely on the ingenuity of this ground-breaking technology and its wide range of possibilities. Little or no consideration is being given to the adverse effects of the seemingly unstoppable advancement of AM technology and unrestricted access to 3D-printing techniques. The wide acceptance and rapid spread of this technology have made 3D printers increasingly openly accessible, and low-cost desktop printing, with capability to reproduce 3D objects from medical prostheses to weapons, is rapidly increasing in availability to the public. This section brings to light some conceivable downsides and challenges of this impending development. Issues discussed include regulation gaps in manufacturing, loopholes in safety and national security, and the need for curbing those problems that can be contained, or otherwise adapting to the eventualities that lie beyond control.

8.2   Operational aspects of additive manufacturing

AM, which is also known as 3D printing (3DP), uses a systematic, layer-by-layer approach in fabricating objects of various complexity, size, and material. From the simple, home-built Legos to an industrial-scale printed car (Local Motors, 2016), almost any design that can be conceived can be made. This systematic approach allows for rapid prototyping (Campbell et al., 2012) and mass customization of products, as different designs can be made to meet individual specifications. Unlike traditional processes like injection molding, milling, and casting, which use a top-bottom approach, AM such as electron beam melting, selective laser sintering (SLS), and MultiJet Modeling involves the sequential deposition of individual layers from the bottom to top. Thus, AM does not require the removal of some parts of the finished product through tooling in order to conform to initial designs (Huang et al., 2013). Traditional manufacturing processes on the other hand are subtractive in nature, which involves the removal of up to 98% of the material depending on the complexity and geometries of the designs (Allen, 2006; Petrick and Simpson, 2013). This leads to waste, which has both economic and environmental consequences (Despeisse and Ford, 2015). The operational process of a typical AM is detailed in Figure 8.1.

8.2.1   3D computer-aided design model

The AM process starts with the design and modeling of the desired finished product. The time required to complete the design depends on the geometry and the complexity of the project and the experience of the designer or the design team. Simple geometries like a water bottle can be designed in less than 10 minutes, while complicated designs such as a robotic arm used as prosthetics can take up to three days. The design stage is very crucial to the AM process and the quality of the final product; therefore, adequate planning is highly recommended when embarking on a new project. Any error will result in an undesired finished product and as such will render the whole process a waste of time and resources.

Typically, the design process utilizes computer-aided design (CAD) or 3D modeling software to create a digitalized version of the product. Examples of software used include

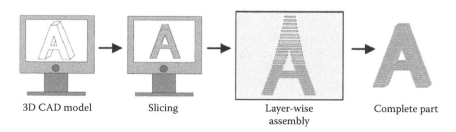

| 3D CAD model | Slicing | Layer-wise assembly | Complete part |

Figure 8.1 The 3D *additive manufacturing* process. (Courtesy of European Space Agency.)

AutoCAD, TinkerCAD, Repetier, SketchUp, Blender, and MeshLab, all of which are free and are easy to use depending on the designer's familiarity. The digital file created can be easily shared or saved in a repository for future use. Many websites offer a collection of different designs, contributed by other users in the 3DP community, from which new designs can be built. Thingiverse, YouMagine, PinShape, GrabCAD, and 3Dagogo are examples of such repositories. The ubiquitous presence of designs and the relative ease of reverse engineering of designs present a challenge in intellectual property (IP) protection (Piller et al., 2014). This is a threat and avenues to mitigate that such must be explored. As it was the case with the protection of contents in the music industry, this threat could turn to opportunities.

In this digital age, customization and flexibility are important drivers of current and future technologies (White and Lynskey, 2013). AM technology is positioned for success, in that it allows for the design and fabrication of products from the comfort of one's home and the relative ease of customizing designs to meet customers' specifications. Like the printing technology, the ability to design and print photos from individual homes and offices opened up the technology to billions of users and changed the whole market paradigm. Apart from opening up the market to more users, the ability to use AM at home and in school provides children and students a valuable hands-on familiarization experience with the technology, thereby raising a generation of early hobbyist and users, who are important to the future of the technology. Kids and students can play with the design software, and can be allowed to explore imaginative and creative designs.

During the industrial revolution, mass production was the key to cost reduction and the ability to reach many consumers. However, in this present age, mass customization is the driver to derive consumer satisfaction (White and Lynskey, 2013). The ability to design anything using CAD software allows for ease of customization. For example, in 2014, Local Motors® demonstrated in front of a live audience the printing of a 3D car at the International Manufacturing Technology Show (IMTS) in Chicago (Ulanoff, 2014). This accomplishment is an important milestone in the customization of products and the ability to print locally. Now, companies like Local Motors can set up their factories locally to meet the customization needs of the local customers. The turnaround time in production which is around one day will make this technology to have more patronage. Also, with the increase of wearables through internet of things (IoT), it is now possible to further customize the electronics of your cars with features that are tailored for your specific needs (IBM, 2016). While customization in traditional manufacturing method only attracted high-end, deep-pocketed users, with AM it will now be possible to reach both low- and medium-end customers, hence making the technology more affordable.

8.2.2   *Slicing*

After the model is designed, the next step is to slice the CAD file in a form that allows for it to be printed layer-by-layer. This is compatible with the AM technology, since it prints individual layers in order to fabricate the object. The chosen direction of slice is crucial as it affects the strength of the object, build time, and the visual quality (Hildebrand et al., 2013; Wang et al., 2015). For simple 2D objects this is trivial, but with increasing layer thickness and complexity, the direction at which the model is sliced becomes more important. Intuitively, a particular slice angle is selected—usually the angle orthogonal to the direction of print. However, from the works of Kristian Hildebrand et al., slicing in only one angle might be trivial but does not always give the best results. It was observed that the optimized direction of slice is not a single, mono direction but a combination of

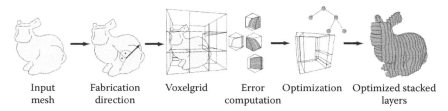

| Input mesh | Fabrication direction | Voxelgrid | Error computation | Optimization | Optimized stacked layers |

Figure 8.2 Overview of the slice angle optimization. (From Hildebrand, K. et al., *Comput. Graph.*, 37, 669–675, 2013.)

different slice angles. The process of optimizing the slice angles is shown in Figure 8.2. It starts with the model as input mesh, which is then sliced in three different directions, whose base vectors are perpendicular to one another. A voxelization process is then performed to determine the slicing error in each direction from which the optimized slicing angles are selected. Slicing the object in this manner ensured a faster build time and a reduction in the error of the finished object. It is important to note that if the directions are not orthogonal to one another, it makes it difficult to fuse the different parts together. Apart from reducing the geometric errors of the final print, optimizing slice directions also helps to improve the flexibility of printing to reduce built time, especially in cases where the shape is larger than the build volume of the printing instrument. Other slicing optimization work was done by W. Wang et al., in which they developed a model to reduce manufacturing time. These optimization models are embedded in software used for developing models in AM applications. Example of such software includes CADfix.

8.2.3 Layer-wise assembly

The layer-by-layer assembly depends on the type of materials to be printed. For polymers, MultiJet Modeling is commonly used, whereas metallic objects are printed with electron beam melting and SLS methods. The printing of transparent materials like glasses can be printed using a material extrusion printer (Klein et al., 2015). MultiJet Modeling uses an inkjet nozzle to print molten polymeric materials. As one layer hardens, the subsequent layer is deposited such that it forms a fuse and melt together. SLS involves the use of high-power laser to melt and sinter lasers in order to bind them with subsequent layers. The laser is usually a carbon dioxide laser beam. Electron beam melting is similar to SLS, however it uses high electron beam instead of laser beam. The high electron beam is powered by a high voltage, typically 30–60 kV in a high vacuum chamber. Technologies such as 3DP involve the use of binders, which are sprayed onto a bed to bind the materials. This is commonly used to handle a variety of polymers (Halloran et al., 2011). Fused deposition modeling (FDM) works usually with plastics such as polycarbonates (PC) and styrene. The print head melts the material and then extrude it in thickness of around 0.25 mm. Usually, this process requires finishing which may increase the time required to finish the printing (Wong and Hernandez, 2012).

The mechanical, optical, and morphological properties of the printed material are dependent on the processing conditions such as temperature, pressure, flow rate, viscosity, layer height, and feed rate (Gu et al., 2012). Hence for each project, it is important to optimize these parameters in order to achieve the best results (Gu, 2015).

8.2.4   Finishing

Depending on the accuracy of the slicing and printing processes, an additional finishing process might be necessary (Wong and Hernandez, 2012). After the printing process, surface roughages are removed for aesthetic purposes. The material removed in this process is negligible when compared to the amount of materials removed via tooling in traditional manufacturing systems. Also, at this stage, it is possible to enhance the finish of the object by the application of paints and coatings.

8.2.5   Prospects of additive manufacturing technology

Additive manufacturing systems have been used by aerospace manufacturers since its beginnings in the '80s. But in the past few years, rapid advancements in AM technology have brought about a notable rise to applications of the technology in the aerospace industry. AM was formerly one of the prototyping technologies in aerospace manufacturing (Markets, 2014). However, as recent developments suggest, additive manufacturing has the potential to transform the production of aerospace and defense components, and its prospects in these industries are already growing fast. For instance, Airbus is exploring 90 separate cases where AM might be applied on its next generation commercial aircraft. GE Aviation is also set to manufacture up to 100,000 parts with AM by 2020. (GE, 2016)." The aerospace industry expects to derive value from additive manufacturing have been identified, which are: reduction of lead times, reduction of component weight, reduction of production and operational costs, and reduction of the negative environmental impacts of production (Markets, 2014).

8.3   A review on the challenges posed by regulatory gaps in additive manufacturing

AM techniques offer a higher degree of creative flexibility, allowing the use of multiple materials in the course of construction, as well as the ability to print multiple colors and color combinations simultaneously. Parts can now be created with complex geometries and shapes, which in many cases are impossible to create without 3DP. However, these alluring capabilities also bring with them grave concerns that seem to be going unheeded, and data suggest that those risks could end up costing several industries billions of dollars (Brugger, 2014).

3DP has been described as a *disruptive technology* (Campbell et al., 2011), because, due to a reduction in cost and the development of direct metal technologies, we are able to visualize a disruption in the manner in which products are being made in virtually all industries—architecture, consumer products, construction, industrial design, automotive, aerospace, food, engineering, biotechnology, and fashion. However, the process of 3DP is not the real disruption. It is the fact that anyone is free to own a 3D printer and create seemingly anything imaginable, from human bones to product parts. Consumers are now having access to 3D systems at retail stores allowing them to create their own products. Small, low-cost 3D printers are becoming available to be purchased for home use (Zurich, 2015).

In essence, anyone today can begin creating and selling a variety of products even from the comfort of their homes. The introduction of a turnkey solution to manufacturing coupled with a growing freedom of use forms the basis of the concerns raised.

8.3.1 Safety—Does lighter equal safer?

In aerospace manufacturing research, there is the desire to drive down the cost and weight of aircraft and to improve economy and design aesthetics. There also remains a requirement to adhere to stringent Federal Aviation Administration (FAA) regulatory and compliance standards. Complex AM processes must therefore be developed to meet the industry's stringent requirements and to ensure that products can achieve the robust performance levels established by traditional manufacturing methods. This could pose a challenge in that achieving these standards with AM may just be more cumbersome and introduce more undesirable alterations than with traditional manufacturing. Also, the research into lighter weight-printed products may require materials such as plastics and nanofibers that are new to a manufacturer. Contaminated, defective, or incorrect materials may result in a faulty product. Eventually, the materials used may even create an overall greater failure risk than those presented by the 3D printer itself (Campbell et al., 2011).

The concern is therefore how printed products will perform over time, the consistency of their quality, and the types and safety of materials used with this technology, especially with very large-scale additive-manufactured products. Thus, there is a need for an introduction of government regulation and inspection to launch quality and safety standards. For its part, the FAA says that it is making efforts to understand the implications of 3DP in the aerospace industry (Long, 2015).

8.3.2 Quality assurance and quality control gaps

New 3DP design freedoms makes simpler, lower-cost design, and assembly possible, and this means that many tools can be created with 3DP much faster than with traditional manufacturing methods. However, a very crucial issue, perhaps the most concerning, with AM is the lack of regulatory oversight for this process, since much of it takes place outside of a traditional mass production factory and not subjected to inspection from agencies (Long, 2015). Even individuals with printers at home can, with relative ease, put a variety of products in the marketplace, without the standard quality assurance/control regulatory oversight that is imbedded in traditional manufacturing. Therefore, attention needs to be paid that the zeal, particularly of unlicensed individuals, to rapidly roll out mass quantities of a printable part could leave a big gap for production of substandard items, leading to part failure and endangering the lives/health of end users.

On another note, even while printing under proper licensing and regulatory conditions, a defective product could inevitably come out of a 3D printer. Because of the multiple contributors to the production process—the printer manufacturer, software designer, materials supplier, distributor, and retailer—identifying who is liable for the failure will be a challenge.

8.3.3 National security

Another pressing concern with open access to 3DP technology is the ability for anyone, anywhere to eventually have the means of creating a weapon such as a firearm. At the present time, it may be easier for an individual with criminal intentions to obtain a weapon illegally via other means; however, with the advancement of AM technology, and new composite materials being fabricated over the next decade, could this issue become more pronounced? A more troubling prospect involves the technology being used to render detection of weapons and nuclear proliferation more difficult, which by itself makes the case for understanding the possible uses of the technology. According to New York representative Steve Israel in 1988, when the Undetectable Firearms Act was passed, the notion of a 3D-printed plastic firearm slipped through metal detectors and onto planes in secure environments was thought fiction. The problem is today a reality.

Currently in some countries like the United States, there are laws requiring a permit in order to purchase a firearm. While this serves to impede unlawful possession of such weapons, this restriction may lose effectiveness if the freedom of ownership of a 3D printer is free-for-all. Thus, for instance, a simple CAD file could be downloaded, and a gun could be fabricated within hours. Such weapons will become cheaply available nearly to everyone. Therefore, there will be a need in the very near future for government to come up with means of governing the possession of 3D printers, or otherwise restricting the kind of items that can be printed, while at the same time ensuring that citizens are not deprived of fundamental rights to freedom (Freedman and Lind, 2012).

3DP technology offers the ability to produce a wide range of objects that cannot be controlled yet, and as noted in a white paper released from the National Defense University (McNulty et al., 2012), there are definitely national security risks that need to be analyzed in the near future, and addressing criminal and legal concerns will require active cooperation across multiple agencies in the national security community.

8.3.4 Intellectual property and digital piracy

The digitization of physical artifacts allows for global sharing and distribution of designed solutions. It lowers the barriers to manufacturing and allows everyone to become an entrepreneur (Campbell et al., 2011). Open-source 3DP technology, however, also increases the risk of design theft as an original software file could easily be used to produce counterfeit products. A vast majority of the current digital software recipes are unpatented, allowing them to be copied and sold by anyone. Expensive designer objects can also be reverse-engineered and sold at a cheaper price. For product managers, this can mean an increased opportunity for counterfeit products to enter the marketplace (Long, 2015). While there have not been a tremendous number of IP issues involving 3DP yet, it could become a major problem in the near future. As more and more 3D models of products are being sold online, an entire underground market for these files will certainly emerge, and billions of dollars will be lost due to file sharing (Krassenstein, 2015).

According to new research from Gartner, the negative ramifications of 3DP to businesses, particularly those that rely on licensing deals and IP to generate revenue, are going to become a seriously expensive problem in the next few years (Brugger, 2014).

Gartner has said that companies may lose at least $100 billion in four years to licensing or IP owners. This potential digital piracy situation is comparable to the way the internet challenged the movie and music industries for copyrights, trademarks, and illegal downloads. Moreover, the current IP legislation does not explicitly regulate 3DP and will have to rush to catch up with the change in the business market that will be brought about by this technology.

8.4 Conclusion

AM is a prospective game changer with implications and opportunities that affect not just the aerospace or Department of Defense (DoD) alone but the economy as a whole. Some liken it to the *next industrial revolution*. Its ability to print complex geometries without tooling and the flexibility to print structures locally allow for mass customization, which is a key driver in new and future technologies. For aerospace, 3DP extends beyond aircraft manufacturing into ground support systems and repair. Original equipment manufacturers (OEMs) and defense contractors are growing their use of 3DP for a wide range of parts, extending usage into production of airborne parts and complete assemblies.

AM has already impacted a variety of industries and has the potential to present legal and economic issues with its strong benefits. But as history shows, a rapid introduction and adoption of a new process like this often bring with it a number of hitches that can result in grave economic, environmental, and even human loss. Because of its remarkable ability to produce a wide variety of objects, AM also can have a significant national security implications and much more complicated production scenario than the business and manufacturing world typically encounters. Therefore, to fully harness the present-day benefits and future potential of 3DP technology, it is highly necessary and wise to carefully assess the multiple potential risks both for today as well as potentially unknown risks that will continue to evolve, as the technology advances in its strides to revolutionize the face of manufacturing.

References

Allen, J. (2006). *An Investigation into the Comparative Costs of Additive Manufacture vs. Machine from Solid for Aero Engine Parts.* Derby: Rolls-Royce plc., Manufacturing Technology. Retrieved from http://dtic.mil/dtic/tr/fulltext/u2/a521730.pdf.

Brugger, T. (2014, February 5). *The Dark Side of 3D Printing.* Retrieved from AOL news: http://www.aol.com/article/2014/02/05/the-moral-dilemma-of-3d-printing/20823680/.

Campbell, I., Bourell, D., & Gibson, I. (2012). Additive manufacturing: Rapid prototyping comes of age. *Rapid Prototyping Journal, 18,* 255–258.

Campbell, T., Williams, C., Ivanova, O., and Garrett, B. (2011). *Could 3D Printing Change the World: Technologies, Potential, and Implications of Additive Manufacturing.* Washington, DC: Atlantic Council.

Déspeisse, M., and Ford. S. (2015, June). The role of additive manufacturing in improving resource efficiency and sustainability. *Centre for Technology Management working paper series*, pp. 1–9.

GE. (2016). *Additive Manufacturing.* Retrieved from GE Imagination at work: www.ge.com.

Gu, D. (2015). Laser additive manufacturing (AM): Classification, processing philosophy, and metallurgical mechanisms. In D. Gu (Ed.), *Laser Additive Manufacturing of High-Performance Materials* (pp. 15–71). Berlin, Germany: Springer.

Gu, D. D., Meiners, W., Wissenbach, K., and Poprawe, R. (2012). Laser additive manufacturing of metallic components: Materials, processes and mechanisms. *International Materials Reviews, 3*(57), 133–164.

Halloran, J. W., Tomeckova, V., Gentry, S., Das, S., Cilino, P., Yuan, D., Guo, R., et al. (2011). Photopolymerization of powder suspensions for shaping ceramics. *Journal of the European Ceramic Society, 31,* 2613–2619.

Hildebrand, K., Bickel, B., and Alexa, M. (2013). Orthogonal slicing for additive manufacturing. *Computers & Graphics, 37,* 669–675.

Huang, S. H., Liu, P., Mokasdar, A., and Hou, L. (2013). Additive manufacturing and its societal impact: A literature review. *The International Journal of Advanced Manufacturing Technology, 67,* 1191–1203.

IBM. (2016, June 16). *Local Motors Debuts "Olli", the First Self-driving Vehicle to Tap the Power of IBM Watson.* Retrieved from IBM Press Release: http://www-03.ibm.com/press/us/en/pressrelease/49957.wss.

Klein, J., Stern, M., Franchin, G., Kayser, M., Inamura, C., Dave, S., Weaver, J. C., Houk, P., Colombo, P., Yang, M., and Oxman, N. (2015). Additive manufacturing of optically transparent glass. *3D Printing and Additive Manufacturing, 2,* 92–105.

Krassenstein, B. (2015, July 16). *3 Dangers Society Faces from 3D Printing*. Retrieved from 3DPrint.com: https://3dprint.com/81526/3d-print-dangers/.

Lind, M., and Freedman, J. (2012). *Value Added: America's Manufacturing Future*. Washington D.C.: New American Foundation.

Local Motors. (2016, June 19). Retrieved from local motors: https://localmotors.com/3d-printed-car/.

Long, E. (2015, February 16). *Additive Manufacturing Creates Tremendous Opportunities, as well as Risks*. Retrieved from 3D Print.com: https://3dprint.com/44105/3d-printing-risks/.

McNulty, C. M., Arnas, N., and Campbell, T. A. (2012, September). *Toward the Printed World: Additive Manufacturing and Implications for National Security*. Washington, DC: National Defense University, Institute for National Strategic Studies.

Petrick, I. J., and Simpson, T. W. (2013). 3D printing disrupts manufacturing: How economies of one create new rules of competition. *Research-Technology Management, 6*, 12–16.

Piller F.T., Weller C., Kleer R. (2015) Business Models with Additive Manufacturing—Opportunities and Challenges from the Perspective of Economics and Management. In: Brecher C. (Eds.), *Advances in Production Technology. Lecture Notes in Production Engineering*. Springer, Cham.

SmarTech Markets. (2014). *Additive Manufacturing in Aerospace: Strategic Implications*. http://www.smartechpublishing.com

Ulanoff, L. (2014). *World's First 3D Printed Car Took Years to Design, But Only 44 Hours to Print* [Online]. Mashable: Mashable. Available: http://mashable.com/2014/09/16/first-3dprinted-car/[Accessed 5 June 2016].

Wang, W., Chao, H., Tong, J., Yang, Z., Tong, X., Li, H., Liu, X., and Liu, L. (2015). Saliency-preserving slicing optimization for effective 3D printing. *Computer Graphics Forum, 34*, 148–160.

White, G., and Lynskey, D. (2013). *Economic Analysis of Additive Manufacturing for Final Products: An Industrial Approach*. University of Pittsburgh, mimco.

Wong, K. V., and Hernandez, A. (2012). A review of additive manufacturing. *International Scholarly Research Network (ISRN), Mechanical Engineering, 2012*, 1–10.

chapter nine

Additive manufacturing and its implications for military ethics

John Mark Mattox

Contents

Abstract: Recent advances in the so-called *additive manufacturing* pose significant, new challenges, in scope if not in kind, for military ethicists. While the problem of dual-use technologies (i.e., technologies that can be used for both good and malevolent purposes) is not new, the possibility of the rapid, uncontrolled replication of highly sophisticated tools of violent action—tools that heretofore have been largely inaccessible to laymen—could vastly expand the number of persons able to commit violent acts, or even wage war, far beyond traditional boundaries. This article explains the nature of additive manufacturing and identifies the challenges it poses for militaries and governments with the de facto responsibility to keep war-making tools out of the wrong hands. In light of the industrial revolution occasioned by the advent of additive manufacturing and the revolution in military affairs that it portends, it proposes a research agenda for military ethicists. In particular, it argues that military ethicists must now expand their scope of inquiry in a way that accords due prominence to the nexus between these technologies and jus ante bellum issues of conflict avoidance.

Keywords: Additive manufacturing, *jus ante bellum*, industrial revolution, revolution in military affairs, weapons of mass destruction

9.1   Introduction

Whenever a copy is made of anything, there arises a set of easily predictable normative questions, to include the following:

- What should one be allowed or forbidden to copy?
- Who should or should not be allowed to make, see, or otherwise use the copy?
- For what purposes should or should not the copy be allowed to be used?

Indeed, there exist entire bodies of national and international laws, in the form of copyright and patent laws and treaties, that deal with these and similar issues. Hence, the ethical issues involved may seem to be essentially settled matters and, to that extent, uninteresting.

However, with the advent of the most recent copying technologies, ethicists of all kinds, to include military ethicists, must now seriously consider the ramifications of what it means to copy things that, in the wrong hands, could prove monumentally disastrous. It is the ethical challenge associated with the emerging technology known as *additive manufacturing*.

9.2   A new industrial revolution

In order to appreciate the magnitude of the ethical issues potentially associated with additive manufacturing, it is first necessary to appreciate the magnitude of the difference between additive manufacturing and traditional manufacturing techniques. For several millennia, and in increasingly refined form, traditional manufacturing has relied principally on two major kinds of processes: *subtractive manufacturing* and *forming* (Williams and Campbell 2011).

- Subtractive manufacturing involves starting with a large mass of some medium— say, a block of steel or aluminum—and removing portions of that mass until the desired shape is all that remains—similar to what happens when a sculptor chisels away at a rough-hewn block of stone until all that remains is the smooth image of a human figure.
- *Forming*, the other traditional approach to manufacturing, is accomplished through processes such as *casting*, *stamping*, or *injection molding*.
- *Casting* involves pouring a molten liquid into a mold that contains a hollow cavity of the desired shape, and then allowing the liquid to solidify. This technique is used frequently to make complex shapes—for example, an automobile engine block—that would be difficult or uneconomical to make by other methods (see DeGarmo et al. 2003, p. 277).
- *Stamping* includes a range of manufacturing processes in which a medium is manipulated to produce a new shape or design. For example, the metal kitchen sink found in many western homes was once a flat piece of metal, which was bent into the shape of a sink and into which holes were then punched to accommodate the drain and faucets. Coins are another example of something produced by stamping.
- *Injection molding* involves the creation of a mold that can be separated into two parts, into which a liquid medium is injected and from which, after the liquid solidifies, the mold can be removed. For example, many plastic objects—like resealable food containers or cell phone cases—are formed in this way.

In contrast to these traditional manufacturing methods, additive manufacturing actually *adds* materials together so as to form a larger object. This technique starts with a three-dimensional (3D) computer-aided design (CAD) model and uses a computer algorithm to *slice* the 3D model into very thin cross sections. The computer images are then sent to a 3D printer, which builds the object one very thin layer at a time.

The implications associated with the ability to construct things in this manner are nothing less than staggering. Additive manufacturing has already been incorporated, in one form or another, into large segments of the manufacturing economy to include consumer products, motor vehicles, medicine and dentistry, industrial and business machines, and aerospace. It has also found its way into architecture, academe, government, and military. As one observer has noted, the first industrial revolution harnessed steam power and mechanization and defined the nineteenth century; the second industrial revolution combined standardization and mass production to define the twentieth century; additive manufacturing marks the beginning of the third industrial revolution and is already defining the twenty-first century (Pierpoint 2012). Indeed, there appears to be no theoretical limit to the array of objects that can be produced using additive manufacturing—even objects that there is no physical way to create by any other means. This includes, incredibly, the possibility of fabricating replacement human body organs or parts on demand using the genetic specifications of the intended recipient, so that the recipient's body will not reject it. The only limit to the size of the object to be printed in three dimensions is the size of the 3D printer. To date, there exist 3D printers large enough to *print* whole car bodies or large building construction components. Conversely, microscale objects have also been printed in 3D. For example, additive manufacturing technology has been used by a Japanese university to produce the world's smallest sculpture—10-μm long and 7-μm high—the size of a red blood cell (Iga and Kokubun 2006, p. 175).

Likewise, there seems to be no theoretical limit to the media that can be used in additive manufacturing processes. Although, *there are only a few multimaterial printers currently available*, still a much broader array of manufacturing possibilities clearly is feasible, and there is no reason to believe that the coming decades will not include breakthroughs in additive manufacturing that would astonish laypersons living today. Moreover, because additive manufacturing represents not merely an advance but indeed a quantum leap in the development of industrial processes, the industrial revolution of which it is the foundation brings with it the potential for a revolution in military affairs in which laymen acquire the ability to produce highly sophisticated items for violent or even war-making purposes as readily as they could produce an utterly benign item for routine household use.

9.3   Implications for military ethics

However, even if one grants that additive manufacturing likely represents the advent of a new industrial revolution, one may be tempted to conclude that, ethically speaking, there really is nothing new here. After all, traditional technologies, to include both subtractive manufacturing and forming, have been used both for peaceful and violent, good and evil purposes for millennia, and so the fact that one and the same technology can be put to contrary purposes hardly seems, *prima facie*, to generate a new or even broadened set of moral–philosophical concerns. However, precisely because additive manufacturing introduces the prospect of small groups or even individuals operating on small budgets being enabled to produce war-making material that, up to the present, only governments or large commercial firms possessed the wherewithal to produce, the vast scope of the regulatory problem presented by additive manufacturing is unprecedented in the history of technology. Indeed, one is reminded of words frequently attributed to Joseph Stalin: "Quantity has a quality all its own." The idea of ubiquitously available weaponry previously available only to nation–states has profound implications on a host of accounts, including military ethics. Consider the vast extent of the problem set as described in the following sections.

9.3.1   Small arms

At first blush, small arms might seem to be unlikely candidates for additive manufacturing, since militaries typically purchase items like these in large quantities, and at least at present, additive manufacturing technologies are not optimized for mass production. That intuition, however, may be irrelevant in the case of extranational paramilitaries and terrorist organizations. What a perfect way to replicate weaponry that cannot be traced by serial number or by any other traditional accounting means! As the sophistication of the technology and its ability to produce small arms in large quantities mature, paramilitary or terrorist organizations could look far more like regular armies than they now do because of their ability to equip themselves. Whether these regular army-like organizations would feel themselves bound by long-standing normative undertakings relative to the conduct of war is questionable indeed. However, the problem of regulating small arms is not restricted to organizations. Already, individuals are able to produce restricted automatic weapon components using additive manufacturing—an issue that has not gone unnoticed in the current gun control debate in the United States (Farivar 2013). That same manufacturing capability could be employed by individual uniformed military personnel, particularly those in specialized units, to equip themselves at will with personal kit that could transgress a legal or moral line. Couple these concerns with what appears to be an insatiable global small arms market and a commercial business ethic that is all too often trumped altogether by the profit motive, and what results is a milieu in which the problem of illicitly copied small arms may become as ubiquitous as the problem of illicitly copied music for iPods. While the proliferation of pirated music may not be something that the military ethicist spends a lot of time thinking about, the completely uncontrolled and unlimited ability to produce small arms on demand is something the military ethicist ignores at the risk of significant consequences.

9.3.2   Major military hardware

Additive manufacturing is ideal for the reproduction of complex and even custom geometries and low-volume production replacement parts, and it can produce these shapes in a way that greatly reduces assembly requirements. Take, for example, a duct for an F-18 fighter jet. The conventional duct fabricated using traditional processes consists of 16 parts, including nuts and bolts, to form an assembly that still requires being glued together. Using additive manufacturing technologies, the same duct can be made as one complete piece— no assembly required. Soon, major components of military hardware also will be produced in this way. In 2012, it was reported that "Airbus is planning to print out a wing for an airliner in the near future. They are not planning to print out a plan for the wing or a picture of the wing. They are going to print the actual wing itself—a wing that will someday be part of an aircraft flying at 500 miles per hour at 40,000 feet altitude" (Pierpoint 2012).

   The ability to manufacture repair parts—including major assemblies—in this way means that a whole host of issues relative to foreign military sales and export regimens for replacement parts for the so-called *major-end items* of military equipment (such as fighter jets and tanks) must be carefully rethought. Why? Because, hitherto, a certain modicum of control could be exerted upon purchasers of major end items by regulating their access to the specialized replacement parts required by these systems. This limitation has long served both the political and moral purposes of regulating purchasing nations' bellicose conduct and of keeping otherwise unobtainable military hardware out of malevolent hands. However, as the ability of purchasers to create their own replacement parts dramatically increases, the ability to regulate violent action by regulating the supply of replacement parts commensurately decreases. This creates a greatly expanded array not only of political problems but also of moral ones, because it requires that nations make new normative decisions about, and weigh more carefully than ever before, the question of to whom and under what circumstances it is morally permissible to sell military equipment abroad and when and to whom it is not. Failure to do so might result in enabling nonstate actors to wage war on a much larger scale than they otherwise would have been able to do, because of their enhanced ability to supply their own replacement parts.

9.3.3   New concepts in war-making equipment

An even bigger problem than the prospect of additive manufacturing technologies making replacement parts readily available, or even that those technologies could be used to replicate the major end items themselves, is that additive manufacturing could render altogether obsolete the big, bulky machines of the kind that have typified military equipment from the Renaissance to the end of the twentieth century. For example, additive manufacturing is already well on its way to producing, *in toto*, unmanned aerial vehicles of the kind that have proliferated in the first decade of the twenty-first century and which have, in their own right, become objects of *jus in bello* inquiry. In 2011, a university in the United Kingdom designed an entire unmanned aerial vehicle, printed it using additive manufacturing technology, and flew it—all in one week (Tate 2011).

9.3.4   Catastrophic weapons

The problem likely to be of greatest concern, however, is that which ongoing advances in additive manufacturing could pose relative to the production of weapons of mass destruction. Up until now, chemical and biological weapon productions often have required expansive industrial complexes. However, emerging advances in nanotechnology coupled with emerging additive manufacturing technologies may make the production of chemical or biological warfare agents using microreactors possible. It could be, therefore, that additive manufacturing technologies may be just the solution to the production of highly toxic substances, very small quantities of which could produce large numbers of human fatalities. Nuclear weapon production also historically has required vast industrial complexes. (Indeed, one of the secret cities associated with the Manhattan Project, which produced the world's first nuclear weapon, consumed 10% of all electricity produced in the United States at the time.) Although up until now, additive manufacturing processes have not involved heavy metals like uranium or plutonium, if (or when) the technological barriers to their use in additive manufacturing are breached, one cannot discount the prospect of nuclear weapons' components being manufactured by something that looks far more like a *Mom and Pop* operation with a fancy copying machine rather than by something that looks anything remotely like the enormous industrial centers associated with nuclear weapons production by the superpowers during the Cold War. Indeed, one can barely imagine more catastrophic consequences than those that attend the prospect of additive manufacturing technologies being applied to produce weapons of mass destruction. They do not require production in quantity; indeed, the ability to produce them at all is sufficient to shift balances of military and political powers globally. Imagine additive manufacturing technology is applied to weapons of mass destruction in the hands of persons like the 9/11 terrorists. This could mean that highly sophisticated technological means to perpetrate not merely monstrous criminal acts, but indeed acts of *war* could find their way into the hands of terrorist cells bent on wholesale destruction. One might be tempted to dismiss the problem by observing that would-be mass destroyers would have to obtain special nuclear materials—weapon-grade uranium or plutonium—before they could manufacture a device capable of producing a nuclear yield. While this is true, it is also the case that these materials exist in abundance in the aftermath of the Cold War, and that some nation–states continue to manufacture them. Moreover, some possessors of these materials have either questionable ties to terrorists or are suspected in their ability to maintain strict account-ability of, or control over, their weapon-grade nuclear materials. Similarly worrisome is the specter of disposable bioreactors, kits for conducting gene-splicing experiments, or chemical microreactors—all of which are part and parcel of the unfolding drama of addi-tive manufacturing, and which, in the wrong hands, could allow private citizens to manu-facture inherently dangerous materials in the privacy of their own homes.

9.4   New imperatives for military ethics

The challenges posed by additive manufacturing have implications for all aspects of military ethical inquiry. Let us begin, for example, with *jus in bello*—the most thoroughly developed domain of the military ethicist. While the world has had—particularly in the last decade—extensive experience in dealing with nonstate actors whose organizations have looked and acted very much like armies, these *armies* have had to rely on sympa-thetic entities to supply them with manufactured military equipment. What if they were able to make their own equipment—not crude devices like roadside bombs but highly sophisticated devices of the kind traditionally available only to nation–states? Such a capability would further blur the already blurred distinction between combatant and common criminal, and along with it, questions surrounding which rules of engagement, which standards of evidence governing targeting criteria, and which safeguards on civil rights should be applied to whom. For example, is an additive manufacturing capability in the possession of entities sympathetic to perpetrators of violent action a proper mili-tary target? Should persons who use the equipment, even if not for military purposes, be regarded as combatants?

With respect to *jus ad bellum*, if advances in additive manufacturing enable the waging of *de facto* war (as opposed to, say, merely terrorist action) by nonstate entities, the whole question of what constitutes a war and the moral boundaries for responding to it with military as opposed to law enforcement means may have to be reconceived. In a world full of nonstate entities able to self-equip for war, what new kinds of aggrievements, if any, might constitute *just cause* to go to war against them? What might count as a *last resort*? In what way and to what extent should nation–states be held accountable for the malevolent acts of *de facto* armies operating within their boundaries but independent of state support? Of course, questions like these are not, *prima facie*, novel. However, they heretofore have occurred in a context in which the role of nonstate actors was considerably more periph-eral than it may become with the aid of the emerging technologies described herein.

Given the proposition that, while it is good to mitigate the evil effects of war, it is far better to avoid them altogether; the military ethicist of the twenty-first century might, in fact, find the intellectual tools of the profession best applied to *jus ante bellum*. Until recent history, this domain could be left to diplomats, lawyers, and policy makers of different sorts, and, of course, these will continue to play a vital role. However, in a world where wars with nonstate entities frequently occur but are rarely declared, and in which these entities can equip themselves with high-tech military hardware, moral structures more thoroughly focused on war prevention rather than war regulation may become essential. That is not by any means to suggest that *jus ad bellum* and *jus in bello* considerations would become obsolete, but merely to acknowledge that, perhaps far more than any time in the past, *jus ante bellum* considerations calculated to prevent violent conflict could assume an unprecedented level of prominence. Indeed, it may be argued that the challenge presented by additive manufacturing provides precisely the impetus needed to extend *jus ante bellum* discussions into a broader framework that includes not only anticipations of war but also war and violence prevention. This is so because, clearly, the issues involved in the military application of additive manufacturing technologies—especially as they could one day apply to weapons of mass destruction—invite an ethical discussion on how to restrict easy access to these tools (and thereby restrict easy access to the incentives) to engage in violent action in the first instance, whether in the form of traditional warfare or terrorist action. *Jus ad bellum* discourse has always observed a strong presumption against war—that wars, if they are to be fought at all, should be fought only as a last resort. The logical extension of that premise is that if the set of circumstances that constitute justified resort to war *can be* constrained, then morality requires that they *should be* constrained. In that respect, the advances now evident in additive manufacturing technologies provide ample justification for an expanded view of *jus ante bellum* that includes questions on how to constrain access to emerging technologies applied to violent purposes. In short, while, until the present day, a sharp, perhaps even exclusive, focus on *jus in bello* as traditionally conceived may have been appropriate for military ethicists, problems of the kind now posed by additive manufacturing may make it such that it is no longer sufficient for military ethicists to focus exclusively on dissecting questions about who can or cannot kill whom when. On the other hand, if the military ethics community is willing to expand the scope of its discourse beyond its present circumscription, it may find itself uniquely positioned to provide an essential intellectual service to the emerging world of twenty-first century conflict and violence.

In particular, military ethicists might profitably seek a voice in the formulation of new global norms. Since it may be both physically impossible and, indeed, undesirable, to prevent altogether the proliferation of these emerging technologies—especially since, when used properly, they offer tremendous benefits to humankind—it may be that the best safeguard against their malevolent use is the establishment of new international norms. A host of international undertakings (e.g., the Nuclear Nonproliferation Treaty, the Chemical Weapons Treaty, the Biological and Toxins Weapon Convention, and the Conventional Forces in Europe Treaty) represent well-established attempts to regulate access to weaponry. However, such regulatory efforts historically have focused upon nation–states. Since additive manufacturing technology could open the door to the possibility of small groups or individuals availing themselves of weapons that states are not allowed to possess (or if allowed to possess, only within strict parameters), the result could require a radical reconceptualization of what *warfare* entails. For example, if an act of war traditionally associated with nation–states is committed instead, and in similar magnitude, by a self-arming nonstate entity, should it be treated as an act of war—and if so, to what end? (Note that a low-tech and small-scale version of what the future could bring is already evident in the challenges associated with fighting a *war* with the Taliban in

Afghanistan.) Regardless of the form that new international understandings to govern the coming revolutions in manufacturing and military affairs might take, the formulation of appropriate norms is a matter in which military ethicists should seek to have a voice. In a related vein, it may be necessary for military ethics to re-tool its thinking with respect to the problem of terrorism and of what constitutes a terrorist. Not only could additive manufacturing make guns, bullets, bombs, and so on cheaper and more accessible, it also could facilitate their disguise, such that improvised explosive devices (IEDs) could be made to look identical to nonweapons (Williams and Campbell 2011); terrorists could altogether lose their dependency upon sympathetic nation–states for their high-tech-manufactured war material, and the entire complexion of organized warfare could become radically different.

9.5 A proposed research agenda for military ethics

Much of the history of ethics is a history of reacting to extant dilemmas. While this is understandable, the perils of the twenty-first century are such that it may no longer be sufficient for ethicists to content themselves with what might be termed *reactive reflection*. In light of the possibilities presented by additive manufacturing, it now becomes necessary for ethicists to become *proactively* reflective and to consider dilemmas to which technology will surely give birth with ever-diminishing lead time. The slate of military ethical problems associated with this rapidly emerging and truly transformative technology suggests a research agenda that includes a host of specific questions, of which the following are representatives:

- What morally justifiable approaches can be taken perhaps along the lines of the proposed Fissile Material Cutoff Treaty, to restrict access either to stockpiled special nuclear materials or the production of these materials, or to precursors of chemical and biological weapons? In light of the emerging threat, it may be that states unable or unwilling to prevent access to these materials by small groups or individuals should be held accountable—if necessary, with the appropriate application of force. If so, an expanded understanding of acceptable *jus ad bellum* parameters may be required.
- In a related vein, what justifications can be given for violent interventions beyond the boundaries of one's own nation–state's borders, to include preemptive action, to prevent the use of emerging technologies for malevolent purposes—particularly as they might be brought to bear with respect to weapons of mass destruction? Implicit in this question as well is a requirement to explore expanded understandings of *jus ad bellum*.
- What moral justification can be given to insist that some entities are entitled access to special nuclear materials and that others are not? While this has long been an issue, emerging additive manufacturing technologies could add new impetus to it. Already, states such as Iran and North Korea appeal to moral arguments to justify possession of special nuclear materials outside the established international legal frameworks. The ability of small groups or individuals to possess these materials would fundamentally transform the nature and scope of the regulatory mechanisms required.
- To what extent can persons be morally required to surrender claims to rights of privacy in order to ensure that their use of additive manufacturing technologies is not malevolent? The enormous complications associated with this question are reflected in the perennial gun control debate in the United States. One can easily imagine

the following variation on an already famous argument: "Additive manufacturing—even if used by private persons to produce deadly weapons—does not kill people; people kill people." Instead of the current preoccupation with the question "What should private individuals be allowed to buy?," the more pressing normative question might become "What should private individuals be allowed to *make*?"

- Indeed, is there any such thing as a right to privacy when it comes to the production of weapons of mass destruction? (An interesting question—especially in light of the fact that both nation–states and nonstate actors have always shrouded their production of such weapons in utmost secrecy).
- How should additive manufacturing as it relates to the manufacture of large and small conventional war material be regulated?

9.6   Conclusion

These are not easy questions. Moreover, the ethical problems that they present—not to mention those persons whose actions give rise to these problems—will not wait for ethicists to decide when it is professionally interesting or intellectually stimulating to address them. These problems are quickly heading our way. Indeed, additive manufacturing technologies are already being used to produce out of plastic fully functioning weapons and weapon components that are presently made out of metal. Just over the temporal horizon lies the prospect of manufacturing small arms and many other weapon-related items out of multiple media simultaneously. Only a bit farther over the horizon lies the production of not only repair parts, but also whole major end items themselves.

Past discussions over the control of military wherewithal have not necessarily been informed by voices from the military ethics community. However, the community can no longer responsibly avoid participation in that discussion—especially in light of the prediction that, in a future day, individuals will have 3D printers in their homes and, instead of having manufactured items shipped to them, they simply will download a design file and manufacture the item themselves with a 3D printer on their kitchen counter (Pierpoint 2012). If questions related to the possibility of laypersons being empowered to create weapons and wage unrestricted war are not proper questions for military ethicists, then there *are no* proper questions for military ethicists.

The sweeping technological revolution on our immediate horizon could change war fundamentally, and the revolution is occurring so rapidly that, unless we are vigilant, we could find ourselves without the ethical tools necessary to deal with the change. The impending changes will profoundly affect both soldiers on the battlefield and private citizens attacked as noncombatants. Indeed, unless these ethical problems are dealt with, every additive manufacturing problem may become a military ethics problem.

Acknowledgments

The author wishes to acknowledge his debt of gratitude to his colleagues, Dr. Christopher Williams and Dr. Thomas Campbell of the Virginia Polytechnic Institute and State University for introducing him to the world of emerging manufacturing technologies as well as for providing insights without which this paper would not have been possible. Thanks are also due to the referees for this journal. All views expressed herein are the responsibilities of the author and do not necessarily reflect the position of any U.S. government entity.

References

DeGarmo, E. Paul, Black, J. T. and Kohser, Ronald A. (2003) *Materials and Processes in Manufacturing*, 9th edn. (Hoboken: Wiley).

Farivar, Cyrus (2013) "Download This Gun": 3D-printed Semi-Automatic Fires over 600 Rounds, accessed September 23, 2013, available at: http://arstechnica.com/tech-policy/2013/03/download-this-gun-3d-printed-semi-automatic-fires-over-600-rounds.

Iga, Kenichi and Kokubun, Yasuo (Eds) (2006) *Encyclopedic Handbook of Integrated Optics*, s.v. "Nanophotonics" by Satoshi Kawata, Yasushi Inouye and Hong-Bo Sun (Boca Raton, FL: CRC Press).

Pierpoint, Paul (2012) Why Industry Needs to Embrace Additive Manufacturing and the New Industrial Revolution, accessed September 23, 2013, available at: http://www.lvb.com/article/20120517/LVB01/120519887/0/FRONTPAGE/Why-industry-needs-to-embrace-additive-manufacturing-and-the-new-Industrial-Revolution.

Tate, Paul (2011) Additive Manufacturing: Can It Bring Manufacturing Home? accessed May 31, 2012, available at: http://www.manufacturing-executive.com/community/leader-ship_dialogues/game-changing_technologies/blog/2011/12/02/additive-manufacturing-can-it-bring-manufacturing-home.

Williams, Christopher and Campbell, Thomas (2011) Additive Manufacturing (3D Printing): State of the Art, Potential and Implications, lecture given at National Defense University, Washington, DC, August 19, 2011.

chapter ten

Additive manufacturing technologies
State of the art and trends

Julien Gardan

Contents

Abstract: The rapid prototyping has been developed from the 1980s to produce models and prototypes until the technologies evolution today. Nowadays, these technologies have other names such as 3D printing or additive manufacturing, and so forth, but they all have the same origins from rapid prototyping. The design and manufacturing process stood the same until new requirements such as a better integration on production line, a largest series of manufacturing, or the reduced weight of products due to heavy costs of machines and materials. The ability to produce complex geometries allows proposing of design and manufacturing solutions in the industrial field in order to be ever more effective. The additive manufacturing (AM) technology develops rapidly with news solutions and markets, which sometimes need to demonstrate their reliability. The community needs to survey some evolutions such as the new exchange format, the faster 3D-printing systems, the advanced numerical simulation, or the emergence of new use. This review is addressed to persons who wish to have a global view on the AM and to improve their understanding. We propose to review the different AM technologies and the new trends to get a global overview through the engineering and manufacturing process. This article describes the engineering and manufacturing cycle with the 3D-model management and the most recent technologies from the evolution of AM. Finally, the use of AM resulted in new trends that are exposed below with the description of some new economic activities.

Keywords: additive manufacturing, rapid prototyping; manufacturing technologies, bioprinting, design process, topological optimization

10.1   Additive manufacturing

The first method to create a three-dimensional object layer-by-layer using computer-aided design (CAD) was rapid prototyping, developed in the 1980s to produce models and prototype parts. The main advantage of the AM is its ability to create almost any possible shape, and this capacity is run by the layer-by-layer manufacturing. AM technology is most commonly used for modeling, prototyping, tooling through an exclusive machine, or 3D printer. AM is largely used for manufacturing short-term prototypes, but it is also used for small-scale series production and tooling applications (rapid tooling) (Stampfl and Hatzenbichler 2014). This technology was created to help and support the engineers in their conceptualization. Among the major advances that this process presented to product development are the time and cost reduction, human interaction, and consequently the product cycle development (Ashley 1991). It also provides the possibility to create almost any shape to validate functionality and aestheticism. Those shapes could indeed be very difficult to manufacture with other processes (e.g., milling, molding). The complex geometries or the curved surfaces needed have to be maintained with a support material. The feedback has a great influence on the quality or effectiveness of the manufactured model. From one technology to another, the manufacture direction, the model orientation, and the material behavior are important to get an accurate model and an efficient production (Beaman et al. 1997). Nowadays, these technologies have other names such as 3D printing, and so forth, but they all have the same origins from rapid prototyping. The demand of AM machines is increasingly growing since the 1990s (Wohlers 2012). Due to the evolution of rapid prototyping technologies, it has become possible to obtain parts representative of a mass production within a very short time (Bernard and Fischer 2002). AM perfectly fits into the numerical design and manufacturing chain. Thus, AM is very complementary with the reverse engineering to reproduce or repair a model.

Many rapid prototyping technologies have appeared on the market (Wang and Zhang 2012) based on the same layers' manufacturing approach. AM or 3D printing have strongly been developed and currently proposed several solutions. Use of AM leads to new practices in different domains, which push the manufacturer to adapt. The constant improvement of AM technologies also leads to new solutions driven by very strong demand (Wohlers and Caffrey 2015). Use and evolution change gradually the product life cycle in order to reducing the manufacturing cost and time while increasing reliability. Thus, this survey is motivated by the need to get an overview of AM and its emerging domains. This chapter proposes to realize a technologic review of manufacturing processes followed by their illustrative scheme. The Additive Manufacturing principles have been classified by manufacturing technologies to explain them. First of all, we will describe the design process before the technologies' description while involving some industrial and academic trends.

10.2 Engineering and manufacturing process

10.2.1 Rapid prototyping cycle

The stages involved to the product design and the rapid prototyping show that the cycle development is specific (Figure 10.1). These rapid prototyping processes generally consist of a substance, such as fluids, waxes, powders, or laminates, which serve as basis for model construction as well as sophisticated computer-automated equipment to control the processing techniques such as deposition, sintering, lasing, and so on (Dolenc 1994). There are two possibilities to start an AM cycle, begin with a virtual model or a physical model. The

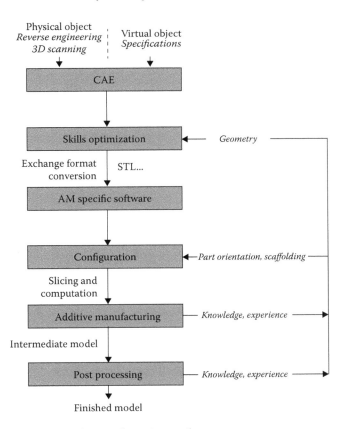

Figure 10.1 AM engineering and manufacturing cycle.

virtual model created by a CAD software can be either a surface or a solid model (Chua et al. 2010). On the other hand, 3D data from the physical model are not at all straightforward, and it requires data acquisition through a method known as a reverse engineering (Chua et al. 2010). The process begins with a 3D model in CAD software before converting it in STL format file. This format is treated by specific software, own to the AM technology, which cuts the piece in slices to get a new file containing the information for each layer. The specific software generates the hold to maintain the complex geometries automatically with sometimes the possibility to choose some parameters. We can decompose the engineering and manufacturing cycle (Figure 10.1) following these steps:

- Part design in CAD or reverse engineering by 3D scanning
- Skills optimization in CAE (computer-aided engineering) to adapt the part to the manufacturing technology chosen
- Conversion of part geometry in exchange format (STL)
- Exchange file implementation into the specific software of the AM machine
- Configuration and orientation of the set (parts and supports)
- Slicing of the part by the specific software
- Computation and layers manufacturing
- Postprocessing

This new file is often proprietary of the machine manufacturer. Rapid manufacturing machine implement the last file to realize the layer-by-layer manufacturing. The operator has to prepare the machine with its raw material (powder, resin cartridge(s), polymer spool(s), etc.) and the manufacturing source (laser, printing head(s), binder cartridge(s), etc.). For the manufacturing, the support material maintains the external and internal surfaces to keep a steady geometry with a structure using scaffolding. In most cases, the support material is cleaned during the finishing (e.g., MJM Technology) or recycled during the postprocessing (e.g., SLS, SLM, CJD/3DP technologies). This step depends on the complex geometry fabricated and if there is need an additional hold resulting in a loss of material. Some technologies allow extracting of the main material, thanks to holes inside closed geometry. The postprocessing step sometimes includes a hardening or infiltration of the main material to obtain the final piece. Several manufacturing constraints require a feedback while involving rules to get a precise and compliant model.

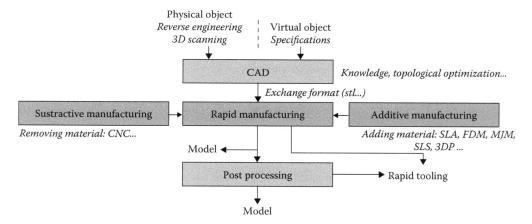

Figure 10.2 Rapid prototyping process.

Rapid prototyping techniques are classified in two categories: subtractive, and additive (Figure 10.2). Subtractive technologies work by removing raw material out of a workpiece until the desired shape is obtained. They include cutting (laser cutting or water-jet cutting) and machining (lathing and milling). Conversely, the additive technologies work by adding of the raw material.

Modeling is a very important step in AM because it shapes the product geometry, but it also must take in account some knowledge since the experiments and equipment are costly. Various potential empirical modeling techniques coexist so that the choice of an appropriate modeling technique for a given AM process can be made. To develop models based on only given data, several well-known statistical methods such as regression analysis or response surface methodology can be applied (Garg et al. 2014a). The formulation of the physics-based models requires in-depth understanding of the process and is not an easy task in the presence of partial information about the process (Vijayaraghavan et al. 2014). Few research studies have been conducted to improve the prediction ability of the genetic programming (GP) and the multigene genetic programming (MGGP) models by hybridizing them with the other potential computational intelligence methods such as artificial neural network (ANN), fuzzy logic, M5′ regression trees, and support vector regression (Garg and Tai 2013; Garg et al. 2014c). MGGP is the most popular variant of GP used recently. Those approaches provide a model in the form of a mathematical equation reflecting the relationship between the mechanical behaviors and the given input parameters. The performance of ANN is found to be better than those of GP and regression, showing the effectiveness of ANN in predicting the performance characteristics of prototype (Garg et al. 2014b).

10.2.2   *Exchange format*

The STL (Standard Tessellation Language) file format was created by 3D systems in 1987 and became a standard for the AM. It offers the advantage of being easily generated by all CAD software. The STL file creation process mainly converts the continuous geometry in the CAD file into a header, small triangles, or coordinates triplet list of x, y, and z coordinates and the normal vector to the triangles. This process is inaccurate, and the smaller the triangles are, the closer to reality it is (Noorani 2006). Each facet is uniquely identified by a normal vector and three vertices (Figure 10.3). The facets define the surfaces of a 3D object. Each facet is part of the boundary between the interior and the exterior of the object, and each triangle facet must share two vertices with each of its adjacent triangles (Chen et al. 1999). The surface creation can generate errors

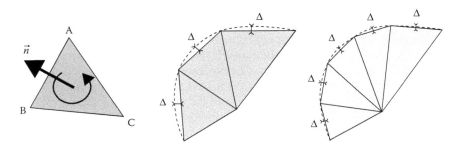

Figure 10.3 STL triangle, A, B, C vertices, and its normal vector n/Facets increment.

because of holes or intersecting triangles, and it is sometimes necessary to repair the STL model. The slicing process also introduces inaccuracy to the file, because here the algorithm replaces the continuous contour with discrete stair steps (Iancu et al. 2010). Edges are added after the slicing process. Today, the computation data and the mesh generation are no longer an obstacle to process models. The computer power used is sufficient to get a refined STL file with many triangles. More the 3D model refined is high, the clearer the details are, and the bigger the file size is.

10.2.3 Economy and users

According to the 2014 Wohlers Report, consumers of 3D printers are classified as those that cost less than $5,000 (Wohlers and Caffrey 2014). The Cornell University and the University of Bath have designed the first open-source 3D printers, which are widely recognized in the area: Fab@home and RepRap (Malone and Lipson 2007; Sells et al. 2010). The entered range 3D printers are predominantly based on fused deposition modeling (FDM) technology, but more recently machines derived from stereolithography (SL) have entered the market due to expiring patents. It is typical to demonstrate that low-cost machines have a low performance. For example, the FDM consumer technology suffers from anisotropic mechanical properties (Es-Said et al. 2000) and a limited selection of thermoplastic materials. A FDM professional printer costs between $10,000 and $300,000. Typical laser and electron beam-based systems can cost anywhere between $500,000 and $1 M (3D Printer Landscape: The View from 30,000 Feet > ENGINEERING.com n.d.). Although these machines are typically high in performance, they come at a high cost. The commercial 3D printers that use more advanced techniques to print objects are usually equipped with proprietary software, which slice the 3D model and command the machine. Companies that sell professional 3D printers include 3D Systems (which bought ZCorporation since 2011), Stratasys (which merged with Objet since 2013), Solido LTD, Voxeljet, and ExOne. Lipson and Kurman (2010) reported that both Hewlett Packard and Xerox "are investing in 3D printing research and technology development."

Each AM technologies have manufacturing constraints linked by printing technology, used material, and expected functions (aesthetic, mechanical, use, etc.). Areas of interest which have used 3D printing to create objects include aeronautics, architecture, automotive industries, art, dentistry, fashion, food, jewelry, medicine, pharmaceuticals, robotics, and toys (Bourell et al. 2009). Automotive manufacturers exploited the technology because of the ability to help new products get quickly to the market and in a predictable manner. Aerospace companies are interested in these technologies because of the ability to realize highly complex and high-performance products. Integrating mechanical functionality, eliminating assembly features, and making it possible to create internal functionality like cooling channels (Pelaingre et al. 2003), internal honeycomb style structures, new topological optimization structure (Schneider et al. 2013), and so on, combine to create lightweight structures. Medical industries are particularly interested in AM technology because of the ease in which 3D medical imaging data can be converted into solid objects. In this way, devices can be customized to suit the needs of an individual patient (Campbell et al. 2012). Thus, each AM technology have advantages and disadvantages for own applications, and we propose to review them. Authors have chosen to classify the technologies according to hardening system or melting system which are characterized by a laser, a flashing source, an extrusion, or a jetting.

10.3   AM technologies

10.3.1   Laser technologies

SLA—Stereolithography is the first of the technologies developed originally and simultaneously in France (CNRS-July 84. French Patent N' 84 11 241) and in the United States (U.V.P-C. HULL August, 84. USA Patent N 45 75 330) to tackle rapid prototyping bottlenecks, as well as faster and better design needs (CAD-induced) (Jacobs 1992). In 1986, 3D Systems was found by Chuck Hull to commercialize this process. Photolithographic systems build shapes using light to selectively solidify photosensitive resins. The laser lithography approach is currently one of the most used AM technologies. Models are defined by scanning a laser beam over a photopolymer surface (Figure 10.4). For a few years, researchers have developed techniques to apply SLA to directly make ceramics. Ceramic powder (silica and alumina) is dispersed in a fluid UV curable monomer to prepare a ceramic–UV curable monomer suspension (Griffith et al. 1995; Griffith and Halloran 1996). The building process is the same as conventional SLA, and the monomer solution is cured forming a ceramic–polymer composite layer. The prototypes have higher stiffness than a standard workpiece, and their temperature resistance is over 200°C. A higher resolution machine has been developed and called microstereolithography, and it can print a layer with thickness of less than 10 μm (Halloran et al. 2011). The microstereolithography shares the same principle with its macro-scale counterpart but in different dimensions. In microstereolithography, an UV laser beam is focused to 1–2 μm to solidify a thin layer of 1–10 μm in thickness. Submicron resolution of the x–y–z translation stages and the fine UV beam spot enable precise fabrication of real 3D complex microstructures (Zhang et al. 1999).

SLM—Selective laser melting - the system starts by applying a thin layer of the powder material spread by a roller on the building platform. A powerful laser beam then fuses the powder at exactly the points defined by the computer-generated component design data. The platform is then lowered and another layer of powder is applied (Figure 10.4). Once again, the material is fused so as to bond with the layer below at the predefined points (EOS n.d.). During the process, successive layers of metal powder are fully melted and consolidated on top of each other. Today, the 3D-printer manufacturers propose machines with powerful double or multilaser technology with layers from 75 to 150 μm in thickness.

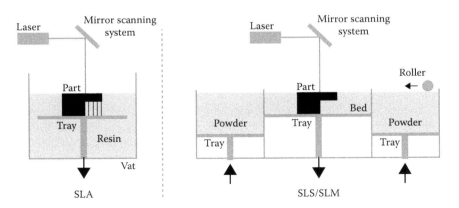

Figure 10.4 Laser technologies.

The material types that can be processed include steel, stainless steel, cobalt chrome, titanium, and aluminum. Electron beam melting (EBM—developed by Arcam [www. arcam.com] in 2001) is a powder process which distinguishes by its superior accuracy and high-power electron beam (up to 3000 W while maintaining a scan speed) that generates the energy needed for high melting capacity and high productivity.

SLS—Selective laser sintering - use a high-power laser to fuse small particles (polyamide, steel, titanium, alloys, ceramic powders, etc.). SLS was invented by Dr. Joe Beaman and Dr. Carl Robert Deckard (Beaman and Deckard 1988) in 1988. As the SLM, the powder bed is lowered by one layer thickness, a new layer of powder is applied on top, and the process is repeated until the model is completed. But what sets sintering apart from melting is that the sintering processes do not fully melt the powder but heat it to the point that the powder can fuse together on a molecular level. The latest SLS machines (www.3dsystems.com) offer laser powers from 30 W to 200 W in a CO_2 chamber controlled (in range ProX and sPro). The porosity of the material can be controlled. This porosity requests a posttreatment by infiltration to harden the final model like the bronze use to the steel. The SLS prototypes have a greater dimensional accuracy than the PolyJet and 3DP models (Ibrahim et al. 2009).

DMLS—Direct metal laser sintering - is similar to SLS with some differences. The technology is a powder bed fusion process by melting the metal powder locally using the focused laser beam. A product is manufactured layer-by-layer along the Z axis, and the powder is deposited via a scraper moving in the XY plane. The DMLS process from EOS© is well established for the net shape fabrication of prototype and short series tooling for plastic injection moulding. The first generation of EOS machine includes a 200-W laser source, when the second generation (EOSINT M 280) was launched with a 400-W fiber laser (www. eos.info). The trend shows an increase in laser power and also an increase in work chamber. DMLS often refers to the process that is applied to metal alloys for the manufacturing of direct parts in the industry including aerospace, dental, medical, and other industry that have small to medium size, highly complex parts, and the tooling industry to make direct tooling inserts. Today, recent developments in the powders coupled with the durability of the materials are extending its reach to the direct manufacturing of functional prototypes for powder metallurgical and cast components (Hänninen 2001). Support structures are required for most geometry because the powder alone is not sufficient to hold in place the liquid phase created when the laser is scanning the powder. The rapid manufacturing of parts by the DMLS process requires the use of a powder, which is composed of two types of particles. One type has a low melting point, and the other has a high melting point. The high melting point particles generate a solid matrix, whereas the particles with the low melting point bind the matrix after being melted by the laser energy input (Pessard et al. 2008).

10.3.2   *Flash technology*

In order to reduce lead time and increase in build speed, a new technology has emerged as a derivative from SLA. On the same principle proposed by Pomerantz (Barequet et al. 1996; Bieber et al. 1990), a photomask system (masking technology) to produce 3D models, the DLP—digital light processing, also known as FTI—film transfer imaging, use the UV photopolymerized materials. A film is coated in resin which is then cured by a UV flash of light from a projector for each slice of product (Figure 10.5). Unlike the 3D laser printer, the DLP projector projects the entire layer and not only of lines or points. This method allows building much quicker than other methods of rapid prototyping by substituting scanning time of a laser. With SLA, the part descends downward into the resin, whereas

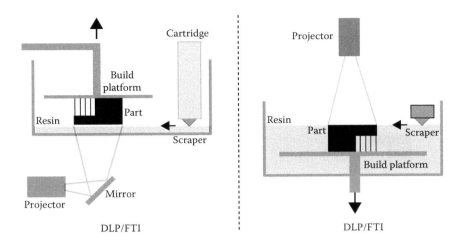

Figure 10.5 Flash technology (masking).

it is pulled upward out of the resin in a DLP printer. SLA process is gentler on the forming implant than the DLP process because, in DLP, the part must attach much more firmly to the build platform to prevent damage when newly formed layers are peeled from the basement plate after each exposure (Dean et al. 2012). The building platform can be angled upward and the light source down in some masking machines (e.g., Phidias technologies with Prodways 3D printer). The DLP technology is known for its high resolution, typically able to reach a layer thickness of down to 30 μm.

A new innovation in mask-image-projection based on the SLA process has been developed to produce objects with digital materials (Zhou et al. 2013). The proposed approach is based on projecting mask images with a new two-channel system design which reduces the separation force between a cured layer and the resin vat. The fabrication results demonstrate that the developed dual-material process can successfully produce 3D objects with spatial control over placement of both material and structure.

Close to DLP principle, the Continuous Liquid Interface Production (CLIP) is a new type of AM that uses photopolymerization working continuously, thanks to a projector and the ability to control oxygen levels through-out an oxygen-permeable membrane. This latter process is 30 times faster than the SLS or the MultiJet Modeling (MJM) (DeSimone 2015).

10.3.3   Extrusion technologies

FDM—Fused deposition modeling - is a layer AM process that uses a thermoplastic filament by fused depositing. FDM is trademarked by Stratasys Inc in the late 1980s, and the equivalent term is fused filament fabrication (FFF). The filament is extruded through a nozzle to print one cross section of an object and then moves up vertically to repeat the process for a new layer (Figure 10.6). The most used materials in FDM are ABS, PLA, and PC (polycarbonate), but you can find out new blends containing wood and stone as well as filaments with rubbery characteristics. Compared to ABS, PLA responds differently to moisture, to ageing UV with a discoloration, and to withdrawal of material. To predict the mechanical behavior of FDM parts, it is critical to understand the material properties of the raw FDM process material, and the effect that FDM build parameters have on anisotropic material properties (Ahn et al. 2002). The support

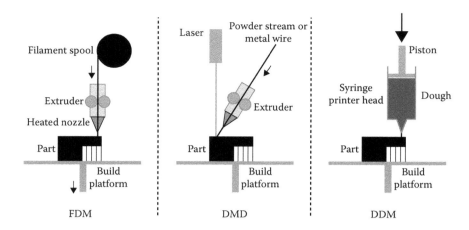

Figure 10.6 Extrusion technologies.

material is often made of another material and is detachable or soluble from the actual part at the end of the manufacturing process (except for the low-cost solutions, which use the same raw material). The disadvantages are that the resolution on the z-axis is low compared to other AM process (0.25 mm); so if a smooth surface is needed, a finishing process is required, and it is a slow process which sometimes takes days to build large complex parts (Wong and Hernandez 2012). FDM technology is the most popular desktop 3D printers and the less-expensive professional printers. The FDM technology was invented in the 1980s by Scott Crump (1992, 1994).

DED—Directed energy deposition - covers a range of terminology: laser engineered net shaping (LENS), directed light fabrication (IFF—ion fusion formation), direct metal deposition (DMD), and 3D laser cladding (Figure 10.6). It is a more complex printing process commonly used to repair or add additional material to existing components (Gibson et al. 2010). LENS is used to melt the surface of the target point, while a stream of powdered metal is delivered onto the small targeted point. IFF melts a metal wire or powder with a plasma-welding torch to form an object. This is a near-net-shape manufacturing process that uses a very hot ionized gas to deposit a metal in small amounts. DMD melts metal wire by electron beam as feedstock used to form an object within a vacuum chamber. The objects created in DED can be larger, even up to several feet long.

DDM—Dough deposition modeling - groups the marginal processes which file different doughs (Figure 10.6). Some technologies based on FDM printers use a syringe to deposit a dough material like silicone, food dough, chocolate, and so on.

A syringe-based extrusion tool which uses a linear stepper motor to control the syringe plunger position (Malone and Lipson 2007). The medical research uses the deposition of biomaterial and cells to realize a tissue structure. It presents a novel method for the deposition of biopolymers in high-resolution structures, using a pressure-activated microsyringe (Vozzi et al. 2002). Other works show applications to extrude a bio-based material to reconstitute a model and preserve the ecological environment. Experimentation uses a piston and 3D printer head adapted on a computer numerical control (CNC) machine to deposit a pulpwood based on wood flour to create a reconstituted wood product (Gardan and Roucoules 2011; Gardan 2014).

10.3.4   Jet technologies

MJM—MultiJet Modeling - deposits droplets of photopolymer materials with MultiJets on a building platform in ultra-thin layers, until the part is completed. Two different photopolymer materials are used for building, one for the actual model and another gel-like material for supporting (Singh et al. 2010). The photopolymer layers are cured by UV lamps, and a gel-like polymer supports the complexity of geometry in wrapping it (Figure 10.7). The soluble gel-like polymer (support material) is then removed by a water jet. The PolyJet technique reproduced details more accurately with a very good surface finish (Ibrahim et al. 2009) and smoothness. The accuracy of a PolyJet machine can reach thickness from 50 to 25 μm, besides the parts have a higher resolution. It is also known as *Thermojet*; some systems can produce wax models in jetting tiny droplets of melted liquid material which cool and harden on an impact to form the solid object.

3DP—three-dimensional printing, also known as CJP - Color Jet Printing, combines powders and binders. 3DP has been developed by the MIT. Each layer is created by spreading a thin powder layer with a roller, and the powder is selectively linked by inkjet printing of a binder. The build tray goes down to create the next layer (Figure 10.7). The thickness of layers is about 90–200 μm. This process has been used to fabricate numerous metal, ceramic, silica, and polymeric components of any geometry for a wide array of applications (Moon et al. 2001). Other powders have been tested to realize green products in wood (Gardan and Roucoules 2011). 3DP can print in multicolor directly into the part during the build process from a color cartridge. The final model is extracted from the powder bed to realize infiltration with liquid glue. The infiltrate improves the color definition and the mechanical behaviors. 3DP can provide architects a useful tool to quickly create a realistic model.

Prometal is a 3D-printing process to build rapid tools and dies. This is a powder-based process, in which stainless steel is used. The printing process occurs when a liquid binder is spurt out in jets to steel powder (Wong and Hernandez 2012). A final treatment is required to solidify the part like sintering, infiltration, and finishing processes.

A process developed at the University of Texas Arlington, is known as liquid metal jetting (LMJ). LMJ involves the jetting of molten metal in a process similar to inkjet printing, whereby individual molten droplets are ejected and connected to each other (Priest et al. 1997). This process is not commercially available yet.

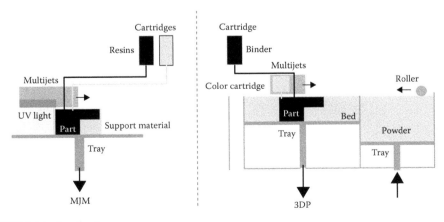

Figure 10.7 Jet technologies.

10.3.5 *Lamination and cutting technologies*

LOM—Laminated object manufacturing - is a rapid prototyping process where a part is sequentially built from layers of paper. The process consists of the thermal adhesive bonding and laser patterning of uniformly thick paper layers (Paul and Voorakarnam 2001). The system includes an *x–y* plotter device positioned above a work table vertically movable. The *x–y* plotter device includes a forming tool to create a layer from a sheet of material positioned on the work table. The layers are bonded to each other with heat-sensitive adhesives provided on one side thereof. A bonding tool or fuser is mounted to translate across the work table and to apply a lamination force and heat to each of the layers (Diamond et al. 1998). The layers are superimposed to give the final object, and the layer resolution is defined by the thickness of the paper sheet. 3D printers can print in full colors (Mcor Technologies).

Stratoconception is a rapid prototyping process with layers of sheets. It consists of the decomposition of a model by calculating a set of elementary layers called *strata* and by placing reinforcing pieces and inserts in strata. The elementary layer is displayed and manufactured by rapid milling or laser cutting. The strata are assembled with inserts to rebuild the final object (Houtmann et al. 2009). This process is useful, thanks to milling of a low-cost sheet in raw material (wood, MDF, PVC, aluminum, etc.).

10.3.6 *Discussion*

When you find out the AM technologies and you can use some of them, experts know that several manufacturing constraints and mechanical behaviors bring complications. For example, the powder technology leads to extract the final product outside of a power bed before cleaning it and often to applying a posttreatment. Moreover, the manufacturing orientation of the model influences the quality of geometry because of material gradient and the manufacturing direction. The part orientation can deeply modify the planarity, the circularity, and the surface accuracy. You have the same constraints with other technologies as the 3DP or DED. The internal structure of product due to the material orientation, the manufacturing technology, and its manufacturing by layers generates use constraints which need to be integrated. We can quote in a nonexhaustive list the anisotropy for the part made by FDM, the crack propagation for powdered parts, and the ageing UV for the models in photopolymers. You can find out the accuracy of layer thickness of some AM machines from manufacturer sources on the Table 10.1. From a 3D printer to another,

Table 10.1 Accuracy of AM technologies announced by manufacturers (2014)

AM technology	Layer thickness (µm)
SLA	<10 µm
MicroSLA	From 1 to 10 µm
SLM	From 75 to 150 µm
SLS	From 25 to 92 µm
DLP/FTI	From 30 to 100 µm
FDM	From 100 to 250 µm
DDM	From to 1000 µm
MJM	From 16 to 30 µm
3DP	From 100 to 4 µm

designer does not answer to the same need, and accuracy is often decisive to get a reliable product or a functional mechanism. Furthermore, the posttreatment, postmachining, or postfinishing are often required to get a finished product. The recycling and the raw material cost have to be taken into account too. To sum up, a set of stages are to define in upstream to assess the AM technology implications. The incrementation of experience greatly improves the engineering and manufacturing process.

The expiring patents open the market for other manufacturers proposing of new machines. Since February 2014, a major patent related to SLS expired (apparatus for producing parts by selective sintering n.d.). New technologies resulting from expiring patents appear with the solutions proposed by the companies: DWS Systems (Italy) or Formlabs (USA) (www.dwssystems.com and www.formlabs.com). Similar to the FDM fallen in the public domain in 2009, the SLS will perhaps know the same situation in leading a significant decline of prices with more manufacturers and more solutions.

10.4 New trends in additive manufacturing

10.4.1 Biomedical

3D printing applied to medical has appeared for some years through different applications. The organ transplantation sector has difficulties and the organ printing by jet based on 3D tissue engineering offers a possible solution. Some research defines organ printing as a rapid prototyping computer-aided 3D-printing technology based on using layer-by-layer deposition of cell and/or cell aggregates into a 3D gel with sequential maturation of the printed construct into perfused and vascularized living tissue or organ (Mironov et al. 2003). The bioprinting is an attractive method to create tissues and organs at hospitals. The success of an implantation depends on compatible materials. We can find a variety of biomaterials such as curable synthetic polymers, synthetic gels, and naturally derived hydrogels (Skardal and Atala 2014). Prosthetic is the first biomedical area which has used the 3D printing, and it presents several successes. We can quote a patient's skull anatomy reproduced via 3D printing for presurgical use in manual implant design and production (Dean et al. 2003), and the enhancement of the fixation stability of the custom made total hip prostheses and restore the original biomechanical characteristics of the joint (Rahmati et al. 2012). Several applications combine some degradable or allogeneic scaffolding with cellular bioprinting to create customized biologic prosthetics that have the great potential to serve as transplantable replacement tissue (Giovinco et al. 2012; Mannoor et al. 2013; Xu et al. 2013). New articles showed that the medical 3D-printing market might reach 983.2 million dollars by the year 2020 (Meticulous Research 2015).

10.4.2 Building construction

Projects for home construction through 3D printing are emerging such as the Shanghai WinSun Decoration Engineering Company (WinSun n.d.). This company can print the basic components separately before assembling them on site. These concrete houses are built in one day by 3D printing, and their construction costs about 3800 $. The 3D printer developed by the Chinese group is much larger than a conventional system and uses the same DDM technology. It has demonstrated the capabilities of 3D printer by rapidly constructing ten (10) houses in less than 24 hours. We do not have any pictures yet, but the manufacturer talks about a printer measuring 6.6 m (22 ft) tall, 10 m (33 ft) wide, and 32 m (105 ft) long. The building industry introduced a vocabulary such as rapid construction or rapid building.

10.4.3   Enriched exchange format

The use of the STL format limits the exchange of trades data (§2.2). If the STL format allows exporting from a surfacing model toward the specific software, the designer needs to insert rules in upstream work in CAD. The emergence of more enriched new exchange format appears such as the additive manufacturing file format (AMF) with important parameters (<material>, <composite>, <metadata>, etc.) or the STL 2.0 (Hiller and Lipson 2009). Alternative file format exports are also required to support depiction of complex organic geometry, while allowing multiple-material and mono/multicolor capabilities; the development of STL 2.0 or AMF is promising, particularly for the composition of complex geometries and multiple materials (Paterson et al. 2012). The article shows that we need to transfer more trades data to the AM machine through an enriched exchange format. The standard ISO/ASTM 52915:2013 *Standard specification for additive manufacturing file format (AMF) Version 1.1* (International Organization for Standardization: www.iso.org) describes a framework for an interchange format to address the current and future needs of AM technology.

10.4.4   Manufacturing interoperability

The manufacturing units and the small size of AM build tray complicate the production line. Industrials seek to reduce the lead time and increase in build speed, but a lot of AM technologies are not adapted. The interoperability is little studied by 3D-printer manufacturers. Reflecting the strategy of some companies like ExOne or Voxeljet, the professional 3D printers can be combined to the production line and offer the largest printers on the world market for 3D printing of sand and metal materials. Announced as a new industrial revolution (Berman 2012), the AM technologies will make the difference when it will be interoperable with the set of manufacturing process. Development orientations show that the new 3D printers will be more integrated inside the production line with the automation and the connectivity with the digital chain. A recent example is the emergence of hybrid system combining the 3D printing by laser deposition of metals (DMD) and the CNC machining through the *LASERTEC additive manufacturing* (en.dmgmori.com) solution proposed by DMG MORI©, which accelerates the realization of the finished product.

10.4.5   Rapid tooling

In order to reduce the time and cost of molds' fabrication, AM is used to develop and manufacture systems of rapid tooling. Powder-based sintering processes are now able to produce metal moulds that can withstand a few thousand cycles of injection molding (Nagahanumaiah et al. 2008). AM technologies propose to manufacture sand molds for the casting. A method to produce a lost mold for casting is used with the *thermojet* technology by wax. We saw that some powder technologies could realize sand molds for casting (Voxeljet, ExOne). Other approaches ally the AM technology and the topological optimization to realize a rapid tooling and to use less material while keeping its properties (Schneider et al. 2015). The layers manufacturing is able to improve a product or a tooling by inserting new methods as cooling channels or sensors. For example, an injection mold manufactured by a stratoconception and after assembly of strata, cooling channels are provided in the various interstratum planes for allowing a fluid to pass through the part (Pelaingre et al. 2003). You must perceive that this type of method can improve the behavior of a molded part by adjusting the location of the cooling channels to a specific geometry.

10.4.6 Topological optimization

Another challenge is to reduce weight and decrease the material used while keeping the product functions (mechanical, use…). Moreover, the main and support material can be expensive in the AM technology. Topology optimization is a mathematical approach that optimizes material layout within a given design space for a given set of loads and boundary conditions, so that the resulting layout meets a prescribed set of performance targets (Bendsoe and Sigmund 2003). Using topology optimization, engineers can find the best concept design that meets the design requirements. Any complex geometry is feasible in AM, the topological optimization implementation of a model leads to a new internal structure while maintaining conditions (mechanical, design shape, functions, etc.). Topologically, optimized parts have been created with internal geometry, using a narrow-waited structure that avoids the need for building supports (Galantucci et al. 2008). This method also creates new shapes of products.

10.4.7 Standards in additive manufacturing

AM technology standards are intended to endorse the knowledge of the industry, help stimulate research and encourage the implementation of technology. The standards define terminology, measure the performance of different production processes, ensure the quality of the end products, and specify procedures for the calibration of AM machines (ASTM n.d.). Several major standards were created very recently by the International Organization for Standardization (ISO) (International Organization for Standardization: www.iso.org); we can mention the main ones:

- *ISO 17296-2:2015, AM—General principles—Part 2*: Overview of process categories and feedstock. It describes the process fundamentals of AM with the existing processes and the different types of materials used.
- *ISO 17296-3:2014, AM—General principles—Part 3*: Main characteristics and corresponding test methods: It covers the principal requirements applied to testing with the main quality characteristics of parts, the appropriate test procedures, and the recommendations.
- *ISO/ASTM DIS 20195, Standard Practice—Guide for Design for AM*: It is being developed since 2015 and will bring together good practices in design for getting a reliable product.

You can also find other standards specifying the terminology in AM or the requirements for purchased AM parts.

10.5 Conclusions

This article aims to review the different technologies in AM before presenting other trends. Its redaction is motivated by a continuous monitoring of latest technologies and applications to better follow the thematic in full growth. After seeing the 3D model management and current technologies in AM, this chapter has proposed to survey the progression of some domains like the biomedical, the building, or the numerical simulation. A large number of additive processes are now available, but each AM technology has these own manufacturing constraints depending on material behavior and the solidification system by layers.

The weight reduction aims are required by the market constraints and the user-centered design. The realization of complex geometries opens many possibilities of new functionalities or manufacturing methods improving our perception of the product, component, or tooling design. The topological optimization is adapted to propose new complex geometry, which is easier to produce by AM with weight gain while keeping a high mechanical behavior. New developments show interest in the integration of the AM directly in the production environment. Thus, the hybrid solutions present an immediate solution to a productivity demand based on known machines, and we can also observe an increase in volume of fabrication chambers and laser powers in powder technologies.

The biomedical 3D printing is an emerging technology to construct artificial tissues and organs, which is currently feasible, fast evolving, and predicted to be a major technology in tissue engineering. We have seen that the AM market is increasing, but also that the trades' feedbacks increase too. This review has identified some trends, but we can specify that 3D printers are specializing for specific applications and that the modeling is decisive to improve the product and propose new ways to optimize productivity. AM opens new possibilities with the rapid construction in the civil engineering or the expiring patents.

New international standards are being published to endorse the knowledge of the industry, help stimulate research and encourage the implementation of technology. Also, the knowledge in manufacturing solutions, in materials developed, and in modeling techniques has to be capitalized to accelerate the future developments and understand the new complex environment, which is gradually taking shape.

Disclosure statement

No potential conflict of interest was reported by the author.

Nomenclature

AM   Additive manufacturing
3DP   Three-dimensional printing
CJP   Color jet printing
CLIP   Continuous liquid interface production
CNC   Computer numerical control
DMD   Direct metal deposition
DED   Directed energy deposition
DLP   Digital light processing
DLMS   Direct metal laser sintering
DDM   Dough deposition modeling
EBM   Electron beam manufacturing
FDM   Fused deposition modeling
FTI   Film transfer imaging
IFF   Ion fusion formation
LENS   Laser-engineered net shaping
LMJ   Liquid metal jet
LOM   Laminated object manufacturing
MJM   Multi jet modeling
SLA   Stereolithography
SLM   Selective laser melting
SLS   Selective laser sintering

ORCID

Julien Gardan http://orcid.org/0000-0001-6521-4526

References

Ahn, S.-H., M. Montero, D. Odell, S. Roundy, and P. K. Wright. 2002. Anisotropic material properties of fused deposition modeling ABS. *Rapid Prototyping Journal* 8: 248–257.

Apparatus for producing parts by selective sintering. n.d. Accessed January 5, 2015. http://www.google.com/patents/US5597589.

Ashley, S. 1991. Rapid prototyping systems. *Mechanical Engineering* 113 (4): 34.

ASTM. n.d. http://www.astm.org/Standards/additive-manufacturing-technology-standards.html.

Barequet, G., B. Ben-Ezra, Y. Dollberg, S. Gilad, M. Katz, I. Pomerantz, Y. Sheinman, and M. Katz. 1996. *Three Dimensional Modeling Apparatus*. *Google Patents*. Accessed December 16, 2014. http://www.google.com/patents/US5519816.

Beaman, J. J., and C. R. Deckard. 1988. *Selective Laser Sintering with Assisted Powder Handling*. Patent US 4938816 A.

Beaman, J. J., J. W. Barlow, D. L. Bourell, R. H. Crawford, H. L. Marcus, and K. P. McAlea. 1997. *Solid Freeform Fabrication: A New Direction in Manufacturing*, Vol. 2061, pp. 25–49. Norwell, MA: Kluwer Academic Publisher.

Bendsoe, M. P., and O. Sigmund. 2003. *Topology Optimization: Theory, Methods and Applications*. Berlin, Germany: Springer.

Berman, B. 2012. 3-D Printing: The New Industrial Revolution. *Business Horizons* 55: 155–162.

Bernard, A., and A. Fischer. 2002. New trends in rapid product development. *CIRP Annals—Manufacturing Technology* 51: 635–652. doi:10.1016/S0007-8506(07)61704-1.

Bieber, A., J. Cohen-Sabban, J. Kamir, M. Katz, M. Nagler, and I. Pomerantz. 1990. *Three Dimensional Modelling Apparatus*. Google Patents. Accessed December 16, 2014. http://www.google.com/patents/US4961154.

Bourell, D. L., M. C. Leu, and D. W. Rosen. 2009. *Roadmap for Additive Manufacturing: Identifying the Future of Freeform Processing*. Austin, TX: University of Texas.

Campbell, I., D. Bourell, and I. Gibson. 2012. Additive manufacturing: Rapid prototyping comes of age. *Rapid Prototyping Journal* 18: 255–258.

Chen, Y. H., C. T. Ng, and Y. Z. Wang. 1999. Generation of an STL file from 3D measurement data with user-controlled data reduction. *The International Journal of Advanced Manufacturing Technology* 15: 127–131. doi:10.1007/s001700050049.

Chua, C., K. Leong, and C. Lim. 2010. *Rapid Prototyping: Principles and Applications*. 3rd ed. Singapore: World Science Publisher.

Crump, S. S. 1992. *Apparatus and Method for Creating Three-dimensional Objects*. Google Patents. Accessed December 17, 2014. http://www.google.com/patents/US5121329.

Crump, S. S. 1994. *Modeling Apparatus for Three-dimensional Objects*. Google Patents. Accessed December 17, 2014. http://www.google.com/patents/US5340433.

3D Printer Landscape: The View from 30,000 Feet > ENGINEERING.com. n.d. Accessed December 12, 2014. http://www.engineering.com/3DPrinting/3DPrintingArticles/ArticleID/3916/3DPrinterLandscapeTheView-from30000Feet.aspx.

Dean, D., K.-J. Min, and A. Bond. 2003. Computer aided design of large-format prefabricated cranial plates. *Journal of Craniofacial Surgery* 14: 819–832.

Dean, D., J. Wallace, A. Siblani, M. O. Wang, K. Kim, A. G. Mikos, and J. P. Fisher. 2012. continuous digital light processing (CDLP): Highly accurate additive manufacturing of tissue engineered bone scaffolds. *Virtual and Physical Prototyping* 7: 13–24.

DeSimone, J. M. 2015. *Continuous Liquid Interphase Printing*. US20150097315 A1. http://www.google.com/patents/US20150097315.

C. R. Deckard / P. F. McClure by Nova Automation Corp., Austin, TX; 1988

Diamond, M. N., E. Dvorskiy, M. Feygin, and A. Shkolnik. 1998. *Laminated Object Manufacturing System*. Google Patents. Accessed December 19, 2014. http://www.google.com/patents/US5730817.

Dolenc, A. 1994. *Overview of Rapid Prototyping Technologies in Manufacturing.* Helsinki, Finland: Helsinki University of Technology National Technical Information Service.

EOS. n.d. http://www.eos.info/additive_manufacturing/for_technology_interested.

Es-Said, O. S., J. Foyos, R. Noorani, M. Mendelson, R. Marloth, and B. A. Pregger. 2000. Effect of layer orientation on mechanical properties of rapid prototyped samples. *Materials and Manufacturing Processes* 15: 107–122.

Galantucci, L. M., F. Lavecchia, and G. Percoco. 2008. Study of compression properties of topologically optimized fdm made structured parts. *CIRP Annals—Manufacturing Technology* 57: 243–246.

Gardan, J. *Rapid Prototyping System of an Object by Material Extrusion—Système De Prototypage Rapide D'un Objet Par Extrusion De Matière.* FR 3002179 (A1), 2014. http://fr.espacenet.com/publicationDetails/biblio?FT=D&date=20140822&DB=fr.espacenet.com&locale=fr_FR&CC=FR&NR=3002179A1&KC=A1.

Gardan, J., and L. Roucoules. 2011. Characterization of beech wood pulp towards sustainable rapid prototyping. *International Journal of Rapid ManufacturingIJRapidM Inderscience* 2: 215–233. doi:10.1504/IJRAPIDM.2011.044700.

Garg, A., and K. Tai. 2013. Selection of a robust experimental design for the effective modeling of nonlinear systems using genetic programming. *IEEE Symposium on Computer Intelligent Data Mining CIDM 2013,* 287–292, July 21, 2015. http://ieeexplore.ieee.org/xpls/abs_all.jsp?arnumber=6597249.

Garg, A., K. Tai, and M. M. Savalani. 2014a. Formulation of bead width model of an SLM prototype using modified multi-gene genetic programming approach. *The International Journal of Advanced Manufacturing Technology* 73: 375–388.

Garg, A., K. Tai, and M. M. Savalani. 2014b. State-of-the-art in empirical modelling of rapid prototyping processes. *Rapid Prototyping Journal* 20: 164–178. doi:10.1108/RPJ-08-2012-0072.

Garg, A., V. Vijayaraghavan, S. S. Mahapatra, K. Tai, and C. H. Wong. 2014c. Performance evaluation of microbial fuel cell by artificial intelligence methods. *Expert Systems with Applications* 41: 1389–1399.

Gibson, I., D. W. Rosen, and B. Stucker. 2010. *Additive Manufacturing Technologies: Rapid Prototyping to Direct Digital Manufacturing.* Springer, US, Springer E-Books, Imprint: Springer, Springer E-Books, Boston, MA: Etats-Unis.

Giovinco, N. A., S. P. Dunn, L. Dowling, C. Smith, L. Trowell, J. A. Ruch, et al. 2012. A novel combination of printed 3-dimensional anatomic templates and computer-assisted surgical simulation for virtual preoperative planning in charcot foot reconstruction. *The Journal of Foot and Ankle Surgery* 51: 387–393.

Griffith, M. L., and J. W. Halloran. 1996. freeform fabrication of ceramics via stereolithography. *Journal of the American Ceramic Society* 79: 2601–2608.

Griffith, M. L., T.-M. Chu, W. C. Wagner, and J. W. Halloran. 1995. Ceramic stereolithography for investment casting and biomedical applications. *Proceedings of Solid Free Fabrication Symposium,* Austin, TX: The University of Texas, pp. 31–38.

Halloran, J. W., V. Tomeckova, S. Gentry, S. Das, P. Cilino, D. Yuan, et al. 2011. Photopolymerization of powder suspensions for shaping ceramics. *Journal of the European Ceramic Society* 31: 2613–2619.

Hänninen, J. 2001. DMLS Moves from rapid tooling to rapid manufacturing. *Metal Powder Report* 56: 24–29. doi:10.1016/S0026-0657(01)80515-4.

Hiller, J., and H. Lipson. 2009. STL 2.0: A Proposal for a universal multi-material additive manufacturing file format. *Proceedings of Solid Free. Fabrication Symposium.* SFF'09 Austin, TX, pp. 266–278, February 16, 2014. http://utwired.engr.utexas.edu/lff/symposium/proceedingsArchive/pubs/Manuscripts/2009/2009-23-Hiller.pdf.

Houtmann, Y., B. Delebecque, and C. Barlier. 2009. Adaptive local slicing in stratoconception by using critical points. *Advances in Production Engineering & Management Journal* 4: 59–68.

Iancu, C., D. Iancu, and A. Stăncioiu. 2010. From Cad model to 3D print via "STL" file format. *Fiability Durability/Fiabilitate Si Durabilitate* 1 (5): 73–81.

Ibrahim, D., T. L. Broilo, C. Heitz, M. G. de Oliveira, H. W. de Oliveira, S. M. W. Nobre, et al. 2009. Dimensional error of selective laser sintering, three-dimensional Printing and PolyJet™ models in the reproduction of mandibular anatomy. *Journal of Cranio-Maxillofacial Surgery* 37: 167–173. doi:10.1016/j.jcms.2008.10.008.

Jacobs, P. F. 1992. Rapid Prototyping & Manufacturing: Fundamentals of Stereolithography. *Society of Manufacturing Engineers*. Accessed December 15, 2014. http://books.google.fr/books?hl=fr&lr=&id=HvcN0w1VyxwC&oi=fnd&pg=PA1&dq=stereolithography&ots=SsD1QU3VZH&sig=X5BtKIq39e_gaufmkWRyqJHXD64.

Lipson, H., and M. Kurman. 2010. *Factory@ Home: The Emerging Economy of Personal Fabrication.* Representative Committee US Official Science Technology Policy. Accessed December 12, 2014. http://risti.kaist.ac.kr/wp-content/uploads/2013/08/Factory-at-Home-The-Emerging-Economy-of-Personal-Fabrication.pdf.

Malone, E., and H. Lipson. 2007. Fab@Home: The personal desktop fabricator kit. *Rapid Prototyping Journal* 13 (4): 245–255.

Mannoor, M. S., Z. Jiang, T. James, Y. L. Kong, K. A. Malatesta, W. O. Soboyejo, N. Verma, D. H. Gracias, and M. C. McAlpine. 2013. 3D printed bionic ears. *Nano Letters* 13: 2634–2639.

Meticulous Research. 2015. Global Medical 3D Printing Market to Reach $983.2 Million by the Year 2020. *Meticulous Research*. http://www.pr.com/press-release/608468.

Mironov, V., T. Boland, T. Trusk, G. Forgacs, and R. R. Markwald. 2003. Organ printing: Computer-aided jet-based 3D tissue engineering. *Trends in Biotechnology* 21: 157–161.

Moon, J., A. C. Caballero, L. Hozer, Y.-M. Chiang, and M. J. Cima. 2001. Fabrication of functionally graded reaction infiltrated SiC–Si composite by three-dimensional printing (3DP™) process. *Materials Science and Engineering: A* 298: 110–119.

Noorani, R. 2006. *Rapid Prototyping: Principles and Applications*. Hoboken, NJ: Wiley.

Paterson, A. M., R. J. Bibb, and R. I. Campbell. 2012. Evaluation of a Digitised Splinting Approach with Multiple-material Functionality using Additive Manufacturing Technologies. *Proceedings of 23rd Annual International Solid Freeform Fabrication Symposium—An Additive Manufacturing Conference*, pp. 6–8, December 19, 2014. http://utwired.engr.utexas.edu/lff/symposium/proceedingsarchive/pubs/manuscripts/2012/2012-50-paterson.pdf.

Paul, B. K., and V. Voorakarnam. 2001. Effect of layer thickness and orientation angle on surface roughness in laminated object manufacturing. *Journal of Manufacturing Processes* 3: 94–101.

Pelaingre, C., L. Velnom, C. Barlier, and C. Levaillant. 2003. A Cooling Channels Innovating Design Method for Rapid Tooling in Thermoplastic Injection Molding. *1st Conference on Advanced Research in Virtual and Rapid Prototyping Proceedings*, Leiria, Portugal.

Pessard, E., P. Mognol, J.-Y. Hascoët, and C. Gerometta. 2008. Complex cast parts with rapid tooling: Rapid manufacturing point of view. *The International Journal of Advanced Manufacturing Technology* 39: 898–904.

Priest, J. W., C. Smith, and P. DuBois. 1997. Liquid Metal Jetting for Printing Metal Parts. *Solid Freeform Fabrication Proceeding*. Austin, TX: University of Texas, December 12, 2014. http://utwired.engr.utexas.edu/lff/symposium/proceedingsArchive/pubs/Manuscripts/1997/1997-01-Priest.pdf.

Rahmati, S., F. Abbaszadeh, and F. Farahmand. 2012. An improved methodology for design of custom-made hip prostheses to be fabricated using additive manufacturing technologies. *Rapid Prototyping Journal* 18: 389–400.

Schneider, A., J. Gardan, N. Gardan. 2015. Material and process characterization for coupling topological optimization to additive manufacturing. *Computer-Aided Design and Applications* 1–11. doi:10.1080/16864360.2015.1059192.

Sells, E., Z. Smith, S. Bailard, A. Bowyer, and V. Olliver. 2010. RepRap: The replicating rapid prototyper: Maximizing customizability by breeding the means of production. In *Handbook of Research in Mass Customization and Personalization*, edited by F. T. Piller, and M. M. Tseng, Vol. 1, pp. 568–580. World Scientific, Singapore.

SLS was invented by Dr. Joe Beaman and Dr. Carl Robert Deckard (Beaman and Deckard 1988) in 1988. https://www.google.com/patents/US4938816

Singh, R., V. Singh, and M. S. Saini. 2010. Experimental Investigations for Statistically Controlled Rapid Moulding Solution of Plastics using Polyjet Printing. *ASME 2010 International Mechanical Engineering Congress and Exposition*, pp. 1049–1053. American Society of Mechanical Engineers, December 19, 2014. http://proceedings.asmedigitalcollection.asme.org/proceeding.aspx?articleid=1616009.

Skardal, A., and A. Atala. 2014. Biomaterials for integration with 3-D bioprinting. *Annals of Biomedical Engineering* 43 (3): 730–746.

Stampfl, J., and M. Hatzenbichler. 2014. Additive Manufacturing Technologies. *CIRP Encyclopedia of Production Engineering*, Springer, pp. 20–27, December 11, 2014. http://link.springer.com/content/pdf/10.1007/978-3-642-20617-7_6492.pdf.

Nagahanumaiah, K., B. Subburaj, and B. Ravi. 2008. Computer aided rapid tooling process selection and manufacturability evaluation for injection mold development. *Computers in Industry* 59 (2-3): 262–276.

Vijayaraghavan, V., A. Garg, J. S. L. Lam, B. Panda, and S. S. Mahapatra. 2014. Process characterisation of 3D-printed FDM components using improved evolutionary computational approach. *International Journal of Advanced Manufacturing Technology* 78: 781–793. doi:10.1007/s00170-014-6679-5.

Vozzi, G., A. Previti, D. De Rossi, and A. Ahluwalia. 2002. Microsyringe-based deposition of two-dimensional and three-dimensional polymer Scaffolds with a well-defined geometry for application to tissue engineering. *Tissue Engineering* 8: 1089–1098.

Wang, H., and H. G. Zhang. 2012. State of the art in rapid prototyping. *Advanced Materials Research* 549: 1046–1050.

WinSun. n.d. http://www.yhbm.com.

Wohlers, T. 2012. *Wohlers Report 2012: Additive Manufacturing and 3D Printing State of the Industry Annual Worldwide Progress Report*. Wohlers Associates, Inc., Fort Collins, CO.

Wohlers, T. T., & Caffrey, T. (2015). *Wohlers Report 2015: 3D Printing and Additive Manufacturing State of the Industry Annual Worldwide Progress Report*. Wohlers Associates, Inc., Fort Collins, CO.

Wong, K. V., and A. Hernandez. 2012. A Review of additive manufacturing. *International Scholarly Research Network Mechanical Engineering* 2012: 1–10.

Xu, T., W. Zhao, J.-M. Zhu, M. Z. Albanna, J. J. Yoo, and A. Atala. 2013. Complex heterogeneous tissue constructs containing multiple cell types prepared by inkjet printing technology. *Biomaterials* 34: 130–139.

Zhang, X., X. N. Jiang, and C. Sun. 1999. Micro-stereolithography of polymeric and ceramic microstructures. *Sensors and Actuators A: Physical* 77: 149–156.

Zhou, C., Y. Chen, Z. Yang, and B. Khoshnevis. 2013. Digital material fabrication using mask-image-projection-based stereolithography. *Rapid Prototyping Journal* 19: 153–165.

chapter eleven

A new global approach to design for additive manufacturing

R. Ponche, J. Y. Hascoet, O. Kerbrat, and P. Mognol

Contents

Nowadays, due to rapid prototyping processes improvements, a functional part can be built directly through additive manufacturing (AM). It is now accepted that these new processes can increase productivity while enabling a mass and cost reduction and an increase of the parts functionality. However, in order to achieve this, new design methods have to be developed to take into account the specificities of these processes, with the design for additive manufacturing (DFAM) concept. In this context, a methodology to obtain a suitable design of parts built through AM is proposed; both design requirements and manufacturing constraints are taken into account.

11.1 Introduction

Recent progress has permitted transition from rapid prototyping to AM. Indeed, today, with this kind of manufacturing process, not only prototypes can be produced but also real functional parts in current materials including metals, polymers, and ceramics can be produced [1]. Because AM for production eliminates the need of tooling and can generate free forms, many of the current restrictions of design for manufacturing (DFM) and assembly are no longer valid [2]. However, whatever the technology used [3], as in all the manufacturing processes, the AM ones have characteristics and specificities of their own which may have an impact on the manufactured parts' quality. In order to utilize the AM possibilities in the best way in terms of design and to ensure the quality of the produced parts, a global numerical chain which allows moving from functional specifications of a part to its manufacturing must be defined (Figure 11.1). The purpose of this numerical chain [4] is to reach a global process control from knowledge of process obtained from experimentations, measurements, and simulations. Among the prerequisite to achieving such numerical chain, a DFM [5] approach is required which allows the AM processes' capabilities to be taken into account and limits directly from the design stage.

11.1.1 Design for additive manufacturing

Several works have been carried out concerning the classical DFM approach [6] for AM. By concerning manufacturability estimation, manufacturing cost and time have been analyzed [7,8] according to the manufacturing sequence. Similarly, the relationship between parts surfaces quality and manufacturing sequence has been studied [9,10]. From these different specific works, [11] proposed a methodology to map parts in relation to its manufacturability. Based only on geometrical analysis, these studies are limited because they do not take into account the physical phenomena that occur during the manufacturing

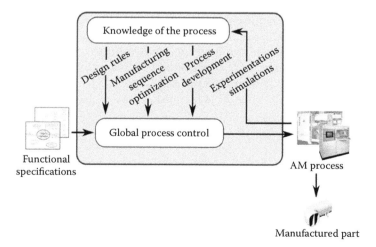

Figure 11.1 Global numerical chain concept.

process [12,13], which may have an impact on parts quality [14] and therefore on their manufacturing time and cost. Concerning manufacturability improvement, there have been very few studies reported on AM processes. General build guidelines have been established [15,16] and a methodology proposed by [17] enables modification of some non-critical geometric features. Moreover, a method for providing mesostructures within a part so as to achieve improved functional requirements part has been established [18]. Here again, the purpose is to minimize fabrication cost and time without a real awareness of the process planning, which do not allow to guarantee the parts quality expected. In addition to these limitations, classical DFM approaches may restrict the new perspectives of design opened up by the AM processes.

11.1.2   Partial approach versus global approach

Indeed, all these works enable to determine and to improve the manufacturability of a part from its computer aided design (CAD) model for a given AM process. Because they start from an initial geometry (given by the initial CAD model), these analysis can be qualified as *partial approaches* (Figure 11.2). In this case, it is difficult to determine the real optimized characteristics for a given AM process while fulfilling original functional specifications. Indeed, the initial CAD model was thought to be manufactured by an initial manufacturing process often very different from an AM process (e.g., machining which is the most often used); moreover the proposed modifications are local, and the result is never far from the initial design. The CAD model that is obtained is thus never really designed for the AM process that is chosen. On the contrary, a *global approach* (Figure 11.3) starts directly from both the chosen manufacturing process characteristics and the functional specifications of the parts to design. Designers can thus determine the geometry which optimizes the use of the chosen AM process characteristics while meeting the functional specifications. The purpose is not to limit geometry by an initial idea of the part shapes but to define it only from the manufacturing process and the functional specifications. This new way of thinking is in opposite with traditional DFM methodologies. However, capitalization of the entire knowledge about the manufacturing processes is needed. A beginning of a global approach, based on topology optimization with manufacturing constraints, has been applied to casting process [19].

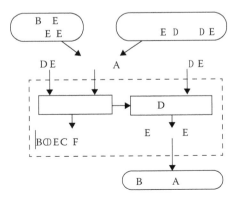

Figure 11.2 Partial approach.

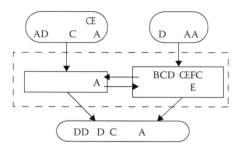

Figure 11.3 Global approach.

11.1.3   Scope of the paper

Although the global approach seems very interesting, no existing work has been carried out in this way in AM. Because these processes are quite new, still little known and very different from the other manufacturing processes, the psychological inertia phenomena may prevent the designer from utilizing all their capabilities in the best way [20].

Moreover, most of the AM processes are based on a layer-by-layer manufacturing where the material is locally merged, thanks to a local moving energy source (usually a laser or an electron bean) which follows a programmed manufacturing trajectory. It entails that the characteristics of the manufactured volumes, in terms of microstructure [21], geometry [14], and manufacturing time [10] depend first on the manufacturing direction (MD) and secondly on the manufacturing trajectories (MT). In addition to the consideration of the processes' characteristics, the choice of the *MD* and the *MT* according to functional specifications is thus the key of a global DFAM which would facilitate designers to explore new design spaces. That is why a new methodology which starts directly from both the functional specifications and the process characteristics is proposed in this paper.

First, in Section 11.2, the required data are presented. Then in Section 11.3, the different steps of the methodology, based on the choice of *MD*, are explained. The methodology has been applied on a part manufactured by a powder-based metal deposition (PBMD) process [22]; this constitutes the fourth section.

11.2   Proposed requirements for a global DFAM

In this section, requirements for a global DFAM are presented. They constitute the required data for the proposed methodology.

11.2.1   Functional specifications

The global purpose is to propose a structured approach which would help the designer to integrate the knowledge of the chosen AM process in his design to meet the functional specifications. The functional specifications can be detailed as follows:

- *Functional surfaces* (*FS*): type, dimensions, and position.
- Dimensional and geometrical specifications linked to the *FS*.
- *Mechanical requirements*: They depend on the chosen material characteristics.
- *Empty volumes*: dimensions and position. They correspond to the volumes which must not contain material, due to the assembly constraints of the designed part into the system to which it belongs.

11.2.2 Context

The study context is an influential factor. Because it can be translated into a concrete objective in terms of mass, cost, or manufacturing time, it has thus to be taken into account too.

11.2.3 Manufacturing characteristics

Manufacturing characteristics are linked to one another, and they cannot be seen separately; a global view which draws upon all the knowledge and experience of the community is consequently needed. The main characteristics of the manufacturing machines which must be taken into account are

- Kinematics
- Maximal and minimal dimensions
- Capability in terms of accuracy
- Required accessibility

But the physical phenomena that are involved in the manufacturing process and which are decisive in terms of final properties of parts, are also linked to the manufacturing sequence.

11.2.4 Finishing process characteristics

Similarly, if specifications (geometrical and dimensional) cannot be directly reached by the AM process that is chosen, a finishing process is needed. Because it can influence the final geometry, particularly in terms of overthickness and required accessibility, it has to be taken into account.

11.3 The proposed design methodology

From this data, a structured methodology can help designers in taking into account the manufacturing constraints while suggesting him an appropriate design for AM. The methodology is presented in Figure 11.4. It is divided into three main steps which enable to include gradually manufacturing knowledge in the shapes and the volumes of the parts to be designed. The first step is a global analysis which allows delimiting the design problem in terms of geometrical dimensions in relation to the dimensional characteristics of the AM process. The second one allows to fulfil the dimensional and geometrical specifications in relation to the AM process capability and the finishing process characteristics. Finally, the third step allows to fulfil the physical and assembly requirements in relation to the capability of the AM process.

11.3.1 Step 1: Analysis

The *FS* are obtained from the functional analysis of the product and are given by the designer. The first methodology step enables to find out if all the surfaces can be merged with one another by the chosen AM process into a single part. A first geometrical analysis is carried out; it takes into account the maximal and minimal dimensions, which can be obtained by the chosen manufacturing machine. If the dimensions are not suitable, the product has to be modified or divided into different parts by the designer, and the functional specifications of these new parts are then studied.

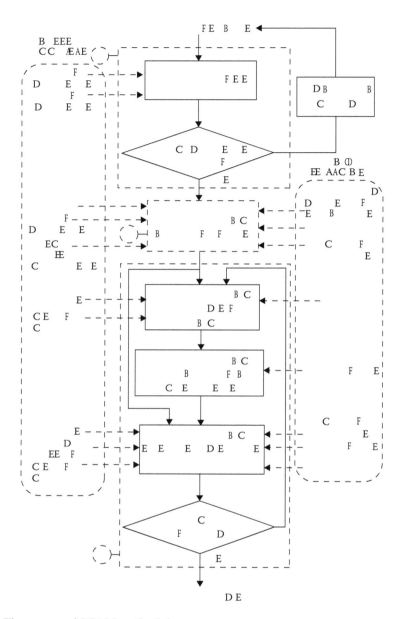

Figure 11.4 The proposed DFAM methodology.

11.3.2 Step 2: Determination of the functional volumes

The functional volumes (*FV*) are defined directly from the *FS* on to which a thickness is added. Indeed, only the tolerances in the normal direction of the surfaces are significant in terms of functionality. The others are initially ignored. The thickness, denoted *E*, depends on

- Dimensional accuracy of the AM process denoted *a*.
- Tolerances linked to the *FS*, denoted *p*.

For each *FS*, there are two different possibilities: first, if a finishing step is not needed ($p \geq a$), thickness is determined from Equation 11.1 which ensures the functional minimal thickness and from Equation 11.2 which ensures the compatibility with the AM process.

$$E \geq t + \frac{a}{2} \tag{11.1}$$

$$E = n.d - (n-1).d.\alpha \tag{11.2}$$

where:
   t is the minimal thickness corresponding to the local mechanical requirements.
   d is the minimal dimension that can be obtained by the chosen AM process.
   α is the overlap between two adjacent paths.
   n is a positive integer.

Equation 11.1 is illustrated in Figure 11.5a. If a finishing step is needed ($p < a$), then the thickness is determined from Equations 11.2 through 11.4 (Figure 11.5a).

$$E \geq t + \frac{a}{2} + e_{\min} \tag{11.3}$$

$$E \leq t + \frac{a}{2} + e_{\max} \tag{11.4}$$

where $e_{\min}$ and $e_{\max}$ are the minimal and maximal overthicknesses, which depend on the finishing process and the surfaces geometry. There can be different values of the parameters *d*, α, and *a* (according to the *MD*).

11.3.3   Step 3: Determination of the linking volumes

The purpose of this step is to merge the *FV* to define the volumes of the part while taking into account at best both design requirements and manufacturing constraints. In the

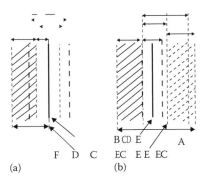

(a)   (b)

Figure 11.5 Definitions of the thickness: (a) without finishing and (b) with finishing.

case of AM, volumes are usually obtained layer-by-layer. It involves that their geometries strongly depend on the *MD*. Indeed, the choices of the *MD* have a direct influence on material quantity (need of supports), build time [23], and mechanical properties [24]. Moreover, the *MT* defining the energy source path during the process have a strong impact on the physical phenomena that occur during the manufacturing process and therefore on the final part quality [25]. That is why, the determination of the *MD* and *MT* is at the center of the methodology. The linking volume (*LV*) definition is divided into four steps. The first step is to determine the most critical *MD*, which can be characterized by the shape or the number of *FV* which can be manufactured in the same way. It is carried out according to the study context and the capability of the process given by its kinematics and the physical phenomena involved. The second step is, in the chosen *MD*, to merge the selected *FV* which can be manufactured from the same substrate. It is carried out according to the empty volumes, which must not be, in the *MD*, between two *FV* manufactured from a same substrate. In the third step, the substrates and supports' shapes are determined according to

- The selected *FV*.
- The other *FV*.
- The kinematics of the process.
- The accessibility required by the process.
- The mechanical requirements.
- The empty volumes.
- The study context.
- The physical phenomena that occur during the process.

The latter being strongly linked to the type and the shape of the *MT* that are used, it is essential to select suitable *MT* among all the possibilities to control them and to guaranty the expected geometrical quality. A classification of the different *MT* that are possible has been done (Figure 11.6) to describe and parameterize each one of them.

The last step is to check if all the volumes are merged. If it is not the case, the first three steps are repeated while taking into account the *FS* that have not yet been analyzed and

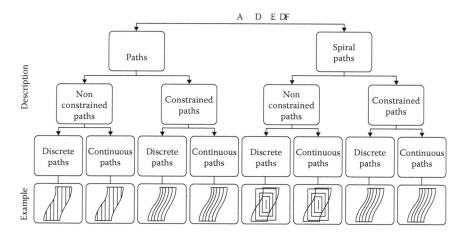

Figure 11.6 Classification of the manufacturing trajectories (*MT*).

the substrates obtained previously, and so on, until all the *FV* are merged. In the end, the process is complete, and an appropriate design for the chosen AM process is obtained. Indeed, the AM process specifications are taken into account step by step in parallel with the functional requirements. This ensures that the most possible process-related knowledge is taken into account to obtain the final shapes of the studied part.

11.4   Example

The proposed methodology has been applied to a case of a robot hinge in stainless steel (Figure 11.7). The study input data are detailed, and the three steps of the DFAM methodology are illustrated in this section.

11.4.1   Input data

11.4.1.1   Functional specifications

The case is composed of 20 *FS*: four bearing holders (hollow cylinders) and 16 flat surfaces, which are shown with their nominal dimensions in black in Figure 11.8. To enable the assembly of the case with the other parts of the robot, some empty volumes are defined. They are represented by the transparent volumes.

   The functional analysis of the robot has enabled to determine the geometrical and dimensional specifications linked with each surfaces. An extract is shown in Figure 11.9. The mechanical requirements are translated into a final minimal thickness *t* for each surface. It is 5 mm for the hollow cylinders and 3 mm for the based planes.

11.4.1.2   Context

Because of the robotic context, the global objective is to minimize the mass of the studied part.

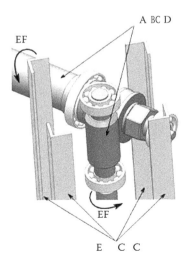

Figure 11.7 Global view of the studied system.

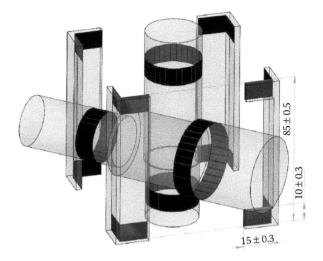

Figure 11.8 The *FS* and the empty volumes of the part to design.

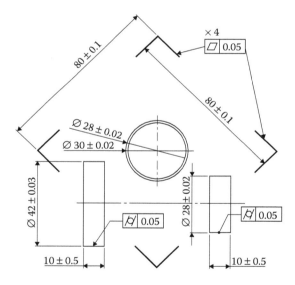

Figure 11.9 An extract of the specifications taken from the functional analysis of the case.

11.4.2   *Manufacturing characteristics*

11.4.2.1   *The additive manufacturing process*

The AM process chosen to manufacture this case is the construction laser additive direct or direct laser additive construction (CLAD) process (Figure 11.10). It is a PBMD process, based on the 3D layer-by-layer deposition of laser-melted powders. Its main characteristics are presented in Table 11.1.

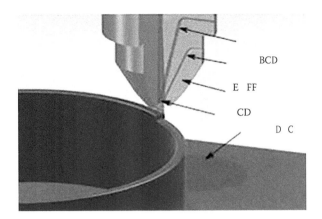

Figure 11.10 Clad process.

Table 11.1 CLAD process characteristics

Kinematics	Required accessibility (mm)	Maximal dimensions (mm)	d (mm)	a (mm)	a (mm)
5 axis	0.60	500*560*700	0.8	0.3	0.2

In this example, the assumptions that parameters d, a, and α are equal in all the directions are made.

The constraints due to the physical phenomena linked with the process (in particular, the thermal phenomena) are considered in ways: first, the substrates' thickness must be at least equal to the thicknesses of the volumes that it enables to manufacture. Second, because the discontinuities in the *MT* may generate an unwanted variation of the process parameters and also a gap between the designed and the manufactured part, the geometry will be chosen to limit them.

11.4.2.2   The finishing process

The finishing process that is chosen is high-speed milling; in view of the specification related to the *FS*, this choice involves a minimal and a maximal overthicknesses of 0.5 mm and 1 mm, respectively.

11.4.3   Step 1: Analysis

According to the geometrical analysis of the FS (Figures 11.8 and 11.9) and to the maximal and minimal dimensions imposed by the manufacturing process (Table 11.1), all the dimensions are compatible, and all the FS can be merged in only one part.

11.4.4   Step 2: Determination of the functional volumes

The thicknesses E are determined from the different parameters of Equations 11.2 and 11.3; for each *FS*, the results are given in Table 11.2. All the FV are shown in Figure 11.11a.

Table 11.2 Definition of the *FS* thicknesses

Surface type	t (mm)	a (mm)	e_{min} (mm)	e_{max} (mm)	d (mm)	E (mm)
Cylinder	5	0.2	0.5	1	0.8	5.9
Flat	3	0.2	0.5	1	0.8	3.6

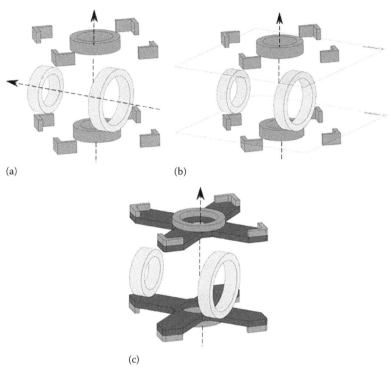

Figure 11.11 Definition of the *LV* in Z_1 when Z_1 is favored (a) the *FV*, (b) position of the substrates, and (c) the LV geometry in Z_1.

11.4.5   Step 3: Determination of the linking volumes

The *MD* are determined, due to the robotic context, to minimize the mass of the part and therefore the support structures' quantity. A geometrical analysis of the *FV* enables to determine two *MD*: Z_1 and Z_2 (Figure 11.11a). Z_1 enables to manufacture heighten *FV* (in dark gray), and Z_2 enables to obtain the two last (in light grey). If one or other of the *MD* is favored (which means that it is analysed in first), the final part geometry could not be the same. Both cases will be therefore analyzed, and the geometry which represented the best way of satisfying the study context will be selected. In case 1, Z_1 is favored whereas in case 2, Z_2 is favored.

11.4.5.1   Case 1: When Z_1 is favored

LVs in Z_1 To minimize supports in the empty volumes shown in Figure 11.8 and thus to minimize the finishing operations, all the selected *FV* linked to Z_1 must not be manufactured together but from two different substrates. Their positions are determined (Figure 11.11b) to minimize supports and time of finishing and to guarantee accessibility for the powder feed nozzle.

The selected *FV* geometry is locally modified in relation to the value of e_{max} and e_{min} to minimize the discontinuity of the *MT*. In particular, in the case of the raster discrete paths, it involves to define a radius *R* (Figure 11.12), which is given by Equation 11.5:

$$R = \frac{(2).(e_{max} - e_{min})}{(2) - 1} \tag{11.5}$$

Then the geometry of the substrates is defined according to the *FV* position and their thickness (in black in Figure 11.11c). Because all *FV* are not merged, a second *MD* is analyzed.

LVs in Z_2 As previously discussed, because of the empty volumes, the two cylinders linked to Z_2 cannot be manufactured together. It involves two substrates (Figure 11.13a). Moreover, the space between these *FV* and those already analyzed (those linked to Z_1) being lower than the required one by the powder feed nozzle of the CLAD machine (Table 11.1), the accessibility requirements are not satisfied, and the *MD* Z_2 cannot be used; the process is repeated one more time.

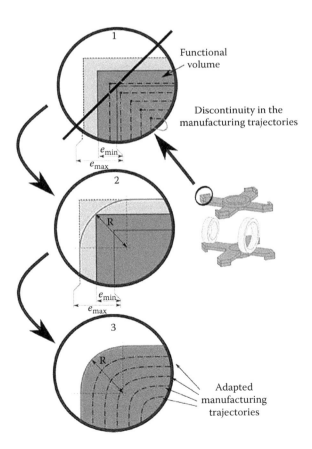

Figure 11.12 Local geometrical modifications to avoid the *MT* discontinuities.

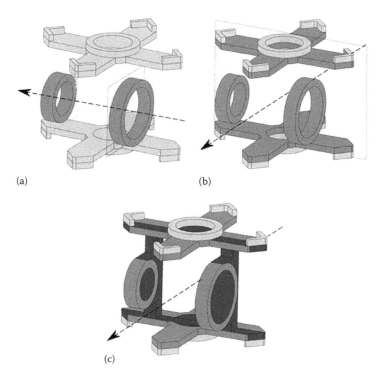

(a) (b)

(c)

Figure 11.13 Definition of the LV in Z_2 and Z_3, when Z_1 is favored (a) position of the substrates in Z_2, (b) position of the substract in Z_3, and (c) geometry of the substrate and the supports in Z_3.

LVs in Z_3 A third MD Z_3 is determined from the previous substrates (linked to Z_1) and FV (Figure 11.13b). The substrate and supports linked to Z_3 are determined, as previously, due to the context, to minimize the material quantity and to avoid, as much as possible, the empty volumes. The substrate dimensions are given by the volumes that it merges. The supports geometry are simply defined by the orthogonal projection of the volumes linked to Z_3 onto the substrate (Figure 11.13c). Finally, all the FV are merged, the design process is then complete, and the final blank part is obtained.

11.4.5.2 Case 2: When Z_2 is favored

From the FV in Figure 11.11a, the same reasoning is applied, starting with the analyzed Z_2.

LVs in Z_2 In the same way as above, because of the empty volumes, the two cylinders linked to Z_2 cannot be manufactured together. Two substrates whose geometry is shown in black in Figure 11.14a are therefore defined.

LVs in Z_1 Similarly, all the FV linked to Z_1 cannot be manufactured together. First, as in the case where Z_1 was favored, to minimize supports in the empty volumes, a minimum of two separated substrates is needed. Moreover, because of the manufacturing of the cylinders linked to Z_2, the accessibility required by the powder feed nozzle involves that each one of these two substrates has to be subdivided once again into tree-separated substrates. In the same way as previously, the geometry of the FV linked to Z_1 is locally modified in relation to the value of e_{max} and e_{min}, and then the substrates' geometry is defined (in black in the Figure 11.14b).

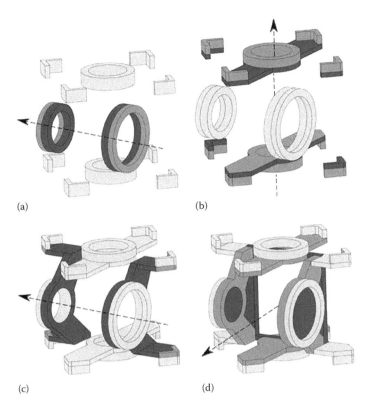

(a)   (b)

(c)   (d)

Figure 11.14 Definition of the different LV, when Z_2 is favored (a) the LV geometry in Z_2, (b) the LV geometry in Z_1, (c) the LV geometry in Z_2 after it is reanalyzed and (d) the Orthogonal projection to Z_3.

New LVs in Z_2 As all the FV are not merged, another MD should be, thus, analyzed. Because of the accessibility requirement, all the substrates previously obtained cannot be merged to one another following only one MD. Z_2, being already selected and allowing to merge again several volumes of the part, is reanalyzed. The result is shown in Figure 11.14c.

LVs in Z_3 Because all the FV are still not merged, the design process is repeated once again. All the previous substrates (defined in Z_2 and Z_1) can be merged into a third MD: Z_3. The substrate position and geometry are determined; then the supports are defined by the orthogonal projection of the volumes linked to Z_3 onto the substrate (Figure 11.14d).

Finally, all the FV are merged; the design process is thus complete, and the final blank part is obtained.

11.4.6   Final result

Thanks to the proposed methodology, each shape of the part has been designed in order to utilize the CLAD® process characteristics and capabilities to fulfil the study functional specifications while taking into account its general context. In the initial stage, from the geometrical analysis of the FS and according to the objective of minimizing the final part mass, two MDs have been selected (Z_1 and Z_2). Because the favoring of the one or other may have an impact on the final part geometry, the two cases have been studied. The

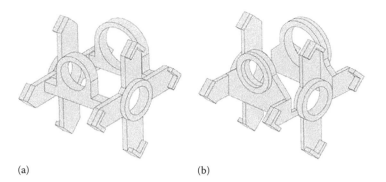

(a)   (b)

Figure 11.15 The final geometry: (a) when Z_1 is favored and (b) when Z_2 is favored.

Table 11.3 Characteristics of the two prosed solutions

	When Z_1 is favored	When Z_2 is favored
Blank part mass (g)	775.9	627.5
Finished part mass (g)	548.6	499.0
Maximal von-Mises stress (MPa)	15.3	22.6
Maximal displacement (mm)	0.041	0.015

final results, obtained after considering the finishing operation which enables to meet the required specifications (geometrical and dimensional) and to remove the material in the empty volumes, are shown in Figure 11.15. A finite element method analysis has been done to simulate the mechanical behavior of the proposed designs under the robot normal condition of use. Because, the results are quite close to each other compared with the material limits, they do not really enable to make a choice between both proposed solutions. However, as it is shown in Table 11.3, the favoring of Z_2 over Z_1 results in a lower mass of the blank part (around 19%) and of the finished part (around 9%). It allows, therefore, a lower manufacturing cost and to meet the global objective given by the robotic context of the study better. This solution is thus finally selected.

11.5   Conclusion

This paper described the initial stage of a promising research project which deals with a global DFAM approach. A new methodology is proposed to obtain an appropriate design for AM processes. In contrary to the classical DFM approaches, to prevent the psychological inertia phenomena which may limit the design innovation and to best utilize the AM processes' capabilities, the proposed methodology starts directly from both functional specifications and AM processes' characteristics. The required data for such a global DFAM approach have been presented. Then the three steps of the methodology allowing to reach to take into account all of them has been detailed and illustrated by a case study taken from the robotic field.

Further research will be conducted to optimize the methodology in particular regarding the local optimization of the shapes and the internal structures of the LVs in terms of functionality as it is done, for example, by [26]. In parallel, new criteria of choice for

MT will be developed to always adapt the local geometry more regarding the physical phenomena which occur during the manufacturing process.

References

1. P. Mognol, P. Muller, and J. Y. Hascoet. A novel approach to produce functionally graded materials for additive manufacturing. In *Proceedings of the Conference on Advanced Research in Virtual and Rapid Prototyping*, CRC Press, Leiria, Portugal, 2011.
2. E. Pessard, P. Mognol, J. Y. Hascoët, and C. Gerometta. Complex cast parts with rapid tooling: Rapid manufacturing point of view. *The International Journal of Advanced Manufacturing Technology*, 39(9–10):898–904, 2007.
3. T. Wohlers. *Wohlers report: Additive Manufacturing State of the Industry*. Wohlers associates, Fort Collins, CO, 2010.
4. R. Bonnard, P. Mognol, and J. Y. Hascoët. A new digital chain for additive manufacturing processes. *Virtual and Physical Prototyping*, 5(2):75–88, 2010.
5. S. K. Gupta, W. C. Regli, D. Das, and D. S. Nau. Automated manufacturability analysis: A survey. *Research in Engineering Design*, 9(3):168–190, 1997.
6. S. A. Shukor, and D. A. Axinte. Manufacturability analysis system: Issues and future trends. *International Journal of Production Research*, 47(5):1369–1390, 2009.
7. P. Alexander. Part orientation and build cost determination in layered manufacturing. *Computed-Aided Design*, 30(5):343–356, 1998.
8. M. Ruffo, C. Tuck, and R. Hague. Cost estimation for rapid manufacturing laser sintering production for low to medium volumes. *Journal of Engineering Manufacture*, 220(9):1417–1427, 2006.
9. R. Arni, and S. K. Gupta. Manufacturability analysis of flatness tolerances in solid freeform fabrication. *Journal of Mechanical Design*, 123(1):148, 2001.
10. M. Ancǎu, and C. Caizar. The computation of Pareto-optimal set in multicriterial optimization of rapid prototyping processes. *Computers & Industrial Engineering*, 58(4):696–708, 2010.
11. O. Kerbrat, P. Mognol, and J. Y. Hascoet. Manufacturability analysis to combine additive and subtractive processes. *Rapid Prototyping Journal*, 16(1):63–72, 2010.
12. X. He, and J. Mazumder. Transport phenomena during direct metal deposition. *Journal of Applied Physics*, 101(5):053113, 2007.
13. S. Y. Wen, Y. C. Shin, J. Y. Murthy, and P. E. Sojka. Modeling of coaxial powder flow for the laser direct deposition process. *International Journal of Heat and Mass Transfer*, 52(25–26):5867–5877, 2009.
14. M. Alimardani, E. Toyserkani, and J. Huissoon. A 3D dynamic numerical approach for temperature and thermal stress distributions in multilayer laser solid freeform fabrication process. *Optics and Lasers in Engineering*, 45(12):1115–1130, 2007.
15. S. Filippi, and I. Cristofolini. The Design Guidelines (DGLs), a knowledge-based system for industrial design developed accordingly to ISO-GPS (Geometrical Product Specifications) concepts. *Research in Engineering Design*, 18(1):1–19, 2007.
16. G. A. Teitelbaum. *Proposed Build Guidelines for Use in Fused Deposition Modeling to Reduce Build Time and Material Volume*. Master of Science in Mechanical Engineering, University of Maryland, College park, MD 2009.
17. S. Sambu, Y. Chen, and D. W. Rosen. Geometric tailoring: A design for manufacturing method for rapid prototyping and rapid tooling. *Journal of Mechanical Design*, 126(4):571–580, 2004.
18. D. W. Rosen. Computer-aided design for additive manufacturing of cellular structures. *Computer-Aided Design & Applications*, 4(5):585–594, 2007.
19. L. Harzheim, and G. Graf. A review of optimization of cast parts using topology optimization part II. *Structural and Multidisciplinary Optimization*, 31(5):388–399, 2005.
20. R. Hague. Unlocking the design potential of rapid manufacturing. In *Rapid Manufacturing: An Industrial Revolution for the Digital Age*, Loughborough University, Loughborough, UK, 2006.
21. L. Costa, R. Vilar, T. Reti, and A. Deus. Rapid tooling by laser powder deposition: Process simulation using finite element analysis. *Acta Materialia*, 53(14):3987–3999, 2005.

22. J. Ruan, T. E. Sparks, Z. Fan, J. K. Stroble, and A. Panackal. A review of layer based manufacturing processes for metals. In *Proceedings of the Seventeenth Solid Freeform Fabrication Symposium,* Austin, Texas, USA, pp. 233–245, 2006.

23. P. Singh, and D. Dutta. Multi-Direction slicing for layered manufacturing. *Journal of Computing and Information Science in Engineering,* 1(2):129, 2001.

24. B. Caulfield, P. E. McHugh, and S. Lohfeld. Dependence of mechanical properties of polyamide components on build parameters in the SLS process. *Journal of Materials Processing Technology,* 182(1–3):477–488, 2007.

25. E. Foroozmehr, and R. Kovacevic. Effect of path planning on the laser powder deposition process: Thermal and structural evaluation. *The International Journal of Advanced Manufacturing Technology,* 51(5–8):659–669, 2010.

26. H. De Amorim Almeida, and P. Jorge Da Silva Bártolo. Virtual topological optimisation of scaffolds for rapid prototyping. *Medical Engineering & Physics,* 32(7):775–82, 2010.

chapter twelve

A new methodological framework for design for additive manufacturing*

Martin Kumke, Hagen Watschke, and Thomas Vietor

Contents

12.1 Introduction

Additive manufacturing (AM) extends the spectrum of conventional manufacturing processes, since parts or complete products are produced by adding material in layers in contrast to subtracting it. AM opens opportunities for innovative designs and

---

* The chapter is reprinted with permission from Kumke et al., A new methodological framework for design for additive manufacturing, *Virtual and Physical Prototyping*, 11(1), 2016. http://dx.doi.org/10.1080/17452759.2016.1139377.

advances in product performance, for example, through geometric freedom and highly integrated structures, which are impossible with machine tooling. Moreover, customization of parts can be realized economically, since AM does not need product-specific tools (Rosen 2014, Wohlers 2014). Up until recently, AM application was limited to prototyping. Due to improvements in accuracy and mechanical material properties, tools and even end-use products can be increasingly manufactured directly from digital models (Campbell et al. 2012).

There are several limits/obstacles for industrial application of AM, and design engineers' lack of experience and knowledge is just one of them. Some methodologies and appropriate tools that support design engineers in taking the specifications of AM into account have been proposed in previous research. However, these are usually tailored to a specific design stage and/or limited to certain AM processes. In particular, there are no interfaces between existing approaches. Methodologies for a continuous support in the entire design process are scarcely available (Laverne et al. 2014). In addition, the applicability of existing design methods and tools to the development of AM parts has not been examined.

Methods for the integration of different considerations into the design process are subsumed under the term *design for X* (DFX). Several DFX strategies aim at a product's simplification (Kuo et al. 2001). For example, guidelines for *design for assembly* (DFA) and *design for manufacturing* (DFM)—subsumed under the term *design for manufacture and assembly* (DFMA)—are widely available (Bralla 1999, Boothroyd et al. 2011). The term DFM has been transferred to AM and is then called *design for additive manufacturing* (DFAM). The purpose of DfAM is defined as a "synthesis of shapes, sizes, geometric mesostructures, and material compositions and microstructures to best utilise manufacturing process capabilities to achieve desired performance" (Rosen 2007b) and even "to maximize product performance" (Gibson et al. 2015).

The objective of this paper is to develop a new methodological DfAM framework, which integrates existing tools, provides continuous support for design engineers to fully exploit AM-specific potentials for new product generations, and facilitates AM-conformal designs. The focus of this paper is on industrialized AM processes that are suitable for creating end-use products and mechanical parts, although the findings might be applicable to prototyping-only technologies and other part categories as well. First, we analyze the existing DfAM approaches comprehensively and classify them into distinct categories. Next, we relate them to a general design process. In this regard, we shall point out particular limitations of previous research. Finally, we propose a new modular DfAM framework based on both AM-specific and general design processes, integrate existing methods and tools into the framework, and provide a concept for using the framework. We conclude our paper with a summary and recommendations for future research.

12.2   Review and classification of DfAM research

The term DfAM is far from being used consistently among researchers. Although the first DfAM approaches primarily focus on the investigation of design potentials and product optimization opportunities created by AM in contrast to design restrictions imposed by traditional manufacturing technologies, others regard DfAM as tools or systems supporting designers at creating AM-conformal designs. Still others use the term in the context of continuous design methodologies providing systematic guidance through the development process of AM products. For the introduction of a continuous DfAM framework, all aspects mentioned above must be included into the definition of the generic term DfAM.

Two literature classifications based on reviews of DfAM approaches have been published. Yang and Zhao (2015) distinguished between *general design guidelines, modified conventional design theory and methodology for AM*, and *design for additive manufacturing*. Laverne et al. (2014) used *DfAM for concept assessment* and *DfAM for decision making* as high-level categories and broke down the latter into the subcategories *guidelines, product properties, design optimization*, and *geometrical validation*. Although both classifications provide insights into the different DfAM types, they have two major drawbacks in common: first, in many cases, the proposed categories are not mutually exclusive, preventing an unambiguous assignment of new approaches into the classification. Second, there is no clear distinction between general process-focused approaches and approaches specifically developed for the design engineer within the design process.

Hence, we propose a new classification represented in Figure 12.1 which distinguishes between *DfAM in the strict sense* and *DfAM in the broad sense*. *DfAM in the strict sense* includes approaches concerning the actual design process, for example, guidelines and methodologies supporting design engineers at their key tasks of creating products, which utilize AM design potentials and adhere to AM design rules. *DfAM in the broad sense* contains additional approaches beyond the core design process. These include upstream, downstream, and other generic DfAM-related activities carried out in new product development processes. These activities must be included in a comprehensive DfAM definition since many approaches use design-based decision criteria, and the outcomes of the activities directly influence the design process, for example, by selecting part candidates and AM processes.

We emphasize that activities concerning the manufacturing process itself are not part of our DfAM definition, since they are carried out under the responsibility of the manufacturing specialist instead of the design engineer. These activities also include process-planning steps such as decisions on build orientation which have, of course, a strong influence on part quality. However, information like this should be already taken into account in the design phase, for example, in AM design rules.

Previous research is presented in the following sections whose structure is based on the new classification.

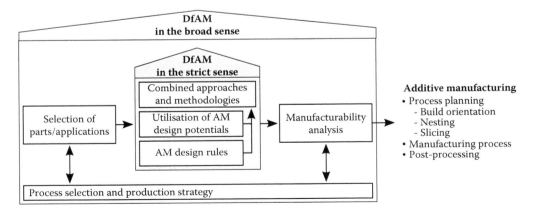

Figure 12.1 Classification of previous DfAM approaches.

12.2.1   DfAM in the strict sense

This category comprises approaches tailored to the core design process. AM design rules (Section 12.2.1.1) constitute the basic level of *DfAM in the strict sense*, since they ensure the creation of AM-producible designs. However, they do not necessarily take into account the unique capabilities of AM which can be indispensable for optimizing a product in terms of its performance or cost-efficient production. Thus, the (systematic) utilization of AM design potentials (Section 12.2.1.2) forms the superordinate level. Only if a product design obeys both levels, it can be referred to as truly DfAM optimized. Some approaches thus combine design rules and design potentials or provide comprehensive DfAM methodologies (Section 12.2.1.3).

12.2.1.1   AM design rules

Although AM provides huge design potentials, geometric freedom is not unlimited. New restrictions arise from the technological principle itself, the processed material, or even the machine. Design engineers have to be aware of the design rules to ensure manufacturability. Similar to conventional manufacturing processes, rules have been developed for various AM technologies and range from general qualitative guidelines, such as build orientation, to specific quantitative limitations, such as minimal wall thickness. In the literature, design rule catalogs can be found particularly for selective laser melting (SLM), selective laser sintering (SLS), and fused deposition modeling (FDM).

Thomas' (2009) research focused on the geometric limits imposed by SLM and was based on series of experiments. He found various quantitative constraints for geometric elements, for example, radii and minimum gap features, as well as general recommendations for high-quality results, for example, surface roughness as a function of build orientation.

Seepersad et al. (2012), Wegner and Witt (2012), and Gerber and Barnard (2008) investigated the limitations of SLS regarding minimal sizes of geometric features such as holes, cylinders, walls, and graven fonts depending on their orientation. Additionally, Wegener and Witt analyzed the durability of functionally integrated parts, for example, hinges and snap-fits.

Adam and Zimmer (2014) conducted experiments on SLM, SLS, and FDM machines based on test specimens with predefined standard elements, which include basic geometric elements, element transitions, and aggregated structures (spatial arrangements of basic elements and their transitions). They developed a comprehensive catalog applicable to all three technologies. They pointed out, however, that numerical values are only valid for the respective boundary conditions (i.e., machine, material, parameter set, layer thickness, etc.). Kranz et al. (2015) compiled a design rule catalog specifically for the SLM-based production of TiAl6V4 parts.

Further design rule collections were, for example, published by Hochschule Bremen (2008) for FDM as well as by various machine manufacturers and AM service providers. In addition, design rules are increasingly finding their way into engineering standards and guidelines, for example, VDI 3405 Part 3 (VDI 2015).

12.2.1.2   Utilization of AM potentials

AM provides design engineers with an immense new geometric freedom making conventional guidelines of DFMA obsolete. The elimination of manufacturing constraints can be used, for example, to improve product performance, reduce assembly cost, or realize innovative designer items that are impossible to manufacture with other technologies. Hague et al. (2003, 2004) were among the first who pointed out these implications on design. Becker et al. (2005) summarized the opportunities in a list of general AM design suggestions.

Although the need to systematically support engineers in utilizing AM design potentials was identified early, previous research has primarily focused on case studies of exemplary parts whose design benefited from AM. Among others, these include topology optimization to achieve high-strength lightweight designs (Watts and Hague 2006), conformal cooling ducts (Petrovic et al. 2011), personalized medical parts as an example of mass customization (Eyers and Dotchev 2010), and parts consolidation enabled by undercuts (Becker et al. 2005). Most of the case studies discuss parts with very specific requirements that cannot be transferred to other product categories. In addition, the usual purpose is to improve an existing product in contrast to a new product development based on requirements. Broad descriptions of design potentials and successful case studies have also been included in educational textbooks (Gebhardt 2011, Gibson et al. 2015).

Little research has been conducted on the systematization and simultaneous methodical utilization of more than one AM design potential. Burton (2005) introduced a questionnaire approach: Based on the responses to questions in different design areas, he suggests part redesigns to exploit AM potentials, for example, through part consolidation. Bin Maidin et al. (2012) built on this approach and developed a digital design feature database which provides a higher number of features and an easier access. Doubrovski et al. (2012) further extended the idea and suggested a collaboratively edited knowledge database similar to a wiki.

12.2.1.3 *Combined approaches and methodologies*

Although AM design rule collections ensure manufacturability and approaches related to AM design potentials foster creativity and the development of innovative solutions, the isolated application of both aspects can prevent the creation of optimal AM products. Therefore, some researchers systematically incorporate both design rules and design potentials into their approaches. However, they usually do not build directly upon the available tools and methods illustrated in the previous two sections.

Ponche et al. (2012) adopted a more global approach in view of not limiting design freedom by an initial computer-aided design (CAD) model. The new aspect of this methodology is to define a part's design from its functional specifications and process restrictions (particularly manufacturing direction and manufacturing trajectories) instead of using an initial CAD model for an AM-specific improvement. Ponche et al. (2014) designed a methodology which optimizes the manufacturing process through process simulation. Although their methodology is based on directed energy deposition (DED), similar approaches can be suitable for other AM technologies. The optimization is split up into three steps and covers part orientation, functional optimization, and paths optimization to balance functional requirements and process specifications. With an improved paths generation depending on process parameters and part geometry, it is possible to minimize the gap between the virtual model and the manufactured part. A similar function-based method is proposed by Vayre et al. (2012).

Leary et al. (2014) combined AM design rules and topology optimization. They showed that the theoretically optimal topology can be modified to ensure manufacturability without any additional support structures. They also identified the optimal build orientation by assessing manufacturing time and component mass. Emmelmann et al. (2011) and Kranz

(2014) also used topology optimization, but for bionics-inspired lightweight design for SLM with TiAl6V4. They argued that a systematic design process is indispensable, in particular, to guide inexperienced designers through the concept phase and to manufacture fully optimized structures. In addition, they embedded the part optimization process in combination with design rules and a bionic catalogue into the framework of guideline VDI 2221. Tang et al. (2015) developed a method for the creation of lattice structures which are generated within specified functional volumes. Afterwards, the distribution of the lattice struts' thicknesses are optimized in consideration of AM design rules.

AM has been identified as a key enabler for mass customization, the objective of satisfying individual customer needs with mass production efficiency (Pine II 1993). Research on design for mass customization has emphasized the importance of product families/platforms, modularization, and the involvement of customers in the design process. A possible design strategy for realizing mass customization is the concept of concurrent engineering proposed by Tseng and Jiao (1998). AM-enabled mass customization particularly includes personalized medical products, such as hearing aids, which require advanced 3D-scanning technologies (Eyers and Dotchev 2010) and approaches for customer codesign of aesthetically appealing consumer products (Ariadi et al. 2012). Although DfAM and mass customization are closely related, only few approaches are targeting the intersection of these two fields directly (Gibson et al. 2015). Specific approaches were proposed by Tuck et al. (2008) who provided a method for customized aircraft seats (using 3D scanning, reverse engineering, and AM) and by Ko et al. (2015) who developed a method entitled customized design for additive manufacturing (CDFAM), which is based on a formal representation of design knowledge.

Some researchers integrate different DfAM aspects into comprehensive methodologies. Rosen (2007a, b) introduced a CAD system for DfAM which is particularly designed for the utilization of mesostructured materials. It contains a mapping between process, structure, property, and behavior, incorporating both geometry and material of an AM product. Rodrigue and Rivette (2010) proposed a methodology which starts with parts consolidation enabled by AM design freedoms based on an existing assembly concept. In the subsequent steps, it contains several options for function optimization. Boyard et al. (2013) built on this approach, but focused on the abstract formulation of functional specifications. They used a standard design methodology and maintained the conventional differentiation between DFA and DFM. The distinctive characteristic is that DFA and DFM are carried out simultaneously instead of successively. Based on an extensive literature review on DfAM methodologies, Yang and Zhao (2015) proposed their own design method focusing on the downstream design stage. Its core process contains steps for function integration and structure optimization of an initial CAD model (part redesign). An European Union project on standardization in additive manufacturing (SASAM) proposed a design strategy draft which incorporates the whole design process from task to final part design. Details are provided for the identification of the general AM potential and AM process selection (Verquin et al. 2014).

12.2.2   DfAM in the broad sense

This category incorporates further DfAM-related approaches around *DfAM in the strict sense*, which is included in this category as its center. Further approaches can be generic DfAM-related activities such as process selection (Section 12.2.2.1), upstream activities such as the selection of AM parts/applications (Section 12.2.2.2), and downstream activities such as manufacturability analyses (Section 12.2.2.3).

12.2.2.1 Process selection and production strategy

In the context of AM, various decision support systems have been developed. The first category consists of database-supported selection tools designed for choosing the most suitable process or machine, in particular for rapid prototyping applications, for evaluating AM's suitability for a certain application, or for a systematic assessment of an economic manufacturability based on quantitative requirements, for example, accuracy or build speed (Campbell and Bernie 1996, Bibb et al. 1999, Kaschka and Auerbach 2000, Byun and Lee 2005, Venkata Rao and Padmanabhan 2007, Kirchner 2011, Zhang et al. 2014b). Some of these approaches include sophisticated and detailed analyses. For example, Munguía et al. (2010) used an advice system based on artificial intelligence to compare additive and conventional manufacturing in order to recommend optimal production parameters.

The second category aims at choosing an optimal production strategy depending on product requirements and process limitations. For example, Achillas et al. (2014) considered the complete supply chain in a scenario-based framework. By means of a decision support system including alternative available processes, the optimal production strategy is chosen. Merkt et al. (2012) introduced an integrative technology evaluation model that includes several levels for economic and technology analyses to assess the competitiveness of AM in comparison to other manufacturing processes. They also analyzed the interaction between product and process innovations to include the potential of AM for improved products. A typical method for a quantitative evaluation is the calculation of part complexity factors.

12.2.2.2 Selection of parts/applications

The selection of suitable parts or applications is an upstream activity performed before the actual design process. Due to the novelty of AM for designers and producers, the selection of appropriate candidates remains a challenge.

On the strategic level, Conner et al. (2014) identified three main product attributes or criteria: complexity, customization, and volume. These part properties are structured in a three-axis model containing eight areas for which different manufacturing strategies can be specified.

On the component level, Klahn et al. (2014, 2015) used four decision criteria to identify parts of a product for an AM-conformal redesign, namely integrated designs, individualization, lightweight design, and efficient design. Lindemann et al. (2015) proposed a methodology for a selection of AM part candidates based on a workshop concept applicable by both AM novices and AM experts. Its core is a trade-off methodology matrix which includes part candidates in its columns and decision criteria in its rows. Its purpose is a screening of parts and whether the AM of those parts enables benefits. Selected parts are then redesigned in the following steps.

12.2.2.3 Manufacturability analysis

In addition to the consideration of design rules already in the design phase to create AM-producible parts, some approaches analyze the finished design solution with regard to manufacturability. Kerbrat et al. (2010, 2011) designed a multiprocess strategy at the component level by combining conventional manufacturing processes and AM. Depending on manufacturability indexes calculated by design parameters, a CAD model is divided into a modular structure, whose parts are manufactured separately and then assembled. Zhang et al. (2014a) developed a two-level methodology for design evaluation and a better understanding of process characteristics. First, the general manufacturability of a given design solution is analyzed, and the process parameters are set up. Second, component-specific aspects like build orientation and slicing strategy are defined. These last steps, however, do not belong to our DfAM definition (see Figure 12.1). The initial design solution may be revised based on the evaluation results.

12.3   Discussion of DfAM research

In this chapter, we first relate previous DfAM approaches to general design methodologies (Section 12.3.1). Second, we critically evaluate the approaches with regard to their limitations (Section 12.3.2). We then deduct requirements for a new DfAM framework (Section 12.3.3).

12.3.1   DfAM approaches in the context of general design methodologies

The term DfAM is, as mentioned in Section 12.1, basically a modification of the original term DFM. DFM activities are primarily carried out in the embodiment and detail design phase (Boothroyd et al. 2011). To some degree, AM design rules (Section 12.2.1.1) are comparable to conventional DFM/DFX guidelines. Our literature review, however, shows that even *DfAM in the strict sense* is by no means at all limited to embodiment and detail design, but also strongly influences conceptual design, for example, by employing function structure analyses and parts consolidation methods (Rodrigue and Rivette 2010, Bin Maidin et al. 2012, Boyard et al. 2013). DFA, in contrast, is considered primarily in the conceptual design phase (Boothroyd et al. 2011).

Consequently, DfAM clearly is a concept or idea related to the whole product development process and provides approaches for all of its phases. We therefore propose a combination of the existing DfAM approaches with general design methodologies.

Researchers have developed many general design methodologies which serve as structured guidelines for design engineers. These methodologies are usually independent of specific products and manufacturing processes. Tomiyama et al. (2009) provided a detailed review on design methodologies. One of the best known methodologies is guideline VDI 2221 (VDI 1993), which is widely recognized both in research and practice. Its core element is a process chart represented in Figure 12.2. Similar to other methodologies, for example, Pahl et al. (2007), the process is divided into four phases. In *phase* 1, the problem or task is clarified. *Phase* 2 deals with the conceptual design including function structures, basic solution principles, and modular structures. The result of *phase* 2 is a product concept. It is concretized and refined in *phase* 3 (embodiment design), which contains the largest part of actual design engineering work. In the following detail design in *phase* 4, the exact part characteristics (e.g., surface qualities and dimensions) are defined and documented.

In order to precisely relate DfAM to general design methodologies, we create a matrix which contains existing approaches of *DfAM in the strict sense* in its rows and the four VDI 2221 phases in its columns, thereby allowing a direct matching of both. This positioning represented in Table 12.1 shows the respective phases that every DfAM approach presented in the literature review is covering. "X" denotes phases that an approach covers comprehensively, for example, by developing new DfAM-specific support tools for this phase. "(X)" denotes phases that an approach covers partly, that is, it adopts existing support tools for this phase or its new contributions to this phase are rudimentary.

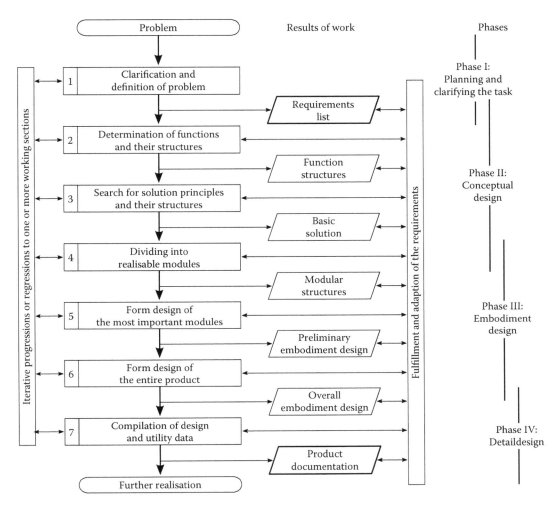

Figure 12.2 General procedure of systematic development and design according to VDI 2221 (VDI 1993). (Note: Reproduced with the permission of the Verein Deutscher Ingenieure e. V.)

12.3.2 Limitations of previous DfAM research

Literature review and the analysis of DfAM in the context of VDI 2221 expose two *overarching or primary limitations* of the existing DfAM approaches which provide, at the same time, promising research opportunities as follows:

- *Missing integration into common framework*: Although DfAM by quite a few is understood as a concept covering all design phases from product requirement/idea to design solution, no continuous method or framework in the style of VDI 2221 is available for DfAM. In particular, all of the presented approaches improve the utilization of AM only in their specific manner and facilitate the application of additively manufactured parts in the end-use products. However, design engineers are currently not

Table 12.1 Positioning of DfAM in the strict sense approaches in VDI 2221 ("X" denotes comprehensive coverage of the phase, "(X)" denotes a partly coverage of the phase)

	Phase	Clarification and definition of problem I	Conceptual design II	Embodiment design III	Detail design IV
AM design rules	Thomas (2009)			X	X
	Seepersad et al. (2012)			X	X
	Wagner and Witt (2012)		(X)	X	X
	Gerber and Barnard (2008)			X	X
	Adam and Zimmer (2014)			X	X
	Kranz et al. (2015)			X	X
	Hochschule Bremen (2008)			X	X
	VDI (2015)			X	
Utilization of AM design potentials	Hague et al. (2003), Hague et al. (2004)		(X)	(X)	
	Becker et al. (2005)		X	(X)	
	Burton (2005)	(X)	X		
	Bin Maidin et al. (2012)	(X)	X		
	Doubrovski et al. (2012)	(X)	X		
Combined approaches and methodologies	Ponche et al. (2012)			X	(X)
	Ponche et al. (2014)			X	(X)
	Vayre et al. (2012)			X	
	Leary et al. (2014)			X	
	Emmelmann et al. (2011)		(X)	X	
	Kranz (2014)	(X)	(X)	X	
	Tang et al. (2015)		(X)	X	(X)
	Ko et al. (2015)	(X)	(X)	(X)	(X)
	Tuck et al. (2008)			X	X
	Rosen (2007a)			X	(X)
	Rosen (2007b)		(X)	X	(X)
	Rodrique and Rivette (2010)			X	
	Boyard et al. (2013)		X	(X)	
	Yang and Zhao (2015)		(X)	(X)	
	Verquin et al. (2014)	(X)	(X)	(X)	(X)

provided with a methodical AM product development process guiding them from product idea to detail design.

- *Independence of DfAM approaches*: Previous research is fragmented; almost all DfAM approaches are developed independently and do not build on each other. For instance, most combined approaches (Section 12.2.1.3) do not utilize the existing knowledge from AM design rules (Section 12.2.1.1) and concepts for utilizing AM potentials

(Section 12.2.1.2). Even though there are approaches that include a continuous DfAM rudimentarily based on a standardized product development process model (Kranz 2014), there are basically no interfaces between the existing DfAM elements.

In addition, the existing DfAM research possesses the following *inherent or secondary limitations*:

- *Limited universal validity of AM design rules*: The interactions between process parameters and part-specific properties are highly complex, making predictions by means of process simulation difficult (Ponche et al. 2014). Thus, the validity of design rules is restricted to a specific physical principle, material, and machine class, for example, Adam and Zimmer (2014) and Kranz et al. (2015). Although rules of such kind seem appropriate for AM-experienced design engineers, only few general guidelines are available to introduce the new principles and restrictions to AM-inexperienced design engineers.
- *Focus on utilization of single AM potentials*: AM is often used for the optimization of one specific design objective, for example, weight reduction achieved by topology optimization (Emmelmann et al. 2011), or a decreased number of assembly operations (Boyard et al. 2013). Although a selectively increased product performance justifies the use of these methodologies, many additional AM design potentials oftentimes remain untapped, especially for completely new designs. Approaches based on functional specifications due to product requirements instead of an initial CAD model (Ponche et al. 2012) are promising exceptions for simulation-driven design.
- *Disproportional attention to innovative designs*: Very few methodologies are available to inspire new designs and product innovations in the conceptual phase of the product development process (Bin Maidin et al. 2012, Doubrovski et al. 2012). Therefore, a systematic utilization of AM potentials is limited by conceptual and cognitive barriers as a result of conventional process restrictions that have to be completely disregarded to exploit the full design potential of AM (Seepersad 2014). For this reason, specific geometries caused by manufacturability limitations should not be taken into account too early in AM product development processes.

12.3.3 Requirements for a new DfAM framework

The last section demonstrates the need for a new framework or methodology integrating previously independent DfAM approaches and tools. The goal is to guide DfAM novices and experts through the development process, to provide them with the right tools at the right time, and thereby to facilitate the development of truly DfAM-optimized products.

In some previous research papers, requirements referring to a continuous DfAM methodology have been formulated, and important specifications, for example, increased support in early design stages, have been pointed out (Laverne et al. 2014, Yang and Zhao 2015). However, none of them provides an own detailed concept to thoroughly fulfill these challenging requirements. Their solutions' usually low degree of detail is exemplarily shown in Figure 12.3. In addition, the idea of integrating and

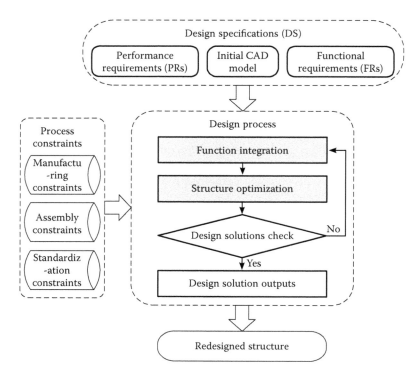

Figure 12.3 AM-enabled design method proposed by Yang and Zhao (2015). (With kind permission from Springer Science + Business Media, Yang, S. and Zhao, Y.F., *Int. J. Adv. Manuf. Technol.*, 80, 327–342, 2015, p. 339, Fig. 18.)

utilizing the existing DfAM approaches has not been included in these methodological concepts.

Therefore, the central objective is to turn an established framework and a uniform interface for the connection of individual approaches with their different advantages and abstraction levels into a holistic tool and to add missing modules to ensure a systematic exploitation of AM potentials and an easy adaption to further advancements in AM technologies. We formulate the requirements for our new DfAM framework as follows:

- *Comprehensiveness*: Similar to VDI 2221, the framework must provide continuous support in all design stages from task clarification to detail design.
- *Modularity*: The framework must be based on a modular structure to ensure that individual approaches and tools can be integrated easily into the overall framework and that their strengths are capitalized. In addition, modularity enables an easy updation of the framework in case of new technologies or redefined requirements.
- *Guidance, ease of use, and abstraction level*: The framework must serve as a guideline through the development process, useful for both DfAM novices and experts and valid for different types of products. Skipping or tailoring modules based on the individual application must therefore be easily feasible. The kind and abstraction level of support modules must thus be adjustable to respective stages of the design process and the level of knowledge of the product as well as the design engineer's experiences and skills.

- *Goal orientation*: AM-specific design potentials must be easily identified, and their interrelations for a systematic application/utilization in the design process must be ensured to achieve concrete goals/product improvements.
- *Degree of novelty*: The differentiation between new (innovative) designs and redesigns (routine-customized designs) has to be reflected in the procedure.

12.4 Proposal of a new DfAM framework

Based on the requirements compiled in the previous chapter, we develop a new DfAM framework particularly alleviating the primary limitations of previous DfAM research. Our proposal is based on three main ideas: First, we present the DfAM framework itself which is derived from VDI 2221 (Section 12.4.1). Second, we integrate the existing general design methods as well as the existing DfAM methods into the framework by means of different integration types (Section 12.4.2). Third, we provide a concept for using the framework based on specific criteria such as the product design's degree of novelty (Section 12.4.3). Our overall approach is visualized in Figure 12.4.

12.4.1 DfAM framework

The derivation of a new framework constitutes the first step in the development of a continuous DfAM. The framework represented in Figure 12.5 is based on the general VDI 2221 process model and adheres to the traditional subdivision into the phases *planning and clarifying the task, conceptual design, embodiment design,* and *detail design* (see Section 12.3.1). This helps design engineers to familiarize themselves quickly with the new methodology as they can build on existing process knowledge. However, the last two phases are consolidated in our framework because they are increasingly blended due to CAD utilization and iterations between these phases (Ehrlenspiel and Meerkamm 2013).

The standard VDI procedure is adapted to the distinctive characteristics of AM by including additional AM-driven steps at certain points in the process, for example, a new decision gate after the conceptual phase due to the currently limited applicability of AM for end products. Other traditional steps are modified or abbreviated. The framework modules are numbered from 1 to 9; many of them comprise several submodules (e.g., module 3a). Modularity allows an easy integration of various existing methods and tools (see Section 12.4.2). The modules can be compared to the seven steps of VDI 2221; the submodules are more specific and provide detail support.

In module 1, the product requirements list is compiled. However, there are different alternatives to enter the DfAM framework, for example, a preliminary parts selection (see Section 12.2.2.1). The possible ways to start the process are described in Section 12.4.3 together with various options for using the framework in practice.

The conceptual design phase (modules 2 and 3) starts with the determination of function structures based on product requirements. In the following step, basic solution ideas are developed. This step particularly focuses on the systematic utilization of AM design potentials, for example, through association aids like bionics catalogs creating the awareness for AM potentials. We emphasize the iterative character of creating and revising ideas both within this phase and the following phases. The result of the conceptual design phase is one or more conceptual models depending on the product type.

Based on the conceptual model(s), a modular product structure is developed. This is the first step of a decision gate containing modules 4 to 6. The division into realizable modules considers general AM restrictions (e.g., maximum part size) as well as other

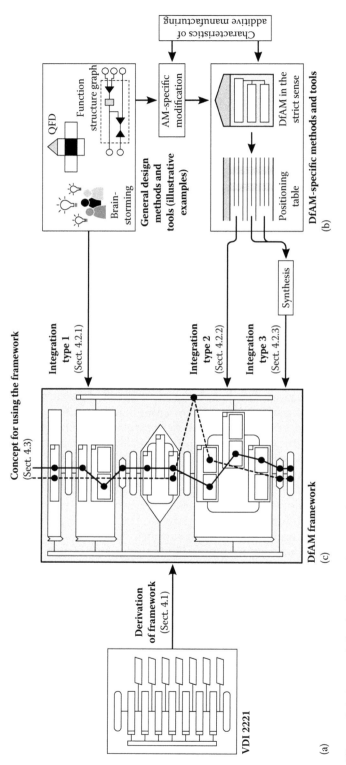

Figure 12.4 Approach for developing the new DfAM framework consisting of (a) framework derivation, (b) integration types, and (c) a concept for using the framework.

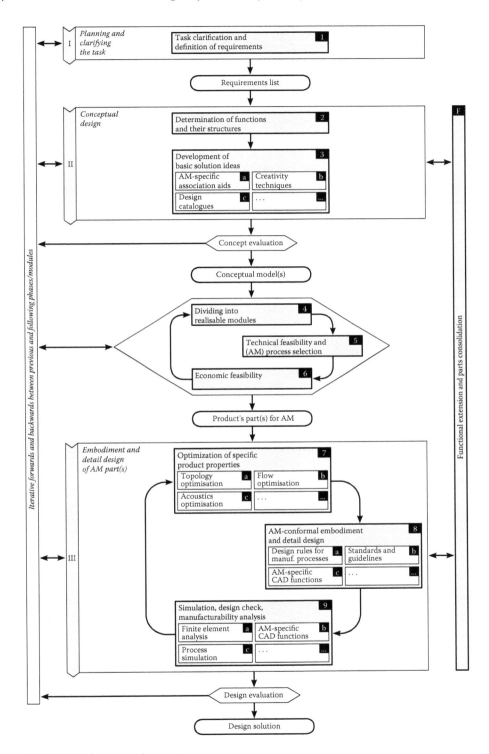

Figure 12.5 DfAM framework overview.

manufacturing technologies which may be more suitable for specific parts of the product. Afterwards, technical and economic feasibilities are checked. Criteria for technical feasibility evaluation include, for instance, the comparison of available AM material properties with product requirements or available build chamber sizes. Economic feasibility is evaluated with the help of several criteria including fundamental production characteristics (e.g., lot size) or product complexity. An early economic evaluation is helpful in order to avoid unnecessary efforts in the subsequent steps if profitability seems far from being attained. Hybrid manufacturing strategies, that is, a combination of additive and conventional manufacturing may also be suggested in the iterative process of the decision gate. Only those parts of the product which are to be additively manufactured are subjects of the following phase of the DfAM framework. Parts produced with conventional manufacturing technologies will be developed further with the help of other design methodologies.

In the embodiment and detail design phase (modules 7–9), the design solution is completed. In the first step, specific product properties may be optimized, for example, by topology optimization. Afterwards, the actual AM-conformal design is created, for example, with the help of an AM design rule collection or AM-specific CAD functions. It may be necessary to take conventional DFX methods into account as well, for example, DFM guidelines for subsequent processes such as milling. The last step contains submodules for simulation, design check, and manufacturability analysis. Several iterations may be necessary to achieve the final design solution.

Module F (right-side bar) deals with functional extension and parts consolidation, which are major AM design potentials. It is relevant for both conceptual and embodiment/detail design phases, since functional integration is not limited to one of these phases.

Following the structure of VDI 2221, the framework contains a left-side bar for iterative forward and backward between phases and modules. Although the main path suggests a linear process flow, today's design processes require high degrees of flexibility. Information flow to the left-side bar is either triggered at the evaluation gates or through special advices within the framework modules depending on the design goals.

12.4.2   Concept for the integration of existing tools and methods

The basis for the integration and utilization of existing methods and tools is the modularity of the framework. We distinguish three types of integration (see Figure 12.4). Integration type 1 focuses on the inclusion of general design methods and tools which are also helpful in a product development with AM. In contrast, integration type 2 contains existing DfAM-specific tools and methods. Integration type 3 finally tends on a consolidation of related DfAM approaches to provide a more general and/or wider support for the design engineer. The different types of integration are described and exemplified in the following subsections.

12.4.2.1   Integration type 1: Direct integration of general design methods and tools

Integration type 1 is based on the idea that there is no necessity to create AM-specific tools for each and every step of the framework. Many proven general design methods and tools developed in the past are independent of manufacturing processing and can thus be used for DfAM as well. Particularly in the conceptual phase of the process, tools of general application can be even more suitable since they provide a broad scope for solutions. These include, but are not limited to, methods such as quality function deployment, theory of inventive problem solving, brainstorming, function structure graphs, catalogs of solution principles, concepts for mass customization, and so on (Osborn 1963, Mizuno and Akao 1994, Tseng and Jiao 1998, Altshuller 1999, Roth 2002). Moreover, conventional DFMA guidelines (Bralla 1999, Boothroyd et al. 2011) must be taken into account if other manufacturing technologies are involved in the process chain, for example, machining for surface finishing.

12.4.2.2   Integration type 2: Direct integration of DfAM-specific approaches

Integration type 2 describes DfAM-specific approaches that can be integrated directly into the framework without major adaptations. Such methods and tools generally provide very specific support at single steps or particular tasks. Integration type 2 approaches are either independent developments essentially on the basis of AM characteristics or AM-specific modifications of general design methods and tools.

Approaches essentially developed on the basis of AM characteristics include, for example, the DfAM feature database of Bin Maidin et al. (2012), which provides support in developing basic solution ideas. It can be used in module 3a of the conceptual design phase to seize a suggestion of design features according to selective AM-related design objectives such as improved functionality or parts consolidation. Due to the integration into the framework, the tool fosters the selection of suitable AM benefits/potentials based on product requirements. Afterwards, the results of the inspiration can be concretized with additional tools in the embodiment and detail design phase to provide the design engineer with a continuous support. For this reason, the database already contains information about limitations of the design features like machine or material.

Furthermore, many existing DfAM approaches are based both on conventional design methods and AM characteristics. By means of a transfer of AM-enabled design opportunities to general design methods, modifications of these are established, for example, by Boyard et al. (2013) and by Rodrigue and Rivette (2010). Both are aimed at parts consolidation and the resultant reduction of part count. The method of parts consolidation proposed by Boyard et al. (2013) is based on the principle of abstraction and a parallelization of DFA and DFM. Rodrigue and Rivette (2010) developed a method to reduce the number of joints between parts systematically using a flowchart and the shape complexity potential of AM processes. Function structure graphs and flowcharts are both typical tools adopted from conventional design methodologies.

Another example for integration type 2 is the use of proven optimization methods such as DfAM-specific topology optimization approaches provided by, for example, Emmelmann et al. (2011) and Leary et al. (2014). These are integrated into module 7a of the framework.

12.4.2.3   Integration type 3: Synthesis of similar DfAM approaches

Integration type 3 proposes a synthesis of similar DfAM approaches with the same or comparable objective. AM design rules, for example, are widely available for several processes and materials as well as in different levels of detail.

To support engineers optimally in AM-conformal design for variable design tasks, they should be provided with guidelines and rules independent of the AM machine as it is rarely well defined prior to the design process. Moreover, AM-inexperienced design engineers rather need general guidelines to familiarize themselves with the most important aspects of AM-conformal design (e.g., the principle of support structures). Experienced

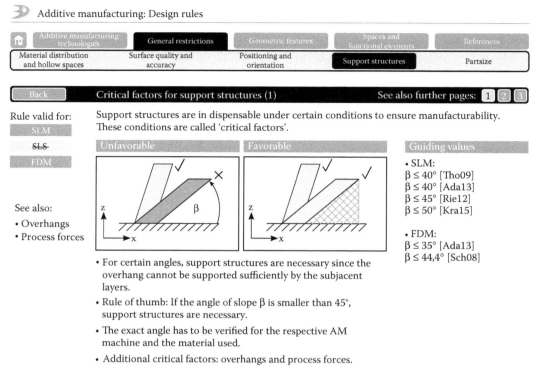

Figure 12.6 Interactive system for AM design rules (example page.).

design engineers, on the other hand, need quantitative rules of thumb for specific geometric features (e.g., minimal wall thickness). We therefore integrated all design rules currently available for SLM, SLS, and FDM into a comprehensive catalog incorporating both general AM characteristics and detailed restrictions. To facilitate its application, we created an interactive user interface (an example page is shown in Figure 12.6). It provides users with a structured access to various types of rules. Each rule is illustrated in a standard format including a precise description, its validity for specific AM processes, graphics for unfavorable and favorable design solutions, and guiding values. Although the latter differ depending on the source, users get a reasonable feeling for quantitative limitations. Moreover, links to related rules are provided on each page.

12.4.3 Using the framework

The third element of our approach is a concept for using the framework in the context of product development in practice. Only in this way, the framework serves as an actual support tool for the design engineer. We propose different paths through the framework depending on (1) the product's degree of novelty, (2) the respective user's experience with AM, and (3) the main design goal as follows:

1. Before starting the actual process, the design's degree of novelty is determined, since it influences the use of specific modules. For new designs, on the one hand, it is usually essential to go through the whole process. Redesigns, on the other hand,

are usually based on predetermined function structures and initial CAD models. Hence, they require less work in the conceptual design phase. It has to be noted, however, that the conceptual design phase should not be skipped particularly for redesigns of assemblies. Redesigns can be converted into new designs if unexpected AM potentials are discovered in the course of the structured process, for example, parts consolidation potentials enabled by AM. Even for redesigns of single parts, it can be helpful to analyze functional extension potentials.

2. User experience with AM is another criterion used to decide whether specific modules are helpful or not. An AM novice needs basic information about the capabilities of AM processes (e.g., the general idea of incorporating lattice structures), whereas an AM expert is rather interested in quantitative values (e.g., minimum and maximum cell sizes of lattice structures).

3. The main design goal also influences the utilization of specific parts of the framework. We suggest determining one or more design goals, for example, lightweight design, at the beginning of the process. Based on this decision, it is possible to offer goal-oriented methods and tools (e.g., topology optimization) or to disable irrelevant modules.

We therefore suggest predefining the applicability of each module, submodule, and tool in terms of these criteria. The concept is schematically shown in Table 12.2. Module 2, for example, is only applicable to new designs, but not influenced by other criteria. Module 3 is applicable to every design task, but its submodules and tools are task specific. The definition of applicability leads to an automatic generation of paths through the framework based on the given setting.

Table 12.2 Schematic concept for the applicability of DfAM framework elements in dependence of given boundary conditions ("X" denotes applicability)

| | Degree of novelty | | | User experience with AM | | Design goal | |
| | Redesign | | | | | | |
	Single part	Assembly	New design	Novice	Expert	Light weight	Parts consolidation
Module 2: determination of functions and their structures			X	X	X	X	X
Module 3: development of basic solution ideas	X	X	X	X	X	X	X
Submodule 3a: AM-specific association aids	X	X	X	X		X	X
Submodule 3b: design catalogs	X	X	X	X	X	X	X
Tool 3b1	X	X	X	X	X	X	
Tool 3b2		X		X	X		X

Although it is possible to use the framework modules individually or start the process at any module, we always suggest compiling at least a requirements list and using a pre-defined standard path through the framework based on distinctive criteria as this greatly improves the usability of the proposed concept.

An example case showing the practical usage of the framework is presented in Figure 12.7. The case study is about the redesign of a housing cover assembly; the user is an AM novice, and the goal is lightweight design. After compiling the requirements list, the user enters the conceptual design phase. He is provided with ideas for AM-specific features based on the design goal, that is, lightweight design enabled by lattice structures. Since the product comprises several separate parts, the user is redirected to module F (functional

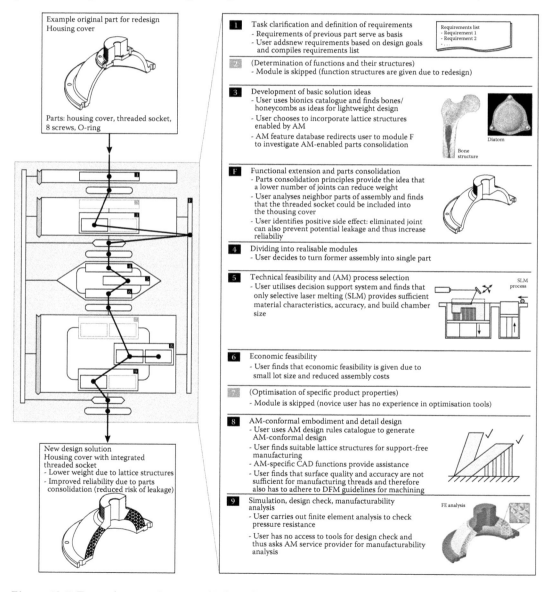

Figure 12.7 Exemplary application of DfAM framework for part redesign.

extension and parts consolidation) where he gains ideas for reducing part count. The result of the conceptual phase is a new conceptual model of the product. In the following decision gate, the user decides to turn the assembly into a single part based on the ideas gained in module F. The user then identifies SLM as the most suitable technology supported by an AM process selection system. Due to the small lot size, economic feasibility is given. In the following embodiment and detail design phase, the user is provided with specific design rules and CAD functions in the CAE environment to create an AM-conformal design with regard to support-free lattice structures and necessary postprocessing steps. The user is also advised of carrying out finite element (FE) and manufacturability analyses. In addition to weight reduction achieved by incorporating lattice structures, part count is reduced from 10 to 1 which leads to lower assembly costs and higher reliability.

Each module has specific input data, provides appropriate methods and tools, and generates output data based on both inputs and utilized tools. The input–output relations of the information flow between the modules are thus flexible and strongly dependent on the boundary conditions. In order to demonstrate the input–output relations in more detail, we exemplarily highlight them for module 3 (development of basic solution ideas):

- *Inputs*: Existing product geometry, requirements list, and boundary conditions (product's degree of novelty, design goal, user's experience with AM)
- *Methods and tools (recommendation based on boundary conditions)*: Bionics catalog for lightweight design, AM feature database, and brainstorming
- *Outputs*: Redirection to module F for parts consolidation and revised conceptual model

The case study reveals that even AM-inexperienced design engineers can create AM-optimal products if they are provided with customized guidance through the design process and if selected existing tools and methods are utilized at the correct point in time. The combination of its higher degree of detail within the modules and the comprehensive coverage of the whole design process make the framework unique. In contrast, previously published DfAM methods (e.g., Figure 12.3) usually have a lower degree of detail and do not integrate other established general and AM-specific design tools and methods.

12.5   Conclusion and future research

In this paper, previous research on DfAM is reviewed and classified. Limitations particularly exist in the conjunction of different approaches and in the continuous methodical support of design engineers from product idea to design solution. The existing DfAM approaches were basically developed independently, and they are not integrated into a common DfAM framework to support all design phases. Based on general design methodologies (e.g., VDI 2221), a new DfAM framework is proposed which provides design engineers with structured guidelines to fully exploit AM potentials, for example, by identifying paths to innovative AM-conformal redesigns in contrast to simple part modifications. Its core advantages over previous approaches are comprehensiveness and modularity, which allow an easy integration of the existing DfAM tools and methods into the correct design phases and facilitate a goal-oriented utilization of AM design potentials. In addition, a concept for using the DfAM framework is proposed which ensures specific support and guidance for the design engineer by considering user experience (AM knowledge), the design's degree of novelty, and design goals. The applicability of the framework is shown by an exemplary case study.

Nonetheless, an extensive validation of the proposed DfAM framework cannot be realized on the basis of one example part in this paper. However, the integrated methods and tools are established and therefore validated individually. Thus, the quality of the framework depends on its contents.

Future research will concentrate on the elaboration of individual modules to provide continuous specific support in all design phases, including the typically indispensable postprocessing steps for AM parts. Due to the modular structure of the new DfAM framework, these can be easily integrated into the overall structure. The interfaces between the modules have to been defined and standardized. Finally, the framework can also provide the architecture for DfAM-specific software tools. For example, these can be designed for supporting design engineers in selecting the appropriate process modules and design tools according to the design goal or the user's AM experience. In addition, DfAM software based on the proposed framework can include structured databases of the existing AM example parts serving as association aids or the option to generate standardized scorecards for part candidates based on technical and economic feasibility evaluations. The modularity of the framework allows the inclusion of conventional software tools which can be easily replaced, once AM-specific solutions are available. Software implementations could thus greatly improve the applicability of the framework.

Disclosure statement

No potential conflict of interest was reported by the authors.

References

Achillas, C. et al., 2014. A methodological framework for the inclusion of modern additive manufacturing into the pro duction portfolio of a focused factory. *Journal of Manufacturing Systems*, 37 (1), 328–339.

Adam, G.A.O., and Zimmer, D., 2014. Design for additive manufacturing—Element transitions and aggregated structures. *CIRP Journal of Manufacturing Science and Technology*, 7 (1), 20–28.

Altshuller, G., 1999. *The Innovation Algorithm: TRIZ, Systematic Innovation, and Technical Creativity.* Worcester, MA: Technical Innovation Center.

Ariadi, Y. et al., 2012. Combining additive manufacturing with computer aided consumer design. In *Solid Freeform Fabrication Symposium*. Austin, TX, pp. 238–249. Available from: http://sffsymposium.engr.utexas.edu/Manuscripts/2012/2012-17-Ariadi.pdf (Accessed February 08, 2017).

Becker, R., Grzesiak, A., and Henning, A., 2005. Rethink assembly design. *Assembly Automation*, 25 (4), 262–266.

Bibb, R. et al., 1999. Development of a rapid prototyping design advice system. *Journal of Intelligent Manufacturing*, 10 (3–4), 331–339.

Bin Maidin, S., Campbell, I., and Pei, E., 2012. Development of a design feature database to support design for additive manufacturing. *Assembly Automation*, 32 (3), pp. 235–244.

Boothroyd, G., Dewhurst, P., and Knight, W.A., 2011. *Product Design for Manufacture and Assembly*. Boca Raton, FL: CRC Press.

Boyard, N. et al., 2013. A design methodology for parts using additive manufacturing. *In*: P.J. Bártolo et al., eds. *High Value Manufacturing*. Boca Raton, FL: CRC Press, pp. 399–404.

Bralla, J.G., 1999. *Design for Manufacturability Handbook*. New York: McGraw-Hill.

Burton, M.J., 2005. *Design for Rapid Manufacture: Developing an Appropriate Knowledge Transfer Tool for Industrial Designers.* Thesis (PhD). Leicestershire, UK: Loughborough University.

Byun, H.S., and Lee, K.H., 2005. A decision support system for the selection of a rapid prototyping process using the modified TOPSIS method. *The International Journal of Advanced Manufacturing Technology*, 26 (11–12), 1338–1347.

Campbell, I., Bourell, D.L., and Gibson, I., 2012. Additive manufacturing: Rapid prototyping comes of age. *Rapid Prototyping Journal*, 18 (4), 255–258.

Campbell, R.I., and Bernie, M.R.N., 1996. Creating a database of rapid prototyping system capabilities. *Journal of Materials Processing Technology*, 61 (1–2), 163–167.

Conner, B.P. et al., 2014. Making sense of 3-D printing: Creating a map of additive manufacturing products and services. *Additive Manufacturing*, 1–4, 64–76.

Doubrovski, E.L., Verlinden, J.C., and Horvath, I., 2012. First steps towards collaboratively edited design for additive manufacturing knowledge. In *Solid Freeform Fabrication Symposium*. Austin, TX, pp. 891–901. Available from: http://sffsymposium.engr.utexas.edu/Manuscripts/2012/2012-68-Doubrovski.pdf (Accessed February 08, 2017).

Ehrlenspiel, K., and Meerkamm, H., 2013. *Integrierte Produktentwicklung: Denkabläufe, Methodeneinsatz, Zusammenarbeit*. Munich, Germany: Hanser.

Emmelmann, C. et al., 2011. Laser additive manufacturing and bionics: Redefining lightweight design. *Physics Procedia*, 12, Part A, 364–368.

Eyers, D., and Dotchev, K., 2010. Technology review for mass customisation using rapid manufacturing. *Assembly Automation*, 30 (1), 39–46.

Gebhardt, A., 2011. *Understanding Additive Manufacturing: Rapid Prototyping, Rapid Tooling, Rapid Manufacturing*. Munich, Germany: Hanser.

Gerber, G.F., and Barnard, L.J., 2008. Designing for laser sintering. *Journal for New Generation Sciences*, 6 (2), 47–59.

Gibson, I., Rosen, D.W., and Stucker, B., 2015. *Additive Manufacturing Technologies: 3D Printing, Rapid Prototyping and Direct Digital Manufacturing*. New York: Springer.

Hague, R., Mansour, S., and Saleh, N., 2003. Design opportunities with rapid manufacturing. *Assembly Automation*, 23 (4), 346–356.

Hague, R., Masood, S., and Saleh, N., 2004. Material and design considerations for rapid manufacturing. *International Journal of Production Research*, 42 (22), 4691–4708.

Hochschule Bremen. 2008. *Design guidelines for rapid prototyping: Konstruktionsrichtlinie für ein fertigungsgerechtes Gestalten anhand des Fused Deposition Modeling mit Dimension SST 768*. Available from: http://homepages.hs-bremen.de/~dhennigs/SERP%20-%20Downloads/Gestaltungsrichtlinien%20DIMENSION% 20768%20SST.pdf (Accessed October 24, 2015).

Kaschka, U., and Auerbach, P., 2000. Selection and evaluation of rapid tooling process chains with protool. *Rapid Prototyping Journal*, 6 (1), 60–66.

Kerbrat, O., Mognol, P., and Hascoet, J.Y., 2010. Manufacturability analysis to combine additive and subtractive processes. *Rapid Prototyping Journal*, 16 (1), 63–72.

Kerbrat, O., Mognol, P., and Hascoët, J.Y., 2011. A new DFM approach to combine machining and additive manufacturing. *Computers in Industry*, 62 (7), 684–692.

Kirchner, K., 2011. *Entwicklung eines Informationssystems für den effizienten Einsatz generativer Fertigungsverfahren im Produktentwicklungsprozess*. Munich, Germany: Dr. Hut.

Klahn, C., Leutenecker, B., and Meboldt, M., 2014. Design for additive manufacturing—Supporting the substitution of components in series products. *Procedia CIRP*, 21, 138–143.

Klahn, C., Leutenecker, B., and Meboldt, M., 2015. Design strategies for the process of additive manufacturing. *Procedia CIRP*, 36, 230–235.

Ko, H., Moon, S.K., and Hwang, J., 2015. Design for additive manufacturing in customized products. *International Journal of Precision Engineering and Manufacturing*, 16 (11), 2369–2375.

Kranz, J., 2014. Design for manufacturing approach for laser additive manufactured bionic lightweight structures in TiAl6V4. *Rapid Tech*. Erfurt, Germany: Desotron.

Kranz, J., Herzog, D., and Emmelmann, C., 2015. Design guidelines for laser additive manufacturing of lightweight structures in TiAl6V4. *Journal of Laser Applications*, 27 (S1), S14001–14016.

Kuo, T.C., Huang, S.H., and Zhang, H.C., 2001. Design for manufacture and design for 'X': Concepts, applications, and perspectives. *Computers and Industrial Engineering*, 41 (3), 241–260.

Laverne, F. et al., 2014. *DfAM in the Design Process: A Proposal of Classification to Foster Early Design Stages*. Sibenik, Croatia: Confere.

Leary, M. et al., 2014. Optimal topology for additive manufacture: A method for enabling additive manufacture of support-free optimal structures. *Materials & Design*, 63, 678–690.

Lindemann, C. et al., 2015. Towards a sustainable and economic selection of part candidates for additive manufacturing. *Rapid Prototyping Journal*, 21 (2), 216–227.

Merkt, S. et al., 2012. Geometric complexity analysis in an integrative technology evaluation model (ITEM) for selective laser melting (SLM). *South African Journal of Industrial Engineering*, 23 (2), 97–105.

Mizuno, S., and Akao, Y., eds., 1994. *QFD: The Customer-Driven Approach to Quality Planning and Deployment*. Tokyo, Japan: Asian Productivity Organization.

Munguía, J. et al., 2010. Development of an AI-based rapid manufacturing advice system. *International Journal of Production Research*, 48 (8), 2261–2278.

Osborn, A.F., 1963. *Applied Imagination: Principles and Procedures of Creative Problem-Solving*. 3rd ed. New York: Scribner.

Pahl, G. et al., 2007. *Engineering Design: A Systematic Approach*. 3rd ed. London: Springer.

Petrovic, V. et al., 2011. Additive layered manufacturing: Sectors of industrial application shown through case studies. *International Journal of Production Research*, 49 (4), 1061–1079.

Pine II, B.J., 1993. *Mass Customization: The New Frontier in Business Competition*. Boston, MA: Harvard Business School Press.

Ponche, R. et al., 2012. A new global approach to design for additive manufacturing. *Virtual and Physical Prototyping*, 7 (2), 93–105.

Ponche, R. et al., 2014. A novel methodology of design for additive manufacturing applied to additive laser manufacturing process. *Robotics and Computer-Integrated Manufacturing*, 30 (4), 389–398.

Rodrigue, H., and Rivette, M., 2010. An assembly-level design for additive manufacturing methodology. Bordeaux: *Proceedings of IDMME—Virtual Concept*.

Rosen, D.W., 2007a. Computer-aided design for additive manufacturing of cellular structures. *Computer-Aided Design & Applications*, 4 (5), 585–594.

Rosen, D.W., 2007b. Design for additive manufacturing: A method to explore unexplored regions of the design space. In: *Solid Freeform Fabrication Symposium*. Austin, TX, pp. 402–415. Available from: http://sffsymposium.engr.utexas.edu/Manuscripts/2007/2007-34-Rosen.pdf (Accessed February 08, 2017).

Rosen, D.W., 2014. Research supporting principles for design for additive manufacturing. *Virtual and Physical Prototyping*, 9 (4), 225–232.

Roth, K., 2002. Design catalogues and their usage. *In*: A. Chakrabarti, ed. *Engineering Design Synthesis: Understanding, Approaches and Tools*. London: Springer, 121–129.

Seepersad, C.C., 2014. Challenges and opportunities in design for additive manufacturing. *3D Printing and Additive Manufacturing*, 1 (1), 10–13.

Seepersad, C.C. et al., 2012. A designer's guide for dimensioning and tolerancing SLS parts. In: *Solid Freeform Fabrication Symposium*. Austin, TX, pp. 921–931. Available from: http://sffsymposium.engr.utexas.edu/Manuscripts/2012/2012-70-Seepersad.pdf (Accessed February 08, 2017).

Tang, Y., Kurtz, A., and Zhao, Y.F., 2015. Bidirectional evolutionary structural optimization (BESO) based design method for lattice structure to be fabricated by additive manufacturing. *Computer-Aided Design*, 69, 91–101.

Thomas, D., 2009. *The Development of Design Rules for Selective Laser Melting*. Thesis (PhD). *Cardiff*: University of Wales.

Tomiyama, T. et al., 2009. Design methodologies: Industrial and educational applications. *CIRP Annals—Manufacturing Technology*, 58, 543–565.

Tseng, M.M., and Jiao, J., 1998. Concurrent design for mass customization. *Business Process Management Journal*, 4 (1), 10–24.

Tuck, C.J. et al., 2008. Rapid manufacturing facilitated customization. *International Journal of Computer Integrated Manufacturing*, 21 (3), 245–258.

Vayre, B., Vignat, F., and Villeneuve, F., 2012. Designing for additive manufacturing. *Procedia CIRP*, 3, 632–637.

VDI. 1993. *VDI 2221: Systematic Approach to the Development and Design of Technical Systems and Products*. Berlin, Germany: Beuth.

VDI. 2015. *VDI 3405 Part 3: Additive Manufacturing Processes, Rapid Manufacturing—Design Rules for Part Production Using Laser Sintering and Laser Beam Melting*. Berlin, Germany: Beuth.

Venkata Rao, R., and Padmanabhan, K.K., 2007. Rapid prototyping process selection using graph theory and matrix approach. *Journal of Materials Processing Technology*, 194, 81–88.

Verquin, B. et al., 2014. *SASAM (support action for standardisation in additive manufacturing): Guidelines for the development of the EU standards in additive manufacturing.* Available from: http://www.sasam.eu/index.php/downloads/send/3-deliverables-public/3-d3–3-first-and-second-draft-guidelines.html (Accessed October 24, 2015).

Watts, D.M., and Hague, R.J., 2006. Exploiting the design freedom of RM. In: *Solid Freeform Fabrication Symposium.* Austin, TX, pp. 656–667. Available from: http://sffsymposium.engr.utexas.edu/Manuscripts/2006/2006-57-Watts.pdf (Accessed February 08, 2017).

Wegner, A., and Witt, G., 2012. Design rules for laser sintering. *Journal of Plastics Technology*, 8 (3), 253–277.

Wohlers, T., 2014. *Wohlers Report 2014: 3D Printing and Additive Manufacturing State of the Industry—Annual Worldwide Progress Report.* Fort Collins, CO: Wohlers Associates.

Yang, S., and Zhao, Y.F., 2015. Additive manufacturing-enabled design theory and methodology: A critical review. *The International Journal of Advanced Manufacturing Technology*, 80 (1), 327–342.

Zhang, Y. et al., 2014a. Evaluating the design for additive manufacturing: A process planning perspective. *Procedia CIRP*, 21, 144–150.

Zhang, Y., Xu, Y., and Bernard, A., 2014b. A new decision support method for the selection of RP process: Knowledge value measuring. *International Journal of Computer Integrated Manufacturing*, 27 (8), 747–758.

section two

Technical section

chapter thirteen

Development and implementation of metals additive manufacturing

Ian D. Harris

Contents

Additive manufacturing (AM) technologies build near-net shape components one layer at a time using data from 3D computer-aided design (CAD) models. AM technologies are the result of evolution of work in 3D printing and stereolithography (the STL files used to convert 3D CAD to layers for building parts come from stereolithography terminology) and could revolutionize many sectors of the U.S. manufacturing by reducing component lead time, cost, material waste, energy usage, and carbon footprint. In addition, AM has the potential to enable novel product designs that could not be fabricated using conventional subtractive processes and to extend the life of in-service parts through innovative repair methodologies.

The opportunities for the offshore oil and gas industry largely remain to be identified but are considered to involve combined functionality, functionally gradient materials, and embedded sensors for structural health or other monitoring functions.

13.1 Definition of additive manufacturing

AM has grown from the early days of rapid prototyping and as a dynamic field of study has acquired a great deal of related terminology. The ASTM F-42 committee was recently formed to standardize AM terminology and develop industry standards. According to their first standard, ASTM F2792-10, AM is defined as:

> *The process of joining materials to make objects from 3D model data, usually layer upon layer, as opposed to subtractive manufacturing technologies.*

A key point in this definition is that in order to qualify as AM under the ASTM definition, the 3D model data controlled by a computer must be used as a design precursor. Simply adding or building up material is not included in this process definition. The notion of

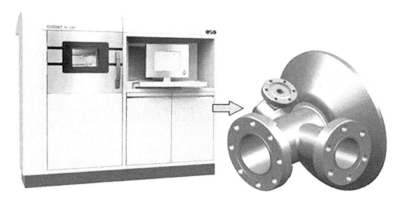

Figure 13.1 The promise of AM.

automation and control using software is an essential distinction in assessing what does and does not fall under the technology of AM. There are many related terms used to describe AM, and common synonyms include additive fabrication, additive layer manufacturing, direct digital manufacturing, and free-form fabrication (FFF).

Within the last 20 years, AM has evolved from simple 3D printers used for rapid prototyping in nonstructural resins to sophisticated rapid manufacturing that can be used to create parts directly without the use of tooling. Most work to date has been conducted using plastics, but a significant effort is now focused on metals.

In principle, a designer can engineer a part using 3D model data in a CAD program and simply email the file to a local manufacturer who can then return the part in a few days. This vision of a new paradigm of mass customized manufacturing is driving much of the excitement in this growing field (Figure 13.1).

13.1.1   *Current landscape in additive manufacturing*

The global market for AM exceeded $1 billion in 2009 with direct revenues for systems and materials sales of over $500 million.[1] Ninety percent of the AM machines sold are 3D printers for making polymer-based parts and models.

In addition to market growth, the visibility of AM technology and industry is increasing. In early 2010, a group of companies led by Materialise formed a group to conduct collective marketing for AM.[2] The cover story for a recent issue of the UK magazine *The Economist* addressed the potential of AM as a revolutionary manufacturing technology.[3]

Although a majority of the current global activity in AM is using polymer-based systems, there has been a significant activity and interest in fabrication of metallic parts. This is of interest because of the possibility for direct fabrication of net or near-net shape components without the need for tooling and with minimal or no machining. There has been particular interest in aerospace and biomedical industries owing to the possibility for high-performance parts with reduced overall cost. The opportunity for such in the oil and gas industry is only just now being explored.

Researchers and industry leaders in the European Union (EU) have identified AM as a key emerging technology.[4] Teaming relationships have been formed between university, industry, and government entities within and across countries. The overall level of activity and infrastructure in the EU is greater to that of the United States in this key area. Several large cooperative projects have been funded, worth of millions of dollars across Europe, including the rapid production of large aerospace components (Rapolac) and the custom fit project[5] for mass-customized consumer and medical project manufacturing. Though much of the original research developing these technologies was done in the United States, much of the subsequent development has been done elsewhere, particularly in Europe.

In 2009, a workshop was held in the United States to form a roadmap for research in AM for the next 10–12 years.[6] The workshop focused on identifying possibilities for development in design, process modeling and control, materials, biomedical applications, energy and sustainability, education, and efforts at development in the overall AM community. The overall assessment was that there are many opportunities for these technologies if investments are made to continue to advance the state-of-the-art. A key recommendation of the report was the establishment of a National Test Bed Center (NTBC) that could leverage equipment and human resources in future research and to demonstrate the concept of cyber-enabled manufacturing research.

Based on results from the roadmap developed in 2009, EWI organized an additive manufacturing consortium (AMC) to bring together key partners in the U.S. AM community. The AMC now consists of 22 industrial members and partner organizations, representing both large and small industry members, government agencies and other partner organizations, and key universities active in the field of AM research. The main goal of the AMC is to advance the manufacturing readiness of AM technologies and to advocate on a national basis for investment in AM to move these technologies into the mainstream of manufacturing technology from their present emerging position. The highest rated technical need is to produce mechanical property data suites for qualification of combinations of the many processes and materials of interest.

13.2   *Technologies for additive manufacturing*

The two main components of any metal AM process are the type of raw material input and the energy source used to form the part. Three main technology categories of AM are considered: powder bed, laser powder injection, and FFF systems that do not use lasers.

The powder bed systems are used in enclosed chambers, and energy is supplied by either a laser or an electron beam to melt the powder in a powder bed to form the desired shape, Figure 13.2. In laser powder injection, a powder nozzle adds material, and a laser beam melts the powder. The free-form processes are a broader category, and the types addressed here include electron beam deposition of metal wire, arc deposition of powder and wire, and ultrasonic consolidation of metal layers.

Each of the processes has its own unique characteristics for speed of manufacturing, postdeposition treatment required, porosity, and level of impurities in the as-built part. In each technology category and for each individual manufacturing process, there are tradeoffs between build rate and maximum build size with surface quality and between deposition of excess material and overall deposition accuracy, Figure 13.3.

For plastics, work on automated near-net-shape AM of components dates back to the 1980s. Work on metals is more recent, and by far the bulk of metal AM research has focused on fusion processes, where successive layers of metal are deposited by melting. Several energy sources (e.g., laser, electron beam, arc) and material forms (e.g., metal powders, wires) have been employed. Powder bed processing has dominated this research over the last decade. Powder bed processing includes variants integrating electron beam or laser power systems. Commercial systems have been introduced, which are

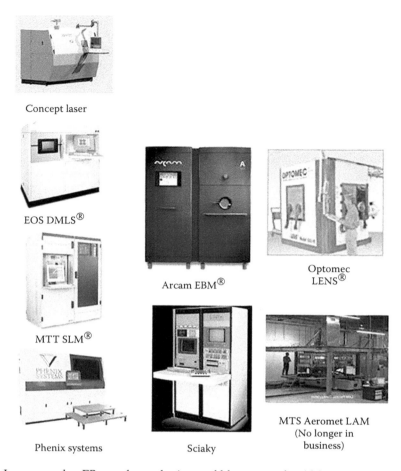

Figure 13.2 Laser powder, EB powder and wire, and blown powder AM systems.

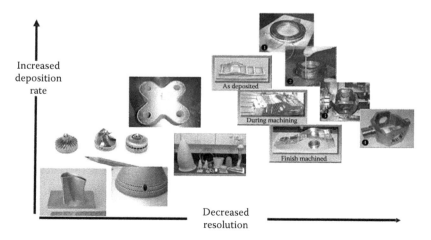

Figure 13.3 Trade-off of part resolution and deposition rate for small and large parts.

capable of producing components of limited size. End users have recognized a number of limitations of this technology, including inconsistent results and poor productivity. More recently, considerable research has been expended on the so-called FFF approaches. Such processing has incorporated a range of power input types (electron beam, laser, gas tungsten arc, plasma arc) to allow development and manufacture of a broader range of product sizes at higher deposition rates. Early precursor work using robotic arc welding, Figures 13.4 and 13.5, achieved similar results but was not truly an AM process by definition.

Figure 13.4 Large component made with robotic arc welding.

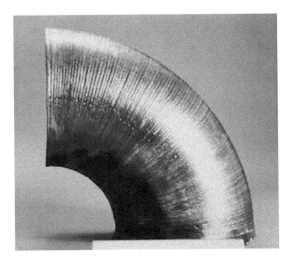

Figure 13.5 Large pipe elbow made with robotic arc welding.

13.2.1 The state of the art in AM

The tangible efforts in research are also aligned with the practical realities of cost, and it is known that the present embodiment of metals AM for combat aircraft now meets cost requirements compared to conventional manufacturing technologies, even if a simple approach to replace a casting or forging is taken, as long as the deposition rate of the process is high enough. In short, the clear buy-to-fly ratio advantages of metals AM reducing from 20:1 or 12:1 to even 2:1 can still be insufficient, if the deposition rate of the manufacturing technology is not high enough, and cost components of finishing technologies are high, affecting the overall cost. One result is that the clear direction is for AM to address a *mind to fly* approach embodying the unique advantages of AM to produce parts that are otherwise unmanufacturable for one or more reasons, along with integrated rather than parasitic functionality and including functionally gradient metals. The latter offers considerable potential for transition pieces between joints in otherwise dissimilar materials.

Much of the emphasis of AM research around the world including the United States is focused on laser and electron beam processes, especially laser and electron beam powder bed processes such as DMLS and EBM through commercialized equipment such as EOS M280, Arcam A2, and electron beam with wire addition, known as EBFFF, respectively. These powder bed systems are closed architectures and are made in Germany and Sweden, respectively.

A notable exception is an emerging solid-state technology known as ultrasonic AM (or ultrasonic consolidation). This technology employs ultrasonic vibrations and pressure to deposit successive layers of metal strips. Work to date has been primarily with soft aluminum alloys. Inconsistent bond quality and poor through-thickness properties have limited application for higher strength alloys, such as titanium.

In addition to AMC leadership and capability in many emerging AM processes for metals, EWI has developed technology thrusts in ultrasonic additive manufacturing (UAM) based initially on work with Solidica. This technology development has increased available power by fivefold from 2 kW to 10 kW and down force tenfold from 500 lbs to 5000 lbs. This capability, developed by EWI and funding through the Ohio Department of Development (ODOD) and an industry consortium including Boeing and GE (both AMC members), has delivered a new very high power UAM (VHP UAM) capability to the EWI shop floor and enabled a new company, Fabrisonic (jointly owned by EWI and Solidica), to be spun off, providing new jobs in Ohio based on the U.S. manufacturing of custom VHP UAM machines for industry and low-cost machines for innovative university research. EWI was recently awarded a project under the Ohio Third Frontier Program to develop and build a laser-based system to augment VHP UAM bond integrity with the aim to substantially increase the technology readiness of this emerging solid-state AM technology.

EWI alone and in conjunction with AMC partners are working in most applicable research areas for processes, sensors, and modeling. Active areas include modeling of residual stresses and distortion, and multifunctional and multimaterial structure modeling for hypersonic vehicle structures with embedded functionality rather than parasitic functionality. Residual stress and distortion modeling have been used to predict the effect of continuous and semi-continuous AM for airfoil fabrication and repair. Simulations accounted for combined thermal and resulting distortion prediction and validation. In terms of sensors and sensor fusion, EWI and its AMC partners are active in visual, thermal, laser-based, and other sensor technologies and are specialized in novel sensor integration.

EWI conducted over $500K of research into arc-based titanium AM and titanium armor cladding for a Prime that is part of AMC. This work demonstrated a deposition

Figure 13.6 A large ground vehicle control arm made by robotic hot wire GTAW, The Additive Manufacturing Consortium (AMC).

rate of over 10 lbs/hour in Ti-6Al-4V using hot wire GTAW in air using a trail shield, Figure 13.6. The development work showed that the Ti-6Al-4V deposits produced met requirements regarding oxygen levels, thus opening up the possibility of production of large titanium parts without the constraints of a small powder bed working enveloped or a vacuum chamber and concomitant consequences including specialty overalloyed wire to compensate for aluminum vaporization in a vacuum chamber.

By combining the attributes of EWI with the collective human and equipment capabilities of the AMC members and partners, a distributed capability presently called the National Test Bed Center (NTBC) is in place for metals AM and joining processes that also include considerable expertise in NDE processes and AM part finishing by machining, grinding, and other surface-finishing technologies.

The AMC, formed in 2010 and led by EWI, is a consortium of 22 organizations in industry, government, and academia which was formed to address needs in metals AM on behalf of U.S. manufacturing competitiveness for the advancement of the manufacturing readiness of this disruptive manufacturing technology. AMC meets quarterly to network, exchange information on public domain research conducted, discuss plans and implementation for program definition for lead candidate technology and data development associated with ultimate process qualification. AMC members including EWI are represented on ASTM F42, the international standards committee for development of AM standards.

AMC members include both multinational corporations and small businesses, Boeing, Lockheed Martin, General Dynamics, GE, Rolls-Royce, Morris Technologies, Applied Optimization, B6Sigma, and EWI. Government agency partners are U.S. Army (Picatinny, Benet), U.S. Air Force (WPAFB), U.S. Navy (NAVAIR), NIST, and NASA, and key universities active in the AM research field, namely the Ohio State University (OSU), University of Texas, El Paso (UTEP), University of Louisville (UofL), North Carolina State University (NCSU), and South Dakota School of Mines and Technology (SDSMT). Other partners include Lawrence Livermore National Laboratory (LLNL), TechSolve, NCMS, and SAE RTAM.

In short, the AMC partnership operated by EWI combines industry (both large and small), government, and university expertise and knowledgebase along with relationships with other consortia to provide a very comprehensive national capability to advance

AM, and other associated technologies in an integrated open manufacturing basis for the nation, jobs, and national competitive advantage in advanced manufacturing. These organizations, most of which are also EWI member companies, together possess a thorough knowledge of the AM landscape and are actively researching in the field.

13.2.2 AMC goals accomplished in year 1

1. Obtain broad industry and government support—achieved for A and D, reaching out to oil and gas, power, and heavy fabrication community.
2. Organize *National Test Bed Center* research partners network—in place with extensive equipment and staff resource capabilities.
3. Identify technology priorities and create development plan—priorities identified for Ti- and Ni-based alloys. $60M of proposals developed to government agencies.
4. Conduct state-of-the-art review of metal AM technology—complete.
5. Establish a database for collecting metal AM property information—will use MMPDS.

13.2.3 The future of AM

There is a rich landscape of available technologies and materials for metals AM. Parts in titanium alloys, nickel alloys, high-grade stainless steels, and many others are being produced using lasers, electron beam, and arc techniques with a variety of consumable forms. This is a dynamic, constantly evolving field with many researchers and industrial users continually improving the state-of-the-art while moving to develop and qualify combinations of material and process for commercial exploitation. Currently, the number of commercially made parts is low because of the high-performance demands and associated costs for industry to qualify parts. Parts qualification costs relate to the demands of the applications and market environments.

Using AM to fabricate metal parts opens the possibility for reducing material usage that could enable overall reduction in cost and greenhouse gas emissions related to manufacturing. Promising case studies have been undertaken, and there are a number of ongoing studies investigating how AM can enable *green manufacturing*.

The opportunities for the offshore oil and gas industry largely remain to be identified but are considered to involve combined functionality and functionally gradient materials. For the aerospace industry, this could lead to a reduction of required raw materials used to fabricate an in-service component, which is known as the *buy-to-fly* ratio. AM could also lead to new innovations for lightweight structures that could see application in unmanned aerial vehicles.

Applications where legacy parts are still necessary for operation and fabricators are no longer in business, which could use AM to create parts direct from a CAD file. For the medical industry, AM is already leading to a revolution in customized medicine where dental implants, orthopedics, and hearing aids are manufactured to fit an individual's unique physiology.

The *digital thread* (fully computerized design to production using computer-based technical data packages) and *moving manufacturing to the left* (integrated computer modeling of materials, process, distortion, metallurgical, and mechanical properties) are visions implicit in the future of open architecture advanced manufacturing for AM.

The most exciting possibilities for AM are for unique applications that could not be fabricated using standard machining practices. Examples include tailored medical implants that can be built with the exact bodily geometry output using an MRI or advanced turbine blades with application-specific cooling channel designs. As a fundamental enabling technology, novel applications that are just beginning to be imagined could be built. Novel functional gradient materials could be generated using these techniques that could enable entirely new applications.

However, there are a limited number of technologies commercially available, and there is a great deal of work to be done on ruggedizing these processes for commercial scale manufacturing. In particular, the larger scale FFF technologies, though their fundamental technologies are commercially deployed in many industries, are at a lower stage of manufacturing readiness as compared to the powder bed or powder-injected laser approaches when it comes to AM part production. Closed-loop feedback control sensing systems and intelligent feed forward schemes will need to be developed and integrated into systems to better control the manufacturing cycle. Currently, part properties and quality can vary from machine to machine for a given material and technology. In addition, new methodologies for nondestructive evaluation need to be developed as many of the microstructures formed present inspection challenges.

Metals AM is still a relatively new and immature technology, and there is a need for understanding the basic science of each particular AM process as most of the processing parameters to this point have been empirically derived. In particular, there is a need to understand the material microstructure resulting from a particular thermal processing cycle. There have been many studies on individual processes and resulting properties, but there is still a need for a comprehensive material property database and testing methodology to be developed. Many studies have been carried out for tensile strength and elongation as a function of material compositions. Further studies of the effect of processing parameters on dynamic loading in high and low cycle fatigue and impact toughness, creep, and other situations will be important to fully understand the performance of AM parts in service-like conditions. The newly formed ASTM F-42 committee is working to write standards that address a wide array of these needs, and there is much work still to be done.

Mechanical properties of parts can vary greatly depending on the process used, parameters of the individual process, loading direction, and postfabrication heat and surface treatments.

Furthermore, different part geometries require special design considerations such as supports and heat sinks that ensure built parts to maintain geometric accuracy. And depending on the technology used, the deposition path can affect the final properties.

To date, there has been a relatively large body of work and focus on Ti-6Al-4V but not as much on other alloys and metals. This is understandable given its high cost and utility in high-value aerospace and medical applications. There is a rich landscape of other high-value applications requiring metal alloys that include nickel, aluminum, and refractory metals that could be manufactured using AM that heretofore have not been extensively investigated.

AM of metals is opening up new possibilities for lower cost manufacturing and novel-integrated parts designs that cannot be made using current technology. This is generating a great deal of enthusiasm around the world for future high-value manufacturing applications. Currently, there are niche applications; particularly in the medical field and to a lesser extent aerospace where parts made using plastics AM and some metals are being put into initial evaluation of service. To meet the full potential of these processes, continued development to *productionize* the machines for full manufacturing readiness and further understanding of the materials properties are essential. With the pace of advancement, this key-emerging field is poised to grow rapidly over the coming years.

13.3   Summary

In summary, AM represents a whole new paradigm and range of opportunities for design, functionality, and cost. The AM field represents an exciting and rapidly emerging industry:

1. AM for metals is rapidly developing through a range of powder bed and wire-fed technologies.
2. AMC is poised for growth into many manufacturing sectors.
3. AMC offers collaboration for development of metals AM manufacturing readiness using laser, EB, arc, and other processes—consortium has 22 members and partners, and welcomes more.
4. Looking for potential applications within the oil and gas market that fits one or more of the following scenarios and opportunities:
   • Nominally *unmanufacturable* components
   • High added value, long lead time items
   • Adding features to low-yield castings and forgings
   • Repair applications

References

1. Wohler, T., Wohlers Report 2010, Wohlers Associates, Ft. Collins, CO, 2010. http://www.pdfslibforme.com/wohlers-report.pdf (accessed on January 22, 2017).
2. Materialise.com press release, Jan 2010. http://www.materialise.com/ (accessed on January 22, 2017).
3. The Economist Journal, Cover Story (2011), Print Me a Stradivarius, *The Economist*, Feb. 10, 2011. http://www.economist.com/node/18114327 (accessed on January 22, 2017).
4. Beaman, J. J., Atwood, C., Bergman, T. L., Bourell, D. L., Hollister, S., and Rosen, D., WTEC Panel Report on Additive/Subtractive Manufacturing Research in Europe, World Technology Evaluation Center, Baltimore, 2003.
5. Custom-fit.org; European Commission online, http://ec.europa.eu/research/infocentre/article_en.cfm?id=/research/headlines/news/article_09_02_11_en.html&item=&artid= (accessed on January 22,2017).
6. Bourell, D. L., Leu, M. C., and Rosen, D. W. eds., *Roadmap for Additive Manufacturing, Identifying the Future of Freeform Processing*, The University of Texas at Austin Laboratory for Freeform Fabrication, Austin, TX, 2009.

chapter fourteen

Selective laser melting (SLM) of Ni-based superalloys:

A mechanics of materials review

Sanna F. Siddiqui, Abiodun A. Fasoro, and Ali P. Gordon

Contents

Abstract: Additive manufacturing (AM) of high-resolution components for applications in the defense, aerospace, power generation, propulsion, and biomedical industries has led to accelerated design and production schedules of these parts as well as geometric flexibility, leading to the development of more complex component designs. AM technologies use a digital solid model (e.g., Computer-aided design, CAD) to develop a component layer-by-layer, with layer thickness varying depending upon the material used during manufacturing. Among the most commonly used AM processes are the powder-bed technologies, which include selective laser sintering (SLS), direct metal laser sintering (DMLS), electron beam melting (EBM), and selective laser melting (SLM), from which a variety of alloys such as Inconel 718 (IN718), Stainless Steel 316L and Ti-6Al-4V can be produced. The SLM process involves selectively melting metal powder by a high-energy laser within an inert gas (nitrogen or argon) environment. The focus of this review is on understanding the effect of SLM processing parameters on the microstructure and mechanical properties of Ni-based superalloys commonly used in the manufacturing of components subjected to elevated temperatures. The main goal here is to highlight the developments and to reflect on further research needed to better understand the SLM AM process. In addition to this review, this paper will present simulated elastoplastic tensile stress–strain response of SLM Inconel 718 and heat-treated SLM Inconel 718 exhibiting anisotropy, based upon mechanical properties found in literature. It will also discuss the viability of using the resulting tensile stress–strain response trends observed for SLM Inconel 718 and heat-treated SLM Inconel 718, to model a first-order approximation of the elastic behavior (e.g., Young's Modulus) response with build orientation.

Keywords: Constitutive modeling, Surface roughness, Residual stress, Fatigue, Creep, Ramberg–Osgood

14.1 Introduction

Research in the area of AM technology has intensified over the last decade as a result of component development flexibility and rapid production capabilities afforded by additive manufacturing (AM) [1]. 3D printers began to appear in the late 1980s with the introduction of stereolithography followed by fused deposition modeling (FDM) and SLS in the early 1990s [2]. SLM was introduced in the 1990s, followed by the introduction of DMLS and EBM technologies in the 2000s [2]. AM can be characterized into powder bed, powder feed, or wire feed systems [3], of which the powder bed systems are comprised of SLS, DMLS, SLM, and EBM [3,4].

The laser-sintering process involves selective melting or fusing of powder particles, one layer at a time within a gas chamber environment to develop a component [5]. Distinctions between SLM and EBM include the power source (fibre laser versus electron beam), pre-heating build plate temperature, build chamber (inert gas versus vacuum), beam-focusing lenses, and so on [6]. The EOSINT M 280 model based upon DMLS is shown in Figure 14.1. This particular device has a (250 × 250 × 325 mm) printing space suitable for the printing of small to moderately sized components that can range from *tooling inserts* to complex geometries such as *free-form surfaces, deep slots,* and *coolant ducts* as suggested by EOS [7].

AM system manufacturers include but are not limited to: EOS, 3D Systems, Renishaw, and ARCAM; and guidance is available to understand the steps involved in the 3D-printing process [8]. Literature is also available for individuals interested in building their own 3D-printing device [9], as well as reducing cost associated with 3D printing, through the use of open-source software and low-cost hardware [10]. Applications of additive-manufactured components manufactured in industry include but are not limited to NASA's SLM 3D Rocket Injector [67] and a model of GE's DMLS GEnx Jet Engine [68].

AM offers several advantages over conventional manufacturing processes. These advantages include reduced part volume, tailoring of localized mechanical properties, the realization of more complex designs not achievable with cast or subtractive-machined materials, and so on. The performance (strength, ductility, weight, etc.) of Ni-based superalloys (e.g., Inconel 718, Inconel 625, Inconel 939, IN738LC, CMSX486 and CM247LC) may be improved through AM as compared with conventionally developed materials. Of the Inconel® family of metals, Inconel 718, Inconel 625, Inconel 939, and Inconel 738LC are among the most commonly used in high-temperature applications because of their strength at high temperatures, corrosion

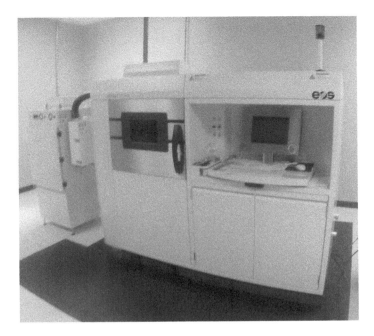

Figure 14.1 The EOS M280 DMLS machine at Central State University (CSU), Wilberforce, OH.

resistance, machinability, weldability, and so on. These classes of specialty Ni-based alloys will be the focus of this review article, with emphasis on Inconel 718.

A number of concepts are germane to the mechanics of AM Inconel. This review continues with an overview of the SLM processing parameters and subsequent optimization of these parameters with respect to porosity, cracking density, surface roughness, and residual stress of as-built components. It also presents the impact of heat treatment (HT) on the performance of these components, which is provided in Section 14.2. The influence of processing parameters on microstructural characteristics of AM Inconel is described in Section 14.3. In Section 14.4, the mechanical properties such as tensile, fatigue, and creep performance of AM Inconel are discussed. Constitutive and life models that have been developed for SLM Inconel, including a first-order approximation of Young's Modulus and stress–strain response with build orientation, from tensile properties provided in literature are presented in Section 14.5. Finally, Section 14.6 will close with a discussion of the ongoing challenges observed in this field of study, where further research and development can enable SLM of Inconel parts to become a more ubiquitous manufacturing method for critical components subjected to elevated temperatures.

14.2 Selective laser melting

The SLM process as illustrated in Figure 14.2a occurs in an ambience of argon or nitrogen gas, in which an Ytterbium (Yb) fibre laser is passed through a beam scanner, which selectively melts metal powder located within a powder bed on a substrate plate/build platform. The powder for each layer is supplied from a powder storage system via a roller or rake. For a given candidate 3D component, its associated digital solid model analogy (e.g., STL) file is sliced into 2D layers, from which instructions typically in the form of G-Code are used to guide the laser scanner during the building process. As

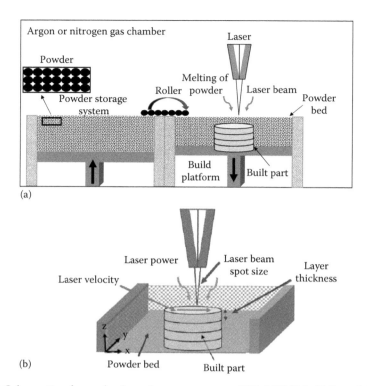

Figure 14.2 (a) Schematic of powder laser beam processes (SLM, DMLS, SLS) and (b) powder bed processing parameters.

each layer is formed, the build platform shifts incrementally downward to allow for development of the next layer. The common laser beam scanning orientations include scanning along the *y*-axis direction, scanning along the *x*-axis direction, or using island/sector scanning strategies [11,12]. Island/sector scanning comprises of dividing the powder layer into islands with specific areas, for example, 2×2 mm^2, from which the laser beam scans using an input scanning orientation. A schematic of a few of the common scanning orientations on each powder layer are as shown in Figure 14.3, which include unidirectional, bidirectional (zig-zag), and reverse bidirectional orientations [1,13,14]. For unidirectional scanning, the laser scanner scans along the powder layer in one direction, for example, along the *x*-direction as shown in Figure 14.3. Bidirectional scanning involves repeated back and forth scanning also known as zig-zag scanning across the powder layer, and reverse bidirectional scanning involves repeated back and forth scanning along both the *x* and *y* directions. Figure 14.3 depicts these scanning orientations along subsequent powder layers.

There are a number of processing parameters that can be varied to influence the physical, mechanical, and thermal properties, cost, speed, and quality of a candidate AM component. The combination of processing parameters coupled with post-processing directly influences the microstructure (e.g., grain size, porosity, cracking density), and mechanical properties, such as tensile, fatigue, and creep properties. The next section will review SLM-processing parameters (i.e., laser power, scan speed, etc.) and their impact on energy density, porosity, cracking density, balling, surface roughness, and residual stress of as-built

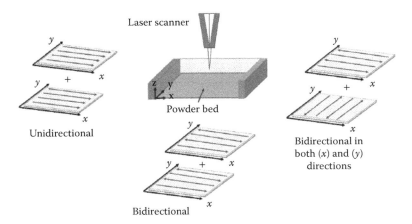

Figure 14.3 Schematic of most commonly used scanning strategies on powder layer (Unidirectional, bidirectional, and bidirectional in both x and y directions).

components. It will also reflect upon processing parameter optimization used to improve these factors, which are inherent to the SLM process.

14.2.1 *Processing parameters and optimization*

Research in SLM and other AM technologies of Ni-based alloys aims to reduce production time and cost, in addition to advancing geometric design, while at least maintaining, if not exceeding, the current mechanical properties of as-cast parts [15]. For a given service condition, it is critical to optimize SLM-processing parameters to achieve desired performance such as strength, ductility, fatigue, creep, and so on. Studies often use the analysis of variance (ANOVA) statistical approach to determine optimal processing parameter sets for AM [16]. Primarily, processing parameters include the powder layer thickness (t), scanning velocity (V), scanning method, laser power (P), laser spot size or focus diameter (δ), and hatch spacing (d). A schematic representation of these parameters is presented in Figure 14.2b, which depicts the typical powder bed AM process as a laser beam scanner with a fixed beam spot size, power, and velocity, is used to develop a component through layer-wise deposition. Depending upon the material used for developing an AM part, these processing parameters are optimized. A comparison of varying process parameters used across literature on SLM Inconel is presented in Table 14.1. It is evident from Table 14.1 that a wide variation exists across processing parameters used for SLM Inconel manufacturing, which most likely is a result of varying systems used to manufacture these parts. Most studies have used Argon as the gas chamber environment for Inconel alloys. Layer thickness across studies has ranged from 20 μm to 60 μm, with laser power ranging mainly from 100 W to 200 W.

The energy density (E) incorporates many of these parameters (i.e., laser power, scanning velocity, scan spacing, focus diameter, hatch spacing, layer thickness), where E_v is the volumetric energy density (J/mm³) function [4,14,17] and E_p is the planar/two-dimensional (J/mm²) energy density function [18].

$$E_v = \frac{P}{Vdt} \tag{14.1}$$

Table 14.1 Comparison of processing parameters used in various studies on Inconel across literature

Material (average particle size)	Laser power, P	Layer thickness, t	Diameter of laser beam/ spot size, δ	Hatch spacing, d	Scanning speed, V	Geometry and orientations	Chamber environment	Equipment	Source
IN718 (5–50 μm)	100 W	50 μm	180 μm	160 μm	85.7 mm/s	Cylindrical/0°, 90°, 45° and 45° × 45° (tensile direction with respect to build direction)	Argon	SLM Realizer II 250 (MPC-HEK)	Chlebus et al. [15]
IN718 (17 μm)	200 W	~50 μm	100 μm	—	800 or 1200 mm/s (in Argon or Nitrogen)	Cylindrical/ vertical and horizontal to build orientation	Argon or Nitrogen	EOS M270 SLM	Amato et al. [42]
IN718 (<50 μm)	170 W	20 μm	100 μm	—	25 m/min	Tensile Samples	Argon	LSNF-I (Self-developed SLM)	Wang et al. [45]
IN718	275 W	50 μm	81 μm	120 μm	805 mm/s	Mechanical test Specimens	Argon	SLM 280HL	Popovich et al. [51]
IN625 (20 μm)	200 W	—	100 μm	—	800 and 1200 mm/s	Cylindrical/ perpendicular to build direction (xy-axis) and parallel to build direction (z axis)	Nitrogen	EOS M270 SLM	Amato et al. [49]
IN939 (30 μm)	400 W	30 μm	—	0.15 to 0.12 mm dependent on which part of sample was manufactured	540 to 620 mm/s	Dogbone/0° and 90° with respect to build platform	Argon	SLM 250HL (SLM-Solutions GmbH)	Kanagarajah et al. [46]
IN738LC	50–100 W	20–60 μm	—	—	50–500 mm/s	Cylindrical/xy specimen and z specimen	Argon	Concept Laser *LaserCusing M1* machine	Rickenbacher et al. [55]

Most commonly, the two-dimensional energy density equation is employed because powder layer thickness is maintained at a constant value.

$$E_p = \frac{P}{V\delta} \tag{14.2}$$

An optimum energy density has been shown to be favorable in allowing complete consolidation of the material, thus reducing the porosity and presence of cracks, yielding a generally accepted optimal density of greater than 99.5% purity for SLM-manufactured parts [4,15]; however, application of too much energy density in DMLS and EBM Inconel 718 build components was suggested to lead to delamination [19]. Image analysis of Ni-based superalloys (i.e., CM247LC, CMSX486, IN625 and IN718) manufactured using varying processing conditions revealed that an energy density of approximately 1.7 J/mm² served as the boundary between low and high void area percentage obtained for these alloys [20]. No direct correlation between energy density and cracking density was acquired [20]. Other research has determined through an ANOVA response surface model that a reduction in cracking density is achievable by increasing the scan speed and decreasing the laser power for SLM-manufactured CMSX486 [16]. This was suggested to reduce the specific energy input from the laser beam, as a result lowering residual stresses, which are linked to the initiation of cracks [16]. Yet, an increase in void percentage was observed by increasing the scan speed and decreasing the laser power through use of an ANOVA response surface model [16]. For DMLS Inconel 625, it has been shown that an increase in scan speed results in a decrease in density and dimensional accuracy of the part, when other parameters are held constant (hatch thickness, beam diameter, and laser power) [21].

The island scanning size has been shown to affect the relative density of SLM IN718 built parts [22]. A direct relation between island scanning size and energy density, in which enlarging the island scanning sizes from 2 × 2 mm² to 5 × 5 mm² or 7 × 7 mm² resulted in an increase in the relative density from 98.67% to 99.10% has been achieved [22]. It was found that the most pores and cracks were evident using a small island scan size of 2 × 2 mm² [22].

The layer thickness has also been shown to impact the microstructure and mechanical properties of SLM parts. Overlap of melt arc pools has been found in vertical cross section (*YZ* plane) of SLM IN625 for a layer thickness of 20 μm, whereas less overlap of melt arc pools was found for a layer thickness of 40 μm, at constant laser power and scan speed of 195 W and 800 mm/s, respectively [23]. This further suggests that a layer thickness of 20 μm will provide improved mechanical properties and part density [23].

An optimum energy density is necessary to avoid an unfavorable phenomena known as balling, in which the molten powder does not fully *wet* the substrate as a result of surface tension, thus causing the liquid melt pool to break down into spheres [24,25,26]. The balling phenomena results in loss of densification of SLM-manufactured part as well as changes in surface roughness [17,24,25]. The phenomena of balling can be classified into two types: spherical and ellipsoidal, from which it was found that ellipsoidal balling impacts the quality of SLM-manufactured parts considerably [27]. It has been demonstrated that peak intensities as high as 2 × 10³ W/mm², which is achieved by operating the laser in pulse mode, completely removes the balling phenomena [25]. This is due to the *evaporation effect* [25], in which evaporated particles develop a recoil pressure on the molten pool that assists in flattening the melt layer [28]. Research indicates that a reduction in balling can be achieved with the use of high peak powers, since increase in recoil pressures flattens and increases the wettability of the melt pool [24,25].

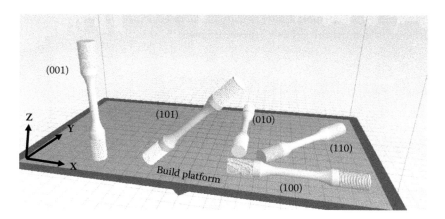

Figure 14.4 Build orientations in Miller Indices for tensile samples in MakerBot© used to observe anisotropic behavior of selective laser melted samples.

Other parameters that also influence the performance of SLM material include the island scanning strategy, powder particle size, build orientation, post-build HT processing, preheating of the substrate plate, chamber environment, and so on. Varying build orientations of tension samples including (001), (100), (110), (010) and (101) are presented in Figure 14.4, with the z-axis representing the build direction and *xy* plane representing the build platform. The build orientations in Figure 14.4 are presented without depicting an attachment of the build plate to the component, which is often removed through the electrical discharge-machining process. The build orientation alters the grain orientation and consequently the mechanical properties of AM parts, which is further discussed in Sections 14.4 and 14.5 of this review.

14.2.2   Surface roughness

Surface roughness in SLM is attributed to the *stair-stepping* effect or ridges that develop across the surface of SLM-manufactured parts as a result of the layer-by-layer processing, which is intrinsic to AM [29,30]. Average surface roughness, R_a, is defined by Equation 14.3 in which $f(x)$ is a function relating the distance between the measurement location at the surface and the reference centerline; here l is the distance over which the surface roughness is being measured [29,30]. Figure 14.5b presents a schematic depicting this surface roughness relationship.

$$R_a = \frac{1}{l} \int_0^l |f(x)| dx \qquad (14.3)$$

A schematic representing the *stair-stepping* effect is presented in Figure 14.5a. From Figure 14.5a, an enlarged view of the layer-wise distribution from a cross section of a tension sample reveals the *staircase* effect with change in geometry of the sample. Partially bonded particles on top surfaces have also been shown to contribute to the overall surface roughness, in addition to the *stair-stepping* profile [30]. Reduction in the surface roughness of SLM-manufactured parts by optimization of process parameters is necessary in order to improve the life and performance of these parts. Fatigue crack initiation phenomena due to high surface roughness in these as-manufactured SLM parts can be relieved by mechanical post-processing techniques (e.g., polishing). Although AM has allowed for

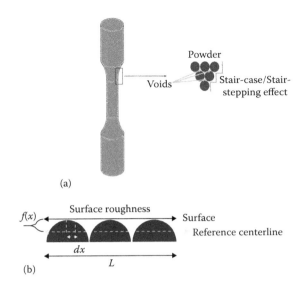

Figure 14.5 (a) Schematic of stair-case/stair-stepping effect and (b) schematic to determine surface roughness.

greater flexibility in geometric design of cooling channels in turbine blades, the surface roughness present within these internal channels may limit the aerodynamic performance of these blades. Studies on DMLS Inconel 718 have shown variation in pressure loss and friction factor for different channel sections and build orientations [31,32].

Before optimization of SLM process parameters, a response surface needs to be developed from which optimization can be achieved to minimize R_a without the need for mechanical post-processing techniques. Representative surface roughness values found for SLM Inconel 625 ranged between approximately 4 μm to 37 μm [24], whereas the mean R_a for creep-fatigue testing purposes should be less than 0.2 μm [33]. In a laser solid forming (LSF) investigation, overlap rate was found to affect the flatness of Inconel 718 produced parts [34]. For SLM Inconel 625, high peak power was shown to reduce both top and side surface roughnesses [24]. This study found that a reduction in top surface roughness can be achieved by an increase in the repetition rate and overlap, with a decrease in the scan speed [24]; however, a decrease in the repetition rate and overlap with an increase in the scan speed was found to reduce side surface roughness [24]. This variation in methods to reduce both top and side surface roughness is suggested to be attributed to surface tension forces [24]. A top and side surface roughness for SLM IN625 of 9 μm and 10 μm, respectively, has been achieved at a scanning speed of 400 mm/min, 0.7 J pulse energy, and 40 Hz repetition rate [24]. For laser-assisted machining of IN718, it was found that surface roughness reduced with an increase in the feed up to 0.25 mm/rev [35].

14.2.3   Residual stress

The rapid and repeated melting and solidification of the layer-wise deposition involved in the fabrication of SLM parts causes a large thermal gradient to develop across the material, which gives rise to high residual stresses that can lead to geometric distortions of as-built/manufactured parts [11,36]. This thermal gradient mechanism is represented in Figure 14.6, which shows the development of a component layer-by-layer, with heating and melting

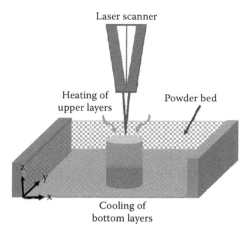

Figure 14.6 Schematic of thermal gradient mechanism (TGM) that gives rise to high residual stress variation in additive-manufactured parts.

of top layers as well as cooling and solidification of bottom layers, thus causing a thermal gradient to exist across the component. Studies have measured residual stress profiles in additively manufactured Inconel 718 through neutron diffraction [19,36], synchrotron X-ray diffraction, [37] and Vickers' hardness measurements [22,34,38].

A variety of relationships can be used to obtain residual stress values from Vickers' hardness measurements [38,39,40]. It is assumed that the uniaxial σ-ε curve of the material obeys the power law function and that the residual stress is in a state of equal-biaxial condition [38].

$$\sigma = K(\varepsilon_{pl})^n \tag{14.4}$$

For the power law function in Equation 14.4, ε_{pl} is the plastic strain, K is the Ramberg–Osgood strain-hardening coefficient, and n is the Ramberg–Osgood strain-hardening exponent [41]. Here, the term K is typically slightly above the tensile strength of the material. The relationship between microhardness and flow stress is as follows:

$$H = C\sigma(\varepsilon_{repr} + \varepsilon_{res}) \tag{14.5}$$

The microhardness, H, is related to the flow stress, $(\varepsilon_{repr} + \varepsilon_{res})$, in which ε_{repr} and ε_{res} are the representative effective plastic strain and effective von-Mises residual plastic strain, respectively, and C is a constant. These constants are presented in Equation 14.5. The relation between the contact areas from the indenter and residual stress can be determined next.

$$c^2 = c_0^2 - 0.32\ln[1 + \sigma_{res}/\sigma(\varepsilon_{res})] \tag{14.6}$$

Equation 14.6 can be used to determine the residual stress σ_{res} from the ratio of the real contact area, A, to the nominal contact area, A_{nominal}, defined as c^2.

$$A_{\text{nominal}} = \left[\frac{(L_1 + L_2)}{2}\right]^2 \tag{14.7}$$

The nominal contact area is determined from the length of the diagonal lines L_1 and L_2 of the indenter, as depicted in Equation 14.7. From these sets of equations, C, ε_{repr}, and c_0^2 have been found to be 3, 0.08, and 1, respectively [38,40]. These equations have been used to obtain a direct solution for σ_{res}, by first determining ε_{res} from Vickers' Hardness value H [34,38].

Optimization of processing parameters can assist in reducing residual stress. A study by Lu and coauthors [22] presented the effect of island scan size on residual stress behavior in SLM IN718. It was shown that 2×2 mm^2 island scan size yields the lowest residual stress, but it was further suggested that the 5×5 mm^2 island scan size would be better, because the 2×2 mm^2 may have had stress release because of crack prevalence, causing it to indicate a lower residual stress [22]. Cao and coauthors [34] have observed the effect of varying overlap rate, ranging from 20% to 50% on residual stress profile for LSF Inconel 718, showing wider variation in residual stress with an increase in the overlap rate [34]. Higher residual stress was found to occur in overlap areas rather than inner pass areas [34].

Post-processing in the form of HT is often used to reduce or even completely eliminate the residual stress present within SLM parts. Vacuum-annealed HT of SLM iron has shown both a decrease in microhardness and relaxation of tensile residual stresses [38].

14.3   *Microstructural influence*

Inconel manufactured through traditional casting methods is used in critical components due to its high strength, excellent oxidation, corrosion resistance, and its resistance to creep and fatigue at elevated temperatures. Inconel is a trademark name for Ni–Fe–Cr austenitic (γ)-based superalloy that exhibits three main precipitation phases: γ' Ni$_3$(Al, Ti, Nb) with a cubic L1$_2$ crystal structure, γ'' (Ni$_3$Nb) with a body-centered tetragonal (BCT) DO$_{22}$ crystal structure, and δ (Ni$_3$Nb) with an orthorhombic DO$_a$ crystal structure, in addition to the presence of Lave phases [42]. The chemical composition of select Inconel alloys can be found in Table 14.2. High concentrations of Ni are present in all alloys shown. A comparison of these alloys reveals low quantities of Co in IN718 and IN625 ($\leq$1.0 wt.%), with the highest quantity of Co in IN939 (19 wt.%). Fe is shown to be present only in IN718 and IN625. Cr is present in all alloys, with the lowest concentration in IN738LC (15.88 wt.%) and the highest concentrations in IN625 (20–23 wt.%) and IN939 (22.4 wt.%), followed by IN718 (17–21 wt.%). The highest concentration of Mo is observed in IN625 (8–10 wt.%), with lower quantities present in IN718 (2.8–3.3 wt.%) and IN738LC (1.75 wt.%). Highest concentration of Nb and Ta can be found in IN718 (4.75–5.5 wt.%). Ti quantity is highest in IN939 (3.7 wt.%), and Al quantity is highest in IN738LC (3.51 wt.%). The presence of W can be found in IN939 (2 wt.%) and IN738LC (2.62 wt.%) only. Traces of other elements in these alloys can also be observed in Table 14.2. The chemical composition of these alloys may vary by manufacturer.

The microstructure of SLM Inconel parts contributes significantly to their observed mechanical properties. Multiple studies [15,42,43] have shown a characteristic dendritic microstructure in additively manufactured IN718 components: Arc lines representing melt pools were observed parallel to the build direction, and a series of elongated vectors/tracks representing the repeated laser scanner movement is observed perpendicular to the build direction as depicted in Figure 14.7. These columnar dendrites have been observed to grow epitaxially along the (100) crystallographic plane [42,44], as well as in the (200) direction [42,43]. The characteristic dendritic microstructure has been shown to disappear with post-build HT [45]. Recrystallization is often exhibited in SLM parts that have been heat treated [46,47,48]. Recrystallization for Inconel 718 occurs during the annealing process at temperatures above 1100°C [15,42,47] and results in the strengthening phases γ' distributed within a

Table 14.2 Chemical composition of selected Inconel alloys (wt.%). Note: Chemical composition may vary by manufacturer/study

Material	Cr	Co	Nb	Mo	Ti	Al	C	Mn	Si	Fe	W	Ta	Zr	B	Ni	P	S	Cu
IN718[a]	17–21	≤1.0	4.75–5.5 (with Ta)	2.8–3.3	0.65–1.15	0.20–0.80	≤0.08	≤0.35	≤0.35	b*	—	Included with Nb	—	≤0.006	50–55 (with Co)	≤0.015	≤0.015	≤0.30
IN625[b]	20–23	≤1.0	3.15–4.15 (with Ta)	8–10	≤0.40	≤0.40	≤0.10	≤0.50	≤0.50	≤5.0	—	Included with Nb	—	—	58	≤0.015	≤0.015	—
IN939[c]	22.4	19	1	—	3.7	1.9	0.15	—	—	—	2	1.4	0.1	0.01	b*	—	—	—
IN738LC[d]	15.88	8.3	0.9	1.75	3.31	3.51	0.1	—	—	—	2.62	1.9	—	0.011	b*	—	—	—

Source:

[a] Special Metals Corporation, Inconel alloy 718. Publication No. SMC-045, Huntington, WV, 1-28, 2007.

[b] Special Metals Corporation, Inconel alloy 625, Huntington, WV, 1-18, 2013.

[c] Kanagarajah, P. et al., *Mat. Sci. Eng A.*, 588, 188–195, 2013; Miskovic, Z. et al., *Vacuum.*, 43, 709–711, 1992.

[d] Kunze, K. et al., *Mat. Sci. Eng A.*, 620, 213–222, 2015; Rickenbacher, L. et al., *Rapid Prototyping J.*, 19(4), 282–290, 2013; Daubenspeck, B.R., Gordon, A.P., *Journal of Engineering Materials and Technology*, 133(2), 021023, 2010.

b* [Balance element].

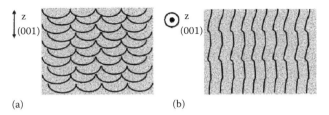

Figure 14.7 Schematic of a generic alloy made by SLM (not to scale): (a) Melt arc pools can be observed parallel to the build direction and (b) laser scan tracks can be observed perpendicular to the build direction. Note: Dendritic microstructure is not depicted.

field of γ', with δ precipitate occurring at recrystallized grain boundaries [42]. The strengthening precipitation phases γ and γ' are observed to be spherical/cuboidal and lenticular in shape, respectively [42]. Inconel 625 SLM has been found to exhibit columnar arrays of γ'' [49]. Brittle phases (e.g., Lave phases) as well as high concentrations of Niobium (Nb) and Molybdenum (Mb) elements occur in overlap areas between adjacent laser tracks and inter-dendritic locations [15]. Locations with rich concentrations of the elements Nb and Mb lead to crack initiation and propagation [44]. Furthermore, the characteristic melt pool boundaries (MPB) evident in SLM microstructures, (e.g., *track-track* and *layer-layer*) which are a result of molten pool overlap, can serve as initiation sites for crack development [50].

The epitaxial grain structure contributes to the strong bonding behavior observed between SLM layers [46]. The average grain size found for SLM IN939 sample oriented parallel to the build platform was 35 microns versus 30 microns for the sample oriented perpendicular to the build platform [46]. The average grain size in LSF Inconel 718 subjected to a temperature of 1100°C for 60 minutes was found to decrease with an increase in the overlap rate [34].

Porosity is a common problem that occurs in the manufacturing of SLM parts. There are two types of commonly observed pores: spherical and irregular [13,14]. The voids present between these pores serve as initiation sites for failure mechanisms (e.g., cracking) to occur. A schematic of these voids is presented in Figure 14.5a. Use of plasma rotation electrode preparation (PREP) powders showed reduction in pores and microcracks in laser rapid forming (LRF) of IN718 [44]. To achieve a consistent powder distribution for SLM manufactured samples, it is important to ensure that the Hausner ratio H_r is less than 1.25 [18].

$$H_r = \frac{\rho_T}{\rho_a} \tag{14.8}$$

The Hausner ratio is a measure of powder flowability and is the ratio of tapped density "ρ_T" to apparent powder density "ρ_a," as shown in Equation 14.8 [18].

14.4 Mechanical properties influence

The anisotropic microstructural behavior of SLM parts contributes significantly to the mechanical strength observed for these components. It has been found that SLM Ni-based superalloys have a higher yield and ultimate tensile strength than Ni-based superalloys made through conventional methods. These mechanical properties in addition to ductility, Young's Modulus, fatigue, and creep behavior of SLM Ni-based superalloys will be presented in the sections to follow.

14.4.1   Tensile

A variety of tensile tests have been carried out for SLM Inconel. Tables 14.3 through 14.5 present a summary of SLM Inconel-based studies for samples oriented with respect to build direction. Highlighted within these tables are the 0.2% yield strength, tensile strength, elongation

Table 14.3 Tensile properties for varying orientated SLM Inconel 718 samples, tested at ambient temperature

Orientation (Millers Indices)	Young's Modulus, E (GPa)	0.2% Yield strength, σ_{ys} (MPa)	Tensile strength, Sut (MPa)	Elongation, EL (%) (strain at failure)	Source
L – (001)	$162 \pm 18^{\S}$	$572 \pm 44^{\S}$	$904 \pm 22^{\S}$	$19 \pm 4^{\S}$	Chlebus
	163 ± 30^{a}	1074 ± 42^{a}	1320 ± 6^{a}	19 ± 2^{a}	et al. [15]
L – (001)	—	850^{b}	1140^{b}	28^{b}	Amato
	—	880^{b*}	1140^{b*}	30^{b*}	et al. [42]
L – vertical	—	$737 \pm 4^{\S}$	$1010 \pm 10^{\S}$	$20.6 \pm 2.1^{\S}$	Strobner
	—	1136 ± 16^{c}	1357 ± 5^{c}	13.6 ± 0.2^{c}	et al. [43]
	—	1186 ± 23^{d}	1387 ± 12^{d}	17.4 ± 0.4^{d}	
T – (010)	$193 \pm 24^{\S}$	$643 \pm 63^{\S}$	$991 \pm 62^{\S}$	$13 \pm 6^{\S}$	Chlebus
	199 ± 15^{a}	1159 ± 32^{a}	1377 ± 66^{a}	8 ± 6^{a}	et al. [15]
T – (100)	—	890^{b}	1200^{b}	28^{b}	Amato
	—	930^{b*}	1120^{b*}	27^{b*}	et al. [42]
T – horizontal (length parallel with substrate surface)	$204^{\S}$	$889–907^{\S}$	$1137–1148^{\S}$	$19.2–25.9^{\S}$	Wang
	201^{f}	$1102–1161^{f}$	$1280–1358^{f}$	$10–22^{f}$	et al. [45]
T – horizontal	—	$816 \pm 24^{\S}$	$1085 \pm 11^{\S}$	$19.1 \pm 0.7^{\S}$	Strobner
	—	1227 ± 1^{c}	1447 ± 10^{c}	10.1 ± 0.6^{c}	et al. [43]
	—	1222 ± 26^{d}	1417 ± 4^{d}	15.9 ± 1.0^{d}	
D – (011)–45°	$200 \pm 23^{\S}$	$590 \pm 15^{\S}$	$954 \pm 10^{\S}$	$20 \pm 1^{\S}$	Chlebus
	188 ± 19^{a}	1152 ± 24^{a}	1371 ± 5^{a}	15 ± 5^{a}	et al. [15]
(111)– 45°×45°	$208 \pm 48^{\S}$	$723 \pm 55^{\S}$	$1117 \pm 45^{\S}$	$16 \pm 3^{\S}$	Chlebus
	209 ± 44^{a}	1241 ± 68^{a}	1457 ± 55^{a}	14 ± 5^{a}	et al. [15]
N/A	—	$569–646^{\S}$ (Yield strength)	$851–1002^{\S}$	$9.8–31.7^{\S}$	Popovich et al. [51]
	—	1160^{e} (Yield strength)	1350^{e}	17.6^{e}	

[a] Heat-treated (solution treatment at 1100°C for 1 hour [water cooling] + age hardening at 720°C for 8 hours [furnace cooling at rate of 100°C/hour] to 620°C for 10 hours [air cooling]) (Chlebus, E. et al., *Mat. Sci. Eng. A*, 639, 647–655, 2015).

[b] HIP + annealed for argon chamber environment (Amato, K. N. et al., *Acta Mater.*, 60, 2229–2239, 2012).

[b*] HIP + annealed for nitrogen chamber environment (Amato, K. N. et al., *Acta Mater.*, 60, 2229–2239, 2012).

[c] Heat treatment (AMS 5662 standard-solution annealing at 980°C for 1 hour [air cooling] + age hardening at 760°C for 10 hours [furnace cooling for 2 hours] to 650°C for 8 hours) (Strøßner, J. et al., *Adv. Eng. Mater.*, 1–7, 2015).

[d] Heat treatment (AMS 5664 standard [homogenization at 1065°C for 1 hour] + AMS 5662 standard) (Strøßner, J. et al., *Adv. Eng. Mater.*, 1–7, 2015).

[e] Heat treatment (homogenization at 1065°C for 1 hour)+ aging (hold for 10 hours at 760°C, cool to 650°C–2 hours, hold at 650°C for 8 hours) (Popovich, A. A. et al., *Key Eng. Mat.*, 651–653, 665–670, 2015).

[f] Heat treatment (solution treatment at 980°C for 1 hour with air cooling) + (double aging at 720°C for 8 hours with furnace cooling and 620°C for 8 hours with air cooling) (Wang, Z. et al., *J. Alloy. Compd.*, 513, 518–523, 2012.)

[§] As-built;—Not reported; L—Longitudinal, T—Transverse, D—Diagonal.

Table 14.4 Tensile properties for varying orientated SLM IN738LC samples, tested at ambient temperature

Orientation (Millers Indices)	Young's Modulus, E (GPa)	0.2% Yield strength, σ_{ys} (MPa)	Tensile strength, Sut (MPa)	Elongation, EL (%) (strain at failure)	Source
(001)	158 ± 3[§]	786 ± 4[§]	1162 ± 35[§]	11.2 ± 1.9[§]	Rickenbacher et al. [55]
	158 ± 3[a]	—	—	—	Kunze et al. [52]
(100)	233 ± 9[§]	933 ± 8[§]	1184 ± 112[§]	8.4 ± 4.6[§]	Rickenbacher et al. [55]
	237 ± 7[a]	—	—	—	Kunze et al. [52]

[a] Heat-treated (HIP at 1180°C for 4 hours, standard solution (SHT) at 1120°C for 2 hours and precipitation hardening (PHT) at 850°C for 20 hours) (Kunze, K. et al., *Mat. Sci. Eng. A*, 620, 213–222, 2015).
[§] As-built;—Not reported.

Table 14.5 Tensile properties at ambient temperature for heat-treated SLM IN625 of varying orientated samples

Orientation (Millers Indices)	0.2% Yield strength, σ_{ys} (MPa)	Tensile strength, Sut (MPa)	Elongation EL (%) (strain at failure)	Vicker's hardness (HV) (GPa)	Source
(001) SLM (Z)+HIP[a]	360	880	58	2.9	Amato et al. [49]
SLM (xy)+HIP[a]	380	900	58	3.4	Amato et al. [49]

[a] HIP at 1120°C for 4 hours in argon at 0.1 GPa pressure (Amato, K. N. et al., *J. Mater. Sci. Res.*, 1(2), 3–41, 2012).

(strain at failure), and Young's Modulus for studies with and without HT or HIP (hot isostatic pressing). A comparison of these mechanical properties across most studies on SLM IN718, as displayed in Table 14.3, reveal an increase in the 0.2% yield strength and tensile strength with HT as opposed to a decrease in the elongation percentage or ductility with HT. Also evident is minimal difference in mechanical properties with change in gas chamber environment for Argon or Nitrogen from Amato et al. 2012 study [42]. Since varying processing parameters, equipment used for manufacturing, and HTs are used for SLM IN718, tensile properties vary across studies. However, a general range of values can be determined from Table 14.3 based upon build orientation. The impact of HT of SLM IN718 has been found to confer room temperature mechanical properties similar to hot-rolled IN718 along with high ductility [51]. From Table 14.4, which presents the tensile properties of SLM IN738LC, results from Rickenbacher et al. 2013 [55] reveal an increase in the Young's Modulus, 0.2% yield strength and tensile strength for samples manufactured along the x-axis direction (100); however, a large ductility or elongation is observed for samples manufactured along the z-axis direction (001). For HIP SLM IN625 results displayed in Table 14.5, higher 0.2% yield strength, tensile strength, and Vickers hardness are observed for samples manufactured perpendicular to the build direction (xy) as opposed to parallel to the build direction (z). In contrast, the elongation is the same for both orientations (xy) and (z). Table 14.6 presents conventional cast properties of Inconel alloys to serve as a comparison with results found for SLM/SLM-heat treated Inconel.

Table 14.6 Tensile properties of conventionally cast Inconel

Material	Young's Modulus, E (GPa)	0.2% Yield strength, σ_{ys} (MPa)	Tensile strength, Sut (MPa)	Elongation, EL (%) (strain at failure)	Source
IN718(Cast AMS 5383)	—	758	862	5	Qi et al. [62]
IN939	—	500–800 (Yield Strengths)	750–950	—	Kanagarajah et al. [46]
IN738 LC	200	765	945	7.5	Rickenbacher et al. [55]; (ALSTOM (Switzerland) Ltd, 2013)

Anisotropic tensile ductility behavior has been observed in horizontal versus vertically built specimens, because of slipping of *track–track* MPB in horizontal specimens versus slipping of both *track–track* and *layer–layer* MPB in vertical specimens [50].

A study on SLM IN738LC has modeled variation in Young's Modulus using the single crystal elastic tensor and has found that minimum Young's Modulus occurs parallel to the build direction as well as along the laser scanning direction, which was set at 45° or the $x+y$ diagonal for this study [52]. Furthermore, the maximum Young's Modulus was predicted to occur at approximately 55° from the build direction, in the x–z plane [52]. A methodology for interpolation of the Young's Modulus across various orientations, assuming a transversely isotropic material, is provided in Section 14.5 based on experimental findings on SLM IN718 from [15].

14.4.2 Fatigue

An understanding of the fatigue life of AM components as compared to conventionally manufactured parts is important, most especially because of the inherent surface roughness that is present within the as-built components. The presence of surface roughness allows for crack initiation to begin with cyclic loading [53]. Cold rolled and DMLS Ni-718 bending fatigue life was found to be within the range of 2×10^5 and 2×10^6 cycles to failure [54]. A study on aged SLM IN939 has found that fatigue life was reduced as when compared with fatigue life at room temperature [46]. This reduction was attributed to precipitate formation/brittleness, resulting in higher sensitivity to crack initiation [46]. A summary of fatigue results presented by P. Kanagarajah and coauthors is presented in Table 14.7. Here, it can be observed that both at room temperature and 750°C, the fatigue life is higher for as-built SLM IN939, as compared with aged SLM IN939 [46]. In contrast, as-cast IN939 has a lower fatigue life at room temperature and 750°C, then cast aged IN939 [46].

14.4.3 Creep

Creep rupture strength behavior of SLM Ni-base superalloys is important for their use in high-temperature applications. The high cooling rates associated with the SLM process results in a fine grain microstructure of SLM components. A study on SLM IN738LC has shown that the build orientation as well as the grain size and microstructural orientation contribute to the creep properties observed for these materials [55]. This study found that a specimen built perpendicular to the build direction (*xy specimen*) had inferior creep

Table 14.7 Fatigue life properties of Inconel

Material	Source	Total strain amplitude	Strain rate	R	Fatigue life, N_f
SLM-processed *As-built* IN939	Kanagarajah et al. [46]	0.50%	6×10^{-3}	−1	4702 (Room temp.)/209 (750°C)
SLM-processed *aged*[a] IN939	Kanagarajah et al. [46]	0.50%	6×10^{-3}	−1	1598 (Room temp.)/73 (750°C)
Cast, as-cast IN939	Kanagarajah et al. [46]	0.50%	6×10^{-3}	−1	313 (Room temp)/230 (750°C)
Cast, aged IN939	Kanagarajah et al. [46]	0.50%	6×10^{-3}	−1	2677 (Room temp)/272 (750°C)

Source: Kanagarajah, P. et al., *Mat. Sci. Eng. A*, 588, 188–195, 2013; Miskovic, Z. et al., *Vacuum*, 43, 709–711, 1992.

[a] SLM-processed, aged (solution annealing for 4 hours at 1160°C + single stage ageing for 16 hours at 850°C).

properties as compared with a specimen built parallel to the build direction (*z specimen*) [55]; however, when compared with cast IN738LC, the specimen built parallel to the build direction (*z specimen*) exhibited creep rupture strengths that were in the lower band of cast IN738LC [55]. This was confirmed by Kunze and colleagues who found similar anisotropic creep behavior in IN738LC samples, suggesting that this behavior may be attributed to the orientation of the crystals, build orientation variation of the Young's Modulus, application of stress loading parallel versus transverse to the columnar grains in the build direction for *z-specimens* and *xy specimens*, respectively [52]. Other factors include the fine grain microstructure produced by SLM processing as well as the arrangement of precipitates around grain boundaries, in addition to γ′ precipitate size and morphology [52,55].

14.5   Constitutive modeling

In order to validate and support the experimental mechanical performance results of a material, it is important to develop constitutive models. This section begins with an overview of constitutive models appropriated for additively manufactured Ni-based superalloys. The emphasis is placed on the continuum-scale where the micromechanical response of the material is of interest. The following sections will present approximations of Young's Modulus variation with build direction and the stress–strain response of SLM Inconel 718 *as-built* and *heat-treated* from tensile properties presented by Chlebus et al. [15].

14.5.1   Constitutive models

Finite element (FE) simulation of the formation of a solid individual layer formation on a powder bed through the SLM process has observed a large tensile stress σ_y occurrence between solidified tracks, suggesting that initiation of cracks may occur in this region [56]. Correction factors have been used to fine-tune recent FEM simulation of IN718 laser-melting pool geometries, with the aim to reflect experimental findings [57]. The importance of FEM models in considering both rapid heating and cooling of the melt pool and its impact on phase transformations has been emphasized for SLM IN625 [23]. Studies have also simulated residual stress development in IN718 during the EBM build process through computational modeling [58], as well as simulating the effect of electron beam current on resulting spatial temperature variations across EBM built IN718 [59]. One study, which developed a thermal expansion model on residual stress development in powder-bed direct laser deposition (PD DLD) Ni-Superalloy C263 suggested that to understand residual stress development, one must consider both the thermal gradient within the built part as well the thermal strain incompatibility with the base plate on which the part was built [60].

14.5.2 Young's modulus orientation dependence

It is well understood that SLM Ni-based superalloys exhibit anisotropic behavior, which is most pronounced perpendicular and parallel to the build direction (z-axis), suggesting that this material exhibits transversely isotropic behavior. As such, when applied to SLM IN718, it has been assumed that the z-orientation corresponds to the build direction with the isotropic plane being the xy plane.

$$
\mathbf{H} = \begin{bmatrix}
\dfrac{1}{E_T} & \dfrac{-v_{TT}}{E_T} & \dfrac{-v_{LT}}{E_L} & 0 & 0 & 0 \\[2ex]
\dfrac{-v_{TT}}{E_T} & \dfrac{1}{E_T} & \dfrac{-v_{LT}}{E_L} & 0 & 0 & 0 \\[2ex]
\dfrac{-v_{TL}}{E_T} & \dfrac{-v_{TL}}{E_T} & \dfrac{1}{E_L} & 0 & 0 & 0 \\[2ex]
0 & 0 & 0 & \dfrac{1}{G_{LT}} & 0 & 0 \\[2ex]
0 & 0 & 0 & 0 & \dfrac{1}{G_{LT}} & 0 \\[2ex]
0 & 0 & 0 & 0 & 0 & 2\left(\dfrac{1+v_{TT}}{E_T}\right)
\end{bmatrix}
\tag{14.9}
$$

In order to model the elastic response of SLM IN718, the elastic compliance matrix [**H**] must be determined, for which transversely isotropic materials are defined by five independent elastic constants: E_T, E_L, v_{TT}, and v_{TL}, G_{TL} [69,70]. E represents the Young's modulus, v is Poisson's ratio, G is the shear modulus, L represents the grain orientation along the longitudinal (z-direction), and T represents the grain orientation along the transverse (x or y directions) [69].

$$
E(\theta) = \left[\frac{1}{E_L}\cos^4\theta + \left(\frac{4}{E_{45}} - \frac{1}{E_L} - \frac{1}{E_T}\right)\sin^2\theta\cos^2\theta + \frac{1}{E_T}\sin^4\theta\right]^{-1}
\tag{14.10}
$$

The Young's Modulus can be found within the LT-plane through use of Equation 14.10 [69], in which E_L is the Young's Modulus in the z-direction (parallel to build axis), E_T is the Young's Modulus in the x- or y-direction (perpendicular to the build axis), and E_{45} is the Young's Modulus 45° from the z-direction (build axis) either along the zx-plane or the zy-plane. It is important to note that θ is referenced from the z-direction (L) build axis. For this analysis, E_T is the Young's Modulus in the y-direction (perpendicular to the build axis), and E_{45} is the Young's Modulus, 45° from the z-direction (build axis) along the zy-plane as provided by Chlebus et al. [15]. Figure 14.8 presents Young's Modulus variation with angle from z-axis build axis (L) using Equation 14.10, in which anisotropic behavior can be observed. Figure 14.8 is plotted from 0 to $\pi/2$ radians from the

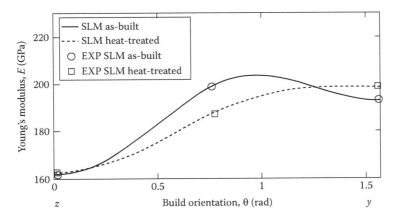

Figure 14.8 Young's modulus for varying build orientations measured from z-axis in *zy*-plane for SLM IN718 as-built and heat-treated, simulated based upon experimental data (Young's moduli) provided in Chlebus et al. 2015.

z-direction in the *zy*-plane and includes both the variation for *as-built* and *heat-treated* SLM IN718. Using the main experimental Young's Modulus values provided in [15], which are indicated by *EXP* for SLM *as-built* and *heat-treated* IN718 in Figure 14.8, it can be seen that $E(0) < E(\pi/2) < E(\pi/4)$ for *as-built* SLM IN718, as previously found by Chlebus et al. [15]. Also evident from Figure 14.8 is the reduction, but not removal, of texture observed for *heat-treated* SLM IN718 as opposed to *as-built* SLM IN718, as discussed in Chlebus et al. study [15]. Furthermore, it can be observed that the peak elastic modulus within the *zy*-plane varies for *as-built* SLM IN718 and *heat-treated* SLM IN718. For *heat-treated* SLM IN718, the peak elastic modulus of 199 GPa is shown to occur perpendicular to the build direction, for this case, along the *y*-direction (010). For *as-built* SLM IN718, the peak elastic modulus of 203.696 GPa is found to occur at ~0.97 radians or ~55.5769°, which is approximately 10° above the bias orientation of 45°. Furthermore, there are two observed intersection points for both the *as-built* SLM IN718 and *heat-treated* SLM IN718 Young's Modulus variation with orientation plots. This is found to occur at ~1.257 radians or ~72° from the build direction (z-axis) and ~0.16 radians or ~9.167° from the build direction (z-axis). This suggests that there are two build orientations for which the Young's Modulus in the *zy*-plane will not vary regardless of HT postprocessing technique, for the HT applied in this study [15] (solution treatment at 1100°C for 1 hour [water cooling] and age hardening at 720°C for 8 hours [furnace cooling at 100°C/hour] to 620°C for 10 hours [air cooling]). These results are simulated based on experimental findings from Chlebus et al. [15].

14.5.3   *Tensile stress–strain response*

Figure 14.9 presents the tensile stress–strain response of varying build orientations for SLM IN718 *as-built* and *heat-treated* from tensile data provided in Chlebus et al. (2015) publication. It is important to note that only the main values provided from the tensile data are included, although error range of these values is listed in Chlebus et al. (2015). These orientations include (001), (010), (011), and (111). The HT applied in Chlebus et al. (2015) study was as discussed in Section 14.5.2.

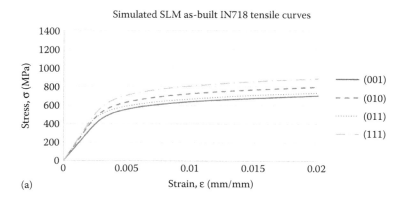

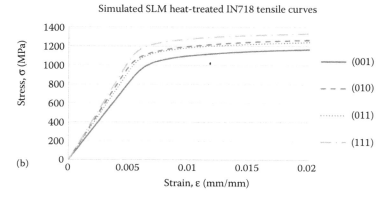

Figure 14.9 Simulated stress-strain behavior through use of the Ramberg–Osgood model, for varying build orientations of SLM IN718 (a) as-built, and (b) heat-treated based upon experimental data (0.2% yield strengths and ultimate tensile strengths) provided in Chlebus et al. 2015.

$$\varepsilon = \left(\frac{\sigma}{E}\right) + \left(\frac{\sigma}{K}\right)^{\frac{1}{n}} \tag{14.11}$$

These plots were developed using the Ramberg–Osgood Model in Equation 14.11, from which the monotonic strength coefficient K and strain-hardening exponent n were determined between the provided 0.2% yield strength and ultimate tensile strength for each build orientation in Chlebus et al. publication [15], through application of Equation 14.12. The tensile plots are simulated and shown up to a strain range of 0.02 mm/mm.

$$\sigma = K\left(\varepsilon_p\right)^n \tag{14.12}$$

A comparison of both tensile plots for *as-built* and *heat-treated* SLM IN718 in Figure 14.9 reveals that the lowest strength is observed for (001) direction, and the highest strength is observed for the (111) direction, as presented in Chlebus et al. (2015). With the applied HT, there is an increase in the strength observed for all build orientations.

14.6   Conclusions and ongoing challenges

This review has presented a compilation of literature on SLM Inconel Superalloys, with a focus on Inconel 718, as well as simulated models of tensile stress–strain curves and Young's Modulus variation for varying build directions, with and without HT, from experimental data presented in Chlebus et al. (2015) study on SLM IN718. It is well understood that these materials exhibit anisotropic mechanical behavior, which has been observed parallel and perpendicular to the build direction, as a result of anisotropic microstructural behavior. SLM manufacturing produces a fine grain microstructure within Inconel alloys that exhibit a dendritic microstructure. A cross section of the microstructure reveals melt arc pools parallel to the build direction and laser scan tracks perpendicular to the build direction. The presence of texture has been observed in IN718 along the (100) and (200) directions, as well as brittle phases and elements (i.e., Laves, Nb, Mb) in between laser tracks and interdendritic microstructure, which can lead to crack initiation.

Anisotropy has also been observed with fatigue and creep performance of these materials. Mechanical properties (Young's modulus, yield strength, and ultimate tensile strength) have been observed to be lower parallel to the build direction as when compared with perpendicular to the build direction. Our simulated model of Young's Modulus variation with orientation based upon experimental findings in Chlebus et al. (2015) study has assumed transversely isotropic material behavior, with xy being the plane of isotropy. Transversely isotropic behavior has been assumed previously for a SLM Hast-X weld-like structure [61]. Also presented in this review, was simulated tensile stress–strain curves based upon the Ramberg–Osgood model for varying build orientations with strength coefficient K and strain-hardening exponent n modeled using power law between 0.2% yield strength and ultimate tensile strength experimental results presented in Chlebus et al. (2015) study. Mitigation of the imminent effects of the SLM process including porosity, cracking, surface roughness, and residual stress can be achieved through adjustments in processing parameters during manufacturing as well as postprocessing processes (i.e., HT, polishing, build direction, prebuild base plate temperature). Furthermore, a consistency among the processing parameters that achieve mechanical performance equivalent to or surpassing that of conventional manufacturing techniques across manufacturers of SLM 3D printers, will allow for a more thorough understanding of how these materials behave with optimized or standard processing parameters.

Acknowledgments

This material is based upon work supported by the National Science Foundation Graduate Research Fellowship Program under Grant No. (1144246) and the Zonta International Amelia Earhart Fellowship awarded to Sanna F. Siddiqui. Abiodun A. Fasoro would like to acknowledge the financial support of the Airforce Research Collaboration Program (RCP) through AFRC/UTC/Clarkson Aerospace. Ali P. Gordon is thankful for the research sponsorship of United Technology Corporation (UTC), Dayton, OH.

References

1. Kruth, J.P., Badrossamay, M., Yasa, E., Deckers, J., Thijs, L., Humbeeck, J.V., 2010. Part and material properties in selective laser melting of metals. *Proceedings of the 16th International Symposium on Electromachining* (ISEM XV1), Shanghai Jiao Tong University Press, Shanghai, China.
2. Wohlers, T., Gornet, T., 2014. History of additive manufacturing. Wohlers Report 2014, 1–34, Wohlers Associates, Inc., Fort Collins, CO.

3. Frazier, W. E., 2014. Metal additive manufacturing: A review. *Journal of Materials Engineering and Performance*. 23(6), 1917–1928.
4. Gu, D. D., Meiners, W., Wissenbach, K., Poprawe, R., 2012. Laser additive manufacturing of metallic components: Materials, processes and mechanisms. *International Materials Reviews*. 57(3), 133–164.
5. ASTM Standard, 2012. *Designation: F2792–12a: Standard Terminology for Additive Manufacturing Technologies*, ASTM International, West Conshohocken, PA.
6. Sing, S. L., An, J., Yeong, W. Y., Wiria, F. E., 2015. Laser and electron-beam powder-bed additive manufacturing of metallic implants: A review on processes, materials and designs. *Journal of Orthopaedic Research*. doi:10.1002/jor.23075.
7. EOS e-Manufacturing Solutions. EOSINT M280. Website: http://www.eos.info/systems_solutions/metal/systems_equipment/eosint_m280 (accessed on December, 2015).
8. Utela, B., Storti, D., Anderson, R., Ganter, M., 2009. A review of process development steps for new material systems in three dimensional printing (3DP). *Journal of Manufacturing Processes*, 10, 96–104.
9. Kelly J. F., *Hood-Daniel, P., 2011. Printing in Plastic: Build Your Own 3D Printer*, Apress, New York.
10. Tucker, C. S., Saint John, D. B., Behoora, I., Marcireau, A., 2014. Open Source 3D Scanning and Printing for Design Capture and Realization. *ASME 2014 International Design Engineering Technical Conferences and Computers and Information in Engineering Conference*, 1B. 34th Computers and Information in Engineering Conference, Buffalo, NY, August, pp. 17–20, 2014.
11. Mercelis, P.; Kruth, J.P., 2006. Residual stresses in selective laser sintering and selective laser melting. *Rapid Prototyping Journal*, 12(5), 254–265.
12. Carter, L.N., Martin, C., Withers, P.J., Attallah, M.M., 2014. The influence of the laser scan strategy on grain structure and cracking behaviour in SLM powder-bed fabricated nickel superalloy. *Journal of Alloys and Compounds*, 615, 338–347.
13. Thijs, L., Verhaeghe, F., Craeghs, T., Humbeeck, J. V., Kruth, J. P., 2010. A study of the microstructural evolution during selective laser melting of Ti-6Al-4V. *Acta Materialia*, 58, 3303–3312.
14. Song, B., Zhao, X., Li, S., Han, C., Wei, Q., Wen, S., Liu, J., Shi, Y., 2015. Differences in microstructure and properties between selective laser melting and traditional manufacturing for fabrication of metal parts: A review. *Frontiers of Mechanical Engineering*, 10(2), 111–125.
15. Chlebus, E., Gruber, K., Kuznicka, B., Kurzac, J., Kurzynowski, T., 2015. Effect of heat treatment on the microstructure and mechanical properties of Inconel 718 processed by selective laser melting. *Materials Science & Engineering A*, 639, 647–655.
16. Carter, L. N., Essa, K., Attallah, M. M., 2015. Optimisation of selective laser melting for a high temperature Ni-superalloy. *Rapid Prototyping Journal*, 21(4), 423–432.
17. Jia, Q., Gu, D., 2014. Selective laser melting additive manufacturing of Inconel 718 superalloy parts: Densification, microstructure and properties. *Journal of Alloys and Compounds*, 585, 713–721.
18. Liu, B., Wildman, R., Tuck, C., Ashcroft, I., Hague, R., 2011. Investigation the effect of particle size distribution on processing parameters optimisation in selective laser melting process, *International Solid Freeform Fabrication Symposium: An Additive Manufacturing Conference*, University of Texas at Austin, Austin, TX, pp. 227–238.
19. Sochalski-Kolbus, L. M., Payzant, E. A., Cornwell, P. A., Watkins, T. R., Babu, S. S., Dehoff, R. R., Lorenz, M., Ovchinnikova, O., Duty, C., 2015. Comparison of residual stresses in inconel 718 simple parts made by electron beam melting and direct laser metal sintering. *Metallurgical and Materials Transactions A*, 46A, 1419–1432.
20. Carter, L. N., Wang, X., Read, N., Khan, R., Segerra, M. A., Essa, K., Attallah, M. M., 2016. Process optimisation of selective laser melting using energy density model for nickel-based superalloys. *Materials Science and Technology*, 32(7), 657–661. doi: 10.1179/1743284715Y.0000000108.
21. Sateesh, N. H., Mohan Kumar, G. C., Krishna, P., Srinivasa, C. K., Vinod, A. R., 2014. Microstructure and mechanical characterization of laser sintered Inconel-625 superalloy. *Procedia Materials Science*, 5, 772–779.
22. Lu, Y., Wu, S., Gan, Y., Huang, T., Yang, C., Junjie, L., Lin, J., 2015. Study on the microstructure, mechanical property and residual stress of SLM Inconel-718 alloy manufactured by differing island scanning strategy. *Optics & Laser Technology*, 75, 197–206.

23. Anam, M. A., Pal, D., Stucker, B., 2013. *Modeling and Experimental Validation of Nickel-based Super Alloy (Inconel 625) Made Using Selective Laser Melting*, J. B. Speed School of Engineering, University of Louisville, Louisville, KY.

24. Mumtaz, K., Hopkinson, N., 2009. Top surface and side roughness of Inconel 625 parts processed using selective laser melting. *Rapid Prototyping Journal*, 15(2), 96–103.

25. Kruth, J. P., Froyen, L., Vaerenbergh, J. V., Mercelis, P., Rombouts, M., Lauwers, B., 2004. Selective laser melting of iron-based powder. *Journal of Materials Processing Technology*, 149, 616–622.

26. Kruth, J.P., Mercelis, P., Vaerenbergh, J. V., Froyen, L., Rombouts, M., 2005. Binding mechanisms in selective laser sintering and selective laser melting. *Rapid Prototyping Journal*, 11/1, 26–36.

27. Li, R., Liu J., Yusheng, S., Wang, L., Jiang, W., 2012. Balling behavior of stainless steel and nickel powder during selective laser melting process. *International Journal of Advanced Manufacturing Technology*, 59, 1025–1035.

28. Morgan, R., Sutcliffe, C. J., O'Neill, W., 2004. Density analysis of direct metal laser re-melted 316L stainless steel cubic primitives. *Journal of Materials Science*, 39, 1195–1205.

29. Turner, B. N., Gold, S. A., 2015. A review of melt extrusion additive manufacturing processes: II. Materials, dimensional accuracy, and surface roughness. *Rapid Prototyping Journal*, 21(3), 250–261.

30. Strano, G., Hao, L., Everson, R. M., Evans, K. E., 2013. Surface roughness analysis, modelling and prediction in selective laser melting. *Journal of Materials Processing Technology*, 213, 589–597.

31. Snyder, J. C., Stimpson, C. K., Thole, K. A., Mongillo, D., 2015. *Build Direction Effects on Additively Manufactured Channels*. Proceedings of ASME Turbo Expo 2015: Turbine Technical Conference and Exposition, June 15–19, 2015, Montreal, Canada, pp. 1–10.

32. Stimpson, C. K., Snyder, J. C., Thole, K. A., Mongillo, D., 2015. *Roughness Effects on Flow and Heat Transfer for Additively Manufactured Channels*. Proceedings of ASME Turbo Expo 2015: Turbine Technical Conference and Exposition, June 15–19, 2015, Montreal, Canada, pp. 1–13.

33. American Society for Testing and Materials (ASTM), 2013. E2714-13 Standard test method for creep-fatigue testing, ASTM International, West Conshohocken, PA. https://doi.org/10.1520/E2714.

34. Cao, J., Liu, F., Lin, X., Huang, C., Chen, J., Huang, W., 2013. Effect of overlap rate on recrystallization behaviors of Laser Solid Formed Inconel 718 superalloy. *Optics & Laser Technology*, 45, 228–235.

35. Lonikar, K. V., 2015. Laser Assisted Machining (LAM) of Inconel 718 with Thermal Modeling and Analysis of Process Parameters. *International Journal of Engineering Research & Technology*, 4(07), 308–312.

36. Watkins, T., Bilheux, H., An, K., Payzant, A., Dehoff, R., Duty, C., Peter, W., Blue, C., Brice, C., 2013. Neutron Characterization for Additive Manufacturing. *Advanced Materials & Processes*, 171(3), 23–26.

37. Idell, Y., Campbell, C., Levine, L., Zhang, F., Olson, G., Snyder D., 2015. Characterization of nickel based superalloys processed through direct metal laser sintering technique of additive manufacturing. *Microscopy and microanalysis*, 21(3), 465–466.

38. Song, B., Dong, S., Liu, Q., Liao, H., Coddet, C., 2014. Vacuum heat treatment of iron parts produced by selective laser melting: Microstructure, residual stress and tensile behavior. *Materials and Design*, 54, 727–733.

39. Carlsson S., Larsson, P. L., 2001. On the determination of residual stress and strain fields by sharp indentation testing. Part I: Theoretical and numerical analysis. *Acta Materialia*, 49, 2179–2191.

40. Carlsson S., Larsson, P. L., 2001. On the determination of residual stress and strain fields by sharp indentation testing. Part II: Experimental investigation. *Acta Materialia*, 49, 2193–2203.

41. Stephens, R. I., Fatemi, A., Stephens, R. R., Fuchs, H. O., 2001. *Metal Fatigue in Engineering*. 2nd Ed. John Wiley & Sons, New York.

42. Amato, K. N., Gaytan, S. M., Murr, L. E., Martinez, E., Shindo, P. W., Hernandez, J., Collins, S., Medina, F., 2012. Microstructures and mechanical behavior of Inconel 718 fabricated by selective laser melting. *Acta Materialia*, 60, 2229–2239.

43. Strøßner, J., Terock, M., Glatzel, U., 2015. Mechanical and Microstructural Investigation of Nickel-Based Superalloy IN718 Manufactured by Selective Laser Melting (SLM). *Advanced Engineering Materials*, 17(8), 1099–1105.

44. Zhao, X., Chen, J., Lin, X., Huang, W., 2008. Study on microstructure and mechanical properties of laser rapid forming Inconel 718. *Materials Science and Engineering A*, 478, 119–124.

45. Wang, Z., Guan, K., Gao, M., Li, X., Chen, X., Zeng, X., 2012. The microstructure and mechanical properties of deposited-IN718 by selective laser melting. *Journal of Alloys and Compounds*, 513, 518–523.

46. Kanagarajah, P., Brenne, F., Niendorf, T., Maier, H.J., 2013. Inconel 939 processed by selective laser melting: Effect of microstructure and temperature on the mechanical properties under static and cyclic loading. *Materials Science & Engineering A*, 588, 188–195.

47. Liu, F., Lin, X., Yang, G., Song, M., Chen, J., Huang, W., 2011. Microstructure and residual stress of laser rapid formed Inconel 718 nickel-base superalloy. *Optics & Laser Technology*, 43, 208–213.

48. Liu, F., Lin, X., Huang, C., Song, M., Chen, J., Huang, W., 2011. The effect of laser scanning path on microstructures and mechanical properties of laser solid formed nickel-base superalloy Inconel 718. *Journal of Alloys and Compounds*, 509, 4505–4509.

49. Amato, K. N., Hernandez, J., Murr, L. E., Martinez, E., Gaytan, S. M., Shindo, P. W., Collins, S., 2012. Comparison of Microstructures and Properties for a Ni-Base Superalloy (Alloy 625) Fabricated by Electron and Laser Beam Melting. *Journal of Materials Science Research*, 1(2), 3–41.

50. Shifeng, W., Shuai, L., Qingsong, W., Yan, C., Sheng, Z., Yusheng, S., 2014. Effect of molten pool boundaries on the mechanical properties of selective laser melting parts. *Journal of Materials Processing Technology*, 214, 2660–2667.

51. Popovich, A. A., Sufiiarov, V. S., Polozov, I. A., Borisov, E. V., 2015. Microstructure and mechanical properties of Inconel 718 produced by SLM and subsequent heat treatment. *Key Engineering Materials*, 651–653, 665–670.

52. Kunze, K., Etter, T., Grasslin, J., Shklover, V., 2015. Texture, anisotropy in microstructure and mechanical properties of IN738LC alloy processed by selective laser melting (SLM). *Materials Science & Engineering A*, 620, 213–222.

53. Huang, Y., Leu, M. C., 2013. *Frontiers of Additive Manufacturing Research and Education: An NSF Additive Manufacturing Workshop Report- July 11 and 12*, University of Florida, Gainesville, FL.

54. Scott-Emuakpor, O., Schwartz, J., George, T., Holycross, C., Cross, C., Slater, J., 2015. Bending fatigue life characterisation of direct metal laser sintering nickel alloy 718. *Fatigue & Fracture of Engineering Materials & Structures*, 38(9), 1105–1117.

55. Rickenbacher, L., Etter, T., Hovel, S., Wegener, K., 2013. High temperature material properties of IN738LC processed by selective laser melting (SLM) technology. *Rapid Prototyping Journal*, 19(4), 282–290.

56. Osakada, K., Shiomi, M., 2006. Flexible manufacturing of metallic products by selective laser melting of powder. *International Journal of Machine Tools & Manufacture*, 46, 1188–1193.

57. Romano, J., Ladani, L., Sadowski, M., 2016. Laser additive melting and solidification of Inconel 718: Finite element simulation and experiment. *Journal of Materials*, 68(3), 967–977.

58. Prabhakar, P., Sames, W.J., Dehoff, R., Babu, S.S., 2015. Computational modeling of residual stress formation during the electron beam melting process for Inconel 718. *Additive Manufacturing*, 7, 83–91.

59. Sames, W.J., Unocic, K.A., Dehoff, R.R., Lolla T., Babu, S.S., 2014. Thermal effects on microstructural heterogeneity of Inconel 718 materials fabricated by electron beam melting. *Journal of Materials Research*, 29, (17), 1920–1930.

60. Song, X., Xie, M., Hofmann, F., Illston, T., Connolley, T., Reinhard, C., Atwood, R. C., Conner, L., Drakopoulos, M., Frampton, L., Korsunsky, A. M., 2015. Residual stresses and microstructure in Powder Bed Direct Laser Deposition (PB DLD) Samples. *The International Journal of Material Forming*, 8, 245–254.

61. Brodin, H., Andersson, O., Johansson, S., 2013. *Mechanical Behavior and Microstructure Correlation in a Selective Laser Melted Superalloy*. Proceedings of ASME Turbo Expo 2013: Turbine Technical Conference and Exposition, June 3–7, 2013, San Antonio, TX, pp. 1–7.

62. Qi, H., Azer, M., Ritter, A., 2009. Studies of standard heat treatment effects on microstructure and mechanical properties of laser net shape manufactured Inconel 718. *Metallurgical and Materials Transactions A*. 40(10), 2410–2422.
63. Mišković, Z., Jovanović, M., Gligić, M., Lukić, B., 1992. Microstructural investigation of IN 939 superalloy. *Vacuum*, 43(5–7), 709–711.
64. Special Metals Corporation, 2007. Inconel alloy 718. Publication No. SMC-045, Huntington, WV, 1–28.
65. Special Metals Corporation, 2013. Inconel alloy 625, Huntington, WV, 1–18.
66. Daubenspeck B.R.; Gordon, A.P., 2010. Extrapolation techniques for very low cycle fatigue behavior of a Ni-base superalloy. *Journal of Engineering Materials and Technology*, 133(2), 021023.
67. Kraft, R., McMahan, T., Henry, K. 2013. NASA Tests Limits of 3-D Printing with Powerful Rocket Engine Check. National Aeronautics and Space Administration (NASA). http://www.nasa.gov/exploration/systems/sls/3d-printed-rocket-injector.html (accessed on December 2015).
68. General Electric. Additive Manufacturing. http://www.ge.com/stories/advanced-manufacturing (accessed on December 2015).
69. Bouchenot, T., Gordon, A.P., Shinde, S., Gravett, P., 2014. Approach for stabilized Peak/Valley stress modeling of Non-Isothermal fatigue of a DS Ni-Base superalloy. *Materials Performance and Characterization*, 3(2), pp. 16–43, doi:10.1520/MPC20130070.ISSN 2165-3992.
70. Moore, Z. J., Neu, R.W., 2011. Fatigue life modeling of anisotropic materials using a multiaxial notch analysis. *Journal of Engineering Materials and Technology*, 133, 031001-1–031001-1.

chapter fifteen

A review on powder bed fusion technology of metal additive manufacturing

*Valmik Bhavar, Prakash Kattire, Vinaykumar Patil,
Shreyans Khot, Kiran Gujar, and Rajkumar Singh*

Contents

Abstract: Additive manufacturing (AM) is a novel method of manufacturing parts directly from digital model by using layer-by-layer material build-up approach. This tool-less manufacturing method can produce fully dense metallic parts in short time with high precision. Features of AM like freedom of part design, part complexity, light weighting, part consolidation, and design for function are garnering particular interests in metal AM for aerospace, oil and gas, marine and automobile applications.

Powder bed fusion, in which each powder bed layer is selectively fused using energy source like laser or electron beam, is the most promising AM technology that can be used for manufacturing small, low volume, and complex metallic parts. This review presents evolution, current status, and challenges of powder bed fusion technology. It also compares laser and electron beam-based technologies in terms of performance characteristics of each process, advantages/disadvantages, materials, and applications.

15.1   Introduction

Additive manufacturing (AM), also known as 3D printing, is a process of joining materials to make objects from 3D model data, usually layer upon layer, as opposed to subtractive manufacturing methodologies [1]. This tool-less manufacturing approach can give industry new design flexibility, reduce energy use, and shorten time to market [2]. Main applications of AM include rapid prototyping, rapid tooling, direct part production, and part repairing of plastic, metal, ceramic, and composite materials [3]. Recent advancements in computation power of electronics, material and modeling science, and advantages offered by AM technology have shifted focus of AM from rapid prototyping to direct part production of metallic parts [4].

The two main parameters of any metal AM process are types of input raw material and energy source used to form the part [5]. Input raw material can be used in the form of metal powder or wire, whereas laser/electron beam or arc can be used as an energy source as shown in Figure 15.1. AM machine requires computer-aided design (CAD) model of the part in the .stl (stereolithography) file format. Specialized slicing software then slices this model into a number of cross-sectional layers. AM machine builds these layers one by one to manufacture complete part [6]. Thickness of these layers depends on the type of raw material and the AM process used to manufacture the given part. Every AM-manufactured part has inherent staircase-like surface finish due to layer-by-layer build up approach.

Metal AM processes can be broadly classified into two major groups, powder bed fusion (PBF) based technologies and directed energy deposition (DED) based technologies. Both of these technologies can be further classified based on the type of energy source used. In PBF-based technologies, thermal energy selectively fuses regions of powder bed [1]. Selective laser sintering/melting (SLS/SLM), laser cusing, and electron beam melting (EBM) are main representative processes of PBF-based technologies. In DED-based technologies focused, thermal energy is used to fuse materials (powder or wire form) by melting as they are being deposited [1]. Laser engineered net shaping (LENS), direct metal deposition (DMD), electron beam free form fabrication (EBFFF), and arc-based AM are some of the popular DED-based technologies.

This paper describes laser and electron beam-based PBF technologies of metal AM and their applications.

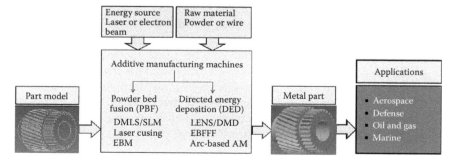

Figure 15.1 Common metal additive manufacturing process. (From Santosa, E.C. et al., *Rapid Manufacturing of Metal Components by Laser Forming*, Department of Mechanical Science and Bioengineering, Osaka University, Machikaneyama, Toyonaka, Osaka, pp. 560–8531; Horn, T.J. and Harrysson, O.L.A., *Sci. Prog.*, 95, 255–282, 2012.)

15.2 Powder bed fusion additive manufacturing

The powder bed is in inert atmosphere or partial vacuum to provide shielding of the molten metal. An energy source (laser or electron beam) is used to scan each layer of the already spread powder to selectively melt the material according to the part cross section obtained from the digital part model. When one layer has been scanned, the piston of building chamber goes downward, and the piston of the powder chamber goes upward by defined layer thickness. Coating mechanism or roller deposits powder across build chamber, which is again scanned by the energy source. This cycle is repeated layer-by-layer, until the complete part is formed. The end result of this process is powder cake, and the part is not visible until excess powder is removed. Build time required to complete a part in PBF-based processes is more as compared to DED technologies, but higher complexity and better surface finish can be achieved which requires minimum postprocessing. Several parts can be built together so that build chamber can be fully utilized [7,9]. Schematic of the PBF technology is shown in Figure 15.2.

These processes inherently require support (of same material as part) to avoid collapse of molten materials in case of overhanging surfaces, dissipating heat, and preventing distortions. Supports can be generated and modified as per part requirement during preprocessing phase, and the same has to be removed by mechanical treatment during postprocessing phase [9]. After support removal, part may undergo postprocessing treatments like shot peening, polishing, machining, and heat treatment depending on the requirement. Some critical components may even require hot isostatic pressing (HIP) to ensure part density.

SLS or direct metal laser sintering (DMLS) from EOS, SLM from Renishaw and SLM solution, and laser cusing from Concept Laser are some of the popular PBF-based technologies which use laser as energy source, whereas EBM from ARCAM is PBF-based technology which uses electron beam as energy source.

15.2.1 Laser-based systems (DMLS/SLM/laser cusing)

SLS is the first among many similar processes like DMLS, SLM, and laser cusing. SLS can be defined as PBF process used to produce objects from powdered materials using one or more lasers to selectively fuse or melt the particles at the surface, layer-by-layer, in an enclosed chamber [1]. It was developed by Carl Deckard in 1986 and commercialized by

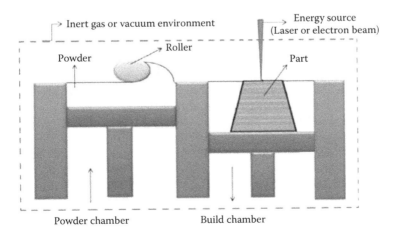

Figure 15.2 Powder bed fusion process.

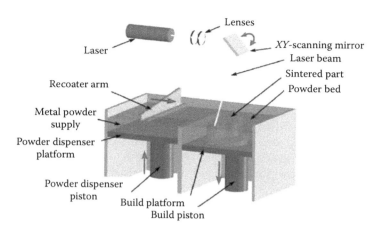

Figure 15.3 Laser-based powder bed fusion technology. (From R. Udroiu, *Powder Bed Additive manufacturing Systems and its Applications, Academic Journal Of Manufacturing Engineering*, Vol. 10, Issue 4/2012.)

DTM Corporation in 1992 [10]. The first commercial metal-sintering machine EOSINT M250 was introduced in 1995 by the EOS from Germany [11]. SLM is an advanced form of the SLS process, where full melting of the powder bed particles takes place using one or more lasers. It was developed by Fockele and Schwarze (F and S) in cooperation with the Fraunhofer institute of laser technology in 1999 and then commercialized with MCP Realizer250 machine by MCP HEK Gmbh (now SLM Solutions Gmbh) in 2004 [12] (Figure 15.3).

Laser cusing is similar to SLM process, where laser is used to fuse each powder bed layer as per required cross section to build the complete part in the enclosed chamber. It was commercialized by Concept Laser GmbH (Germany) in 2004. The term laser cusing comes from letter "C" (concept) and the word fusing. The special feature of laser cusing machine is the stochastic exposure strategy based on the island principle. Each layer of the required cross section is divided into a number of segments called *islands*, which are selected stochastically during scanning. This strategy ensures thermal equilibrium on the surface and reduces the component stresses [14].

Most of these systems use one fiber laser of 200 W to 1 KW capacity to selectively fuse the powder bed layer. The build chamber is provided with inert atmosphere of argon gas for reactive materials and nitrogen gas for nonreactive materials. Power of laser source, scan speed, hatch distance between laser tracks, and the thickness of powdered layer are the main processing parameters of these processes [15]. Layer thickness of 20–100 μm can be used depending on the material. All of these processes can manufacture fully dense metallic parts from wide range of metal alloys like titanium alloys, inconel alloys, cobalt chrome, aluminium alloys, stainless steels and tool steels.

Most of the laser-based PBF systems have low build rates of 5–20 cm³/hour, and maximum part size that can be produced (build volume) is limited to 250 × 250 × 325 mm³, which increases part cost and limits its use only for the small-sized parts. So in recent years, the machine manufactures and the research institutes are focusing on expanding the capabilities of their machines by increasing the build rates and the build volumes. SLM solution from Germany has launched SLM500 HL machine in 2012, which uses double beam technology to increase the build rate up to 35 cm³/hour and has a build volume of 500 × 350 × 300 mm³. Two sets of lasers are used in this machine, each set having two lasers (400 W and 1000 W). This means four lasers scan the powder layer simultaneously [16]. EOS

Table 15.1 Features of different PBF based AM machines

Manufacturer	Model	Energy source	Build volume (mm)	Build rate (cm^3/hr)	Scan speed (m/s)
EOS [17]	EOS M400	1 KW fiber laser	400 × 400 × 400	NA	7
EOS [17]	EOS M280/ M290	200 W or 400 W fiber laser	250 × 250 × 325	5–20	7
SLM solution [16]	SLM 500HL	2 × 400 W lasers and 2 × 1000 W lasers (optional)	500 × 280 × 325	Up to 70	Up to 15
SLM solution [16]	SLM 280	400 W and 1000 W (optional) lasers	280 × 280 × 350	20–35	Up to 15
Concept laser [18]	M2 cusing	200 W (cw) or 400 W (cw)	250 × 250 × 80	2–20	7
Renishaw [19]	SLM 250	200 W or 400 W fiber laser	250 × 250 × 300	5–20	7
ARCAM [20]	Q 10	3000 W electron beam	200 × 200 × 180	Up to 80	1000 max
ARCAM [20]	Q 20	3000 W electron beam	350 × 380 mm (Ø×H)	Up to 80	1000 max

from Germany has just launched (2013) EOSINT M400 machine, which is having a build volume of 400 × 400 × 400 mm^3 and uses one 1 KW fiber laser to increase build rate [17]. Concept Laser and Fraunhofer Institute for Laser Technology (ILT) have developed largest AM machine for metals (X line 1000R) with build volume of 630 × 400 × 500 mm^3 and build rate up to 100 cm^3/hour [18]. Details of the process capabilities in terms of build volumes build rates and scan speeds of some PBF machine models are given in the Table 15.1.

15.2.2   Electron beam melting

EBM is another PBF-based AM process in which electron beam is used to selectively fuse powder bed layer in vacuum chamber. It was commercialized by ARCAM from Sweden in 1997. EBM process is similar to the SLM with the only difference being its energy source used to fuse powder bed layers: here, an electron beam is used instead of the laser [9]. In EBM, a heated tungsten filament emits electrons at high speed which are then controlled by two magnetic fields, focus coil, and deflection coil as shown in Figure 15.4a. Focus coil acts as a magnetic lens and focuses the beam into desired diameter up to 0.1 mm, whereas deflection coil deflects the focused beam at required point to scan the layer of powder bed [21,22]. When high speed electrons hit the powder bed, their kinetic energy gets converted into thermal energy which melts the powder [10]. Each powder bed layer is scanned in two stages, the preheating stage and the melting stage. In preheating stage, a high current beam with a high scanning speed is used to preheat the powder layer (up to 1.4–0.6 Tm) in multiple passes. In melting stage, a low current beam with a low scanning speed is used to melt the powder [23]. When scanning of one layer is completed, table is lowered, another powder layer is spread, and the process repeats till required component is formed. The entire EBM process takes place under high vacuum of 10^{-4} to 10^{-5} mbar. The helium gas supply during the melting further reduces the vacuum pressure, which allows part cooling and provides beam stability [21,23]. It also has multibeam feature which converts electron beam into several individual beams which can heat, sinter, or melt powder bed layer [24].

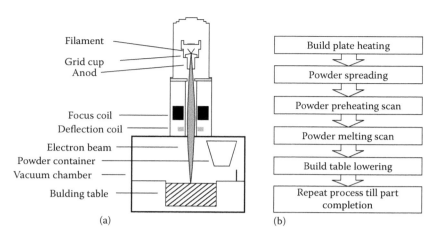

Figure 15.4 Electron beam-based powder bed fusion technology (EBM). (a) Schematic of EBM process (From Gong, X. et al., *Review on Powder-Based Electron Beam Additive Manufacturing Technology*, Mechanical Engineering Department, The University of Alabama Tuscaloosa, Tuscaloosa, AL, 2012. and (b) Steps in EBM process (From Wooten, J. and Dennies, D.P., *Electron Beam Melting Manufacturing for Production Hardware*, SAE International, Cobo Center, Detroit, MI, 2008.)

ARCAM EBM system uses high power electron beam of 3000 W capacity to melt powder bed layers. Electron beam power, current, diameter of focus, powder preheat temperature, and layer thickness are main processing parameters of the EBM. Layer thickness of 50–200 μm is typically used in this process [21]. EBM systems can work with wide range of materials like titanium alloys (Ti-6Al-4V, Ti-6Al-4V EI), cobalt chrome, titanium aluminide, inconel (625 and 718), stainless steels, tool steels, copper, aluminum alloys, beryllium, and so on [13].

15.3   Comparison between SLM and EBM

As compared to the SLM system, the EBM has higher build rates (up to 80 cm³/hour because of the high energy density and high scanning speeds) but inferior dimensional and surface finish qualities [24]. In both the SLM and EBM processes, because of rapid heating and cooling of the powder layer, residual stresses are developed. In EBM, high build chamber temperature (typically 700°C–900°C) is maintained by preheating the powder bed layer. This preheating reduces the thermal gradient in the powder bed and the scanned layer, which reduces residual stresses in the part and eliminates postheat treatment required. Preheating also holds powder particles together, which can act as supports for overhanging structural members. So, supports required in the EBM are only for heat conduction and not for structural support. This reduces the number of supports required and allows manufacturing of more complex geometries. Powder preheating feature is available in very few laser-based systems where it is achieved by platform heating. In addition, entire EBM process takes place under vacuum, since it is necessary for the quality of the electron beam [26]. Vacuum environment reduces thermal convection, thermal gradients, and contamination and oxidation of parts like titanium alloys [24]. In SLM, part manufacturing takes place under argon gas environment for reactive materials to avoid contamination and oxidation, whereas nonreactive materials can be processed under nitrogen environment. So it can be expected that EBM-manufactured parts have lower oxygen content than SLM-manufactured parts [27]. In spite of having these advantages, EBM is not as popular

Table 15.2 Characteristic features of SLM and EBM

	SLM	EBM
Power source	One or more fiber lasers of 200–1000 W	High power electron beam of 3000 W
Build chamber environment	Argon or Nitrogen	Vacuum/He bleed
Method of powder preheating	Platform heating	Preheat scanning
Powder preheating temperature (°C) [16,17]	100–200	700–900
Maximum available build volume (mm)	500 × 350 × 300	350 × 380(Ø × H)
Maximum build rate (cm³/hr)	20–35	80
Layer thickness (μm)	20–100	50–200
Melt pool size (mm)	0.1–0.5	0.2–1.2
Surface finish [9] (Ra)	4–11	50–200
Geometric tolerance (mm) [14]	±0.05–0.1	0.2–1.2
Minimum feature size (μm) [28]	40–200	100

as SLM because of its higher machine cost, low accuracy, and nonavailability of large buildup volumes. Characteristic features of SLM and EBM are summarized in Table 15.2.

15.3.1   *Applications of metal additive manufacturing*

Metal AM started to gain attention in aerospace, oil and gas, marine, automobile, manufacturing tools, and medical applications because of the advantages offered by this process. First, it can reduce buy-to-fly ratio considerably, which is the ratio of input material weight to final part weight. For conventional manufacturing processes, buy-to-fly ratio for aerospace engine and structural components can be as high as 10:1 and 20:1, respectively. AM can produce near-net shape using layer-by-layer addition of materials as per requirement, which can reduce buy-to-fly ratio up to 1:1 [7,29]. AM can produce highly complex parts and provide freedom in part design. It can be used for structural optimization using finite element analysis (FEA) to get benefit in terms of light weighting. It can also produce lattice structures with low density, high strength, good energy absorption, and good thermal properties which can be used for light weighting and better heat dissipation in applications like heat exchangers in aerospace, automobile, and computer industries [30]. Figure 15.5 shows conventional and optimized door bracket manufactured by AM, where internal *bamboo* structure is used for light weighting [31].

Figure 15.5 Conventional and optimized part by additive manufacturing. (From Additive Manufacturing, White paper, January 2013. Available online at http://www.qmisolutions.com.au/accessed on 5th April 2014.)

Conventionally manufactured part may require a number of different manufacturing processes like casting, rolling, forging, machining, drilling, welding, and so on, whereas same part can be produced using AM which eliminates required tooling and produces part in single processing step. Every part manufactured by AM can be unique and produced in very short time, which enables mass customization [32].

AM also reduces assembly requirements by integrating number of parts required in assembly into a single part. It reduces overall weight, decreases manufacturing time, reduces number of manufacturing processes required, reduces cost and material requirements, and optimizes required mechanical properties [29]. General Electric (GE) has integrated fuel nozzle assembly of 20 small parts into single fuel nozzle part of cobalt chrome material (Figure 15.6), which is under testing phase. It is 25% lighter and five times more durable than conventional assembly [34]. Oak Ridge National Laboratories has manufactured lightweight, compact underwater robotic system (hydraulic manifold) in which the robot base, hydraulic reservoir, and accumulator are integrated into a single lightweight structure [35].

AM can enhance part performance and add value to the product as parts can be designed for function. Injection molding tools can be provided with conformal cooling channels (Figure 15.7), which increases cooling efficiency and reduces the cycle

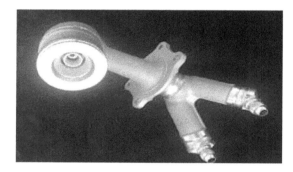

Figure 15.6 GE's additively manufactured fuel nozzle. (From GE capital, *Additive Manufacturing Redefining What's Possible*, Fall 2013. QMISOLUTIONS, Additive manufacturing, white paper, January 2013.)

Figure 15.7 Conformal cooling channel insert. (From GE capital, *Additive Manufacturing Redefining What's Possible*, Fall 2013. QMISOLUTIONS, Additive manufacturing, white paper, 2013.)

Figure 15.8 **(See color insert.)** Lattice structure for light weighting. (From Vaerenbergh, J.V. et al., *Application Specific AM Technology Development for High-End Mechatronic Systems*, LayerWise NVLeuven, Belgium, 2014.)

time. Different types of lattice structures can be used to achieve unique properties like improved heat dissipation, structures with negative poisons ratio, and improved energy absorption characteristics [29,37]. A negative Poisson's ratio increases impact resistance, fracture toughness, and shear resistance [21] (Figure 15.8).

15.3.2   Summary and challenges of PBF technologies

Though significant progress and technological advancement have been made by the PBF AM technology, performances in terms of speed, accuracy, process control, and cost effectiveness still need to improve. Knowledge of processing–structure–property relationship for existing materials is required to predict part performance. In-process quality monitoring and closed-loop control systems are required to improve the consistency, repeatability, and uniformity across machines. Closed loop melt pool temperature control system has proven its significance in deposition-based LENS process by maintaining desired quality of part. Such close loop melt pool temperature control system has still remained as a challenge for SLM systems. Early stage defect detection through in-process quality monitoring could save required raw material and manufacturing time [39]. Better understandings of physical and metallurgical mechanisms responsible for variation in properties are required for predictive process modeling. AM part cost is still on the higher side for some applications. Increasing build speeds of AM machines, designing more complex geometries, and reducing assembly requirements could reduce required part cost and widen application areas of AM in future.

Abbreviations

The following symbols are used in this paper:

AM: Additive manufacturing
CAD: Computer-aided design
DED: Directed energy deposition
SLS: Selective laser sintering

DMLS: Direct metal laser sintering
SLM: Selective laser melting
EBM: Electron beam melting
DMD: Direct metal deposition
EBFFF: Electron beam free-form fabrication
LENS: Laser-engineered net shaping
PBF: Powder bed fusion

References

1. F-2792- 12a, *Standard Terminology for Additive Manufacturing Technologies.* ASTM International, West Conshohocken, PA,2012.
2. *Additive Manufacturing: Pursuing the Promise, advanced manufacturing Office,* U.S. Department of Energy, DOE/EE-0776, August 2012.
3. T. Wohlers, *Additive Manufacturing State of the Industry,* Wohlers report 2010, Wohlers Associates, Fort Collins, CO.
4. A. Koptyug, L. E. Rännar, M. Bäckström, S. F. Franzén, DDS Dérand, Additive Manufacturing Technology Applications Targeting Practical Surgery, *International Journal of Life Science and Medical Research,* 3(1) 15-24, Sweden 2013.
5. I. D. Harris, PhD Director, AMC, EWI, Columbus, OH, *Development and Implementation of Metals Additive Manufacturing,* DOT International, New Orleans, LA, 2011.
6. B. Stucker, Additive Manufacturing Technologies: Technology Introduction and Business Implications, *Frontiers-of-engineering-reports-on-leading-edge-engineering-from-the 2011 Symposium,* pp. 5–14, The National Academic Press, Washington, DC.
7. C. L. English, S. K. Tewari, D. H. Abbott, *An Overview of Ni Base Additive Fabrication Technologies for Aerospace Applications (Preprint),* GE Aviation, Cincinnati, OH, March 2011.
8. E. C. Santosa, M. Shiomia, K. Osakadaa, T. Laoui, *Rapid Manufacturing of Metal Components by Laser Forming,* Department of Mechanical Science and Bioengineering, Osaka University, Machikaneyama, Toyonaka, Japan, pp. 560–8531.
9. B. Vayre, F. Vignat, F. Villeneuve, Metallic additive manufacturing: State-of-the-art review and prospects, Mechanics & Industry, 13(2), 89–96, 2012.
10. J. Ruan, T. E. Sparks, Z. Fan, J. K. Stroble, A. Panackal, F. Liou, *A Review of Layer Based Manufacturing Processes for Metals,* Department of Mechanical and Aerospace Engineering, University of Missouri, Rolla, MO.
11. M. Shellabear, O. Nyrhilä, *DMLS—Development History and State of the Art,* 1EOS GmbH Electro Optical Systems, Germany; 2EOS Finland, Finland.
12. D. Thomas, *The Development of Design Rules for Selective Laser Melting,* PhD Thesis, University of Wales Institute, Cardiff, 2009.
13. R. Udroiu, Powder bed additive manufacturing systems and its applications, academic. *Journal of Manufacturing Engineering,* 10(4), 122–129, 2012.
14. M. Aliakbari, *Additive Manufacturing: State-of-the-Art, Capabilities, and Sample Applications with Cost Analysis,* Master of Science Thesis, Department of Industrial Production, KTH, 2012.
15. A. B. Spierings, K. Wegener, G. Levy, *Designing Material Properties Locally with Additive Manufacturing Technology SLM,* INSPIRE AG–Institute for Rapid Product Development IRPD, St. Gallen, Switzerland, 2012.
16. (2014) SLM Solution website, available at www.slm-solution.com, accessed 5th April 2014.
17. (2014) EOS website, available at www.eos.info, accessed 5th April 2014.
18. (2014) Concept Laser website, available at www.concept-laser.de, accessed 5th April 2014.
19. (2014) Renishaw website, available at www.renishaw.co.in, accessed 5th April 2014.
20. (2014) Arcam website, available at www.arcam.com, accessed 5th April 2014.
21. X. Gong, T. Anderson, K. Chou, *Review on Powder-Based Electron Beam Additive Manufacturing Technology,* Mechanical Engineering Department, The University of Alabama Tuscaloosa, Tuscaloosa, AL, 2012.

22. J. Hiemenz, *EBM offers a new alternative for producing titanium parts and prototypes, Time-Compression Technologies*, Vol 14, Rapid News Publications, UK, 2006.
23. L. E. Murr, E. Martinez, K. N. Amato, S. M. Gaytan, J. Hernandez, D. A. Ramirez, P. W. Shindo, F. Medina, R. B. Wicker, *Fabrication of Metal and Alloy Components by Additive Manufacturing: Examples of 3D Materials Science,* Department of Metallurgical and Materials Engineering, The University of Texas, El Paso, TX, 2012.
24. B. Vayre, F. Vignat, F. Villeneuve, Identification on some design key parameters for additive manufacturing: application on electron beam melting, Forty Sixth CIRP, Conference on Manufacturing Systems, Sesimbra, Portugal, 29–31 May, Vol. 7, pp. 264–269, 2013.
25. J. Wooten, D. P. Dennies, *Electron Beam Melting Manufacturing for Production Hardware,* SAE International, Cobo Center, Detroit, MI, 2008.
26. M. Koikea, K. Martineza, L. Guoa, G. Chahine, R. Kovacevic, T. Okabe, Evaluation of Titanium alloy fabricated using electron beam melting system for dental applications, *Journal of Materials Processing Technology*, 211(8), 1400–1408, 2011.
27. L. Loeber, S. Biamino, U. Ackelid, S. Sabbadini, P. Epicoco, P. Fino, J. Eckert, *Comparison of Selective Laser And Electron Beam Melted Titanium Aluminides,* Online available at https://sffsymposium. engr.utexas.edu/Manuscripts/2011/2011-43-Loeber.pdf, accessed on 5th April 2014.
28. (2014) additively website, learn-about/3d-printing-technologies, available at www.additively. com, accessed April 2014.
29. Horn, T. J., Harrysson, O. L. A., Overview of current additive manufacturing technologies and selected applications, Science Progress, 95(3), 255–282, 2012.
30. S. L. Campanelli, N. Contuzzi, A. Angelastro, A. D. Ludovico, Capabilities and performances of the selective laser melting process, *New Trends in Technologies: Devices, Computer, Communication and Industrial Systems, Sciyo,* Scottsdale, AZ, 2010.
31. QMISOLUTIONS, *Additive Manufacturing*, White paper, January 2013. Available online at http://www.qmisolutions.com.au/accessed on 5th April 2014.
32. J. Geraedts, E. Doubrovski, J. Verlinden, M. Stellingwerff, *Three Views on Additive Manufacturing: Business, Research, and Education,* Proceedings of TMCE, 2012.
33. GE capital, *Additive Manufacturing Redefining What's Possible,* Fall 2013. QMISOLUTIONS, *Additive manufacturing,* white paper, January 2013.
34. MIT Technology Review, *2013 Breakthrough Technology: Additive Manufacturing,* April 23, 2013, online available at www.technologyreview.com, accessed on 5th April 2014.
35. T. Wohlers, *Additive Manufacturing at Work,* April 2012, http://www.sme.org/manufacturingengineering, accessed on 5th April, 2014.
36. S. Kumar, S. Pityana, *Laser-Based Additive Manufacturing of Metals,* Council for Scientific and Industrial Research, National Laser Centre, Pretoria 0001, South Africa.
37. M. Thymianidis, C. Achillas, D. Tzetzis, E. Iakovou, Modern additive manufacturing technologies: an upto-date synthesis and impact on supply chain design, *2nd International Conference on Supply Chains 2nd Olympus ICSC,* Katerini, 5–6 October, 2012.
38. J. V. Vaerenbergh, L. Thijs, W. V. der Perre, T. D. Bruyne, *Application Specific AM Technology Development for High-End Mechatronic Systems,* LayerWise NVLeuven, Belgium, 2014.
39. J. Scott, N. Gupta, C. Weber, S. Newsome, *Additive Manufacturing: Status and Opportunities,* Wohlers Associates, Washington, DC, pp. 1–29, 2012.

chapter sixteen

Additive manufacturing of titanium alloys

B. Dutta and Francis H. Froes

Contents

Titanium alloys are among the most important advanced materials and are key to improved performance in both aerospace and terrestrial systems due to an excellent combination of specific mechanical properties and outstanding corrosion behavior.[1]

The high cost of producing conventional titanium components has spurred numerous investigations into potentially lower cost processes, including powder metallurgy (PM) near-net-shape techniques such as additive manufacturing (AM).[1] This article reviews AM with an emphasis on the *work horse* titanium alloy, Ti-6Al-4V. AM is receiving a significant attention from numerous organizations including the U.S. Navy, as its envisions use aboard carriers with parts able to be rapidly fabricated for immediate use by battle groups.[1] Various approaches to AM, along with examples of components made by different AM processes, are presented. The microstructures and mechanical properties of Ti-6Al-4V produced by AM are also discussed and compared well with cast and wrought products. Finally, the economic advantages of AM compared to conventional processing are presented.

16.1   Additive manufacturing overview

All AM technologies are based on the principle of slicing a solid model into multiple layers and building the part up layer-by-layer following the sliced model data. Following ASTM classification, AM technologies for metals can be broadly classified into two categories: directed energy deposition and powder bed fusion (PBF) (Table 16.1). Several technologies fall under each category as branded by different manufacturers. While powder bed fusion technologies enable construction of complex features, hollow cooling passages, and high precision parts; they are limited by the build envelope, single material per build, and

Table 16.1 Additive manufacturing technologies for processing

AM category	Technology	Company	Description
Directed Energy Deposition (DED)	Direct Metal Deposition (DMD)	DM3D Technology LLC (formerly POM Group)	Laser and metal powder used for melting and depositing with a patented closed loop process
	Laser Engineered Net Shaping (LENS)	Optomec Inc.	Laser and metal powder used for melting and depositing
	Direct Manufacturing (DM)	Sciaky Inc.	Electron Beam and metal wire used for melting and depositing
Powder Bed Fusion (PBF)	Selective Laser Sintering (SlS)	3D Systems Corp. (acquired Phenix Systems)	Laser and metal powder used for sintering and bonding
	Direct Metal Laser	EOS GmbH	Laser and metal powder used for sintering, melting, and bonding
	Laser Melting (LM)	Renishaw Inc.	Laser and metal powder used for melting and bonding
	Laser Melting (SLM)	SLM Solutions GmbH	Laser and metal powder used for melting and bonding
	LaserCUSING	Concept Laser GmbH	Laser and metal powder used for melting and bonding
	Electron Beam Melting (EBM)	Arccam AB GmbH	Electron beam and metal power used for melting and bonding

horizontal layer construction ability. In comparison, directed energy deposition technologies offer larger build envelopes and higher deposition rates, but their ability to construct hollow cooling passages and finer geometry is limited. Direct metal deposition (DMD) and laser engineered net shaping (LENS) technology can deposit multiple materials in a single build and add metal to existing parts.

Commercially available AM technologies melt powder or wire feedstock using either laser or electron beam heat sources. Laser-based systems operate under inert atmosphere (for titanium processing) in contrast to the vacuum environment of electron beam systems. Although vacuum systems are more expensive, they offer low residual stress compared to laser-based systems, and electron-beam-processed parts can be used without stress-relieving operations. Heat source effects on mechanical properties are discussed in more detail further in this article.

16.2   Powder bed fusion

PBF technologies place a layer of metal powder on the build platform, and then the powder is scanned with a heat source, such as a laser or electron beam, to either partially or completely melt the powder in the path of the beam and resolidify and bind it together as it cools (ASTM specification F2924-12a and -13 for Ti-6Al-4V and Ti-6Al-4VELI grades, respectively). Layer-by-layer tool path tracing is governed by the computer-aided design (CAD) data of the part being built. Figure 16.1 shows a schematic explaining the steps involved in this process:

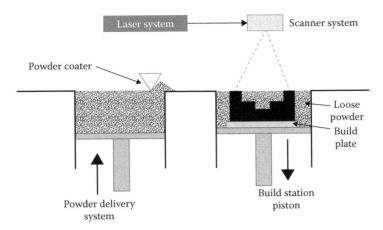

Figure 16.1 Schematic showing powder bed fusion technology. (Courtesy of Jim Sears.)

- A substrate is fixed on the build platform.
- The build chamber is filled with inert gas (for laser processing) or evacuated (for electron beam processing) to reduce the chambers oxygen level to the desired level.
- A thin layer of metal powder (20–200 mm thick, depending on the technology and equipment) is placed on the substrate and leveled to a predetermined thickness.
- The laser or electron beam scans the powder bed surface following the tool path pre-calculated from the CAD data of the component being built.
- This process is repeated or the following layers until the build is complete.

16.3   Directed energy deposition

Directed energy deposition technologies work by injecting material into a melt pool rather than scanning a powder bed (AMS specification 4999A for Ti-6Al-4V). Figure 16.2 shows a schematic of Direct Metal Deposition (DMD) technology (laser-based metal deposition). Steps for the directed energy deposition process include the following:

- A substrate or existing part is placed on the work table.
- Similar to PBF, the machine chamber is closed and filled with inert gas (for laser processing) or evacuated (for electron beam processing) to reduce the chambers' oxygen level to the desired level (AMS 4999A specifies below 1200 ppm). The DMD process offers local shielding and does not require an inert gas chamber for less reactive metals than titanium, such as steels, Ni alloys, and Co alloys.
- At the start of the cycle, the process nozzle with a concentric laser or electron beam is focused on the part surface to create a melt pool. Material delivery involves powder traveling through a coaxial nozzle (laser) or through a metal wire with a side delivery (electron beam). The nozzle moves at constant speed and follows a predetermined tool path created from the CAD data. As the nozzle (tooltip) moves away, the melt pool solidifies and forms a metal layer.
- Successive layers follow the same principle and build up the part layer-by-layer.

Table 16.2 provides a comparison of capabilities, benefits, and limitations of various AM technologies used for producing titanium parts.[3–6]

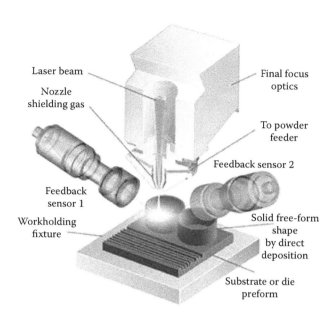

Figure 16.2 Schematic showing direct metal deposition technology. (Courtesy of DM3D Technology.)

Table 16.2 Comparison of various additive manufacturing technologies

Item/Process	Laser-based PBF (DMLS)	Electron beam-based PBF (EBM)	Laser-based directed energy deposition (DMD)
Build envelope	Limited	Limited	Large and flexible
Beam size	Small, 0.1–0.5 mm	Small, 0.2–1 mm	Large, can vary from 2–4 mm
Layer thickness	Small, 50–100 μm	Small, 100 μm	Large, 500–1000 μm
Build rate	Low, cclh	Low, 55–80 cclh	High, 16–320 cclh
Surface finish	Very good, Ra 9/12 μm, Rz 35f40 um	Good, Ra 25/35 μm	Coarse, Ra 20–50 μm, Rz 150–300 um depends on beam size
Residual stress	High	Minimal	High
Heat treatment	Stress relief required, HIP'ing preferred	Stress relief not required, HIP'ing may or may not be performed	Stress relief required, HIP'ing preferred
Chemistry	ELI grade possible, negligible loss of elements	ELI grade possible, loss of Al needs to be compensated for in powder chemistry	ELI grade possible, negligible. loss of elements
Build capability	Complex geometry possible with very high resolution. Capable of building hollow channels	Complex geometry possible with good resolution. Capable of building hollow channels	Relatively simple geometry with less resolution. limited capability for hollow channels

(Continued)

Table 16.2 (Continued) Comparison of various additive manufacturing technologies

Item/Process	Laser-based PBF (DMLS)	Electron beam-based PBF (EBM)	Laser-based directed energy deposition (DMD)
Repair/ Remanufacture	Possible only in limited applications (requires horizontal plane to begin remanufacturing)	Not possible	Possible; capable of adding metal on 30 surfaces under 5 + 1-axis configuration, making Repair solutions attractive
Feature/metal addition on existing parts	Not possible	Not possible	Possible; depending on dimensions, ID cladding is also possible
Multi-material build or hard coating	Not possible	Not possible	Possible

16.4   Titanium AM applications

Extensive exploration regarding use of AM titanium parts in aerospace and medical applications is underway. Other potential AM applications include the chemical and defense industries (Defense News, 2013). Although PBF technologies are suitable for smaller, complex geometries with hollow unsupported passages/structures, directed energy deposition is better suited for larger parts with coarser features requiring higher deposition rates.

Use of finer powder grains combined with smaller laser/electron beam size achieves a superior surface finish on the as-built parts from the PBF technologies compared to directed energy deposition technologies. However, the majority of AM parts require finish machining for most applications. Beyond building new parts, the ability of directed energy technologies to add metal onto existing parts makes it possible to apply protective surface coatings, remanufacture and repair damaged parts, and reconfigure or add features to the existing parts.

16.5   Complex geometry

A small beam size and small layer thickness, along with support of the powder bed, allow PBF technologies (such as electron beam melting, EBM; direct metal laser sintering, DMLS; or selective laser sintering, SLS) to produce complex geometries with high precision and unsupported structures. Figure 16.3 shows an example of a hydraulic manifold mount for an underwater manipulator built using EBM technology. Building the integrated mount and manifold with internal passageways in a single operation eliminates fabrication of multiple parts and costs much less. A quality surface finish eliminates the need to machine finish all surfaces except seal surfaces and threading of screw holes. Generally, the PBF technique achieves a better surface finish than DED approach, though demanding applications such as aerospace require finish machining. Figure 16.4 shows a biomedical implant built with a Ti-6Al-4V alloy using DMLS technology. These technologies also make it possible to build patient-specific custom implants.

Figure 16.3 Hydraulic manifold built using EBM technology. The part was built at Oak Ridge National Laboratory (ORNL). (Courtesy of ORNL.)

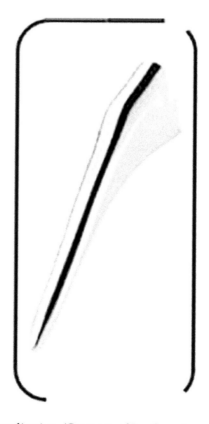

Figure 16.4 Medical implant application. (Courtesy of Jim Sears.)

16.6 Adding features to the existing parts

Directed energy deposition technologies, such as DMD and/or LENS, can add metal to 3D surfaces to allow additional features to be added to existing parts and/or blanks, which is not possible using the PBF approach. Adding features to a forged or cast preform, as opposed to machining such features, can result in the most cost-effective manufacturing process, where a significant reduction of preform size and weight can be achieved by eliminating the need for a machining allowance. Examples include various casings and housings in jet engines, where flanges and bosses can be added on cast or forged cylindrical preforms. To illustrate, Figure 16.5 shows a feature added to a titanium fan casing for an aerospace engine.

16.7 Remanufacturing

One of the application areas best suited to directed energy deposition techniques is remanufacturing and repair of damages, worn, or corroded parts. Due to the ability to add metal to select locations on 3D surfaces, these technologies can be used to rebuild lost material on various components.[7,8] Closed-loop technologies, such as DMD, achieve a minimum heat-affected zone (HAZ) in the repaired part, which helps retain its integrity.

Figure 16.6 shows cross-section microstructures of the DMD area of a remanufactured turbine blade. Excellent process control during DMD leads to a fully dense microstructure as observed in the vertical cross section. A layer thickness of roughly 0.1–0.2 mm was applied, and a minimal HAZ occurs in the as-deposited blade. The DMD vision system plays a significant role in this type of remanufacturing application. An integrated, calibrated vision system allows automatic identification of part location in the machine coordinate system, resulting in precise processing. Other titanium components that can be repaired using DMD include housings, bearings, casing flanges, and landing gears.

Figure 16.5 Fan case produced by adding features with AM (laser-aided directed energy deposition) to a forged perform. (Courtesy of Jim Sears.)

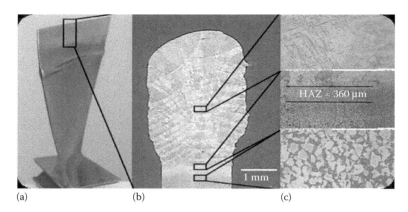

(a)   (b)   (c)

Figure 16.6 DMD repair of turbine components. (a) Repaired vane, (b) macrocross section, and (c) microstructures (top to bottom shows the clad, interface, and base material). (Courtesy of DM3D technology.)

16.8   Microstructure and mechanical properties

The aerospace materials' specification SAE AMS4999A covers Titanium Alloy Direct Products Ti-6Al-4V Annealed. This calls for a postbuild annealing treatment of 1025°F (550°C). If a hot isostatic pressing (HIP) treatment is used, it should be at no less than 14.5 ksi (100 MPa) within the 1650°F–1750°F (899°C–954°C) temperature range for 2–4 hours followed by a slow cool below 800°F (437°C). Minimum tensile properties should be ultimate tensile strength (UTS) 124–129 ksi (855–889 MPa, depending on direction), YS 110–116 ksi (758–800 MPa), and elongation of 6%.[9]

Typical microstructures of as-built material using the DMD process and after subsequent HIPing and aging are shown in Figure 16.7. The as-built microstructure shows the typical martensitic structure expected for Ti-6Al-4V cooled rapidly from the beta phase field, whereas the HIPd and aged material shows the expected grain boundary of alpha and intergranular coarse alpha plates. This microstructural transition from as-deposited to the HIPd-aged condition is also reflected through their tensile properties. Although tensile strength and yield strength are a little lower after HIPing and aging, ductility improves significantly as a result of the microstructure changing from martensitic to a transformed beta (precipitated alpha) structure. The as-built electron beam-processed material contains a similar microstructure, though martensite is replaced by a lamellar alpha phase.

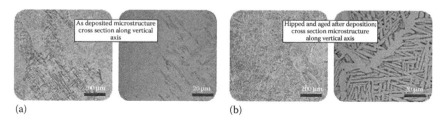

(a)   (b)

Figure 16.7 Microstructure of DMD-built Ti-6Al-4V (a) before and (b) after HIP'ing. (Courtesy of DM3D technology.)

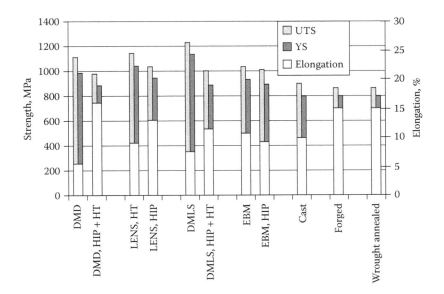

Figure 16.8 Tensile strength, yield strength, and elongation of Ti-6Al-4V alloy built using various AM processes: DMD (direct metal deposition) (From Titanium Alloy Direct Deposited Products Ti-6Al-4V Annealed. SAE Aerospace Material Specification (AMS) 4999A, www.sae.org/technical/standards/AMS4999A, Sept. 2009), LENS (laser-engineered net shaping) (Private communication, DM3D Technology, www.optomec.com/Additive-Manufacturing-Technology/Laser-Additive-Manufacturing, July 2013), DMLS (direct metal laser sintering) (From www.morristech.com/Docs/Ti64ELI%20DataSheet.pdf), EB (electron beam melting) (From www.morristech.com/Docs/Ti64ELI%20DataSheet.pdf), HIP (hot isostatic pressing), and HT (heat treatment).

Tensile properties of Ti-6Al-4V fabricated by a number of AM techniques are shown in Figure 16.8. All processes achieve strength levels superior or comparable to conventional material (cast, forged, and wrought annealed). The as-built materials in laser-based processes such as DMD, LENS, and DMLS exhibit less ductility due to formation of the martensite phase. However, ductility can be improved through subsequent HIPing and/or heat treatment (HT). As a result of reduced residual stress, EBM-processed Ti-6Al-4V achieves greater ductility when compared to laser-processed Ti-6Al-4V. Fatigue properties were tested using many different cycles. In general, as-built Ti-6-4 offers fatigue resistance similar to cast and wrought material, even without hot isostatic pressing (HIP) treatment, as shown in Figure 16.9.

16.9   Additive manufacturing economics

Among the main benefits of PBF processes is their ability to create hollow structures and, therefore, to achieve weight savings. The aerospace industry, where weight savings can make significant impacts, is actively looking into AM processes. A case study involving a seat buckle for commercial passenger jets is a prime example of this capability.[12]

A lightweight seat buckle with hollow structures was designed based on an extensive finite element analysis study to ensure adequate strength against shock loading. The part was produced using a DMLS Ti-6Al-4V alloy. Replacing a conventional steel buckle with a hollow AM titanium buckle achieves weight savings of 85 g per buckle, a 55% weight reduction. By applying this across, an Airbus A380 with 853 seats results in weight savings of 72.5 kg. According to the project sponsor, Technology Strategy Board, United Kingdom, this

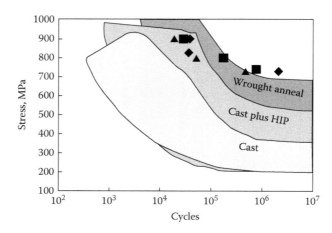

Figure 16.9 Comparison of room temperature fatigue properties of Ti-6Al-4V processed by AM versus conventionally fabricated Ti-6Al-4V. ■, ♦, and ▲ represent properties in the three orthogonal directions, x, y, and z, respectively. (Courtesy of EADS/Jim Sears.)

weight saving translates to 3.3 million liters of fuel savings over the life of the aircraft, equivalent to $3 million, whereas the cost of making the buckles using DMLS is only $256,000.

The direct manufacturing ability of AM technologies also helps to reduce manufacturing costs in the case of high buy-to-fly ratio parts. For example, researchers at Oak Ridge National Laboratory built a Ti-6A-4V bleed air leak detect (BALD) bracket for the joint strike fighter (JSF) engine using EBM technology (Figure 16.10).[13] Traditional manufacturing from wrought Ti-6Al-4V plate costs almost $1000/lb due to a high (33:1) buy-to-fly ratio as opposed to just over 1:1 for the AM-built part. Estimated savings through AM is approximately 50%.

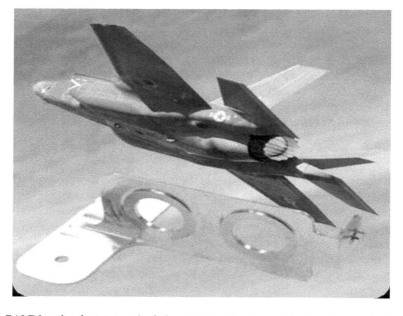

Figure 16.10 BALD bracket for joint strike fighter (JSF) built using EBM technology. (Courtesy of ORNL.)

Direct deposition techniques such as DMD cannot only be used to create parts, but these technologies can also be used for remanufacturing, repair, and/or feature building on the existing parts. Damaged aerospace titanium components such as bearing housings, flanges, fan blades, casings, vanes, and landing gears can be rebuilt using these technologies at 20–40% of the cost of new parts.[10] Worn flanges in jet engine casings have been rebuilt using DMD at less than half of the cost of new parts. Extensive work is also underway to investigate the feasibility of using such technologies to salvage components that are mismachined during conventional manufacturing. Successful realization of these efforts will have a significant impact on the titanium-manufacturing industry. Although most of the commercial activities in the AM industry are concentrated in the United States and Europe, significant efforts are underway in other parts of the world as well, including China.[14]

16.10   Conclusions

Significant advances in AM technologies over the past few years have led to the production of fully functional parts using titanium and its alloys. Although PBF technologies offer the ability to build hollow near-net shapes with finer resolution, directed energy-based technologies offer the ability to add features to the existing parts and to remanufacture and repair damaged parts, besides building parts directly from CAD data. Most studies reveal that the properties of AM material are as good as, or superior to, conventionally fabricated titanium alloys. Matching the correct AM technology to the application, along with proper design optimization, can achieve significant savings by reducing both weight and scrap. The aerospace and medical industries have so far been the largest users of titanium AM materials, while other industries, such as automotive, are beginning to exploit the benefits of AM titanium alloys as well.

Acknowledgments

The authors recognize input for this article and useful discussions with Jim Sears, Ryan DeHoff, Richard Grylls, Jessica Nehro, Anders Hultman, Scott Thompson, Laura Kinkopf, Michael Cloran, David Whittaker, Karl D'Ambrosio, and Ma Quin.

For more information: Bhaskar Dutta is the chief operating officer at DM3D Technology LLC, 2350 Pontiac Rd., Auburn Hills, MI 48326, 248/409-7900, bdutta@dm3dtech.com, www.dm3dtech.com.

References

1. F.H. (Sam) Froes, Powder Metallurgy of Titanium Alloys, *Advances in Powder Metallurgy.* Eds. Isaac Chang and Yuyuan Zhao, Woodhead Publishing, Philadelphia, p 202, 2013.
2. *Defense News*, 24, June 10, 2013.
3. www.eos.info/additive_manufacturing/for_technology_interested.
4. www.arcam.com/technology/additive-manufacturing.
5. www.dm3dtech.com/index.php/expertise-innovations/experticeandinnovations-dmddtechnology.
6. www.morristech.com/Docs/Ti64ELI%20DataSheet.pdf.
7. B. Dutta, et al., *Adv. Mater. & Processes*, 33–36, May 2011.
8. B. Dutta, et al., Rapid manufacturing and remanufacturing of DoD components using direct metal deposition, *The AMMTIAC Quarterly*, 6(2), 5–9, 2011.

9. Bhaskar Dutta, Harshad Natu, and Jyoti Mazumder, Near net shape repair and remanufacturing of high value compo- nents using DMD, TMS Proceedings, Vol 1: Fabrication, Materials, Processing and Properties, 131–138, 2009.
10. Titanium Alloy Direct Deposited Products Ti-6Al-4V Annealed. SAE Aerospace Material Specification (AMS) 4999A, Sept. 2009, www.sae.org/technical/standards/AMS4999A.
11. Private communication, DM3D Technology, July 2013. www.optomec.com/Additive-Manufacturing-Technology/Laser-Additive-Manufacturing.
12. www.manufacturingthefuture.co.uk/_resources/case-studies/TSB-AirlineBuckle.pdf.
13. Ryan Dehoff, et al., *Adv. Mater. & Processes,* 171(3), 19–22, 2013.
14. Private communication, Ma Quin, July 2013.

chapter seventeen

Ultrasonic additive manufacturing

Paul J. Wolcott and Marcelo J. Dapino

Contents

17.1 Introduction

Worldwide sales of additive manufacturing products and services are estimated to reach $11 billion in 2021, up from $2 billion in 2012 [1]. Despite this anticipated level of growth, the value of additive manufacturing to industries that rely on mass manufacturing is unclear. As a relatively new technology, additive manufacturing has not yet reached the levels of throughput, cost effectiveness, and standardization required for implementation in industry sectors such as automotive and electronics. Ultrasonic additive manufacturing (UAM), a niche technology within the additive manufacturing area, offers the manufacturing industry a different approach for creating lightweight metal-based structures incorporating dissimilar metals, nonmetallic materials, smart materials, and intricate features that are difficult to produce through conventional means. In-depth research is needed to address the challenges posed by this relatively new technology and to move UAM from the laboratory to practical applications. This chapter discusses those challenges and presents research efforts conducted by the authors and other researchers to understand the process.

UAM, a solid-state 3D-printing technology based on traditional ultrasonic metal welding, [2] makes it possible to fabricate metal structures from foil stock. The fundamental principle of UAM operation is the layering of foils through solid-state metal welding to achieve fully dense, gapless 3D parts. In a broad sense, UAM can also be used as a joining technology to integrate dissimilar metals, seam welding of metallic sheets, [3] or as a cladding technology by which high-value materials are layered over a bulk substrate. UAM structures can also incorporate dissimilar metals along with embedded features such as reinforcement fibers, smart sensors and actuators, and heat-wicking materials. In addition, UAM has been shown to address various traditional joining and manufacturing needs such as joining of metals to nonmetals, provided suitable joint configurations are developed. Figure 17.1 illustrates possible uses for UAM.

A key benefit of UAM as a technology for 3D printing of metals and joining of dissimilar materials is that process temperatures are low, typically less than one half of the melting temperature of aluminum alloys [4,5]. The low thermal loading inhibits the formation of brittle intermetallics with the subsequent advantage of not altering the microstructure of the constituent metals. Low operating temperatures also can limit corrosion through mitigation of electrochemical reactions. Further, finished parts suffer no heat-induced distortion, and hence no remedial machining is required to bring parts to their intended dimensions.

In the UAM process, a sonotrode driven by one or more piezoelectric transducers imparts ultrasonic vibrations to a metal foil, creating a scrubbing action and plastic deformation between the foil and the material to which it is being welded (Figure 17.2a). The vibration frequency is nominally 20 kHz on most systems.

The scrubbing action displaces surface oxides and contaminants and collapses asperities, exposing nascent surfaces that instantaneously form a metallurgical bond under a compressive force. The first layer is welded onto a metal baseplate, which is used to support the build. By welding a succession of tapes, first side by side and then one on top of one another, a three-dimensional metal part can be fabricated [2]. Periodic machining with a computer numerical control (CNC) stage (Figure 17.2b) or laser-etching system allows

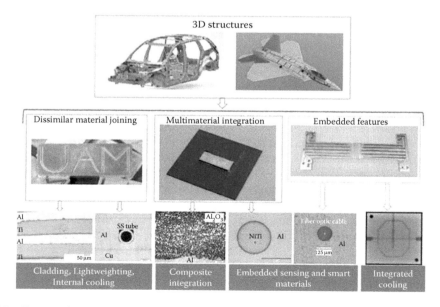

Figure 17.1 Potential UAM applications and capabilities. Low UAM process temperatures allow joining of dissimilar metals without the formation of brittle intermetallics and the integration of temperature-sensitive components, smart materials, cooling channels, organic polymers, and electronics into metal matrices.

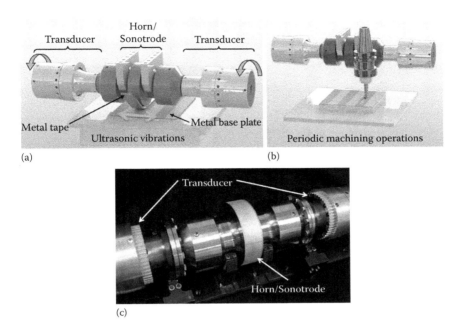

Figure 17.2 (a) Schematic of a UAM welder, which utilizes ultrasonic vibrations and pressure to join foil stock to a baseplate or other foils. The process is solid state, implying that no melting is present; (b) the process usually features a CNC mill for conducting subtractive operations; and (c) image of a welder outside of the machine indicating the transducers and sonotrode. The welder is acoustically tuned, so that the sonotrode resonates at 20 kHz (nominal).

for selective material removal and *in situ* machining to final dimensions. Depending on the part being built, additive and subtractive processes are repeated in various sequences until a solid component has been created or material has been added to a component. The subtractive processes are also utilized to create internal channels for thermal management, to align fibers within the matrix, or for surface texturing of embedded fibers. Upon completion of the build, the baseplate and build can be separated by conventional or electrical discharge machining (EDM) if this material is not desired in the final design.

The most advanced UAM systems [6] deliver 9 kW of ultrasonic power to the weld interface, which improves the strength and quality of UAM builds, greatly enhances the ability to weld dissimilar materials, and enables the construction of previously unfeasible adaptive structures (Figure 17.3). This is illustrated in Figure 17.4, where the metallurgical section of an aluminum 3003 build using a 1 kW UAM system shows gaps, in contrast to a build made with a 9 kW UAM system which shows no gaps despite the material used being aluminum 6061. Even though a low-void content does not guarantee high mechanical strength, [7] obtaining gapless builds is a necessary condition for optimizing the strength of UAM components.

Fundamental investigations in the field of UAM are aimed at creating an exact understanding of the process and to develop experimental approaches and models to describe the relationship between process conditions and build properties. As with ultrasonic metal welding, the main control variables for the UAM process include weld speed (or time), down force (or pressure), and vibration amplitude [2,8]. Ultrasonic vibration frequency is fixed at the designed resonance frequency of the sonotrode. For some materials, a heat

Figure 17.3 **(See color insert.)** State of the art UAM system featuring a 9 kW welder and 20 kN of normal force. This amount of power makes it possible to make high quality, gapless metallic parts.

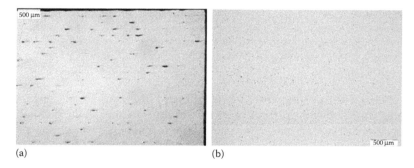

(a)   (b)

Figure 17.4 (a) Metallurgical section of aluminum 3003 build showing gaps left by 1 kW UAM system and (b) metallurgical section of gapless aluminum 6061 build fabricated with the 9 kW UAM system shown in Figure 17.3.

plate is utilized to further intensify softening and enhance weldability. However, quality, void-free welds can be achieved for many aluminum alloys with no additional heating, as shown by way of example for Al 6061 [9].

UAM joining of relatively soft alloys including Al 3003 and Cu 1100 has been extensively studied [7–12]. These alloys are known to react well to the ultrasonic metal-welding process and have been shown to be compatible with the UAM process. More recent work performed on 9 kW systems has successfully demonstrated the fabrication of Al 6061 [9] builds and Al/Ti composites [13]. However, how harder materials such as iron alloys respond to the UAM process is not yet fully understood, in part due to current difficulties in fabricating iron-based UAM builds. Since these alloys do not deform easily, higher normal force and ultrasonic power are needed to achieve successful UAM joints. The increased mechanical rubbing, combined with an affinity to the steel sonotrode, leads to foils welding to the sonotrode, known as nuggets. This technological barrier may be addressed by brazing certain ceramic materials to the sonotrode, reducing the affinity for bonding to the sonotrode, and therefore increasing the likelihood of generating viable welds.

17.1.1 UAM microstructure

The microstructure of UAM built structures has been extensively investigated using optical microscopy, scanning electron microscopy (SEM), focused ion beam (FIB), and transmission electron microscopy (TEM) techniques [11,14–16]. Such investigations have shown that within approximately 15 μm of the bond interface region, an area of small recrystallized grains exists [11,15,17]. Figure 17.5 shows an electron backscatter diffraction image of the bond interface in a UAM sample. Within the bond interface, small, mostly equiaxed grains are observed, whereas the bulk of the foils shows an elongated microstructure from the original rolling texture. Similarly, the polar mapping of the grain orientations shows a distinct texture in the bulk due to rolling, whereas the interface region is much more equiaxed. TEM measurements showing this effect have been presented by Johnson [14]. These observations indicate that a small recrystallized zone exists within approximately 15 μm of the bond interface, whereas the remainder of the bulk material has the same microstructure as the as-received foils.

The equiaxed grains at the bond interface indicate that a highly localized process of recrystallization has occurred due to deformation and limited heating during processing [10]. The equiaxed grain structure within the interface has a 111 <110> shear texture [15]. A shear texture of this type is expected to have developed through the scrubbing action of the sonotrode and the deformation of microasperities at the interface of each material.

A bond theory of the UAM process is developed based on these findings, which is described as follows. The scrubbing action delivered to the bonding interfaces via the sonotrode creates plastic deformation, collapses asperities, removes surface oxides and contaminants from the faying surfaces, and creates nascent metal surfaces that instantaneously bond under sufficient normal force. Figure 17.6 illustrates this bond progression. In steps 1–3, microasperities are present or formed on the surface of the foil. Dynamic recrystallization occurs through the deformation and heating of this surface after sonotrode contact. During steps 4–6, an additional tape is being welded. During the weld process, dynamic recrystallization occurs by shear deformation of the microasperities on the material surfaces. Therefore in the interface region, a fine-equiaxed grain structure is formed in the microasperity locations. The interface region then expands by static recrystallization,

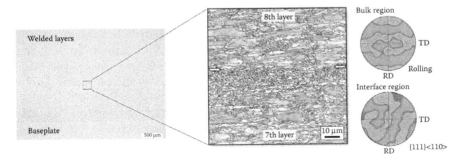

Figure 17.5 Electron backscatter diffraction (**EBSD**) image showing grain maps colored by inverse pole figure along with pole figure indicating strong rolling texture in bulk region and recrystallized texture in interface region. (From H. Fujii, M. Sriraman, and S. Babu, *Metallurgical and Materials Transactions A*, **42A**, 4045–4055, 2011.)

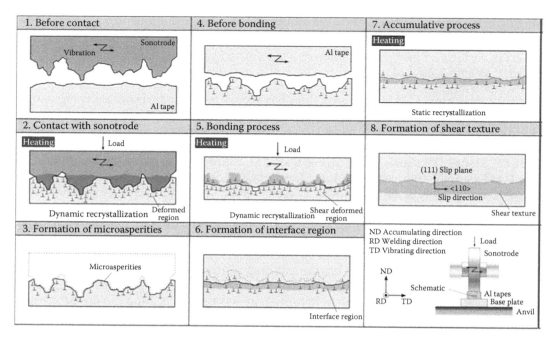

Figure 17.6 Schematic illustration of the microstructural evolution of the UAM process. Steps 1–3 show formation of microasperities due to plastic deformation of the top surface due to sonotrode contact, steps 4–6 show the bond formation process, and steps 7–8 show the effects of cumulative loading. (From H. Fujii, M. Sriraman, and S. Babu, *Metallurgical and Materials Transactions A*, **42A**, pp. 4045–4055, 2011.)

whereas additional tapes are subsequently welded, as shown in steps 7–8. Further, a shear texture is developed during this cumulative process which is oriented with the vibrating direction.

17.2   Example builds and components

Due to the low temperatures involved, the UAM process is a proven method for creating unique components including smart structures, thermal management devices, parts with embedded fibers, selectively reinforced parts, and dissimilar material joints. The temperatures of the UAM process are on the order of 150°C, well below the critical temperatures of smart materials, enabling incorporation into metallic structures without degradation of their properties as in fusion-based processes [5,17]. An example build incorporating Galfenol into aluminum is shown in Figure 17.7. Galfenol, an alloy of iron and gallium, exhibits moderately high magnetostriction (magnetic field-induced strain) and magneto-elasticity, whereby the material changes its magnetization when stressed. These responses are used to design sensors and actuators with fast dynamic response, few to no moving parts, and compact operation. Galfenol withstands combined mechanical loads, tension, and shear, and it can be machined and formed using conventional means. When implemented into aluminum structures via UAM, the Galfenol element is shielded from outside factors and the resulting robust composites can be used as contact-less sensors, electrically tunable variable resonators, and solid-state actuators. Figure 17.8a and b shows model calculations for Galfenol composites, where the effect of Galfenol volume fraction on the normalized natural frequency is shown as a function of magnetic field. Figure 17.8c

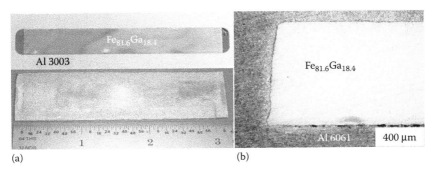

Figure 17.7 (a) Galfenol composite before and after UAM integration and (b) optical microscopy image of a Galfenol composite's cross section.

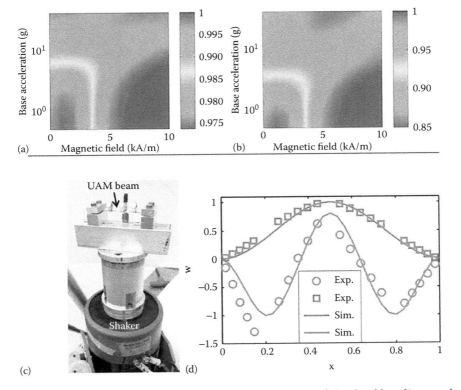

Figure 17.8 (a) Model results for normalized resonant frequency of the third bending mode at 10% Galfenol volume fraction, (b) 46% Galfenol volume fraction, (c) image of setup for testing Galfenol–aluminum beam, and (d) comparison of mode shapes from model and experimental results.

and d shows the test setup and performance of a Galfenol–aluminum composite under a mechanical load showing comparisons of experimental results with model simulations for the third bending mode. Testing was conducted using a mechanical shaker to induce specified vibration modes and measured with a laser vibrometer.

Another smart material system that can be integrated into metal matrix composites with UAM is Ni–Ti. Shape memory Ni–Ti, or Nitinol, exhibits large strains under thermal

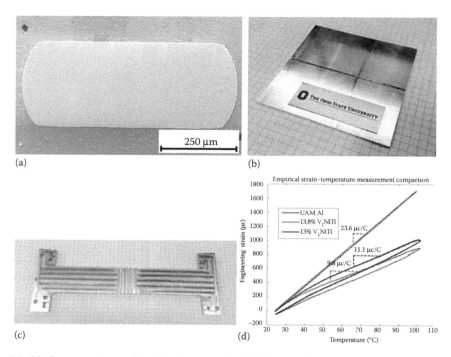

Figure 17.9 (a) Cross section of Ni–Ti wires embedded into Al 6061, (b) solid-state Ni–Ti hinge, (c) Ni–Ti beam for coefficient of thermal expansion mitigation, and (d) performance of Ni–Ti composite during thermal loading showing decreased thermal expansion with higher volume fraction.

loading due to a solid-state phase change to and from a memorized shape. The material is able to withstand very large elastic strains and actuate under thermal loads. These characteristics make Ni–Ti a suitable material for small, solid-state hinges and can also be used in the creation of composites with low coefficient of thermal expansion. Examples of Ni–Ti composites are shown in Figure 17.9. The cross section in Figure 17.9a shows the complete integration of a Ni–Ti ribbon into an aluminum matrix. A solid-state hinge is shown in Figure 17.9b, whereas the beam shown in Figure 17.9c is an aluminum and Ni–Ti composite which exhibits decreased thermal expansion when heated. Results of testing this composite are presented in Figure 17.9d showing the change in thermal expansion with Ni–Ti volume fraction. As the composite heats up, the aluminum matrix expands, whereas the Ni–Ti elements contract as they undergo a phase change from martensite to austenite, therefore leading to a net decrease in thermal expansion for the overall composite structure.

In addition to metallic smart materials, polymers and fiber optics can be integrated into composites using the UAM process. Figure 17.10 shows beta-phase polyvinylidene fluoride (PVDF), an active polymer with extremely high-frequency response and sensitivity, embedded in aluminum for purposes of integrated impact detection. When the surface of the aluminum plate is struck by an object, the stress that propagates through the structure creates a polarization change in the PVDF element that results in a voltage across thin electrodes deposited on the polymer. An insulating layer such as Kapton is used to prevent electrical conductivity between the electrodes and aluminum matrix. The frequency response of commercial PVDF embedded via UAM has been shown to be in the MHz range [18].

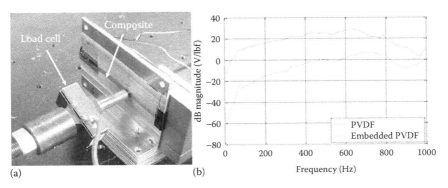

(a)   (b)

Figure 17.10 (a) Impact detection concept using PVDF embedded into aluminum using UAM and (b) response of PVDF sensor in air and embedded. The amplitude difference is caused by attenuation of the applied stress field due to damping in the host material.

(a)   (b)   (c)

Figure 17.11 (a) Cross section of fiber optic wire embedded into Al 6061, (b) image of fiber Bragg grating embedded into aluminum and strain gage attached to surface, and (c) comparison of responses of fiber optic sensor and strain gage under a step load input.

Implementation of fiber optics into components can be done for *in situ* health monitoring without affecting the structural properties of the component being monitored. An example build with an embedded fiber optic sensor is shown in Figure 17.11. The sensor is a fiber Bragg grating which detects very small displacements through frequency shifts in monochromatic light waves passing through the sensor. The fiber is completely encapsulated in the surrounding aluminum matrix without suffering degradation of mechanical or optical properties. Comparison of the sensor performance to a strain gage is shown in Figure 17.11c. The two signals track extremely closer to one another, indicating the sensor is unaffected due to the integration into the aluminum matrix. Use of these types of devices can allow for longer part duration, as the component performance can be monitored throughout its lifetime.

State of the art UAM systems are built within a CNC machining framework, controlling table motion and maintaining the ability to perform machining operations. The ability to conduct welding and milling operations in tandem allows for creation of components with unique channel geometries built into the structure. An example device is shown in Figure 17.12, in which channels traverse throughout the build. The image is an X-ray of the build showing the channels as darker areas on the image. UAM thus makes it possible to create unique thermal management devices with cooling channels that are difficult to achieve using other processes, in addition to dissimilar materials with suitable thermal properties.

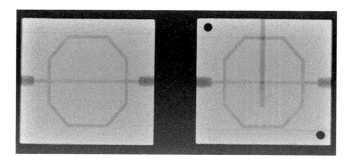

Figure 17.12 X-ray of UAM build with internal channels for thermal management.

(a)   (b)

Figure 17.13 (a) X-ray of UAM component with a complex array of cooling channels in aluminum and (b) cross section of channels traversing an aluminum build. (Courtesy of Fabrisonic LLC.)

Additional examples of embedded features are shown in Figure 17.13. Figure 17.13a shows an X-ray of a device with conformal cooling channels incorporated throughout the structure, whereas Figure 17.13b shows a part with many channels traversing through the aluminum structure. These types of components can be used for highly efficient, localized cooling in applications for power generation, electronics, manufacturing, and other industries.

The typical joint configuration in the UAM process uses thin foil, on the order of 0.005 in. thick, welded onto a baseplate or previous foil layers. However, recent developments have enabled the joining of aluminum sheet material 0.076 in. thick using a scarf joint geometry [3]. The concept uses UAM equipment to create a seam weld of two aluminum sheets, illustrated in Figure 17.14a. By utilizing a scarf joint configuration and welding on both sides of the sheet, seamless joints can be achieved with properties similar to bulk material. A schematic of the configuration along with a cross section from a joint is shown in Figure 17.14b, where no voids are apparent. This design effectively demonstrates that the UAM process can be used for joining of sheet material with properties matching bulk material, resulting in parts with no protrusions into flow fields exhibited by mechanical fasteners such as rivets. A key advantage of this approach over friction stir welding is the lower capital cost involved.

Similar to the concepts introduced with smart materials, passive materials can be incorporated into metal matrix structures as well. Passive fibers such as carbon fiber can be incorporated into aluminum utilizing the UAM process. Examples are shown in Figure 17.15. In Figure 17.15a, a carbon fiber composite is shown integrated with aluminum, while carbon nanopaper embedded into an aluminum matrix is shown in Figure 17.15b.

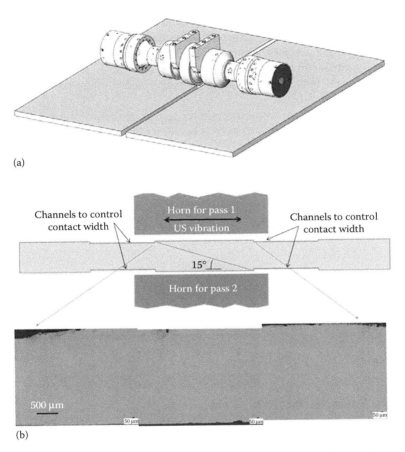

(a)

Channels to control contact width

Horn for pass 1
US vibration

Channels to control contact width

15°

Horn for pass 2

500 μm

50 μm

50 μm

50 μm

(b)

Figure 17.14 (a) UAM-based approach for butt joining sheet material and (b) schematic and cross section of a viable joint created using the process.

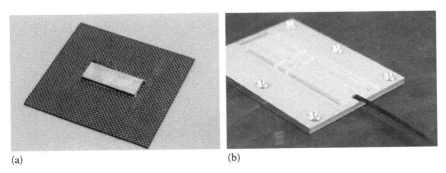

(a)

(b)

Figure 17.15 (a) Carbon fiber incorporated with aluminum build using UAM and (b) carbon nano-paper embedded into an aluminum matrix using the UAM process.

The integration of passive fibers or wires can be used for selective reinforcement. High stress areas of components can be reinforced, whereas areas of little to no load can remain thin for lightweighting purposes. One such material used for localized reinforcement is MetPreg, a metal matrix composite prepreg made of an aluminum matrix and continuous alumina ceramic fibers. This material can be readily welded using the UAM process, with examples shown in Figure 17.16.

Other examples of reinforcement include the incorporation of high-strength steels into an aluminum matrix. A cross section from a build with a stainless steel mesh embedded into aluminum is shown in Figure 17.17.

Similar to the MetPreg composites, this construct can allow for selective property control. Properties such as stiffness and strength can be controlled within a single component, allowing gradients of functionality where specific needs are required.

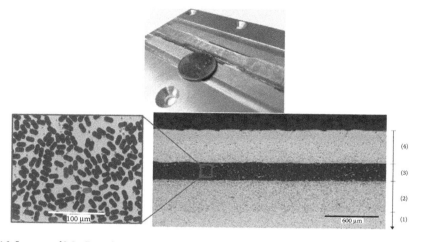

Figure 17.16 Image of MetPreg build in aluminum and cross section of a UAM build using MetPreg, a metal matrix composite prepreg. Sections on the macrosection are described as (1) baseplate material, (2) four layers of Al 6061 tape, (3) one layer of MetPreg, and (4) four layers of Al 6061 tape.

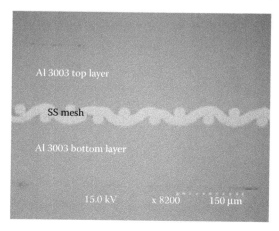

Figure 17.17 Cross section of UAM build with stainless steel mesh embedded for improved strength and stiffness. (From G.D. Janaki Ram, C. Robinson, Y. Yang, and B.E. Stucker, *Rapid Prototyping Journal*, **13**(4), 226–235, 2007.)

The UAM process is a proven method of creating dissimilar material joints. It was established before in this article that because the process operates at low temperatures compared to fusion-based methods, diffusion is inhibited, therefore preventing the formation of brittle intermetallic layers at the bond interface. This key property of the UAM process has motivated trials on new material combinations, demonstrating that numerous material combinations are possible and many more are expected to be viable. Example uses of dissimilar joints include lightweight armors, cladding of high-value materials onto standard materials, and materials with tailored properties. A laminate of aluminum and titanium is shown in Figure 17.18 for a ballistic armor application. This construct enables high ballistic resistance while reducing weight overall compared to thick steel armors.

An example of cladding is presented in Figure 17.19, where aluminum foils are welded onto a stainless steel substrate. These types of joints are of growing importance in the integration of lightweight aluminum components into larger substructures. Although aluminum and stainless steel are highlighted, many other combinations have been proven, including Al/Cu, Al/Zn, Cu/Ni, Al/Ag, Cu/Ag, Mo/Al, Ta/Al, and Ni/stainless steel [19–22].

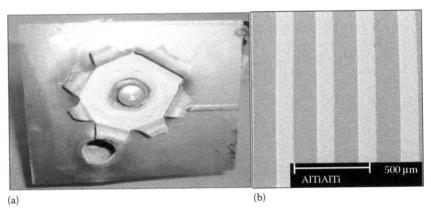

(a) (b)

Figure 17.18 (a) Al–Ti laminate armor created using UAM and tested via ballistic impact and (b) cross section of Al–Ti laminate armor. (Courtesy of Fabrisonic LLC.)

Figure 17.19 Image of aluminum foils welded onto a stainless steel substrate. (Courtesy of Fabrisonic LLC.)

17.3 UAM process property relationships in aluminum alloys

17.3.1 Process parameter optimization

Process parameters that are commonly controlled to affect the build quality of the UAM process are weld force, weld speed, weld amplitude, and baseplate temperature. These parameters must be optimized in order to create components with maximum build strength. In order to determine optimal process parameters, a study was performed joining Al 6061-H18 foil using the 9 kW UAM process [9]. A design of experiments (DOE) approach was used such that multiple processing parameters could be explored simultaneously. The study uses fully work-hardened Al 6061 which was purchased in the annealed heat treatment and fully work-hardened cold working condition, known hereafter as Al 6061-H18. Al 6061 was chosen due to its frequent use in industry and strong compatibility with UAM. Samples were manufactured on a Fabrisonic SonicLayer 4000 9 kW UAM machine. The machine is fully automated and includes CNC and laser-machining capabilities to complement the additive ultrasonic welding stage.

The DOE approach determines the optimal process parameters through mechanical strength testing of multiple build strips manufactured with varying process parameters. Build strips were generated for the DOE following a Taguchi L18 design matrix by varying the temperature, weld force, weld amplitude, and weld rate. The design matrix is shown in Table 17.1 with the 1, 2, and 3 designations indicating the low, medium, and high levels, respectively, for each of the parameters within a treatment combination. This type of experimental design allows for investigation of the effect that each parameter has on mechanical strength in a minimal number of experimental runs. The exact levels for each

Table 17.1 Taguchi L18 orthogonal array including 18 treatment combinations and three levels (low, medium, and high) for each of the four parameters investigated: temperature, weld force, weld amplitude, and weld rate

Treatment combination	Temperature	Weld force	Amplitude	Weld rate
1	1	1	1	1
2	1	1	2	2
3	1	1	3	3
4	1	2	1	1
5	1	2	2	2
6	1	2	3	3
7	1	3	1	2
8	1	3	2	3
9	1	3	3	1
10	2	1	1	3
11	2	1	2	1
12	2	1	3	2
13	2	2	1	2
14	2	2	2	3
15	2	2	3	1
16	2	3	1	3
17	2	3	2	1
18	2	3	3	2

of the parameters were determined from a pilot study, which established the build envelope of parameter levels for the study. These parameters are given in Table 17.2. The lower
limit signifies levels of parameters where welds could not occur, whereas the upper limit
indicates when the foils would weld to the sonotrode as opposed to the previous layer.
This DOE methodology has been applied to the UAM process to determine optimal process parameters in 1 kW UAM for Al 3003, titanium to Al 3003, and stainless steel alloys,
proving an effective method of determining the best process parameters for mechanical
strength [7,19,23].

Weld strips were built following the Taguchi matrix on four baseplates, with nine strips
welded onto each plate in 15/16 in. (23.81 mm) wide strips. Temperature was held constant
during welding at either room temperature or 200°F, whereas the location on the plate was
randomized for each parameter set. Push-pin tests were conducted after the samples were
built. Further details on push-pin testing were provided by Zhang et al. [24]; subsequent
studies in which this type of testing was directly applied to UAM include, for instance,
Truog [25]. For the tests, 20 layers were welded onto a 12.7 mm (0.5 in.) thick baseplate with
the weld strips built such that four test specimens could be machined from each strip. An
example baseplate with UAM welds is shown in Figure 17.20. Utilization of solid baseplate
material in the sample designs reduced the required number of layers, thus expediting the
testing. Push-pin testing was conducted using a Gleeble thermal–mechanical system, where
a pin was pressed into the sample while load and displacement of the frame were recorded.

Results from a representative push-pin treatment combination for a good bond are
shown in Figure 17.21. A poor bond implies that the failure is predominately driven by
delamination between layers, whereas a good bond implies that the failure is predominately driven by tensile failure of the layers. In the latter case, the pin presses through the
layers forcing a failure which propagates through the foils, rather than by delamination.

Following mechanical testing, an analysis of variance (ANOVA) was performed on
the measured data. The ANOVA is used to examine three or more variables, in this case
the four parameters listed in Table 17.2, for statistical significance within a process. Main
effects' plots are then used to indicate the optimal levels of the parameters for mechanical
strength. The ANOVA uses a generalized linear model with four main effects, with the
model equation given by

Figure 17.20 Test strips from push-pin sample manufacturing.

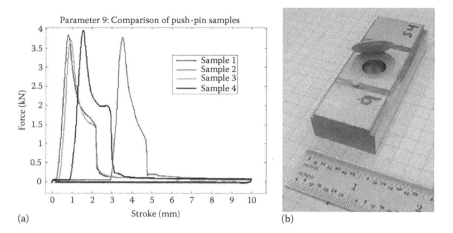

Figure 17.21 (a) Push-pin results for parameter set 9 representing good interlaminar failure and (b) sample built with parameter set 9.

Table 17.2 Parameter levels for each of the DOE treatment combinations

Parameter	Level 1	Level 2	Level 3
Temperature	22.2°C (72°F)	93.3°C (200°F)	–
Force	4000 N	5000 N	6000 N
Amplitude	28.28 µm	30.47 µm	32.76 µm
Speed	84.6 mm/sec (200 in./min)	95.2 mm/sec (225 in./min)	105.8 mm/sec (250 in./min)

$$Y_{ijkl} = \mu + \alpha_i + \beta_j + \gamma_k + \delta_l + \varepsilon_{ijkl} \tag{17.1}$$

This linear equation models the dependence of the response variable, Y_{ijkl}, on the levels of the treatment factors [26]. In Equation 17.1, μ is the overall average of the response variable (in this case, push-pin strength), and α_i, β_j, γ_k, and δ_l represent the effects of each of the process parameters on the mean response. In this case, α_i is the effect of temperature at the ith level on the response, while the other factors are fixed. Similarly, β_j, γ_k, and δ_l represent the effects of weld force, amplitude, and weld rate at the jth, kth, and lth levels, respectively, while the other factors are fixed. The error variable, ε_{ijkl}, is a random variable with normal distribution and zero mean, which denotes any variation in the response unaccounted for by the main four process parameters.

ANOVA results are given in Table 17.3 using the area under the force-displacement curve representing mechanical work as the response variable. In a statistical analysis, the p value represents the probability of obtaining a test at least as extreme as the one observed, assuming that the null hypothesis of no trend or no effect is true; p values less than 0.05 were chosen to indicate that a particular source of variation is statistically significant in the process. This means that a source of variation has a 95% likelihood of being a statistically significant influence on the process. In this case, amplitude and speed are considered significant with p values lesser than 0.001 and 0.007, respectively. Both temperature and force have p values greater than 0.05 and are therefore considered statistically insignificant.

Table 17.3 ANOVA results for push-pin testing using mechanical work as the response variable

Source	DF	Adj SS	Adj MS	F-ratio	p-value
Temperature	1	0.4018	0.4018	0.94	0.337
Weld force	2	0.3689	0.1845	0.43	0.652
Amplitude	2	19.1955	9.5977	22.39	<0.001
Weld speed	2	4.5869	2.2934	5.35	0.007
Error	64	27.4299	0.4286	–	–
Total	71	51.9830	–	–	–

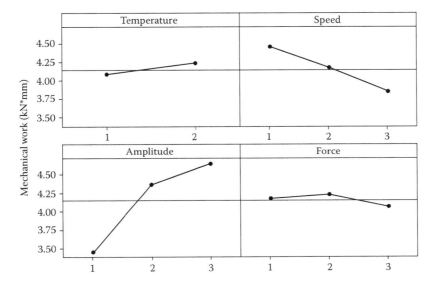

Figure 17.22 Main effects plot of push-pin test for each factor.

Main effects' plots shown in Figure 17.22 visually confirm the ANOVA results. The amplitude plot shows a significant increase in mechanical work with increasing amplitude, whereas the mechanical work decreases as the speed increases. By comparison, the temperature and force plots indicate very little change in response depending on their level. These results indicate that higher mechanical strengths are achieved with increases in amplitude, decreases in speed, and are not dependent on temperature and force within the levels tested in this study.

As seen in these results, amplitude is the driving and most sensitive variable for bond quality. This result is expected to be from an enhanced scrubbing action which more effectively disperses oxides and contaminants away from the interface, in turn improving the strength of the interface by increasing the density of metallic bonding. This trend cannot necessarily be extrapolated because defects may be introduced within the structure at higher amplitudes; yet this variable appears to have a critical correlation with the mechanical strength of UAM builds within the levels tested.

Speed was also found to have a statistically significant effect on strength. A slower speed allows additional time for scrubbing of the interface and therefore increased ultrasonic energy supplied to the interface for welding. As a result, enhanced dispersion of oxides and contaminants at the interface can be achieved by decreasing the weld speed. Similar to the amplitude observation, it is not known if there is a point of diminishing returns for decreases in speed.

Table 17.4 Optimal weld parameters for Al 6061-H18 as determined by analysis of push-pin tests. The optimization was performed on a Fabrisonic SonicLayer 4000 system; these results do not necessarily apply to other UAM systems

Parameter	Level
Temperature	RT to 93.3°C (200°F)
Force	4000–6000 N
Amplitude	32.8 μm (70%)
Speed	84.6 mm/sec (200 in./min)

Based on this work, optimal weld parameters for Al 6061-H18 foil material in the 9 kW UAM process are presented in Table 17.4.

17.3.2   Process improvements studies

17.3.2.1   Foil overlap and stacking

To create builds wider than 1 in., the UAM process requires abutting foil tapes next to one another, which in turn creates a source for void formation, as shown by Obielodan et al. (2010) [27]. Overlapping of tapes can minimize or prevent void formation at these abutments, but the build surface becomes less uniform due to accumulation of material at the seam locations. This effect is illustrated in Figure 17.23 [44]. Another production factor that must be addressed is the stacking sequence, or stagger, of layers as they progress higher in the build. If the tapes are all aligned at the same location, the possibility of voids running through a single area greatly increases. Therefore moving the seam location in a brick-like fashion to create a less direct crack path, should a crack develop, is ideal. Many methods of stagger can be implemented, but typical patterns use an ordered, or random layup, as shown in Figure 17.23.

UAM builds were constructed to investigate the effect of tape to tape overlap and stacking sequence on strength. Al 6061-H18 foils that were 0.006 in. (0.1524 mm) thick and 1 in. (25.4 mm) wide were used, built onto an Al 6061-T6 baseplate. The weld parameters used for the builds follow previously optimized parameters by Wolcott et al. [9] for Al 6061 presented in Table 17.4. All builds were performed at room temperature using a sonotrode with a roughness of 7 μm R_a.

Two build plates were used to investigate each of the effects with each of the sample sets summarized in Table 17.5. In plate 1, the stacking remained constant, while the tape to tape overlap (α in Figure 17.23) was varied from 0.0015 in. (0.038 mm) to 0.0045 in. (0.1143 mm). In the SonicLayer system, this is achieved using a constant 1 in. (25.4 mm) wide tape and setting the tape width to varying levels. The specified tape overlaps were selected at levels which provide overlap while minimizing ash, or excess material, at the abutment points. The ordered stacking sequence followed a 0,1,0,–1... pattern as shown in Figure 17.23, with the amount of stagger given by β. In plate 2, the tape overlap was held constant while the stacking sequence was varied. Samples A and B were built using similar 0,1,0,–1,0 ordered stacking sequences with varying amounts of stagger. Build C had random stacking with a maximum stagger value of β = 0.3 in. (7.62 mm). Build D had ordered stacking with 50% stagger from tape to tape. Stacking sequences were selected such that both randomized and ordered sequences were investigated, and the entire design space of stagger values was covered.

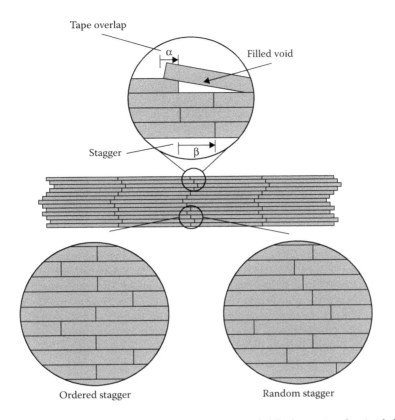

Figure 17.23 Tape overlap and stagger with potential void filled in via plastic deformation of the tape and schematics for ordered and random stagger sequences. (From P. Wolcott, A. Hehr, C. Pawlowski, and M. Dapino, *Journal of Materials Processing Technology*, **233**, 44–52, 2016.)

Table 17.5 Overlap and stacking sequence prescribed for each sample set

Sample set	Overlap (α in in.)	Stacking sequence (β in in.)
1A	0.0015	Ordered with $\beta = 0.15$
1B	0.0025	Ordered with $\beta = 0.15$
1C	0.0035	Ordered with $\beta = 0.15$
1D	0.0045	Ordered with $\beta = 0.15$
2A	0.003	Ordered with $\beta = 0.1$
2B	0.003	Ordered with $\beta = 0.15$
2C	0.003	Random with $\beta = 0.3$ at maximum
2D	0.003	Ordered with $\beta = 0.497$ (50% stagger)

Tensile samples were sectioned from the blocks using the CNC stage on the machine following ASTM subsize tensile sample dimensions [28] such that testing occurred across the various tape interfaces, transverse to the welding direction. Tensile tests were performed on a 22 kip (98.7 kN) Interlaken 3300 test frame, using a displacement rate of 0.05 in./min (1.27 mm/min) while recording the load to failure.

Table 17.6 Ultimate tensile strength results for overlap and stagger study builds

	UTS (MPa)							
	1A	1B	1C	1D	2A	2B	2C	2D
	133.1	121.7	202.3	227.1	225.5	185.1	221.7	184.2
	117.7	129.8	214.7	178.8	177.5	196.3	223.9	163.7
	124	144.2	211.4	228.2	185.1	167.7	222	153.6
Mean	124.9	131.9	209.5	211.4	196	183	222.5	167.2
St Dev	7.7	11.4	6.4	28.2	25.8	14.4	1.2	15.6

The test results are summarized in Table 17.6. From the results of plate 1, there is a clear delineation between the first two samples (A, B) and the last two samples (C, D), with samples C and D producing strengths of approximately 210 MPa on average versus 125 MPa for samples A and B. Results of test data from plate 2 indicate that sample C resulted in the highest tensile strength, 222.5 MPa on average, compared to the other samples.

Based on these results, it is recommended that for UAM block builds, a tape to tape overlap of at least 0.0035 in. (0.0635 mm) should be used and that the stacking sequence follows a random stacking with maximum stagger of 0.3 in. (7.62 mm). The optimal overlap value of at least 0.0035 in. is approximately half the height of the original foil thickness (0.006 in.). This could represent a threshold value whereby the plastic deformation of the foil is able to completely fill the void at the abutting area. Further work using varying thickness foils could be performed to test this hypothesis.

Randomized stacking is shown to produce the highest strength results. It is hypothesized that this is due to the configuration producing a more tortuous crack propagation path. If the abutting points are assumed to be the crack initiation points, failure in the ordered structure would require transmission through only a single layer at a time. In the randomized case, instances of the failure would have to traverse multiple layers, leading to a more complete crack arresting mechanism.

Tape to tape overlap results are consistent with those found by Obielodan et al. [27] in 1 kW UAM, who recommended overlaps of at least 0.00275 in. (0.07 mm). Stacking sequences recommended by Obielodan et al. use a 50% stagger; however, only two stacking methods were investigated, while the study presented here investigated four separate stacking sequences. Of note, these recommendations are based on the testing performed here. A globally optimal value may be achieved through further optimization of these parameters. The recommended stacking sequence proposed is based on the findings from sample 2C, indicating that randomized stacking should be used. However, the magnitude of the proposed stagger may not scale in taller builds where tape flash at the build edge creates areas of poor support leading to inconsistent welds at the build edge. This effect can propagate inward as a build progresses higher, making further welds near the edges difficult. In these instances, a random stacking pattern with smaller stagger should be used.

17.3.2.2   *Effects of surface roughness*

Periodic flattening passes conducted throughout a build using the CNC stage in state of the art UAM systems can remove excess material at seams due to tape overlap. However, inhomogeneities are created within the build due to the smoothly machined surface following these flattening passes. Consequently, it is necessary to understand how welding onto smooth and textured surfaces affects bond quality.

To study this effect, builds were conducted with welds onto smooth, freshly machined surfaces and welds onto roughened surfaces. Flattening passes were performed using a carbide insert shell mill within the CNC stage of the SonicLayer 4000 UAM machine. Roughened surface samples were built onto surfaces which were textured by vibrating the sonotrode at a low amplitude, similar to weld operations. The roughness of the machined surface was 0.12 μm R_a, and the roughened surface was 5.7 μm R_a, measured with a Mitutoyo mechanical probe profilometer. Each build consisted of 20 total layers, such that five flattened surfaces were introduced into each build. All builds were constructed on a 0.5 in (12.7 mm) thick Al 6061-T6 baseplate with Al 6061-H18 foils 1 in. (25.4 mm) wide and 0.006 in. (0.1524 mm) thick. Sample strength was measured via push-pin testing to compare delamination strength and resistance.

Push-pin test results are provided in Table 17.7; of note, one textured sample was damaged during test setup and was not tested. The mean peak force for the textured samples (4.42 kN) is similar to the nontextured samples (4.45 kN). However, there is a measurable difference in how the two types of samples are measured for mechanical work. The textured samples exhibit an average push-out energy of 5.51 kN-mm compared to 4.75 kN-mm for the nontextured samples, indicating that a larger amount of energy is necessary to produce failure in the textured material.

The improvements in weld properties due to surface texturing are believed to originate at the weld interface. It is hypothesized that the increased surface roughness after texturing enhances asperity deformation during welding leading to increases in plastic deformation, oxide dispersal, mixing, and the driving force for dynamic recrystallization. In combination, these factors increase the potential for grain growth across the bond interface leading to improved metallurgical bonding [11]. To further investigate this phenomenon, in-depth characterization of the grain structure at the interface will be required.

A second study involving surface roughness was performed to determine the effect of sonotrode roughness on weld quality. In this study, samples were fabricated using sonotrodes of 7 μm and 14 μm R_a roughness, respectively. Both of these sonotrodes were textured with electrical discharge machining to create the desired surface profile. Two 12 in. (30.5 cm) long, 1 in. (25.4 mm) wide build strips, 20 layers tall were constructed using the two different sonotrodes, each with identical weld parameters. The strips were built onto a 0.5 in. thick Al 6061-T6 baseplate, yielding eight total push-pin samples.

Push-pin results comparing the two sonotrode roughnesses are shown in Figure 17.24 and Table 17.8. As seen in the figure and table, the 14 μm samples exhibit a larger peak force until failure and a similar push-out energy compared to the 7 μm samples.

Table 17.7 Push-pin data for textured and nontextured builds

Sample	Textured		Nontextured	
	Peak force (kN)	Energy (kN*mm)	Peak force (kN)	Energy (kN*mm)
1	4.66	5.32	4.28	3.82
2	4.18	5.47	4.66	5.50
3	4.42	5.74	4.51	5.58
4	–	–	4.35	4.11
Mean	4.42	5.51	4.45	4.75
St Dev	0.20	0.17	0.15	0.80

Table 17.8 Averaged results of push-pin testing with varying roughness sonotrodes

7 µm Roughness		14 µm Roughness	
Max force (kN)	Energy (kN-mm)	Max force (kN)	Energy (kN-mm)
4.9	7.3	5.8	7.7

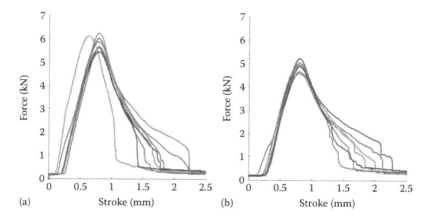

Figure 17.24 Push-pin results: (a) 14 µm R_a and (b) 7 µm R_a.

Based on these measurements, enhanced bond quality can be achieved using a 14 µm R_a roughness sonotrode compared to a 7 µm R_a sonotrode. Li and Soar [29] also noted that rougher sonotrode surfaces produced better bonds when other processing conditions were kept the same. They also reported that rougher surfaces increase the void concentration at weld interfaces. However, interface void presence is essentially nonexistent with 9 kW UAM if appropriate machine settings are used. If this is not the case, for example, for sonotrodes rougher than 14 µm R_a or low weld amplitudes and normal forces, interface voids can still form. Consequently, selecting an optimal surface roughness for a given material or UAM welding application is necessary. Friel et al. [30] discussed this matter as well.

The improved bond quality associated with rougher sonotrodes is likely to originate in the consolidation at the interface, similar to the effect seen in roughened versus smooth surfaces. Due to the creation of larger asperities, more plastic deformation may occur, which would enhance the bonding mechanisms of oxide fracture, dispersal, and increase the driving force for dynamic recrystallization. It is suspected that peak force during push-pin testing is enhanced from the 14 µm sonotrode roughness because it increases the resistance to initial crack formation, whereas push-out energy is largely unaffected because it is a measure of the resistance to crack propagation. Further work using mechanical testing and microscopy is required to understand these failure energy differences.

17.3.2.3   *Effect of heat treatments*

Heat treatments on UAM components have shown an ability to improve mechanical and microstructural properties [9,31]. Because Al 6061 is a heat-treatable alloy, it is necessary to determine the strength improvements achievable on postprocessed components. To test the effects of heat treatments on out-of-plane UAM tensile strength, a 0.8 in. (20.32 mm) tall UAM block was fabricated using Al 6061-H18 foils 1 in. (25.4 mm) wide and 0.006 in. (0.1524 mm) thick. This build was constructed using the weld parameters from Wolcott et al. [9] shown in Table 17.4 with a 7 μm horn and no tape to tape overlaps or flattening passes. From this block, nine cylindrical samples were sectioned using wire electrical discharge machining. These samples were sectioned avoiding any seam locations. Three samples were annealed, three were treated to a T6 condition, and three samples were tested as-built. The specific heat treatment temperatures and settings as described by ASM standards are summarized below.

For annealing (O): Heat to 413°C for 2.5 hours, cool at 1°C/min until 280°C, and then air cool [32]. For T6: Heat to 530°C for 1 hr to solutionize, quench in water, and heat to 160°C for 18 hours [32]. H18: indicates an as-built condition. Following heat treatment, samples were machined via CNC lathe to final dimensions for tensile testing. The dimensions for the specimens are based on ASTM standards [28]. Machined specimens were then tested in tension using a 22 kip (98.7 kN) Interlaken 3300 test frame with displacement rate of 0.05 in./min (1.27 mm/min).

Out-of-plane ultimate tensile strength results are summarized in Table 17.9. Comparisons to initial foil stock in each of the H18, T6, and O conditions are presented. H18 comparisons used as-received stock tested in tension with no heat treating, while foils for the annealed and T6 references were processed using the same heat treatment as the samples from the UAM block build. Of note, the elongation values provided are not exact as they represent deflection of the entire load frame measured by the linear variable differential transformer (LVDT). However, the displacement values can provide useful comparative evaluations. Use of an extensometer was not possible due to the small sample size of the specimens. Results show a significant improvement in strength with heat treatment. The T6-treated samples exhibit strength almost 90% of reference material, while annealed samples exhibit strength nearly the same as reference at 97%.

Overall, these results indicate that the mechanical properties of UAM structures can be enhanced considerably when a postprocess heat treatment is applied. This observation coincides well with the microscopy work of others in Al 3003. In particular, Sojiphan et al. [31] observed that the recrystallized grain structure at weld interfaces in optimized aluminum UAM builds was very stable after heat treating. This stable microstructure results in less defects and defect nucleation sites, which in turn improves mechanical properties. It was also observed that significant recrystallization and grain growth occurred in the bulk weld foil after heat treating. Heat treating also enhances precipitate distribution and concentration in Al 6061. Consequently, strength improvements are suspected to be a combination of improved precipitate density and microstructure stability.

Table 17.9 Comparison of UAM tensile strength for various postprocess heat treatments with comparisons to solid material references

Group	Avg. UTS (MPa)	Avg. Elo. (%)	Ref. UTS (MPa)	Ref. Elo. (%)	UTS (%)	Elo. (%)
H18	135.6	1.4	266.1	3.1	51	45
T6	300.3	13.1	337.3	12.5	89	105
O	117.1	13.7	121.1	18.6	97	74

17.3.3   Elastic modulus

Prior research has shown that the failure strength of UAM parts depends on the testing direction with respect to foil orientation due to the presence of interfacial voids [12]. Thus, it is likely that interfacial voids not only have an effect on failure strength, but also have an effect on the elastic properties of UAM components. Therefore, there is a necessity for research focusing on the measurement of the elastic constants in the three material directions (rolling direction, vibration direction, and transverse direction) and the characterization of how interfacial voids affect these elastic constants.

Elastic constants can be measured by mechanical testing. However, due to the small geometries of typical builds and limited yielding, this approach is difficult. Ultrasonic testing can be used as an alternative to mechanical testing for accurate determination of elastic constants [33,34]. The process uses ultrasonic waves which are transmitted and reflected into the sample to measure the elastic constants from speed of sound measurements and measured or estimated material density. Directions and equations for these computations are shown in Figure 17.25.

Foster et al. [33] investigated UAM samples with 65% and 98% bonded areas. The 65% bonded area case was constructed using the 1 kW UAM process, a known procedure for creating joints that contain voids. The 98% bonded area samples were constructed using 9 kW UAM, which significantly limits voids in the weld zone. Step builds were created for each condition, such that accumulative effects could be examined. An example step build is shown in Figure 17.26. Data for the 65% bonded area sample are shown in Table 17.10. The results are presented in comparison to a solid Al 3003 sample with similar measurements taken. A decrease in the elastic moduli is observed for each of the material directions.

Measurements for the 98% bonded area sample are shown in Table 17.11. The elastic constants for this material are significantly closer to solid material than the 65% bonded

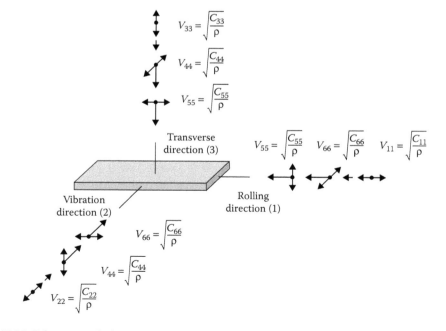

Figure 17.25 Schematic of ultrasonic wave propagation along the three Cartesian directions.

Figure 17.26 Image of step build showing 12 build sections.

Table 17.10 Comparison of elastic constants of 1 kW UAM builds to reference material

	Al 3003-H14 (GPa)	65% Bonded UAM sample	% Difference
C33	108.9	78.2	−28
C44	26.1	23.4	−10
C55	26.1	23.1	−11

Table 17.11 Comparison of elastic constants of 9 kW UAM builds to reference material

	Al 3003-H14 (GPa)	9 kW UAM Al 3003	% Difference
C33	108.7	109.2	0.5
C44	26.1	28.1	7
C55	26.1	28.1	7

area samples. Both the C44 and C55 directions exhibit slightly higher stiffness than the control sample.

It is expected that the lower material stiffness in the 65% bonded sample is due to the presence of voids at the welding interface. These void volumes are filled with no matrix material and thus have negligible mass and strength. As a result, when the material is loaded, the bulk foil portion of the UAM part elastically deforms a small amount, while the interface region under the same load will deform more. This occurs because the load bearing cross-sectional area at the interface is smaller due to the presence of voids for a given load. The combined loading response from the bulk foils and interface region results in an overall greater elastic deformation of the part for a given load. This phenomenon creates a component with an effective stiffness that is lower than the foils used to construct it. The elastic constants measured in samples made by 9 kW UAM were close to those of monolithic aluminum. This is attributed to the low void content in these samples. Therefore, for components with design requirements for stiffness matching solid material, a voidless bond must be achieved.

17.4   Dissimilar material joining in UAM

17.4.1   Al/Ti dissimilar welding

The low density, high conductivity, and high specific strength and stiffness of Al/Ti composites make them attractive for a number of aerospace, electronic, and automotive applications. Despite these benefits, joining aluminum and titanium can be problematic with conventional methods due to large differences in their melting temperatures, thermal conductivity, coefficient of thermal expansion, and crystal structures. Because UAM operates at low temperatures, much of these issues can be overcome, making it an attractive technology for creating Al/Ti joints.

A study was therefore conducted using 9 kW UAM for joining aluminum and titanium [34]. All welds in this study were performed with a Fabrisonic SonicLayer 4000 9 kW UAM system. Aluminum 1100 foils and commercially pure titanium foils of 0.005 in. (0.127 mm) thick were used. During joining, a bilayer arrangement was used where titanium on top of aluminum was welded in one step, which is shown schematically in Figure 17.27. In this arrangement, the sonotrode is in contact with the titanium layer only. All samples were built onto a solid Al 6061-T6 baseplate with the Al 1100/Al 6061 interface as the first layer. The weld parameters used for the joints are shown in Table 17.12.

Figure 17.28 shows an EBSD scan of an as-built Al/Ti UAM build. Results show significant deformation in the aluminum layers at the titanium–aluminum interfaces. The aluminum layers have a nominal thickness of 127 µm prior to welding, which is reduced to approximately 70 µm after the UAM process. By contrast, the titanium layers are nominally 127 µm prior to welding and 125 µm after welding with layers lower in the build showing more grain refinement and deformation than layers further up the build.

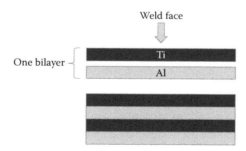

Figure 17.27 Arrangement for Al/Ti bilayers.

Table 17.12 Weld parameters for joining aluminum and titanium

Parameter	Level
Temperature	RT to 93.3°C (200°F)
Force	3500 N
Amplitude	41.55 µm (70%)
Speed	60 in./min (25.4 mm/sec)

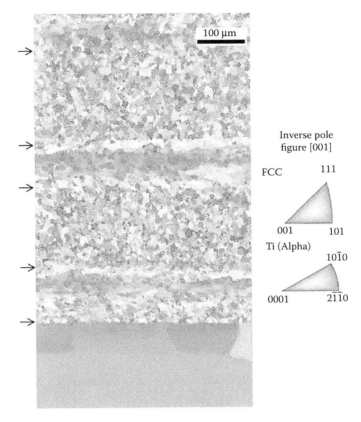

Figure 17.28 Electron backscatter diffraction image of Al/Ti joint. Arrows indicate approximate location of material interfaces.

The microstructure in the titanium layers is unchanged during the welding process, with all deformation and refinement occurring in the softer aluminum layers.

EBSD measurements of an Al/Ti sample heat treated at 600°C for one hour are shown in Figure 17.29. The grain structure in the titanium layers appears unchanged compared to the as-built samples, while the aluminum layers show significant grain growth. In each of the aluminum layers, it appears that the heat treatment has caused preferential grain growth into only a few grains for each layer. Grain growth in the substrate Al 6061 material appears as well, though not to the extent of the growth in Al 1100 layers.

In addition, though not shown in the EBSD results, an intermetallic layer forms at the Al/Ti interfaces. This is an approximately 5 μm thick layer caused by diffusion during the heat treatment process. In addition to microstructural evaluations, mechanical strength was tested via push-pin and shear testing. A summary of the mechanical test results is presented in Table 17.13. The mechanical work, or area under the force-displacement curve, was used as the metric for evaluating the strength of the samples. The heat-treated samples yield much higher values of mechanical work for failure than the as-built samples, roughly 12.7 kN-mm versus 3.5 kN-mm on average. These results indicate that heat treatment significantly increases the mechanical strength of UAM-joined Al/Ti material.

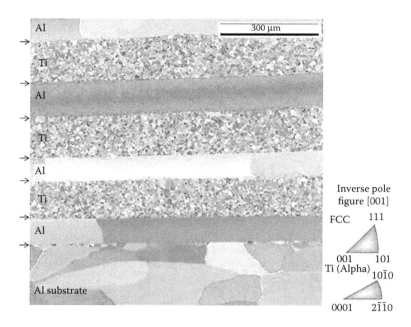

Figure 17.29 Electron backscatter diffraction image of Al/Ti joint after heat treatment. Arrows indicate approximate location of material interfaces.

Table 17.13 Mechanical test results for as-built and heat-treated Al/Ti joints

	Shear strength (MPa)	Pushpin (kN-mm)
As-built	46.3	3.5
Heat-treated	102.4	12.7

Results of shear testing are likewise summarized in Table 17.13 showing the average ultimate shear stress (USS) of the tests. The shear strength of the heat-treated samples exhibits ultimate shear strengths over two times that of the as-built samples with strengths of 102.4 MPa and 46.3 MPa, respectively.

Previous studies have examined various aspects of Al/Ti joining using 1 kW UAM. Using a shear test and 1 kW UAM, Hopkins et al. [19] measured the as-built shear strengths of 63 MPa on average, slightly above the average value of 46.3 MPa for the 9 kW UAM as-built shear strengths measured. Studies by Obielodan et al. [36] using CpTi and Al 3003 suggest the as-built shear strengths of 34 MPa. Following a heat treatment of 480°C for 30 min, shear strengths of 73 MPa were measured while exhibiting diffusion of approximately 5 µm. This diffusion zone was said to provide solid solution strengthening at the interface, not present in the as-built samples. The study presented here demonstrates that shear strengths of 102 MPa on average are possible when using 9 kW UAM and a postprocess heat treatment which generates a similar 5 µm diffusion zone. However, in this case, the diffusion zone is believed to provide a biaxial constraining action at the interface which provides the strengthening. Weld amplitudes of 41.55 µm are expected to increase the plastic deformation at the bond interfaces, thus increasing the driving force for recrystallization at the interface and improving bonding as compared to the studies using 1 kW UAM. The as-built samples in all three cases lack indications of diffusion which, based on results of heat-treated specimens, is necessary for maximizing mechanical strength.

17.4.2   Steel/Ta joining

Tantalum is an attractive material for corrosive environments and nuclear applications, due to its low permeability to radiative species and high melting temperature [37]. Due to cost, it is advantageous to use tantalum clads whenever possible to take advantage of its material properties without encountering exorbitant costs. Therefore, joints of tantalum and steel are required to meet these goals. However, due to significant differences in melting temperature (Ta: 3020°C and Fe: 1538°C), solid-state welding techniques are preferred over fusion-based welding.

An examination of Ta/Steel welds was conducted using 9 kW UAM [38]. Welds of 99.5% tantalum sheet were joined onto a 1010 steel substrate using the 9 kW UAM process. These joints used a single 50 μm thick tantalum layer as a clad onto a 2.5 mm thick steel substrate. Weld parameters for the joints are shown in Table 17.14. Successful joints were examined using electron microscopy and nanoindentation following joining.

An electron microscopy image of a Ta/Steel joint is shown in Figure 17.30. The image shows three distinct areas that are characteristics of the Ta/Steel joints observed. The first

Table 17.14 Weld parameters for joining tantalum to steel

Parameter	Level
Force	7000 N
Amplitude	36 μm
Speed	35.4 in./min (15 mm/sec)

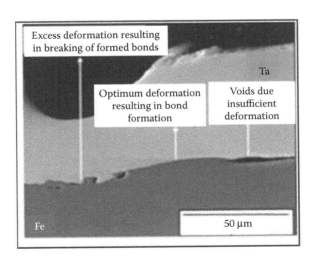

Figure 17.30 Electron microscopy image of Ta/Steel joint showing regions of voids, successful bonding, and areas of excess deformation leading to broken bonds. (From N. Sridharan, M. Norfolk, and S. Babu, *Metallurgical and Materials Transactions A* 47(1), 2517–2518, 2016.)

is a region of voids, where insufficient deformation occurred. Under insufficient deformation, the surface asperities do not collapse, and intimate contact required for bonding does not occur. A second region observed is that of good bonding, characterized by a lack of voids at the interface. The final region observed is excess deformation. In these regions, it is hypothesized that a bond occurs, however continued deformation of the interface leads to breakage of these bonds. This region is characterized by small voids along the interface along with distinct deformation zones that are atypical in the well-bonded areas.

Electron backscatter diffraction results are presented in Figure 17.31 showing the grain map and grain orientation spread (GOS) map at the interface. The interface shows a fine, mostly equiaxed grain structure, while the bulk of the tape suggests little changes to the grain structure. The GOS map in Figure 17.31b shows a gradual increase in the amount of plastic deformation at the interface and into the Ta layer. Within the bond region, the GOS map conveys little plastic deformation indicating that recrystallization has occurred, which is consistent with the equiaxed grain structure found in this region.

This grain structure is similar but not identical to previously measured microstructures for UAM, which indicate dynamic recrystallization at the bond interface [15]. Therefore, a different mechanism must take place to create the microstructural features found in these dissimilar joints featuring body-centered cubic crystal structures. Based on the observed measurements, it is proposed that rotational dynamic recrystallization is occurring. This is a mechanism which has been observed in adiabatic shear bands in Ta alloys where dislocations generated during plastic deformation reach a critical level to form elongated subgrains. These subgrains minimize the strain energy in the lattice, and with continued deformation, eventually break up into equiaxed grains at the interface while continued grain rotation increases the misorientation between grains. This mechanism explains the high-angle grain boundaries observed at the interface. Further deformation would then lead to further refinement and thus the very fine grain structure which is observed.

Nanohardness tests of the tantalum and steel portions of the joint are shown in Figure 17.32. The hardness at the interface is higher than that in the bulk portions of each material. Similarly, the strength decreases for tests further from the interface. Because

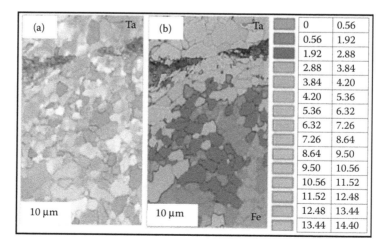

Figure 17.31 Electron backscatter diffraction image of Ta/Steel joint showing (a) grain map and (b) GOS map indicating plastic strain. (From N. Sridharan, M. Norfolk, and S. Babu, *Metallurgical and Materials Transactions A* 47(1), 2517–2518, 2016.)

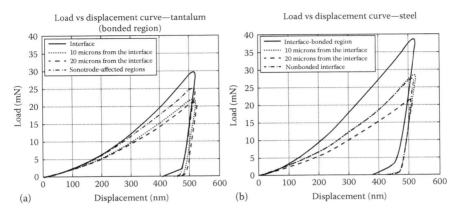

Figure 17.32 Nanohardness testing of (a) tantalum and (b) steel sides of Ta/Steel joint. (From N. Sridharan, M. Norfolk, and S. Babu, *Metallurgical and Materials Transactions A* 47, 1, 2517–2518, 2016.)

higher strength and grain refinement are shown at the interface, it is hypothesized that the strengthening mechanism in this region is due to Hall–Petch strengthening from small grain sizes. Regions close to the interface which show a decrease in hardness are attributed to a decrease in the plastic strain away from the interface. This is generally confirmed in the GOS results, which indicate plastic deformation decreases further from the interface.

17.4.3   Other dissimilar joining

The UAM process is a proven technology for joining a number of other dissimilar material combinations in addition to the Al/Ti and Ta/Steel combinations highlighted. In a study by Truog [25], Al/Cu combinations were proven using the 9 kW UAM process. This work shows that viable Al/Cu welds can be achieved using the UAM process with a cross section from an Al/Cu joint shown in Figure 17.33.

Heat treatments at 350°C for 10 min were shown to significantly improve the bond quality of the joints. Push-pin tests for as-built and heat-treated Al/Cu welds are shown in Figure 17.34. For each welded combination, the joint strength increases following heat treatment. This is consistent with Al/Ti joints which show similar mechanical strength increases following heat treatment, as discussed in Section 17.4.1.

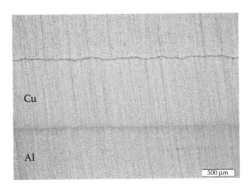

Figure 17.33 Cross section of Al/Cu joint.

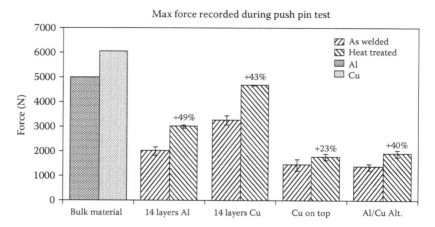

Figure 17.34 Push-pin results for as-welded and heat-treated Al/Cu combinations. (From A. Truog, Bond improvement of al/cu joints created by very high power ultrasonic additive manufacturing, Master's thesis, The Ohio State University, Columbus, OH, 2012.)

Work by Mueller et al. [39] suggests that Al/Cu joints produce very small-scale inter-metallic diffusion. Evidence of diffusion is only observable via TEM. This work also confirms the prevailing bond theory for Al/Cu joints based on the concept of dynamic grain recrystallization at the interface through rearrangement of dislocations.

A study by Gonzalez and Stucker [23] proved that stainless steel 316L joints could be achieved using the UAM process. Their work shows that voidless joints can be achieved using the 1 kW UAM process. A cross section of a successful joint is presented in Figure 17.35, where four foils are welded onto an Al 3003 substrate. Following a DOE study, optimal process parameters for the 1 kW UAM process were identified. A normal force of 1800 N, weld rate of 26 in./min, weld amplitude of 27 μm, and baseplate temperature of 400°F were identified as optimal for achieving successful joints.

In a study by Obielodan et al. [20] combinations of titanium, silver, tantalum, aluminum, molybdenum, stainless steel, nickel, copper, and MetPreg (commercial metal matrix composite) were all proven using the UAM process. Figure 17.36 shows a cross section of a joint containing nickel, copper, and silver foils, on an Al 3003 baseplate. Two layers of silver were welded onto the aluminum baseplate, followed by a layer of copper; then a layer of

Figure 17.35 Cross section of stainless steel 316L foils welded onto an Al 3003 substrate. (From R. Gonzalez and B. Stucker, *Rapid Prototyping Journal* **18**(2), 2012.)

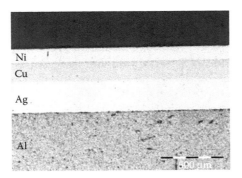

Figure 17.36 Cross section of a multimaterial UAM build including nickel, copper, and silver welded onto an Al 3003 substrate. (From J. Obielodan, A. Ceylan, L. Murr, and B. Stucker, *Rapid Prototyping Journal* **16**(3), pp. 180–188, 2010.)

nickel was welded on the top of the structure to complete the build. This build highlights the ability of the UAM process to join multiple material systems.

These works, while not representing all dissimilar material joining using UAM, shows the extent to which the UAM process can be used for joining dissimilar materials. Due to the low process temperatures, dissimilar joints can be achieved for a number of material combinations and encompassing multiple different crystallographic structures.

17.5   Challenges and future directions

A lack of complete scientific understanding of the UAM process and how it affects build properties limits the quality and size of builds as well as the range of dissimilar material combinations that can be additively welded. The underlying challenge is that no comprehensive models exist to describe the UAM process, specifically models which quantify the energy flow in the welder and how the available scrubbing energy effectively induces plastic deformation and dynamic recrystallization in a build. Efforts to model the process have partially addressed this need, but much progress is needed before process–property relationships can be mathematically described and predicted in UAM with any degree of accuracy. The approach to correlate process settings with build properties currently entails DOE studies, though these approaches typically focus on part strength rather than a full set of properties such as strength, fatigue characteristics, functionality, and cost. Examples of existing design of experiment studies were presented in Section 17.3.1.

As is the case with ultrasonic metal welding, the main control variables for the UAM process include weld speed (or time), down force (or pressure), and vibration amplitude [2,8]. Ultrasonic vibration frequency is fixed at the designed resonance frequency of the sonotrode, which is critical for the successful operation of the process. Ringing of the sonotrode at the correct frequency also represents the focus of control strategies implemented within commercial UAM equipment [6].

The input weld energy can be expressed as a function of the main control variables by assuming that weld energy E_{weld} is imparted into a build as mechanical scrubbing,

$$E_{\text{weld}} = \int P \times dt = \int F_s \times \dot{\delta} \times dt = \frac{1}{V_t} \int F_s \times \omega \times \delta \times dx \tag{17.2}$$

Here, F_s is the scrubbing or shear force at the interface, which is a function of the downforce during initial tape slip (due to sliding friction) and of the vibration amplitude of the sonotrode after slip ceases (due to shear deformation during collapse of asperities). The variable δ is the velocity of the sonotrode vibration or the derivative of the vibration displacement δ_t, and ω is the fixed frequency at which the welder vibrates. The welder amplitude and velocity are both sinusoidal functions since the piezoelectric transducers are supplied with a nominal 20 kHz sinusoidal voltage. The integral (Equation 17.2) is calculated over the amount of time the welder is welding a specific tape area, based on the weld speed V_t at which the welder travels over the surface of a build. The weld energy can be indirectly determined by measuring the electrical current and voltage applied to the transducers and assuming constant energy transfer efficiency in the piezoelectric transducers. This efficiency is estimated to range between 80% and 90% if the energy transfer characteristics of the welder are well understood. The dynamic response of the welder has been characterized and modeled by Hehr and Dapino [40].

The Al 6061 DOE study described in Section 17.3.1 suggests that, for this material, temperature and normal force have no statistically significant effect on build strength, whereas amplitude and weld speed do have a statistically significant effect on build strength. Since normal force contributes to the weld power only when the tape is slipping and does not influence build strength, it can be inferred that the time of frictional slip is small compared to the time of tape sticking. This conclusion is supported by Sriraman et al. [5] who showed that heat generation correlates with plastic deformation (while tape is sticking) and that force has no significant effect on heat generation over a similar range of forces. The effective weld energy (Equation 17.2) is thus largely dominated by sonotrode amplitude and linear weld speed.

For consistent welding throughout a UAM build, it is necessary to maintain a constant amplitude of relative motion between the foil and the workpiece to which it is being welded. As a build is being constructed, however, the mechanical stiffness of the system changes [41]. According to simple beam theory, an increase in build height leads to a decrease in the stiffness of the workpiece. This, in turn, has an effect on the relative motion of the foil and workpiece because the workpiece deflects with the loads from the sonotrode. This is represented schematically in Figure 17.37, where the imparted sonotrode displacement is represented by δ_{limit}, the displacement of the build due to its finite stiffness is δ_E, and the relative displacement available to weld the foil is the difference $\delta_{\text{limit}} - \delta_E$. If the part did not bend due to compliance, δ_E would be zero, and all the imparted sonotrode displacement would be available to weld the foil. Because in practice there is mechanical

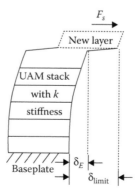

Figure 17.37 Schematic of UAM build undergoing deformation imparted by the shear force applied to the sonotrode (F_s).

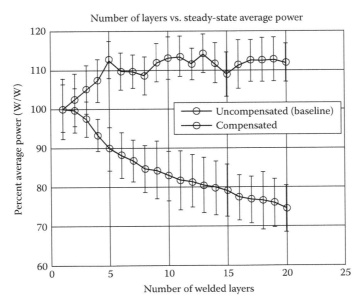

Figure 17.38 Measured power with and without amplitude compensation. The compensated build maintains the power applied to the build within +15%, whereas the uncompensated build shows a decay in input power of 25% by the twentieth layer.

compliance (inverse of stiffness) and a resulting bending motion of the part, the net displacement between the new foil and the rest of the stack is increasingly diminished as part height increases. Since amplitude has been shown to have a significant effect on weld quality, [7,19,42], this decrease in relative motion will typically lead to a degradation of weld quality through the height of the build.

Measurements presented in Figure 17.38 show that the effective weld power can decrease by as much as 25% after only 20 foils have been welded. Manual compensation of the imparted sonotrode amplitude was shown to effectively prevent a decrease in power. These particular measurements show, in fact, a slight increase in imparted weld energy with build height due to overshoot in the manual compensation. Builds were measured without and with power compensation using push-pin tests. Experimental data are shown in Figure 17.39. The samples built with compensated power exhibit failure through all the welded layers compared to more delamination in the uncompensated samples. Additionally, the compensated samples require additional mechanical work to drive the sample to failure and exhibit slightly higher strength.

Microstructural analysis for the fifth and fifteenth tape interfaces for the uncompensated and compensated power samples is shown in Figure 17.40 with electron microscopy images of ion-etched samples. Interfaces 5 and 15 were chosen for comparison because push-pin testing occurs near layer 5 and above. Figure 17.40 shows fine grains at all the interfaces. However, for the fifteenth uncompensated power weld interface, the recrystallized region is narrow, showing little to no mixing in some areas. On the other hand, the fifteenth compensated power interface shows strong mixing and dimensions similar to the fifth power-compensated interface. The difference in mixing and grain refinement originates from the uncompensated build having received less strain energy than the compensated one.

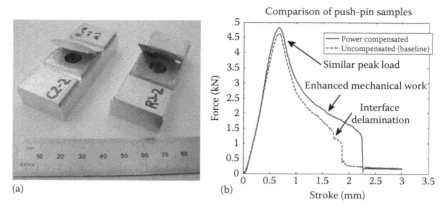

(a)

(b)

Figure 17.39 Comparison between compensated and uncompensated push-pin samples: (a) photo comparing failure behavior and (b) force-displacement plots.

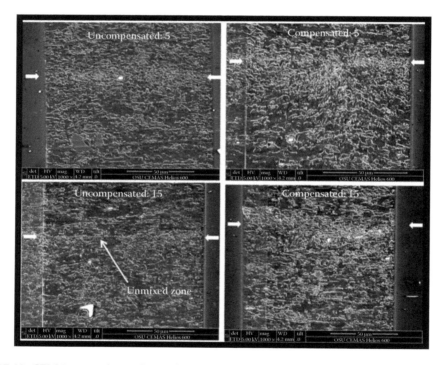

Figure 17.40 SEM image of samples etched with ion beam milling showing interface microstructure at layers 5 and 15 for uncompensated and compensated samples. Arrows indicate the approximate interface region.

The hypothesis of structural compliance impacting ultrasonic weld power and corresponding mechanical strength was tested in this study and found to be influential. It was shown that both mechanical testing and microstructure analysis correlate with variations in weld power input due to poor properties observed when power was not held constant through the UAM process. Consequently, future UAM systems should incorporate means to monitor and control weld power during the UAM process, rather than operating in amplitude control mode. *In situ* power monitoring and control can be used both to ensure part quality and to monitor any degradation in part quality in real time [41].

The decrease in plastic deformation energy ($E_{plastic}$) due to compliance can be analyzed by considering the force and displacement involved in plastically deforming the material:

$$E_{plastic} = F_{plastic} \int_{\delta_E}^{\delta_{limit}} d\delta \tag{17.3}$$

where $F_{plastic}$ is the force at which plastic deformation initiates. The integration limits represent the deformation at which elastic deformation ends (δ_E) and the limit prescribed by the controller (δ_{limit}), as shown in Figure 17.37. If one assumes, for simplicity, that the plastic deformation force is constant, the plastic energy can be written in terms of build stiffness (k) as follows:

$$E_{plastic} = F_{plastic}\delta_{limit} - k\delta_E^2 \tag{17.4}$$

This expression shows that the amount of deflection associated with build stiffness, δ_E, has the effect of reducing the available energy to plastically deform the interface. Exact quantification of each of the terms in (Equation 17.4) requires first-principles models which describe the flow of energy through the welder and into the part, including elastic, plastic, and dissipative terms. This requires understanding of the transfer path of input electrical power into mechanical weld power and the energy of a given weld. Using this methodology, a control scheme can be implemented which can account for structural compliance effects during the build process, improving the consistency of welds throughout.

From a process viewpoint, UAM is not compatible yet with mass manufacturing production lines. The existing UAM equipment was designed for small production batches and one-off parts. That being said, because UAM welders can be treated as another tool in a CNC workflow, the potential exists for integration of UAM welders into mass production settings. For instance, a UAM welder can be attached to an end effector for 3D printing or joining of automotive parts in production settings. Given limitations in throughput, these parts may need to be manufactured offline and brought to the assembly line alongside preassembled subsystems. This is a typical approach in automotive manufacturing; so conceptually, UAM could be incorporated into vehicle assembly lines without significant disruption or retooling.

References

1. Wohlers Report, http://wohlersassociates.com/2013report.htm, 2013.
2. K.F. Graff, Ultrasonic additive manufacturing, *Welding Fundamentals and Processes*, ASM Handbook, 6A, 2011.
3. P.J. Wolcott, C. Pawlowski, L.M. Headings, and M.J. Dapino, Seam welding of aluminum sheet using ultrasonic additive manufacturing system, *Journal of Manufacturing Science and Engineering*, **139**(1), 011010, 2017.
4. C.Y. Kong, R.C. Soar, and P.M. Dickens, Optimum process parameters for ultrasonic consolidation of 3003 aluminum, *Journal of Materials Processing Technology*, **16**(2), 181–187, 2004.
5. M.R. Sriraman, M. Gonser, H.T. Fujii, S.S. Babu, and M. Bloss, Thermal transients during processing of materials by very high power ultrasonic additive manufacturing, *Journal of Materials Processing Technology*, **211**, 2011.
6. www.fabrisonic.com.

7. C.D. Hopkins, P.J. Wolcott, M.J. Dapino, A.G. Truog, S.S. Babu, and S.A. Fernandez, optimizing ultrasonic additive manufactured Al 3003 properties with statistical modeling, *Journal of Engineering Materials and Technology*, **134**(1), 011004, 2012.

8. K. Graff, J. Devine, J. Keltos, and N. Zhou, Ultrasonic welding of metals, in *American Welding Society Welding Handbook*, Miami, American Welding Society, 2000.

9. P. Wolcott, A. Hehr, and M.J. Dapino, Optimized welding parameters for Al 6061 ultrasonic additive manufactured composites, *Journal of Materials Research*, **29**(17), 2014.

10. M.R. Sriraman, S.S. Babu, M. Short, Bonding characteristics during very high power ultrasonic additive manufacturing of copper, *Scripta Materialia*, **62**, 560–563, 2010.

11. R. Dehoff, and S.S. Babu, Characterization of interfacial microstructures in 3003 aluminum alloys blocks fabricated by ultrasonic additive manufacturing, *Acta Materialia*, **58**(13), 4305–4315, 2010.

12. D.E. Schick, R.M. Hahnlen, R. Dehoff, P. Collins, S.S. Babu, M.J. Dapino and J.C. Lippold, Microstructural characterization of binding interfaces in aluminum 3003 blocks fabricated by ultrasonic additive manufacturing, *Welding Journal*, **89**(5), 105s–115s, 2010.

13. P.J. Wolcott, N. Sridharan, S.S. Babu, A. Miriyev, N. Frage, and M.J. Dapino, Characterisation of Al-Ti dissimilar material joints fabricated using ultrasonic additive manufacturing, Science and Technology of Welding and Joining, 21 (2), 114–123, 2016.

14. K. Johnson, *Interlaminar Subgrain Refinement in Ultrasonic Consolidation*. PhD thesis, Loughborough, UK, Loughborough University, 2008.

15. H. Fujii, M. Sriraman, and S. Babu, Quantitative evaluation of bulk and interface microstructures in Al-3003 alloy builds made by very high power ultrasonic additive manufacturing, *Metallurgical and Materials Transactions A*, **42A**, 4045–4055, 2011.

16. E. Mariani and E. Ghassemieh, Microstructure evolution of 6061 O Al alloy during ultrasonic consolidation: An insight from electron backscatter diffraction, *Acta Materialia*, **58**, 2010.

17. S. Shimizu, H. Fujii, Y. Sato, H. Kokawa, M. Sriraman, and S. Babu, Mechanism of weld formation during very-high-power ultrasonic additive manufacturing of Al alloy 6061, *Acta Materialia*, **74**, 2014.

18. L.M. Headings, K. Kotian, and M.J. Dapino, Speed of sound measurement in solids using polyvinylidene fluoride (PVDF) sensors, *Proceedings of ASME Conference on Smart Materials, Adaptive Structures and Intelligent Systems*, SMASIS2013–3206, Snowbird, Utah, September 16–18, 2013.

19. C.D. Hopkins, M.J. Dapino, and S.A. Fernandez, Statistical characterization of ultrasonic additive manufacturing Ti/Al composites, *ASME Journal of Engineering Materials and Technology*, **132**(4), 1–9, 2010.

20. J. Obielodan, A. Ceylan, L. Murr, and B. Stucker, Multi-material bonding in ultrasonic consolidation, *Rapid Prototyping Journal*, **16**(3), 180–188, 2010.

21. I.E. Gunduz, T. Ando, E. Shattuck, P.Y Wong, and C.C. Doumanidis, Enhanced diffusion and phase transformations during ultrasonic welding of zinc and aluminum, *Scripta Materialia*, **52**, 2005.

22. T.H. Kim, J. Yum, S.J. Hu, J.P. Spicer, and J.A. Abell, Process robustness of single lap ultrasonic welding of thin, dissimilar materials, *CIRP Annals—Manufacturing Technology*, **60**, 2011.

23. R. Gonzalez and B. Stucker, Experimental determination of optimum parameters for stainless steel 316L annealed ultrasonic consolidation, *Rapid Prototyping Journal*, **18**(2), 2012.

24. C. Zhang, A. Deceuster, and L. Li, A method for bond strength evaluation for laminated structures with application to ultrasonic consolidation, *Journal of Materials Engineering and Performance* **18**(8), 1124–1132, 2009.

25. A. Truog, Bond improvement of al/cu joints created by very high power ultrasonic additive manufacturing, Master's thesis, The Ohio State University, Columbus, OH, 2012.

26. A. Dean and D. Voss, *Design and Analysis of Experiments*, Springer, New York, 1999.

27. J. Obielodan, G. J. Ram, B. Stucker, and D. Taggart, Minimizing defects between adjacent foils in ultrasonically consolidated parts, *J. Eng. Mater. Technol* 132(1), 011006 (Nov 03, 2009)

28. ASTM E8/E8M-16a, *Standard Test Methods for Tension Testing of Metallic Materials*, ASTM International, West Conshohocken, PA, 2016, www.astm.org

29. D. Li and R. Soar, Influence of sonotrode texture on the performance of an ultrasonic consolidation machine and the interfacial bond strength, *Journal of Materials Processing Technology*, **209**, 1627–1634, 2009.

30. R. Friel, K. Johnson, P. Dickens, and R. Harris, The effect of interface topography for ultrasonic consolidation of aluminum, *Materials Science and Engineering A*, **527**, 4474–4483, 2010.

31. K. Sojiphan, S. Babu, X. Yu, and S. Vogel, Quantitative evaluation of crystallographic texture in aluminum alloy builds fabricated by very high power ultrasonic additive manufacturing, in *Solid Freeform Fabrication Symposium*, Austin, TX, 2012.

32. ASM-International, *Heat Treating—Heat Treating of Aluminum Alloys*, Vol. 4, ASM International, 1991.

33. D. Foster, M. Dapino, and S. Babu, Elastic constants of ultrasonic additive manufactured Al 3003-H18, *Ultrasonics*, **53**, 2013.

34. H. Jeong, D. Hsu, R. Shannon, and P. Liaw, Characterization of anisotropic elastic constants of silicon-carbon particulate reinforced aluminum metal matrix composites, part 1, *Metallurgical and Materials Transactions A*, **25A**, pp. 799–809, 1994.

35. P. Wolcott, N. Sridharan, S. Babu, A. Miriyev, N. Frage, and M. Dapino, Characterisation of Al-Ti dissimilar material joints fabricated using ultrasonic additive manufacturing, *Science and Technology of Welding and Joining*, **21**(2), 114–123, 2015.

36. J. Obielodan, B. Stucker, E. Martinez, J. Martinez, D. Hernandez, D. Ramirez, and L. Murr, Optimization of the shear strengths of ultrasonically consolidated Ti/Al 3003 dual-material structures, *Journal of Materials Processing Technology*, **211**, 2011.

37. R. Balliett, M. Coscia, and F. Hunkeler, Niobium and tantalum in materials selection, *Journal of Metals*, **38**(9), 25–27, 1986.

38. N. Sridharan, M. Norfolk, and S. Babu, Characterization of Steel-Ta dissimilar metal builds made using very high power ultrasonic additive manufacturing (VHP-UAM), *Metallurgical and Materials Transactions A*, **47**(1), 2517–2518, 2016.

39. J. Mueller, J. Gillespie, and S. Advani, Effects of interaction volume on x-ray line-scans across an ultrasonically consolidated aluminum/copper interface, *Scanning*, **35**, 2013.

40. A. Hehr and M.J. Dapino, Dynamics of ultrasonic additive manufacturing, *Ultrasonics*, **73**, 49–66, 2017.

41. A. Hehr, P. Wolcott, and M. Dapino, Effect of weld power and build compliance on ultrasonic consolidation, *Rapid Prototyping*, **22**(2), 377–386, 2016.

42. Y. Yang, G. Janaki Ram, and B. Stucker, An experimental determination of optimum processing parameters for Al/SiC metal matrix composites made using ultrasonic consolidation, *Journal of Engineering Materials and Technology*, **129**, 538, 2007.

43. P. Wolcott, A. Hehr, C. Pawlowski, and M. Dapino, Process improvements and characterization of ultrasonic additive manufactured structures, *Journal of Materials Processing Technology*, **233**, 44–52, 2016.

44. G.D. Janaki Ram, C. Robinson, Y. Yang, and B.E. Stucker, Use of ultrasonic consolidation for fabrication of multi-material structures, *Rapid Prototyping Journal*, **13**(4), 226–235, 2007.

chapter eighteen

Printing components for reciprocating engine applications

Michael D. Kass and Mark W. Noakes

Contents

18.1   Introduction

Engines, no matter what their configuration, can be defined as machines used to convert hydrocarbon fuels into mechanical motion. The generated motion can be a rotational torque (as in a reciprocating engine) or thrust (from a turbine). The combustion of hydrocarbon fuels in an engine produces environments consisting of extremely hot and pressurized gas, which often include highly reactive chemical species, such as carboxylic and sulfuric acids. The materials used in the construction of engines must be able to withstand these extremes in temperature, pressure, and chemistry. Outside of the combustion chamber, engine components are subjected to cyclic loads, friction and wear, and torsion. As a result, metals, especially high-strength steels, are predominantly used in engine construction.

The earliest engines were relatively simple in design and construction. Over time, engine designs became more complex as they improved in efficiency and performance. As component geometries, tighter tolerances, and material integration become more complex, fabrication methods too have evolved. To date, most solid metal engine components are fabricated using technologies suited for rapid mass production; these include casting and machining, thermal treatments, and stamping. Newer designs are necessitating complex geometries that are not as easily machined. This is especially true for interior channeling for better thermal management. Manufacturing processes are becoming more complicated as well; for instance, advanced diesel pistons have cooling channels that necessitate additional fabrication steps such as welding into the overall process. One potential means of meeting these fabrication challenges is additive manufacturing (AM).

AM, also known as three-dimensional (3D) printing, has gained notable attention as a means of rapid prototyping and manufacturing small quantities of specialized components. The term additive manufacturing actually refers to a group of technologies that construct 3D objects through an iterative additive process. In each case, a computer-generated model of the object to be printed is processed to produce a sequence of commands suitable to produce the object by laying down successive layers of the material in solid form. These technologies range from fused deposition modeling for plastics and ceramics to laser, e-beam, ultrasonic, and binder/powder/sintering-based systems for metal production. Materials that have been used in AM to fabricate solid components include polymers (including embedded composites), metals, and (to a much lesser degree) ceramics, and even concrete. AM differs from traditional fabrication methods in that it is typically additive rather than subtractive. As such, the level of final machining and waste is normally much lower than traditional methods. While postmachining and surface finishing of AM objects are common, another key advantage is that complex features may not require additional machining steps typical of traditional manufacturing methods. AM also enables fabrication of objects with complex and minimally accessible internal structures that could not otherwise be implemented [1,2].

In polymer applications, the component is typically built up by physically extruding fluidized material through a nozzle and depositing the material layer by layer. Printing resolution for commercial printers in these applications is typically equal to or better than 0.25 mm (or 0.010 inch). However, at these fine resolutions, printing an object of any appreciable size may take many hours (many plastic printers currently run in the 16 cc/hour range). One attractive aspect of high-resolution polymer printing is that these systems can often be used with little to no postprocessing/machining.

There have been recent developments in big area additive manufacturing (BAAM). These large-scale plastic printers use plastic chips instead of filament and mount an extruder at the build point. Chips may have embedded chopped carbon fiber for added strength. While bead size will vary, one typical example flows a 0.22 inch bead 0.1 inch tall. The flow rate for this particular machine is 70 lbs/hour. Print volume capability is increasing rapidly, but this particular printer has a range of motion of 20 feet long by 8 feet wide by 6 feet high. These printers have been used to print cars and a small technology demonstrator home [3,4].

Metal printers use a wide array of fusing technologies to bond the additive layers— e-beam, laser, ultrasonic, and so on—and the printed media may be either in powder form or as fine sheets (in the case of ultrasonic-based AM). E-beam and laser systems both use powdered print media as the starting material. Between these two types, e-beam units are often preferred because they are much faster than laser-based systems. Laser-based systems, while slower, tend to produce better final surface finish and may more accurately match the computer model of the part printed. The powder is put down in thin layers and then fused (or sintered) by the directed energy from the e-beam (or laser). Each successive powder layer builds on the previous layer, and the final printed object is loosely coated with particles from the powder bed. The particle coating is removed using blast media consisting of the same powder that is used in the print process, so that it may be reclaimed for reuse. Another feature of these types of metal printers is that they often require the use of breakaway supports during the build process. These supports must be mechanically removed after the part has been blasted; *nibs* are left that may require machining depending on the required surface finish of the part. Ultrasonic-based AM uses thin sheets of material ultrasonically welded in a stack. Each layer is fused to the one below it. Ultrasonic AM in its current form is an additive/subtractive form of manufacturing. As the object is built up of stacks, milling is used to cut away portions of the build to shape the part to the model. The end result may require no further postprocessing. They are one of the fastest AM systems with a print rate of up to 100 cc/hour (depending on the geometry of the part printed).

Another process that produces AM metal objects involves laying down metal powder as with other print technologies, but the extrusion process bonds the layers of metal powder together with polymer. After the object is completely printed, it is cured and then fired in an oven to sinter it. Sintering will always leave some voids in the object that can then be back filled with some metal that melts at a lower temperature than the base metal. Bronze-infused stainless steel bushings have been manufactured this way. Copper is currently a major focus of development for this printing process.

To date, e-beam printing has seen the most success with titanium alloys, but other metal alloys are possible. The laser-based systems can do titanium, Inconel, cobalt chrome, and stainless steels but print about an order of magnitude more slowly than the e-beam printers. For both systems, the repertoire of print media material is expanding rapidly. Any list published will be quickly out of date. However expanding the list of materials that can be printed is not a trivial process as the print parameters change based on the material to be printed. Aluminum and copper are more often suited to the ultrasonic AM process, but progress is being made in their use with other print processes.

For automotive engine applications, the emphasis has been on printing components from metal, in particular titanium and aluminum alloys [1,2,5-9]. In many instances, AM has been used to produce prototypes for first-order evaluations. These have included lightweight components to reduce overall engine and vehicle weight and highly complex parts, such as heads and integrated heat exchangers. AM is also being evaluated for its potential to manufacture small unmanned aircraft engines and subsystems. Weight is a critical concern in unmanned aircraft, and AM offers a feasible means of producing engine components out of titanium, which is much lighter than steel. Compared to aluminum, titanium is heavier but much stronger and can withstand higher operating temperatures and pressures. Its thermal conductivity is also much lower than aluminum, meaning that less heat is lost through the walls during combustion. Small engine sizes, such as those used to power small unmanned aerial systems (UASs) have lower bulk combustion and exhaust temperatures due to incomplete combustion of the fuel and heat losses through the cylinder walls and head. In some cases, the exhaust temperatures are less than 250°C, which would enable the use of polymers in some component applications. The ability to manufacture engine blocks and heads out of polymer composites would lead not only to additional reductions in weight but possibly noise as well.

18.2 Reciprocating internal combustion engine environment

In a reciprocating engine, those components directly exposed to the combustion chamber environment are subjected to rapid transients in pressure and to a lesser degree temperature as well. Part of the pressure rise is associated with the compression ratio (which is the ratio of the cylinder volumes at the bottom and top ends of the cycle). The top end of the cycle is usually referred to as top dead center (TDC). For a nominal compression ratio of 10:1, the pressure achieved at the TDC will be close to 25 bar, which approximates the pressure value at the start of combustion. Engines are being designed with even higher compression ratios (approaching 18:1) to achieve higher power outputs and efficiencies. However, most of the pressure rise occurs as a result of the combustion event as the burning fuel and air mixture are converted into gaseous CO, CO_2, H_2O, NO, and NO_2. The sudden increase in heat and pressure drives the piston (to produce work) in very short timescales (milliseconds). During the start of combustion, the cylinder pressure rises rapidly as shown in Figure 18.1 for a typical automotive diesel engine. As the piston approaches the TDC (which corresponds to 0 crank angle degrees), the pressure rise is extremely rapid. For the case illustrated below, a peak pressure of 190 bars occurs shortly after the TDC. The peak pressure in diesel engines can be as high as 200 bar, while spark-ignited gasoline engines have values of around 100 bar. After achieving peak value, the pressure decreases almost as dramatically as it rose. The figure also shows the corresponding flame front temperature; however, the bulk combustion temperature is much lower due to thermal quenching by the unburned fuel–air mixture, radiative losses, and quenching through the cylinder wall, piston, and head. In fact, the typical peak inner wall temperatures for most reciprocating engines achieved are around 400°C.

To get an idea of the transient forces impacting cylinder components, let us consider a four-stroke diesel engine operating at 4000 rpm with pressure fluctuations depicted in Figure 18.1. A pressure fluctuation of ~170 bar occurs at a rate of 33 bar per second. Running at these conditions for 30 minutes means that the cylinder components have experienced 59,400 high pressure fluctuations. The impact of these high-frequency pressure pulses greatly impacts material selection and design. Any in-cylinder components manufactured by AM will need to withstand these peak pressure fluctuations and also maintain integrity while heated to temperatures approaching 400°C. For intake and exhaust valves, additional considerations include dynamic impact forces and wear caused by physical impact of metal-on-metal surfaces as when the valves contact the valve seats.

Other components, such as the upper head assembly and the connections from the piston to the shaft, are not exposed to the extreme environmental conditions of the combustion chamber, yet they are still subject to impact, frictional, and torsional loads. These

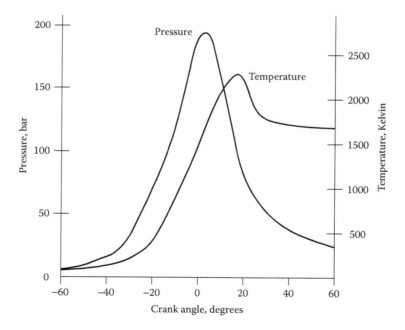

Figure 18.1 In-cylinder pressure trace for a typical diesel engine. The corresponding flame front temperature is also shown for comparison.

forces do impact the material selection and fabrication processes. It is important to note that engine components are machined to very tight tolerances, and AM technologies, by themselves, do not provide components with the necessary geometric and dimensional precision. Some level of machining (or polishing) will be required to finish certain component surfaces regardless of the manufacturing method.

18.3 Additive manufacturing approaches to engine systems

The evolution of engines over time has brought increased geometric complexity to their design, and it is likely that future engines will have even more integrated complexity. It will be necessary for manufacturing technologies to be able meet these needs in a cost-effective manner. AM is potentially an important component of the engine manufacturing process. AM can help meet future engine manufacturing needs by providing near-net shape rapid prototyping, *in situ* sensor integration, and enabling the use of difficult to machine materials (such as titanium). These features are discussed briefly in this section, and a case study utilizing these advantages is presented in the following section.

18.3.1 Production of prototype components

There are several key enabling aspects of AM technologies that make this manufacturing method especially attractive for the production of engine components. While AM is being evaluated as a means of reducing the number of machining steps to lower the overall cost, a more immediate application is the development of prototypes for development applications. To date, traditional manufacturing methods are better suited for mass production, but AM offers a more economical approach to produce near-net shape components by eliminating the need for molds and additional tooling. As such, the most widespread use is the production of prototype components for empirical testing and evaluation. Several UAS engine manufacturers, Ford Motor Company, and Daimler are looking at AM to manufacture heads for some proprietary prototype applications, but otherwise not much information exists in the public literature [7]. The cylinder head of a reciprocating engine has one of the (if not the) most complex geometries of any single engine component. Reciprocating engine heads start out as castings or billets, which are then machined to final dimensions. Machining the internal channels (for fuel and air delivery, spark/glow plug ignition, exhaust removal and cooling changes) is especially difficult and time consuming. The inclusion of channeling, precision tolerances, and polished surface finish translate to high fabrication costs for this component.

In addition to engine heads, AM is also used to produce prototype turbochargers and heat exchangers. These components also have complex geometries and are costly to produce via traditional fabrication methodology. Turbines manufactured using AM are not only being produced as prototypes but also as a production step for large-scale manufacturing. In fact, turbine manufacture via AM is being pursued aggressively by a number of different manufacturers. The majority of these applications are for jet turbines, but there is an interest for automotive turbochargers (including the turbocharger housing) as well. The Swedish sports car manufacturer, Koenigsegg, has used AM to manufacture turbocharger housings and exhaust components out of titanium. The primary motivation for utilizing titanium is for weight reduction [5]. Prototype hydraulic manifolds were produced by the Red Bull racing team to better facilitate coolant flow in the engine. These manifolds are made out of a titanium alloy (using direct metal laser sintering) and contain highly complex channeling [6]. Once again, the primary motivator was to reduce the overall weight of the engine system. The new manifolds produced improved coolant flow characteristics over the stock units, which resulted in lower energy requirements from the engine to move the coolant.

18.3.2   Integration of sensors

AM offers a means of facilitating sensor placement within a larger complex component. These sensors include temperature and pressure measurement for diagnostic and control applications. Modern engines are an example of minimalism in volume and mass. The overall size of automotive and aircraft engines (with respect to power output) have decreased over the years, while, at the same time, the packaging of the subsystems has become more complex and dense. These factors have increased the difficulty of placing instrumentation into the engine and subsystem components. The ability to print in ports and geometries to facilitate monitoring systems and circuitry (especially for small engine sizes) would enable better control and monitoring of engine systems. Another area of potential that opens up with AM is the ability to print in solid state, air gaps, or other channels to facilitate or manipulate heat flow in an engine via a more local (rather than bulk) approach.

18.3.3   Titanium

AM is especially attractive for the manufacture of components from titanium. Titanium is a difficult material to use in the production of solid components, because of its brittle nature and its phase change properties. However, the incorporation of titanium is attractive due to its low density, high strength, and low thermal conductivities and can be used to develop improved engine systems. Titanium is easily fabricated using AM and has been used to manufacture engine components in some small engine applications. Components made out of titanium using AM include turbines, turbochargers, turbocharger housings, heads, pistons, connecting rods, and engine blocks.

18.4   Case study of printed engine components for a small two-stroke engine

In 2013, researchers at Oak Ridge National Laboratory (ORNL) evaluated a two-stroke 4-cc glow-assist engine equipped with a titanium head that was manufactured using AM [8]. The original stock engine head, block, piston, and connecting rod were composed of aluminum, while the cylinder liner and the crankshaft were made of chrome-plated brass and steel, respectively. The study had three primary goals: One was to determine whether an engine component directly exposed to the combustion event would remain intact over a significant exposure period. The second objective was to measure the in-cylinder pressure cycles by printing in a port for sensor placement, and the third goal was to determine whether a head made from titanium would retain more heat inside the combustion chamber, leading to improved efficiency of combustion. The performance of the printed Ti head was evaluated against the stock aluminum head provided with the engine.

An important aspect of this study was to utilize AM capabilities to build in a port to enable placement of a pressure sensor capable of making cycle-resolved pressure measurement. The diameter of the combustion chamber was only 2 cm, with the glow plug occupying 0.45 cm^2 or roughly 14% of the total area. The small available surface area combined with the low head thickness precluded machining a port in the stock head. AM was used to build up additional surface to enable sensor placement. Sensor standoff (not flush) was

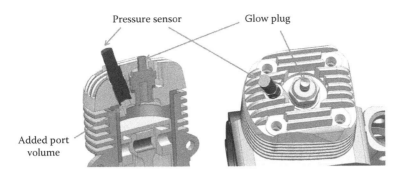

Figure 18.2 Computer-aided drawing showing printed head geometry including pressure sensor placement.

implemented as specified by the sensor vendor. The printed head sensor port added an additional volume (as shown in Figure 18.2). In order to maintain the same compression ratio as with the stock head, the interior volume geometry was adjusted (as shown in the figure) to maintain the same cylinder volume for both head types.

These high-resolution sensors are used to measure the pressure rise during combustion as a function of crank angle. As such, they are a critical tool to characterize combustion in an operating engine. They are normally placed on larger automotive-scale engines by machining a port directly through the head to the combustion chamber.

A photograph showing the outer surfaces of the two heads is shown in Figure 18.3. The printed head, shown in the bottom of the figure, was geometrically similar to the stock

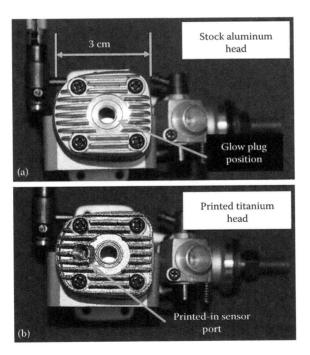

Figure 18.3 Photograph showing the stock aluminum head (a) and the titanium head (b) manufactured via AM.

aluminum head. The surface texture for the AM head is noticeably rougher than the stock aluminum head. Mating surfaces were machined to achieve the desired finish; otherwise, there was minimal finishing of the printed surfaces. The primary physical difference in the titanium head was the inclusion of a port for sensor measurement.

A Kistler Type 6054BR transducer was selected due to its small size for measurement of the in-cylinder pressure. (The sensitivity of this device is better than 1.5%.) Even though this is the smallest sensor available from Kistler for combustion analysis applications, the stock engine head, by itself, did not have enough thickness to allow sensor insertion.

A titanium head was printed, and a port was built in and reinforced to support the sensor. The stock head gasket was used and is composed of aluminum.

The engine was mounted to a motoring dynamometer and successfully operated with the titanium head over a wide range of speed and load conditions. More than 20 hours of total operational time were run with the titanium head with speeds ranging from 3000 to 7000 RPM, and no wear or degradation to the head or other components was observed. It is important to note that the engine was tuned for combustion with aluminum head, which would have higher heat losses and lower combustion temperatures. The utilization of a thermally less conductive material would be expected to advance the start of combustion (SOC) such that the pressure rise would begin before the piston had reached the top dead center. This would reduce the combustion efficiency even though more heat was retained in the cylinder. Unfortunately, it was not possible to tune the engine such that the TDC would occur earlier in the cycle to take advantage of the higher cylinder temperature (and also pressures). In one set of experiments using the titanium head, a preignition phenomenon was observed. This event was characterized by very early combustion phasing, in which combustion was completed 30–40 crank angle degrees before the top dead center. When the engine was disassembled, significant fouling was observed on the glow plug. It is hypothesized that high temperature deposits (caused by burned oil) may have been the source of the preignition event. This behavior was not noted for the aluminum head, nor again on the titanium head once the glow plug was energized. The lesson is that while AM allows for advanced materials to be used as components, the operation will need to be adjusted to compensate for expected combustion changes (even if they improve the performance overall). This will especially be true for materials with different thermal properties from those of the original design.

The engine performance for both heads was evaluated at speeds ranging from 3000 to 7000 rpm. In each case, the speed was controlled by the dynamometer, and the load was matched by adjusting the airflow via the throttle to achieve maximum brake torque (MBT). The measured fuel flow and airflow were used to calculate the relative air/fuel ratio, lambda, for each test condition and are presented in Figure 18.4. As shown, at MBT, the engine was observed to run rich ($\lambda < 1$) for all speeds, although at 6000 RPM, combustion was relatively close to stoichiometric. The implication is that the kinetics (i.e., scavenging) associated with fuel–air mixing was more favorable for this condition than for the other operating points. In addition, the titanium head was observed to produce maximum torque at less-rich fuel–air mixtures than the aluminum head. For 4000, 6000, and 7000 RPM, the increase in lambda with the titanium head was relatively constant and around 5%. Below 4000 RPM, MBT occurred at a near-constant lambda for the aluminum head but at increasingly less-rich conditions for the titanium head. The less-rich operation associated with the titanium head is attributed to the reduced heat transfer for this material. Less heat conduction should translate into higher in-cylinder combustion temperatures and potentially more complete combustion of the fuel.

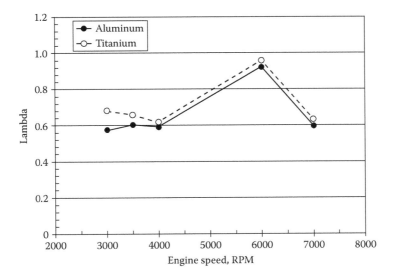

Figure 18.4 Lambda measurements associated with MBT for the printed titanium and stock aluminum heads.

It is important to keep in mind that, in order to achieve MBT with the titanium head, the fuel air mixture was tuned to a less-rich operation than that of the aluminum head. Interestingly, the exhaust temperatures measured for the two head materials were roughly the same. The implication is that titanium allowed less-rich operation at similar combustion temperature, and this effect is attributed to the reduced wall heat losses associated with titanium.

The MBT and efficiency were measured at several speeds from 2000 to 7000 RPM and are shown in Figure 18.5. For each data point, the torque measurements fluctuated by as

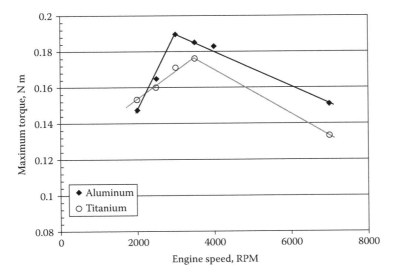

Figure 18.5 MBT map for a 4-cc single cylinder two-stroke engine. The engine was mapped with the stock aluminum head and a titanium head produced via AM.

much as 0.025 N m depending on the operating condition, but definite trends could be ascertained. Engine torque peaked around 3000 to 3500 RPM. In general, the measured torque with the titanium head was similar to, but less than, that measured with the aluminum head. For instance, the shaft torque peaked around 3000 to 3500 RPM for both head materials. Interestingly, at 2000 RPM, the MBT produced by the titanium head slightly exceeded that observed with the aluminum head.

For speeds of 2000 and 2500 RPM, the MBT for both head materials was roughly the same. At slower speeds, the lower heat losses through the cylinder walls accompanying the titanium head reduce quenching and may allow for more complete combustion of the fuel.

The exhaust temperatures at MBT were measured for each operating condition for speeds ranging from 2000 to 7000 RPM. As shown in Figure 18.6, the results for both head materials were similar, although the titanium head did increase the exhaust temperature around 10 degrees for the 3000, 3500, and 7000 RPM operating points. It is important to keep in mind that the air–fuel ratio was optimized for maximum torque for each head material. Therefore, a direct comparison of the exhaust temperatures at matching fuel flow and airflow is not presented. When the air–fuel ratio was maintained for each head material (as was the case for richer operation), the exhaust temperatures were approximately 20–30 degrees higher when the titanium head was used.

The brake specific fuel consumption (BSFC) at MBT was calculated for the operating points between 3000 and 7000 RPM, and the averaged results are shown in Figure 18.7. For speeds between 4000 and 7000 RPM, the engine operation was more efficient with the stock aluminum head, especially at 6000 RPM. The precise reason for the improved aluminum performance is not known, but it is likely associated with changes in the combustion phasing due to higher heat rejection properties of the titanium, and/or the sensor port cavity may be affecting the scavenging and level of mixing between the fuel and air. The reduced piston–cylinder wall friction (expected with the aluminum head) may also be a factor.

For speeds of 4000 and especially 6000 RPM, the BSFC fluctuations were higher with titanium than aluminum. The reason for this variability is unknown but may be related to incomplete scavenging caused by the sensor port. It is interesting to note that the torque

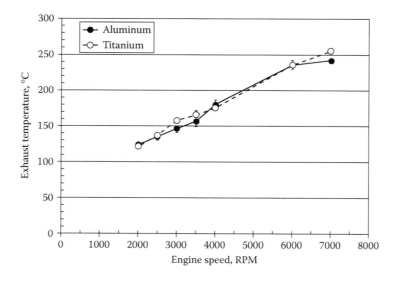

Figure 18.6 Exhaust temperatures associated with MBT for the aluminum and titanium heads.

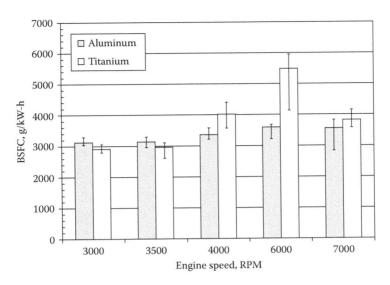

Figure 18.7 BSFC results associated with MBT for each operating point between 3000 and 7000 RPM.

values measured for the titanium head (at these two points) also deviated considerably from the values measured for the aluminum head.

At lower speeds (3000 and 3500 RPM), the BSFC values averaged 7% lower when the titanium head was used. This efficiency improvement is attributed to more complete combustion of the fuel caused by the less-rich operation of the titanium head. Improved combustion performance at slower speeds is important as it may expand the overall operating range of an engine used in UAV propulsion.

18.4.1   Combustion results and analysis

The in-cylinder pressure was measured (using the titanium head) for every 0.5 crank angle degree at MBT for speeds of 4000, 6000, and 7000 RPM. The 300-cycle ensemble-averaged pressure traces are shown in Figure 18.8, where the top dead center corresponds to 180 crank angle degrees. Analysis of the pressure traces shows that the peak cylinder pressure increased with decreasing speed. The crank angle positions for the peaks were 12, 16, and 18 degrees after the top dead center for speeds of 4000, 6000, and 7000 RPM, respectively. Although these peak pressure positions are similar to those for automotive-scale SI and compression ignition engines, the measured peak pressure values are quite low. High heat losses and incomplete combustion may be partially responsible for the relatively low cylinder pressures, but the much lower compression ratio is probably the primary reason.

As shown in Figure 18.8, the pressure rise rate was observed to increase with a decreasing speed; and both, it and the corresponding peak pressures, were caused by the combustion process starting earlier in the cycle as the speed is reduced. As the piston speed is reduced, more time elapses per degree of crank angle, and therefore, more time is available for combustion to initiate and propagate. As the SOC is advanced (occurs earlier in the cycle), the subsequent pressure rise counteracts more strongly against the piston movement resulting in higher pressure rise rates and higher peak pressures.

The inclusion of the pressure sensor enabled the characterization of the combustion process in a very small 4 cc engine. These measurements were not possible without placing

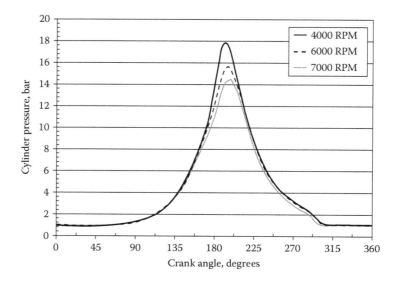

Figure 18.8 Averaged in-cylinder pressure measurements at MBT for speeds of 4000, 6000, and 7000 RPM.

a sensor port on the head. For this particular application, traditional cast and machining processes would have added both cost and time to the effort. Especially, time consumption would have been the costs associated with mold development. AM provided a cost- and time-effective means of producing this prototype head by eliminating the mold development step. Concerns centered on whether printed components were durable enough to survive the pressure, temperature, and chemical extremes of the combustion chamber of reciprocating engines. The titanium head was operated over 20 hours without failure indicating that AM can successfully be used to make components for engine applications.

18.5   Printed component lessons learned

Several factors must be taken into account when producing engine components via AM. One of the most important design criteria is to account for settling or possible warpage following printing, which can occur even for very small components less than 15 cm in length and weighing less than 200 grams. This issue was demonstrated during fabrication of a titanium engine block for the small two-stroke engine described in the previous section. Originally, the block was printed in vertical arrangement starting with the top of the block to the bottom. This arrangement produced a small reduction in the distance from the center of the crankshaft to the top of the cylinder head, which resulted in the piston physically hitting the cylinder head surface. In order to avoid this issue, it was necessary to print the engine block in a horizontal arrangement. Because engine components are held to very tight physical tolerances, factors related to uniformity and physical changes must be considered during the design stage.

Keep in mind that AM is still relatively new in its application to general fabrication, and there are no known published formal design guides to direct its use. This is especially true for printing metal. Machining tolerances must allow for the precision level of the printer. For titanium, enough extra material must be printed to permit proper surface finishing. Holes for threaded fasteners are better left out completely, so that they can be accurately placed without concern for machine tools/drills being misdirected by a slightly off-printed hole. Titanium is difficult to tap, especially for smaller cap screws; if at all possible, small size screws should be avoided. Titanium galls easily and therefore has issues with application in any moving interfaces. Those surfaces should either be coated (nitrides, zirconium, etc.), or a bushing should be used. The ORNL-printed titanium block used a bronze infused stainless steel-printed bushing. Minimum wall thicknesses must be maintained to provide strength and avoid any issues from any defects in the printing process. Generally with machines that exist at the time of this writing, it is best to add material to a wall thickness and to add additional material if that surface will be machined. Depending on the application, 1–2 mm additional material may be required.

In the manufacturing process with the metal powders and e-beam fusing, printed components are normally postprocessed by blasting with the same printed powder, so that the residuals from the printing process may be collected and reused. It can be quite difficult to completely remove all granules of powder from the printed object. It is also possible that particles may later break loose from internal structures and cause scoring on moving parts. This was seen on the small-scale printed engine work. Abrasive grit blasting after blasting to recover the print media may help remove particles on inner surfaces that might otherwise not break loose until the part is in use. An ultrasonic bath has also been found useful. Some benefit has also been found from extrude honing on some types of parts.

Engineers and designers spend their early career learning how to design parts that can be fabricated with traditional machining techniques. This must be unlearned at least partially in designing for fabrication with AM, or the maximum benefits will not be achieved. Interestingly, students and less experienced engineers will adapt much more quickly than the most experienced staff. Expect the adaptation to be a process that will take time before an organization can realize the full potential that AM has to offer.

18.6   Summary

AM has been demonstrated to facilitate rapid prototyping of engine components and has value in the creation of small production lots of customized parts, especially when those parts include complex features that would be difficult or impossible to achieve with conventional fabrication techniques. These AM techniques were successfully used to facilitate investigation of the science of small-scale combustion and are ready to use for large-scale engine parts. While some of the materials may be difficult to work, the benefits to research are obvious.

References

1. Ford, Sharon. "Additive Manufacturing Technology: Potential Implications for U.S. Manufacturing Competitiveness." *Journal of International Commerce and Economics*. Published electronically September 2014. http://www.usitc.gov/journals.
2. William E. Frazier, Metal Additive Manufacturing: A Review, *Journals of Materials Engineering and Performance* (2014) 23: 1917–1928.
3. Curran, S., Chambon, P., Lind, R., Love, L. et al., "Big Area Additive Manufacturing and Hardware-in-the-Loop for Rapid Vehicle Powertrain Prototyping: A Case Study on the Development of a 3-D-Printed Shelby Cobra," SAE Technical Paper 2016-01- 0328,2016, doi:10.4271/2016-01-0328.

4. AMIE Demonstration Project. Retrieved from http://web.ornl.gov/sci/eere/amie/ on January 19,2017.

5. D. Eylon, F.H. Froes, R.W. Gardiner, Developments in Titanium Alloy Casting Technology, in *Titanium Technology: Present Status and Future Trends*, Titanium Development Association, Dayton, OH (now The International Titanium Association, Editors F.H. Froes, D. Eylon and H.B. Bomberger, 1985, pp. 35–48.).

6. Beiker Kair, A. and Sofos, K. Additive Manufacturing and Production of Metallic Parts in Automotive Industry/A Case Study on Technical, Economic and Environmental Sustainability Aspects. Masters Thesis in Production Engineering and Management. Kth Royal Institute of Technology, Stockholm, Sweden, June 2014.

7. John Newman, 3D Printing Applications, Koenigsegg Harnesses Additive Manufacturing for the One:1, April 2014, http://www.rapidreadytech.com/2014/04/koenigsegg-harnesses-additive-manufacturing-for-the-one1/.

8. Cooper, D.E., Stanford, M., Kibble, K.A., Gibbons, G.J., Additive Manufacturing for Product Improvement at Red Bull Technology, *Materials and Design* (2012). doi:10.1016/j.matdes.2012.05.017.

9. Michael D. Kass, Mark W. Noakes, Brian Kaul, Dean Edwards, Timothy Theiss, Lonnie Love, Ryan Dehoff, John Thomas, *Experimental Evaluation of a 4-cc Glow-ignition Single Cylinder Two-Stroke Engine*, SAE Paper No. 2014-01-1673 April 2014.

chapter nineteen

Developing practical additive manufacturing design methods

David Liu, Alan Jennings, Kevin D. Rekedal,
David Walker, and Hayden K. Richards

Contents

Additive manufacturing (AM) has developed past specialty prototyping to having commercially feasible production parts. The increased production cost can be offset by smaller logistical requirements, better performance, or incorporating multiple functions into a single part reducing part count. AM allows for creating parts that would be impossible or impractical for traditional manufacturing methods. This chapter presents how AM can support military applications to illustrate the principles for applying it to other industries. The focus of the chapter is first on material characterization to address differences from traditionally machined parts and second on advanced design methods such as topology optimization. Aerospace applications are presented to demonstrate applications for the use of topology optimization in AM.

19.1   Motivation

Starting from 3D plastic printers, AM machines have come a long way to delivering production-quality metal parts. In the metal-printing process, metal powder is melted layer-by-layer building up a part via selective laser melting (SLM) or electron beam melting (EBM) in a single direction. To print in any direction, material deposition processes (MDP) can lay down material along the path of the tool. Recently, more aerospace materials, including inconel, stainless steel, aluminum, and titanium have become more readily available. Although additive techniques can construct the bulk shape, tolerance and surfaces finish specifications typically require machining. Support structures' removal may also need machining for removal. In most cases, the finished products or components will also need heat treatments to relieve AM-related build stresses [1].

Since the beginning of warfare, logistics has been a central cornerstone to successful operation. Over the last decade of war in both Iraq and Afghanistan, logistics has driven the need for the United States to have strategic bases in the region with a sizable in-theater footprint keeping our aircraft in the fight. Aircraft battle damage repair (ABDR) replaces or repairs parts to provide a temporary fix. After the ABDR's flight limit, a permanent repair or inspection will be performed [2]. Fully stocking every part is infeasible, so being able to print complicated geometries would prevent long downtimes waiting for replacements from CONUS. For safety critical parts, they must have an engineering determination for the part life; otherwise the repair could risk damage to an expensive airframe or even the life of the aircrew.

Topology optimization applies optimization to finite element analysis to determine the best placement of material [3]. These designs typically have a much more organic look as the material branches to directly support the load on the part shown in Figure 19.1. Traditionally, these designs were discarded as impossible or impractical to manufacture. AM makes them feasible and beneficial since they require less material, so they are lighter and print faster. Whereas ABDR is primarily about replicating parts, topology optimization tools aid design and analysis, whether in-theater or during acquisition.

This chapter presents work done to help establish material properties and design methodology. These are two tools necessary for an engineer to make good design decisions. The goal of this chapter is that an engineer who is exploring AM applications can understand some of the challenges and opportunities of additive-manufactured parts. To this end, principles and examples are provided. The reader can then apply them to the domain of interest. If the reader wishes to develop the field, the chapter will give an introduction for exploring the references provided.

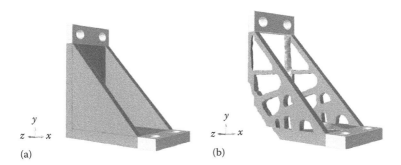

Figure 19.1 From the valid design area (a), material is removed to support a load. Because material is placed along the load paths, the optimized design (b) is stiffer than a tradition design with the same mass.

19.2   Fatigue characterization

19.2.1   Test samples

In order to determine the hours for flight limits, characterization of AM aerospace materials is required. To this end, fatigue tests of Ti-6AL-4v (Ti64) samples created with SLM were performed. The methodology presented in this section is representative of other material testing, whereas there is a very long history of properties for conventionally formed materials, there is a great deal work required for characterization of additive-manufactured materials in order to establish empirical factors needed for determining performance. Some standards are currently in place but are not yet accepted by industry or fully justified by data.

SLM uses a finer powder than EBM and with layer thicknesses of 30–60 microns. The ASTM E8/E8M standard specimen illustrated in Figure 19.2 took roughly 500 layers to reach a thickness of 20 mm. Due to layer thickness, stair-step ridges can form on shallow curves in the print direction, as shown in Figure 19.3, whereas the opposite side is cleaned by a wire EDM cut. By having both features on the test specimen, results will show preferential crack formation if surfaces on ABDR parts are unfinished [4].

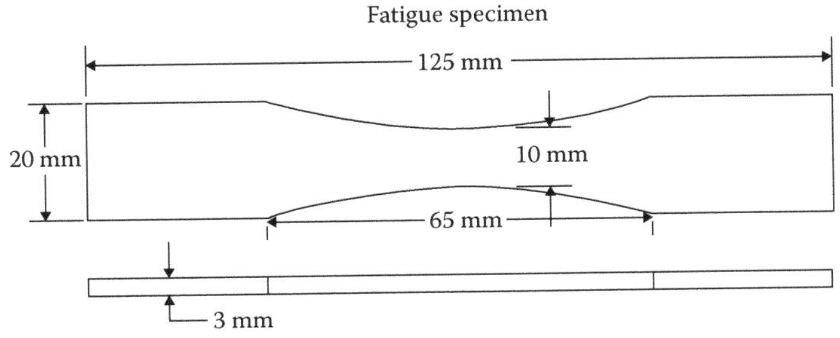

Fatigue specimen

125 mm

20 mm

10 mm

65 mm

3 mm

Figure 19.2 Schematic of test specimens for fatigue testing in accordance with ASTM E8/E8M standards.

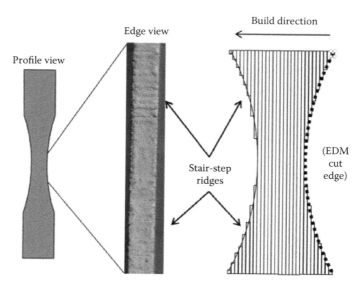

Figure 19.3 Stair-step effect in inclined surfaces of additively manufactured parts for selected laser-melted Ti-64 test specimens.

19.2.2   Fatigue life

Fatigue curves have two regions. For high-stress fatigue, a lower stress (S) results in prolonged life (N). Once the stress drops below the fatigue limit, life is dramatically increased. AM parts do follow the expected knee curve, as shown in the stress–life (S–N) curve on Figure 19.4. Nevertheless, more data collection is required on how fatigue changes with test parameters (e.g., round vs. flat samples, R: stress ratios) along with manufacturing parameters (e.g., powder size, production time, build direction) [4].

Researchers have used various treatments on AM-produced parts to alter the high-cycle fatigue properties. In particular, Figure 19.4 illustrates the effects between a heat treatment and a hot isostatic pressing (HIP) treatment on Ti64. Although the fatigue life is similar for both treatments, the high stress fatigue region of the S–N curve shows a 100% increase at 300 MPa and a 60% increase at 500 MPa in the average number of cycles to failure.

A finished surface also provides an increase in the average number of cycles to failure. A life increase is expected due to the decrease of crack initiation sites. For the HIP-treated specimens, a few samples were machined on the edge to determine the importance of postprocessing, which may have limited availability in-theater. In comparison of the results from Figure 19.4, the data show that the effects of machined edges have an 80% increase at 300 MPa and 75% increase at 500 MPa in the average number of cycles to failure.

19.2.3   Crack behavior

Surface examinations were also performed on the fracture surface. For Ti64, the typical fracture surface is shown in Figure 19.5. The cracks originated at the as-built edge and would propagate toward the EDM side of the specimen. The location of the crack initiation region was expected there, since the rough surface provided ample crack initiation sites. The crack propagation region is shown with both the crack propagation surface and fatigue striations. The fast fracture region is identified by the ductile dimpling.

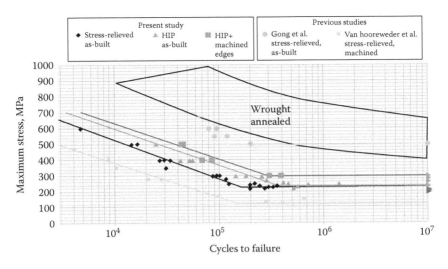

Figure 19.4 The left side of the fatigue curves shows how life is limited by the stress. The right side shows the upper limit of stress for high-life products.

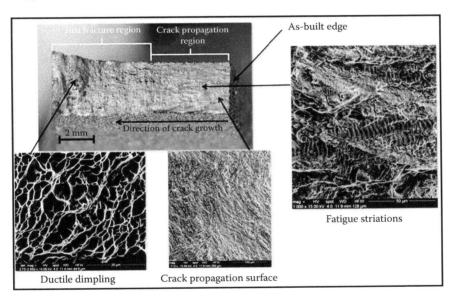

Figure 19.5 Fracture surface examinations for AM Ti64 specimen in both the fast fracture and crack propagation regions.

19.2.4 Summary

The data presented show that the AM Ti64 material did not have a high-cycle fatigue lifetime as high as wrought Ti64. However, the study showed that HIP treatments can increase the high-stress fatigue region for AM Ti64. The ability to predict fatigue properties will allow ABDR engineers to better estimate component lifetimes for the repair. Additionally, the understanding of treatments allows engineers to design to a desired life for a given part. This example shows one of the many possible studies of heat treatment effects. Due to the limited market share of additive-manufactured parts, it will take time to gather the data to fully characterize treatments.

But as more companies develop additive-manufactured parts, companies will develop sets of best practices to improve product performance.

19.3   Topology optimization

AM lends itself well for topology optimization. The focus on topology optimization is on choosing where the material should exist, or conversely where material needs removal. At the end of the optimization, the finite amount of material allowed is distributed offering the best result.

With any design, manufacturing considerations are incorporated in determining superior designs. Standard material sizes, access to cavities, fixturing, stamping or mold directions, and rigidity for cutting operations are prime considerations for mass production where manufacturing cost drives the design. Additive-manufactured parts do have design considerations as well, such as the requirement to clean support material, wall thicknesses limits, slope limitations, and rate of cross sectional change, but these limitations have little to do with the overall shape of the part. With AM, designs requiring extensive welding are as simple to create as conventional designs. Specifically, the *organic* branching structure seen from topology optimization has the same manufacturing burden as conventional cast or machined designs, except with lower mass and/or greater stiffness. Getting a new sense of intuition for the component design is at the heart of applying topology optimization for AM. Whereas the focus for conventional design is on removing material from standard shapes, the focus for AM is on where to place material.

Topology optimization consists of determining the connectedness or shape of the structure. Traditional designs are based on traditional practices which consider manufacturing. By employing topology optimization early in the design process, an intuition is developed for the ideal shape. Because this design is optimal according to the criteria used, it can serve as a baseline for further development. Topology optimization is particularly important for additively manufactured parts to help open and direct the design process.

The focus of this section is to describe the principles of topology optimization as applicable to AM. If the reader is more interested in the calculations and theory of topology optimization, [8] is recommended and can be considered as a reference for much of this section. Finite element analysis is used to evaluate how individual locations influence the broader behavior. With typical optimizations, the designer must understand how cost/fitness and constraints are defined whereas the optimization process is typically treated as a black-box. Topology optimization software is readily available, and commercial products are presented. With the foundation of topology optimization, other optimization techniques, for example, plate thickness and shape optimization, are compared and contrasted with topology optimization. Refining topology optimization results with other methods is suggested. The section concludes with application and limitations of applying topology optimization to component design.

19.3.1   The topology optimization process

19.3.1.1   Finite element analysis

At the heart of the topology, optimization process is finite element analysis. Structural problems are approximated by *small* elements which locally match the structural behavior of solids, plates, beams, or other fundamental units. These elements are then connected to add the stiffness of individual elements, $[k_i] \in \mathbb{R}^{m \times m}$, which are combined

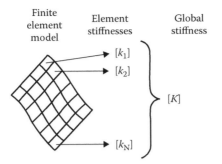

Figure 19.6 The finite element process consists of adding local stiffness matrices together to form a global stiffness matrix.

to approximate the stiffness of the whole structure, $[K] \in \mathbb{R}^{M \times M}$, where m is the number of nodes in element i, and M is the total number of nodes in the structure (illustrated in Figure 19.6).

The finite element method has established itself as sufficiently efficient and accurate despite errors. These errors can arise due to assumptions made when deriving the elements or limitations of the resolution of the results. Good modeling practices limit the extent of these errors. Many times finite element analysis is the only way to simulate the behavior of components offering no analytical, continuous solution.

The finite element method uses a linear approximation of the stiffness,

$$F = [K]X \qquad (19.1)$$

where $F \in \mathbb{R}^M$ are the applied forces and $X \in \mathbb{R}^M$ are the displacements at the nodes. The displacement is found by inverting the stiffness,

$$X = [K]^{-1}F \qquad (19.2)$$

which is typically accomplished via numeric methods. Constraints are required to specify the known location of nodes. These fix the structure and identify the location and magnitude of the reaction forces. Only the unconstrained segment of the structure is used for solving the displacements.

Linearity means the gradient of displacement is based on the stiffness matrix. The gradient assumption significantly simplifies the calculations for optimization. Though the application of topology optimization to nonlinear problems is possible, in principle, the computational complexities limit practical application.

19.3.1.2   Optimization parameterization

The topology of the structure is optimized by adjusting the magnitude of the stiffness of individual elements, $\alpha_i^\gamma [k_i]$, where α is the adjustment, and γ is the penalty factor. The mesh of elements provides a raster representation across the structure. The local stiffness adjustment is traditionally considered as adjusting the density, α, at the selected location. A penalty is applied to the scaling, for example, $\gamma = 2$ or 3, so a half-dense element may only have a quarter or an eighth of the stiffness as a fully dense element. Therefore partially dense elements are less effective for the same amount of mass. Optimizing the designs tends to results with a binary mixture of voids and fully dense material. The magnitude of the penalty is used to adjust the push to a final, binary result for binary results the optimization.

19.3.1.3 *Optimization objective*

The optimization problem requires an objective. Minimizing compliance is a popular objective because it is mathematically convenient and has practical applications. The actual expression for compliance minimization is the strain energy:

$$U = F^T X$$
$$= F^T [K]^{-1} F$$

(19.3)

Because the loading is the same throughout the optimization, F scales the objective by a constant amount. More importantly, it selects the directions relevant to the optimization. This means the structure is optimized for the given loading condition. The optimization step uses the stiffness matrix along with the current displacements to determine where to increase or decrease the element density. Globally defined quantities, such as vibration modes, or results at a specific location, such as the deflection at the tip or an element's stress, also provide convenient gradients. This is because the global results are condensed to a scalar value by well-conditioned functions. Results at unspecified locations, such as the maximum deflection of the structure, are based on using the extremum in a set. As the location of the extremum changes from one point to another, the gradient will change. This results in a poorly conditioned objective that may have many local minima.

19.3.1.4 *Optimization constraints*

Constraints are typically required for optimizations. The design variable, α, is given an upper bound of fully dense, 1, and a small, positive lower bound, for example, 10^{-3}, to prevent singularities in the stiffness matrix. If the local stiffness went to exactly zero, then nodes could have no connection to the structure and have an unbounded displacement. The total amount of material in the final design is limited by a volume fraction. If different materials are used, a total mass constraint is more appropriate and is straightforward to implement.

Deflection limits and stress are also commonly used for constraints. Limits over a set are defined as an individual constraint for each point considered. If the set is large, then the computational complexity can quickly grow. Therefore constraints should be limited to locations where they are likely active, rather than being applied to the entire model. Note that these are optimization constraints, not the boundary condition constraints for the finite element problem.

19.3.1.5 *Optimization methodology*

The optimization process is based on direct numeric optimization using Lagrange multipliers. The specifics of the numeric algorithm do not help in understanding how to employ topology optimization and are not formally given here. The principle of a direct optimization is the candidate solution, the current iteration's local densities, are incrementally changed to improve the solution. The incremental change is based on the gradients of the objective and constraints at the current candidate solution. Lagrange multipliers provide a mathematical method for converting constraints into an augmentation of the objective. Each constraint is multiplied by a Lagrange multiplier and added to the objective. When the constraint is active, the constraint is exactly zero, and it does not affect the objective, but it does influx the incremental change so that the constraint is not violated. When the constraint is satisfied, then the multiplier is zero, and there is no impact to the objective or the incremental change. When the constraint is violated, then neither the multiplier nor constraint is zero, and they impact the objective causing the incremental change to reduce the violation. The finite element model is used by the software to specialize the analysis and accelerate performance over general purpose optimization software.

For a true linear system, only one step is needed for a direct optimization. This is because the gradient is constant and provides the exact direction to the optimum. However, adjustments are made to the linear analysis to improve the quality of the output. The next section will explain some of the issues arising in topology optimization and processes to mitigate their effect. These nonlinear steps mean the local gradients are not representative of the global gradients. The incremental changes should be small enough, so the gradient changes are minor in each iteration.

19.3.2 Issue arising in topology optimization

19.3.2.1 Existence of a solution
Although it may seem obvious that a solution should exist, there is a possibility that a true solution does not exist. This does not mean a good solution does not exist; rather there is a possibility that any given design allows for improvement. This is the case when the optimized structure is singular. For example, the bending stiffness of a rectangular cross section increases as the thickness decreases and the height increases. The ideal solution is an infinitely tall, infinitely thin member; any finite design can be improved by making another adjustment. Obviously this structure is very prone to buckling, so one method to prevent this singularity is to optimize a combination of compliance for multiple loading cases. In practice, the mesh size chosen will limit how close the singularity can be approached resulting in the closest solution to the singularity the mesh allows. When solutions appear to approach singularities, the parameterization and loading conditions should be reviewed to determine if modifications are required.

19.3.2.2 Uniqueness of solutions
Some loading conditions do not provide a unique solution. For example, deflection due to axial loading is dependent on the total cross-section area and is not dependent on the shape of the cross section. This is a case where the solution is not driven to any particular configuration. Typically the penalty drives the solution to a binary material-void result, which will prevent these insensitive results.

A more common occurrence is the concern of local minimum. Nonlinearities of the optimization can cause multiple local optimums. Local optimums are better than all contiguous surrounding points (i.e., all designs with a small change from the current iteration), so there are no infinitesimal steps which could improve the solution. Results should be considered as one possible result, but it is difficult to prove they are the best of all possible local optimums.

19.3.2.3 Model limitations
The basis of topology optimization rests on using finite element analysis to determine the structural response. All the assumptions from the modeling process are inherited and may not be valid throughout the optimization. Finite elements are based on elastic theory of materials and typically have isometric, homogeneity, and dominate length direction assumptions. Though at the microscopic level, the grain structure of additively

manufactured materials is not isometric, at the macroscopic level, the linear elastic properties are typically isometric, which are used to derive the element stiffness. Manufacturer methods, for example, laser powder deposition, are being developed for blending materials but are not sufficiently developed to address design considerations. Since linear analysis is used, all higher order effects and geometric nonlinearities are neglected.

The most significant modeling concern is that the finite element model does not capture localized behavior. It is well known that loadings applied at individual points will produce unrealistic stresses. Idealizing transitions as sharp edges will also not accurately model stress at the given location, although results are affected minimally a few elements away [6]. Local behavior is a concern when the topology optimization produces sharp local features. In the extreme is the checker-board pattern where elements alternate fully dense to void. Checker boarding occurs with plate elements due to the idealized corner connections of the plate elements (e.g., Figure 19.7). In finite element models, they do not shear but transmit loads through at a single, infinitesimal point. This is often the mathematically correct optimum. Similar problems occur based on the other nuances of the mesh used for analysis. In reality, there is a single physical optimum, and filtering methods are used to reduce the effect of the mesh. The density result, or its sensitivity, is convoluted with a smoothing filter larger than the element size, so the geometry of an individual element does not influence the result.

Recall densities are constrained from reaching zero with a small, positive lower bound. The density lower bound was added, so that there is always some stiffness connecting every node, preventing infinite displacements. The singularity in the stiffness matrix would be an invalid result. The nonzero amount of material still counts against the total mass or volume of the design, so the total mass of the *void* area should be checked at the end of the project to see if it was negligible. In the rare case it is significant, the design space should be reduced to more closely match the presumed ideal design. The final result will then have a larger volume fraction with the void having a smaller effect. Previous results could be used to provide intuition of how the design space can be trimmed.

19.3.2.4 Computational limitations

Large finite element models are computationally challenging in certain circumstances, even for linear analysis; therefore, there is some upper limit on computation for topology optimization. The number of iterations required depends on the level of fidelity required and similarity of the design space to the result. Each iteration involves solving for displacements and sensitivities. Typically 10–100's of iterations are used along with the convergence criteria. The computation time of a single analysis scaled by an estimated number of iterations can

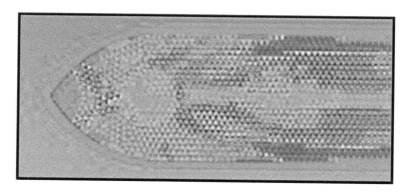

Figure 19.7 A checkerboard pattern is seen here due to not applying filtering constraints.

be used to determining the order of magnitude for processing time. Note the optimization is an iterative process, which is inherently serial in nature. Parallel processing may reduce the individual iteration time but cannot be used to run multiple iterations in parallel.

19.3.2.5   Available optimization software

There are a number of commercial off-the-shelf products for topology optimization. Altair's Hyperworks have placed design optimization at the forefront of the OptiStrut code. Add-ons are available for most major structural modeling programs, such as MSC NASTRAN Optimization Module, and the GENESIS Topology for Ansys Mechanical (GTAM) extension. Topology optimization is incorporated into CFD* (e.g., the TOSCA Fluid Optimization add-on for Ansys Fluent) or multiphysics packages, such as COMSOL's Optimization module. For basic 2D designs, there are open-source options, such as the 99-lines-of-code program for MATLAB [3]. These are products that have matured over more than a decade and fit many industry needs.

19.3.3   Similar optimizations

Topology optimization's focus is determining the connectedness of the structure. Other optimizations have a different focus and are beneficial later in the design phase. This section briefly introduces the concepts of other structural optimizations where the result is conducive to AM. Methods begin with the greatest similarity to topology optimization.

19.3.3.1   Plate thickness optimization

The design variable for plate thickness is the thickness for plate elements. Thicknesses are determined for independent groups (e.g., determining thickness for each arm of a bracket) or for each individual element (e.g., determining where a cover should be reinforced). At the extreme, plate thickness optimization is analogous to topology optimization, where thickness is equated to density. It is important to understand the different perspectives though. Topology optimization considers removing material and dividing the space into material and void through penalty functions. Plate thickness optimization is more focused on reinforcing regions and has a direct analogy to the final structure, opposed to the non-physical partial-densities of topology optimization. Large transitions from thin to thick do violate the plate assumptions, so results should be smoothed for final analysis. Variable thickness plates are ideally suited for AM.

19.3.3.2   Shape optimization

Shape optimization is based on finding the best parameters for a parameterized geometry. Typical applications include sizing members, changing cross-sectional shapes, moving reinforcing spars, or changing the path of extensions. The mesh used for finite element analysis is tied to the parameterized geometry. As a section extends, nodes are shifted according to the original distribution, and elements are connected to the same nodes at their new location. The challenge is to determine a parameterization allowing for adjustment without resulting in a poor mesh. If features cross, then the mesh is likely entangled and will give an erroneous result. For this reason, shape optimization must have an existing design which is then refined and adjusted, opposed to considering large, structural changes to the design. Since the finite element model is based on the design geometry, a final validation of the design is automatically done as part of the shape optimization,

---

* Computational Fluid Dynamics

whereas a topology optimization requires another analysis after clearing partial densities and accounting for manufacturing constraints. The primary reason for shape optimization is it has near infinite resolution as shapes are adjusted continuously. On the other hand, topology optimization is limited by the raster representation set by the computation time and results in noticeably blocky results.

19.3.3.3   Truss optimization

Truss optimization, like topology optimization, optimizes where members are placed to connect together to form a structure. Though truss optimization is similar to topology optimization in purpose, they are very different in method. Discrete members are sized and connected in truss optimization, while topology optimization uses a raster representation. The discrete connections result in a much more complex optimization space. While topology optimization follows gradients to a strong local maximum, truss optimization is strongly tied to the initial configuration. A solution may assume all possible connections initially and then may enlarge or remove them. A factorial number of connections are possible, making this method only suitable for small problems.

Truss optimization has been used successfully in many architectural applications, where many members are straight and standardized. For additive-manufactured designs, topology optimization is recommended for most applications followed by a shape optimization. This recommendation is based on topology optimization-combined exploration of the connectedness, sizing of members, and the amount of reinforcement at joints. The truss optimization may approximate curved beams, but the topology could estimate them directly.

19.3.3.4   Heuristics methods

Heuristics are developed by applying intuition, rather than proofs, to solving a problem. They are very popular in cases where a reasonable solution is quickly determined, but the optimal solution is infeasible or impossible to find. Examples of heuristic optimizations include genetic algorithms, sampling/voting methods, particle-swarms, or other biomimicry. There is a strong allure to using these methods since they can offer a global search to avoid locally optimal results. However, in practice they are typically ad hoc and require high-level application-specific knowledge, at which point the designer probably has an intuition of the ideal result without an analysis. Despite the goal of global optimization, the tuning parameters for the method often limit a global search in practice. Therefore, topology optimization is again recommended due to its maturity and completeness for most applications.

19.3.4   Applications conducive to optimization

Many structural parts are rated in terms of load or deflection. These were the earliest of the objectives implemented for use in topology optimization. Since then, the developments of other objectives to provide more comprehensive part design are utilized. Each new objective requires an efficient method of determining its gradient with respect to the design variables. Structural objectives are available based on deflection, stress, or vibration. Thermal objectives are also utilized for temperatures or flux. Topology optimizations for computational fluid simulations are not currently available, since the blocky edges and partial densities interfere with the computational fluid dynamic simulations.

Often, improving one objective deteriorates performance of other metrics. Multiple objectives are a way of combining multiple objectives in a weighted sum for a single combined objective. Each objective is based on the same or different loading conditions. Since a weighted sum of the individual objectives is used, the gradient is simply the weighted sum of the individual gradients; so computational complexity does not scale exponentially. This linear multiobjective optimization results in more robust designs. For a minimization, poor performance in any individual objective will adversely affect the entire result.

19.3.5 Additive manufacturing constraints

Though manufacturing constraints are limited for AM, there are still some affecting the final design. Some constraints are incorporated into the final design, but others are outside the scope of the current tools.

19.3.5.1 Supported constraints

As was mentioned before, topology optimization seeks solutions partitioned into fully dense material or void. In the extreme, the members are infinitesimally thin, until the material is effectively partially dense from a macroscopic view. Current AM equipment is not able to make high-quality partial dense material and have a minimum member size, on the order of 1–4 millimeters. As was mentioned in the previous section, filtering connections on the length scale of the element size is a good practice to prevent checkerboards and other mesh-dependent results. Manufacturing constraints should dictate the size of the filter window. Note that this is a process to ameliorate the occurrence of small members but is not a hard constraint on the optimization. It is possible that this filtering can cause large areas of partial density.

One suggestion for interpreting partially dense results is to use a lattice with equivalent density for the region. This is more heuristic, since lattices are anisotropic and introduce geometric nonlinearities at a smaller deflection scale. Anisotropic stiffness is important for designing lattices for stiffening, while geometric nonlinearities allow for absorbing impact energy. Accounting for the physically realizable or nonlinear stiffness matrices would add significant complexity to the optimization. Despite the limitations of the model, lattice structures are worth considering for evaluation against only using large members.

19.3.5.2 Unsupported constraints

Other manufacturing conditions are more complicated to relate to the optimization and are currently not supported by commercial programs. Whereas plastic 3D printing uses support material under all deposited material, power-bed metal can build overhangs up to a point. Very shallow overhangs do not yet have the material above them, which will support the final product during the build and as a result can sag and not properly fuse. Long-thin columns are also prone to build failure by catching the spreading plate. While some of these limitations are remedied by adding support material, it is not always allowable due to access to the area for cleaning.

The build process for AM involves localized heating to fuse the material together which may or may not cool prior to the heating caused by the layer above it. When the cross section of the region changes, the cooling rate in the specific area will also change. This transition region will have latent heat stresses which may cause failure at lower loads than expected. These stresses are not accounted for in the optimization process. In general, more research is required to understand build quality before incorporating it into the optimization. The current practice is to use an AM design expert to adjust the design and to determine good build directions.

19.3.6   Finishing the design

Results from topology optimization should be first interpreted into binary design, possibly including a printable lattice. Then the design will require analysis, and if desired, refinement with shape optimization. Results are then compared for consistency with the topology optimization analysis; otherwise, the optimization process may have exploited modeling assumptions which are not realistic. Final designs should be reviewed and adjusted for manufacturing considerations, including residual stresses and finishing operations. A complete design review is then done to ensure other concerns were not created, such as fatigue, dynamic loading, or buckling failures, especially if these were not considered in the optimization process.

19.4   Optimization of an aircraft wing

The idea of topology optimization for aerospace applications is not new. In fact, Airbus has used topology optimization in the design of their structural components in their commercial aircraft fleet [8]. With that said, one interesting research avenue is looking at the possible gains from applying topology optimization for AM components in aerospace applications [6].

Simulated aerodynamic loading was applied on a wing rib shown in Figure 19.8a [9]. A topology optimization was then done to the wing rib with the loads found in the aerodynamic simulations [7]. The resultant-optimized structure, shown in Figure 19.8b, is not typical due to the complicated geometry. Whereas the first design can be manufactured via drilling operations, the optimized rib would require an individualized die, which is very expensive for a custom or prototype part.

The design illustrated in Figure 19.8b is then further expanded to the entire wing as shown in Figure 19.9. Now a complex geometry is printed together as a single part. Not only does this reduces the part count, but also provides near-ideal connections at the spar-rib and rib-skin joints while reducing assembly man-hours. A distinct advantage of topology optimization is the placement of material only in locations to benefit the stiffness. As such, the printed-optimized structure has a mass savings of 15% over the original design.

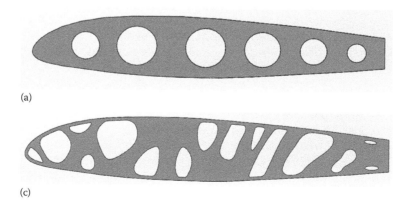

(a)

(c)

Figure 19.8 A wing rib is optimized from the initial design for better stiffness for the aerodynamic loads: (a) baseline design and (b) topology-optimized design.

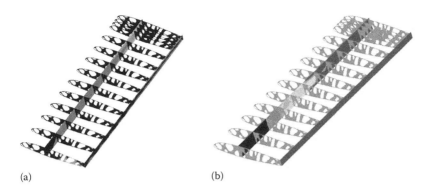

(a)   (b)

Figure 19.9 Optimized ribs can be used to reduce wing mass. The same model can be used to optimize the spar. Skin is included in analysis but is not shown, so the internal results are visible. (a) Stress results for uniform spar and (b) element thickness for optimized spar.

By assuming each rib has the same thickness, further optimization was implemented on the entire wing structure. In this case, a size optimization was performed on the main spar located at the center of the wing. The size optimization determined the required thickness of the main spare from root to tip based on the local strain. Figure 19.9b shows the result of the size optimization. As expected, the region of higher strain is located toward the root of the wing, thus resulting in a thicker main spar at the root. Some cross sections are not uniform with a thinning in the middle. This new design further increased the manufacturing complexity of the entire wing structure since the thickness of members is allowed to vary. As a result, this design is difficult to manufacture via traditional techniques. However, these varying geometries are simple to build in 3D printing.

The current design still does not fully leverage the capabilities inherent in AM. In the proposed design, nearly all components of the wing and current manufacturing techniques would likely suffice. To further exploit AM technology in design, consider a structure which not only supports the design load but also provides another purpose. In such instances, the multipurpose structure shown in Figure 19.10 incorporates both the fuel tank and the wing structure.

The use of multipurpose structures is well established in commercial products most notably in the motorcycle industry. However, with the aid of AM, the practice of using multipurpose structure is now more likely as the complicated design will not increase the cost of production significantly. In the baseline design, the fuel tank was a separate component which does not significantly contribute to the overall strength of the wing. In the new design illustrated in Figure 19.10, the tank is now part of the structure and helps support the aerodynamic loads. The shape and location of the tank were also determined using topology optimization and not determined heuristically.

Once AM is considered, there is no longer a reason to require ribs. An optimization was done for the entire interior of the wing to determine what the ideal stiffener layout would look like. Results considering only solid elements yielded poor performance with many partial density elements in the results without a clear support structure. The partial density elements (tetrahedrons) were then converted to a lattice structure. Each edge of the bars was created into a bar element. The thickness of each bar was then optimized. Despite a very large number of optimization parameters (>90 k), the analysis was reasonably fast (<3 hr).

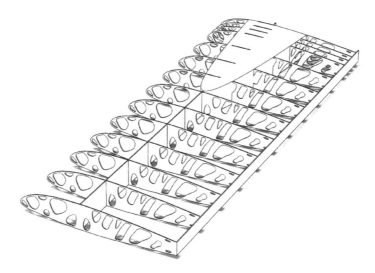

Figure 19.10 Illustration of multipurpose structures for a topology-optimized wing structure where the fuel tank now supports structural loading and is located at the root of the wing. The skin of the aircraft is hidden to illustrate the internal structures.

Results show branching lines of lattice with a reinforcement of the skin near the root shown in Figure 19.11. The branching lattice lines are able to reach a wide area for support with relatively low density. Solid elements alone would result in a support as thick as the size of the tetrahedron, which is much thicker than the ribs considered before. This would result in worse structural efficiency, but computation limitations would never allow for a tetrahedron mesh on the order of the thickness of sheet metal. The lattice structure allows for better structural efficiency. The skin reinforcement is acting as the

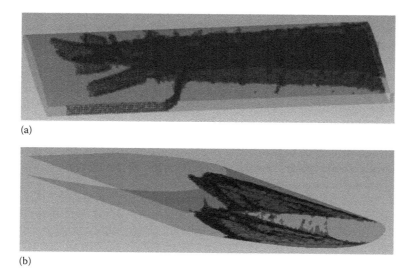

(a)

(b)

Figure 19.11 Optimization of solid core wing into lattice. (a) Full resulting structure and solid and (b) structure with lattice hidden for visibility elements.

main spar. The skin reinforcement is able to spread and approach the ideal I-beam shape of thin, broad flanges, and thin web. In fact, there is a web-like, vertical support near the limits of the reinforcement. These were below the threshold, so they were replaced with lattice structure but are more reinforced than surrounding lattice. The shape of this spar-like structure was used in the inspiration of the integrated fuel tank. If the fuel tank is integrated into the structure, then its mass would not be parasitic to the structure, rather it would allow for a lighter wing with equivalent compliance.

As these results show, AM opens the door for topology optimization through complex, custom part design. Topology optimization focuses the design of AM parts to loading conditions, so parts can be custom designed for applications. Together they enable stiffer parts, with lower mass, and fewer part counts.

19.5   Optimization of penetration warheads

As potential adversaries have witnessed over the last decade, the United States is very successful in the implementation of precision-guided munitions. In order to defend against these precision strikes, the strategy is to bury vital systems. Penetrating these structural defenses requires the warheads to have very thick walls to withstand the initial impact without crumpling, so it can deliver the required explosives. However, thicker walled warheads have worse fragmentation patterns resulting in poorer performance. The AM-topology optimization solution combines thin walls for better fragmentation performance with internal stiffeners for better impact performance. These integral stiffeners can only be manufactured by AM techniques. A small-scale prototype was designed and built for exploratory testing [10].

With the purpose of creating a more desirable fragmentation pattern, the thickness of the outer wall was reduced by 50%, and this mass is redistributed within the warhead. To maintain penetration capability, topology optimization was used to determine where material was required to support the anticipated loads within the volume of the warhead. Figure 19.12 shows the constraint and body loads applied to the given warhead geometry along with the optimized topology. The load was applied at an angle to simulate an oblique impact, which represents a worst-case scenario. The variations of shade represent the relative material density of each element. From the topology-optimized solution, there are clearly defined truss like structures toward the aft section of the warhead (on right). In the forward region of the topology, the solution shows regions of partial density material. In practice, such partial density materials are not possible within the confines of traditional manufacturing.

The use of AM allows for a compromise between partial dense materials and a solid. In this case, the partial dense region was filled with lattice structure. Stretch-dominated lattice structures were preferred due to their stiffness and strength characteristics for a given mass. The pillar kagome lattice structure was used in this build and is illustrated in Figure 19.13. There are many variations of stretch-dominated lattice, which may have

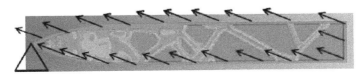

Figure 19.12 Diagram of loads and constraints applied to a penetration warhead. The inner region within the warhead represents density of material required to support the loads, whereas the outer boundary represents the loading.

Figure 19.13 Pillar kagome structure has pillars in the four corners and intersections at the face and cube center. No angles with respect to the horizontal are less than 45 degrees, which means print support is not required.

provided greater support; the pillar kagome solution was selected based on printability. Other stretch-dominated structures have overhang angles which exceed the capability of the AM printer. The near-horizontal lattice members in other structures would collapse during the build due to poor support.

The layout of the internal structure, see Figure 19.14a, is inspired from a 2D topology optimization of a 20± oblique impact. This result shows both regions of high density (solid trusses) and partial density (lattice). A truss structure is drafted to follow dominant loading lines. This truss structure is used as radial ribs, so the warhead is axially symmetric, Figure 19.14b. A lattice structure is placed for the partial density material in the tip. The lattice provides a broad, semi-stiff support, which is very efficient for the tip to resist buckling or compressing on impact. With minor changes to accommodate manufacturing angle limitations and to move the center of mass forward, the final design is shown in Figure 19.15.

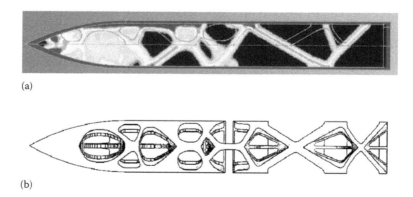

(a)

(b)

Figure 19.14 Topology optimization inspired internal structure design. (a) Internal structure design overlaid on optimization results and (b) rib-based design of internal structure.

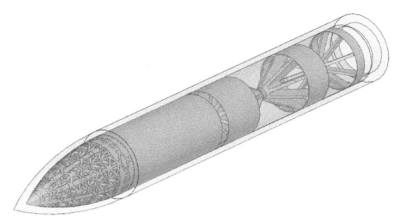

Figure 19.15 3D solid model of the topology-optimized penetration warhead with lattice support structures at the nose. The outer jacket of the warhead was made transparent to show the inner design structure.

The AM topology-optimized warhead testing was conducted at the Air Force Research Laboratory. An AM warhead of equivalent mass and geometry without topology optimization was also tested. The geometry of the *standard warhead* had a baseline wall thickness similar to current penetrative warhead designs. The *optimized warheads* were compared to the standard design to determine the effectiveness of penetration. In the first test series, both the standard and optimized impacted the concrete target slab at approximately the same velocity and the same angle of obliquity. After impact, the depth of penetration was measured for both warheads.

Both the standard and optimized warheads penetrated 60% of their body length. Image of the warhead striking the concrete target is shown in Figure 19.16. From both the penetration depth and the recovered warhead, shown in Figure 19.17, the tests suggested the optimized warhead performed similarly compared to the standard warhead.

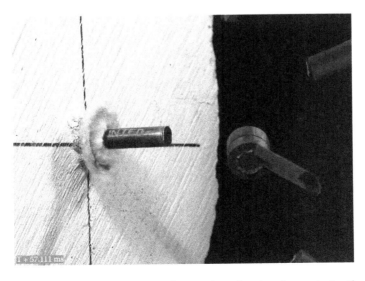

Figure 19.16 Still image of the additively manufactured warhead as it penetrates the concrete target.

Figure 19.17 Additively manufactured warhead after it was recovered from concrete impact testing.

19.6   Summary

The focus of this chapter has been on the potential for AM for solving today's and tomorrow's challenges. The ability to produce many different replacement parts in theater will significantly decrease the Department of Defense's extensive logistical supply lines. Similar applications can be realized for just-in-time manufacturing or product customization. Accurate characterization of material properties, failure criteria, and design guidelines is needed to reduce the qualification process, especially for safety-critical parts.

Additive-manufactured parts are not limited to manufacturing restrictions like those in the original part. Topology optimization allows for getting an intuition for the best allocation of material to satisfy performance requirements. Due to the pervasiveness of engineering best practices based on traditional manufacturing constraints, there is a need for these *clean sheet* design methods. As performance requirements increase, optimized designs are necessary to maintain competitiveness.

AM has a bright future, and there is a significant investment in developing the technology. This drive is no longer primarily supported by research institutes; rather it is supported by commercial companies as they use AM for a competitive advantage. The examples presented in this chapter were chosen to illustrate how AM can benefit the military. While these applications are directed to the military, the principles described in the chapter should help with agile development and supply and better structural performance.

Topology optimization methodologies are largely incorporated into commercial software packages and do not require an expert for its use in design. These tools are well developed and continue to keep pace as the manufacturing techniques improve. However, there is still a need for expert evaluation. Until material behavior is better understood, part qualification is still required. As production ramps up, better understanding of defects will be developed as statistical metrics have sufficient data. These relate not just to the material, but the build parameters, especially part placement and orientation. Current design tools do not incorporate these considerations.

While AM is cost competitive for customized production, it is not cost competitive with standard product designs. For this to take place, a single additively manufactured part should combine multiple parts and offer better performance. The cost of the eliminated parts should be combined with the base part's cost along with any related assembly time when making cost comparisons.

Acknowledgments

The authors would like to thank Mr. Dennis Lindell from the Joint Aircraft Survivability Program (JASP) Office for the support for our research efforts. We would also like to thank Mr. Don Littrell and Dr. Rachel Abrahams from the Munitions Directorate, Air Force

Research Laboratory (AFRL) for their support of the warhead research. In addition, we would like to thank the Materials Directorate of AFRL (Wright-Patterson AFB) for the use of their facilities and expertise.

References

1. I. Gibson, D. W. Rosen and B. Stucker, *Additive Manufacturing Technologies: Rapid Prototyping to Direct Digital Manufacturing*, New York: Springer, 2010.
2. AFMC Handbook, Aircraft Battle Damage Repair Engineering Handbook v2010, Department of Defense Specifications, Wright-Patterson Air Force Base, OH, 2010.
3. O. Sigmund, A 99 line topology optimization code written in Matlab, *Structural and Multidisciplinary Optimization*, 21, 120–127, 2001.
4. K. Rekedal and D. Liu, Fatigue Life of Selective Laser Melted and Hot Isostatically Pressed Ti-6Al-4v Absent of Surface Machining, in *AIAA SciTech*, Kissimmee, FL, 2015.
5. M. P. Bendsoe and O. Sigmund, *Topology Optimization*, New York: Springer, 2003.
6. V. Adams and A. Askenazi, *Building Better Products with Finite Element Analysis*, Santa Fe: OnWord Press, 1999.
7. XFOIL, *Subsonic Airfoil Development System, Software Package, Ver. 6.99*, Cambridge, MA: Massachusetts Institute of Technology, 2000.
8. L. Krog, A. Tucker, A. Rollema and R. Boyd, Topology Optimization of Aircraft Wing Box Ribs, in *Proceedings of the Altair Technology Conference*, Troy, MI, 2004.
9. D. Walker, D. Liu and A. Jennings, Topology Optimization of an Aircraft Wing, in *AIAA SciTech*, Kissimee, FL, 2015.
10. H. Richards and D. Liu, Additively-Manufactured, Lattice-Reinforced Penetrative Warheads, in *AIAA SciTech*, Kissimmee, FL, 2015.

chapter twenty

Optical diagnostics for real-time monitoring and feedback control of metal additive manufacturing processes

Glen P. Perram and Grady T. Phillips

Contents

20.1 Introduction

After two decades of research and development, additive manufacturing (AM) is emerging as a viable commercial manufacturing technology for the production of mechanical parts from three-dimensional (3D) model data. Process advantages of AM over conventional approaches include a potential reduction in energy expenditure by a factor of 4–50,[1] and deposition rates of 1–10 kg/hr conducive for reducing part costs.[2] Because of these advantages, metal-based AM, in particular, has been implemented in automotive and aerospace industries.[3] For example, Aerojet Rocketdyne is now employing AM to build operational RS-68 engines for the Delta 4 rocket and RL-10 upper stage engines for the Atlas 5.[4]

In metal AM, components are built one layer at a time using a metal powder bed, a powder feed, or wire feed process.[5] Processing techniques include selective laser melting (SLM), direct metal laser sintering (DMLS), electron beam free form fabrication (EBFFF), and laser engineered net shaping (LENS).[6–9] AM of metals employing solid free-form fabrication (SFFF) by plasma-transferred arc (PTA) processing may also be used in larger format 3D printing.[10]

To ensure parts produced in an AM process meet production standards, methods for nondestructive inspection (NDI) and real-time process control are needed to assess the physical characteristics of deposited material, as it is formed and before it is encased in an inaccessible location within the part. Diagnostics are sought that can be incorporated into a feedback loop that optimizes build parameters to minimize defects in the construction of high-quality components. To guide the development of AM capabilities, a national roadmap for developing closed-loop control systems was updated in 2013.[3,11,12] The report recognizes that new *in situ* measurement techniques are needed to control microstructure and mesostructure, surface finish, defects, porosity, online convection, powder bed distortion, and other measures of part quality.[3] Process control parameters may include the power density, focus, and dwell time of the laser or e-beam source, powder or wire feed rate, beam scan velocity, hatch spacing, chamber environmental controls (e.g., pressure and oxygen content), and surface thermal controls. Small changes in these processing conditions can lead to large differences in the quality of the fabricated part, providing the sensitivity required for detailed process control.[13]

This chapter describes optical diagnostics currently employed for *in situ*, on-line monitoring of metal AM processes. The chapter also provides brief summaries of related industrial techniques potentially applicable to AM. Finally, the chapter introduces emerging spectral sensing and hyperspectral imagery diagnostics.

20.2   Optical diagnostics employed in metal AM

Optical monitoring of the melt pool characteristics including maximum surface temperature, temperature distribution and gradients, size and shape, and cooling dynamics has been of interest throughout the development of AM process control techniques.[14] Variations in local geometry and sublayer structure drive the need to control laser power or dwell time to accommodate differing thermal transport conditions. In particular, the thermal diffusivity of powder and solid metal in previous layers is dramatic and readily observed with thermal imaging methods.

Thermography, that is infrared thermal imaging, has been employed for process monitoring of SLM.[15] The technique's capability for detecting defects caused by insufficient heat dissipation, porosity, and other irregularities was evaluated in terms of the surface temperature distribution. The heat-affected zone was monitored by analyzing its spatial extent and the temporal evolution. A 640 × 480 pixel camera responsive in the long wave infrared (LWIR) spectral range with a 50 Hz frame rate was used to image a 160 mm × 120 mm surveillance area on a 250 mm × 250 mm build area. Differences were discernible in irradiance profiles as a function of pixel index between a build with an artificial flaw and an unflawed build.

Optical pyrometry usually employs one or more nonimaging detectors to determine surface temperature. For example, a bicolor pyrometer was integrated with a SLM apparatus to monitor surface temperature in the laser impact zone.[16] The sensor was mounted to make observations coaxially with the laser beam to exclude the influence of scanning speed. The pyrometer consisted of two InGaAs photodiodes with two optical bandpass filters transmissive at 1.26 μm ± 100 nm, and 1.4 μm ± 100 nm. The sampling rate was 50 ms. The observation zone was 560 μm. Experiments focused on the observation of the superposition of thermal processes in adjacent tracks due to the laser path passing through the heat-affected zone of the previous track. The pyrometer signal, which varied due to heat loss into the substrate, was shown to have a dependence on hatch distance. This relationship, in turn, was shown to differ according to layer thickness.

The simultaneous employment of two optical techniques has shown some merit. Pyrometry has been combined with visible imagery into a feedback control system for a SLM process.[17] A photodiode and a camera responsive in the 400–900 nm wavelength range were used for monitoring. The CMOS camera had an eight-bit gray value image. Melt pool area, length, and width were correlated to an image gray value. The investigation focused on the influence of local geometry on melt pool size. In particular, when static process parameters were used, local overheating occurred in zones containing overhang structures, producing an increase in the melt pool size which led to dross formation, excessive surface roughness, and differences in local microstructure due to the drastic change in melt pool cooling rate. The conclusion is feedback from the photodiode signal can be used to control laser power to maintain optimal melt pool dimensions.

We recently collected high-speed mid infrared imagery using a FLIR SC6000 InSb ($\lambda = 3 - 5\,\mu m$) camera with 620×512, $15\,\mu m$ pixels while performing SLM of Ti-6Al-4V and Co-Cr-Mo alloy powders. The field of view was limited to 128×160 pixels to increase the framing rate to 420 Hz. The temperature dynamic range was improved for many of the tests by sequencing through four integration times of 0.2, 4, 8, and 12 ms, reducing the effective framing rate to 70–80 Hz. The camera was positioned at 40–54 cm and at 30° relative to the surface normal to view the build surface and monitor surface temperatures. The instrument's field of view with a f = 25 mm lens was $22° \times 17°$, providing a spatial resolution of $513 \times 703\,\mu m$/pixel after accounting for the viewing angle. The camera was calibrated for absolute temperature measurements by observing a large area blackbody at 323 K–873 K in 50 K increments for each integration time. A neutral density filter ($OD = 1.0$) and a band pass filter were employed during testing to prevent detector saturation. For a 0.2 ms integration time, the background signal was ~3,000 counts. Saturation occurred on the 14 bit readout at 16,384 counts. The corresponding temperatures for the background were 621 K and 1102 K. To extend the dynamic range to lower temperatures, the longer integration times of 4, 8, and 12 ms were used in sequential frames of a single movie. The lowest temperature detectable at the background signal is about 475 K. Detection of temperatures higher than 1100 K is not required, despite the high melt temperatures reported in Table 20.1, due to spatial averaging. The standard deviation in temperature across the camera is $T = \pm 0.4\,K$.

An example of the surface temperatures observed from the calibrated FLIR camera is provided in Figure 20.1. The laser at $P = 150\,W$, is scanning at $v = 150$ mm/s diagonally from the bottom left to the top right side of the figure. The total scan distance is 4 cm, and five separate single frames separated by 12.5 ms are coadded to illustrate the laser scan direction. The oxygen concentration has been minimized (0.2%) to produce a better build in Ti-6Al-4V powder. The maximum temperatures of 1000 K are lower than the melt temperature due to the rather large area associated with a single pixel of $513 \times 703\,\mu m$, relative to the laser spot size of $100\,\mu m$. For example, a Gaussian temperature profile with peak temperature of 2000 K and half width at half maximum of $240\,\mu m$ yields an average temperature in a pixel of size $500\,\mu m$ of about $T = 1050\,K$. Deconvolution of the imaging point spread function with a thermal model of the material response to laser heating would enable a full determination of the peak temperature. It would appear that thermal

Table 20.1 Metal bulk properties

Property	Ti-6Al-4V	CoCr
Density, ρ (g/cm³)	4.42	8.30
Specific heat, c (J/g °C)	0.53	0.43
Melting temperature, Tm (K)	1878–1933	1653–1713
Thermal conductivity, k (W/m K)	6.70	12.5
Thermal diffusivity, k (cm²/s)	0.0029	0.0035
Heat of melting, hm (kJ/kg)	360	315

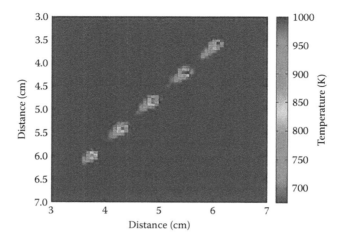

Figure 20.1 Ti surface temperatures with five frames, 125 ms apart, coadded. Processing parameters were v = 150 mm/s, [O_2] = 0.2%, and P = 150 W.

diffusion extending the melt pool by a factor of 2 and 3 larger than the spot size is a necessary assumption for consistency with the present observations.

To examine the spatial distribution of temperatures, a single frame is analyzed in Figure 20.2. The temperature profile along the laser scan and across the track is compared. The leading edge and cross-track dimensions are similar, with a significant decrease in the temperature beyond the central pixel. The temperature remains elevated trailing the laser spot. The characteristic exponential decay time is about 5 ms, considerably longer than the radial diffusion time for the bulk metal. For example, the thermal conductivity of bulk Ti is 13.8 times higher than Ti powder with 63 μm diameter.[18] The influence of particle size can be dramatic. When particle size decreases from 100 to 10 μm, the thermal conductivity can decrease by a factor of 3.[19]

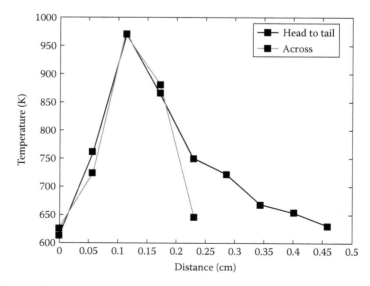

Figure 20.2 FLIR surface temperature along and across the laser track.

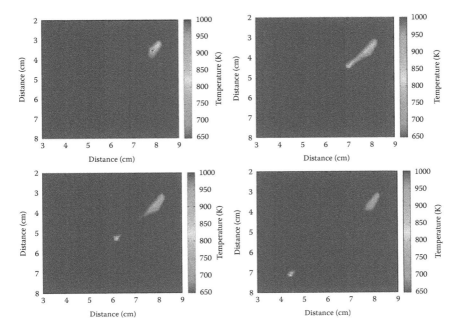

Figure 20.3 Four frames separated by 62.5 ms from FLIR surface temperatures. Only the first 2 cm of the 6 cm scan has unmelted powder in the lower layer.

The decreased thermal diffusivity for the powdered Ti alloy is dramatically demonstrated for multilayer builds. In Figure 20.3, the surface temperatures for a second layer are shown for several points in a single laser scan. Powdered Ti alloy exists below the first 2 cm with previously melted Ti alloy beneath the remaining 4 cm of the scan. Cooling in the regions with powder in the lower layer is dramatically slower. The temperature decays exponentially to background with a decay time of ~25 ms when powder exists in the lower layer, nearly 5 times longer than the 5 ms observed for the metal substrate.

The imaged surface temperature varies by ±5% as the laser spot is scanned at 150 mm/s. Even for conditions with low oxygen and reduced particle ejection, the energy delivered to the surface is likely varying, inducing a fluctuating melt pool size and irregular surface in the build material. A fast control loop to maintain a constant surface temperature would be a key diagnostic.

The temperature maps are also sensitive to gas chamber conditions. For example, in Figure 20.4, the temperature at the laser spot is independent of oxygen concentration, but the temperature remains high behind the laser spot when oxygen is present. Presumably, the exothermicity resulting from the Ti oxidation reaction provides increased heating after the laser spot has translated.

The surface temperature maps provide signatures that are sensitive to laser power, oxygen content, powder size, and sublayer structure. Statistical uncertainty in the measurements has been demonstrated at $\Delta T = \pm 0.4\,K$. Absolute temperatures require deconvolution of the imaging point spread function, which would be aided by a shorter focal length lens.

A brief overview of some basic relationships between typical AM process parameters is presented below in the context of AFIT's work with Ti-6Al-4V and cobalt chromium molybdenum alloys in a powder bed, SLM system. The basic thermal properties of the two alloy powders are provided in Table 20.1. Ti-6Al-4V, grade 5 titanium samples are 6% Al and 4%

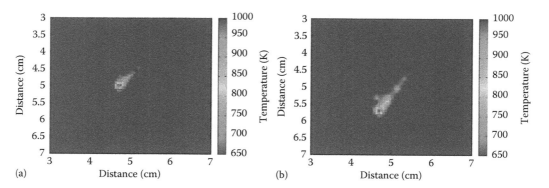

Figure 20.4 SLM of Ti-6Al-4V in an atmosphere of (a) 0.2% O2 and (b) 4.3% O2.

vanadium alloys. The cobalt chromium molybdenum samples in Table 20.1 are 59–65% Cr and 5–7% Mo. Thermal properties depend on particle size, with powders often sifted to provide an average particle diameter of $d = 10-50\,\mu m$. The powder layer thickness is usually >100 μm. Parts may be built on various substrates including aluminum and stainless steel.

The fluence required to melt the surface depends on the thermal properties and the size of the melt pool. For a thin melt pool with no thermal diffusion, the melt fluence is[20]:

$$F_m = \frac{\rho l}{(1-R)}\left[c\left(T_m - T_o\right) + h_m\right]$$
(20.1)

where:

ρ is the mass density
c is the specific heat
l is the thickness of the melt pool
h_m is the latent heat of melting
R is the surface reflectivity

The delivered fluence is expressed as

$$F = \frac{P}{A}\tau = \frac{P}{A}\frac{2w}{v}$$
(20.2)

where:

P is the laser power
A is the beam area (where $A = \pi w^2$)
w is the laser beam diameter
v is the scan rate

The laser dwells on a single spot for a time interval given by $\tau = v/2w$. For a scan rate of 500 mm/s, a laser power of $P = 53\,W$ is required to produce a 0.25 mm melt layer in the Ti alloy powder. For most of the build conditions, the laser power is greater than that required for melting a thin layer, and the size of the melt pool may be somewhat larger than the laser spot size. Melt pool size should be consistent with the selection of hatch spacing between laser scans.

Thermal conductivity will increase the required power if the diffusion time,

$$\tau_d = \frac{w^2}{D} = w^2 / \left(\frac{k}{\rho c} \right) \qquad (20.3)$$

is short compared to the dwell time. For higher scan rates, the time for thermal diffusion is longer than the dwell time, and heating is largely limited to the laser spot.

In comparison, CoCrMo AM processing typically requires more laser power or a lower scan speed. For a scan rate of 500 mm/s, a laser power of $P = 76\,W$ is required to produce a 0.25 mm melt layer in the CoCrMo powder.

In general, melt pool formation with short dwell times requires higher laser irradiance and can induce plume chemistry and particulate production. The surface tension induced in small melt pools can pull the powder into the melt pool, leading to rough surfaces. The melt pool size and its aspect ratio are critical to the surface tension.[21]

Improvements in metal AM could be achieved by further developing strategies to automatically monitor the molten area in various layers, the *in situ* release of stresses induced by temperature gradients, alloy composition, or contaminants.

20.3   *Optical diagnostics employed in related industrial processes*

Optical diagnostics have been utilized for real-time process monitoring in a variety of industrial processes to include laser cladding, laser beam welding, machining processes, and metal emission control. Several comprehensive reviews are available.[22–24] A few approaches relevant to AM are highlighted in this section.

Atomic spectroscopy has been used to estimate penetration depth in laser welding processes based on emissions from the laser–material interaction zone.[25] Spectral line intensities of iron present during the welding of stainless steel plates are used to compute a plasma electron temperature.[26] The variation in electron temperature with penetration depth shows potential as a means of countering laser beam attenuation and defocusing due to the presence of the plasma plume, which can lead to insufficient penetration depth and other weld flaws. The changes in electron temperature observed are partly due to sensor positioning and the fact that the plasma may be partially obscured due to its formation at a given depth within the keyhole and not a true change in plasma plume temperature. Regardless, the observed signal serves as an indicator for process control. It is noted that the optical signals can exhibit a wide background contribution due to thermal blackbody radiation.

The correlation of atomic, gas phase, chemical species to AM process parameters has promise. The technique does not require source illumination, and the spectroscopy is well known. Atomic species likely to be encountered in AM include Fe, Ti, Cr, and Co depending on the alloy of choice.

In other work, a high-power laser welding process was monitored with a high-speed camera equipped with ultraviolet and visible band pass filters.[27] Features such as size and growth direction of the laser-induced plume, the splatter radius, and its ejection direction, gray value, and velocity were shown to be useful indictors of weld quality.

Optical monitoring of laser cladding process may also be transferrable to AM. One technique employs a visible spectrometer to collect emissions from the plasma plume during the laser hot wire cladding of Inconel 625.[28] Emissions from chromium and nickel were used to calculate electron temperatures, which were then correlated with surface appearance, clad dilution, hardness, and microstructure.

A technique shown to detect chromium spectra may also be useful for AM processing of cobalt–chromium–molybdenum alloys, which are of interest to the orthopedic implant industry.[29]

An optical technique devised for surface roughness characterization during a standard machining process might be adapted to AM monitoring.[30] Coherent light scattered from a rough surface becomes less diffused as the surface is smoothed. Specimens of tool steel were illuminated with a He–Ne laser ($\lambda = 628.8$ nm). Light was collected at the reflection angle with a photodiode. The ratio of the specular light intensity to the incident intensity was used to characterize surface roughness. A similar light-scattering diagnostic was modified to include a compressed air source to clear the workpiece surface of debris for accurate measurements.[31] Surface roughness measurements differed by less than 10% when compared to conventional stylus measurements.

Vorburger and Teague have provided a review of optical techniques for surface topography to include specular reflectance, total integrated scatter, diffuseness, angular scattering distributions, speckle, ellipsometry, and interferometry.[23] While some of these techniques may only be applicable postprocessing, it is worthwhile to evaluate their potential for *in situ* measurements.

Finally, many spectroscopic and spectrometric techniques have been employed as online monitoring systems for metal emission control in industrial processes.[22] Many are established laser-based methods. Some techniques can detect chemical species encountered in AM. For example, heavy metals such as Co, Cr, Ni, and V have been investigated with laser-induced plasma spectroscopy (LIPS).[32] The continuous monitoring of Fe in the top gas tube of a blast furnace was accomplished with laser-induced breakdown spectroscopy (LIBS).[33]

20.4   Advanced sensors for laser–material interactions

20.4.1   Fast framing hyperspectral imagery

Fast framing hyperspectral imagery based on imaging fourier transform spectroscopy (IFTS) has recently been developed to monitor combustion events and laser material processing.[34–36] Correlations between IFTS spectral or spatial signatures and manufacturing defects might be used to enhance AM process control. Adding high-resolution spectral information to fast framing imagery may enable monitoring of surface temperatures with high accuracy and mapping the chemical evolution of laser-irradiated surfaces. In the present work, we seek to demonstrate the utility of IFTS high-speed broadband imaging and spectral analysis for monitoring SLM processing of metals and alloys. Extension of the IFTS sensor to metal AM will require reduction of focal distances to achieve increased spatial resolution as well as development of new methods for image interpretation. Our strategy involves complimenting the IFTS with high speed, visible, and infrared focal plane imagers that possess high spatial resolution. Merging data from these sensors and extracting key imagery features that correlate to process control and manufactured part quality will inform development of a lower cost system for SLM processing. This prototype sensor system will likely combine spectrally filtered, low-cost cameras with high-speed point detectors, persevering the most important information from the IFTS data, while greatly reducing the amount of data that must be processed, making real-time SLM process control feasible.

Two versions of an IFTS have been used to study the reactive gas plume and particulates above laser-irradiated metal powders.[37] The first version is the Telops,

Inc. Hyper-Cam IFTS that couples a Michelson interferometer to an infrared camera. A sequence of modulated intensity images corresponding to optical path differences are collected on a focal plane array (FPA) to form an interferogram cube (i.e., an interferogram at each pixel). Fourier transformation of each pixel's interferogram produces a raw hyperspectral image. The maximum optical path difference (MOPD) defines the unapodized spectral resolution of up to 0.25 cm^{-1}.

The second version of IFTS in use is the mid-infrared IFTS using a 320 × 256 pixel, Stirling cooled, InSb FPA responsive in the 1.5–5.5 μm spectral band. The 16-tap InSb array frames quickly at 1.2 kHz for the full 320 × 256 frame size and at 6.0 kHz for a 128 × 128 pixel window. The single pixel instantaneous field of view is 1.42 mrad, providing a spatial resolution of 0.64 mm at the focal distance of 45 cm. The spatial resolution is approximately twice the diffraction limit, with 73% of the energy at 4.3 μm from an on-axis pixel delivered to a single pixel. Variation in spatial resolution across the central 32 × 32 pixels is less than 1%. The point spread function changes by less than 5% along a 5 cm depth of field. A subset of pixels, 128 × 128 being a typical choice in the graphite experiments, can be read out to improve acquisition rate at the expense of a reduced field of view. The rate at which hyperspectral images are acquired depends on spectral resolution, FPA integration time, and the number of pixels in the image.

The MWIR IFTS equipped with a neutral density filter of OD = 1.0 saturates with 65,000 counts at 300 μW/(cm^2·sr·cm^{-1}) and has a background radiance of 6 μW/(cm^2·sr·cm^{-1}). For typical conditions encountered in SLM with gas plume temperatures near 1000 K, the DC component of the interferogram represents 17% of the dynamic range, with the interferometer producing a 20% modulation at zero optical path difference (ZPD). Calibration for absolute radiance with large area and high-temperature blackbody sources has been described previously.[38]

A LWIR version of the IFTS instrument employs a HgCdTe detector array with spectral response in the λ = 7.7–11.8 μm band. This version of the instrument is useful for improving the detection limit at low temperatures and potentially identifying process contaminants. Two internal wide-area blackbody sources (one maintained at a temperature of 303.2 ± 1.5 K, and the other at 333.1 ± 1.7 K) are used to calibrate the LWIR raw spectra.

To evaluate the use of IFTS sensors for monitoring laser–material interactions, we imaged the combustion plumes generated from laser irradiation of graphite targets. Porous graphite targets were irradiated with a 1.07 μm, 10-kW ytterbium fiber laser at irradiances of 0.25–4 kW/cm^2. Emissive plumes from the oxidation of graphite in air were monitored using the MWIR IFTS. Strong spectral emissions of CO and CO_2 were observed in the infrared between 1,900 and 2,400 cm^{-1} at an instrument spectral resolution of 2 cm^{-1}. A homogeneous, single-layer plume, line-by-line radiative transfer model (LBLRTM) was applied to estimate spatial maps of temperature and column densities of CO and CO_2 at a temporal resolution of 0.47 s per hyperspectral data cube. Steady surface temperatures between 1,800 and 2,900 K are achieved after approximately one minute for irradiances of 0.25–1.0 kW/cm^2. A stable, gas phase combustion layer extends from 4 to 12 mm away from the surface, with buoyancy driving a gas flow of ~8 m/s. Plume extent and intensity are greater for the large porosity samples (particle size: 6 mm) than for the small porosity samples (particle size: 0.0102 mm). Steady-state gas temperatures exceeded surface temperatures by up to 400 K. Column densities of up to 1,018 molec/cm^2 were calculated for CO and CO_2 based on the hyperspectral imagery.

A portion of the MWIR IFTS spectrum between 1900 and 2500 cm^{-1}, highlighting the CO_2 antisymmetric stretching band and CO fundamental band emission, is shown in Figure 20.5a.

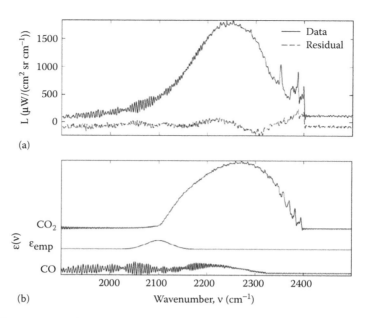

Figure 20.5 (a) Time averaged (41 s ≤ *t* ≤ 120 s) spectral radiance for (*x*, *y*) = (1.4 mm, 35.3 mm) with (…..) fit residuals and (b) spectral basis functions from the LBLRTM and the optional empirical emissivity without atmospheric attenuation.

The spectrum is presented for a single pixel located 1.4 mm from the surface, near the center of the beam at (*x*, *y*) = (1.4 mm, 35.3 mm), and averaged temporally from 41 to 120 s. The CO_2 emission dominates over the weaker CO emission. The hot CO_2 emission is attenuated by the colder CO_2 along the atmospheric path leading to the IFTS. The signal to noise at the peak of the spectrum is 87:1. The imaginary portion of the Fourier transform is not shown but is less than 22 μW/(cm²·sr·cm⁻¹) with little spectral structure. Also shown in Figure 20.5b are the spectral basis functions for CO_2 and CO at 2995 K. The instrument spectral resolution of 2 cm⁻¹ coincides with the CO rotational spacing, giving rise to the structure for CO rotational lines obtained in Figure 20.5b.

To extract emitter column densities and line-of-sight averaged plume temperature from the observed spectra, a LBLRTM was employed.[39] By ignoring scattering and assuming that the plume is in local thermodynamic equilibrium (LTE), the spectral radiance, *L*, can be expressed as

$$L(\nu) = \int \tau_{atm}(\nu')\varepsilon(\nu')B(\nu';T)ILS(\nu-\nu')d\nu' \qquad (20.4)$$

where:

τ_{atm} is the atmospheric transmittance along a 47 cm path between the imaging spectrometer, and the plume ε represents the plume's spectral emissivity

B is the Planck's distribution for blackbody radiation at temperature *T*

ILS is the instrument spectral line shape

An unapodized instrument line shape function, $ILS(\nu) = 2a\,\mathrm{sinc}(2\pi a\nu)$, is used, where *a* = 0.3 cm is the instrument's MOPD. The spectral emissivity is expressed as

$$\varepsilon(\nu) = 1 - \exp\left(-\sum_i l\,\xi_i\,n\,\sigma_i(\nu,T)\right)\tau_p\,\tau_{emp} \qquad (20.5)$$

where:

ξ_i is the ith species volume mixing fraction

l is the path length through the plume which is assumed constant, and approximately the laser beam diameter is 4.46 cm

n is the total gas density

The absorption cross section σ_i for the ith molecular specie is computed using the HITEMP[40] extension to the HITRAN[41] spectral database and includes the temperature-dependent partition function. The particulate transmittance, τ_p, is assumed to be independent of frequency. The emission intensity in the 2400–2500 cm^{-1} range is low and largely due to particulates. With a particulate volume fraction constrained to 0.2%, the modeled intensity in this spectral region was more than four times larger than observed. Even at this high soot fraction, the choice of soot model influences temperature extraction by less than 7 K and column densities by less than 5%.[35] A gray body particulate transmittance is used in the model for simplicity.

Best estimates for the plume concentrations and temperature, averaged along the line of sight, are obtained from a nonlinear fit of Equation 20.4 to the observed spectra, using a Levenberg–Marquardt algorithm. The average fit residual for Figure 20.5a is 100 µW/(cm^2·sr·cm^{-1}) and is only ~10% larger than the instrument noise. The best estimate for the effective gas temperature is 2999 ± 395 K, with a column density of $7.06 \pm 0.06 \times 10^{17}$ molec/cm^2 for CO_2, and $7.44 \pm 0.06 \times 10^{17}$ molec/cm^2 for CO. The statistical error bounds represent the 95% confidence intervals and do not include systematic error bounds associated with the spectral model or the nonuniformity of the plume along the instrument line-of-sight.

The spatial distributions of the column densities for CO and CO_2 and the spatial distribution for the gas-phase plume temperature were computed from the LBLRTM model for the $41\ s \le t \le 120\ s$ time frame, in which the temperature was stable. The results are shown in Figure 20.6. The column densities were significant, ~10^{17} molec/cm^2, and comparable to those observed in laminar flames with optimal combustion efficiency.[36] Statistical uncertainties were small, typically less than 7%. The CO_2 column density is observable over a larger region than the CO column density. At $y = 30$ mm, the CO column density is greater than CO_2 at the surface boundary. The CO_2 column density reaches a maximum further away from the surface at $x = 2.88$ mm. These results are generally consistent with CO production at the surface and gas phase oxidation to CO_2 further into the plume.

The time-averaged gas temperature at steady state between $y = 17.2$ and $y = 51.2$ mm and extending out to $x = 3$ mm is mostly uniform, and the highest temperature region of the plume is at 3,000 K. The plume temperature decreases further from the surface where hot gases rise, driven by buoyancy and the flow rate from a fume hood. As the plume rises, it curves around the top edge of the graphite plate at $y = 60$ mm where it continues in the free stream toward the fume hood. The statistical fit uncertainty in gas temperature is generally small, $\Delta T_g \le 37$ K but can exceed 250 K for $T_g > 3000$ K where the partition function in the spectral database is less certain.

These results represent the first observations of spatially resolved CO and CO_2 column densities and gas temperatures during the laser irradiation of graphite using IFTS. The IFTS imagery produces hyperspectral video cubes every 0.47 s, enabling dynamic monitoring of plume chemistry. Extension of this instrument to AM process control is under

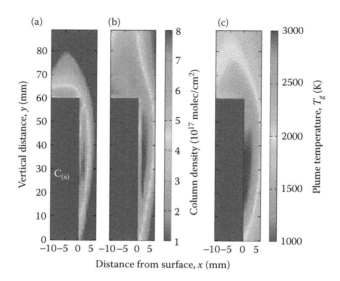

Figure 20.6 Spatial maps at steady state ($41\ s \leq t \leq 120\ s$) of (a) CO column density, (b) CO_2 column density, and (c) gas plume temperature.

development. The fast scan rates encountered in SLM are challenging for this diagnostic and prompt the need for advanced capabilities. The diagnostic offers the potential for real-time control of alloy composition and minimization of contaminants. Extension of this sensor to metal AM will require reduction of focal distances to achieve increased spatial resolution and new methods for image interpretation.

20.4.2 Polarimetry

Polarization techniques may provide enhanced feature contrast compared to a laser-based specular imagery technique.[30,42] A viable approach consists of illuminating the build area with a low-power laser and collecting the reflected radiation from the metal surface with a high frame-rate camera equipped with a polarization filter. A Phantom v12.1 monochrome camera would be a suitable choice for imaging. It is a CMOS-based detector capable of taking 6,242 frames per second (fps) at the full $1,280 \times 800$ pixel array size and up to 1 million fps with reduced resolution.

A schematic of the experimental layout is shown in Figure 20.7. When the polarized beam is reflected from the surface of the build area, it undergoes a change in polarization state which can be correlated to variations in the surface structure. Signals can be interpreted using a Stokes–Mueller formalism.[43]

An alternative approach would involve replacing the visible camera with the LWIR IFTS equipped with a rotatable polarization filter. The spectral data cube rate can be varied from 0.5 to 5 Hz depending on the choices for spatial and spectral resolution. The instrument's rotatable polarizer is a ZnSe window with a 4,000 lines/mm wire grid on its front surface capable of achieving a nominal extinction ratio of 400.

The visible imagery technique offers the greatest simplicity. The Telops camera offers richer data through the simultaneous measurement of spatial and temporal fluctuations in temperature and molecular species concentrations during laser–material interactions.

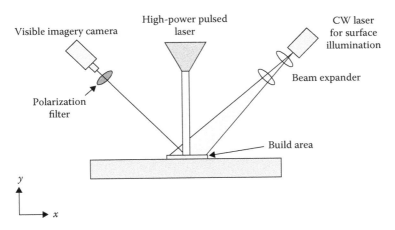

Figure 20.7 Experimental layout for employing a high-speed visible camera to collect polarized images of the build area during the selective laser melting process.

Rough-surface characterization has been explored through polarimetric imaging, that is pixel-based estimation.[44] The purpose was to partition an object according to the reflection properties of its surface as a means of identifying defects. A Stokes polarimeter was used to measure the polarization state of surface-reflected light. The technique was shown to have potential for detecting imperfections on varnished objects. Further work would be needed to assess applicability to AM online monitoring.

The measurement of weld pool surface temperatures based on polarization of state of thermal emission has been explored for a gas tungsten arc welding (GTAW) process.[45] A spectral window in the NIR, insensitive to the arc plasma emissions, was exploited for observations. The refractive index and surface radiance of the liquid steel were determined from the polarization signatures.

The efforts to fuse polarimetric and spectral imagery for object separation have been pursued by the remote sensing and computer vision communities.[46] Although the scene encountered in AM is substantiality smaller in scale and less varied than those in remote sensing environments, the benefits gained from combining the two techniques is worthy of consideration.

Polarimetric techniques have been developed for the measurement of optical and thermophysical properties of metals at the melting point and in the liquid phase.[47,48] Conducting materials are observed during an ohmic-pulsed heating process during which properties can be deduced at temperatures ranging from 1,200 K (at which most metals are in the solid state) up to 5,000 K (the liquid state). The normal spectral emissivity of the materials was determined using a microsecond division-of-amplitude-photopolarimeter (μs-DOAP).[49] Applicability to AM may be possible with modifications. At the very least, results from such pulsed heating experiments can inform interpretation or modeling of results.

20.5   Conclusion

Thermography and pyrometry have been established as viable online monitoring techniques for AM. For improving the effectiveness of these techniques, challenges requiring attention include accurate determination of an imaged object's emissivity for robust thermodynamic temperature calculations, assessment of spectral bandwidth limitations imposed by optical components (e.g., f-theta lenses), particularly in a coaxial scanning system, and mitigation window/viewport contamination by metallic debris.[50]

A number of laser-based techniques currently employed to monitor industrial processes such as laser cladding, laser beam welding, machining processes, and metal emission control could be considered for real-time monitoring of AM processes.

Imaging Fourier transform spectrometry shows strong potential for monitoring spatial and temporal variations in molecular species concentrations and gas-phase plume temperatures. AFIT is exploring this approach through small business technology transfer (STTR) collaborations.

Polarimetry seems to be largely unexplored as a diagnostic for AM at this time. The technique may be applicable for rough surface characterization and melt pool temperature determination. Modifications to current techniques used in remote sensing and the thermophysical characterization of materials may be beneficial for AM monitoring.

20.5.1   Selected definitions

The following definitions are taken from *ASTM F2792—12a, Standard Terminology for Additive Manufacturing:*

> **additive manufacturing (AM), n**—a process of joining materials to make objects from 3D model data, usually layer upon layer, as opposed to subtractive manufacturing methodologies. Synonyms: additive fabrication, additive processes, additive techniques, additive layer manufacturing, layer manufacturing, and freeform fabrication.

> **direct metal laser sintering (DMLS®), n**—a powder bed fusion process used to make metal parts directly from metal powders without intermediate 'green' or 'brown' parts; term denotes metal-based laser-sintering systems from EOS GmbH—Electro Optical Systems. Synonym: direct metal laser melting.

> **laser sintering (LS), n**—a powder bed fusion process used to produce objects from powdered materials using one or more lasers to selectively fuse or melt the particles at the surface, layer-by-layer, in an enclosed chamber.
>
> DISCUSSION—Most LS machines partially or fully melt the materials they process. The word *sintering* is a historical term and a misnomer, as the process typically involves full or partial melting, as opposed to traditional powdered metal sintering using a mold and heat and/or pressure.

> **subtractive manufacturing, n**—making objects by removing of material (milling, drilling, grinding, carving, etc.) from a bulk solid to leave a desired shape, as opposed to additive manufacturing.

20.5.2   Acronyms

AFIT	Air Force Institute of Technology
AM	Additive manufacturing
CMOS	Complementary metal-oxide-semiconductor
DMLS	Direct Metal laser sintering
EBF3	Electron beam free-form fabrication
FLIR	Forward looking infrared
FPA	Focal plane array
GTAW	Gas tungsten arc welding
HgCdTe	Mercury cadmium telleride
IFTS	Imaging fourier transform spectrometer
ILS	Instrument line shape
InSb	Indium antimonide
LBLRTM	Line-by-line radiative transfer model
LIBS	Laser-induced breakdown spectroscopy
LIPS	Laser-induced plasma spectroscopy
LTE	Local thermodynamic equilibrium
LWIR	Long wave infrared
MOPD	Maximum optical path difference
MWIR	Mid wave infrared
NDI	Nondestructive inspection
NIR	Near infrared
OD	Optical density
OPD	Optical path difference
PTA	Plasma-transferred arc
SFFF	Solid free-form fabrication
SLM	Selective laser melting
STTR	Small business technology transfer
ZPD	Zero path difference
μs-DOAP	microsecond division of amplitude photopolarimeter

References

1. Sutherland, J. W., D. P. Adler, K. R. Haapala, and V. Kumar, A comparison of manufacturing and remanufacturing energy intensities with application to diesel engine production, *CIRP Ann. Manuf. Technol.*, 57 (1): 5–8 (2008).
2. C. A. Brice and K. M. Taminger, Additive Manufacturing Workshop, in Anonymous (Melbourne, Australia, 27 June 2011).
3. National Institute of Standards and Technology, Measurement Science Roadmap for Metal-Based Additive Manufacturing Workshop Summary Report, (May 2013).
4. W. Ferster, Aerojet rocketdyne to 3-D print rocket engine parts under air force demo, *Space News* (21 August 2014).
5. Frazier, W. E., Metal additive manufacturing: A review, *J. Mater. Eng. Perform.*, 23 (6): 1917–1928 (2014).
6. Rombouts, M., J. P. Kruth, L. Froyen, and P. Mercelis, Fundamentals of selective laser melting of alloyed steel powders, *CIRP Ann. Manuf. Technol.*, 55 (1): 187–192 (2006).
7. Khaing, M. W., J. Y. H. Fuh, and L. Lu, Direct metal laser sintering for rapid tooling: processing and characterisation of EOS parts, *J. Mater. Process. Technol.*, 113 (1–3): 269–272 (2001).

8. K. Taminger and R. Hafley, Electron Beam Freeform Fabrication: A Rapid Metal Deposition Process, in *3rd Annual Automotive Composites Conference*, 9–10 September 2003, Troy, MI, Anonymous (NASA Langley Research Center; Hampton, VA, United States, Jul 1, 2003).

9. C. Atwood, M. Ensz, D. Greene, M. Griffith, L. Harwell, D. Reckaway, T. Romero, et al., *Laser Engineered Net Shaping (LENS(TM)): A Tool for Direct Fabrication of Metal Parts*. SAND98–2473C, Retrieved from http://www.osti.gov/scitech/servlets/purl/1549(1998).

10. M. Samandi, R. Storm, R. Loutfy, and J. Withers, The Development of Plasma Transferred Arc Solid Free Form Fabrication as a Cost Effective Production Methodology, in *Heat Treating 2003: Proceedings of the 22nd Heat Treating Society Conference and the 2nd International Surface Engineering Congress (ASM International)*, Indiana Convention Center, Indianapolis, Indiana, December 1, pp. 513–519.

11. National Center for Manufacturing Sciences, Industrial Roadmap for the Rapid Prototyping Industry, Report 0199RE98, (1998).

12. D. L. Bourell, M. C. Leu, and D. W. Rosen, Roadmap for Additive Manufacturing: Identifying the Future of Freeform Processing, The University of Texas at Austin, Laboratory for Freeform Fabrication, Advanced Manufacturing Center (2009).

13. Thijs, L., F. Verhaeghe, T. Craeghs, J. V. Humbeeck, and J. Kruth, A study of the microstructural evolution during selective laser melting of Ti–6Al–4V, *Acta Mater.*, 58 (9): 3303–3312 (2010).

14. Bi, G., C. N. Sun, and A. Gasser, Study on influential factors for process monitoring and control in laser aided additive manufacturing, *J. Mater. Proc. Technol.*, 213 (3): 463–468 (2013).

15. H. Krauss, C. Eschey, and M. F. Zaeh, Thermography for monitoring the selective laser melting process, in *Proceedings of Solid Freeform Fabrication*, University of Texas, Austin, TX, Anonymous (2012).

16. Pavlov, M., M. Doubenskaia, and I. Smurov, Pyrometric analysis of thermal processes in SLM technology, *Phys. Procedia*, 5: 523–531 (2010).

17. Craeghs, T., F. Bechmann, S. Berumen, and J. Kruth, Feedback control of Layerwise Laser Melting using optical sensors, *Phys. Procedia*, 5: 505–514 (2010).

18. Fischer, P., V. Romano, H. P. Weber, N. P. Karapatis, E. Boillat, and R. Glardon, Sintering of commercially pure titanium powder with a Nd:YAG laser source, *Acta Mater.*, 51 (6): 1651–1662 (2003).

19. Alkahari, M. R., T. Furumoto, T. Ueda, A. Hosokawa, R. Tanaka, and M. S. A. Aziz, Thermal conductivity of metal powder and consolidated material fabricated via selective laser melting, *Key Eng. Mater.*, 523–524: 244–249 (2012).

20. S. M. Baumann, B. E. Hurst, M. A. Marciniak, and G. P. Perram, Fiber laser heating and penetration of aluminum in shear flow, *Opt. Eng.*, 122510 (2014).

21. Mumtaz, K. and N. Hopkinson, Selective laser melting of Inconel 625 using pulse shaping, *Rapid Prototyping J.*, 16 (4): 248–257 (2010).

22. Monkhouse, P., On-line diagnostic methods for metal species in industrial process gas, *Prog. Energy Combust. Sci.*, 28 (4): 331–381 (2002).

23. Vorburger, T. V. and E. C. Teague, Optical techniques for on-line measurement of surface topography, *Precis. Eng.*, 3 (2): 61–83 (1981).

24. Purtonen, T., A. Kalliosaari, and A. Salminen, Monitoring and adaptive control of laser processes, *Phys. Procedia*, 56: 1218–1231 (2014).

25. Sibillano, T., D. Rizzi, A. Ancona, S. Saludes-Rodil, J. Rodríguez Nieto, H. Chmelíčková, and H. Šebestová, Spectroscopic monitoring of penetration depth in CO2 Nd:YAG and fiber laser welding processes, *J. Mater. Proc. Technol.*, 212 (4): 910–916 (2012).

26. Konuk, A. R., R. G. K. M. Aarts, A. J. H. i. Veld, T. Sibillano, D. Rizzi, and A. Ancona, Process control of stainless steel laser welding using an optical spectroscopic sensor, *Phys. Procedia*, 12: 744–751 (2011).

27. You, D., X. Gao, and S. Katayama, Monitoring of high-power laser welding using high-speed photographing and image processing, *Mech. Syst. Signal Proc.*, 49 (1–2): 39–52 (2014).

28. Liu, S., W. Liu, M. Harooni, J. Ma, and R. Kovacevic, Real-time monitoring of laser hot-wire cladding of Inconel 625, *Opt. Laser Technol.*, 62: 124–134 (2014).

29. S. H. Riza, S. H. Masood, and C. Wen, 10.11—Laser-assisted additive manufacturing for metallic biomedical scaffolds, in *Comprehensive Materials Processing*, S. Hashmi, G. F. Batalha, C. J. Van. Tyne, B. S. Yilbas, ed. (Amsterdam, the Netherlands, Elsevier, 2014), pp. 285–301.

30. Persson, U., In-process measurement of surface roughness using light scattering, *Wear*, 215 (1–2): 54–58 (1998).

31. Tay, C. J., S. H. Wang, C. Quan, and H. M. Shang, In situ surface roughness measurement using a laser scattering method, *Opt. Commun.*, 218 (1–3): 1–10 (2003).

32. Panne, U., R. E. Neuhauser, M. Theisen, H. Fink, and R. Niessner, Analysis of heavy metal aerosols on filters by laser-induced plasma spectroscopy, *Spectrochim. Acta B*, 56 (6): 839–850 (2001).

33. V. Sturm, A. Brysch, R. Noll, H. Brinkmann, R. Schwalbe, K. Mülheims, P. Luoto, et al. Online multi-element analysis of the top gas of a blast furnace by LIBS, in *Progress in analytical chemistry in the steel and metal industries: Seventh International Workshop on Progress in Analytical Chemistry in the Steel and Metal Industries, 16 to 18 May 2006 in Luxembourg*, J. Angeli, ed. (Glückauf, 2006), pp. 183–188.

34. Gross, K. C., K. C. Bradley, and G. P. Perram, Remote Identification and Quantification of Industrial Smokestack Effluents via Imaging Fourier-Transform Spectroscopy, *Environ. Sci. Technol.*, 44 (24): 9390–9397 (2010).

35. Acosta, R. I., K. C. Gross, G. P. Perram, S. M. Johnson, L. Dao, D. F. Medina, R. Roybal, and P. Black, Gas-phase plume from laser-irradiated fiberglass-reinforced polymers via imaging Fourier transform spectroscopy, *Appl. Spectrosc.*, 68 (7): 723–732 (2014).

36. Rhoby, M. R., D. L. Blunck, and K. C. Gross, Mid-IR hyperspectral imaging of laminar flames for 2-D scalar values, *Opt. Express.*, 22 (18): 21600–121617 (2014).

37. G. P. Perram, In-Process Monitoring of Additive Manufacturing, NASA STTR 2014 Phase I Solicitation, STTR_14_P1_140028, Contract No.: NNX14CL86P, (2014).

38. Revercomb, H., H. Buijs, H. Howell, D. LaPorte, W. Smith, and L. Sromovsky, Radiometric calibration of IR Fourier transform spectrometers: solution to a problem with the high-resolution interferometer sounder, *Appl.Opt.*, 27: 3210–3218 (1988).

39. Clough, S. A., M. W. Shephard, E. J. Mlawer, J. S. Delamere, M. J. Iacono, K. Cady-Pereira, S. Boukabara, and P. D. Brown, Atmospheric radiative transfer modeling: a summary of the AER codes, *J. Quant. Spectrosc. Radiat. Transfer*, 91 (2): 233–244 (2005).

40. Rothman, L. S., I. E. Gordon, R. J. Barber, H. Dothe, R. R. Gamache, A. Goldman, V. I. Perevalov, S. A. Tashkun, and J. Tennyson, HITEMP, the high-temperature molecular spectroscopic database, *J. Quant. Spectrosc. Radiat. Transfer*, 111 (15): 2139–2150 (2010).

41. Rothman, L. S., I. E. Gordon, Y. Babikov, A. Barbe, D. Chris Benner, P. F. Bernath, M. Birk, et al. The HITRAN2012 molecular spectroscopic database, *J. Quant. Spectrosc. Radiat. Transfer*, 130: 4–50 (2013).

42. Tyo, J. S., D. L. Goldstein, D. B. Chenault, and J. A. Shaw, Review of passive imaging polarimetry for remote sensing applications, *Appl. Opt.*, 45 (22): 5453–5469 (2006).

43. Garcia-Caurel, E., A. De Martino, J.P. Gaston, and L. Yan, Application of spectroscopic ellipsometry and mueller ellipsometry to optical characterization. *Appl Spectrosc.*, 67 (1): 1–21 (2013).

44. Terrier, P., V. Devalminck, and J. Charbois, Segmentation of rough surfaces using a polarization imaging system, *J. Opt. Soc. Am. A*, 25 (2): 423–430 (2008).

45. N. Coniglio, A. Mathieu, O. Aubreton, and C. Stolz, Weld pool surfaces temperature measurement from polarization state of thermal emission, in *Conférence QIRT 2014, at Bordeaux, France, 7-11 July 2014*.

46. Zhao, Y., L. Zhang, D. Zhang, and Q. Pan, Object separation by polarimetric and spectral imagery fusion, *Comput. Vis Image Und.*, 113 (8): 855–866 (2009).

47. Cezairliyan, A., S. Krishanan, and J. L. McClure, Simultaneous measurements of normal spectral emissivity by spectral radiometry and laser polarimetry at high temperatures in millisecond-resolution pulse-heating experiments: Application to molybdenum and tungsten, *Int. J. Thermophys.*, 17 (6): 1455–1473 (1996).

48. Seifter, A., F. Sachsenhofer, and G. Pottlacher, A fast laser polarimeter improving a microsecond pulse heating system, *Int. J. Thermophys.*, 23 (5): 1267–1280 (2002).
49. Pottlacher, G. and A. Seifter, Microsecond laser polarimetry for emissivity measurements on liquid metals at high temperatures—application to tantalum, *Int. J. Thermophys.*, 23 (5): 1281–1291 (2002).
50. M. Mani, B. Lane, A. Donmez, S. Feng, S. Moylan, and R. Fesperman, Measurement science needs for real-time control of additive manufacturing powder bed fusion processes, NISTIR 8036, (February 2015).

chapter twenty one

3D printed structures for nanoscale research

Tod V. Laurvick

Contents

This chapter discusses using 3D-printed structures to enable nanoscale research. Several industries have utilized 3D printing, and much has been published in the last decade. These fields include medicine, forensics, aeronautics, optics, and microelectronics. The appropriate process to utilize depends on the application, so selection of the best process will be discussed including materials available and what methods these can be deposited with, as well as a discussion of the pros and cons of each process. Two examples of using 3D-printed structures to facilitate nanoscale research are (1) prototyping an extremely low flow pump and (2) macrosized vessels for depositing polystyrene nanospheres for nanoscale lithography (NSL).

3D printing has been used for a wide variety of applications since its invention over a decade ago. In the medical industry, applications including tissues with blood vessels, prosthetics, biosensors, bones, heart valves, synthetic skin, and even replacement organs are all in various stages of development [1–7]. We have the ability to print a wide range of materials and the ability to print metals such as aluminum, stainless steel, and titanium. The use of these materials has raised interest in aviation and aerospace applications, which uses 3D printing for prototyping parts [8]. To fill this demand, companies have risen up to provide this service to a broader range of users [9–11]. The versatility, low cost, and efficiency of this technology have led to it branching into other, unexpected applications ranging from printing concrete blocks for building, tools for astronauts, fully functional automobiles, and even food [12–15]. As new applications are addressed, and the technology changes to meet these demands, this process fuels constant improvements which trickle down to the commercially available printers already available today. This growing market in turn drives better quality at a lower price, making this technology more accessible than it has ever been. While hardware improvements are part of what is driving this growing market, the quality and types of materials used are also steadily improving.

This in turn improves the quality of the finished product and also expands the number of applications for which 3D printing can be used—particularly for a variety of advanced research applications.

21.1   Materials for commercially available printing

If we consider the most inexpensive and readily available 3D-printing processes, plastic is a common material to use. With a variety of manufacturers now producing 3D printers, the printers currently available typically have the ability to use one or two kinds of plastics—PLA and ABS. A third type, TPU is a more recent addition to this list and is slowly working its way into use. These are all polymers, and the exact material properties of any given type will vary depending on the manufacturer and method of fabrication. For example, one prototyping company currently offers five different types of ABS alone. In general however:

> *Polylactic acid (PLA)*—This is the most rigid of the three materials and typically more prone to fracturing rather than plastic deformation when under load. This material also generally exhibits less change in dimensions during printing, making PLA more suitable for size-critical applications (gears, enclosures, circuit mounting, etc.) This material is also biodegradable and thus is used in plastic drinking cups and fibers used in teabags and diapers [16,17].
>
> *Acrylonitrile butadiene styrene (ABS)*—This plastic is typically softer than PLA, exhibiting more flexibility under stress and leading to more plastic deformation before breaking. This material tends to have more variability in its material properties when manufactured and deposited but typically is easier to process. As it has a slightly lower melting point than PLA, it is slower to cool, and thus it remains more pliable during the cooling process. This characteristic makes it a more common choice for injection molding and is commonly used in consumer products such as toys, kitchen appliances, and so on, where durability, strength, and flexibility are more critical than precise mechanical dimensions [18,19].
>
> *Thermoplastic polyurethane elastomers (TPU)*—This most recent addition to the list of printable plastics is being marketed as a flexible printing material, with a consistency very similar to rubber. It is highly durable and resistant to a moderate range of temperatures. While it is suitable for applications like gaskets and tubing, it is already in use in textiles and clothing (clothing, shoes, handbags, etc.) [20,21].

These subtle differences are substantial enough to drive continued commercial availability for all three materials, and these even drive the need within a single printer to utilize all three materials interchangeably. Despite their differences, they do share some common properties. 3D printing of plastic is normally accomplished through a headed extruder, which typically operates between 100°C and 250°C, and all three of these materials are formable in this range. They also can be bonded together using common solvents as a glue (both ABS and PLA are easily bonded using acetone for example), and the surfaces can similarly be smoothed either through sanding or rubbing with these solvents.

There are other 3D-printing techniques which allow for a variety of materials to be deposited. While spools of plastic thread are more common as they are easy to work with, some techniques utilize liquid plastics, or even powders, as a raw material. There are also a variety of deposition techniques. Heated extruders are typically for melting solid plastics, but for liquid raw materials, lasers might be used to selectively cure parts being printed. Ultraviolet-induced chemical reactions are also an option. These other techniques often come with their own benefits and limitations, but where these techniques are cost-justified, commercial availability follows. If we consider just extrusion printing of the three relatively inexpensive and versatile plastics mentioned above, they form a powerful combination of options for use in supporting advanced research.

21.2   Applications

The process of developing 3D-printed test fixtures and equipment starts just as it does with any fabrication technique—design of the prototype. Traditional machine shop methods can be used, which utilize a variety of 3D design software packages. These allow the designer to build a virtual structure to meet the needs of the process being considered. If these parts were to be fabricated using traditional means where bulk materials are cut and shaped through removal processes, then the parts must be designed with this in mind, and these designs must consider how these fabrication operations will or will not work based on the part. Such complex parts may be impossible to design as a single piece due to limited access of machine tools. Fabrication in multiple segments may be an option, but this approach introduces seams which may not be acceptable for some applications. Since 3D printing is an entirely additive process in which the design is broken down virtually into numerous planes and each plane deposited one layer at a time, there is much more flexibility in terms of forming complex shapes and surface in close proximity to other surfaces or even in locations that will eventually be inside other volumes. This means that parts which would otherwise be difficult if not impossible to fabricate with bulk removal may be done easily with 3D printing.

One potential issue arises if part of the final design is unsupported. To avoid material being dropped in mid air, this issue is addressed by building temporary *scaffolds* and *rafts*, and if formed properly, these temporary supporting features can be easily removed later. To illustrate the versatility of this approach, let us review a few specific research projects which used 3D printing to provide unique solutions.

21.3   Nanosphere deposition

Nanospheres have been used for over a decade in a variety of research fields (optics, micro-electronics, etc.). One key feature is their ability to self-assemble into planes of regularly spaced, closely packed monolayers on a scale otherwise difficult or impossible to achieve due to inherent limitations with lithography approaching the size of the wavelength of the light used. One difficulty in using nanospheres is finding a reliable method of forming these monolayers. One method which has recently been investigated involves a process shown in Figure 21.1. Samples which are to receive the monolayer are submerged beneath the surface of a carrier liquid. The nanospheres are then floated on or near the top of the liquid, and then the liquid is slowly drained. As this happens, the spheres are either transferred as a plane of particles to the sample and/or are drawn individually through capillary forces near the edge of the liquid onto the samples and form these regularly patterned monolayers. The few publications which pursue this method however are limited, in part due to the challenges of building a setup capable of accomplishing this. This is where 3D printing comes in.

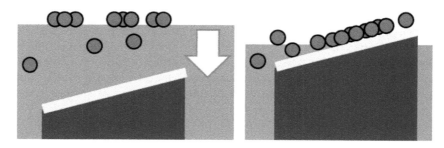

Figure 21.1 Nanosphere deposition through carrier liquid draining.

Before discussing how to design this apparatus, the first step is to review what other researchers have found on the subject to help guide the design. Such research indicates that this process is a function of many factors, including the size of the particles, their composition, the viscosity of the liquid used, the surface concentration of spheres, the angle of the substrate to the surface of the liquid, and the rate at which the liquid is drained [22]. To come up with a successful design, some of these factors can be addressed in what materials are chosen and what processes are followed. The rest of the factors will be addressed within the design of the apparatus itself. 3D printing allows a relatively fast and inexpensive means to test theories, refine the design based on results, and iterate this process until an optimal design is reached.

To do this, several vessels were designed, printed, and evaluated. The first of these is shown in Figure 21.2. This vessel is capable of holding nine samples, with each sample being held at slightly different angles relative to the surface of the liquid, ranging from 5° to 45°. A small ridge along the bottom of each pillar keeps the samples from sliding off, the back surface is textured to aid in drainage and eliminating any adhesion between the back of the substrate and the pillar, and the sides of the vessel are marked to aid in leveling. From this vessel, using a standardized suspension solution and 500 nm polysilicon

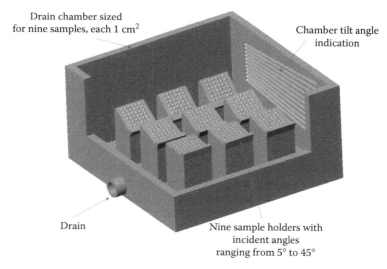

Figure 21.2 First iteration of a vessel for depositing nanospheres onto samples with incident angles between the liquids surface and substrate between 5 and 45 degrees.

nanospheres, it was determined that 15°–25° of inclination provided the optimal coverage. These results were utilized in the next iteration.

Using the results from the first vessel, a second vessel was designed and is shown in Figure 21.3. Previous nanosphere research suggested that sphere size, liquid viscosity, and surface inclination were interrelated [23,24]. It was decided to maintain the same nanosphere size and liquid composition, and thus all the surfaces were inclined to a standard 20 degrees as the results indicated this combination produced promising results. The back texturing was removed as these ledges have some roughness from the printing process itself, and adhesion did not turn out to be a problem. The bottom ledge was notched to aid in draining as the previous straight ledge tended to trap liquid.

It was observed that the next issue which needed to be addressed was how to actually place the nanospheres on the surface. Previous publications addressed this using a slightly inclined glass slide [25]. To attempt this approach, a ledge was added to hold a slide, and this technique could then be used to aid in introducing the nanospheres to the top of the liquid. The next area under investigation was surface concentration, and more specifically if surface area reduction techniques (such as those used with a Langmuir trough) could be included in the design to improve this [26].

The third vessel attempted utilized changeable inserts shown in the upper left corner of Figure 21.4. These inserts were coated with a hydrophobic material and placed around the pillars. These inserts were printed as shown in the figure, but need to be flipped when placed in the vessel so that as the liquid drained, the surface area would decrease. This would then theoretically compress the nanospheres on the surface resulting in more uniform monolayer formation.

After testing this design, these inserts caused too much disruption in the surface, primarily because they were not fixed in place. Also, it was noted that the surface compression made the loading stage unnecessary. The next design iteration would focus on addressing these issues.

The fourth vessel build used a fixed, linear reduction in surface area shown in Figure 21.5. In order to coat this surface in a hydrophobic material (and thereby reduce particle adhesion to the sidewalls), a separate insert was made that was coated separately and then was inserted into the outer shell. While this method did show more orderly nanospheres on the surface of the liquid to transfer these preformed floating monolayers to the substrate,

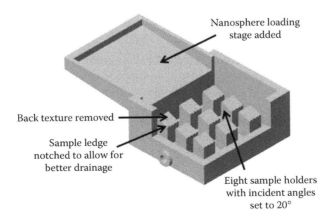

Figure 21.3 Nanosphere deposition vessel with 20 degree incident angle with addition of loading stage.

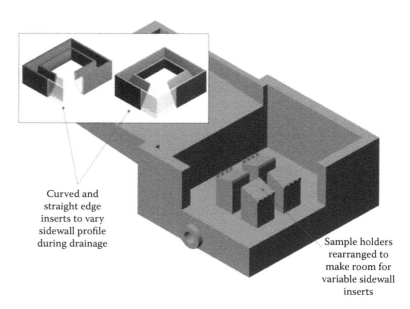

Curved and
straight edge
inserts to vary
sidewall profile
during drainage

Sample holders
rearranged to
make room for
variable sidewall
inserts

Figure 21.4 Nanosphere deposition vessel with inserts that change surface area during deposition process.

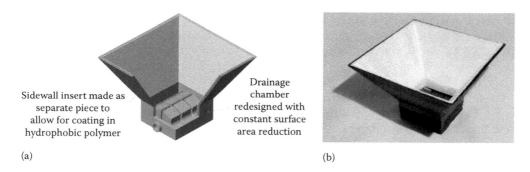

Sidewall insert made as
separate piece to
allow for coating in
hydrophobic polymer

Drainage
chamber
redesigned with
constant surface
area reduction

(a)   (b)

Figure 21.5 Nanosphere deposition vessel with set surface area reduction with hydrophobic coating shows (a) the as-designed vessel and (b) the actual vessel printed.

it appeared to require more precise flow control than what was available. To address this, a solution involving a custom-made pump was devised and will be discussed in the next section as the second 3D printer application. Before that, however, let us discuss one final vessel design.

In processing microelectronic devices, it is often times necessary to work with full-sized wafers—2, 3, or 4 inches in diameter, or even larger. Also, the size of the nanospheres used thus far was limited to 500 nm, and in practice different size particles may be required depending upon the application. To accomplish deposition of different sized nanospheres (which would then require different deposition angles as discussed above), a full wafer vessel was designed and is shown in Figure 21.6. This vessel has a fairly simple center post which is printed in two parts. After printing, embedded magnets would ensure proper positioning is maintained. This positioning creates the desired substrate to liquid surface angle. Depending on the rotation angle of the pillar relative to the vessel, the angle of the

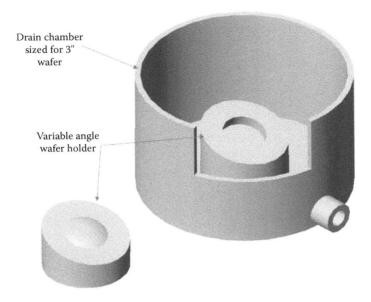

Figure 21.6 Nanosphere deposition vessel for 3 inch circular wafers with variable incident angle center stage.

substrate to the liquids surface will be altered. Thus this single design should be sufficient for a variety of nanosphere particle sizes, suspension liquids, and substrate materials.

For all of these vessels, the typical time to print was between 6 and 12 hours depending on the overall volume and complexity. The resulting structure was also typically fairly resistant to leaks, but in some instances leaks did develop after repeated use. To counter this, acetone could be used to *smooth* the surfaces and help seal these gaps. Treating the surface before use with a sealing material such as silicone sealant could preemptively solve this problem. When considering how well these vessels were able to form monolayers of nanospheres, the results obtained from these first attempts showed improvement when compared to the best obtained published data, and while the results depend greatly on the type of nanospheres, substrate material, nanosphere size, and so on, improvement in the largest observed regularly patterned area appeared to be roughly an order of magnitude better than the best results obtained from literature. As previously mentioned, it may be possible to further improve these results if the drain rate could be more precisely controlled, and for this let us consider the second application to be discussed.

21.4   Low flow, low friction pump

In order to pump liquids containing nanospheres, several factors must be considered. First, the flow rates in question are typically extremely low (~0.05–0.3 mL/min produced the best results), but the way in which this liquid is moved must be done in a manner that does not damage the nanospheres. Hospitals and medical researchers typically use peristaltic pumps in order to avoid cell damage. While these devices work well for biological samples, they are not necessarily ideal for moving fluids with nanospheres. First, the method of transporting the fluid may be harmful to some of the softer materials used to make nanospheres. In the case of harder nanospheres, continual brushing of these hard particles against the soft plastic lines used in peristaltic pumps could potentially be

Inverted gear
component   Mutually geared
components

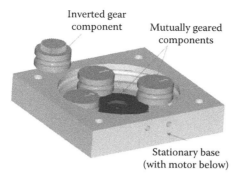

Stationary base
(with motor below)

Figure 21.7 Variation of peristaltic pump for extremely low flow with low impact to particles suspended in solution.

damaging. To address this, another technique was devised which uses 3D printing as a fast, cheap means of building a prototype device to evaluate its potential, with part of this design shown in Figure 21.7.

While plastic parts may wear when exposed to long-term use, several or all of these components could either be regularly replaced or fabricated from sturdier plastics or metal. The resolution is more than sufficient for extremely solid gearing, resulting in parts that fit together with virtually no gear backlash. The design can be optimized with extreme easy with this material, and even can go through several iterations at almost no cost (other than design time and a few hours to print), and then the design can actually be physically tested before more expensive, permanent machined parts are obtained.

For the third application, let us consider an unrelated topic involving the measurement of material properties under specific test conditions.

21.5   Three-dimensional pressure effects testing

Consider an experiment in which the effects of pressure on resistance in a sheet of material are to be investigated. The two-dimensional configuration of this test could be as simple as a thin layer of the material confined between two rigid planes, which is then subjected to a constant uniform pressure, and prepositioned probes at various points on the plane of the material are used to gather the required measurements. To adapt this experiment into three dimensions, 3D printing provides a relatively simple solution. Consider Figure 21.8, which shows a conceptual illustration of the experiment in three dimensions. Inside the shell (A) is a spherical void and inside of this void is a second spherical *eggshell* (B). Sandwiched between these two rigid spherical surfaces is the material to be tested (C). The inner shell is segmented such that a bladder placed inside allows the inner shell to press outward, spherically compressing the layer of material and allowing measurements to be made from the test probes, which access the material through holes formed in the shell during printing (E). These holes are in fixed locations but can be placed in virtually any location and orientation.

To build this, let us first consider the outer shell shown in Figure 21.9. This four-part shell can be printed with the alignment bumps and holes in place as well as the holes for sensor lines, with all sections being printed simultaneously in around 8–10 hours. Next let us consider the inner shell shown in Figure 21.10 below. Initially an eight-segment inner shell (A) was printed for evaluation. However, more uniform pressure may be required, in which case a shell with more segments (B) may be just as easily printed. To print structures such as these, the time and quality may vary slightly

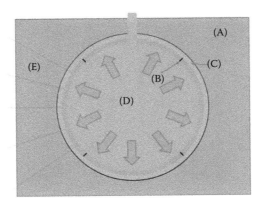

Figure 21.8 Conceptual illustration of device to measure material properties under pressure using a 3D printed outer shell (A) and interior, segmented shell (B) with the material under test (C) under radial, spherical pressure provided by a bladder (D) with test measurements being taken from preformed access holes (E).

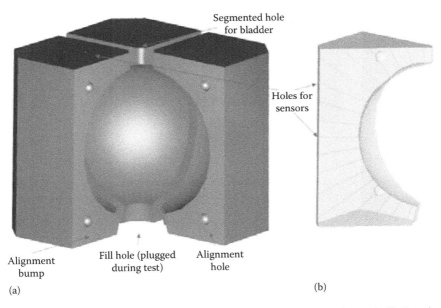

Figure 21.9 Shell for radial pressure testing designed in four segments (three-shell pieces) as shown in (a), with alignment bumps and required access for test setup and operation and (b) shows a cross section of one corner piece, highlighting the holes included for sensors to reach the material under test.

depending on how much supporting structure needs to be applied, but for the eight-segment option, these can be produced in a matter of hours.

Depending on the material to be tested, the surface roughness of contacting shells may be critical. A common, commercially available 3D printer is going to produce surfaces which may not meet this uniformity requirement. In Figure 21.11, the ridges produced can be seen in the reflected light, particularly on the curved surfaces. However, the basic structure is more than rigid enough to withstand sanding and/or solvent smoothing as previously mentioned. While this final processing does add time to the process, the printing itself is autonomous.

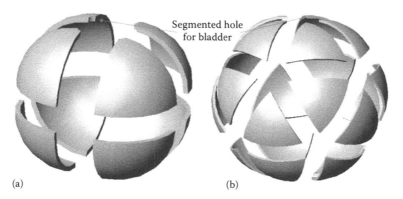

Segmented hole
for bladder

(a)   (b)

Figure 21.10 Interior shells for 3D pressure testing apparatus including both eight-segment interior (a) and 20 segment interior (b) options.

Figure 21.11 Pressure-testing apparatus printed in ABS plastic with eight-segment inner shell and outer shell quarters with alignment holes and test probe holes fabricated during printing.

Contrast this with the time required to produce these through machining a bulk block of material. For the outer shells, this would require several steps for removal of the inside void and would either require an extremely sophisticated, multiaxis tool to access all the various angles required or multiple repositioning of the block during machining. This entire process is what is required for a single quadrant, and since four are required, this must be repeated four times for the four sections. Alternatively, a single piece could be made, and from that single piece, a mold could be fabricated, but even that approach would require the sensor holes to be drilled separately on the all the final molded parts. The inner shells could also be machined, but as both the inner and outer surfaces require machining, each piece would require a separate block of material, incurring considerable waste. Also, in order to hold the part during machining, several supports would need to remain during machining and later removed by hand, incurring more time for every segmented piece. Thus, even a 3D-printed process which requires some labor after printing is still likely to be much faster and less expensive than other means available.

References

1. T. Boland, T. Xu, B. Damon and X. Cui, Application of inkjet printing to tissue engineering, *Biotechnology Journal*, 1(9), 910–917, 2006.
2. P. NG, P. Lee and J. Goh, Prosthetic sockets fabrication using rapid prototyping technology, *Rapid Prototyping Journal*, 8(1), 53–59, 2002.
3. P. Charles, E. Goldman, J. Rangasammy, C. Schauer and M. Chen, Fabrication and characterization of 3D hydrogel microarrays to measure antigenicity and antibody functionality for biosensor applications, *Biosensors and Bioelectronics*, 20(4), 753–764, 2004.
4. B. Leukers, H. Gulkan, S. Irsen, S. Milz, C. Tille, M. Schieker and H. Seitz, Hydroxyapatite scaffolds for bone tissue engineering made by 3D printing, *Journal of Materials Science: Materials in Medicine*, 16(12), 1121–1124, 2005.
5. L. Hockaday, K. Kang, N. Colangelo, P. Cheung, B. Duan, E. Malone, J. Wu, L. Girardi, L. Bonassar and H. Lipson, Rapid 3D printing of anatomically accurate and mechanically heterogeneous aortic valve hydrogel scaffolds, *Biofabrication*, 4(3), 1–12, 2012.
6. W. Lee, J. Debasitis, V. Lee, J. Lee, K. Fischer, K. Edminster, J. Park and S. Yoo, Multi-layer culture of human skin fibroblasts and keratinocytes through three-dimensional freeform fabrication, *Biomaterials*, 30(8), 1587–1595, 2009.
7. C. Schubert, M. van langeveld and L. Donoso, Innovations in 3D printing: a 3D overview from optics to organs, *British Journal of Ophthalmology*, 98(2), pp 159–161, 2013.
8. F. Arcella and F. froes, Producing titanium aerospace components from powder using laser forming, *JOM*, 52(5), 28–30, 2000.
9. Xometry, Rapid Prototypes & Parts, Xometry, [Online]. Available: http://www.xometry.com (Accessed December 8, 2015).
10. RapdiPSI, RapidPSI, 3D Pringing Since 1998, Rapid PSI, [Online]. Available: http://www.rapidpsi.com (Accessed December 8, 2015).
11. stratasys, Stratasys Direct Manufacturing, Stratasys, [Online]. Available: http://www.stratasysdirect.com (Accessed December 8, 2015).
12. D. Sher, Texas Student Builds Concrete 3D Printer with Hopes of House Printing, 3D Printing Industry, December 8, 2015.
13. J. Kell, *Fortune*, This new tool will make astronauts' lives a lot easier, [Online]. Available: http://fortune.com (Accessed October 29, 2015).
14. LM3D, "We're building cars & trust," *Local Motors*, [Online]. Available: http://localmotors.com (Accessed December 8, 2015).
15. T. Halterman, 3D Print, The Voice of 3D Printing Technology, 6 Jan 2015. [Online]. Available: http://3dprint.com (Accessed December 8, 2015).
16. R. Giordano, B. Wu, S. Borland, L. Cima, E. Sachs and M. Cima, "Mechanical properties of dense polylactic acid structures fabricated by three dimensional printing," *Journal of Biomaterials Science, Polymer Edition*, 8(1), 63–75, 1997.
17. Plastic Ingenuity, Plastic ingenuity, thermoforming your vision, PI, [Online]. Available: http://www.plasticingenuity.com (Accessed December 8, 2015).
18. J. Rodruiguez, J. Thomas and J. Renaud, Mechanical behavior of acrylonitrile butadiene stryrene (ABS) fused deposition materials. Experimental investigation, *Rapid Prototyping Journal*, 7(3) 148–158, 2001.
19. ULProspector, Acrylonitrile Butatdiene Styrene (ABS) Plastic, Underwriter Laboratories, [Online]. Available: http://plastic.ulprospector.com (Accessed December 8, 2015).
20. G. Kumar, L. Mahesh and N. Neelakantan, Studies on thermal stability and behaviour of polyacetal and thermoplastic polyurethane elastomer blends, *Polymer International*, 31(3), 283–289, 2007.
21. American Chemistry Council, Thermoplastic polyurethane, ACC, [Online]. Available: http://polyurethane.americanchemistry.com (Accessed December 2015).
22. Y. Wang and Z. Weidong, A Review on Inorganic Nanostructure Self-Assembly, *Journal of Nanoscience and Nanotechnology*, 10, 1–21, 2010.

23. J. Smith, M. Hadley, G. Craig and F. Lowe, Methods and apparatus for fluidic self assembly. US Patent 6623579, September 23, 2003.

24. K. Velikov, F. Durst and O. Velev, Direct observation of the dynamics of latex particles confined inside thinning water-air films, *Langmuir*, 14(5), 1148–1155, 1998.

25. S. Barcelo, S. Lam, G. Gibson, X. Sheng and D. Henze, Nanosphere lithography based technique for fabrication of large area, well ordered metal particle arrays, HP Laboratories, 2012.

26. Q. Guo, X. Teng, S. Ragman and H. Yang, Patterned langmuir-blodgett films of monodisperse nanoparticles of iron oxide using soft lithography, *Journal of the American Chemical Society*, 125(3), 630–631, 2003.

chapter twenty two

Additive manufacturing at the micron scale

Ronald A. Coutu, Jr.

Contents

This chapter presents three micron-scale additive manufacturing techniques based on specialized microelectromechanical systems (MEMS) fabrication processes: bulk micromachining, surface micromachining, and micromolding. Bulk micromachining is a *subtractive* process where portions of a substrate are patterned and etched resulting in 3D-micromachined features. Surface micromaching is an *additive* process where thin film layers are deposited, patterned, and etched resulting in planar surface structures. In micromolding, high-aspect ratio devices' molds (i.e., 100 μm features) are created using thick photosensitive layers that are patterned, developed, and filled using electroplating.

22.1 Introduction

As the size of a structure decreases, traditional macroscale fabrication techniques become more difficult to implement. As a result, in order to fabricate MEMS device transducers, sensors, and optical structures having features in the micron scale, an entirely new set of fabrication techniques is required.

Almost all of the techniques that are used to fabricate MEMS devices require a method of patterning. Following the footsteps of the microelectronics industry, much of this patterning is done using photolithographic methods. In a photolithography process, the substrate is coated with a light-sensitive polymer film called photoresist. When exposed to light, the polymers in the resist become either more or less cross-linked, leading to a change in solubility. These materials are called negative and positive photoresists, respectively. By shining light on the resist through a photolithography mask, only certain areas of the photoresist layer are exposed. These exposed areas are then developed away, resulting in a patterned film on the substrate. This film can then be used in a variety of ways, such as, a protective layer during an oxide etch as in Figure 22.1.

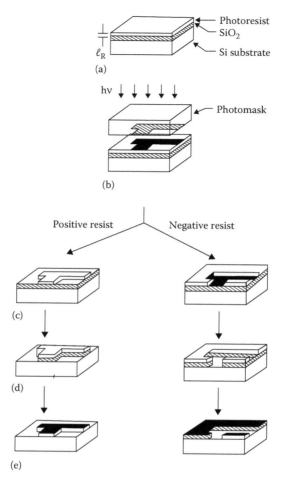

Figure 22.1 Photolithographic patterning of an oxide: (a) Oxidized substrate is coated with photoresist. (b) Resist is selectively exposed through a photomask. (c) Developing creates a pattern. (d) Photoresist acts as a protective layer during etch. (e) The resist is removed with an organic solvent. (From May, G. S. and Sze, S. M. *Fundamentals of Semiconductor Fabrication*, Hoboken, NJ, Wiley, 1998.)

Photolithography is generally classified by the wavelength of light used to expose the resist. The most common of these is ultraviolet (UV) lithography, using wavelengths such as the 365 nm I-line. Since the wavelength of light is directly related to the minimum feature size possible, the industry also uses deep UV sources such as the 248 nm or 193 nm excimer lasers to produce line widths as low as 22 nm [2].

As wavelengths decrease, photon–matter interactions change drastically, leading to increasingly difficult and expensive lithography tools. Lithographic systems using extreme ultraviolet (EUV) or X-ray wavelengths are currently being researched. Alternatively, electrons, which have a significantly smaller effective wavelength, can be used in place of photons. This method, known as electron beam lithography, provides increased resolution past 10 nm at the cost of lower device throughput [1,3].

22.2 Bulk micromachining

Bulk micromachining deals with subtractive processes in which the substrate itself is selectively etched through a variety of processes into the desired structure. These processes include both wet and dry chemical etches, plasma or reactive ion etches, and laser etching [4]. These can be anywhere on a scale from completely isotropic, etching the material equally in all directions, to completely anisotropic, which can create high-aspect ratio structures.

Many wet chemical etches can be used in MEMS fabrication, the selection of which depends on a wide variety of factors such as substrate material, available etch stop layers, etch rates, compatibility with other materials, and mask selectivity [4]. The substrate is generally masked using photolithography, and the sample is immersed into an etchant for a set period of time. If the photoresist is not sufficient to mask the wafer in a particular etchant, a secondary layer such as an oxide may be patterned by the resist to act as a hard mask. Often some form of agitation is used in order to increase the etch rate by speeding the delivery of fresh etchant to and removal of byproducts from the etching locations. By inserting a layer of resistant to a certain etchant into the substrate through methods such as ion implantation, the etch depth can be controlled as shown in Figure 22.2.

Wet chemical etching can also take advantage of the differences in etch rates for different crystallographic directions in many substrates to produce highly predictable anisotropic etches. An example of this is shown in Figure 22.3.

Dry or vapor chemical etches can also be used to etch substrate materials. These processes use reactive chemicals in the vapor phase to produce highly selective isotropic etches. Common vapor phase etchants include xenon di-fluoride (XeF_2) and hydrofluoric acid (HF) [4].

Plasma-based etching similarly uses chemical reactions to etch substrate materials. The main difference is that radio frequency (RF) energy is used to generate plasmas needed to provide the energy to drive the chemical reactions. In addition, plasma-generated ions can be used to physically erode or etch various materials. A variety of plasma etch recipes are available ranging from isotropic to anisotropic depending on the percentage of chemical and physical etching. Depending on the positioning of the ground electrode, this process may be called plasma-enhanced or reactive ion etching (RIE) [4]. In addition, extra RF power supplies (e.g., inductively coupled plasma or ICP) can be used to improve plasma quality and directionality [5]. Deep silicon etching or deep reactive ion etching (DRIE) is a

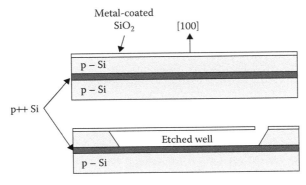

Figure 22.2 Doping layers can act as an etch stop layer in silicon in making the well underneath this cantilever. (From Kovacs, G., *Micromachined Transducers Sourcebook*, Boston, MA, McGraw-Hill, 1998.)

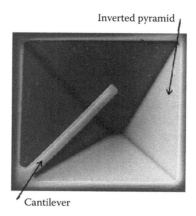

Figure 22.3 The anisotropic wet etch of (100) silicon in KOH produces an inverse pyramidal pit with faces in the (111) direction. The cantilever is formed from silicon dioxide and aluminum, and thus is not etched. (From Vaishnav, U. et al., Micromechanical Components with Novel Properties, in SPIE Symposium on *Smart Materials, Structures and MEMS*, 3321, 287–297, SPIE, 1996.)

variation that uses RIE/ICP etching in concert with sidewall passivation steps to achieve high-aspect ratio anisotropic etches [5].

Lasers can also be used in bulk micromachining by direct material ablation. Since this must be done serially, laser ablation is a slow process. Lower energy lasers can also be used to speed the etch rate of the substrate in the presence of certain gases, but care must be taken with the chemistry of different substrate–gas mixtures. This is called laser-assisted chemical etching (LACE) [4].

A focused ion beam (FIB) can also be used to etch a wide range of materials. Ions are accelerated and focused electrostatically before bombarding the substrate. Upon impact, kinetic energy is transferred from the ions to atoms on the surface of the substrate, some of which are ejected. The minimum resolution of FIB etching is on the nanometer scale [7].

Bulk micromachining also commonly uses wafer bonding to assemble final devices. Multiple wafers that have already been etched appropriately are physically bonded together using a combination of adhesives, elevated temperatures, high pressure, and electrical fields. Methods have also been developed that directly bond surface combinations such as metal and oxide, silicon and silicon, or oxide and oxide without the use of an adhesive. This is basis for silicon-on-insulator (SOI) micromachining that is discussed in Section 22.5.

22.3 *Surface micromachining*

Surface micromachining includes processes where thin films are deposited, patterned, and etched on top of a substrate to produce planar structures with features on the micron scale. A variety of films can be added, such as metals, dielectrics, and polymers. With a wider variety of materials and processes, more intricate devices can be made with surface micromachining as compared to bulk micromachining.

An important consideration when depositing thin films is the resulting film conformance or step coverage. This affects the consistency of film thickness when the height of the underlying layer changes. For instance, a process with vertically directional step coverage will result in almost no deposition on the side walls of the underlying layer, but

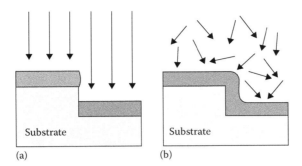

Figure 22.4 Step coverage: (a) Vertical directional processes do not coat side walls and (b) non-directional processes produce conformal coverage. (From Kovacs, G., *Micromachined Transducers Sourcebook*, Boston, MA, McGraw-Hill, 1998.)

a nondirectional process will have conformal side walls on approximately the same order of thickness as the rest of the device. This is illustrated in Figure 22.4.

One method of physical vapor deposition (PVD) is evaporation, where a material is heated under vacuum until melting. The evaporated material travels directly outward from the heating area until they arrive at the surface of the sample, where they condense and adhere. By placing the substrate a certain distance away from the heating element, highly directional coverage can be obtained. Commonly used methods to heat and evaporate materials are thermal or resistive, RF or a scanned electron beam [4].

The lift-off process is a commonly used patterning technique used in conjunction with evaporation. Generally, two photoresist layers are deposited onto the substrate: a lower resist (LOR) and an upper imaging resist [8], although in some cases only the imaging resist is required if the layer is thick enough [4]. During the process, the LOR is over exposed leading to an undercut photoresist ledge. When a material is deposited, the undercut prevents it from forming a continuous sheet along the side walls. Thus, only the metal deposited directly on the substrate is attached, and removal of the photoresist layers with an organic solvent will similarly remove the remainder of the metal. This process is described visually in Figure 22.5.

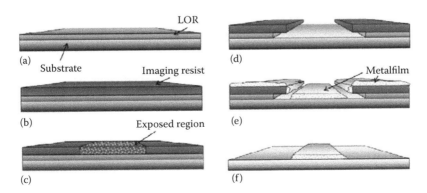

Figure 22.5 Metal lift-off process: (a) coat and soft-bake LOR. (b) Coat and soft-bake imaging resist. (c) Expose imaging resist. (d) Develop imaging resist and LOR. (e) Deposit film. (f) Lift-off resists and unwanted metal. (From Microchem, *LOR and PMGI Resists Data Sheet*, 2009.)

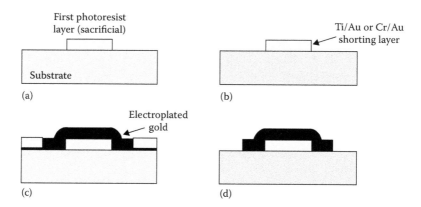

Figure 22.6 Electroplating process over a sacrificial layer: (a) pattern sacrificial photoresist using standard lithography, (b) Deposit shorting layer, (c) Perform masking lithography and electroplating, and (d) Remove photoresist and etch-shorting layer. (From Kovacs, G., *Micromachined Transducers Sourcebook*, Boston, MA, McGraw-Hill, 1998.)

Sputtering is another PVD method where direct current (DC)- or RF-generated plasmas are used to accelerate ions toward target materials. Upon impact, these ions exchange momentum with the target material resulting in target material erosion. Plasma sputtering yields high quality, conformal thin films that typically have higher adhesion than evaporated films. Due to their high conformance, sputtered films are generally patterned using the etch-back process where the films are deposited first, patterned as in Figure 22.1 and then etched through the photolithographic pattern.

Direct deposition of a material from a reactive vapor to the substrate is also possible in a process known as chemical vapor deposition (CVD). Advantages can include high film uniformity and conformal step coverage [3]. CVD usually requires elevated temperatures unless energy is supplied using a plasma as in plasma enhanced (PECVD) [5]. Other variants of CVD include low-pressure CVD (LPCVD), atmospheric pressure CVD, and metal organic CVD (MOCVD) [1].

Sacrificial layers are commonly used in MEMS fabrication. These space-holding layers are selectively removed and replaced with air gaps or voids when the device is released. Figure 22.6 illustrates an example of MEMS fabrication where a sacrificial photoresist is used to make an electroplated fixed–fixed beam. One common difficulty when removing sacrificial layers is stiction, where surface tension from a liquid solvent or etchant produces large attractive forces between nearby surfaces causing them to permanently stick together. This can be avoided by using super-critical CO_2 drying or dry etching [4].

22.4   Microforming

Surface micromachined structures have limited aspect ratios due to the relatively thin layers that are used during fabrication. In microforming, however, devices are built on top of the substrate like surface micromachining but with much larger aspect ratios. One microforming process, called LIGA [Lithography, Galvanoformung (electroplating) and Abformung (molding)], uses thick photoresists as molds which are then filled in with electroplated films [4]. These devices typically have thicknesses ranging from 50 to 500 μm [4].

Electroplating is a common technique for depositing metals into microformed molds, especially when thicker layers are required. The areas to be electroplated are held at a negative voltage compared to an electrode, and both are immersed in a solution containing reducible metal ions [4]. The electrons provided by the voltage allow the ions to be reduced at the wafer, thus depositing onto the surface. A metallic seed material is deposited over the entire wafer, prior to electroplating, for electrical contact. This layer is then selectively protected by the photoresist mold and allows electroplating only into specific exposed regions. After plating, the seed layer can be etched away to remove electrical conductivity between structures, as shown in Figure 22.6 [3].

22.5   Silicon-on-insulator micromachining

SOI wafers consist of a thin oxide layer sandwiched between a thick silicon substrate (handle wafer) and a top layer of bonded silicon (device layer). There are two main manufacturing methods for creating SOI wafers: separation by implantation of oxygen (SIMOX) and wafer bonding. SIMOX uses high-energy ion implantation to place oxygen atoms well below the surface of a silicon wafer. The wafer is then annealed at high temperature to produce a buried oxide layer below the silicon surface [4]. In silicon wafer-to-wafer bonding, a silicon wafer is oxidized to form a SiO_2 layer. A second silicon wafer is then placed in contact with this oxidized surface. The two wafers are annealed at high temperature to form a bond. After bonding, the top silicon layer is thinned by chemical etching, until it reaches the desired device level thickness [4]. This type of SOI wafer is illustrated in Figure 22.7.

In SOI micromaching, photolithography is used to pattern the device layer followed by silicon etching to form the devices. Once the structures are outlined in the device layer, they are then released by etching the SiO_2 and removing the handle wafer.

22.6   Multilevel micromachining

Multilevel micromaching is a hybrid microfabrication process where subsequent material layers are patterned and deposited in a similar fashion to surface micromachining or microforming. Unlike surface micromachining, however, that is typically restricted to 2–3 thin film mechanical layers [9], multilevel micromaching utilizes high aspect ratio (HAR) layers with offset sacrificial layers resulting in true 3D structures and devices on micron/millimeter scale [10,11].

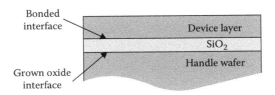

Figure 22.7 Bonded silicon-on-insulator (SOI) wafer. SiO_2 is thermally grown on the handle wafer, which is then bonded to the device or mechanical layer.

22.7   Conclusions

In this section, five advanced MEMS fabrication processes were presented and discussed. Each of these processes represents a foundational technology for additive manufacturing. In addition, key microelectronics fabrication processes were presented because they enable these micron scale additive manufacturing approaches. These processes can be used to create stand-alone 3D structures or critical micron sized parts for other macrosized devices.

References

1. G. S. May and S. M. Sze, *Fundamentals of Semiconductor Fabrication*. Hoboken, NJ: Wiley, 1998.
2. B. Hardwidge, IBM and AMD create first 22 nm SRAM cell, *Custom PC*, 2008.
3. R. C. Jaeger, *Introduction to Microelectronic Fabrication*. Upper Saddle River, NJ: Prentice Hall, 2002.
4. G. Kovacs, *Micromachined Transducers Sourcebook*. Boston, MA: McGraw-Hill, 1998.
5. M. Madou, *Fundamentals of Microfabrication*, 2nd Edition. Boca Raton, FL: CRC Press, 2002.
6. U. Vaishnav, P. Apte, S. Lokhre, V. Palkar, and S. Pattalwar, Micromechanical Components with Novel Properties, in SPIE Symposium on *Smart Materials, Structures and MEMS*, vol. 3321, pp. 287–297, SPIE, 1996.
7. Zeiss, *Focused Ion Beam Essential Specifications: NVision* 40, 2009.
8. Microchem, *LOR and PMGI Resists Data Sheet*, 2009.
9. MEMSCap website, *PolyMUMPs Process* (http://www.memscap.com/products/mumps/polymumps), accessed on December 14, 2015.
10. EFAB website (http://www.isi.edu/efab), accessed on December 14, 2015.
11. Microfabrica website (http://www.microfabrica.com/technology.html), accessed on December 14, 2015.

chapter twenty three

Computer modeling of sol–gel thin-film deposition using finite element analysis

Alex Li

Contents

This chapter explores the use of a multiphysics approach to model the sol–gel thin-film deposition, which is a versatile wet chemistry method for thin-film deposition and surface medication and may play an important role in additive manufacturing processes. First, we briefly introduce the sol–gel process through two representative chemical reactions: hydrolysis and condensation of alkoxides. Second, we discuss thin-film deposition using the sol–gel process, including heat treatment of the sol–gel thin films using laser processing and indirect micro-oven approach. Finally, we describe our ongoing effort on computer modeling of thin-film deposition using finite element analysis.

23.1   General route for sol–gel processing

The sol–gel thin-film deposition usually starts with the preparation of a solution by means of hydrolysis and condensation of the precursor chemical compounds, for example, metal alkoxides. The sol–gel solution is then deposited on a substrate surface to form a thin film using different methods, such as spin- and dip-coating techniques. The thin-film deposition process can be repeated for multiple times to achieve the desired thickness or material properties. Heat treatments using laser or other methods are often required to convert the precursor sol–gel films into fully dense coatings. Multifunctional materials can be produced using different catalysts and template molecules to modify the pore structure and size distribution, hybridizing molecules with different functions, and processing the materials under different conditions. For instance, the sol–gel process of silicon alkoxide includes, in general, the following chemical reactions. The silicon alkoxides, for example, $Si(OR)_4$, are hydrolyzed in water with the aid of acidic or base catalysts (Reaction [1]).

$$Si(OR)_4 + H_2O ----- (RO)_3 Si(OH) + ROH \qquad (23.1)$$

The partially or fully hydrolyzed metal alkoxides are condensed to form a network (1D, 2D, or 3D), depending upon the properties of chemicals, experimental conditions, and chemical compositions of the system.

$$\left(\mathrm{RO}\right)_3 \mathrm{Si}\left(\mathrm{OH}\right) + \left(\mathrm{HO}\right)\mathrm{Si}\left(\mathrm{OR}\right)_3 \ ---- \ \left(\mathrm{RO}\right)_3 \mathrm{Si} - \mathrm{O} - \mathrm{Si}\left(\mathrm{OR}\right)_3 + \mathrm{H}_2\mathrm{O} \qquad (23.2)$$

or

$$\left(\mathrm{RO}\right)_3 \mathrm{Si}\left(\mathrm{OH}\right) + \left(\mathrm{RO}\right)\mathrm{Si}\left(\mathrm{OR}\right)_3 \ ---- \ \left(\mathrm{RO}\right)_3 \mathrm{Si} - \mathrm{O} - \mathrm{Si}\left(\mathrm{OR}\right)_3 + \mathrm{ROH} \qquad (23.3)$$

These reactions follow a S_N2 mechanism in which the nucleophilic addition of negatively charged $HO^{\delta-}$ groups onto positively charged metal atoms $M^{\delta+}$ which leads to an increase of the coordination number of the metal atom during a transition state. The chemical reactivity of metal alkoxides toward hydrolysis and condensation depends mainly on the positive charge of the metal atom and its ability to increase its coordination number. Therefore, it is possible to control the rates of chemical reactions involved in the sol–gel process by catalysis (acid or base), leading to different polymeric structures. Acid-catalyzed condensation often leads to chain polymers, whereas the base-catalyzed condensation leads to more isolated nanoparticle structures. In addition, solvent, oxidation state of the metal atom, and steric hindrance of alkoxide group also have a significant influence on the hydrolysis and condensation reactions. Some metal alkoxides, for example, $Ti(OR)_4$, are highly reactive with water, resulting in precipitates. The reaction rates of the alkoxides can be modified by introducing other chemicals such as carboxylic acids or β-diketones. These bidentate ligands behave either as bridging or chelating groups to reduce the reactivity and functionality.

23.2   *Laser processing of sol–gel coatings*

Laser processing is a unique technique for surface treatment. Temperature profile can be controlled by the laser technique (laser power, pulse or chopping frequency, and translation rate) depending on the thermal properties of the materials involved, so that a high temperature required for coating or surface processing can be achieved at the very surface layer while the thermal penetrating depth into the substrate or bulk materials is minimized to submicrometer range. This technique is particularly useful for processing high-temperature coatings deposited on low-temperature substrates. Laser processing of thin coating materials generally uses two different methods. One is the *direct writing* approach in which the sample is exposed to laser energy, and the heating comes from the absorption of the laser energy by the coating or/and by the substrate. The spatial resolution achievable is mostly limited to the diffraction-limited width of the laser wavelength. The *direct writing* approach is simple, but for application to weakly absorbing materials, a high-energy laser is required. The other approach usually referred as *indirect writing* involves depositing a layer of light-absorbing materials (usually metallic alloys like Au/Pd) on the coatings. The overlayer absorbs the laser energy causing localized heating at the interface between the light absorber and the coating sample. Some fine and complex patterns can be obtained by defining different overlayer structures using standard photolithographic techniques. However, one of the drawbacks of such approach is the possibility of introducing contamination in the sample from the overlayer materials. For some optical applications, the optical transparency of the coating materials is critical. We have previously introduced a laser-processing method which is similar to the *indirect writing* technique, except that a thin layer of air gap is inserted between the coating sample and the light absorbing materials, as shown in Figure 23.1. Instead of using metallic alloys like Au/Pd, high-temperature materials such as SiC were used as the light-absorbing materials. Our processing technique integrates optical sol–gel coatings having tailored properties of porosity, surface area, and surface affinity with substrate materials used for integrated and fiber optic infrared technologies such as optical sensors. Multiple depositions of preceramic polymers are usually required to achieve

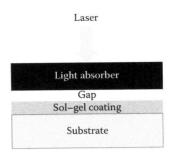

Figure 23.1 Indirect laser-processing method for curing porous coatings.

acceptable properties and performances for advanced composite materials. The deposition process involves several physical and chemical phenomena such as fluid flow, diffusion, and reaction. Traditional methods produce films/coatings by repeating the deposition process for multiple times using the same precursor solution. The substrate structures and properties (e.g., the pore size, the surface wettability, and the substrate permeability) may change when loaded with the infiltrated polymers. Such changes may have a great impact on diffusion and reaction, and thermal transport in the successive filler material, which in turn can significantly affect interfacial strength of multilayer structures. In this work, we explore the use of computer simulation to gain an insight into some important additive manufacturing parameters that are usually difficult to obtain by means of experiment. Our computer simulation focuses upon fluid flow, energy balance in thermally activated chemical reactions, heat and mass transport, thermomechanical stress in the porous structure.

23.3   Computer modeling of thin-film processing and characterization

The model system that we use to simulate the sol–gel deposition on curved surfaces consists of two-layered fiber mesh networks. The networks are embedded in a rectangular container through which the precursor sol–gel solution flows from one to the opposite side of the container under a constant pressure. The mass transport of the sol–gel solution in the model system is treated in the regime of laminar flow. The adsorption and desorption of the sol–gel precursor chemical are assumed to be governed by a thermally activated first-order chemical reaction. The conversion of physically adsorbed precursor molecules to chemically adsorbed ones is also treated to be a thermally activated first-order chemical reaction. In this finite element analysis (FEA) simulation, we study spatial and temporal distributions of the precursor chemical concentration in the sol–gel solution and on the surface of the fiber fabrics (physically and chemically adsorbed), as the precursor chemical is introduced into the model system with different distributions (e.g., Gaussian and rectangular profiles).

Surface reactions: The precursor chemical undergoes an adsorption and desorption process as it flows through the porous network.

$$P + S \underset{k_{\text{des}}}{\overset{k_{\text{ads}}}{\rightleftarrows}} PS \tag{23.4}$$

$$PS \underset{k_{\text{bs}}}{\overset{k_{\text{sb}}}{\rightleftarrows}} PS_0 \tag{23.5}$$

Some of the adsorbed precursor chemical can be chemically bonded to the surface in a reversible fashion. The rate of adsorption is given by a first-order reaction, $r_{ads} = k_{ads}C_P$, where C_P is the concentration of physically adsorbed precursor chemical P. The rate of desorption is given by a first-order reaction, $r_{des} = k_{des}C_{PS}$, where C_{PS} is the concentration of the physically adsorbed precursor chemical P. Part of the physically adsorbed chemical may be converted to be chemically adsorbed chemical, as described by Equation 23.5. The net rate of deposition of the precursor chemical P on the surface is given by, $r_{sb} = -k_{bs}[PS] + k_{sb}[PS_0]$.

The net increase in the concentration of the deposited chemical P is equal to

$$\frac{dC_{AS}}{dt} = r_{ads} - r_{des} - r_{sb}$$

The mass transport of the precursor chemical in the sol–gel solution is described by

$$\frac{\partial C_A}{\partial t} + \nabla \cdot (-D_A \nabla C_A) + u \cdot \nabla C_P = 0 \tag{23.6}$$

where:

C_p is the concentration of the precursor chemical in the solution
D_P is the coefficient of diffusion of the chemical P
u is the flow velocity of the chemical P

The precursor solution enters at one end of the channel and exits at the other end. The time-dependent distribution of the chemical P at the inlet boundary is simulated by a Gaussian pulse having a peak concentration of G_{max} and deviation of σ. The boundary condition at the outlet surface is given by $n \cdot (-D_P \nabla C_P) = 0$. The diffusion of the chemical on the surface is neglected for simplicity.

The laminar flow of the sol–gel solution in the channel is described by the Navier–Stokes equations:

$$\rho u \cdot \nabla u = \nabla \cdot \left[-pI + \eta \left(\nabla u + (\nabla u)^T \right) - (2\eta / 3)(\nabla \cdot u)I \right]$$

$$\nabla \cdot (\rho u) = 0 \tag{23.7}$$

where:

ρ is the density
η is the viscosity
p is the pressure

The calculated flow field is used as input to the mass transport in the solution. The boundary conditions are as follows:

$$u \cdot n = u_0, \text{ at the inlet}$$

$$u = 0, \text{ at the walls of surfaces}$$

$$p = p_0, \text{ at the outlet}$$

The pressure at the outlet is equal to the ambient atmosphere pressure.

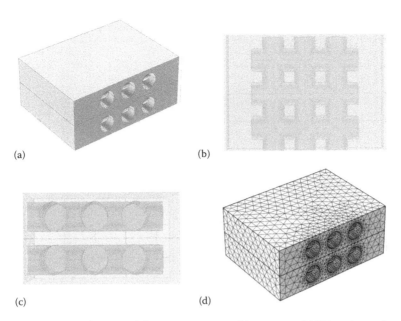

(a)

(b)

(c)

(d)

Figure 23.2 The geometry of the model system measured in meters: (a) 3D rectangular sample with two layers of embedded fiber fabrics; (b) Top view (3D perspective) of the 3D rectangular sample with embedded fiber network; (c) Side view (the yz plane, 3D perspective) of the 3D rectangular sample with embedded fiber network; and (d) Mesh structures of the 3D rectangular sample with embedded fiber network.

The geometry of the model material system is shown in Figure 23.2. The sol–gel solution enters at the left side of the channel in the yz plane. The sol–gel solution exits at the right side of the channel in the yz plane.

The concentration distribution of the precursor deposition (e.g., chemically adsorbed) on the fiber surfaces at a selected time is shown in Figure 23.3. It is clearly seen that the surface coverage is not uniform, highly dependent of locations. The concentration distribution of the physically adsorbed precursor chemical on the fiber surfaces at the same time is shown in Figure 23.4. It appears that these two types of surface adsorptions are complementary to each other. For future study, we may explore different parameters, including initial concentration, flow rate and pressure, diameter and spacing of the fibers to further understand the process of sol–gel deposition.

The velocity of the sol–gel solution increases significantly as the channel narrows in the regions between the different layers of the fiber fabrics, which may have important implications about the rate of adsorption and desorption on the fiber surfaces (Figure 23.5).

It is worth mentioning that at the steady state, the overall pressure distribution is independent of time with large gradients near the regions occupied by the fiber fabrics, as shown in Figure 23.6.

The changes in concentration of the precursor chemical at different locations in the front and rear central, and the front and rear wall regions are illustrated in Figure 23.7. The chemical depositions on the fiber surfaces can be evaluated for different areas and locations of the fiber fabrics.

Time = 45 s   Surface: surface concentration (mol/m²)

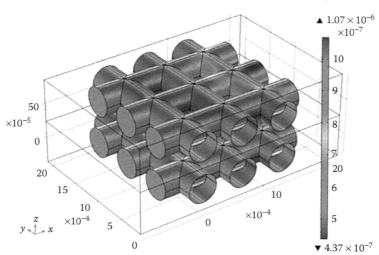

Figure 23.3 **(See color insert.)** The concentration distribution of the chemically adsorbed precursor chemical on the surface of fibers at a selected time.

Time = 45 s   Surface: surface concentration (mol/m²)

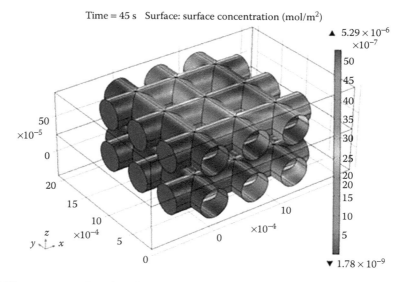

Figure 23.4 The concentration distribution of the physically adsorbed precursor chemical on the surface of fibers at a selected time.

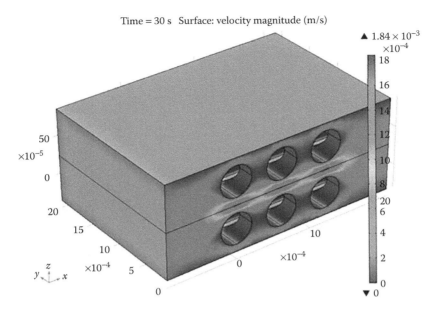

Figure 23.5 The velocity of the precursor solution in the channel at a selected time.

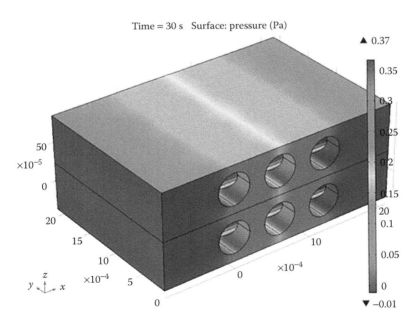

Figure 23.6 **(See color insert.)**   The pressure of the precursor solution in the channel at a selected time.

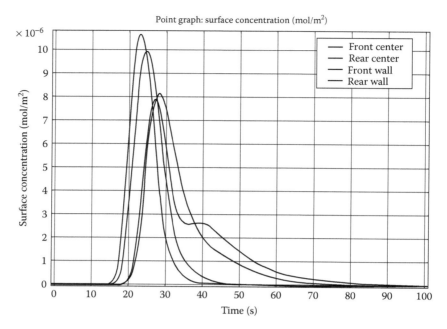

Figure 23.7 The changes in the concentration of the precursor chemical at different locations of the fiber surfaces for different times.

chapter twenty four

Additive manufacturing technology review

From prototyping to production

Larry Dosser, Kevin Hartke, Ron Jacobson, and Sarah Payne

Contents

24.1   Background on direct digital manufacturing

Direct digital manufacturing (DDM) encompasses a broad range of technologies, which can be used to fabricate parts directly from an electronic file as shown in Figure 24.1.

The computer-aided design (CAD) file contains all the necessary geometrical information required to create the part. For the purposes of this program, the DDM processes have been grouped into common categories, which include subtractive, additive, and hybrid technologies. The upstream and downstream processes were also explored as a part of this program, and the relationships are shown in Figure 24.2.

The upstream processes include reverse engineering and CAD and computer-aided manufacturing (CAM) technologies. Reverse engineering is used to gather information from an existing part, which can be translated into CAD data. CAD/CAM

Figure 24.1 Direct digital manufacturing.

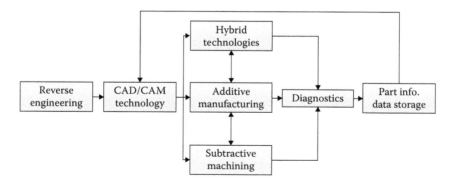

Figure 24.2 Direct digital manufacturing upstream and downstream process.

technology is used to generate digital part information and transform it into machine code. The machine code is loaded into an additive, subtractive, or hybrid process to create the part. The part is then checked using diagnostics including geometrical and nondestructive analysis. Part information data storage was the final area for review, which includes the ability to store detailed part geometry and manufacturing information on the part.

24.2   Data collection methodology

The data collection portion of the project required a variety of methods to gather and collate data. The first step was to create a decision tree (Figure 24.3), which could be used to characterize each process on a common platform. The decision tree allowed all DDM technologies to be characterized using seven key qualifiers.

After the completion of the decision tree, data collection was completed using the following methods:

Internet searches: Each technology was thoroughly searched including equipment manufacturers, forums, and user websites.

Interviews: Key players were interviewed including researchers, users, and equipment manufacturers.

Conferences: Conferences were attended to review the ongoing research in DDM technology.

Industry Reports: Available industry reports were obtained and reviewed.

Survey: A comprehensive survey was completed and sent to researchers, users, and manufacturers in the industry.

24.3   Additive manufacturing

Additive manufacturing (AM) comprises a group of technologies that fabricates parts through a buildup process. All of these technologies start with raw material and a CAD file. AM technologies have progressed from rapid prototyping to functional part fabrication over the past several years. At a high level, these processes appear to be a perfect match for DDM. However, upon detailed investigation, there is much more complexity than which initially meets the eye.

24.3.1   Direct metal laser sintering

The direct metal laser sintering (DMLS) process is used to create metal parts by laser heat source melting metal powder in layer-by-layer part fabrication. The process is contained in an inert gas environment. The part build information is obtained directly by a CAD file. Examples of parts created through this process are shown in Figure 24.4.

The leader in DMLS processing is EOS (www.eos.info), which is a German company. For each EOS company presentation, they have over 900 systems installed worldwide. The systems range in price from $500K to $700K. Other top players in the industry include MTT (www.reinshaw.com) and Concept Laser (www.concept-laser.de). Based on all research, it appears as though all DMLS equipment manufacturers are located in Europe.

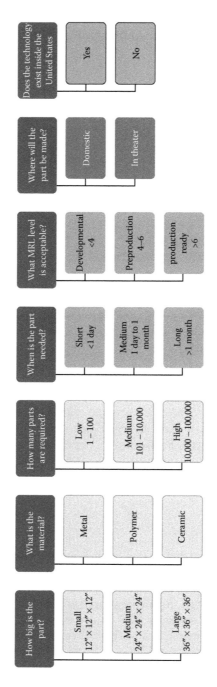

Figure 24.3 Decision tree.

Figure 24.4 Examples of components fabricated using DMLS. (From www.electroptics.com, www. eos.info.)

Part size	Material	Part quantity	Part cycle time	MRL level	Portability	Domestic/ foreign
Small	Metal	Low				
		Medium	Medium	Prepod.	Large size/ high power	Foreign
				Production ready		

Figure 24.5 Decision tree for DMLS.

The technology has been qualified through the decision tree, which is shown in Figure 24.5.

As shown in the decision tree, the technology has limitations to small ($12'' \times 12'' \times 12''$) parts and is only suitable for processing metals. There is also a limited number of metals available including stainless steel and cobalt chrome. The process is being used today to fabricate components. Per an interview with Greg Morris of Morris Technologies, this process requires engineering design knowledge, postprocess thermal treatments, postprocess machining (milling, drilling), and postprocess polishing. Greg estimates that 20% of the process is in the equipment, and the other 80% is in process knowledge and downstream processing.

DMLS is a fast growing technology area in DDM. The technology is currently being used to fabricate dental implants and other medical devices. The main limitations are the size of part, build rate, availability of materials, and in-process monitoring capability.

Based on an interview with Greg Morris of Morris Technology, DMLS is a very capable technology for producing functional parts. Morris technology has 18 DMLS machines and is running CoCr, stainless steel, and titanium alloys. The company is in the process of developing an aluminum alloy suitable for DMLS processing. According to Greg, the market awareness of DMLS has increased rapidly over the past several years. The commercial sector has been much quicker than the government customer to adopt the technology. Greg believes that many of the advances in DMLS are occurring as these commercial applications are being developed, and since this development is privately funded, these advances are not being seen in the public domain. The total process cycle time for a DMLS part is in the range of two weeks. This includes engineering design, CAD/CAM, DMLS processing, postprocess heat treating, machining, and polishing. The DMLS machine time only makes up to 20% of the process, and the remaining time is involved in upstream and downstream processing. Greg predicts the DDM market to be in the range of $9–$10 billion in 10 years and estimates the current market size at $100 million. He also mentioned that flight critical hardware is currently being produced using DMLS, and the mechanical performance is better than wrought material.

24.3.2 Direct metal deposition

Direct metal deposition (DMD) uses a focused laser beam combined with coaxial powder metal delivery to fabricate component. DMD is also known as laser engineered net shaping (LENS) and laser free-form fabrication (LF3). This process is similar to DMLS with the exception of powder delivery through a nozzle for DMD and bed-based for DMLS. DMD is typically used for repair, cladding, and add-on features but can also be used for complete part fabrication (Figure 24.6).

This process has the advantage over DMLS in its ability to produce much larger components. There is also the advantage of being able to mix different metals through the nozzle to create custom alloys. This process is typically used to make near-net shaped components with secondary subtractive operations to finish part fabrication. Figure 24.7 details the decision tree for this technology.

Figure 24.6 Laser engineered net shaping (LENS) in process. (From www.sandia.gov.)

Part size	Material	Part quantity	Part cycle time	MRL level	Portability	Domestic/ foreign
Small	Metal	Low	Short	Develope		Domestic
Medium			Medium	Prepod.	Large size/ high power	
Large						

Figure 24.7 Decision tree for DMD.

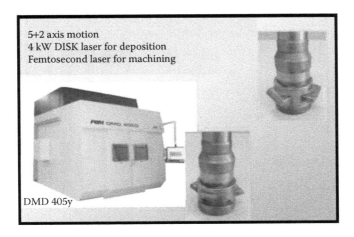

Figure 24.8 POM DMD system combined with subtractive femtosecond laser processing.

POM (www.pomgroup.com) (Figure 24.8) and Optomec (www.optomec.com) are both U.S. based companies developing and selling DMD equipment. Both companies are focusing on large-scale (greater than 12″ × 12″ × 12″) components, which the DMD process is well suited. Fraunhofer has also made great strides using this technology to create near-net shape BLISK components, which is funded through a 10.25 M Euro program call TurPro (Figure 24.9). Using DMD, a single BLISK blade can be fabricated in less than 2 minutes by employing a 10 kW disk laser and coaxial powder delivery system. The GE Research Center in Shanghai has fabricated a 42″ jet engine fan blade using DMD (Figure 24.10).

Deposition rates for the DMD process can be up to 150 mm^3/s based on reports from the Fraunhofer TurPro program. This makes the process industrially viable for large part fabrication.

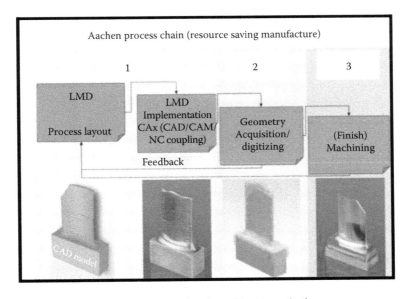

Figure 24.9 BLISK DMD fabrication process developed by Fraunhofer.

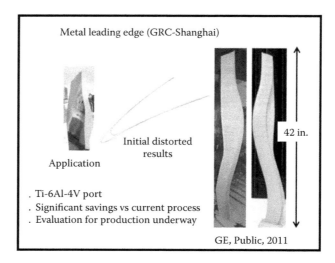

Metal leading edge (GRC-Shanghai)

Application

Initial distorted results

. Ti-6Al-4V port
. Significant savings vs current process
. Evaluation for production underway

42 in.

GE, Public, 2011

Figure 24.10 42″ jet engine fan blade—GE research—Shanghai.

24.3.3 *Electron beam melting*

Electron beam melting (EBM) is a process where an electron beam is used as a heat source to melt metal powder in layer-by-layer part fabrication. Parts are fabricated in a vacuum, and the electron beam heat source can hold the build chamber at an annealing temperature for the duration of the part fabrication cycle. The part build information is obtained directly from a CAD file. An example of parts fabricated via EBM is shown in Figure 24.11.

Figure 24.11 Hip implant components fabricated using EBM. (From www.arcam.com.)

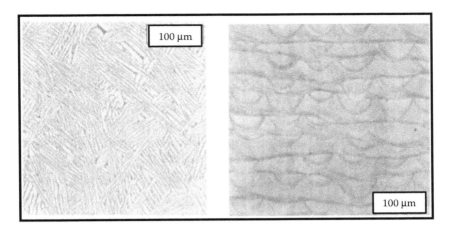

Figure 24.12 EBM Ti-6Al-4V microstructure (a) versus DMLS CoCr microstructure (b).

EBM is very similar to DMLS with the primary difference being the heat source. The ability of the heat source to maintain an annealing temperature during the build is evident in the part microstructure shown in Figure 24.12.

The anisotropic microstructure of the DMLS process requires postprocess heat treating to normalize the material. The EBM process does not require this postprocess heat treating.

The decision tree for EBM is shown in Figure 24.13. The main disadvantage of EBM is the resultant surface finish. The process creates a relatively rough outer surface of the part, which requires postpart polishing or machining.

Based on an interview with Kevin Slattery of Boeing, over 80% of their titanium parts could be fabricated using the Arcam system. The processing time for a baseball size part takes around 15 hours, and the rates would need to increase two to four times to make EBM a cost-effective process.

A total of 50% of the production time is on the machine, and it could take up to 3 hours of postmachining to bring the part into tolerance. Boeing has a bracket produced using EBM, and they found the process to be two to five times cheaper than conventional processing due to the time savings. Overall, EBM has further development to be used across a larger number of applications, but it has been proven effective in a selective number of

Part size	Material	Part quantity	Part cycle time	MRL level	Portability	Domestic/ foreign
Small	Metal	Low				
			Medium		Large size/ high power	Foreign
				Production ready		

Figure 24.13 Decision tree for EBM.

applications. Further area of development include: more qualified raw materials, machine improvements (better uptime and online process monitoring), bigger build chamber, better surface finish, and reduced capital cost.

24.3.4 Electron beam free-form fabrication

Electron beam free form fabrication (EBFFF) is a process, which uses an electron beam as the heat source and an off-axis metal wire to fabricate parts. The process is used primarily for near-net shaped AM and requires postsubtractive processing to fabricate finished parts. The process is specialized for aerospace applications with Sciaky (www.sciaky.com) being the primary developer of the equipment. Figure 24.14 shows a schematic of the process and parts fabricated.

The advantages of EBF3 are the ability to process difficult or specialty alloys in a vacuum into a near-net shape. The process is capable of high deposition rate (need number) when compared to EBM, DMD, and DMLS (Figure 24.15).

Based on an interview with Kevin Slattery of Boeing and Craig Brice of NASA, the EBF3 process provided excellent build rates with typical deposition rates being in the 7–10 lb/hour range. However, the process is only suited for near-net shape fabrication and requires postprocess machining to bring the part into tolerance. However, the process can produce some of the largest parts compared to all of DDM processing with sizes up to 4 ft × 2 ft × 2 ft.

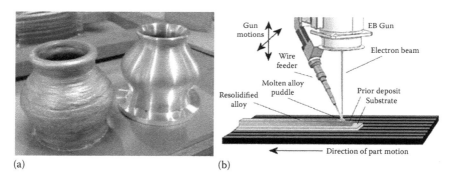

(a)   (b)

Figure 24.14 Pre/post EBF3 Part (a) (From www.nasa.gov) and EBF3 schematic (b). (From www. sciaky.com).

Part size	Material	Part quantity	Part cycle time	MRL level	Portability	Domestic/foreign
Small	Metal	Low			Small size/low power	
			Medium	Preprod.	Large size/high power	Foreign

Figure 24.15 Decision tree for EBF3.

24.3.5 Summary of additive beam-based processes

The four additive beam-based processes including DMLS, DMD, EBM, and EBF3 have distinct advantages and disadvantages. Each technology has enabled the fabrication of metal parts using a CAD file and raw material. The main areas for continued development of these technologies include the following:

- Feature size versus deposition rates
  - Deposition rates need to increase two to three times for full part fabrication while maintaining surface finish and feature size.
  - Limited part size.
- Postprocessing
  - Eliminate the need for support structures and postprocessing.
  - All processes require some or all of the following polishing, machining, and heat treating.
- Characterization of materials
  - Limited material choices.
  - Limited characterization.
  - Fundamental research into new materials made specifically for AM processes.
- Capital equipment
  - Equipment is expensive and has limited number of suppliers.
  - Equipment is not consistent from one machine to the next.
  - Domestic source of equipment.
- Process monitoring
  - Limited development on process control and feedback.
- Process modeling
  - Limited knowledge/development in process modelling.
  - Residual stress/distortion prediction.
- Education
  - *Training*: Common training platform for technicians.
  - *Design*: Improved training of designers on how the technology can be exploited.
  - *Process knowledge*: All current process knowledge is proprietary, and there is no open forum or handbook on how to use the technology.
- Government consortium
  - Europe has multiple consortia working to solve the above issues.
  - US has no funded consortia at this time.

Figure 24.16 shows a qualitative comparison between the four additive beam processes. The size of the data point equates to the size of the part the process can fabricate. The DMLS and EBM processes are well suited for small ($<12'' \times 12'' \times 12''$) parts that require better surface finish with minimal postprocessing. The DMD and EBF3 processes are better suited for medium to large parts ($>12'' \times 12'' \times 12''$) but will require more postprocessing due to lower quality surface finish. Based on the review of all four technologies, it appears that there is an inverse relationship between surface roughness and deposition rate, which equates to a decrease in the part surface finish with an increase in the material deposition rate.

24.3.6 Selective laser sintering

Selective laser sintering (SLS) is a laser-based AM process that is capable of manufacturing parts in metals, polymers, and ceramics in a layer-by-layer buildup. In the case of metals,

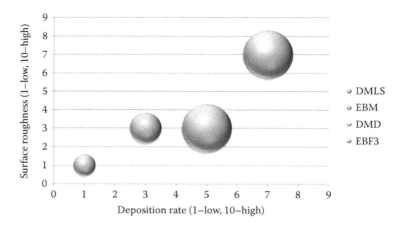

Figure 24.16 Comparison of additive beam processes.

the laser is melting a binder, which holds the powder material together versus fusing the part material directly. This produces a weaker part that requires additional postprocessing and backfilling for metals. For polymers, the laser fuses the polymer together directly. The process is excellent for producing rapid prototypes and functional parts in polymers. Example parts are shown in Figure 24.17.

The main players in this market include EOS (www.eos.info) and 3D Systems (www.3dsystems.com). The technology is being used to manufacture polymer components for the aerospace industry. Based on an interview with Scott Martin of Boeing, SLS is used to fabricate over 80 parts for the F-16. The process is well suited for rapid part manufacturing of polymer parts. However, there are cost considerations when compared to injection molding because of the low cost of the raw material for this process. These cost considerations make the process well suited for low-volume manufacturing. The decision tree results are shown in Figure 24.18.

There are several areas of desired improvement for the SLS process which include improved surface finish, reduced cost of capital, machine to machine consistency, open architecture equipment, formalized test platform, and intelligent feedback and control.

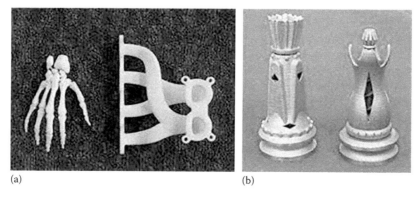

Figure 24.17 Examples of SLS parts in polymer (a) and metal (b). (From www.louisville.edu, www.protocam.com.)

Part size	Material	Part quantity	Part cycle time	MRL level	Portability	Domestic/ foreign
Small	Metal	Low	Short			Domestic
Medium	Polymer	Medium	Medium	Preprod.	Large size/ high power	Foreign
	Ceramic			Production ready		

Figure 24.18 Decision tree for selective laser sintering.

24.3.7 Stereolithography

Stereolithography (SL) is a laser-based process, which can fabricate parts through a layer-by-layer buildup of a liquid photo-curable polymer. The process is typically used only for rapid prototyping and an example part fabricated with this process is shown in Figure 24.19.

The leading company producing SLA equipment is 3D Systems (www.3dsystems. com). The decision tree for this process is shown in Figure 24.20.

24.3.8 Fused deposition modeling

Fused deposition modeling (FDM) uses a polymer filament fed through a head extrusion nozzle to build parts up in a layer-by-layer process. FDM is limited to polymers but can fabricate parts with strength near that of the parent material. Figure 24.21 shows examples of parts fabricated through the FDM process.

The key player in the industry is Stratasys/Hewlett Packard. There is also a movement to produce low-cost FDM equipment through companies such as MakerBot and Bits for Bytes. The low cost of the technology makes it suitable for hobbyist and early adopters. The decision tree for FDM is shown in Figure 24.22.

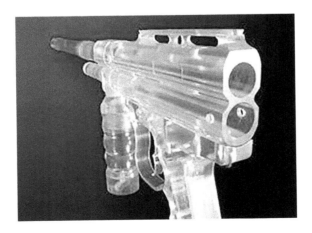

Figure 24.19 Part fabricated using stereolithography.

Part size	Material	Part quantity	Part cycle time	MRL level	Portability	Domestic/ foreign
Small		Low				Domestic
Medium	Polymer		Medium		Large size/ high power	Foreign
				Production ready		

Figure 24.20 Decision tree for stereolithography.

Figure 24.21 Parts fabricated using FDM. (From www.peridotinc.com.)

Part size	Material	Part quantity	Part cycle time	MRL level	Portability	Domestic/ foreign
Small		Low				Domestic
Medium	Polymer		Medium		Large size/ high power	Foreign
				Production ready		

Figure 24.22 Decision tree for fused deposition modeling.

24.3.9   3D Ink-Jet printing

3D ink-jet printing used a multichannel ink-jet print head to deposit a liquid adhesive on a bed of powdered polymer. Another variation of this process deposits tiny droplets of thermoplastic and wax materials directly from the print head. Both variations can be used to fabricate parts in a layer-by-layer buildup process. Examples of parts fabricated using this process are shown in Figure 24.23.

Players in the 3D-printing industry include Objet (www.objet.com) Solidscape (www.solid-scape.com), VoxelJet (www.voxeljet.de), and 3D Systems (www.3dsystems. com). Similar to SLA, the technology is well suited for prototype fabrication; the parts are not robust enough for rugged application. The decision tree for 3D printing is shown in Figure 24.24.

24.3.10   Additive manufacturing summary

The AM industry contains a number of different technologies that continue to progress toward DDM. Figure 24.25 displays a breakdown of the technologies by manufacturing ruggedness (*x*-axis), capital cost (*y*-axis), and part size (bubble size). Based on this review, the technologies can be broken down into two main subgroups. The first subgroup contains technologies that have the capacity to directly produce parts in metals and polymers, which include DMLS, DMD, EBM, EBF3, SLS, and FDM. All of these technologies have the

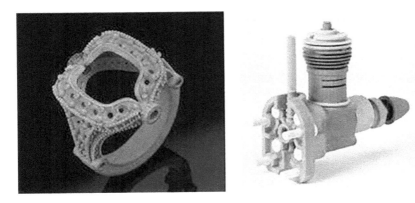

Figure 24.23 Examples of parts fabricated using 3D printing. (From www.objet.com.)

Part size	Material	Part quantity	Part cycle time	MRL level	Portability	Domestic/ foreign
Small		Low	Short		Small size/ low power	Domestic
Medium	Polymer		Medium	Preprod.		Foreign
				Production ready		

Figure 24.24 Decision tree for 3D printing.

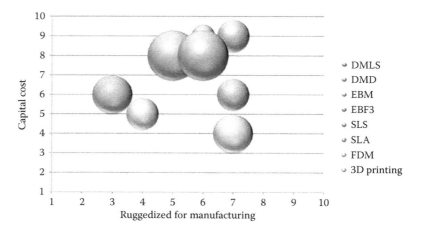

Figure 24.25 Overview of additive manufacturing technologies.

inherent ability to produce usable components directly from a CAD file. The remaining technologies are limited by the material sets and are only suitable for prototype or form models, which include SLA and 3D printing.

In addition to interviews, MLPC (Mound Laser & Photonics Center), Inc also conducted an online survey across academic, industry, equipment, and service providers in the AM industry. The complete survey results are listed in Appendix A.

Survey results for metal parts

- Build speed and surface roughness were each ranked in the top three most important aspects for improvement/development for four of the six AM technologies that build metal parts. In fact, they were ranked most important in two technologies each.
- The results for maximum part size were split. This aspect was ranked in the top three priorities for three of the six technologies but was not of great importance to the respondents of the other three.
- Except for EBM, in which it was ranked as the most important aspect for development, in-process monitoring was ranked in the bottom half for the other technologies.
- Surprisingly, respondents did not seem to place raw material variety very high on their list of priorities of aspects needing improvement.

Survey results for plastic parts

- Three out of the four AM technologies that build polymer parts (SLA, SLS polymers, 3D/inkjet printing, and FDM/FFF) showed finished part strength as being either the first- or second-most important aspect for improvement/development.
- Raw material variety and build speed were other aspects that respondents generally seemed to place emphasis on for improvement/development. These aspects usually fell in the top half of priorities for the different technologies.
- On the other hand, maximum part size and amount of postprocessing required did not generally seem to be of major concern to the respondents. Neither of these aspects fell in the top 50% of choices for needing improvement/development.

Overall, the AM technologies are in an early stage of technical development making a transition from prototyping to production. This transition is occurring in private industry through the design and testing of parts across many industries. There is a significant amount of continued development required for full qualification into critical applications. This transition will occur over the next 10 years as the technical challenges continue to be solved.

24.4   Subtractive manufacturing

Subtractive manufacturing includes all technologies used to remove material to create a part. These processes start with a piece of raw material and a CAD file. These technologies include computer numerical control (CNC) machining, electrical discharge machining (EDM), laser machining, and waterjet machining. Each of these technologies is described in detail in the following sections.

24.4.1   CNC machining

CNC machining uses a computer-controlled motion system combined with a rotating machine tool to fabricate parts. This technology has been used in industrial manufacturing since the 1980s. There are a large number of companies across the world that manufactures CNC equipment. The leaders in this industry include Haas and Morei, and they produce thousands of CNC machines a year.

The decision tree for CNC machining is shown in Figure 24.26.

In addition to large-scale CNC machining which can produce parts in excess of $36'' \times 36'' \times 36''$, a microversion of the technology was also reviewed. MicroCNC machining uses higher resolution motion systems coupled with high (>50,000 RPM) spindle speeds. Examples of parts fabricated with this technology are shown in Figure 24.27.

24.4.2   Waterjet machining

Waterjet machining uses a higher pressure water nozzle combined with an abrasive to cut components. This technology is generally used to cut through a component leaving a kerf of (0.010″–0.015″) behind. The process does not introduce any thermal input into the material and can be used on a wide range of materials. Waterjet machining was developed in

Part size	Material	Part quantity	Part cycle time	MRL level	Portability	Domestic/foreign
Small	Metal				Small size/low power	Domestic
Medium	Polymer	Medium	Medium		Large size/high power	
Large		High	Long	Production ready		

Figure 24.26 Decision tree for CNC machining.

Figure 24.27 Example of microCNC-machined parts.

the 1950s. The process is used across a wide range of industries, and key players providing equipment include Flow and OMAX. The decision tree for waterjet machining is shown in Figure 24.28.

Similar to CNC machining, there is also a microversion of the technology which is capable of producing parts with a kerf width down to 80 microns. Figure 24.29 provides examples of parts cut using microwaterjet.

Part size	Material	Part quantity	Part cycle time	MRL level	Portability	Domestic/ foreign
Small	Metal		Short			Domestic (AJW)
Medium	Polymer	Medium	Medium		Large size/ high power	
Large	Ceramic	High		Production ready		

Figure 24.28 Decision tree for Waterjet.

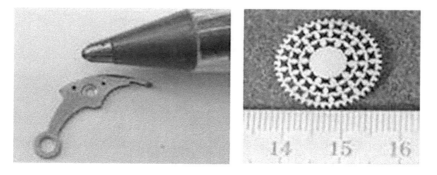

Figure 24.29 Examples of parts fabricated with microwaterjet.

24.4.3   Laser micromachining

Laser micromachining uses pulses from a focused laser beam to ablate small increments of material from a surface. For longer nanosecond pulses, ablation is by melt and evaporation with significant heat transfer. Ultrashort pulses lead to direct ablation and minimal heat transfer. The process works well on most opaque and some transparent materials. The decision tree for laser micromachining is shown in Figure 24.30.

The equipment manufacturers for this technology include Resonetics, 3D Micromac, and JPSA. The technology is used across a wide range of industries including medical device, aerospace, and microelectronics. Figure 24.31 provides examples of features that were fabricated with laser micromachining.

One area of laser micromachining that has been recently developed is ultrafast laser processing. This technology employs picosecond and femtosecond laser pulses to ablate material with minimal or no heat input. Femtosecond laser micromachining was studied as a part of this program due to its potential benefits to the AM process. Figure 24.32 provides examples of surfaces machined using a femtosecond laser.

24.4.4   Subtractive manufacturing conclusions

Subtractive manufacturing processes including CNC, waterjet, and EDM are considered to be industrially hardened and mature technologies. These processes are well suited to support DDM through the ability to selectively remove material. Laser micromachining is a developing technology that has benefits in the area of precision material removal.

Part size	Material	Part quantity	Part cycle time	MRL level	Portability	Domestic/ foreign
Small	Metal	Low				Domestic
	Polymer	Medium	Medium	Preprod.	Large size/ high power	
	Ceramic		Long	Production ready		

Figure 24.30 Laser micromachining decision tree.

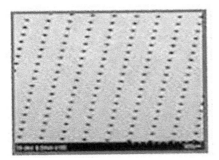 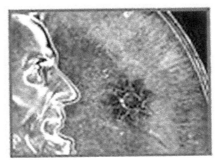

Figure 24.31 Examples of parts fabricated with laser micromachining.

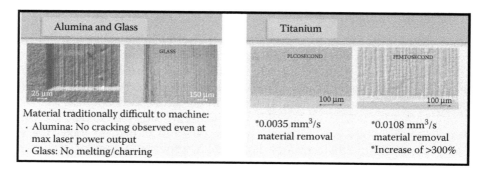

Figure 24.32 Femtosecond laser machining.

Specifically, femtosecond laser micromachining shows promise for material removal without any damage to the part or heat-affected zone.

24.5   *Hybrid manufacturing*

Hybrid manufacturing includes the use of nontraditional methods to fabricate components including ultrasonic consolidation, direct-write electronics, additive/subtractive-combined technologies, and multimaterial processing.

24.5.1   *Ultrasonic consolidation*

Ultrasonic consolidation is a process that uses an ultrasonic welding process to build up a part a layer-by-layer. The process is combined with CNC machining to remove the unwanted material and generate the desired part shape. Solidica (www.solidica.com) is the inventor of ultrasonic consolidation and sells capital equipment to support the technology. A schematic of the process and a sample part is shown in Figure 24.33.

The advantage of this technology is the combination of different types of materials including aluminum and fiber optics. The disadvantage is that the process is limited to

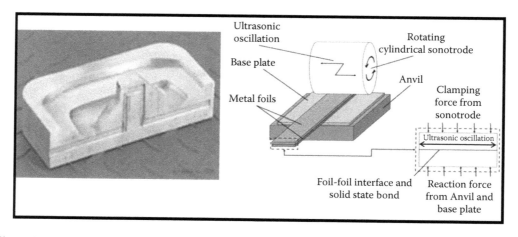

Figure 24.33 Ultrasonic consolidation example part (a) and schematic (b). (From www.solidica.com.)

Part size	Material	Part quantity	Part cycle time	MRL level	Portability	Domestic/ foreign
Small	Metal	Low	Short	Develop.		Domestic
			Medium	Preprod.	Large size/ high power	

Figure 24.34 Decision tree for ultrasonic consolidation.

malleable metals that can be ultrasonically welded. The decision tree for this process is shown in Figure 24.34.

Ultrasonic consolidation is a developing process for hybrid manufacturing and provides a good capability for the combination of malleable materials and embedded electronics and fiber optics.

24.5.2 Direct write electronics

Direct write electronics is not an additive part manufacturing technology but could be used to augment these technologies to incorporate conformal electronics. Companies that produce equipment for this include Optomec (www.optomec.com), Sciperio (www.sciperio.com), and Mesoscribe (www.mesoscribe.com). All of these technologies use a variation of inkjet printing or thermal spray technology to directly apply metal particles to the surface of part, which are fused in process or postprocess.

24.5.3 Additive/Subtractive manufacturing technologies

Additive and subtractive manufacturing technologies are described in detail in the above sections. However, some companies have taken the next step and combined these technologies to take full advantage of both. The POM group has a current program with the US Navy, and the hybrid machine being developed is shown in Figure 24.35. This application uses DMD and dry electrodischarge machining. AM also uses postprocess machining, but this is not typically done on the same machine tool. As these technologies continue to advance, it is likely that a complete integration of the process will be incorporated into one machine tool for maximum efficiency and throughput.

24.5.4 Multimaterial processing

A next generation application of AM technology is the incorporation of multiple materials into one process to create graded material structures. A futuristic vision of this technology would be the ability to make any material through the combination of basic alloying elements into the process. To date, most of the material development has included turning the standard material sets into powders, which are remelted in the additive process. Multimaterial processing offers the next level of development in which two or more of these standard materials are combined in the process. Figure 24.36 provides an example of parts created in the process.

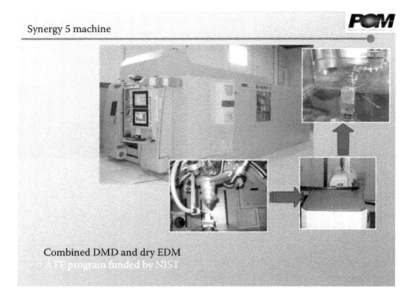

Figure 24.35 POM hybrid DMD and dry EDM machine.

Figure 24.36 Complex geometric parts produced by selective laser melting; a dodecahedron with internal structures (left) and a one dimensional multi-material structure (right)

24.6   Part information data storage

One of the primary values of DDM is the ability to economically make small numbers of any particular part. This makes the paradigm ideal for making replacement parts on an as-needed basis. To facilitate this, it would be valuable to store information needed for the manufacture of any given part directly with the part. This can facilitate finding full details on the part design, dimensions, and fabrication instructions or possibly eliminating the need to look up this information. This can be especially important when replacing parts for vehicles or platforms that are intended to remain in service for long periods of time or are being enabled to continue service through sustainment efforts. In these situations, it is possible that original drawings and specifications may become lost over time with the degradation of records and institutional knowledge. Having key information directly on the part can obviate these concerns.

This section addresses two issues: What information to store with a part, and techniques for storing the information.

24.6.1   Types of information to store

The information of interest to store on a part is that which can be used to guide any DDM technique, or combination of techniques, in accurate reproduction of the part. This may include dimensions, exact material specification, minimum material requirements where a variety of materials may be acceptable, directions for specific fabrication techniques (required equipment, or operating settings, for example), directions for finishing the part (heat treatments, passivation, etc.), and directions for required inspection and validation methods. The total amount of this information may be minimal or quite extensive depending on the part.

There are two basic approaches to associating the above information with the part in question. The first is to comprehensively encode all of the required information directly onto the part. The second is to apply only an identifier number that specifies where the complete set of information can be accessed. (Historically, the latter option, in the form of a producer or supplier specific part number, is the only information typically encoded on a part. A modern alternative to this is the global unique identifier [GUID], described in a subsequent section.)

Each of these methods has advantages and drawbacks. The advantage of comprehensive encoding is that no recourse to other resources is required to begin fabrication. This eliminates dependence on third parties for reliable data storage. It also cuts down on time and expense to access information. If one is attempting to fabricate parts in an environment of active military engagement, there is no dependence on communication lines. However, there are also serious limitations to comprehensive encoding. Many parts simply do not have enough space to hold the necessary information. The physical marking of the information on the part may be inordinately time consuming. Also, damage or wear to the part (which is a near certainty given that the part needs to be replaced) is likely to destroy or efface portions of the information on the part.

Finally, if details of the part are classified or proprietary, it may be undesirable to include the information on the part itself. In general, comprehensive on-part data storage will only be practical for parts that are simple, nonproprietary, and not too small.

Marking of an ID number on the part reintroduces dependence on an external library of part information but addresses all the challenges of comprehensive on-part data storage. An ID number takes up relatively little space and is fast to mark. It can be applied with redundancy to increase the chance that a complete and legible number can be discerned after the part suffers damage or wear. Finally, an ID number removes the primary part information to a point where its proprietary nature can be protected. For most parts and situations, ID numbers will remain the best method to store fabrication information on the part, particularly when implementing the GUID concept discussed below.

24.6.2   GUID

A GUID is the reference number system originally developed to generate unique identifiers in computer software, but the concept is readily adapted for general inventory and database purposes. A GUID is normally represented as a 32-character hexadecimal string (equivalent to a 128-bit binary number).

To use a GUID for part information storage, one must do three things: Generate the GUIDs, apply them permanently to a part, and create and maintain a database that contains the part information associated with each GUID.

Generation of valid GUIDs is trivial. The total number of unique GUIDs ($>10^{38}$) is so large that the probability of random duplication is negligible, even when an enormous number of GUIDs is simultaneously in service. It is literally true that if GUIDs were randomly assigned to every insect on earth[*] (estimated as 10^{19} individuals), the odds that there would be even a single instance of a duplicated GUID is less than 50%. This is so far beyond the number of items to be tracked in any practical database that identifiers can be assigned by any pseudorandom number generator(s) without the concern of confusing parts. The GUIDs can be assigned not just to each type of part, but to each and every *individual* part. Further, because the numbers can be assigned randomly, they do not need to be assigned by central governing body. Any given fabricator of a part can generate a GUID for each part he makes without any fear of a duplicate one already in existence (as long as the generation is done randomly).

Application of a GUID to a part is straightforward, and methods of doing so are discussed in the following sections.

This leaves the issue of creating and maintaining the GUID database. This is a simple, if potentially large, exercise in information technology and data storage, solvable by many providers with off-the-shelf technology and equipment. The main issues include the following:

- *Determining what data will be stored with each GUID*: With sufficient capacity, CAD files to support DDM of each part can be stored in addition to more conventional drawings, specifications, and instructions. Further, since every individual part can have their own GUID, it would be possible to store and update part histories (installation date, last maintenance date, notes taken at last maintenance, hours of cumulative service, etc.).
- *Protocols for accessing the database*: This includes not only the specific technical means of accessing and downloading the information needed to duplicate a part or interest but also the methods for ensuring the security of the information. In some cases, it may be useful to produce and distribute subset databases that can be stored at a local fabrication facility (perhaps in an area of limited or suspected electronic connectivity).
- *Protocols for adding to the database*: Since individual fabricators will be able to generate random GUIDs to cover each part they make, they will need a method for reporting the new GUIDs and part information to the library.

It is beyond the scope of this report to suggest specific methods for setting up such a database.

24.6.3   *Alphanumeric marking*

Alphanumeric marking of a part is the most straight forward way of encoding a GUID on a part. It has the advantage of being readily understood by a human operator but is less

[*] Estimated as 10^{19} individuals by the Smithsonian. (http://www.si.edu/Encyclopedia_SI/nmnh/buginfo/bugnos.htm)

Figure 24.37 Marking ID numbers via dot peening. (Taken from the website of DAPRA, a provider of dot peening equipment, www.dapramarking.com/dot-peen-marking.)

well adapted to optical readers. Marking of parts can be accomplished by conventional engraving techniques, dot peening, or laser marking. Use of ink generally is not advisable due to likelihood of degradation.

The image in Figure 24.37 shows a typical example of how lettering is applied to metal parts using dot peening. An advantage of dot peening on metal is that it introduces compressive stress, which is generally considered to be safer than engraving with respect to the likelihood of reducing the fatigue life of a part. Dot peening also marks deep enough to be legible after substantial wear. A limitation that dot peening is only applicable to materials is ductile and will permanently hold a deformation. Therefore, it is not appropriate for brittle ceramics and may not retain well in some plastics.

Figure 24.38 shows examples of laser marking and micromachining to produce alphanumerics. Advantages of laser marking include speed and the ability to address virtually any material. Laser marking requires only line-of-sight to the mark area and does not access for a physical tool head. Of particular advantage for small parts is the ability to make the font size extremely small.

Figure 24.39 shows, for example, lettering marked into the surface of a penny with characters less than 100 microns tall. This allows for redundant or relatively concealed marking (Figure 24.40).

Part size	Material	Part quantity	Part cycle time	MRL level	Portability	Domestic/foreign
	Metal	Low			Small size/low power	Domestic
Medium		Medium	Medium			Foreign
Large		High		Production ready		

Figure 24.38 Decision tree for dot peen marking.

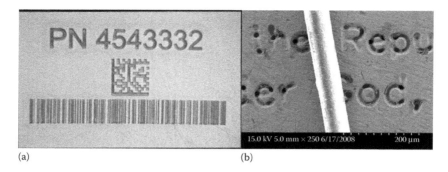

(a)   (b)

Figure 24.39 Marking ID numbers via laser marking (a) or micromachining (b). Fiber shown in (b) for scale is ~60 microns wide. (Courtesy of MLPC, a laser processing company, www.mlpc.com.)

Part size	Material	Part quantity	Part cycle time	MRL level	Portability	Domestic/foreign
Small	Metal	Low	Short		Small size/low power	Domestic
Medium	Polymer	Medium				Foreign
Large	Ceramic	High		Production ready		

Figure 24.40 Decision tree for laser marking.

24.6.4   2D Bar codes

A popular alternative to alphanumerics is barcoding, with 2D (or matrix) barcodes likely to be the most appropriate for most part marking. A 2D barcode encodes information equivalent to alphanumerics as an array of filled and empty cells in a square matrix. Typically, some of the cells are devoted to alignment and registration of the pattern orientation, and the rest are devoted to the actual recorded information. The amount of information that can be stored depends on the size of the matrix. The amount of space that a particular matrix must occupy is limited primarily by the resolution of the reader technology. A common resolution is 0.33 mm/cell, though better can be achieved with high-resolution technologies.

There are a large number of 2D barcode-encoding standards, but public and proprietary. An example of a popular format is the quick response (QR) code. The largest QR codes can store 4000+ alphanumeric characters. They can also be coded with redundancy, up to 30%, by reducing the number of characters. As an example, the QR code shown at right encodes this paragraph.

If this QR code was marked at 0.33 mm resolution, it would fit inside a 23 mm (<1″) square. A QR code that contained only a GUID number would fit in a square of just 8.25 mm on a side.

The physical marking of barcodes on parts can be accomplished by the same techniques of dot peening and laser marking or micromachining described in the previous section. Figure 24.41 shows examples of 2D barcodes marked by dot peening (left) and laser micromachining (right). The contrast for laser marking tends to be better, but both have been shown to be compatible with optical readers.

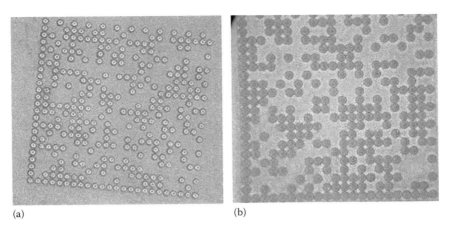

(a)   (b)

Figure 24.41 Examples of 2D barcodes marked by dot peening (a) and laser micromachining (b). Cells are 0.3 mm wide.

The decision tree for 2D barcode marking is essentially the same as those presented for alphanumeric marking. The one essential difference is that 2D barcodes require an optical reader. However, the technology for this is so ubiquitous (readily available as apps on smart phones) that it does not affect the decision tree.

24.6.5   RFID

The possibility of using radio frequency identification (RFID) tags for data storage on parts was investigated but found to be inappropriate. A RFID tag is basically a small antenna with attached integrated circuit chip designed to encode a number and respond to a wireless interrogation device. However, RFID tags cannot be directly produced on a part. They are instead a separate attachment that can become separated from a part. RFID tags typically encode less information than a GUID. Also, they are not very robust. They are more appropriate for inventory and tracking at the warehouse level than for the following individual parts.

24.7   Reverse engineering

Reverse engineering includes any techniques that can be used to gather information from an existing part to inform a DDM process. Ideally, this information can be translated into CAD/CAM format.

24.7.1   Coordinate-measuring machines

A coordinate measuring machine (CMM) is useful for determining precise external dimensions of a complex part. A probe measures the location in 3D space of many representative points on the part surface. The probe can be noncontact (such a laser) or contact (mechanical probe). The accumulated measurements are digitized and form point cloud file that defines the part shape (Figure 24.42).

The largest providers of these machines include Helmel, Trimek, Wenzel, and Zeiss. The main strengths of CMMs are their high accuracy, ability to measure deep slots and pockets, and, if using a contact probe, they are not influenced by surface optical properties like color or transparency. The drawback of a CMM that uses a contact probe is that it is

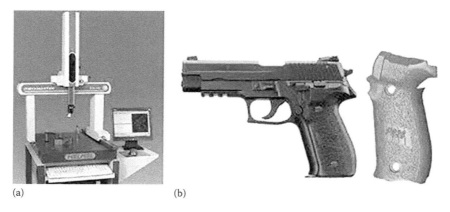

(a)   (b)

Figure 24.42 A table top CMM (a) and example of a point cloud (b) generated for a component of a larger device. (From www.cmmquarterly.com.)

Part size	Material	Part quantity	Part cycle time	MRL level	Portability	Domestic/ foreign
Small	Metal	Low	Short			Domestic
Medium	Polymer				Large size/ high power	Foreign
	Ceramic			Production ready		

Figure 24.43 Decision tree for coordinate measuring machines.

slow for measuring complex surfaces and can deform soft surfaces, leading to an incorrect measurement. The decision tree for CMM is shown in Figure 24.43.

24.7.2   Portable 3D scanning

A portable 3D scanner projects light on the object of interest and then detects the reflection with a camera. Multiple scans are taken from different angles, and point-cloud data are generated to represent the surface (Figure 24.44).

The largest providers of these scanners include Artec, Z Corp, and Creaform. The main strengths of these scanners are their portability and relatively high speed of data collection. The drawback is that internal structures cannot be detected. The decision tree for portable 3D scanners is shown in Figure 24.45.

24.7.3   Laser scanning

A laser scanner reflects laser light from a surface and then generally uses time-of-flight measurements or triangulation to determine the position of points on the surface. As with other optical methods, it generates point-cloud data to represent the surface. Table top and hand-held models are common (Figure 24.46).

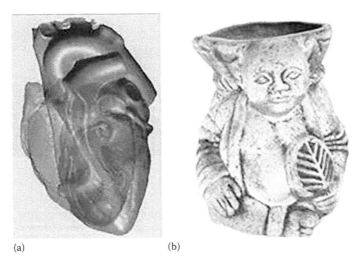

Figure 24.44 (a,b) Examples of complex objects modeled with portable 3D scanners. (a: From www. creaform3d.com and b: From www.artec3d.com.)

Part size	Material	Part quantity	Part cycle time	MRL level	Portability	Domestic/foreign
Small	Metal	Low	Short		Small size/low power	Domestic
Medium	Polymer					Foreign
Large	Ceramic			Production ready		

Figure 24.45 Decision tree for portable 3D scanner.

Figure 24.46 Examples of complex objects measured with laser scanners. (From www.nelpretech. com/reverse_engineering.htm.)

Part size	Material	Part quantity	Part cycle time	MRL level	Portability	Domestic/ foreign
Small	Metal (low density, e.g., aluminum)		Short		Small size/ low power	Domestic
Medium	Polymer	Medium	Medium			
Large	Ceramic	High		Production ready		

Figure 24.47 Decision tree for laser scanners.

The primary domestic providers of laser scanners include NVision and Konica Minolta. The main strengths of laser scanners are fast digitization of large volume parts, combined with good accuracy and resolution. The drawbacks are possible performance limitations on colored or transparent surfaces, lasers' safety cautions, and inability to detect internal structures. The decision tree for laser scanners is shown in Figure 24.47.

24.7.4 Computer tomography X-ray scanning

Computer tomography (CT) takes as series of X-ray scans that map cross-sectional slices of a part and then assembles them into a 3D map. Internal structures (e.g., channels, voids) can be seen, but this ability is limited by density and part thickness. Thus, it works well on plastics and aluminum but not on denser metals. Scan time per part can be relatively long, but information content is high. Output is given in the form of CAD models or blueprints (Figure 24.48).

A main provider of CT scanners is Nel Pre Tech Corp. The main strength of this technique is the ability to detect and model internal structures of a part. However, there are radiation safety cautions to be observed. The decision tree for CT X-ray scanners is shown in Figure 24.49.

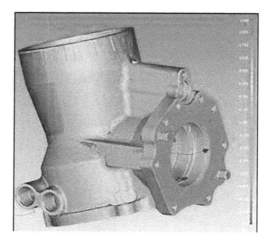

Figure 24.48 A Complex object measured with a scanner for reverse engineering. (From www. nelpretech.com/reverse_engineering.htm.)

Part size	Material	Part quantity	Part cycle time	MRL level	Portability	Domestic/ foreign
Small	Metal (low density, e.g., aluminum)	Low			Small size/ low power	Domestic
Medium	Polymer	Medium	Medium			
			Low	Production ready		

Figure 24.49 Decision tree for CT X-ray scanning.

24.7.5   Industrial computed tomography X-ray scanning

Similar to CT scanning described in the previous section, industrial computed tomography takes a series of X-ray scans of part as it is rotated (see Figure 24.50) rather than cross sections, and then uses digital geometry processing to generate a 3D map. Again, the highlight is the ability to map internal structures along with the outer shape.

Main providers of computed tomography equipment are XViewCT, Zeiss, North Star, and Toshiba. The decision tree for industrial computed tomography X-ray scanning is shown in Figure 24.51.

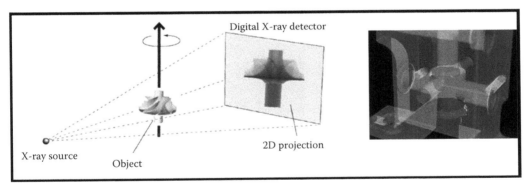

Figure 24.50 Schematic of industrial computed tomography technique (a) and complex object imaged in this way (b). (From www.xviewct.com.)

Part size	Material	Part quantity	Part cycle time	MRL level	Portability	Domestic/ foreign
Small	Metal	Low	Short			Domestic
Medium	Polymer				Large size/ high power	Foreign
	Ceramic			Production ready		

Figure 24.51 Decision tree for industrial computed tomography X-ray scanning.

24.8   *CAD/CAM technologies*

CAD and CAM software were developed in the 1980s and has progressed significantly over the last 30 years. Both pieces of software are vital to the DDM industry. CAD software includes off the shelf programs like AutoCAD, Solidworks, and Pro-Engineer. These programs allow for the 3D design of parts. Figure 24.37 provides an example of a part designed using Solidworks (Figure 24.52).

After the part has been designed in CAD, it needs to go through a postprocessor to be readied for input into the DDM machine. The postprocessing software is being produced by a number of developers including Tesis, Materialise, and Able. The output of the postprocessing software is a common .stl file type. The .stl file format provides all the instructions required to the DDM equipment for fabrication of the part. Each machine platform has proprietary software, which loads the .stl file and allows for a variety of part-processing parameter manipulation.

Figure 24.53 outlines the survey responses regarding software.

Converting solid part files to .stl files was the software task that respondents felt their software programs performed most adequately. Conversely, with the exception of respondents from Academia, there were almost as many people who were dissatisfied with their software *creating necessary hatches and support structures*, as there were people who were satisfied. Overall, it appears that respondents in Academia were somewhat more satisfied than the total average with the performance of their software programs, and respondents in defense were somewhat less satisfied.

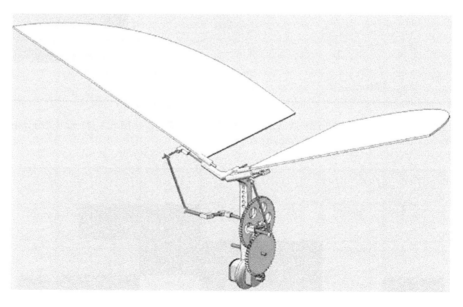

Figure 24.52 Example assembly design in solidworks.

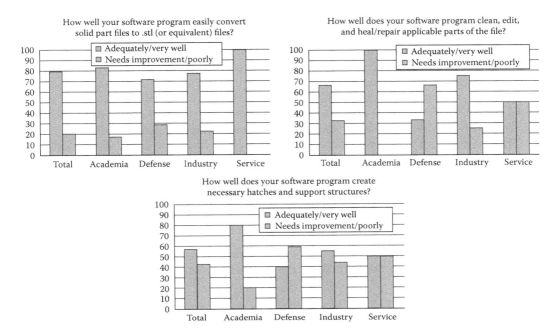

How well your software program easily convert solid part files to .stl (or equivalent) files?

How well does your software program clean, edit, and heal/repair applicable parts of the file?

How well does your software program create necessary hatches and support structures?

Figure 24.53 Online survey responses for DDM software.

24.9   Diagnostics

After part fabrication is complete, diagnostics will be required to verify part integrity. The diagnostics will include both geometrical inspection and scanning of the part for internal defects. Through the online survey, MLPC asked participants to provide feedback on the diagnostic tools used in DDM. The results are shown in Figure 24.54.

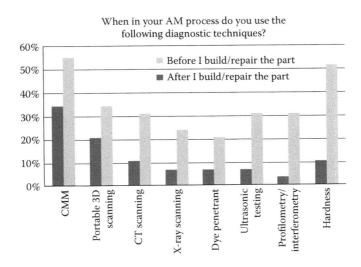

Figure 24.54 Online survey results for diagnostics used in DDM.

Not surprisingly, the proportion of respondents who use a certain diagnostic technology tends to increase as the cost and time associated with performing the diagnostic decrease. So, it is expected that more people use CMM and hardness testing than CT scanning. However, it is interesting to note that while surface roughness was ranked among the highest aspects of AM needing improvement, only ~30% of respondents reported using profilometry/interferometry as a diagnostic. This might indicate that the surface roughness of the AM parts is high enough that the users must perform additional machining and/or polishing anyway, so they do not bother quantifying the initial roughness.

The geometrical inspection would be completed using a coordinate measuring machine (CMM) as shown earlier in Figure 24.42. CMM inspection provides high accuracy measurement (±0.00025″) using a contact probe method of measurements. Companies producing this equipment include Wenzel, Zeiss, and Trimek.

Another method of measurement includes a noncontact method of laser scanning. Laser scanning can be used similar to the CMM as shown in Figure 24.55. The accuracy of a laser scanning system is (±0.0002″).

Key players in the manufacture of laser scanning equipment include NVision and LaserDesign. The final technology that was explored for part diagnostics is computer

Figure 24.55 Laser scanning equipment.

• 3D display

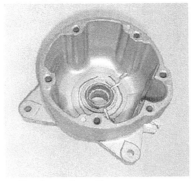

▲ Aluminum die casting

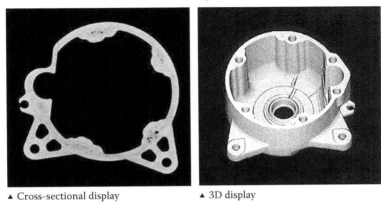

▲ Cross-sectional display   ▲ 3D display

Figure 24.56 Industrial CT scanning of aluminum cast part. (www.toshiba-itc.com.)

tomography X-ray scanning. This technology is important to verifying the absence of defects internal to the part. The technology works by scanning the part with a series of X-rays and reconstructing a 3D map of the part. The downside of this technology is the relatively long scan times (2–3 hours for a small part), but the amount of information gained through this process is important for parts used in critical applications as shown in Figure 24.56.

This technology can be used for both part integrity inspection and reverse engineering. Leading suppliers of equipment in this industry include Zeiss and Toshiba.

Acknowledgments

The contents of this chapter were extracted from *Direct Digital Manufacturing*, Report Number 11-S555-0021-05-C1, a research effort sponsored by the Universal Technology Corporation in Dayton, Ohio.

Bibliography

Bourell, D. L., Leu, M. C., Rosen, D. W. *Roadmap for Additive Manufacturing*, Austin, TX: University of Texas at Austin, 2009.

Frazier, W. E. *Direct Digital Manufacturing of Metallic Components; Vision and Roapmap.* Patuxent River, MD: Naval Air Systems Command, September 2010.

Gibson, I., Rosen, D. W., and Stucker, B. *Additive Manufacturing Technologies Rapid Prototyping to Direct Digital Manufacturing*. New York: Springer, 2010.

Ott, M., Zaeh, F. *Multi-Material Processing in Additive Manufacturing*, Munich, Germany: Institute for Machine Tools and Industrial Management, 2010.

Wohlers, T. T. Wohlers Report 2011: Additive Manufacturing and 3D Printing State of the Industry Annual Worldwide Progress Report. Fort Collins, CO: Wohlers Associates, 2011.

chapter twenty five

Mechanical property optimization of fused deposition modeled polylactic acid components via design of experiments

Jonathan Torres and Ali P. Gordon

Contents

This chapter presents the influences of several production variables on the mechanical properties of specimens manufactured using fused deposition modeling (FDM) with polylactic acid (PLA) as a media and relates the practical and experimental implications of these as related to stiffness, strength, ductility, and generalized loading. A Taguchi orthogonal array test matrix was defined to allow streamlined mechanical testing of several different fabrication settings using a reduced array of experiments. Specimens were manufactured and tested according to ASTM E8/D638 and E399/D5045 standards for tensile and fracture testing. After initial analysis of mechanical properties derived from mechanical tests, analysis of variance (ANOVA) was used to infer optimized production variables for general use and for application/load-specific instances. Production variables are established in order to yield optimized mechanical properties under varying loading types as related to orientation of loading and fabrication. The methodology used exemplifies both the nuances of component production using additive manufacturing methods and shows a refined and streamlined method for determining processing parameter influence on mechanical properties, so as to optimize component production across additive manufacturing technologies.

25.1   Introduction

Rapid prototyping (RP) techniques have been advancing rapidly since the commercialization of the first method additive manufacturing, stereolithography (SL), in the late 1980s, which utilizes a laser to cure successive layers of a liquid polymer into the desired structure (Bártolo 2011; Jacobs 1992; Hopkinson and Dickens 2001; Mellor et al. 2014). Several rapid manufacturing technologies now exist which utilize various techniques and materials to construct components of one layer at a time from a computer-aided design (CAD) file. Systems such as 3D printing (3DP) and selective laser sintering (SLS) use a polymer powder base, which is then joined together layer-by-layer using either an injected binder or fused by laser heating, respectively (Bogue 2013). In FDM the layers are created by the heating and deposition of a thermoplastic filament extruded through a motorized nozzle onto a platform.

It was the expiration of the original patents for FDM in 2009, which made rapid manufacturing technologies widely available to industry and individuals alike (Crump 1992). Rapid manufacturing is now accessible to broad audiences, but the processing-to-property relations are still not well known. Thermoplastic extrusion FDM systems are typically less expensive and safer to maintain as the powders and resins which other systems use can create hazardous environments for hardware and sometimes toxic air conditions for users, along with the high cost of maintaining expensive hardware such as lasers (McMains 2005). These systems were also found to be the most suitable choice for home and small business use, via evaluation of some popular models of each type of technology using performance and cost criteria such as build time, system cost, dimensional accuracy, and material usage and waste (Roberson et al. 2013; Stanek et al. 2012).

As such, with the growing popularity of FDM printer systems for consumer-level use, the mechanical evaluation of components produced in this fashion is of paramount interest. Fused deposition-modeled parts have previously been evaluated for several different parameters, including dimensional accuracy and smoothness, compressive, tensile, flexural, and impact strength (Lee et al. 2007; Panda et al. 2009; Zhang and Peng 2012). Lee et al. (2007) observed that the compressive strengths of FDM parts are 23.6% higher in axially loaded specimens than in transverse specimens, thereby showing the effects of print orientation. Panda and colleagues (2009) showed the effects of processing properties; while decreasing layer thickness raises both tensile and flexural strength, increasing it improves impact strength. Traditionally, however, FDM studies such as those previously cited have focused around acrylonitrile butadiene styrene (ABS) components, rather than the more environmentally friendly and popular PLA, which is used in many of the increasingly popular desktop printers. PLA is a biodegradable thermoplastic polymerized from natural lactic acid from natural sources such as corn (Ashby and Johnson 2013). Some of the characteristics of bulk PLA are given in Table 25.1. Although PLA has a larger strength and lower ductility than the traditional ABS, PLA is a sustainable thermoplastic alternative, which addresses the problem of added waste from end users manufacturing components at home and has similar characteristics as ABS. Parts produced via FDM from PLA have also been of high interest to the medical field, due to the biocompatibility of PLA for use in applications such as tissue engineering and custom-made implants per patient needs (Drummer et al. 2012; Too et al. 2002).

It is evident that there is a need to thoroughly evaluate the properties of PLA components produced via FDM, primarily in strength and fracture characteristics, so that they may continue to be successfully employed for both industrial and general uses. Complications arise; however in testing, these components as several different factors that affect component print quality and strength are affected by the multiple adjustable settings of the FDM machine. It has been shown that factors such as material extrusion temperature, T, component-manufacturing orientation, and layer thickness, δ, strongly affect the strength and durability of components produced via FDM using ABS (Sood et al. 2010). Combined with other commodity desktop printer settings, such as print speed and

Table 25.1 Material properties of bulk polylactic acid as given by manufacturers and literature

Property	Unit	Value
Density, ρ	g/cm^3	1.24
Melting temperature, T_m	°C	130–230
Elongation at break	%	7.0
Elastic modulus, E	MPa	3500
Shear modulus, G	MPa	1287
Poissons ratio, ν	–	0.360
Yield strength, σ_y	MPa	70
Flexural strength, σ_f	MPa	106
Unnotched izod impact	J/m	195
Rockwell hardness	HR	88
Ultimate tensile strength, σ_{uts}	MPa	73

Source: Jamshidian, M. et al., *Comp. Rev. Food Sci.,* 9–5, 552–571, 2010; Bijarimi, M. et al., *Mechanical, Thermal and Morphological Properties of PLA/PP Melt Blends,* International Conference on Agriculture, Chemical and Environmental Sciences, Dubai, pp. 115–117, 2012; Clarinval, A. and Halleux, J., Classification of Biodegradable Polymers, in: Smith, R. (ed.) *Biodegradable Polymers for Industrial Applications.* 1st ed. Taylor & Francis Group, Boca Raton, FL, 2005; Ashby, M. F. and Johnson, K., *Materials and Design: The Art and Science of Material Selection in Product Design,* Butterworth-Heinemann, Oxford, UK, 2013; Henton, D. E. et al., Polylactic Acid Technology, in Mohanty, A., Misra, M., and Drzal, L. (eds.) *Natural Fibers, Biopolymers, and Biocomposites,* Taylor & Francis Group, Boca Raton, FL, 2005; Subhani, A., *Influence of the processes parameters on the properties of the polylactides based bio and eco-biomaterials,* PhD, Institut National Polytechnique de Toulouse, 2011.

infill density, the number of experiments needed to evaluate the effects of each individual parameter on component strength and fracture behavior can quickly escalate. The Taguchi method of design of experiments (DOE) curtails extensive experimentation as encountered when using a full factorial experiment (FFE) (Roy 2001). Using the FFE method, the number of experiments necessary equals the number of levels tested for each factor raised to the number of factors; this means, for example, any arbitrary three-level four-factor experiment would require 3^4, or 81 experiments (Lee and Kuo 2013). The Taguchi method allows for the selection of a partial factorial test matrix to test multiple factors with several levels at once and accounts for interactions between these with a minimum amount of experiments, which can later be analyzed using ANOVA (Roy 2001; Wen et al. 2009). With the three-level four-factor experiment previously mentioned, the L9 matrix can be utilized to reduce the original 81 experiments to just nine experiments, an 89% reduction. The Taguchi method has been shown to be a valid approach for evaluating the effects of the different factors present in FDM and simplifying experimentation while evaluating multiple factor levels and their influence on component performance (Patel et al. 2012; Sood et al. 2011; Zhang and Peng 2012).

ANOVA is a statistical process which evaluates the variance between individuals or groups of individuals in order to assess the effects of treatments (Girden 1992; Montgomery 2012). The use of ANOVA gives a statistical measure, F, which is the ratio of the between-group variance, or variance due to treatments, divided by the error, which is a result of within-group variance (Roberts and Russo 1999). The variance, or mean square, is calculated as the sum of squares divided by the degrees of freedom. The mean square due to treatment can then be divided by the mean square of error to obtain an F value, which is useful in gauging the effect of each condition on the mean (Miller 1997); in this situation, the *F* values can be used to gage the effects of each processing parameter on the selected material properties of interest:

$$F = \frac{\sum_{i=1}^{N}(x_i - \bar{x})^2}{\sum_{i=1}^{n}(n-1)\sigma_x^2} \cdot \frac{N-p}{p-1} \qquad (25.1)$$

where:
  p is the total number of populations
  n is the total number of samples within a population
  N is the total number of observations
  σ_x is the standard deviation of the samples

These results establish a reference guide by which users can more intuitively determine which factors will affect their components and to what extent, thereby allowing them to decide how best to tune their printer for maximum component performance depending on their application. Moreover, the process established herein to determine these properties can be used to optimize collective or individual FDM properties for other platforms and materials, allowing users to optimize for whichever properties are desired, given their own specialized circumstances and equipment.

In this study, DoE is applied to determine an experimental array by which several process settings, or factors, of a commodity FDM device can be evaluated without the need for full factorial testing. Results are analyzed using ANOVA in order to determine factor impact on tensile and fracture characteristics, such as yield strength, σ_y, Young's modulus, E, and critical stress intensity factor, KQ.

25.2 Experimental approach

Drawing out the mechanical properties of the candidate solid, PLA, is accomplished via experimental mechanics of materials. Standard methods, such as fracture toughness testing, tensile testing, and so on, are ubiquitous since they can be applied to monolithics, composites, metals, polymers, and so on. Although the maturity of additive manufacturing materials has yet reached the point where ASTM/ISO protocols have been established for testing, existing methods provide guidelines to establish the fractural and tensile properties of orthotropic PLA FDM components. Standards do exist, however, which set guidelines for general terminology and reporting, such as ISO/ASTM 52921.

Samples were manufactured using a common FDM desktop printer (MakerBot Replicator2), as it represents a commodity device for rapid manufacturing machines. This is a single extruder rapid prototype machine, which uses 1.75 mm PLA filament to produce components via FDM that serves as an ideal representation of desktop FDM printers for small-scale production/prototyping. Temperature, T, print speed, s, infill direction, θ, relative density (or infill %), ρ, and layer thickness, δ, are the most common parameters which may be adjusted based on the object to be printed. The infill direction, which will be defined later, was adjusted using Skeinforge, a software addition which works in conjunction with the proprietary software of the printer and expands the available control level to a more advanced adjustability level.

The number of runs, or experiments, necessary was determined by the use of the Taguchi DoE. These experiments are defined by the unique combination of the settings

Table 25.2 Processing parameters used with range of values possible, minimum increments possible, and actual values used during testing given

Processing parameter	Range possible	Minimum increment	Settings used (Low/high)
Temperature (°C)	Tm-280	1	215/230
Speed (mm/s)	10–200	1	60/120
Infill direction (°)	0–180	N/A	0/90 and 45/135
Relative density/infill (%)	0–100	1	35/100
Layer thickness (mm)	0.10–0.40	0.05	0.1/0.3
Perimeter	Off/On	N/A	Off/On

denoted in Table 25.2. This table denotes the collection of parameters tested, their available ranges, and the values employed in this study. The limitations given for each parameter for the range possible are mostly determined by hardware limitations as defined within the software. The lower temperature boundary, however, is arbitrarily defined at a point below which it is believed that the PLA will not be heated enough to extrude properly. Each run has a high or low setting for each printer process parameter under question, with the different combinations giving a broad spectrum of testing conditions with which the effect of each setting on mechanical properties and loading response can be thoroughly assessed and ranked in importance. The high and low values were based on average slicer settings used for these types of printers, as defined by the manufacturer, with a deviation from the *normal settings* defining the high and low values. For example, the standard setting for layer thickness is designated as 0.2 mm, so a deviation of +/− 0.1 mm was made to determine the high and low values. The printer manufacturer recommends an extrusion temperature of 230°C for all prints, with temperatures higher than this being likely to cause warping of the component. As such, 230°C was chosen as the upper boundary for temperature, and 215°C was chosen as the lower, as this is approximately a 5% difference from the recommended setting which can accurately be achieved by the extruder. The difference between the infill directions of 90°/180° (aligned) and 45°/135° (biased) is shown in Figure 25.1. The 90°/180° orientation is the standard produced by the printer, where the extruded strands are aligned with the axes of print plate, whereas the biased 45°/135° infill direction is the same internal structure but rotated to produce a 45° diagonal version of the aligned structure. Using these two configurations gives two very distinct set of microstructures, which should lead to differences in failure modes between similar runs.

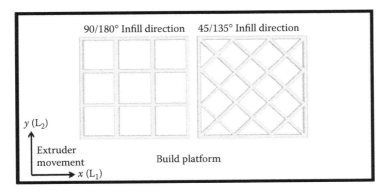

Figure 25.1 Visualization of low (90°/180°) and high (45°/135°) value settings for the infill direction process parameter.

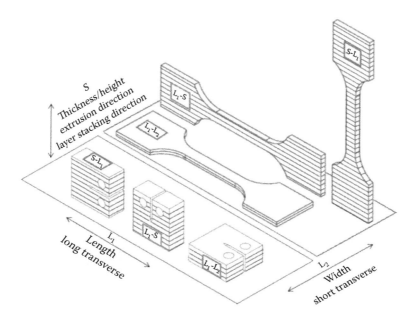

Figure 25.2 Naming convention for the FDM fracture and tensile specimens.

By typical FFE calculation, the experimental array necessary to thoroughly assess the effect of each setting on material properties leads to 2^6 unique experiments; however, while testing for both tensile and fracture properties, the number of experiments doubles. Due to the nature of FDM printing, there exist three orientations in which samples may be printed, as shown in Figure 25.2. This diagram shows both tensile and fracture coupons in the three types of manufacturing orientations possible with this FDM device. Additionally, in order to detect and mitigate the effects of outliers, three samples were tested for each experiment at each orientation. While summing up the total number of tests needed by the FFE format under the given conditions, this leads to a total of 1152 necessary test specimens.

Using the Taguchi method, however, the experimental array is calculated as an orthogonal L8 array leading to eight ensembles of experiments, or runs, necessary for analysis of the six variables chosen, as shown in Table 25.3. Each run is an experiment with a specific combination of high/low values for each setting, as denoted by the requirements

Table 25.3 Runs designed via use of Taguchi L8 test matrix. Each of these runs will be tested with three samples at each of the three orientations of denoted in the text: L_1–L_2, S–L_1, and L_1–S

Run	Temperature, T (°C)	Speed, s (mm/s)	Infill direction, θ (°)	Relative density, ρ (%)	Layer thickness, δ (mm)	Perimeter layer, P
1	215	60	90/180	35	0.1	Off
2	215	60	90/180	100	0.3	On
3	215	120	45/135	35	0.1	On
4	215	120	45/135	100	0.3	Off
5	230	60	45/135	35	0.3	Off
6	230	60	45/135	100	0.1	On
7	230	120	90/180	35	0.3	On
8	230	120	90/180	100	0.1	Off

of the L8 array. With three orientations being tested with three samples for each orientation for both fracture and tensile experiments, the total number of specimens required to be tested becomes 144, an eighth of the 1152 test specimens required by the FFE. Tensile and fracture testing specimens were prepared according to dimensions specified by ASTM standards D638 and D5045 for tensile and fracture testing of plastics, respectively (ASTM 2010; ASTM 2007). Specimen design and dimensions for both tensile and fracture tests are shown in Figure 25.3. Tensile testing specimens were manufactured in the dogbone shape, while fracture tests used the compact tension (CT) specimens. The naming convention was derived from ASTM E399 with modifications made according to the layering orientation (ASTM 2012). Specimens were named by the direction of loading followed by the direction of expected crack propagation or rupture, as shown in Figure 25.2. It should be noted, however, that other permutations exist for loading and cracking directions, which were not chosen for this study as preliminary testing revealed strong similarity to the three chosen due to the symmetrical nature of how the FDM machine prints samples. The material behavior can be classified as a special case of orthotropy with S, L_1, and L_2 being primary stress axes. Properties in L_1 and L_2 (for 90°/180°) are expected to be identical, yet the material is not isotropic in the L_1–L_2 plane. The orientations used (along with those which were not used due to equivalence) are: S–L_1 (S–L_2), L_1– S (L_2–S), and L_1–L_2 (L_2–L_1).

The importance of orientation has been thoroughly documented especially when considering the tensile response of FDM components, as printing the layers such that the direction of tensile loading is along the length of the layers, rather than perpendicular to them, leads to the greatest tensile strength (Ahn et al. 2002; Sood et al. 2010). This is due to the load being applied along the length of the stacked layers, providing the best distribution of loading. Though many of these studies typically concentrate on FDM samples manufactured from ABS thermoplastics, these identical principles apply for other FDM parts from different

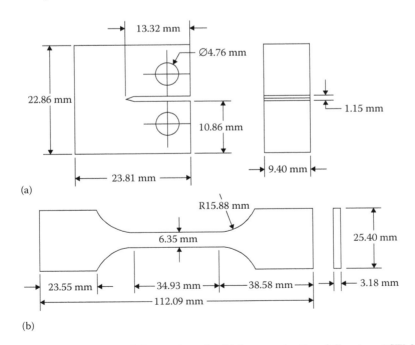

(a)

(b)

Figure 25.3 Specimen designs and dimensions for (a) fracture testing following ASTM D5405 and (b) tensile testing following ASTM D638 guidelines.

materials with similar layering. The fracture properties of FDM components have not been exhaustively investigated, especially as they relate to PLA, as the fracture properties of bulk PLA are not very thoroughly documented. It is important to characterize the influence of the effect of material orientation on both tensile and fracture properties in order to determine which orientation will provide the best results for each situation or when a combination or balance of strengths is desired and also to develop a streamlined optimization process.

During preliminary tests conducted prior to this study, the perimeter, or outer shell, of the object would often display unique behavior different from the infill of the object. Although an object cannot be printed without at least one perimeter layer, the influence of the perimeter layer on mechanical properties should also be explored. As such, for those samples which call for no perimeter layer, the perimeter had to be manually removed, taking care not to damage the internal structure of the components which could affect results. For those tested with the perimeter layer left attached, the printer was set to add two layers, as is the standard setting for this printer. The thickness of these layers was dictated by the thickness setting; thus, if the layer thickness was set to 0.1 mm, the perimeter layers also were printed at 0.1 mm.

Testing procedures followed those set by ASTM standards D638 and D5045 (ASTM 2010; ASTM 2007) in an ambient environment utilizing an electromechanical universal test machine (MTS 1 kN) and TestWorks software; the constant displacement rate was set to 1.524 mm/min (0.001 in/sec) with a data capture frequency of 5 Hz. A direct contact extensometer (MTS model 634.11E-25) was used for extension measurement during tensile testing, as shown in Figure 25.4, and a MTS (model no. 632.02E-20) clip gauge was used to measure crack tip-opening displacement for fracture tests. Due to the flexible nature of thermoplastics, the range of displacements encountered during tests was often beyond the capabilities of the measurement devices used. As such, a simple method was derived to account for the large deflections. For the fracture tests, the clip gage would be removed just before it would reach its limiter at 5 mm, which would have also ended the test prematurely. A correlation was made between the displacement recorded from the clip gage and that exported by the

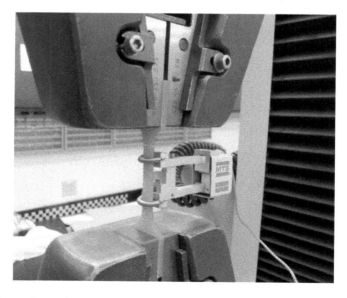

Figure 25.4 Tensile specimen during testing in MTS electromechanical universal test machine attached via mechanical wedge grips (model #M2 0–250 S25) with direct contact extensometer attached.

crosshead; the displacement could then be estimated beyond which the instrument could measure from this correlated equation using the load-line displacement. The same procedure was conducted for the tensile tests; however, the removal of the extensometer was not required midtest. Consequently, a linear calibration was established between extension and clip gage/extensometer displacement to calculate large sample deflections. The collected data were then analyzed to determine the mechanical properties and how the individual printing processes affect them. The test specimens were analyzed to study the rupture modes as they relate to printing orientations and the subsequent effects on material properties.

25.3   Experimental mechanical testing results

25.3.1   Tensile testing

The results acquired via tensile and fracture tests support the well-documented fact that orientation plays a primary role in the performance of FDM-manufactured parts (Ahn et al. 2002; Sood et al. 2010). This applies to all runs conducted, for both fracture and tensile specimens, though the best-performing orientation and run differ for the two specimen types. The extent of this dependence, however, varies from run to run, as shown in Figure 25.5, which displays the stress–strain response of selected tensile runs. Although a general trend

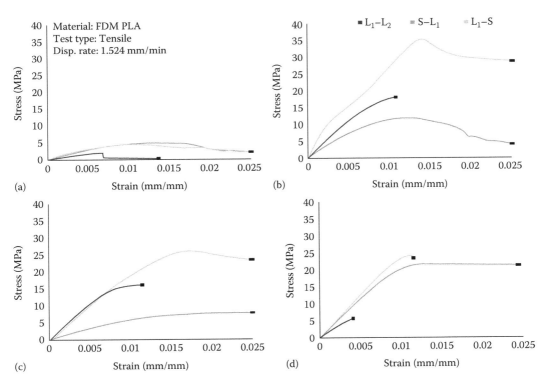

Figure 25.5 Effects of material orientation and process settings on tensile strength of FDM PLA; (a) Run 1: $T = 215°C$, $s = 60$ mm/s, $\theta = 90°/180°$, $\rho = 35\%$, $\delta = 0.1$ mm, $P = $ On, (b) Run 2: $T = 215°C$, $s = 60$ mm/s, $\theta = 90°/180°$, $\rho = 100\%$, $\delta = 0.3$ mm, $P = $ On, (c) Run 4: $T = 215°C$, $s = 120$ mm/s, $\theta = 45°/135°$, $\rho = 100\%$, $\delta = 0.3$ mm, $P = $ Off, and (d) Run 6: $T = 230°C$, $s = 60$ mm/s, $\theta = 45°/135°$, $\rho = 100\%$, $\delta = 0.1$ mm, $P = $ On.

can be observed for each specimen type on the reliance of performance on orientation, the difference in strength between different orientations varies between runs due to the variations in print parameter settings. This can be seen when comparing tensile Run 1 (Figure 25.5a) to Run 2 (Figure 25.5b), in which the main differences are relative density, layer thickness, and the presence of a perimeter layer. In Run 1, the three stress–strain curves are closely grouped together, but in Run 2 there is a clear difference between the three curves, so the effects of print orientation become evident. The clear trend here is that the L_1–S orientation is the least likely to fail for components which will experience a tensile load. Recalling Figure 25.2, it could be expected that L_1–S and L_1–L_2 would display similar response behaviors due to the layout of the layers being along the direction of loading, rather perpendicular to it such as in S–L_1. The stress–strain curves of the different runs in Figure 25.5 show that this is not the case. Due to the nature of the printing process, the L_1–S orientation has more load-bearing fibers, meaning that more fibers print parallel to the direction of the load with very shot perpendicular fibers. The L_1–L_2 samples, then, have numerous long fibers which run perpendicular to the load which will delaminate rather than deform, reducing the total number of load-bearing fibers in the structure and significantly reducing the strength, as shown in Figure 25.6. The difference in the load-bearing area is notable when comparing the lighter portions of each sample, as these are areas which are deformed before failure, rather than separating by delamination or sudden fracture, as is indicated by the darker, transverse strands of Figure 25.6a. Of note in these images is the tendency of the perimeter layers to separate from the bulk of the sample, particularly for L_1–L_2, which decreases load-bearing capabilities. This phenomenon is likely due to the use of the lower temperature setting. Additionally, in Figure 25.5b and c, a behavior is observed outside of the logically expected outcomes in which the samples built in the S–L_1 orientation, which are expected to display the worst tensile response, achieve a higher tensile strength than that achieved in the L_1–L_2 orientation. This is due to the nonuniformity of the rupture mechanism as shown in Figure 25.6a, where it can be seen that separate strands deform independently and to different extents leading to the resulting fracture varying in levels of deformity and strand length, whereas those of Figure 25.6b deformed uniformly and cohesively withstanding a larger amount of force before failure. As such, the interlayer bond strength of the S–L_1 components shown in Figure 25.5b and c exceeds the tensile strength of the corresponding

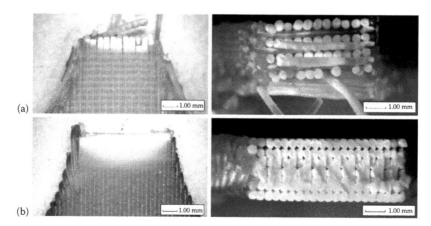

Figure 25.6 Tensile failure mechanisms for (a) L_1–L_2 orientation which displays nonuniform deformation at the rupture site as opposed to, and (b) L_1–S orientation which shows a nearly uniform fracture surface.

L_1–L_2 components due to the individual strands of the structure deforming nonuniformly from one another, rather than as a single cohesive unit.

Further inspection of Figure 25.5 shows that tensile Run 1 conferred the weakest mechanical properties of all runs with yield strength of 4.7 MPa, compared to that achieved by Run 2 at 32 MPa as the highest. Runs 4 (Figure 25.5c) and 6 (Figure 25.5d), on the other hand, performed relatively well in comparison to the remaining runs. Comparing the settings between these runs demonstrates that the low ρ and low δ facilitate weak mechanical properties. This implies that the highest values for tensile properties such as yield strength are closely connected to 100% relative density, ρ, and 0.3 mm layer thickness, δ, settings. Run 4 yielded slightly higher ut and σ_{uts} values than Run 6, and this is likely attributed to having both high density and layer thickness, such as in Run 2; these values being lower than that of Run 2 can be attributed to either utilizing the alternate infill direction of 45°/135° making the strands less resistant to the tensile load due to their offset orientation or the lack of a perimeter layer. The properties yielded by Run 6 are close to those of Runs 2 and 4 despite the lower 0.1 mm δ, but a higher T of 230°C as opposed to the lower setting of 215°C for Runs 1–4. From this, it is observed that for tensile samples, the most important settings are high density and high layer thickness. High temperature and the presence of a perimeter layer are also suspects of impacting tensile response from the aforementioned observations. For the L_1–S direction, the presence of a perimeter layer could simply be advantageous as it provides extra layers to reinforce the structure, and this could be part of the difference in the performances of Runs 2 and 4, such as the yield strengths being 32 and 23 MPa, respectively. Also important to note from Figure 25.5 is that for all three orientations of most runs, the peak stress seems to occur at around 0.01 to 0.015 mm/mm, though it is slightly lower for some of the runs in the S–L_1 orientation closer to 0.005 mm/mm. This is due to the delamination of layers as the bond between the layers is weaker than that of the layers themselves, so the sample delaminates rather than deforming.

Examining the individual mechanical properties yielded by each run and comparing them by orientation allows for more thorough analysis. Figure 25.7 displays the ultimate tensile strength of the samples, σ_{uts}; the yield strength, σ_y; the elastic modulus, E; and the modulus of toughness, u_t.

The modulus of toughness is calculated as

$$u_t = \int_0^{\sigma_{uts}} \varepsilon\,d\sigma \tag{25.2}$$

The toughness was calculated up to the point of σ_{uts} rather than to rupture as is traditionally done, due to the vast differences in material response between the different orientations and runs beyond this point. Since some samples within a run set ruptured at the point of σ_{uts} and some experienced large plastic deformations before rupture, this calculation helps to eliminate some of the behavior discrepancies between samples. This measure is not done to penalize or underestimate samples which may have performed well, rather to eliminate inconsistencies which arise in testing due to the unpredictable deformation behavior beyond the point of σ_{uts}, which is also often the furthest point of usability of a component, and thus the point of greatest interest to most users.

Runs 1 and 3 both have a combination of low T, δ, and ρ and display poor tensile properties; this is likely due to the combination of low density and thickness, though the low temperature likely reduces their performance further. Inspection of Run 2 in the L_1–S orientation, which clearly outperforms all other run/orientation combinations in terms of load support with yield strength of 32 MPa, combined with the aforementioned observation

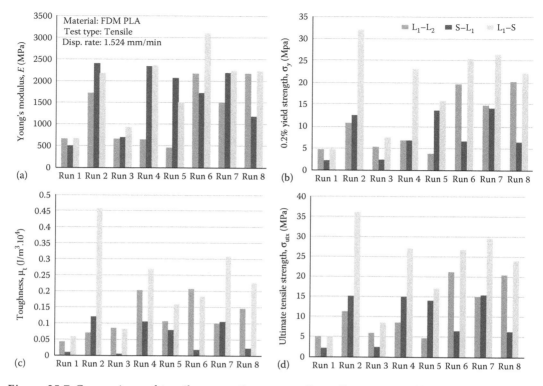

Figure 25.7 Comparison of tensile properties across all tensile test runs: (a) elastic modulus, *E*, (b) 0.2% yield strength, σ_y, (c) modulus of toughness, *ut*, and (d) ultimate tensile strength, σ_{uts}.

leads to the conclusion that both high density and layer thickness are required for good tensile response. Runs 6 and 7, with similar though slightly lower yield strengths of 25 and 26 MPa, respectively, and which do not have both high density and layer thickness, rather the combination of either 100% density and 0.1 mm layer thickness (Run 6) or 35% density and 0.3 mm layer thickness (Run 7) plus high temperature (230°C), show that it is important to have both of these settings on high to achieve maximum tensile strength. This is also an indication that high temperature provides a significant strength increase over samples produced at the lower temperature, as successive layers will develop more cohesive properties. Moreover, these results indicate that low performance actually comes from the combination of 35% relative density and the low layer thickness setting of 0.1 mm, as this is what yields poor, low-control print quality such as that seen in the microscopy of the inner structure of samples from Run 1 which bears this combination of attributes, as shown in Figure 25.8. This microstructure shows uneven thickness and spacing throughout the print, rather than consistent overlapping strands which add strength to the structure.

Although tensile properties favor the settings for Run 2 in the L_1–S orientation, which resulted in the highest values for yield strength and ultimate tensile strength, Runs 4 and 6 resulted in higher *E* values. This is attributed to one of the samples in each of these run/ orientation combinations having a very different response from the other two samples, as shown in Table 25.4. Run 2 has one sample with a significantly lower response than the other two samples, which are much closer together, reducing the average. Runs 4 and 6, on the other hand, both have one sample with a significantly higher response. Although all three samples from Run 6 have a highly varied response, sample 1 is much higher than the other

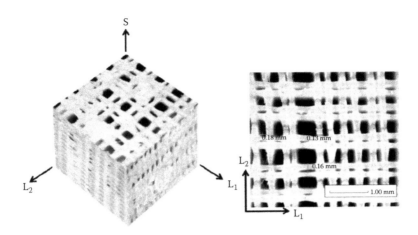

Figure 25.8 Microstructure of low density, low thickness sample from run 1 ($T = 215°C$, $s = 60$ mm/s, $\theta = 90/180$, $\rho = 35\%$, $\delta = 0.1$ mm, $P =$ off).

Table 25.4 Calculation of elastic modulus for all tensile samples printed in the L_1–S orientation for Runs 2, 4, and 6 highlighting outliers and noting their effects on the average values for each run

Run #	Sample 1 (GPa)	Sample 2 (GPa)	Sample 3 (GPa)	Average with outlier (GPa)	Average without outlier (GPa)
Run 2 ($T = 215°C$, $s = 60$ mm/s, $\theta = 90°/180°$, $\rho = 100\%$, $\delta = 0.3$ mm, $P =$ On)	2.42	2.63	1.51	2.19	2.53
Run 4 ($T = 215°C$, $s = 120$ mm/s, $\theta = 45°/135°$, $\rho = 100\%$, $\delta = 0.3$ mm, $P =$ Off)	2.82	2.10	2.15	2.36	2.12
Run 6 ($T = 230°C$, $s = 60$ mm/s, $\theta = 45°/135°$, $\rho = 100\%$, $\delta = 0.1$ mm, $P =$ On)	3.89	2.86	2.56	3.10	2.71

two, this being considered the outlier. The existence of these outliers is considered to be responsible for the variation in responses from the trend which exists for all other properties. Likely, having a larger number of samples would place the values for these much closer together, eliminating the large levels of statistical variance. Omitting these outlier results in average values which are much closer together than those calculated using all three samples, though Run 6 still has the highest value, likely due to the use of a higher temperature.

25.3.2 Fracture testing

The dependence of performance on orientation is as evident in the fracture testing results, as it was in the tensile-testing results. Figure 25.9 shows the load versus displacement behavior of the fracture samples for the same four runs as displayed for the tensile samples for a direct comparison. This time, however, the L_1–L_2 orientation is clearly the best performing orientation in terms of mechanical properties related to the fracture tests, as opposed to the L_1–S orientation which outperformed the others in the tensile tests. This is due to the nature of the fracture test and the sample orientation. Due to the material orientation, when the test is conducted the crack must propagate through multiple

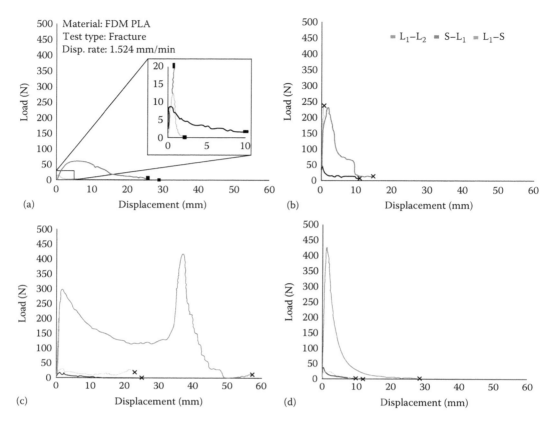

Figure 25.9 Effects of material orientation and process settings on fracture strength of FDM PLA; (a) Run 1: $T = 215°C$, $s = 60$ mm/s, $\theta = 90°/180°$, $\rho = 35\%$, $\delta = 0.1$ mm, $P =$ On, (b) Run 2: $T = 215°C$, $s = 60$ mm/s, $\theta = 90°/180°$, $\rho = 100\%$, $\delta = 0.3$ mm, $P =$ On, (c) Run 4: $T = 215°C$, $s = 120$ mm/s, $\theta = 45°/135°$, $\rho = 100\%$, $\delta = 0.3$ mm, $P =$ Off, and (d) Run 6: $T = 230°C$, $s = 60$ mm/s, $\theta = 45°/135°$, $\rho = 100\%$, $\delta = 0.1$ mm, $P =$ On.

layers at once, rather than one layer at a time as would be the case for the L_1–S samples. As the crack tip advances in the L_1–L_2 samples, it encounters multiple continuous layers throughout the entire sample, as opposed to having to tear through a single layer at a time in the L_1–S samples, which may lead to delamination of the printed layers once the crack completely severs through one of the layers. This can be seen in Figure 25.10c, where the sample undergoes out-of-plane cracking because the energy needed to transversely rupture fibers at the crack tip is too great. The L_1–S sample depicted delaminates at the mounting points of the sample after tearing through only a few layers, rather than tearing through the entire structure, as the L_1–L_2 sample does. Although failure occurs in the expected direction for the S–L_1 sample, Figure 25.10b, inspection of the fracture surface reveals that this sample also delaminates, this time along the thinnest part of the component at the tip of the crack.

Fracture behavior and the effects of production variables can be determined by examining both the load-displacement response of each of the runs as well as the resulting mechanical properties. Examining the curves in Figure 25.9d reveals that Run 6 yields the highest fracture response, and as was the case in the stress–strain curves from the tensile experiment, Run 1 yields the least load support, failing at significantly lower load levels

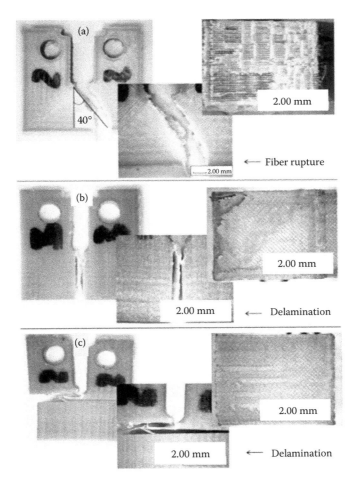

Figure 25.10 Fracture mechanisms of select samples in the (a) L_1–L_2 orientation, (b) S–L_1 orientation, and (c) L_1–S orientation.

than other runs. This can be further corroborated by inspection of the individual mechanical properties yielded by each run compared by orientation in Figure 25.11. This figure shows the critical stress intensity factor, K_Q; the ultimate load, P_{ult}; the fracture energy, U_{TF}; and the strength ratio, R_{sc}. The critical stress intensity factor was calculated according to (ASTM 2012):

$$K_Q = \frac{P_Q}{B\sqrt{W}} f\left(\frac{a}{W}\right)$$ (25.3)

where:
   B is the specimen thickness
   W is the specimen width
   a is the initial crack length

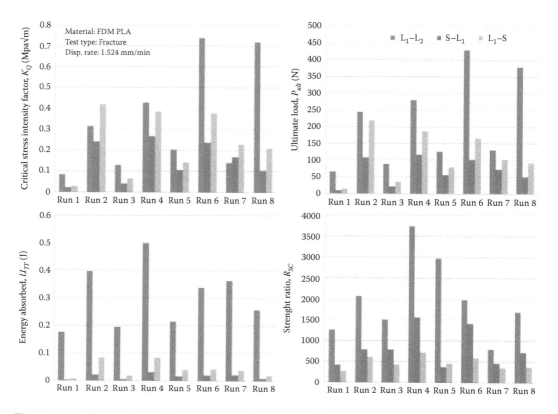

Figure 25.11 Comparison of fracture properties across all fracture test runs: (a) critical stress intensity factor, K_Q; (b) ultimate load, P_{ult}; (c) fracture energy absorbed, U_{TF}; and (d) strength ratio, R_{SC}.

$$f\left(\frac{a}{W}\right) = \frac{\left(2 + \dfrac{a}{W}\right)\left(4.65\dfrac{a}{W} - 13.32\dfrac{a^2}{W^2} + 14.72\dfrac{a^3}{W^3} - 5.6\dfrac{a^4}{W^4}\right)}{\left(1 - \dfrac{a}{W}\right)^{3/2}} \tag{25.4}$$

The requirement for the stress intensity factor, K_Q, to be used as the fracture toughness, KI, for the material is

$$B.a \geq 2.5\left(\frac{K_Q}{\sigma_y}\right)^2 \tag{25.5}$$

This requirement is met by all configurations except Run 4 of the L_1–L_2 orientation. An additional requirement is that the ratio of *Pult/PQ* does not exceed 1.10 in order for *KI* to be valid. All runs from the L_1–L_2 orientation do not meet this requirement, thus the strength ratio, R_{sc}, is calculated according to (ASTM 2012):

$$R_{sc} = \frac{2P_{max}(2W + a)}{B(W - a)^2\sigma_y} \tag{25.6}$$

Here, the strength ratio is a unitless description of material toughness. The combination of settings for Run 4 ($T = 215°C$, $s = 120$ mm/s, $\theta = 45°/135°$, $\rho = 100\%$, $\delta = 0.3$ mm, $P =$ Off) seems to yield the highest toughness when calculated in this manner; however, using samples of the same size, KI was calculated as if all of these validity factors were met. These KI values should not be cited as direct figures of correctly calculated fracture toughness, as only some of the runs met the aforementioned conditions. The values calculated serve as a way to assess the general behavior of each sample and the trends or effects which arise from the varying run settings. As such, in this document, the critical stress intensity factor will be primarily discussed, though the toughness may be referred to as well; both cases will refer to the same value of K_Q as calculated above.

Comparing the performance of the individual runs for specific fracture properties, namely maximum load and critical stress intensity factor, further supports the observation that Run 6 ($T = 230°C$, $s = 60$ mm/s, $\theta = 45°/135°$, $\rho = 100\%$, $\delta = 0.1$ mm, $P =$ On) had the highest performance. However, the graphs in Figure 25.10 also reveal that the Run 8 settings achieved comparable results in terms of these properties. This likeness is associated with the high density (100%) and low layer thickness (0.1 mm) prints, which develops numerous thin layers across the crack tip which must be simultaneously broken in order to facilitate brittle fracture. Any one of these thin layers could also contribute to the out-of-plane crack path deflection, further strengthening the structure. The difference between the two, although slight, is attributed to either the presence of a perimeter layer for Run 6, which adds additional bulk and crack resistance to the samples, or to a difference in infill direction, θ, as the $45°/135°$ orientation of layers provides more divisions or gaps throughout the layers which the growing crack will encounter. These very small gaps between filaments act as grain boundaries or lattice imperfections do within a crystalline material, deflecting crack growth and requiring more energy for crack propagation. Additionally, the strands printed in the $90°/180°$ configuration are either parallel or perpendicular to the crack tip, offering less resistance than samples with the $45°/135°$ orientation which are at an angle to the crack tip, causing it to deflect the growth of its direction each time it encounters a strand at a different angle, as displayed in Figure 25.12. The difference in crack growth shows that the direction of the strands works to divert the crack, not allowing it to grow directly in the expected direction as it does when using the $90°/180°$ setting but forcing it to grow diagonally.

25.3.3   Discussion

As the previous two sections suggest, there are various points which can be discerned from the individual tensile and fracture results to suggest ways to strengthen samples for either situation. Some key observations can also be made regarding what attributes will weaken PLA FDM structures in general; regardless of orientation or loading mode, the structure is expected to endure. Although within each run there is typically a clear difference between orientations, the combination of 0.1 mm layer thickness and 35% relative density, of Runs 1 and 3, is consistently a low performer regardless of which property is being examined or which orientation is being tested. Although there is still some deviation between orientations, it is not nearly as pronounced as in the other runs. This was attributed to the combination of low layer thickness and low infill density, as it is the major commonality between Runs 1 and 3, which differentiates them from the other runs. Logically, this would be due to the low level of infill, creating a largely porous structure with lots of divisions or faults due to the low layer thickness increasing the number of layers. While this still applies, inspecting the microstructure of the samples reveals that

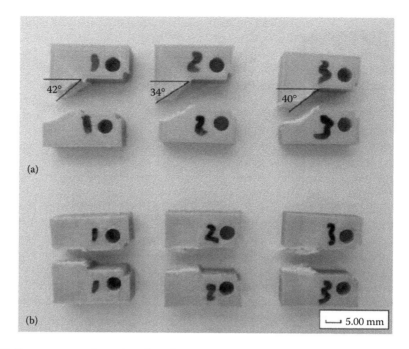

Figure 25.12 Fracture samples printed in the L_1–L_2 orientation showing crack propagation following infill direction of 45°/135° (a), with the angles of initial fracture denoted, and 90°/180° (b).

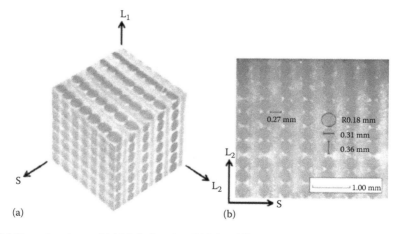

Figure 25.13 Microstructure of (a) high density (100% infill) and (b) high thickness (0.3 mm) sample from run 2.

the combination of low density and low layer thickness creates numerous flaws within the materials' microstructural fiber patterns. The very thin strands that the extruder is attempting to weave together are poorly controlled at such a small layer thickness and diameter with large gaps in between each strand. The result is illustrated in Figure 25.8, which shows the microstructure of samples with 35% relative density and 0.1 mm layer thickness (Run 1), as compared to those in Figure 25.13, which have 100% density and

0.3 mm layer thickness (Run 2). The Run 1 samples with low ρ and low δ, as has been previously noted, have very disorganized, unevenly spaced strands leading to numerous flaws within the structure. Comparatively, the samples which have 100% infill (from Run 2) have an extremely well-organized and evenly spaced pattern of fibers, with everything cohesively aligned and cosupporting. When the strands are unaligned (Run 1), sagging can occur as there is a minimal support for each strand, further causing distortions within the fiber lattice. All of these flaws will contribute to further weaken the samples, mostly negating the advantage that a specific orientation may have in each experiment.

25.4 ANOVA

Further analysis of findings and results based on ANOVA F scores calculated using algorithms embedded into workbooks which utilize Equation 25.1 facilitates ranking the influence level of each setting on each property over all runs. Tables 25.5 through 25.7 show the influence of each parameter on each mechanical property for L_1–S, L_1–L_2, and S–L_1 orientations, respectively. These are ranked in order of influence with first being most influential and sixth being the least. The symbols +, −, and 0 have been assigned to denote which value the setting should be set to for optimizing that property: + for high, − for low, or 0 for null meaning that for that particular mechanical property, the process and the resulting property value are insensitive to the setting used for the process variable.

These tables show that regardless of orientation, density is by far the most influential setting, with high density always being better than low density for both tensile and fracture samples. Intuitively, this is attributed simply to the fact that a component under loading which has more material over which to distribute that loading will be more resistant to failure, thus enduring higher loading levels and increasing the values of mechanical properties such as yield strength. Fracture strength will increase due to the same reason, as so much open space within the microstructure of the component means that a crack only needs to travel through a very finite amount of material before the component fails; in this case, only 35% of the space beyond the perimeter layers is actually filled with material and provides fracture resistance.

Layer thickness comes in at the second most influential setting, with the high value being favored for the L_1–S and S–L_1 orientations. However, inspection of the properties in Table 25.6 for L_1–L_2, which is the favored orientation for fracture samples, reveals that low layer thickness is preferential, corroborating the findings deciphered from Figures 25.9 and 25.10. Therefore, it can firmly be said that the lower layer thickness which produces a larger number of layers is favorable for crack resistance, raising fracture toughness or critical stress intensity factor, and the higher layer thickness which produces larger layers with less faults results in higher tensile strength. Another interesting fact to consider is that the lower layer thickness setting produces much smaller gaps between the strands of PLA than the higher layer thickness, as shown in Figure 25.14, which shows the structure of a Run 8 sample with 100% relative density and 0.1 mm layer thickness. This means that most of the area that the crack must travel is filled in but still has plenty of gaps to divert or stall crack growth, resulting in higher fracture toughness. Comparing the microstructures of fracture samples from Runs 2 and 8 in Figures 25.13 and 25.14, both of which have full density but with a 0.2 mm difference in layer thickness, shows that the structure within the component changes significantly as the layer thickness is varied. The sample with the higher layer thickness (Run 2) has symmetrical, even spacing on both sides of the individual strands, while the lower layer thickness sample (Run 8) only has gaps on the right side of the strands, so that it not only reduces the amounts of gaps but also fuses the two

Table 25.5 ANOVA ranking of process parameter influence on material properties for L1–S orientation

L$_1$–S

Material property	Ranking by influence on property					
	1st	2nd	3rd	4th	5th	6th
Young's modulus, E	Density+	Temperature+	Perimeter+	Thickness+	Infill direction+	Speed+
Ultimate tensile Strength, σ_{uts}	Density+	Thickness+	Perimeter+	Temperature+	Infill direction+	Speed0
0.2% Yield strength, σ_y	Density+	Thickness+	Perimeter+	Temperature+	Infill direction−	Speed0
Modulus of toughness, u_t	Thickness+	Density+	Infill Direction−	Perimeter+	Speed+	Temperature+
Ultimate load, P_{ult}	Density+	Thickness+	Perimeter+	Speed−	Infill direction0	Temperature0
Fracture energy, U_{TF}	Thickness+	Density+	Temperature−	Infill Direction+	Perimeter+	Speed−
Critical stress intensity factor, K_Q	Density+	Thickness+	Perimeter+	Speed−	Temperature0	Infill Direction0

Table 25.6 ANOVA ranking of process parameter influence on material properties for L$_1$–L$_2$ orientation

L$_1$–L$_2$

Material property	Ranking by influence on property					
	1st	2nd	3rd	4th	5th	6th
Young's modulus, E	Density+	Temperature+	Infill direction−	Perimeter+	Thickness−	Speed0
Ultimate tensile strength, σ_{uts}	Thickness−	Infill direction−	Density+	Temperature+	Speed−	Perimeter0
0.2% Yield strength, σ_y	Temperature+	Density+	Perimeter+	Infill direction−	Thickness−	Speed+
Modulus of toughness, u_t	Density+	Infill Direction+	Temperature+	Speed+	Perimeter+	Thickness0
Ultimate load, P_{ult}	Density+	Temperature+	Thickness−	Infill direction+	Perimeter+	Speed0
Fracture energy, U_{TF}	Density+	Thickness+	Speed+	Temperature0	Perimeter0	Infill direction0
Critical stress intensity factor, K_Q	Density+	Temperature+	Thickness−	Infill direction+	Perimeter−	Speed+

Table 25.7 ANOVA ranking of process parameter influence on material properties for S–L$_1$ orientation

S–L$_1$ Material Property	Ranking by influence on property					
	1st	2nd	3rd	4th	5th	6th
Young's modulus, E	Thickness+	Density+	Temperature+	Perimeter+	Infill Direction+	Speed+
Ultimate tensile strength, σ_{uts}	Thickness+	Density+	Temperature+	Perimeter+	Speed0	Infill Direction0
0.2% Yield strength, σ_y	Thickness+	Temperature+	Density+	Speed+	Infill Direction0	Perimeter0
Modulus of toughness u_t	Thickness+	Density+	Infill direction+	Perimeter+	Temperature+	Speed+
Ultimate load, P_{ult}	Density+	Thickness+	Perimeter+	Infill direction+	Temperature+	Speed0
Fracture energy, U_{TF}	Thickness+	Density+	Infill direction+	Perimeter0	Speed0	Temperature0
Critical stress intensity factor, K_Q	Density+	Thickness+	Perimeter+	Infill direction+	Temperature+	Speed+

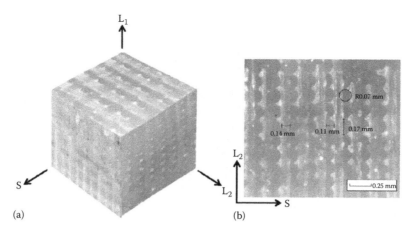

Figure 25.14 Microstructure of (a) high density (100% infill) and (b) low thickness (0.1 mm) sample from run 8.

directions of strands together more continuously. Thus, while the thicker layers of the Run 2 sample with a layer thickness of 0.3 mm produce stronger strands which support a higher tensile load, the thinner 0.1 mm layers seem to result in less overall porosity within the structure and more continuously bonded layers, as the heat retention in this structure should be higher than in the thicker 0.3 mm layer structure due to the reduced air flow within the structure due to decreased gap size. This, with the existence of regularly occurring small gaps to divert crack growth, in turn raises the critical stress intensity factor of the samples with 0.1 mm layer thickness as compared to the 0.3 mm thick layers.

Temperature seems to be the third most important factor; in cases when both infill density and layer thickness are not identified as the two most influential factors, it is typically because temperature is taking one of these two spots, with the high value nearly always favored. This is likely due to the fact that the 230°C setting of the extruder puts the PLA at a temperature close to but below its melting point, which has been reported to be between 130°C–230°C, depending on various structural properties of the material and its overall composition (Henton et al. 2005). This puts the extruded PLA in a semimolten state which improves malleability and adhesion to previous layers, as opposed to the lower 215°C setting. This setting, although still allowing printing, causes the extruded PLA to adhere less to previously printed layers, so that each subsequent layer is still stacked on top of the previous layer but with minimal bonding, thereby decreasing component strength. It is evident, however, that although the extruder head temperature is set to 230°C, the PLA itself is being extruded somewhere below that temperature, and the PLA's actual temperature will vary, given the differences in printer design and heating properties. It can be concluded then that the ideal temperature setting for printing should be at or around the melting point of the media being used as this should place it at a temporarily semimolten state, which will improve printability and component strength. Thus, this setting may have to be altered for each specific printer type depending on design and the heat properties of both the materials used in construction of the printer and the printing media. Although the 230°C setting could be said to be a good set point for most FDM PLA printers, the dependency of this setting on printer models may dictate some manipulation of the setting up or down. Though the trend for near-melting point operating temperature still stands, there will be instances where printers with high heat transfer capabilities will cause the PLA to be heated excessively past the semimolten state and enter a molten state, which will prevent it from forming correctly or clogging the extruder. The heat retention properties of PLA may also cause warping in the structure in this case, thus endangering both physical and aesthetic properties of the printed component.

Perimeter ranks at fourth most influential processing parameter, closely following temperature. As may be deduced intuitively, perimeter is better on than off, given that the extra, fully dense layers will add strength regardless of the other setting values. As mentioned in the experimental section, this is not actually a setting that can be turned off as of yet, but this setting was tested to gage the effects of the perimeter layer and how high of influence it is. It is especially relevant in the L_1–S orientation, ranking at third place for all but one of the material properties. It is of varying importance in the other two orientations, though overall it comes in at fourth and nearly always favors the *on* setting. The most important deduction that can be made here, however, is that if a user wants to use a very low density or infill setting so as to save materials, they could increase the number of perimeter layers in order to increase the overall strength of the component. The perimeter layer will also affect the way the component fractures, as the perimeter layers do not follow the direction of the infill, rather continuously wrap around the contour of the component, which would increase its strength and resistance to fracture regardless of the direction of loading. This will also vary depending on the number of perimeter layers, with more layers contributing a greater effect. It is important to note that comparison of the response curves of individual samples within a run set for those which underwent manual removal of the perimeter layers (Runs 1, 4, 5, and 8) do not show any abnormal behavior beyond that which is expected or shown by samples which were unaltered. This is due to the careful removal of the perimeter layers while preserving microstructural integrity during sample preparation in order to verify the effects of the presence of the perimeter layers.

According to the ANOVA tables, speed and infill direction do not seem to affect the mechanical properties much regardless of orientation. The distribution of which value is favored for speed varies almost evenly across the orientations, though it is mostly ranked rather low when compared to other printer parameters, which means the value that is ultimately chosen will have little or no effect on end results. Since the overall majority of the rankings indicates a null or high value, however, the recommendation can be made to use the high value for most situations in the interest of economizing time, especially when making full-density prints. This will alleviate some of the time added using the high density setting. However, speed may also be important for aesthetic factors, as a slower speed will boost surface quality as it is connected to a higher resolution, though this is at the sacrifice of time. The decision to raise or lower speed should be decided by the user based on their rating of importance for surface quality. Of note, the fact that the highest majority of properties in the ANOVA table favored a null value suggests that a value in between the two speeds tested is best for general-purpose use. Given the speeds tested of 60 mm/s and 120 mm/s, this suggests that 90 mm/s would yield favorable results across all situations.

As stated previously, it is intuitive that infill direction affects fracture behavior as longitudinal fibers are barriers to crack propagation and deflect crack growth, in much of the same manner as grain boundaries and lattice imperfections. It is due to this that the ANOVA tables (Tables 25.5 through 25.7) show that infill direction predominantly favors the high value (45°/135°) setting for all fracture properties across all orientations, with the addition of a few null values. For the tensile properties, on the other hand, the S–L_1 orientation shows a null/positive split, the L_1–L_2 orientation shows a slightly higher favor to the low value (90°/180°), while the L_1–S orientation shows a split between the low and high values. According to the ANOVA tables, the tensile properties tend to favor the 90°/180° configuration overall, indicating that minimally larger values for properties such as yield strength and Young's modulus would result from using this infill direction as opposed to the 45°/135° configuration. In a generalized view of this setting across both experiments, however, there is a greater favor toward the high value setting with more occurrences of the null value than the negative value. In application, this means that the 45°/135° setting is the more reliable one to use as a default, though tensile loading is of significantly greater concern, then the 90°/180° may be the better default. Alternately stated, this parameter setting is based on the situation, especially given the large number of null values. This is a low-ranking setting in order of importance, however, so choosing either setting as permanent should have little impact even if the situation could be said to call for the other setting.

Comparing the results of these experiments to the bulk material properties as given in literature shows favorable results. From Table 25.8, it can be seen that many of the largest values from the experimental results of the FDM PLA fall within the ranges given for bulk material from the various sources under which they are listed. The maximum values achieved for each mechanical property are recorded against bulk properties along with the run and orientation which yielded that result. The large variations within some of the values quoted in the literature stem from the different methods which exist to produce the polymer chains of PLA from lactic acid, two structures of which are noted below the column for Subhani (2011). The D and L subscripts denote which lactic acid isomer is used in synthesizing the PLA compound, the D and L subscripts being an indicator of the spatial configurations of the atoms in the lactide isomer (Meislich 2010). It has been shown that the combination of poly(L-latic acid) and poly(D-Lactic acid) enantiomers produces a stereocomplex with increased crystallization and varied material properties, such as a higher *Tm*, which will vary depending on the ratio of the mixture, thus giving the wide range of material properties of Table 25.8 (Yamane and Sasai 2003; Garlotta 2001). It is important to note that when considering fracture toughness or critical stress intensity factor, only the

Table 25.8 Comparison of select bulk properties of PLA as reported by literature versus the maximum values achieved experimentally with FDM-manufactured samples

Mechanical property	Mechanical properties of monolithic PLA						
	Ashby and Johnson (2013)	Henton et al. (2005)	Subhani (2011)		As tested		
			P(L) LA[a]	P(D,L) LA[b]	Maximum value	Run[c3]	Orientation
Young's modulus, E (GPa)	3.45–3.8	3.31–3.86	2.7–4.1	1–3.5	3.1	6	L–S_1
Yield strength, σ_y (MPa)	48–69	110.3–144.8	15.5–150	27.6–50	32	2	L–S_1
Toughness, K_l (MPa·√m)	0.7–1.1	–	–	–	0.738	6	L–$L_{1\,2}$

[a] $P_{(L)}LA$—PLA formed exclusively from L-lactides (Garlotta, D., *J. Poly. Environ.*, 9–2, 63–84, 2001).
[b] $P_{(D,L)}LA$—PLA formed from the combination of D- and L-lactides to form a stereocomplex with higher *Tm* and differing material properties (Garlotta, D., *J. Poly. Environ.*, 9–2, 63–84, 2001; Subhani, A., *Influence of the processes parameters on the properties of the polylactides based bio and eco-biomaterials*, PhD, Institut National Polytechnique de Toulouse, 2011).
[c3] Properties of Run 2: $T = 215°C$, $s = 60$ mm/s, $\theta = 90°/180°$, $\rho = 100\%$, $\delta = 0.3$ mm, $P =$ On and Run 6: $T = 230°C$, $s = 60$ mm/s, $\theta = 45°/135°$, $\rho = 100\%$, $\delta = 0.1$ mm, $P =$ On.

tests on the L_1–L_2 specimens closely followed linear elastic facture mechanics (LEFM) conditions as denoted in ASTM standard E399 where crack growth is nominally transverse to the applied load, tearing through the layers. This is why this orientation yields the greatest fracture results, as the other two will delaminate or shear before tearing.

25.5   Optimization

Some generalizations can be made about which process variables should be used depending on the application emphasis and for a general basis according to the results of the runs tested here. For situations where the component will experience tensile loading and tensile properties are of high importance, the settings of Run 2 ($T = 215°C$, $s = 60$ mm/s, $\theta = 90°/180°$, $\rho = 100\%$, $\delta = 0.3$ mm, $P = $ On) in the L_1–S orientation yield the most appropriate approximation of ideal settings. This combination yielded the highest yield stress of 32 MPa and ultimate tensile stress of 36 MPa while also giving a comparatively good critical stress intensity factor of 0.421 MPa$\sqrt{m}$, the third highest result overall and the highest value achieved outside of the L_1–L_2 orientation. This may also be applicable for situations where a generally high strength component is desired, and the possibility of sudden failure by delamination or shearing is not an issue or concern.

When an emphasis on fracture properties is desired with consideration being given to maintaining good tensile properties, Run 6 ($T = 230°C$, $s = 60$ mm/s, $\theta = 45°/135°$, $\rho = 100\%$, $\delta = 0.1$ mm, $P = $ On) in the L_1–L_2 direction will give the best results, yielding a critical stress intensity factor of 0.721 MPa$\sqrt{m}$. These settings are best for manufacturing a component which is slow to fracture, as warning of failure will be given by the evidence of crack propagation. This comes at some sacrifice to tensile properties, as the yield strength was lowered to 20 MPa as compared to the maximum of 32 MPa achieved with the Run 2 settings, and thus is more appropriate for components which will not undergo particularly high tensile loading, but may be best for those that will endure low-level cyclic loading.

Run 4 ($T = 215°C$, $s = 120$ mm/s, $\theta = 45/135°$, $\rho = 100\%$, $\delta = 0.3$ mm, $P = $ Off) consistently yields medium to high results for all properties, though this is dependent on the orientation which is best for each particular test. This presents a good combination of default settings for general purpose use as long as care is taken to choose the appropriate orientation for the component and its intended use. The main implication is that the optimal direction for each test is the orientation which yields the largest cross-sectional area per printed layer.

Given these generalized results based on the tested run combinations, compounded with the results from the ANOVA, an optimization of settings can be deduced which should yield high performance across all properties for use on a regular basis for manufacturing various components. These settings, as shown in Table 25.9, may be used for any type of generalized loading, where users are not concerned specifically with tensile or fracture properties, but have some concerns over the general material properties and strengths of the components being manufactured. Although the orientation is denoted as L_1–S, this will be at the discretion of the designer to define and decide upon, as it will be dependent on the expected loading situation and direction and the chosen priority of which type of failure to design against, tensile or fracture. In denoting this setting as L_1–S, the recommendation is being made to define and design for tensile loading. Also given in this table are a summary of the settings when an emphasis is made on tensile or fracture specific material properties. As noted previously, these will closely resemble the settings of Runs 2 and 6 for tensile and fracture, respectively, with slight modifications to boost material properties based on the ANOVA table results. These changes include increasing the temperature from 215°C to 230°C for the tensile settings and noting that speed is given in ranges for both

Table 25.9 Optimization of settings of processing parameters based on loading situation

Processing parameter	Loading situation		
	General	Tensile	Fracture
Relative density, ρ	100%	100%	100%
Thickness, δ	0.30 mm	0.30 mm	0.10 mm
Temperature, T	230°C	230°C	230°C
Perimeter, P	On	On	On
Speed, s	90 mm/s	60–90 mm/s	90–120 mm/s
Infill direction, θ	45°/135°	90°/180°	45°/135°
Orientation	L_1–S[a]	L_1–S	L_1–L_2

[a] User-defined

tensile and fracture due to the results of the ANOVA tables. These showed that although a median speed of 90 mm/s is generally best for all situations, the tensile properties showed a slight favor to the slower 60 mm/s speed, and the fracture properties showed that the higher 120 mm/s speed may be used without any negative effects on properties, giving slight favor to that setting, especially when time savings are considered.

25.6   Conclusions

On the broad scale, a joint characterization–optimization method was developed on the basis of standard methods. The adaptability of the approach to a wider range of materials is evident.

DoE was used to construct a set of experiments by which the effects of FDM printer settings on tensile and fracture properties of components produced via FDM using PLA could be explored. The settings of the printer adjusted were the layer thickness, density or infill percentage, extrusion temperature, speed, infill direction, and component orientation. Each of these settings was assigned a high and low level to be tested at to determine their effect and the best level for each setting. The orientation was tested by printing samples in the three orientations possible while printing via FDM. Tensile and fracture specimens were manufactured and tested according to the ASTM standards D638 and D5045. Test results were then analyzed using ANOVA to determine the influence of each setting.

Through tensile and fracture testing of FDM-printed samples, guidelines have been established for FDM printers employing PLA print media. The variable settings can, therefore, be prescribed based on application. Recommendations have been made for both tensile and fracture applications, as well as a generalized combination of parameters which can be chosen for generic applications which may not necessarily be constructed for a single loading situation. This combination yields consistent medium to high values for all properties tested in comparison to other situations which may yield higher values for either fracture or tensile properties at some sacrifice to the other.

Though the given settings represent the best overall combination as given by test results and ANOVA influence rankings, some of these could be changed due to user or situational preference. A lower layer thickness and slower speed will result in a higher resolution with an improved surface finish when aesthetics are important. Layer thickness could then be lowered with little concern to decreased strength, as has been previously shown. If there is a desire to reduce material consumption, relative density could be lowered, perhaps to a medium setting of ~70%. Though this would decrease strength, an increase in the number of perimeter layers could be used to reduce the negative effect, thereby decreasing the amount of hollow space. For components which will experience negligible mechanical loading, relative density may be sacrificed to the lower setting as strength will not be an issue, and an increase in perimeter layers could be used to prevent it from being too fragile, so that mishandling the component would not cause damage. Additionally, the temperatures studied here are specific to the model printer utilized, with the low setting pertaining to around 95% of the manufacturer-recommended temperature. Due to its impact on material properties, future studies should take care to analyze the optimal temperature setting of the printer in question in the manner shown here with a wider and more refined temperature range, so as to effectively gauge the effects of changing extrusion temperatures.

The suggested settings for tensile, fracture, or general use will allow users of desktop FDM printers to produce components in which they can be confident of their desired performance. A methodology has been suggested which can be used to find the most important processing parameters and their settings in order to identify which parameters will yield optimal material properties with a minimized mechanical test matrix.

References

Ahn, S.H., Montero, M., Odell, D., Roundy, S., and Wright, P. K. 2002. Anisotropic material properties of fused deposition modeling ABS. *Rapid Prototyping Journal,* 8–4, 248–257.

Ashby, M. F., and Johnson, K. 2013. *Materials and Design: The Art and Science of Material Selection in Product Design,* Oxford, UK, Butterworth-Heinemann.

Astm. 2007. D5045, 1999(2007), *Standard Test Methods for Plane-Strain Fracture Toughness and Strain Energy Release Rate of Plastic Materials.* West Conshohocken, PA: ASTM International.

Astm. 2010. D638 (2010) *Standard Test Method for Tensile Properties of Plastics.* West Conshohocken, PA: ASTM International.

Astm. 2012. E399-12e3 *Standard Test Method for Plane Strain Fracture Toughness of Metallic Materials.* West Conshohocken, PA: ASTM International.

Bártolo, P. J. 2011. *Stereolithography: Materials, Processes and Applications,* New York: Springer.

Bijarimi, M., Ahmad, S., and Rasid, R. 2012. *Mechanical, Thermal and Morphological Properties of PLA/PP Melt Blends.* Dubai: International Conference on Agriculture, Chemical and Environmental Sciences, pp. 115–117.

Bogue, R. 2013. 3D printing: The dawn of a new era in manufacturing? *Assembly Automation,* 33–4, 307–311.

Clarinval, A., and Halleux, J. 2005. Classification of biodegradable polymers. *In:* Smith, R. (ed.) *Biodegradable Polymers for Industrial Applications.* 1 ed. Boca Raton, FL: Taylor & Francis Group.

Crump, S. S. 1992. Apparatus and method for creating three-dimensional objects. U.S. Patent 5, 121, 329.

Drummer, D., Cifuentes-Cuéllar, S., and Rietzel, D. 2012. Suitability of PLA/TCP for fused deposition modeling. *Rapid Prototyping Journal,* 18–6, 500–507.

Garlotta, D. 2001. A Literature review of poly(Lactic Acid). *Journal of Polymers and the Environment,* 9–2, 63–84.

Girden, E. R. 1992. *ANOVA: Repeated Measures,* Newbury Park, CA: SAGE Publications.

Henton, D. E., Gruber, P., Lunt, J., and Randall, J. 2005. Polylactic Acid Technology. *In:* Mohanty, A., Misra, M. and Drzal, L. (eds.) *Natural Fibers, Biopolymers, and Biocomposites.* Boca Raton, FL: Taylor & Francis Group.

Hopkinson, N., and Dickens, P. 2001. Rapid prototyping for direct manufacture. *Rapid Prototyping Journal,* 7–4, 197–202.

Jacobs, P. F. 1992. *Rapid Prototyping & Manufacturing: Fundamentals of Stereolithography,* Dearborn, MI: Society of Manufacturing Engineers.

Jamshidian, M., Tehrany, E. A., Imran, M., Jacquot, M., and Desobry, S. 2010. Poly-Lactic acid: Production, applications, nanocomposites, and release studies. *Comprehensive Reviews in Food Science and Food Safety*, 9–5, 552–571.

Lee, C.-T., and Kuo, H.-C. 2013. A Novel algorithm with orthogonal arrays for the global optimization of design of experiments. *Applied Mathematics & Information Sciences*, 7–3, 1151–1156.

Lee, C., Kim, S., Kim, H., and Ahn, S. 2007. Measurement of anisotropic compressive strength of rapid prototyping parts. *Journal Materials Processing Technology*, 187–188, 627–630.

Mcmains, S. 2005. Layered manufacturing technologies. *Communications of the ACM*, 48–6, 50–56.

Meislich, H. 2010. *Schaum's Outline of Organic Chemistry*, New York: McGraw-Hill. Mellor, S., Hao, L., and Zhang, D. 2014. Additive manufacturing: A framework for implementation. *International. Journal of Production Economics*, 149–0, 194–201.

Miller, R. G. 1997. *Beyond ANOVA: Basics of Applied Statistics*, Boca Raton, Florida: Chapman & Hall/CRC.

Montgomery, D. C. 2012. *Design and Analysis of Experiments*, New York, NY: John Wiley & Sons.

Panda, S. K., Padhee, S., Sood, A. K., and Mahapatra, S. 2009. Optimization of fused deposition modelling (FDM) process parameters using bacterial foraging technique. *Intelligent Information Management*, 1–2, 89–97.

Patel, J. P., Patel, C. P., and Patel, U. J. 2012. A review on various approach for process parameter optimization of fused deposition modeling (FDM) process and taguchi approach for optimization. *International Journal of Engineering Research and Applications*, 2–2, 361–365.

Roberson, D. A., Espalin, D., and Wicker, R. B. 2013. 3D printer selection: A decision-making evaluation and ranking model. *Virtual and Physical Prototyping*, 8–3, 201–212.

Roberts, M. J., and Russo, R. 1999. *A Student's Guide to Analysis of Variance*, London, UK: Routledge.

Roy, R. K. 2001. *Design of experiments using the Taguchi approach: 16 steps to product and process improvement*, New York, NY: John Wiley & Sons.

Sood, A. K., Chaturvedi, V., Datta, S., and Mahapatra, S. S. 2011. Optimization of process parameters in fused deposition modeling using weighted principal component analysis. *Journal of Advanced Manufacturing Systems*, 10–02, 241–259.

Sood, A. K., Ohdar, R., and Mahapatra, S. 2010. Parametric appraisal of mechanical property of fused deposition modelling processed parts. *Materials and Design*, 31–1, 287–295.

Stanek, M., Manas, D., Manas, M., Navratil, J., Kyas, K., Senkerik, V., and Skrobak, A. 2012. Comparison of different rapid prototyping methods. *International Journal of Mathematics and Computers in Simulation*, 6–6, 550–557.

Subhani, A. 2011. *Influence of the Processes Parameters on the Properties of the Polylactides Based Bio and Eco-Biomaterials, Tolouse, France*. Ph. D., Institut National Polytechnique de Toulouse.

Too, M., Leong, K., Chua, C., Du, Z., Yang, S., Cheah, C., and Ho, S. 2002. Investigation of 3D Non-Random porous structures by fused deposition modelling. *The International Journal of Advanced Manufacturing Technology*, 19–3, 217–223.

Wen, J.L., Yang, Y.K., and Jeng, M.C. 2009. Optimization of die casting conditions for wear properties of alloy AZ91D components using the Taguchi method and design of experiments analysis. *The International Journal of Advanced Manufacturing Technology*, 41–5–6, 430–439.

Yamane, H., and Sasai, K. 2003. Effect of the addition of poly(d-lactic acid) on the thermal property of poly(l-lactic acid). *Polymer*, 44–8, 2569–2575.

Zhang, J. W., and Peng, A. H. 2012. Process-parameter optimization for fused deposition modeling based on taguchi method. *Advanced Materials Research*, 538–541–1, 444–447.

Figure 2.1 3D Systems' ProX 400 is capable of printing in more than a dozen alloys, including stainless steel, aluminum, cobalt chrome, titanium, and maraging steel. (Courtesy of 3D systems.)

Figure 2.2 Six materials for 3D printing. New material choices are contributing to the growth of additive manufacturing. (Courtesy of 3D Systems.)

Figure 2.5 With a 3D printer, you can prototype a design and hold the results in your hand in just days. (Courtesy of Stratasys.)

Figure 4.2 3D-printed titanium component (on right) with nine-piece welded steel version. (Courtesy of General Dynamics Land Systems.)

Figure 4.3 3D-printed titanium preform before final machining. (Courtesy of General Dynamics Land Systems.)

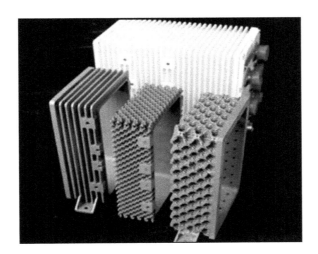

Figure 4.4 3D-printed concepts with *cooling fins* version. (Courtesy of General Dynamics Land Systems.)

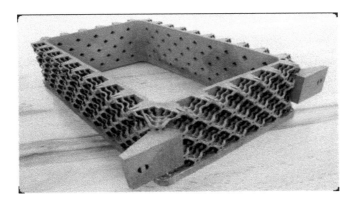

Figure 4.5 3D-printed titanium preform before final machining. (Courtesy of General Dynamics Land Systems.)

Figure 4.7 On-demand spare parts will dramatically enhance the ability to quickly service combat vehicles. (Courtesy of General Dynamics Land Systems.)

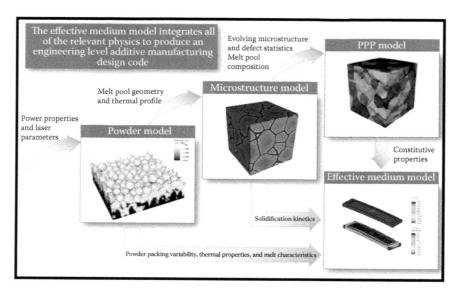

Figure 6.2 LLNL modeling strategy for additive manufacturing. (From https://acamm.llnl.gov/)

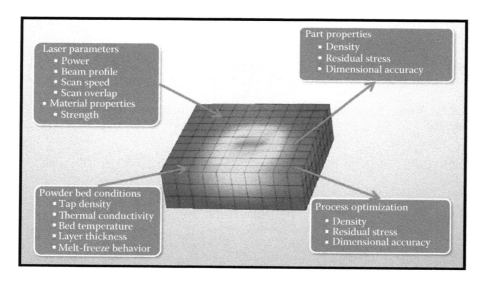

Figure 6.3 LLNL-effective medium model. (From https://acamm.llnl.gov/)

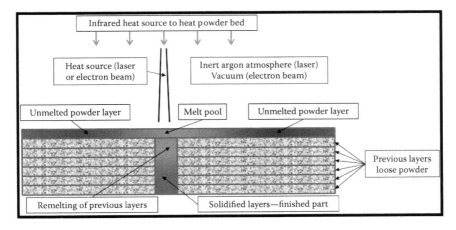

Figure 7.2 Model formulation.

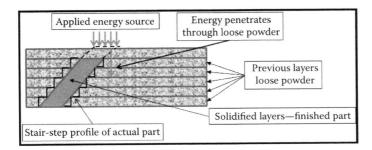

Figure 7.3 The energy has to be distributed through layers of powder beneath the area where it is applied.

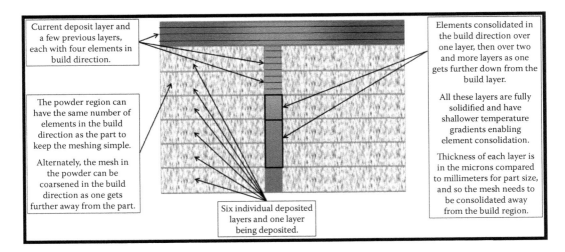

Figure 7.6 Meshing strategy in the build direction.

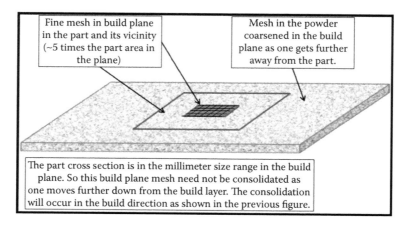

Fine mesh in build plane in the part and its vicinity (~5 times the part area in the plane)

Mesh in the powder coarsened in the build plane as one gets further away from the part.

The part cross section is in the millimeter size range in the build plane. So this build plane mesh need not be consolidated as one moves further down from the build layer. The consolidation will occur in the build direction as shown in the previous figure.

Figure 7.7 Meshing strategy in the build plane.

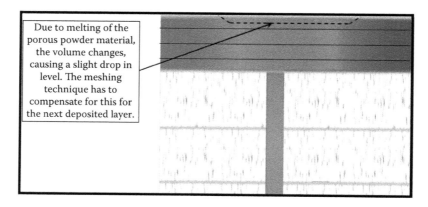

Due to melting of the porous powder material, the volume changes, causing a slight drop in level. The meshing technique has to compensate for this for the next deposited layer.

Figure 7.8 Mesh adjustment due to melting of powder layer.

Figure 15.8 Lattice structure for light weighting. (From Vaerenbergh, J.V. et al., *Application Specific AM Technology Development for High-End Mechatronic Systems*, LayerWise NVLeuven, Belgium, 2014.)

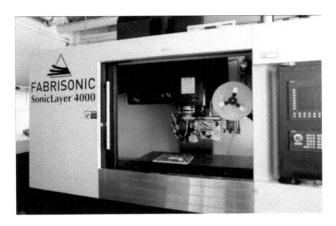

Figure 17.3 State of the art UAM system featuring a 9 kW welder and 20 kN of normal force. This amount of power makes it possible to make high quality, gapless metallic parts.

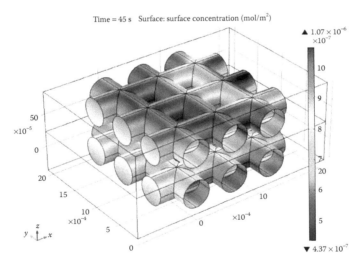

Figure 23.3 The concentration distribution of the chemically adsorbed precursor chemical on the surface of fibers at a selected time.

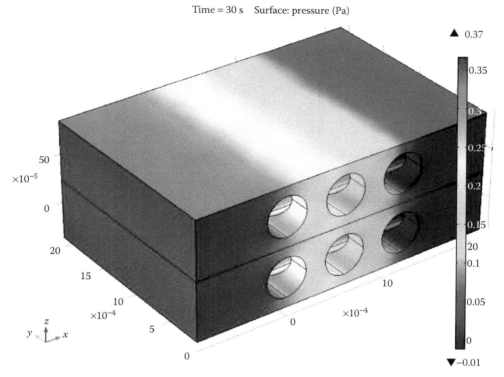

Time = 30 s   Surface: pressure (Pa)

▲ 0.37

0.35

0.3

0.25

0.2

0.15

0.1

0.05

0

▼ −0.01

50
×10⁻⁵

0

20

15

10
×10⁻⁴

5

0

20

10

0

×10⁻⁴

z

y   x

Figure 23.6 The pressure of the precursor solution in the channel at a selected time.

(a)

(b)

Figure 28.8 Surface measurement devices: (a) microscope and (b) surface profile gage.

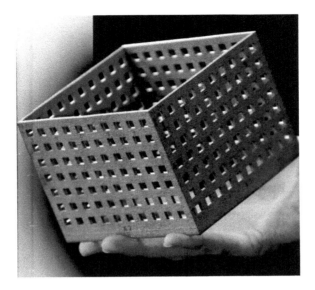

Figure 39.1 A Perforated metal box produced by an Arcam 3D printer. This detailed *calibration* part illustrates some of the versatility of 3D printing. (Courtesy of Jason Richards.)

Figure 39.2 3D printing can build products from a variety of materials for products ranging from heavy equipment to biomedical implants. (Courtesy of Jason Richards.)

Figure 39.3 ORNL scientist Chad Duty removes a finished part from a production-grade 3D printer. (Courtesy of Jason Richards.)

Figure 40.1 MOST Prusa Mendel RepRap.

Figure 40.2 Customized MOST Delta RepRap.

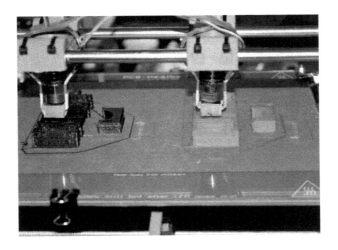

Figure 40.3 Double Whammy Prusa modification.

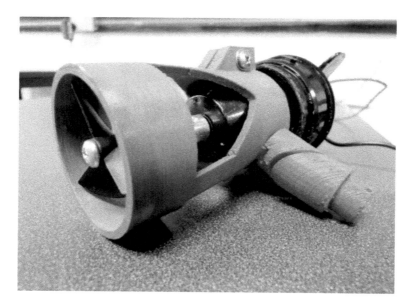

Figure 40.4 High school student ROV project design.

Figure 42.7 ProJet™ 1500 printer, opened to view printing bed (right).

chapter twenty six

Laser powder-bed fusion additive manufacturing of metals; physics, computational, and materials challenges

Wayne E. King, Andrew T. Anderson, R. M. Ferencz, N. E. Hodge,
C. Kamath, Saad A. Khairallah, and Alexander M. Rubenchik

Contents

The production of metal parts via laser powder-bed fusion additive manufacturing is growing exponentially. However, the transition of this technology from production of prototypes to production of critical parts is hindered by a lack of confidence in the quality of the part. Confidence can be established via a fundamental understanding of the physics of the process. It is generally accepted that this understanding will be increasingly achieved through modeling and simulation. However, there are significant physics, computational, and materials challenges stemming from the broad range of length and timescales and temperature ranges associated with the process. In this chapter, we review the current state of the art and describe the challenges that need to be met to achieve the desired fundamental understanding of the physics of the process.*

26.1   Introduction

26.1.1   Laser powder-bed fusion

Metal AM is "the process of joining materials to make objects from 3D computer-aided design (CAD) model data, usually layer upon layer, as opposed to subtractive manufacturing technologies."[1] Metal AM has a number of modalities, including material extrusion,

* King, W.E., Anderson, A.T., Ferencz, R.M., Hodge, N.E., Kamath, C., Khairallah, S.A., Rubenchilk, A.M., 2015. Laser powder bed fusion additive manufacturing of metals; physics, computational, and materials challanges. Applied Physics Reviews 2, 041304 DOI. doi:http://dx.doi.org/10.1063/1.4937809.

material jetting, material droplet printing, binder jetting, sheet lamination, powder bed fusion (PBF), and directed energy deposition.[2] Most current metal AM systems are of the PBF type.[2] In the powder-bed fusion process, thin layers of powder are applied to a build plate, and an energy source (a laser or electron beam) is used to fuse the powder at locations specified by the model of desired geometry. When one layer is completed, a new layer of powder is applied and the process is repeated until a three-dimensional (3D) part is produced. The PBF process is alternatively known as selective laser sintering (SLS), selective laser melting (SLM), direct metal laser sintering (DMLS), direct metal laser melting (DMLM), and electron beam melting (EBM).[3] Current metal PBF AM systems tend to use melting as opposed to sintering to build full-density parts.

Metal LPBF AM systems have designs similar to those illustrated in Figure 26.1.[4] They are composed of powder delivery and energy delivery systems. The powder delivery system comprises a piston to supply powder, a coater to create the powder layer, and a piston that holds the fabricated part. The energy delivery system is made up of a laser (usually a single-mode continuous wave Ytterbium fiber laser operating at 1075 nm wavelength) and a scanner system with optics that enable the delivery of a focused spot to all points of the build platform. A flow of gas (usually nitrogen or argon) passes over the powder bed with the intention to (a) protect the part from oxygen and (b) to clear any *spatter* and metal fumes that are created from the laser path. Some systems have an *in situ* process monitoring capability that can image the melt pool using a high-speed camera or a temperature sensor that is in line with the laser system.[5]

During production, the laser executes a scanning or exposure strategy. The strategies associated with the laser path are characterized by the length, direction, and separation (hatch spacing) of neighboring scan vectors. A detailed discussion of scanning strategies is beyond the scope of this chapter, but a list of scanning strategies has been compiled by Yasa.[6] Scanning strategies can affect the properties of the part

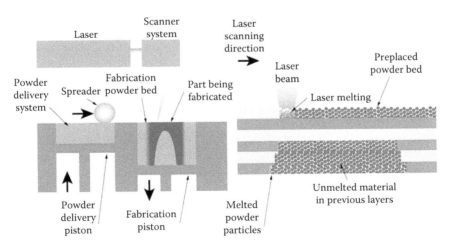

Figure 26.1 This figure provides a schematic overview of the select laser melt process both at the machine and powder scales. (Reprinted from Knowledge Based Process Planning and Design for Additive Layer Manufacturing [KARMA], funded by the European Commission, 7th Research Program, Detailed Report on Laser Cusing, SLA, SLS and Electron Beam Melting (Including Technical, Economical and Safety Features) (Valencia (ESPA~NA), Report No. DL 1.1, 2011.)

including density, mechanical properties, and residual stress. Residual stress is one of several important material responses that need to be optimized for laser-based additively manufactured parts. A part can be fabricated within tolerances only to have residual stress-induced distortions had put the part out of tolerance when removed from the build plate. Residual stresses can also cause a part's connection to support structures to fail or result in surface deformations that can damage the coater or inhibit the coater's motion.

26.1.2 The role of high performance computing for selective laser melting modeling and simulation

In recent years, the state of the art in metal PBF AM has improved to the point where it is transforming from a rapid-prototyping technology to a production technology. Parts can be fabricated at near full density (99.5+%) with mechanical properties that are similar to conventionally produced metals. Potential applications are broad, increasing, and particularly notable in the medical[2] and aerospace sectors.[7-9] Despite this progress, one of the most serious hurdles to the broad adoption of AM of metals is the qualification of additively manufactured parts. Some 47% of manufacturers surveyed indicated that uncertain quality of the final product was a barrier to adoption of AM.[8]

A physical understanding of the metal PBF process can provide insights into performance margins, uncertainties in those margins, and their sensitivities to process parameters. Thus, a physical understanding of the process is an essential element of part qualification. Such an understanding should also enable increased control of the process, which in turn improves the likelihood of producing qualified parts. Modeling and simulation of the AM process provide a mechanism to develop this understanding. Several roadmaps for AM have highlighted the needs for and benefits of a process modeling and simulation capability.[3,10-12] These will only be fully realized through leveraging the scale of modeling achievable through high-performance computing.

Although the PBF process is conceptually simple, the underlying physics is complex and covers a broad range of time and length scales. Laser beams and powder layer thicknesses are ~10 s of μm and laser speeds are ~1 m/s. On the other hand, parts are many cubic centimeters in dimension and build times can be hours, days, even weeks. Further, the process involves around 130 parameters that could affect the quality of the final part.[13] Parameters such as the laser power, speed, and beam size control the length, width, and depth of the melt pool. The geometry of the melt pool is important as its width and depth can affect part density, and length can affect the microstructure through the cooling rate. Generally speaking, it is desirable to maintain a constant or controlled melt-pool geometry during a build. However, because the thermal boundary conditions change as a function of the part geometry, the parameters required to achieve desired melt-pool characteristics will also be a function of geometry. In current PBF systems, geometry-specific parameters can be entered for geometries such as the core, skin, and downward facing surfaces. But, achieving controlled melt-pool characteristics throughout a part requires voxel-by-voxel control of the parameters. *In situ* sensors and feedback schemes aid such control.[14-16] Feedback works best when the parameters are close to the optimal for the given geometry. This is particularly the case for the high laser speeds involved in metal PBF where the time constant for the response of the melt pool to changes in power or speed can be relatively slow. Achieving optimized input parameters is referred to as a *priori*[17] or *intelligent feed forward*[10,18] control.

One system manufacturer is implementing a geometry-dependent scanning (or exposure) strategy.[19] Modeling and simulation combined with high-performance computing optimization (solving the inverse problem) have the potential to provide the next step in such voxel-by-voxel control of the process.

A number of papers have had significant impact (as measured by an average of ≥5 citations/year) in the field of modeling and simulation of the PBF AM process. Williams and Deckard recognized the need for process modeling and simulation in the early days of polymer PBF.[20] In the case of metals, contributions with significant impact include thermal models of the process,[21–24] thermomechanical models of the process,[25,26] residual stress modeling,[27] and laser–powder interactions.[28–30]

26.1.3   Outline

In this chapter, we give a brief review of recent progresses in developing physics-based models for the metal PBF. We first discuss the fundamental aspects of melting of the metal powder. We then discuss a model at the scale of the powder. This model is used to simulate the melting of powder and its resulting densification. It resolves individual powder particles in 3D. The laser–material interaction is treated via ray tracing and a physics-based absorption model. It models melting of the powder, flow and convection of the liquid, and behavior of trapped gases. It covers timescales of fractions of a second and length scales of fractions of a millimeter. We also discuss a model at the scale of the part that is used to computationally build a complete part, and predict properties such as residual stress in 3D. It treats the powder as a lower-density, low-strength solid. The laser–material interaction is treated using an energy source term. The part-scale model represents melting, solidification, and includes strength; it can be readily extended to include solid-state phase transformations for future material systems of interest. It covers timescales to hours and length scales to centimeters. We discuss the role of data mining and uncertainty quantification in the modeling and simulation process and describe its future applications.

26.2   Fundamental aspects of melting

The selective laser melting (SLM) process includes a variety of physical effects with huge disparities in temporal and spatial scales, making comprehensive first-principles modeling practically impossible. However, the disparity in scales enables the use of simplified models for aspects of the process. A simulation at the scale of the powder would consider the laser interaction with the powder, powder melting, and evolution of the melt (see Section 26.3). A simulation at the scale of the part would take into account laser heating and melting treated as a thermal source, part shape, and laser scan strategies, and would be able to calculate the residual stresses (see Section 26.4). The ranges of applicability of the simulations can overlap, opening the possibility for the mutual code validation.

Modeling of the SLM process has some similarities with modeling of welding, but with two significant differences. First, in SLM we must be able to model the new physics associated with the interaction of the laser with the metal powder, including radiation absorption and scattering, powder melting, and melt wetting. The second is a possible significant simplification of the description. It is clear that additively manufactured material quality degrades when the energy deposited exceeds the threshold for keyhole-mode melting.[31] This means that modeling of the SLM process does not need to include the plasma formation description, the radiation interaction with the vapor, and the variety of the interface instabilities that are observed in keyhole mode.

26.2.1   Numerical modeling of powder absorptivity

An important component of metal AM process modeling efforts is the description of the absorption of laser light by the metal powder and the spatial distribution of the absorbed energy. Direct measurements of the absorption are quite difficult.[32] Also, it is problematic to make use of measurements obtained without detailed specifications of the experiment, because the absorption depends on the powder material, the distribution of particle sizes, the spatial distribution of the particles, and the laser beam size and profile. Thus, it is not sufficient to know only the results for one particular powder of a given material and for a particular beam. Similarly, the spatial distribution of absorbed energy is difficult to obtain experimentally. These considerations reinforce the usefulness of absorption calculations. A commonly used laser absorption model proposed by Gusarov et al.[33] assumes diffusive radiation transport in the powder. The model can be applicable to a ceramic powder or to a thick, high-porosity metal powder. This assumption, however, is not applicable for the thin (a few powder particles thick), low-porosity metal powder layers used in the selective laser melting (SLM) process. As we shall see, in this case most of the energy is absorbed at the surface of the top layer, and the absorption is highly nonuniform even on the scale of individual powder particles. This situation is inconsistent with a diffusion model. Also, Gusarov et al.[33] assume volumetric deposition of the energy instead of surface deposition. In a typical experimental situation, the diffusion time a^2/D is longer or comparable with dwell time a/u. Here, a is a powder particle size, D is material thermal diffusivity, and u is scan speed. In reality, the laser deposits the energy on the surface of the particle changing the melt dynamics in comparison with the volume deposition. Physically, the powder is an assembly of metal particles, taken here to be spheres, with sizes appreciably larger than the laser wavelength (taken as about 1 µm) and with a complex refractive index appropriate to the material and the wavelength. It is natural to use ray tracing to calculate the powder absorption. This has previously been considered, for example, in Wang et al.[34] but the angular and polarization dependence of the absorption of incident rays was neglected. Boley et al.[35] reported the results of comprehensive absorption modeling, including all the effects mentioned above. A challenge was the problem of tracing rays within an assembly of thousands of objects, while keeping track of the angle, polarization, power, and reflection/refraction of individual rays. However, this issue has long been considered, and commercial software is available for handling it. Boley et al.[35] used the FRED[36] code, a multipurpose optics code widely used in optical design and analysis.

To begin the calculations, we consider a powder consisting of spheres of a single size that are densely packed in a hexagonal structure. Six materials (Ag, Al, Au, Cu, stainless steel [SS], and Ti) are considered. We first study the overall absorptivity of such a powder, by assuming a uniform beam of large width compared with the particle size, so that the absorption is nearly independent of the beam position. The refractive indices near 1 µm were taken from a data compilation.[37] The results are summarized in Table 26.1. Most important for each metal is the total absorptivity by the spheres and the substrate (column 9). This is to be compared with the absorptivity of the metal at normal incidence on a flat surface (column 4), and the average absorptivity of an isolated sphere illuminated by a uniform beam (column 5). The calculations show that the resulting powder absorptivity is significantly higher than the absorptivity of a flat surface or of a single, isolated sphere, thus confirming the important role of multiple scattering, as illustrated in Figure 26.2.

Table 26.1 Absorptivity calculated for a number of materials and material configuration

(1)	(2)	(3)	(4)	(5)	(6)	(7)	(8)	(9)	(10)
Material	Re(n)	Im(n)	α(flat surface)	α(isolated sphere)	α(top layer)	α(bottom layer)	α(substrate)	α(spheres + substrate)	α(spheres + substrate)/α(flat surface)
Ag	0.23	7.09	0.018	0.020	0.072	0.047	0.010	0.13	7.2
Al	1.244	10	0.047	0.056	0.15	0.063	0.011	0.22	4.7
Au	0.278	7.20	0.021	0.024	0.081	0.050	0.011	0.14	6.7
Cu	0.35	6.97	0.028	0.032	0.101	0.055	0.011	0.17	6.1
SS	3.27	4.48	0.34	0.36	0.53	0.062	0.013	0.60	1.7
Ti	3.45	4	0.38	0.40	0.56	0.062	0.014	0.64	1.7

Note: α denotes the absorptivity.

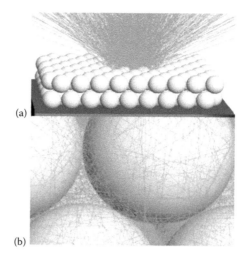

(a)

(b)

Figure 26.2 (a) Typical rays during illumination of the ideal array and (b) details of ray trajectories in (a), showing multiple scattering from spheres.

A ray can scatter repeatedly, leading to additional absorption relative to the case of a flat surface. Thus the relative increase in absorptivity is higher for highly reflective metals (Ag, Al, Au, and Cu) than for moderately absorbing metals (SS and Ti). In the former case, this ratio (column 10) varies from 4.7 to 7.2, whereas in the latter case the ratio is 1.7. Note that most of the power is absorbed in the top layer of the spheres (column 6). Little more than 1% of the power penetrates beneath the two layers to the substrate (column 8).

More generally, one is interested in not only the total absorbed power but also the spatial distribution of the absorbed power. In some AM machines, the laser beam size is roughly comparable to the powder particle size. Here we consider a powder with spheres of radius 10 μm and a beam having a $1/e^2$ radius of 24 μm.

Figure 26.3 shows the distribution of absorbed irradiance along the top layer of an array of SS spheres as the beam is rapidly scanned across the array. This distribution was

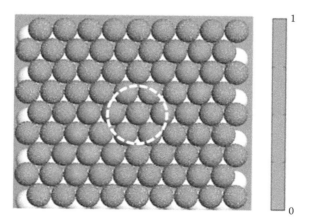

Figure 26.3 Irradiances (arbitrary scale) for 61 successive beam positions, from lower left to upper right, in steps of 2 μm. The irradiances pertain to the spherical surfaces. A sample beam spot ($1/e^2$ radius) is shown.

obtained by calculating the absorbed irradiance pattern at a number of points along the path and plotting the sum as a function of position. It gives a qualitative picture of the absorbed irradiance on a timescale short compared to thermal times, that is, for a sufficiently fast scanning speed. We see that the scattered light is well confined and that the typical absorption area is comparable to the beam area.

The absorptivity can be sensitive to the beam size, and fluctuations of the absorptivity are smoothed with increasing beam size. For the example presented in Figure 26.3, the absorptivity fluctuates along the scan by about 20% and the distribution of absorbed power in a single sphere is very nonuniform.

Real powder is different from the monosized powder considered above. A realistic powder has a distribution of sizes and a nonuniform geometrical arrangement, generally with porosity greater than that of an ideal array. To generate the powder geometry, Boley et al.[35] used a particle-packing algorithm[38] similar to that of the rain model for random deposition.[39] The algorithm randomly places powder particles with a specified distribution of sizes on a powder bed until the first contact with other particles or with the substrate. If the contact is with a particle, the particle is randomly perturbed, in an effort to minimize the potential energy. To simulate the removal of extra powder by a coater blade, the algorithm inserts a plane at a specified distance from the substrate and removes all particles intersected by the plane or situated above it. It should be noted that discrete element method (DEM) modeling is also being used by some investigators to understand the packing of the powder layers.[40–42]

Boley et al.[35] discussed two different types of powders. The first, shown in Figure 26.4, mimics the powder used in the Concept Laser metal additive manufacturing machine (www.concept-laser.de/en/home.html). The powder has a Gaussian distribution of radii, with an average radius of 13.5 μm, a full width at half-maximum equal to 2.3 μm, radial cutoffs at 8.5 μm and 21.5 μm, and a powder layer thickness of 43 μm.[31] In the absorption calculations, the path of the beam extends along the length of the powder bed, as shown in the figure. The calculated absorption for SS for a 1mm laser beam path is shown in Figure 26.5. Local variations in the powder structure give rise to sizeable fluctuations in the absorption. The fluctuations occur on a scale of about 100 μm, which is much larger than the typical sphere size. The mechanism for the fluctuations can be seen in the two insets in Figure 26.5a and b. In Figure 26.5a, the incident beam has mainly struck small spheres, with larger spheres on the periphery. This results in multiple reflections and an increased absorption. In Figure 26.5b, on the other hand, much of the incident power has reached the substrate, producing fewer reflections and a decreased absorption. For the second example of a powder, shown in Figure 26.6[35] consider a bimodal distribution characterized by a 7:1 ratio of radii and a volume fraction of small spheres equal to 20%, as discussed in Kelkar et al.[43] This powder was chosen because of its high density, or low porosity. Following Kelkar et al.[43] we consider a large-sphere radius of 42 μm and a powder thickness of 50 μm.

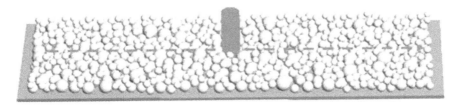

Figure 26.4 Powder with a Gaussian distribution of sizes. The length of the bed is about 1100 μm, and the beam path is indicated.

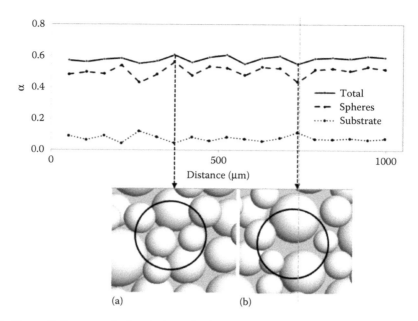

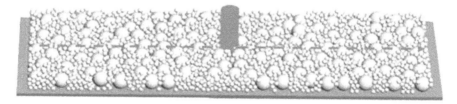

(a)   (b)

Figure 26.5 Absorptivity *a* as calculated along the beam path for the Gaussian powder of Figure 26.7. The material is stainless steel. The insets show the powder and incident beam size ($1/e^2$) at locations with high absorption (a) and low absorption (b)

Figure 26.6 Powder with a bimodal distribution of sizes. The powder bed and the beam are as in Figure 26.4.

Figure 26.7 shows the calculated absorption for SS along a 1 mm laser beam path. In this configuration, holes in the powder layer are practically absent. The absorption minima correspond to situations when the beam mainly strikes a large sphere, with much of the light directly reflected (Figure 26.7a). The largest absorption occurs when the beam strikes a local assembly of small spheres, as seen in Figure 26.7b. The difference between these two cases lies in the ratio of the beam size to the size of the irradiated spheres, with a larger ratio offering more opportunity for multiple reflections. As in the previous case, the absorption fluctuates on a distance scale larger than a particle size, or about 100 μm.

Parenthetically, it should be noted that the problem of a powder structure producing a maximum density has been investigated in a number of studies, for example, see Hopkins et al.[44]

Powder packing with density more than 80% of the bulk material was demonstrated computationally for complex powder size distributions but it may not be practical.

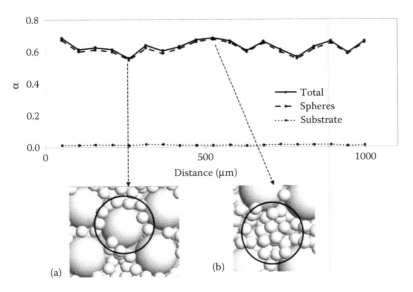

Figure 26.7 Absorptivity a as calculated along the beam path for the bimodal powder of Figure 26.5 (stainless steel). The insets show the powder and incident beam size at locations with low absorption (a) and high absorption (b).

Table 26.2 Total absorptivity for selected materials

Material	Ideal array (Table 26.1)	Gaussian array	Bimodal array
Ag	0.13	0.081	0.14
Au	0.14	0.093	0.16
SS	0.60	0.58	0.63

Returning to the Gaussian and bimodal powders, let us compare the overall results with those for the hexagonal powder array of Section 26.2.1. The results are summarized in Table 26.2, which demonstrates that a change in the powder structure can noticeably affect the absorptivity. For a moderately absorbing metal such as SS, the difference is not large, about a few percent. As a consequence, the absorptivities of the SS and titanium are not very sensitive to powder structure and powder feed system. On the other hand, for highly reflective metals such as Ag and Au, the variation can be nearly a factor of two. In these cases, multiple scattering is very important, and the powder configuration and size distribution affect the total absorptivity.

26.2.2   Direct absorptivity measurements

There are many reasons to do direct absorptivity measurements, even in the presence of detailed absorptivity simulations: the powder particle shape can differ from ideal spheres, (see Figure 26.8), the real powder structure in an experiment can differ from that produced by the numerical model, surface oxides can affect the absorptivity, and the refractive index of the alloy materials can be very different from the pure metal measurements.[45]

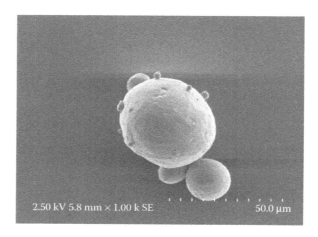

2.50 kV 5.8 mm × 1.00 k SE   50.0 μm

Figure 26.8 Scanning electron microscope image of the real, stainless steel powder, with rough surface and agglomeration with small particles.

As an example, the Al absorptivity for 1 μm light according to Palik[37] is about 5%. The real measurements of the bulk Al give absorptivity of about 20%. This is partially explained by the effect of the oxide layer and partially by the surface roughness.[45,46]

As a result, there is increasing demand for a simple compact system for fast measurements of the temperature dependence of the laser absorptivity up to and including the molten state. Existing systems, for example, see Tolchko et al.[32] and McVey et al.[47] measure the reflected light from the powder with the help of an integrating sphere and are typically complex and expensive. The distribution of the scattered light is broad and even the small absorption in the integrating sphere coating can affect the result. Calorimetry has also been used to measure the absorbed energy for a moving beam melting the powder layer. But in this case, most of the energy was absorbed by the melt not the powder, and the losses due to the radiative and convective transport were unaccounted, effectively increasing the absorptivity.[48] Recently,[49] a simple calorimetric scheme for direct absorptivity measurements had been proposed. The scheme of the measurements is presented in Figure 26.9. A thin layer of powder is placed on a thin disk made from refractory metal. A laser or diode array beam uniformly irradiates the thermally isolated disk. The temperature increase is measured by thermocouples underneath the disk. The disk holder is designed such that it does not significantly absorb radiation nor affect the temperature distribution in the target. The input heating is selected to be slow as compared to the rate of thermal diffusion, resulting in a uniform temperature through the powder and substrate. The temperature across the face of the disk will be uniform due to the uniform nature of the laser irradiation.

Consider a thin layer of powder with thickness d_1 on a flat disk substrate of refractory metal with thickness d_2 and radius R uniformly illuminated by light with intensity I. For absorptivity of powder (or melt), assuming uniform temperature throughout the disk, the temperature evolution is

$$\rho_1 c_1 d_1 + \rho_2 c_2 d_2 \frac{dT}{dt} = A(T) - Q(T) \qquad (26.1)$$

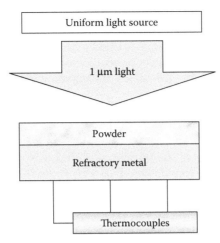

Figure 26.9 Diagram of the measurement scheme. A thin layer of powder is placed on a thin disk made from refractory metal and is uniformly irradiated by 1 μm laser light sources. Temperature is measured by thermocouples attached to the bottom of the disk.

where:
   $A(T)$ is the absorptivity
   $Q(T)$ is the thermal losses including convective and radiative losses
   ρ is the density
   c is the specific heat
   d is the thickness

Subscript 1 denotes powder and 2 denotes the substrate. Consider a flat top of a finite duration heating pulse. A typical temperature history is presented in Figure 26.10 and comprises two phases: heating and cooling. First, we consider the temperature evolution during the cooling phase, when $I = 0$ in order to determine the convective and radiative losses $Q(T)$ for known heat capacities and material densities. Next, we will find the temperature dependent absorptivity $A(T)$ considering the temperature evolution during the heating phase. The missing piece in

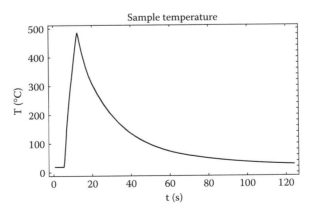

Figure 26.10 Sample data from thermocouples attached to the refractory disk showing the temperature variation during the heating and cooling periods.

this scheme is the measurement of the powder density (porosity). This problem can be solved through a special target design. The target disk with diameter d has a rim with height h to determine the powder thickness. The disk is filled with powder and a blade or roller removes the extra material, mimicking the powder deposition of commercially available AM systems. If we multiply Equation 16.1 by the disk area S, the equation can be rewritten as

$$\left(m_1 c_1 + m_2 c_2\right) \frac{dT}{dt} = A(T)P - Q(T)S \qquad (26.2)$$

Here, m_1 and m_2 are the masses of the powder and disk, respectively. $P = IS$ represents the total power incident on disk. Weighing the disk with and without powder gives the powder weight needed to calculate the absorptivity from Equation 16.2. A similar setup was used in a previous study to measure the absorptivity of solid metal, see Rubenchik et al.[46] where more details of physical effects related to the experiment can be found. Measurements[49] were done for 316L SS, Ti-6Al-4V, and 99.9% purity Al powder. The density and heat capacity of Ta as function of temperature were taken from Bodryokov.[50] For SS, Ti alloy, and Al, we used the density and heat capacities from Mills.[51] The results of the measurements are presented in Figure 26.11a–c. The SS powder (Figure 26.5a) was the same as used in

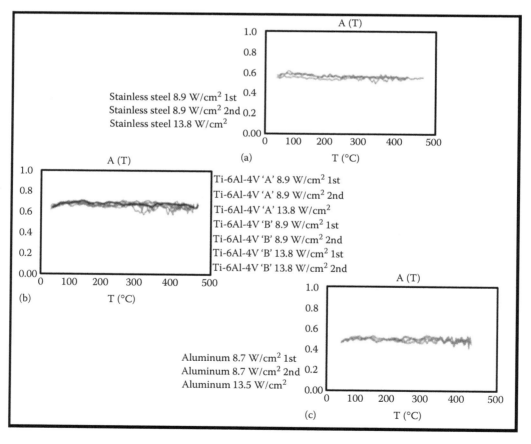

Figure 26.11 Measured absorptivity data for (a) stainless steel 316L, (b) Ti-6Al-4V, and (c) 99.9% purity Al (Goodfellow Al006031).

experiments carried out in a Concept Laser AM machine. The powder has a Gaussian size distribution with average radius 13.5 μm, a full width at half-maximum of 2.3 μm, and radial cutoffs at 8.5 μm and 21.5 μm.[31] After the first measurement, the sample was allowed to cool and the measurement was repeated (blue lines). Some small difference in results had been observed, probably due to powder reconfiguration driven by thermal expansion. Performing the measurement at two laser intensities gave consistent results, suggesting that the absorptivity is independent of the heating rate.

For Ti-6Al-4V (Figure 26.11b), measurements are from powders from two different suppliers. The powders have different particle size distributions (with the same average diameter ~27 μm, same as for SS) and they behaved differently when spread across the target disk. One powder was more cohesive than the other, tending to stick to the coater blade and roller to form clusters. Al powder of 99.9% purity was supplied by Goodfellow (Goodfellow Al006031).

The measurements presented here were done at temperatures up to 500°C. At higher temperatures, oxidation becomes important and the material changes color.[46] In a typical AM process, the melting takes place in an Ar environment, and we plan to make our high temperature measurements under similar conditions. The use of a Ta disk presents a possibility to go above the melting points of most materials of interest and measure the absorptivity of the melt.

Let us compare the results with recent, first-principles modeling of laser absorption in powder using the ray tracing code presented above. It was demonstrated that due to multiple scattering, the powder absorptivity is greatly increased in comparison to flat surface absorptivity. The absorption for the metals with high absorptivity (SS, Ti) is practically independent of powder structure. For SS, the calculated results presented in Tables 26.1 and 26.2 give 60% absorptivity for monosized hexagonally packed powder, and 58% for powder with experimentally measured size distributions packed according to the *rain drop* method.[38] Experimental measurements are consistent with these calculations.

The insensitivity of absorption to the powder structure may explain the independence of Ti-6Al-4V absorption on powder type. The absorptivity value for Ti alloy in our measurements is about 70%, somewhat higher than predicted by the modeling value of _65% (Table 26.1). One possible explanation is that the calculations in Boley et al.[35] used the refractive index for the pure Ti, which can differ from that of Ti-6Al-4V.

Calculated values for Al are very different from the measurements (Figure 26.11c). They suggest that the oxide layer and the structure of the surface are important. For a flat surface, the observed absorptivity of Al is more than 20% for 1 μm light, much higher than the 5% value predicted using the textbook refractive index (see discussion in Rubenchik et al.[46]). The increase in powder absorptivity in comparison with a solid material is consistent with numerical results.[35]

26.3 Modeling at the scale of the powder

26.3.1 Description and purpose

The powder-scale model uses the input laser beam characteristics to transform a particle bed through the dynamics of the molten state into solidified material. The model is initialized with powder particles of the desired size distribution and layer thickness, often on a uniform substrate, or on a previously processed layer. The combined thermal and hydrodynamic simulations model the appropriate distribution of the laser's energy, as it interacts with the powder particles, the substrate, and the melt pool. The deposited energy from the laser heats the powder above the melting point, where the model coalesces the particles into a melt pool that flows under the influence of surface tension and vaporization recoil. The powder model tracks the various modes of heat loss, including conduction and evaporation, until the melted material solidifies onto the existing substrate. The contributions of the powder model to the overall AM modeling effort come in four areas: (1) laser interaction with the powder bed, (2) powder response, (3) melt-pool characterization, and (4) the build quality metrics of surface finish and final part density.

The total energy absorbed from the laser is an important integrated quantity that must be provided by the powder model. Another important quantity, the net energy deposited into the part, accounting for losses including evaporation and thermal radiation, will also come from the powder model simulation results. The net deposited energy plays an important role in the part-scale model simulations, particularly as the details of individual layers are abstracted away for computational efficiency. The powder model has the capability to include effects of the laser beam geometry, including spot size and shape, and various options for the distribution of power within the beam, such as Gaussian, top hat, or *donut*.

Through modeling of various arrangements of individual powder particles, the powder-scale simulations are used to determine their integrated effects. The particle size distribution is used by the powder model to initialize the geometry, thereby affecting a number of model outputs, including the powder bed packing density and the effective thermal conductivity of the unconsolidated powder bed. The thickness of the powder layer has significant effect on the art quality, including the obtainable density and surface roughness, which can be investigated with a powder-scale model.

The powder model includes the formation, evolution, and eventual solidification of the melt pool. Single-track[52] parameters such as width, height, and depth can be compared with experimental data for validation of the model. The powder model can determine the uniformity of the track/bead, which is useful for creating maps of optimal process parameters. As part of the hydrothermal calculations of the melt-pool motion and solidification, the powder model can generate temperature-time history data for use in models of microstructure evolution.

The role of the powder model also includes build quality measures such as surface roughness and obtainable density. Multitrack simulations will give the solidified shape of many overlapping or overlaying melt tracks, giving the roughness of top, bottom, and side-facing surfaces of the part. These simulations can also be used to study the formation of voids in the final part structure. The powder model may be used to investigate mitigation strategies to improve these quantities by such techniques as laser power or speed variations.

26.3.2 *Physics representation*

26.3.2.1 *Included physics*

The powder model begins with the three-dimensional geometry of a random powder layer on a substrate. A preprocessing program generates the locations of nonoverlapping spheres with the desired particle-size distribution in a randomly packed arrangement. The program fits in spheres until the desired density fraction is reached, usually near the typical experimental value of about 55%. The powder particles are overlaid on a uniform background mesh, replacing the background *void material* with metal for finite elements (FEs) within the preprocessing program's defined spherical shapes. The background mesh is fine enough, typically about 3 microns, to resolve the individual powder particles.

Several approaches are possible for modeling the interaction of the laser beam with the powder bed. Some approaches (e.g., see Gusarov et al.[53]) utilize the methods of radiation transport to analyze the absorption and scattering of the laser beam within a 50% dense packing of uniform spheres. These methods determine an energy deposition profile that is not concentrated at the surface of the powder bed, but rather is distributed into the depth of the powder layer with a roughly exponential fall off. The depth of deposition is determined by the packing fraction, the powder bed depth, and the absorptivity of the metal particles. This energy deposition profile is moved with the laser scanning velocity at a fixed height corresponding to the nominal powder bed depth. These methods are most applicable to applications such as SLS, where the particles for the most part retain their original geometric arrangement for the duration of the laser irradiation.

For SLM applications, the powder particles rapidly melt and begin to consolidate well within the laser beam spot, so that a more dynamic laser deposition model is required. An approach is needed that will deposit energy on the powder particles and melt surface, dynamically following the melt-pool evolution. Powder-scale models that include the recoil pressure from evaporation of the metal show significant depression of the melt surface under the laser beam, to below the original substrate level. Energy deposition models that provide fixed deposition versus depth profiles then have almost no metal remaining in the deposition volume.

Perhaps ideally, the laser deposition would be modeled with a ray-tracing algorithm that would operate in an integrated fashion with the calculation of the hydrodynamic motion of the melted particles. Such an approach would properly distribute the laser energy scattered between particles near the leading edge of the beam, while accounting for deposition on the melt-pool surface, and any absorption of laser energy reflected off a dynamically changing melt pool surface. This method is computationally challenging due to the complex, rapidly changing powder and melt surface topology.

For laser deposition, we used a ray-tracing model that does not take into account the multiple reflections. For optimal processing conditions, the laser beam must melt the powder layer and some depth of substrate to provide good bonding of the new layer. When the laser interacts with the powder particles, the particles are practically thermally isolated from each other and the melting is rapid. When the laser starts to melt the substrate, the thermal conduction losses through the substrate slow down the rate of melting. In an optimal processing regime, the substrate under the laser spot will be melted to a depth comparable with the thickness of the new layer. From the above arguments it follows that in the optimal regime, the powder particles must be melted near the leading edge of the laser spot and most of the spot intersects a smooth melted surface. The results of the modeling are consistent with the above arguments. Relatively few of the incident rays would hit the metal after reflection. From this pattern it is clear that the multiple reflections play a limited role and can be disregarded.

Another effect we do not take into account is the interaction of the laser with the evaporated plume. It is usually assumed that for intensities less than 100 MW/cm^2 the laser absorption in the evaporated plume and laser-produced plasma is unimportant. For optimal processing, we want to model the laser intensity $I < 10$ MW/cm^2. Although the plasma produced as a result of this interaction can be useful as the diagnostic tool, it will not affect the energy balance or the material flow.

To account for laser light reflected away from the metal surfaces, a constant absorptivity value $A = 0.3$ was adopted in calculations. This value is a conservative lower limit, for example, the absorptivity at room temperature is ~0.34 and the absorptivity of the melt is higher. The somewhat lower value is chosen to account for the decrease in the effective absorptivity of the sloped surfaces.

The powder model tracks the energy content of the metal throughout the FE mesh. Local temperatures are derived from the energy based on the heat capacity and the latent heat of the material, which are included in a tabular equation of state (EOS). The model also computes the density changes due to thermal expansion using this same EOS. Melting of an alloy occurs over a temperature range, and not at a particular melting temperature, and this is also included in the EOS treatment. What is not included is the hydrodynamic effect of such a *mushy zone* between solidus and liquidus, where solid and liquid phases are mixed. The model instead decreases the material strength linearly from room temperature down to zero at the liquidus temperature. One of the primary drivers for consolidation of the melted particles and subsequent motion of the melt pool is surface tension. The powder model includes an algorithm that identifies the material boundaries between the metal and background void. Based on information on the surface locations in adjoining elements, the model determines local curvature of the metal surface and applies the temperature dependent surface tension force to the appropriate nodes. As there is a large curvature at a *neck* where a particle first contacts an edge of the melt pool, this is a mechanism for drawing powder particles out of the bed and into the melt pool. Gravity effects are also included in the powder model, even though these are overwhelmed by the surface tension forces.

Marangoni convection is driven by the surface temperature gradient between regions of high and low temperature on the surface of the melt pool. For many materials, surface tension decreases as the temperature increases, leading to a flow away from the melt surface closest to the laser spot. Other drivers of melt-flow motion are the inflow of newly melted material and curvature-driven surface tension. The Plateau–Rayleigh instability in a long, cylindrical melt bead can cause a pinching-off of some sections of the pool from others. The strong curvature of the melt pool near the laser spot draws melt flow back into this region.

Eventually the melt must solidify, so the powder model includes several modes of heat loss from the melt and heated solid. One primary loss mechanism is through thermal conduction, largely to the substrate. This energy loss is computed as part of the thermomechanical solution at every time step of the simulation. The model also includes conduction through the adjacent powder bed, though this effect is limited by the poor effective thermal conductivity of the bed, only about an order of magnitude higher than the gas used as the backfill atmosphere.[54] The model uses a boundary condition on the bottom of the substrate that approximates the response of a semi-infinite slab, reducing the thickness of substrate that must be modeled. A thermal radiation loss is also computed for the top surface of the melt pool, based on the usual T4 relation, modified by an effective emissivity of the liquid metal. The model also includes an energy loss due to evaporation, which will be discussed below.

As the melt cools, its motion is eventually stopped by the strength terms in the material model turning back on when the temperature falls below the melt temperature. As the material solidifies, the temperature and thermal shrinkage are tracked, as the energy is brought lower through the thermal loss mechanisms discussed. Again, no *mushy zone* is accounted for as the material passes from liquidus to solidus.

26.3.2.2 *Abstracted physics*

Evaporation of metal, particularly under the intense laser spot, is an important part of the dynamics and energy balance of the SLM process. As the mass loss is expected to be small and the processes are dominated by very near-surface effects, the evaporated material is not modeled directly. The powder model does include the effects of evaporation through an abstracted model approach. The two effects that are modeled are the energy loss due to the loss of metal vapor from the modeled system, and the recoil pressure that balances the momentum of the departing vapor.

The theory of rapid vaporization is well established in the literature.[55–57] Adjacent to the surface, there develops a thin layer where the vapor velocity distribution is dominated by the evaporating material, and so is not in translational equilibrium. Within a few mean-free paths, collisions between the vapor molecules establish equilibrium conditions. The gas dynamics model of this thin Knudsen layer employs jump conditions that conserve mass, momentum, and energy.

There are several different treatments for the evaporation rate that boil down to the same exponential dependence on the surface temperature. For example, in Klassen[58] the Clausius–Claperon relation can be used to compute the saturation vapor pressure from the material's latent heat of vaporization, boiling temperature at atmospheric pressure, and the critical temperature. The net mass transport rate, including the effects of condensation, is computed from a local Mach number based on the ambient pressure and the saturation vapor pressure as a function of surface temperature. This result determines the evaporation coefficient, the net fraction of molecules leaving the surface. Finally, the recoil pressure is computed based on a pressure balance across the Knudsen layer using the evaporation coefficient and the saturation vapor pressure.

The powder model makes use of the recoil pressure following the treatment of Anisimov.[59] A table of recoil pressure as a function of surface temperature for a particular material is first created. The model reconstructs the location of the top surface of the melt pool at every cycle using information on the volume fraction of metal in each zone and its neighbors, as discussed above in relation to application of surface tension forces. The recoil pressure forces are added normal to the local interface direction, with a magnitude determined by the local surface temperature.

The mass lost to evaporation is expected to be small, at least for the optimized build conditions for a particular material. Approximately, the mass loss is less than a percent of a single powder particle per millimeter of laser beam travel, so the powder model does not adjust masses to account for the net vaporization rate. However, the energy content of the vaporized material is significant, because the latent heat of vaporization is quite large. Using the computed net vapor flux and the latent heat of vaporization, table of recoil pressure is constructed giving the energy loss rate as a function of surface temperature for the material under study. The thermal solver part of the powder model uses this table as a (negative) source term applied to the surface elements of the melt pool. One great benefit of applying this evaporative cooling to the simulation is to effectively limit the peak temperature under the laser spot to near the boiling point. For example, a calculation of a 400 W laser beam on steel would give a peak temperature near 3000 K, not the 7000 K that is seen without the evaporative loss. An underlying assumption in this evaporation treatment is that the material is pure, or can be effectively modeled with an averaged set of material properties in the case of alloys. It is well known in the laser welding literature that lower vapor pressure constituents can preferentially vaporize, depleting the remaining material in these components. The effect of our approximation is unknown, and will vary by material and processing conditions.

26.3.2.3   Neglected physics

One of the limits of this powder model is that it cannot be used under conditions of intense vaporization, such as those that might be found in keyhole welding.[60] The powder model assumes that there is no interaction of the vaporized material with the incoming laser beam. Doing so would involve computations of laser–plasma interactions and the subsequent reradiation of energy deposited in the plasma back to the workpiece. Although models for these processes are available in the literature, experimental evidence related to our work has indicated that a mode of SLM processing that approaches the keyhole regime is not advantageous to build quality.[60,61]

Another laser-related feature not included in the powder model is a true laser ray-tracing capability. Such a model would be expected to improve calculations of the spatial distribution of the laser energy deposition. This would particularly be true with strongly concave melt-pool surfaces where reflected laser light might deposit energy on the far side of the depression, rather than being lost. Until this capability is fully installed and tested with the powder model, the line of sight laser deposition model will be used, albeit with some needed experimental calibration for total absorbed fraction.

The convective losses to a flowing guard gas in the build chamber are ignored in the current model. Computational limitations, discussed below, permit only small spatial areas to be modeled, on the order of a fraction of a square millimeter over a few dozen milliseconds. During these timescales, evaporation, thermal radiation, and conduction to the substrate dominate the heat loss from the solidifying melt track. The convective losses will be more important when we can scale up to several square millimeters of build area.

The current powder model ignores the gas dynamics of any trapped background gas, instead modeling only the metal and a *void material*. The void disappears if a gap between two powder particles is closed, obviating the need to track the motion of any gas. The strong dynamics of the melt flow driven by Marangoni convection and the recoil pressure suggest that the melt is fairly well mixed, so any trapped gas should be able to escape, given sufficient melt depths. The powder particles used thus far in the modeling have been spherical. The preprocessing program can utilize ellipsoids and spheres, but this work has not yet been done. A more significant omission in the powder particles is ignoring any oxide layer. Although particles tend to melt fairly quickly and so change the surface, the melt flow is largely driven by surface tension. Surface tension can be significantly changed by contamination, but there tends not to be sufficient data to permit addressing this issue.

The powder model does not contain any consideration for the formation of metal grains and for grain growth as the melt pool solidifies. A simple isotropic model adds material strength as it cools below the melt temperature.

26.3.3   Computational challenges

26.3.3.1   Need to approximate some physics

Evaporation is an important phenomenon in SLM, but the dominant length scale is the Knudsen layer, which is much smaller than what can be explicitly modeled with the powder-scale finite element mesh. This situation demands a subscale model of the process that must be precomputed by a separate model of the Knudsen layer flow. Although somewhat cumbersome, the approach of applying pregenerated tables to the simulations is quite effective.

The most significant approximation with the laser deposition portion of the powder model is the assumption of no interaction of the incoming laser with any vaporized material. With this assumption, the computational model does not need to track laser energy absorption, scattering, and subsequent reemission. The first two factors would involve much more sophisticated laser–plasma interaction capabilities, whereas the last factor would require dynamically changing energy fluxes on the neighboring surfaces of the melt pool and powder bed. We are working on validation experiments to determine if this assumption is valid for the preferred range of SLM operating parameters.

26.3.3.2   Need for fine zoning

The powder model needs to resolve the individual particles in the powder bed. There must be adequate resolution for accurate determination of surface shapes for the surface tension computations. For the thermal solution steps, the cavities between the powder particles must be resolved to obtain an accurate model for the thermal conduction through the unconsolidated powder bed. On a coarse mesh, adjacent particles will appear to have a "neck" of connected metal between them, greatly increasing the effective thermal conductivity. These requirements have led to a typical FE size of about 3 μm on a side, for simulating particles of a 27 μm mean diameter.

26.3.3.3   Explicit time marching limits time step

The powder model uses an explicit hydrodynamic formulation for the motion of the powder and the melt. This approach brings with it a limit on stability based on the time required for a sound wave to cross a zone. With a 3 μm zone size and sound speeds of a typical metal, the time-step size cannot exceed about a nanosecond. Thus simulations covering several milliseconds require millions of time steps, leading to long simulation times for detailed models.

One well-known trick to mitigate the effects of this stability restriction is to artificially raise the density of the material under study, thereby reducing the sound speed somewhat, and so increasing the minimum time step. The difficulty with this trick is that the dynamics of the material motion can be affected by this artificial density change. Through various numerical experiments using various levels of density scaling, it was found that at most a factor of three to five in sound speed could be achieved with this approach without adversely affecting the results.

26.3.4   Materials challenges

26.3.4.1   Experimental data required

Powder morphology is needed to properly initialize simulations of powder bed SLM, as these characteristics affect packing density and minimum reasonable powder layer thicknesses. A primary metric needed from the supplier or preferably from direct measurements is the particle size distribution. Direct measurement is important if the excess powder from previous builds is to be reused, because the particle size distribution will evolve through reuse cycles. Another metric of powder morphology is the particle shape. Spheres are of course easiest to model, but any powder with a significant amount of non-spherical particles should have some quantitative measure of the shape to ensure good fidelity in establishing the initial powder bed for a simulation. The effective packing density should be measured in the SLM machine by performing a build of a known size box that is removed from the build chamber with the enclosed unconsolidated powder still in place. Weight measurements before and after removing this powder will provide a good target for the modeled powder bed. Our measurements have shown approximately a 55% packing fraction. This number should probably be treated as an upper bound as the powder layer becomes thin, nearing the maximum particle size.

The powder model has fairly extensive requirements for material property data. The elemental composition of the powder particles must be known, both for the main constituents and for any oxides or other impurities that might be present. Knowledge of the composition is necessary both to select the proper literature values of any properties that will not be measured, and to allow construction of tabular equations of state, to cover the thermodynamic phase space. Density of the material, not just at room temperature, but as a function of temperature up to and beyond the melting point, is required for proper computation of thermal expansion. The melting temperature or solidus and liquidus temperatures for alloys must be known. The latent heat of melting must be known to properly capture the melting rate and particularly the cooling rates as the melt pool solidifies. As evaporation of the metal can occur under the laser spot, the boiling temperature and heat of vaporization must be known for the material. For alloys, this can be an average of the properties of the constituents, or just the values for the major component, depending on the available data. The heat capacity and thermal conductivity of the material must be known from room temperature up to and beyond melt. As much of the dynamics is driven by surface tension, this must be known for the liquid metal. To include the effects of Marangoni convection, the temperature dependence of the surface tension must also be measured. Some measure of the viscosity of the liquid metal should be made, though, because many metals have rather low viscosities; however, this quantity is not as critical for success of the model. Laser absorption properties of the material must be known, both for a packed powder bed and for a solid and a liquid surface.

The laser input parameters must be known for input into the powder model. The laser wavelength is needed to properly assess absorption properties. The laser temporal characteristics must be known, either steady continuous wave (CW) or pulsed duration and repetition rate. The total power must be specified. The beam size is an important measure, and must be accompanied by a good definition of *beam size*, be it $1/e^2$, $D4\sigma$, 95% power, or another measurement standard. The power distribution within the laser must be specified, usually Gaussian, though flat top, donut, and other options are possible.

26.3.4.2 Description of material models

The powder model is based on the ALE3D Multiphysics code to model the heat transfer and material motion. The thermal solution makes use of the temperature dependent thermal conductivity and heat capacity entered as part of the material data. Thermal expansion and response to pressure loading is handled using a tabular equation of state that defines pressure and temperature as functions of density and energy over the range from room temperature to boiling. The strength model is of less importance to the powder model than to the effective medium model, because the primary use of strength is to compute residual stresses in the part-scale model. A standard ALE3D model of a high-deformation rate and temperature dependent strength is used for the simulations.[62]

26.3.5 Application examples

26.3.5.1 Powder bed thermal conductivity

The thermal conductivity of the particle powder bed is computed *on the fly* from first principles. All that is required is the thermal conductivities of the SS material and of air at a given temperature. The powder has lower thermal conductivity than bulk SS. This is because the particles are at point contact and the heat diffusion in gaps between the particles depends strongly on the gas's thermal conductivity, which is lower than that of the metal.[54] As a code validation test, we compute the thermal conductivity of SS powder. We find for a powder packing density of 36%, the ratio of powder thermal conductivity over thermal conductivity of air is 3.0; for 45% it is 4.2; for 55% it is 6.6. These results agree well with the values 3.0, 4.5, and 6.0, respectively, as shown in Figure 26.5 of Rombouts et al.[54]

26.3.5.2   Single track formation

First, we examine a single track simulation to illustrate the effects of melt flow driven by surface tension, including flow instabilities leading to nonuniformity of the final solidified bead. The material is SS, 316L, in a powder with a log-normal distribution about 27 µm. The powder is distributed in a random packing to a depth of 35 µm on a thicker substrate. The laser source is about 1 µm wavelength, 200 W power, 2.0 m/s scanning speed, and a beam size ($D4\sigma$) of 54 µm diameter. The computational domain is 1000 µm long and 300 µm wide. The simulation includes the effects of surface tension and Marangoni convection, but neglects evaporation and recoil pressure for this case.

We find that the surface-tension effects on topology and heat transfer drive the SLM process. As soon as a melt forms, the surface tension acts to decrease the surface energy. Although the viscosity is low, we still consider it whenever surface tension is computed. The model includes gravity; however, surface-tension forces are stronger and the timescales we consider are short, so we do not expect gravity to play a major role.

Our fine-scale approach demonstrates the 3D nature of the SLM process and the influence of the stochastic powder bed. Figures 26.12 and 26.13 show the temperature contour lines on the surface of the stochastic powder bed and inside the substrate, respectively. The black contours surround a region of temperatures higher than 5000 K (this temperature exceeds the boiling point and is addressed below), which indicates the location of the laser spot. The next interesting contour line is the red melt line with a temperature of 1700 K, which surrounds regions of liquid metal. One notices that the red line, that is, liquid melt races ahead of the laser spot. The region that separates the laser spot and the solid particles ahead is quite narrow. These contour lines also indicate that temperature gradients are the

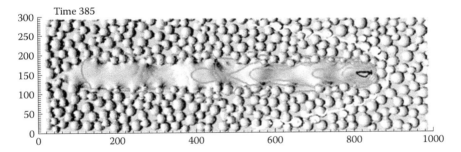

Figure 26.12 3D simulation snapshot shows the temperature distribution on the surface as the laser spot moves to the right. The time is expressed in microseconds, length in micrometers, and the temperature is in Kelvin. The temperature profiles are correlated with the surface topology.

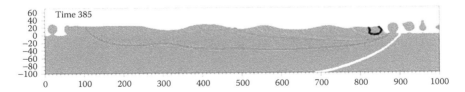

Figure 26.13 This 2D longitudinal slice corresponds to the 3D snapshot in Figure 26.12. The slice is taken in the middle of the melt track. It shows the temperature distribution in the melt pool and substrate. The time is expressed in microseconds, length in micrometers, and temperature in Kelvin. The temperature profiles are correlated with the surface topology.

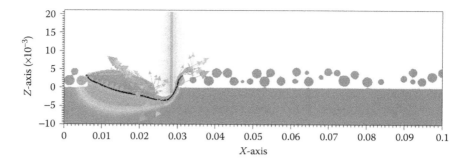

Figure 26.14 This 2D longitudinal slice is taken in the middle of the melt track. Colors indicate the temperature distribution in the melt pool and substrate, with red being melted material, whereas the vectors represent melt velocities. The drawn-in laser beam is approximately in the correct location at this time in the simulation. The *floating* powder particles result from the 2D slice through a random 3D distribution.

strongest near the laser spot and decrease in the back of the flow. This suggests that the Marangoni effect should contribute to the flow and that its effect will be largest close to the laser spot.

One also notices islands of liquid regions at the back of the flow (Figure 26.12). The temperature profiles on the surface and the substrate are intimately connected to the melt topology. These island formations are evidence of the Plateau–Rayleigh instability, which creates peaks and troughs. At the troughs, the melt height is low. It takes less time for the substrate to cool these regions because less liquid is present there. At the peaks, the opposite is true. More liquid means more stored heat and the liquid stays longer. This nonuniformity of surface cooling can be an important feature that the powder-scale model can pass on to the microstructure and part-scale models.

Going beyond what was shown in the previous simulation, the inclusion of recoil pressure in the powder model reveals robust dynamics under the laser spot. Figure 26.14 illustrates the melt-pool dynamics in a slice taken through the center of the laser path, with color indicating temperature and vectors representing the velocity field. Hot droplets are seen ejected from the melt pool, similar to the sparking that is observed experimentally. Spatter of liquid droplets is observed in front of the laser spot. Recoil pressure has a significant effect on the topology of the powder and melt, allowing melt penetration well into the solid substrate. Marangoni convection is moving the surface of the melt away from the laser, but the curvature of the melt surface is strongly pulling the melt back into the depression formed by the laser beam. Though not shown in the figure, inclusion of the evaporative energy loss term decreases the peak temperatures under the laser spot from being in excess of 5000 K previously to about 2700–3000 K, the boiling point of steel.

26.3.6 *Alternate approaches*

26.3.6.1 *Lattice–Boltzmann methods in two dimensions*

Granular or fine-scale models are expected to be computationally demanding. In Klassen[58] and K€orner,[63,64] a mesoscopic simulation of the melting process uses the two-dimensional (2D) lattice Boltzmann method to create a process map. These researchers have included much of the same physics included in the present study, adding more features over time, such as the recent addition of recoil pressure and evaporative losses. These works show the influence of surface tension and the packing density of the powder bed have a significant effect on the melt-pool characteristics. The process maps show regions in scan speed/laser energy space where certain melt bead morphologies are to be expected. Although 2D models are computationally efficient, the SLM process is inherently three-dimensional (3D), requiring the method to employ several approximations to achieve useable results.

26.3.6.2   *Open source models in three dimensions*

Another recent approach applies the computational fluid dynamics toolbox called Open FOAM, with some special purpose routines for SLM, to model the powder bed in three dimensions.[65] The simulations presented in the chapter show a uniformly packed powder bed, which somewhat reduces the stochastic nature of the powder-scale processes. Their flow is driven by surface tension, though Marangoni convection was not included. The model also includes effects of evaporation under the laser beam, resulting in melt surface depression under the laser spot, in general agreement with the present model. A primary disadvantage of their approach is that while the open-source tool is designed to be flexible, it is not well suited to large-scale calculations requiring good parallel scaling for many thousands of processor-hours.

26.3.6.3   *DEM in three dimensions*

The DEM are a natural choice for modeling behavior of particle beds. Some recent work in this area[66] utilizes the DEM method to analyze the selective laser-sintering process, in which particles are heated sufficiently to begin to sinter together, but do not develop into a convecting melt pool. The DEM method is capable of thermomechanical analyses, with some calibrations required for certain parts of the model. The initial conditions in a random powder bed are set by allowing a number of particles to settle on a flat plate under the influence of gravity. For the thermal conduction, the model handles conduction between particles at contact points but must be adjusted to account for the additional gas phase conduction that is usually significant. Ganeriwala et al. described the thermal solution to the laser energy deposition on the layer and the effects of particle size on melt rates. The DEM has promised for selective laser-sintering simulations, but selective laser melting modeling requiring melt-pool dynamics is problematic.

26.4   Modeling at the scale of the part

26.4.1   *Purpose*

The metal AM enterprise needs information and knowledge at the overall scale of the desired part and builds a process to inform many engineering decisions.

Currently, these decisions are primarily informed by past experiences and test fabrications. Ideally, simulation insights would help inform design, process specification and qualification, process monitoring, and part acceptance. Quantities of interest at the part scale include the following:

- Deformations that could halt machine operation or place the completed part outside the desired geometric tolerances
- Residual stresses causing those deformations and/or creating initial conditions detrimental to service-life concerns such as failure and fatigue
- Local effective material properties, or at least indicators of where they might significantly deviate from the nominal properties expected from the process

The ability to reliably predict such responses would aid in adoption of AM technology and speed its ongoing adaptation to new material systems and specific part geometries. These predictions must be attainable in a timely manner with acceptable and assessable computational resources. An eventual goal is fabrication models that are so efficient that they could be evaluated as part of the performance evaluation for a trial design within an automated design optimization process.

26.4.2 *Physics challenges*

SLM is a process calling for multiscale modeling: local ($O(10–100$ μm)) extreme material transformation is taking place over brief time intervals $O(10$ ms) as an overall part $O(10$ cm)^3 is fabricated in a processes lasting O(hours-days). Yet, in-line multiscale material response modeling is little utilized in any application space, let alone AM, due to its extreme computational demands. The very separation of these scales suggests that multiple models can each provide useful insights and build knowledge, leading toward eventual coupling or coordination. In creating tractable simulation approaches for part-scale fabrication, a series of modeling topics must be addressed—or consciously avoided. The eventual strategy decisions must be tested through assessing the ability of the resulting overall model to produce meaningful insights.

The local SLM process is an extreme, thermally driven material transformation, as illustrated in Section III on modeling at the powder scale. At the part scale, one seeks to obscure the details of the local power–laser interaction. Instead, the goal is to capture the aggregate influence of the SLM process on the macroscopic state of the part during and at the completion of its fabrication. By choosing to ignore flow dynamics in the melt pool, the simulation can be cast as the thermomechanical response of a nonlinear solid continuum. Within that perspective, the powder can be represented as a reduced-density, low-strength solid. The deposition of the laser energy into the powder can then be represented by a volumetric energy source term. The spatial distribution derived by Gusarov et al.[67] has been one common choice, even if utilized outside the original assumptions of that analysis. Gusarov introduces a simple knockdown factor to the total nominal laser power to acknowledge the effects of reflected radiation and metal evaporation. Melting can be represented thermally through a latent heat and mechanically as a near-total loss of strength. Some researchers view the only relevant response being the subsequent freezing and choose to simply initialize the active fabrication area at $T_{solidus}$, for example, Zaeh et al.[68] Having the temperature-dependent strength rise as temperature falls below $T_{solidus}$ is currently our only acknowledgement of the complex behavior in the *mushy zone* at the melt-pool boundary.

With an effective medium model such as the one discussed, the geometry of powder particles is not resolved. It is indeed a choice as to what powder volume is directly represented in the computational domain. For true part-scale spatial domains, the common modeling practice to date is to largely ignore the adjacent regions of unmelted powder, at most perhaps representing their thermal interaction with the part through some Neumann boundary condition. Some of the present authors have analyzed Representative Volume Element domains consisting of a cubic millimeter of material.[69] In this case, successive 50 μm layers of powder are initialized and scanned by moving the energy source location. To model the gross loss of porosity due to powder melting, an irreversible *phase strain* was introduced into the thermomechanical constitutive model that is activated during the material's first excursion above $T_{solidus}$. This *phase strain* magnitude was simply assigned to result in a net volume associated with full-density material. If future powder-scale modeling can identify a phenomenological evolution law for porosity, for example, based on the local history of temperature and temperature gradients, then the part-scale model could adaptively assign the appropriate local phase strain or at least output a map of regions likely to have unacceptable porosity.

26.4.3   Computational challenges

The computational challenges of part scale thermomechanical simulations are driven by the disparate spatial scales of the laser energy source and the overall part geometry and compounded by the disparate time scales of local heating versus overall heat transfer and the actual time of fabrication, which is at least hours and often days. This has led most researchers to concentrate on coarse mesh representations capable of capturing overall part deformations while minimizing computational costs for each time step. Others are complementing this with multi-resolution approaches, for example, adaptive mesh refinement or forms of embedded grid, to localize some higher resolution in the vicinity of the active material transformation.

The aggregation of process representation to more computationally tractable length scales reinforces the similarities between SLM and welding: this potentially obviates some of the physical differences already noted between the two. Not surprisingly then, some of the active AM researchers come out of the weld modeling community and are informed by the well-established methodologies represented by standard texts in that field, for example, Goldak[70] and Lindgren.[71] This is particularly so with the related metal AM technology of laser engineered net shaping (LENS) and similar direct metal deposition methods having a larger (wider) active deposition zone, closer to a weld bead. Representative of work in this vein are publications by Michaleris and coworkers, for example, Michaleris et al.[72] and Denlinger et al.[73]

One approach to thermomechanical modeling of SLM fabrication is being pursued in the context of the computational perspective and resources at a national laboratory. Some of the present authors are involved in adapting the in-house, general purpose implicit nonlinear FE code Diablo,[74] capable of effectively utilizing commodity parallel-processing platforms. Early efforts focused on developing SLM modeling and algorithmic approaches in the context of 50 µm layer-resolved simulations for representative volumes comprising 1 mm³.[69] This paper provides a detailed description of the balance laws, boundary conditions, and material models utilized. These coupled thermomechanical simulations utilize the laser deposition model of Gusarov directed in a serpentine pattern with alternating layer orientations. These calculations typically used 32–128 processors simultaneously, eventually taking less than two days. Peak heating and cooling rates of O(105 K) are observed, as also reported in Schilp et al.[75] Importantly, these simulations highlight that it is misleading to think merely in terms of the temperature history of the material in the active powder layer. These simulations clearly show that the material located several or more layers below the active work surface is still undergoing significant temperature excursions, which will contribute to continued evolution of the local microstructure.

26.4.4   Material challenges

With our current thermomechanical modeling strategy, the material response is represented via rather standard heat conduction and J2-plasticity models, parameterized with temperature-dependent properties. As engineering materials are typically not envisioned to have service life at temperatures near $T_{solidus}$, it is not surprising that scant handbook type property data are available in that regime. Thus, to date we have relied on artful interpolations between available elevated temperature properties and melt. The casting literature is another area for us to explore, though the timescale of SLM solidification may not match well with useful representations/correlations established in that field of modeling. Of course none of the thermomechanical responses described says anything about microstructure evolution and resulting service-temperature properties. We envision role of part-scale modeling to be producing histories of temperature, temperature gradients, cooling rates, and so on that would inform a microstructure prediction model.

26.4.5 Application examples

We first consider a case where even looking at the limited domain of a representative volume provides insights into a common SLM challenge: fabrication of downward-facing surfaces. Such *overhang* features often result in an undesirable finish on the underside that could necessitate further machining—if accessible. Figure 26.15 contrasts two build strategies. A common domain is defined: a 1 mm² plan form starting on a build plate shown in gray. For the first six layers, the energy source only scans over the left half. Blue represents unconsolidated powder and red fully transformed material; intermediate colors represent incompletely consolidated material. Starting with the seventh layer, the energy source scans the entire plan form. With the leftward case, which maintains constant laser power, we see the relative insulating properties of powder lead to localization of the energy and deeper penetration of the melt pool. With the rightward case, the laser power is modulated to one-fourth its nominal value as it reaches the right edge of the domain. This produces a transformed overhang region much nearer the desired horizontal surface. Note, however, this simulation also shows that the mitigation strategy should only be utilized during the initial overhang layers. By the third overhang layer, the reduced power is leading to substantially incomplete melting of the powder as evidenced by the green regions.

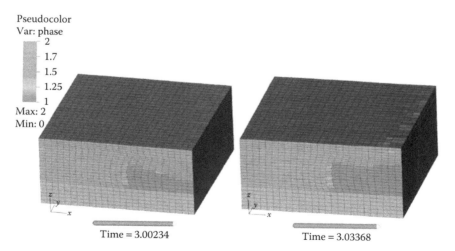

Time = 3.00234   Time = 3.03368

Figure 26.15 A comparison of results from two overhang fabrication scenarios. The domains are 1-mm-square in plan form. The gray base represents the build plate, blue represents untransformed powder, and red fully represents transformed material. The first scenario maintains constant laser power throughout the build scans. The second scenario modulates the laser power whenever the scans extend into the overhang section, leading to a more uniform build thickness.

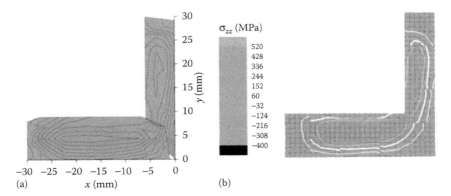

Figure 26.16 A comparison of normal stress at the midplane of a 316L specimen assessed experimentally (a) combining neutron diffraction and digital image correlation[76] to a thermomechanical finite element simulation to a (b) thermomechanical finite element simulation.

Our modeling approach is being revised and extended to address dimensions of real engineered parts. Material addition is modeled via meta-layers comprising the dimension of many physical powder layers (cf. Zaeh et al.[68]); though in our case the energy is deposited in a coarse, serpentine pattern, rather than instantaneously over an entire layer. Early efforts have been encouraging. Figure 26.16 shows a comparison of experimental and simulation results, in particular, contours of normal stress magnitude on the midplane of a 316L specimen 3 cm tall. The experimental characterization, made with the build plate still attached, fused in-volume neutron diffraction measurements with digital image correlation (DIC) on the side-surfaces.[76] The simulation results are plotted with identical color ranges and display the ability of the FE analysis to capture the high compressive stresses in the interior of each arm, balanced by surrounded tensile stress. Further refinements are being pursued, both for physics representation and computational efficiency.

26.4.6   *Alternative approaches*

The majority of academic researchers use commercial computer-aided engineering (CAE) software modeling tools. This reflects both a lack of native software assets within these research teams and a pragmatic recognition that commercial tools are the most likely avenues for subsequent industrial adoption. This choice allows these researchers to leverage software with a rich feature set, yet also places some key restrictions on their modeling approaches. First, having no ability to customize/extend the software, they must rely on the publically documented interfaces allowing specification of simulation components such as user-defined boundary conditions and material models. Furthermore, these software tools typically can still only leverage modest computational resources, thus limiting the size of the computational mesh utilized. Within this context, insightful efforts have been achieved by teams such as Zaeh and coworkers.[68] There, a simple part geometry is accreted through a succession of meta-layers, each 1mm thick, hence representing a collection of roughly 20 actual powder layers. Simultaneous cooling of an entire meta-layer from an assigned temperature of T_{solidus} captures the overall bending behavior induced in the test fabrication, but clearly cannot capture more local behaviors influenced by specifics of the laser scanning strategy. More recently, Zaeh and coworkers have developed a means of abstracting the true laser scan paths to identify areal boxes in the plane of the active meta-layer.[77] This abstraction considers the laser scan strategy on a subset of the true powder layers comprising the meta-layer, arriving at an aggregate heat load applied to successive planar areas. Results for improved fidelity are promising.[75] It is interesting to note that this team often utilizes a *one-way* coupled approach, where a thermal solution for the part formation is precomputed and then utilized by a subsequent mechanics-only simulation of the stress response.

Academic modeling research has not been solely focused on commercial software. One example of this is the work by Stucker and his research team. Pal et al.[78] summarize a series of numerical algorithms they have explored with the goal of significantly reducing the computational cost for FE simulation of the SLM process. For instance, a limited volume of fine mesh near the active melt region incrementally traverses a coarse mesh representation of the entire part volume. They have also explored reusing significant parts of the stiffness matrix to reduce costs associated with numerical linear algebra. The size of the thermal problem is further reduced by representing the coarse-mesh, far-field temperature field through a basis constructed from a small number of eigenvectors. Publications to date have not documented the complete integration of all these numerical technologies, and that process is now being pursued in the context of a commercial start-up. We also note that this team has demonstrated to date perhaps the most complex material constitutive model for SLM applications, a crystal plasticity representation incorporating dislocation density.[79]

26.5 Role of data mining and uncertainty quantification

The use of modeling and simulation to gain insights into the physical processes that govern AM is one step in the process of part qualification. To fully understand the factors that influence part quality and to provide a confidence interval on the properties of a part produced using AM, we also need experiments, data mining, and statistical inference. The role of the experiments is in validating the simulations to ensure that the computer model adequately represents reality.[80] Data mining techniques allow us to extract useful information from both simulations and experiments, providing insights and efficiencies in building parts with desired properties. Statistical inference enables us to reason in the presence of uncertainties; these could be uncertainties in either the experiments or the simulations.

This section provides a glimpse into some of the many ways in which ideas from the multidisciplinary and often overlapping fields of data mining, statistics, and uncertainty analysis can be used in AM. As the application of these ideas in AM is relatively new, this overview is necessarily introductory in nature. It is not intended to be a comprehensive review or comparison of techniques. Instead, we address two questions associated with AM: first, how do we select the different process parameters to build a part, and second, how do we quantify the uncertainties in the properties of a part?

26.5.1 Building additive manufacturing parts with desired properties

Determining the optimal parameters required to create a part with a desired property, such as density >99%, is often challenging, requiring extensive experimentation. This is because the number of parameters that control the AM process is large, numbering over a hundred by some estimates.[13] This optimization process unfortunately has to be repeated with changes in the material or the property being optimized, and changes in the machine parameters, such as the laser power or beam spot size. Modeling and simulation can play an important role in reducing the costs of this process optimization. We next describe briefly several of the current approaches that rely mainly on experimentation and compare them with an approach we have recently proposed that combines experiments with simulations using data mining techniques. We focus on part density, as this is one of the first properties needing to be optimized in building an additively manufactured part. We describe the approach using 316 L SS as an example, though the ideas can be applied to other materials and other properties as well.

26.5.1.1   *Design of computational and physical experiments*

There have been several studies that primarily use experiments to determine the process parameters that result in high-density parts. The approach taken is based on one or more of the following: (i) selecting parameters based on theory where the energy density, defined as a function of laser speed, power, and scan spacing, is restricted to lie within certain predetermined values, (ii) implementing simple single-track experiments to identify parameters that result in a sufficiently deep melt pool, (iii) building small pillars using various combinations of parameters and determining their density, and (iv) combining experiments with ideas from the field of design of experiments.

Single-track experiments[52] are a simple way to determine which combinations of laser power and speed result in melt pools that are deep enough to melt through the powder layer into the substrate. A layer of powder of a specified thickness is spread on a thin plate and several tracks, at varying laser power and speeds, are created. The plate is then cut, the cross section is etched and polished to reveal the melt pool perpendicular to the laser track, and the melt-pool characteristics are obtained, as shown in Figure 26.17. Both the top view of the tracks and the melt-pool characteristics provide useful insights into the surface roughness, the continuity of the track, and the depth and width of the melt pool. With increasing power or reduced speed, the melt pool becomes deeper, as shown in the three examples in Figure 26.17b. For high-density parts, we need to select parameters that locally reduce porosity by ensuring that (i) the powder melts completely, removing any voids in the powder bed and (ii) the process does not enter key-hole mode melting, where the laser can drill deep into the substrate, resulting in vaporization and formation of voids.[31] A simple way to improve the efficiency of the single-track experiments is to use a tilted plate[81] so that several powder layer thicknesses can be evaluated using a single track.

Although a suitable choice of the laser power and speed can ensure sufficient melting locally for a given powder layer thickness, the density of a part is also determined by other processing parameters, including the overlap between adjacent scan lines and the scanning strategy that determines how the area in one layer is scanned and how the scan pattern in one layer is related to the scan pattern in the next layer. A fully experimental approach to study the effect of these parameters was used by Yasa et al.[6,82] who built small pillars using different parameter settings and evaluated their density using the Archimedes method. A slightly different approach was used by Kempen et al.[83] who started with single-track experiments and used the quality of the tracks to identify a process window for building pillars for density evaluation.

As the design space of AM machines has expanded with the use of higher-powered lasers, new scanning strategies, new materials, and new processing techniques, ideas from the field of design of experiments[84,85] have started to play a role in systematic studies to understand the influence of various parameters on properties of parts. This field is relevant to understanding the design space of both experiments and simulations, which are sometimes referred to as computational experiments. It provides guidance on the selection of parameters and their values, and the analysis of the results. For example, Delgado et al.[86] used a full factorial experimental design with three factors (layer thickness, scan speed,

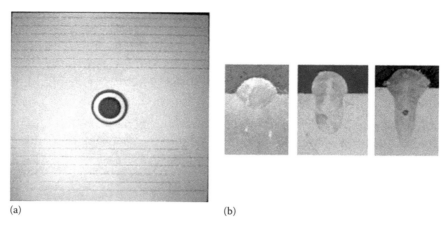

(a)   (b)

Figure 26.17 (a) A small build plate, 40 × 40 mm in dimensions, with 14 single tracks. The plate is tilted, so that when a layer of powder is spread, its thickness is zero at the left edge, and increases linearly to 200 μm at the right edge. Once each track has been created using a specified laser power and speed, the plate can be cut at various powder thickness values to obtain the cross section of the track, as shown in (b) for three sample tracks.

and build direction) and two levels per factor in their study on part quality for a fixed laser power. The outputs of interest were dimensional accuracy, mechanical properties, and surface roughness. The results of the experiments were analyzed using an ANOVA (analysis of variance) approach to understand the effects of various factors on the outputs.

We have recently developed an approach to process optimization for high-density parts that exploits both simulations and experiments by combining the insights from each using data mining and statistical techniques.[87] As multiscale simulations and experiments involving single tracks and pillars can be very expensive, we developed an iterative approach that starts with simple simulations and experiments and uses the results to guide the choice of parameters for more complex simulations and experiments. We first used a very simple and computationally inexpensive, Eagar–Tsai model[88] to explore the design space. This model considers a Gaussian beam on a flat plate to describe conduction-mode laser melting. The temperature distribution is then used to compute the melt-pool width, depth, and length as a function of four input parameters: laser power, laser speed, beam size, and laser absorptivity of the powder. The Eagar–Tsai model does not directly relate the process parameters to the density of a part. Further, it does not consider powder other than the effect of powder on absorptivity, so its results provide only an estimate of the melt-pool characteristics. However, it is a simple model, making it computationally inexpensive. This means that we can sample the input parameter space rather densely to understand how the melt-pool depth and width vary with the four inputs.

26.5.1.2   Sampling strategies

There are many ways in which the design space of input parameters to the simulations and experiments can be sampled. Screening experiments, which are done at the initial stages of a traditional design of experiments endeavor,[89] use a large number of parameters, each sampled at two extreme points that cover the range of each parameter. For d parameters, this results in $2d$ experiments. However, if the range of a parameter is large, sampling at the two extreme values might not be sufficient. If k sample values are used for each parameter, the number of experiments increases exponentially to kd; this can become prohibitively expensive even for moderate values of k and d. Therefore, a screening experiment using just two levels is often used first to identify the important parameters, which are then sampled more frequently. Sampling of the design space is often accompanied by analysis of the results using ANOVA to determine the factors that have an effect on the response, or using response-surface methods, where a first- or second-order model is fit to the data. These parametric response surfaces can also act as surrogates to the data.

The traditional approach to the design of experiments was motivated by physical experiments where it was expected that repeating an experiment would give slightly different results. More recently, similar ideas have been applied to simulations, where repetition usually does not have any effect on the results. The ideas used in the field of DACE (design and analysis of computer experiments) also involve sampling the input parameter space of the simulations and building surrogates that act as predictive models. If the sampling is adequate, the latter can be considered as providing reasonable approximations to the simulation output variables for a specific range of input parameters. These surrogates can be extremely useful in problems when the simulations are computationally very expensive.

As the accuracy of the surrogates depends on how well the input space of parameters is sampled, but the function relating the output to the inputs is unknown, the initial set of samples is usually placed randomly, and additional samples are added as necessary. Using a simple random sampling can result in regions that are under or oversampled, as shown in Figure 26.18 for a 2D domain. To address this, we used stratified sampling, where each of the four input parameters was divided into a number of levels and a point selected randomly in each of the resulting cells. As the range of values of the laser beam size and absorptivity of the powder was small, we used a smaller number of levels for these inputs in comparison with the laser power and speed parameters. Figure 26.18 shows that a stratified sampling approach results in an improved placement of samples relative to a straightforward random sampling.

In our work with 316L SS, we varied the speed from 50 mm/s to 2250 mm/s with 11 levels, the power from 50 W to 400 W using 7 levels, the beam size ($D4\sigma$) from 50 μm to 68 μm using 3 levels, and the laser absorptivity from 0.3 to 0.5 using 2 levels. This resulted in 462 parameter combinations that were input to our simulation. These ranges were selected as follows: The upper bound on the power was set to the peak power of our machine. The lower limit on the speed was set to ensure sufficient melting at the low power values such that the melt-pool depth would be at least 30 μm (the layer thickness selected for our experiments). The upper limit on the speed was estimated at a value that would likely result in a relatively shallow melt pool at the high power value. The lower and upper limits on the beam size were obtained from measurements of the beam size on our machine at focus offsets of 0 mm and 1 mm. By varying the beam size and the absorptivity, we were able to account for possible variations in these parameters over time or with changes in build conditions as we built the parts. A drawback of the stratified random sampling approach is that the number of samples is determined by the number of input parameters and the number of levels in each parameter; it cannot be set to a prespecified value. As mentioned earlier, this number can be quite large. In our work using the Eagar–Tsai model, this was not an issue as the model is computationally inexpensive. However, for more expensive models, where we want to control the number of samples by starting with a small number and incrementally adding new samples, an alternative approach using low-discrepancy sampling is often used. In a low-discrepancy sampling in two (or three) dimensions, the number of sample points falling into an arbitrary subset of the domain is proportional to the area (or volume) of the subset. This essentially results in samples that are randomly

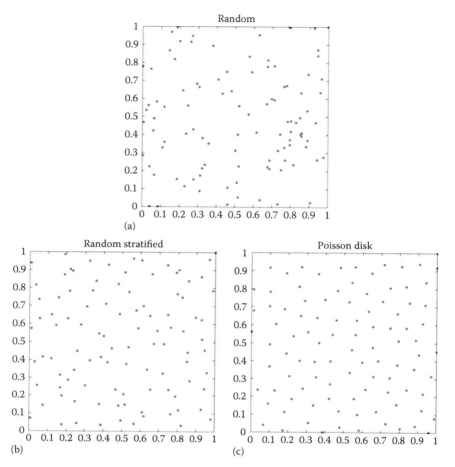

Figure 26.18 A total of 100 sample points distributed randomly in a two-dimensional space using (a) random sampling, (b) random stratified sampling, and (c) Poisson disk sampling.

placed far apart from each other. An example of such a sampling is the Poisson disk sampling, shown in Figure 26.18, where no two points are closer than a prespecified distance.[90]

26.5.1.3 *Feature selection*

In our approach to finding optimal parameters for high part density, we were able to sample the input parameter space of the Eagar–Tsai model quite densely as it had only four input parameters. However, there are more than a hundred variables in AM when we consider material properties, powder bed conditions, laser parameters, and so on. Some of these variables, such as material properties, are fixed for a material, though their values may not be known precisely, or may have a range associated with them. Other variables, such as the laser speed and power, are set during the manufacture of a part. Given this large number of variables, a commonly used class of algorithms in data mining, namely, dimension reduction,[91] become relevant in making the task of process optimization tractable. The dimension of a problem is the number of features or variables describing an experiment or simulation. By reducing the dimension of a problem, we can focus on just the most important variables, making it easier to understand how the outputs, such as melt-pool dimensions, are related to the input variables. Further, as mentioned in Section 26.5.1.2, the number of samples required to fully cover the design space is exponential in the number of dimensions. Therefore, reducing the dimensions is important when the experiments and simulations are expensive.

There are a number of dimension-reduction algorithms, including linear and nonlinear methods that transform the input parameter space into a reduced dimension space, and feature subset selection methods that rank the input variables, or features, in order of importance. The latter are more relevant in the context of AM as we need to set values of specific parameters, not their linear or nonlinear combinations.

One way in which these feature selection techniques can be used is for the identification of the important input parameters in the simulations. In Section 26.5.1.2, we used stratified random sampling to identify the sample points in the four-dimensional space of laser power, laser speed, beam size, and laser absorptivity. We then ran the Eagar–Tsai simulation at these sample points and obtained the melt-pool width, depth, and length. Of these melt-pool characteristics, we are most interested in the depth and the width. The depth indicates if the energy density is sufficient to melt through the powder to the substrate below. The width helps us to determine how far apart the adjacent laser tracks should be to ensure that no unmelted powder is left between the tracks.

We used two feature selection techniques to understand the order of importance of the four input variables in determining the melt-pool depth and width. The correlation-based feature selection (CFS) method[92] is a simple approach that calculates a figure of merit for a feature subset of k features defined as

$$\frac{k\overline{r_{ef}}}{\sqrt{k + k(k-1)\overline{r_{ff}}}}$$

where $\overline{r_{ef}}$ is the average feature-output correlation and $\overline{r_{ff}}$ is the average feature-feature correlation. We use the Pearson correlation coefficient between two vectors, X and Y, defined as

$$\frac{Cov(X,Y)}{\sigma_X \sigma_Y}$$

where $Cov(X,Y)$ is the covariance between the two vectors, and σ_X is the standard deviation of X. A higher value of merit results when the subset of features is such that they have a high correlation with the output and a low correlation among themselves.

The mean-squared error (MSE) method: In the second feature selection method, the features are ranked using the MSE as a measure of the quality of a feature.[93] This metric is used in regression trees (Section 26.5.1.4) to determine which feature to use to split the samples at a node of the tree. Given a numeric feature x, the feature values are first sorted $x_1 < x_2 < \ldots < x_n$. Then, each intermediate value, $(x_1 + x_2)/2$, is proposed as a splitting point, and the samples are split into two, depending on whether the feature value of a sample is less than the splitting point or not. The MSE for a split A is defined as

$$MSE(A) = p_L s(t_L) + p_R s(t_r)$$

where t_L and t_R are the subset of samples that go to the left and right, respectively, by the split based on A, p_L, and p_R are the proportion of samples that go to the left and right, and $s(t)$ is the standard deviation of the $N(t)$ output values, c_i, of samples in the subset t, defined as

$$s(t) = \sqrt{\frac{1}{N(t)} \sum_{i=1}^{N(t)} \left(c_i - \overline{c(t)}\right)^2}$$

where $\overline{c(t)}$ is the mean of the values in the subset t. For each feature, the minimum MSE acrvoss the values of the feature is obtained, and the features are rank ordered by increasing values of their minimum. This method considers a feature to be important if it can split the data set into two, such that the standard deviation of the samples on either side of the split is minimized, that is, the output values are relatively similar on each side. Note that unlike CFS, which considers subsets of features, this method considers each feature individually.

Table 26.3 presents the ordering of subsets of input features by importance for the melt-pool width, length, and depth obtained using the CFS method. A noise feature was added as another input; this is consistently ranked as the least important variable, as might be expected. This table indicates that for the melt-pool depth and width, the single most important input is the speed, whereas the top two most important inputs are the speed and power. In contrast, for the length of the melt pool, the top two most important inputs are power and absorptivity.

Table 26.4 presents the results for the MSE method. These are very similar to the CFS method, with the exception that the beam size is ranked lower than the noise variable for the depth of the melt pool. For all three melt-pool characteristics, the three lowest ranked variables have the MSE value roughly the same, so the corresponding three variables have roughly the same order of importance.

These results indicate that we should focus on the laser power and speed as they are the most important inputs related to the melt-pool depth and width based on the Eagar–Tsai simulations. Although these simple simulations relate just four inputs to the melt-pool characteristics, we expect that as we move to more complex simulations, feature selection and other dimension reduction techniques will become more useful in helping us to focus

Table 26.3 Order of importance of subsets of features using the CFS method. A higher rank indicates a more important input; the best subset of features is the one with the highest ranks

	Speed	Power	Beam size	Absorptivity	Noise
Width	5	4	2	3	1
Length	3	5	2	4	1
Depth	5	4	2	3	1

Table 26.4 Order of importance of subsets of features using the MSE method. A higher rank indicates a more important input

	Speed	Power	Beam size	Absorptivity	Noise
Width	5	4	2	3	1
Length	3	5	2	4	1
Depth	5	4	1	3	2

on the important variables, potentially limiting the number of experiments or simulations required to create parts with desired properties.

26.5.1.4   *Data-driven predictive modeling*

Our simulations using the Eagar–Tsai model provide the melt-pool characteristics at specific input values. These simulation inputs and outputs can also be used to build a data-driven predictive model, or a surrogate, that can be used to predict the output values at other inputs. A simple predictive model is a regression tree,[93] which is similar to a decision tree, but with a continuous instead of a discrete output.

A regression tree is a structure that is either a leaf, indicating a continuous value, or a decision node that specifies some test to be carried out on a feature, with a branch and subtree for each possible outcome of the test. If the feature is continuous, there are two branches, depending on whether the condition being tested is satisfied or not. The decision at each node of the tree is made to reveal the structure in the data.

Regression trees tend to be relatively simple to implement, yield results that can be interpreted, and have built-in dimension reduction. Regression algorithms typically have two phases. In the training phase, the algorithm is *trained* by presenting it with a set of examples with known output values. In the test phase, the model created in the training phase is tested to determine how accurately it performs in predicting the output for known examples that were not used in training. If the results meet expected accuracy, the model can be put into operation to predict the output for a sample point, given its inputs.

The test at each node of a regression tree is determined by examining each feature and finding the split that optimizes an impurity measure. We use the MSE as defined in Section 26.5.1.3, as the impurity measure. The split at each node of the tree is chosen as the one that minimizes MSE across all features for the samples at that node. To avoid splitting the tree too finely, we stop the splitting if the number of samples at a node is less than 5 or the standard deviation of the values of the output variable at a node has dropped to less than 5% of the standard deviation of the output variable of the original data set.

There are different ways in which we can evaluate the accuracy of the regression trees. The first is k runs of m-fold cross validation, where the data are divided randomly into m equal parts, the model is trained on $(m-1)$ parts, and evaluated on the part that is held out. This is repeated for each of the m parts. The process is repeated k times, each with a different random partition of the data. The final accuracy metric is the average of the accuracy for each of the k m parts. We use the relative MSE metric, defined as

$$\frac{\sum_{i=1}^{n} \left(p_i - a_i\right)^2}{\sum_{i=1}^{n} \left(\bar{a} - a_i\right)^2}$$

where p_i and a_i are the predicted and actual values, respectively, of the i-th sample point in the test data consisting of n points, and a is the average of the actual values in the test data. This is essentially the ratio of the variance of the residual to the variance of the target (that is, actual) values and is equal to $(1.0 \geq R^2)$, where R^2 is the coefficient of determination. The second metric is the prediction using a leave-one out (LOO) approach, where a model, which is built using all but one of the sample points, is used to predict the value at the point that is held out. For a data set with N points, this is essentially N-fold cross validation.

A common approach to improving the accuracy of regression algorithms is to use an ensemble, where many models, built from the same training data using randomization, are created.[91] The final prediction is the mean of the prediction from each of the models. In our work, we consider 10 trees in the ensemble, with randomization introduced through sampling. Instead of using all the sample points at a node of the tree to make a split, we use a random subset of the samples, thus making each tree in the ensemble different from the others.

Figure 26.19 shows the accuracy of the regression tree model in predicting the depth using the 462 simulations of the Eagar–Tsai model as the training set. Panels (a) and (b) show the predicted versus actual values using LOO for 1 tree and 10 trees, respectively. We observe that most of the points are near the blue line at 45 degrees (indicating perfect prediction), though the scatter is greater at larger melt-pool depths. The scatter reduces with the use of ensembles as would be expected. Using five runs of five-fold cross validation as the error metric, we obtain a relative MSE of 8% with a single tree and 3.6% with an ensemble of 10 trees.

The regression tree acts as a surrogate for the data from the Eagar–Tsai simulations and can be used to predict the width, depth, and length of the melt pool for a given set of inputs. The inputs for a sample point are used to traverse the tree, following the decision

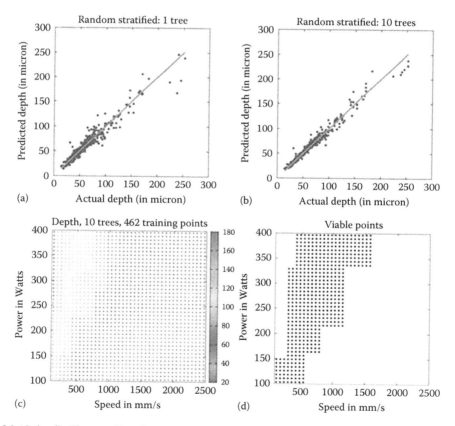

Figure 26.19 (a–d): The predicted versus actual depth using LOO for 1 tree and 10 trees, respectively; the predicted depth on a 40 × 40 grid in the power-speed space and the resulting viable sample points.

at each node, until a leaf node is reached; the predicted value assigned to the sample is the mean of the output values of the training data that end up at that leaf node. Figure 26.19c shows the depth prediction at sample points on a 40×40 grid over the power-speed design space, using a fixed value of $D4\sigma = 52$ μm and absorptivity of 0.4 for 316L SS. The prediction was obtained using the 462 Eagar–Tsai points to build a model with 10 regression trees. Panel (d) shows the viable space in the power-speed plot, where viability is defined as any grid point with predicted depth greater than, or equal to, 60 μm and less than, or equal to, 120 μm. In comparison with the Eagar–Tsai simulations, where each simulation takes approximately 1 min on a laptop, it takes a few microseconds to build the regression tree surrogate from the 462 simulations and practically no time to generate the melt-pool depth for a set of input variables using the surrogate.

26.5.1.5   Example of density optimization

The Eagar–Tsai model, combined with sampling techniques, feature selection, and the building of data-driven predictive models, enables us to determine the melt-pool characteristics for a given power and speed combination. The accuracy of these predictions depends on the number and location of the sample points, the accuracy of the physics model, and the complexity of the function being predicted. The Eagar–Tsai model, being relatively simple, gives us an approximation of the melt-pool characteristics. We use it to determine the viable region of the power-speed space (these being the most important variables) and then select a few points in this region for single-track experiments, as shown in Figure 26.17. Once we know the actual melt-pool characteristics at specific power-speed values, we can use them to identify process parameters for building small, 3D pillars, whose density is measured using the Archimedes method.

Figure 26.20 shows the first set of 24 pillars of 316L SS in powder, along with the density for the first two sets of pillars. Each pillar is $10 \times 10 \times 8$ mm. The rows correspond to different power values, whereas the columns represent different speeds. Having obtained the density estimate for the first set of 24 pillars, we ran another 24 at the same power values, but with the speeds chosen to complete the gaps in the density curves.[87]

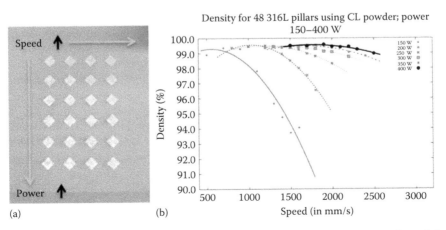

Figure 26.20 (a) A total of 24 pillars from the first set in powder; speed increases from left to right and power increases from top to bottom and (b) the density of pillars from the first two sets, built using 316L powder from Concept Laser (CL) on a CL M2 machine.

We have successfully used this approach to build high density parts for several different materials and powders of different sizes. Although our early work used a tilted plate for single-track experiments, we found that we typically used only one value for powder layer thickness. As a result, it was more efficient to use a flat plate. This also allowed us to double the number of tracks on a plate, leading to better exploration of the power-speed space using experiments.

26.5.2   Uncertainty analysis

There are many ways in which we can use simulations and experiments, both simple and complex ones, in AM. In Subsection 26.5.1, we outlined a process by which we can combine simple simulations with experiments to create an approach that efficiently identifies the process parameters for high-density parts. However, we expect that the conditions under which a part is built will vary even if we set the parameters to certain fixed values. For example, the laser beam size may change as the optics get heated during use, or the porosity of the powder bed may vary depending on the distribution of powder particles in a layer, or the powder size distribution may vary with reuse, or the calibration of the laser power, speed, and beam size may change over time. All these variations will influence the properties of the part being built.

As we start using additively manufactured parts in situations where their failure could have serious consequences, it becomes important to quantify how much variation we can expect in the part resulting from variations in the inputs. This quantification of the uncertainty in the properties of a part, such as its density or dimensions, is an important step in qualification and certification of AM parts. Such issues are not restricted to experiments alone. In the case of simulations, it is not just the input parameters, such as laser power and beam size, that might vary. Other variables in the simulations, such as the material properties, may not be known precisely, or may be known within a certain range. We then need to understand the sensitivity of the simulation outputs to these variations in the inputs and material properties.

One approach to addressing these questions is to build many parts over time with the same set of parameters, and evaluate their properties, or run many simulations, varying the parameters over their expected range, and evaluate the range of output values. This can be prohibitively expensive, especially when the simulations are computationally intensive and measuring the properties of the parts is time consuming and labor intensive. An alternative is to use uncertainty analysis techniques from statistics and machine learning.[80] The application of these ideas in AM is at the very early stages, though they are being applied in other domains. We next present some preliminary thoughts on the ways in which we can address uncertainty issues in AM.

In Section 26.5.1.4, we used a regression tree surrogate model for predicting the depth of the melt pool, with the model built from Eagar–Tsai simulations run at various sample points in the design space. Another class of predictive models, referred to as Gaussian process models[94] provide not only a prediction, but also an associated uncertainty. These techniques can be applied to simulation data, where the simulations run at specific points are used to constrain the uncertainty in the predictions and experimental data, where the experiments, run at specific sample points, also have associated uncertainties that are used to evaluate the uncertainty at new sample points.

The idea of ensembles described in Section 26.5.1.4 can also be used to obtain an idea of the uncertainty due to the surrogate model. As each regression tree in the ensemble will provide a slightly different result, the spread of values will reflect how much we can expect the results to vary as we build regression models with slightly different training data.

26.6   Future applications

26.6.1   Powder model

The powder model is a powerful tool to study and optimize various aspects of the AM process. These include the following:

- Understanding the effects of laser power, speed, beam size, shape, and profile[95–97]
- Understanding the effects of powder size distribution and packing density
- Developing parameters for new materials
- Guiding development of alloys specifically engineered for AM
- Understanding the effects that control surface finish
- Understanding the effects that control *sparks* and *spatter*. (Spatter is molten metal droplets that are ejected from the melt pool.)

Using the powder model is much more economical in terms of time and cost compared with carrying out experiments in an AM system, particularly a commercial system where such studies may be prohibited by the manufacturer. With this capability, it is possible to help define the AM systems of the future.

26.6.2   Effective medium model

Proper effective medium models can be utilized to computationally explore a number of issues regarding part design and fabrication SLM processes:

- Predict the deformations occurring during fabrication and thus evaluate the possibility of process breakdown or out-of-tolerance end product
- Predict the residual stresses from fabrication and thus provide initial conditions for evaluating their impact on a design's intended performance
- Develop parameters that can be used to improve the quality of challenging configurations through examination of representative geometries such as the following:
- Unsupported downward-facing surfaces
- Thin walls
- Horizontal holes
- Vertical holes
- Unsupported overhangs
- Unsupported bridges
- Provide histories of local, configuration-specific temperatures, and temperature gradients to help assess the likely resulting material microstructures
- Permit full initial condition to assess the effects of post processes such as heat treatment
- Provide the basis for reduced order models that can be integrated with control systems on SLM machines. Part-scale modeling would also be the likely point-of-intersection with part design optimization, and one can envision the combining design and process identification under a multiobjective optimization framework.

26.6.3   Solving the inverse problem

Both the powder and effective medium models will be essential elements underpinning the solution of the inverse problem (solving the inverse problem involves "use of the actual results of some measurements of the observable parameters to infer the actual values of the model parameters."[98]). That is, specifying desired properties and using optimization to find the voxel-by-voxel parameters for building a part. Because of the long computational times required to compute a full part at high resolution using the physics-based models described here, an alternate approach is required to control the process better and identify the stable-operating regimes for qualification. This involves approximating the large-scale simulations with low-computational cost surrogates. As PBF is a complex process, it is necessary to identify the important *science* in the regimes of interest. By focusing on the important science and ignoring the less important, we can identify optimized build direction, optimized support structures, and implement an intelligent feed forward capability.

26.7   Summary and conclusions

26.7.1   Powder model

The powder model combines in three dimensions the laser-beam interaction with the powder with the thermomechanical response of the powder bed material through melt and eventual resolidification. The method incorporates as inputs many of the process parameters that may be varied for optimization of SLM for a particular material. The inclusion of many of the important physical processes in the metal's response to the laser beam allows valuable insights to be obtained into the physics of the SLM process. Outputs from the powder model can be used for inputs to other models, for example, temperature histories for microstructural modeling, or the outputs may be directly useable, such as surface roughness estimates. The powder model is an important part of the suite of tools needed to optimize SLM builds and eventually certify the fabricated parts for use.

26.7.2   Effective medium model

Effective medium models provide an abstraction to examine larger-scale phenomenon in a computationally tractable manner that can support the engineering design process in a timely manner. One should anticipate a continuing evolution of approaches, with alternatives building upon general commercial software, adapting in-house tools, and developing dedicated solutions. As analysis of real-world part configurations is the overall objective, and an important consideration will be forging the software links supporting transfers between the geometry specification, the machine scan specification, and the actual simulation. This breadth across the part scale will be complemented by more formal interchanges of information with the powder-scale model and eventual microstructure models, and its integration with formal process and design optimization.

26.7.3   Data mining and uncertainty quantification

Data mining techniques are being used to extract useful information from simulations and experiments, providing insights and efficiencies in building parts with desired properties. These techniques have been used to identify process parameters for high-density parts for a variety of materials. They can be combined with uncertainty analysis to understand how uncertainties in process parameters will affect the properties of a part, or uncertainties in input parameters and material properties used in a simulation will influence the output. When combined with experiments, techniques from data mining and uncertainty analysis will form an integral part of qualification and certification of the AM process.

Acknowledgments

This work was performed under the auspices of the U.S. Department of Energy by Lawrence Livermore National Laboratory (LLNL) under Contract No. DE-AC52-07NA27344. This work was funded by the Laboratory Directed Research and Development Program at LLNL under project tracking code 13-SI-002.

References

1. ASTM International, Standard Terminology for Additive Manufacturing Technologies (ASTM International, West Conshohocken, PA, 2010).
2. T. T. Wohlers, *Wohlers Report 2014 Additive Manufacturing and 3D Printing State of the Industry Annual Worldwide Progress Report* (Wohlers Associates, Fort Collins, CO, 2014).
3. S. Srivatsa, *Additive Manufacturing (AM) Design and Simulation Tools Study* (Air Force Research Laboratory, Wright-Patterson AFB, OH, 45433, 2014).
4. Knowledge Based Process Planning and Design for Additive Layer Manufacturing (KARMA), Detailed report on Laser Cusing, SLA, SLS and Electron Beam Melting (including technical, economical and safety features) (Valencia (ESPA~NA), Report No. DL 1.1, 2011.
5. T. Craeghs, A monitoring system for on-line control of selective laser melting, PhD thesis, Department of Mechanical Engineering, University of Leuven, Heverlee, Belgium, 2012.
6. E. Yasa, Manufacturing by Combining Selective Laser Melting and Selective Laser Erosion/ Laser Re- Melting, **2011D01**, PhD thesis, Faculty of Engineering, Department of Mechanical Engineering, Katholieke Universiteit Leuven, Heverlee (Leuven), 2011.
7. L. Adams, *MICRO Manufacturing* (Don Nelson, Northfield, IL, 2013), Vol. September/October, p. 44.
8. PWC, *3D Printing and the New Shape of Industrial Manufacturing* (PricewaterhouseCoopers LLP, DE, 2014).
9. J. Coykendall, M. Cotteleer, J. Holdowsky, and M. Mahto, *3D Opportunity in Aerospace and Defense* (Deloitte University Press, Washington, DC, 2014).
10. D. L. Bourell, M. C. Leu, and D. W. Rosen, *Roadmap for Additive Manufacturing Identifying the Future of Freeform Processing* (The University of Texas at Austin, Austin TX, 2009).
11. Energetics Incorporated, *Measurement Science Roadmap for Metal-Based Additive Manufacturing* (Energetics Incorporated, Columbia, MD, 2013).
12. R. Berger, *Additive Manufacturing: A Game Changer for the Manufacturing Industry?* (Roland Berger Strategy Consultants GmbH, Munich, Germany, 2013).
13. I. Yadroitsev, *Selective Laser Melting: Direct Manufacturing of 3DObjects by Selective Laser Melting of Metal Powders* (LAP Lambert Academic Publishing, Saarbr€ucken, 2009).
14. J. P. Kruth, P. Mercelis, J. Van Vaerenbergh, and T. Craeghs, *Feedback Control of Selective Laser Melting* (Taylor & Francis Ltd, London, 2008), p. 521.
15. T. Craeghs, F. Bechmann, S. Berumen, and J. P. Kruth, in *Proceedings of the Laser Assisted Net Shape Engineering 6* (LANE 2010), Part 2, edited by M. Schmidt, F. Vollertsen, and M. Geiger (Elsevier Science BV, Amsterdam, the Netherlands, 2010), 5, 505.
16. T. Craeghs, S. Clijsters, J.-P. Kruth, F. Bechmann, and M.-C. Ebert, *Phys. Proc.* 39, 753 (2012).
17. S. Clijsters, T. Craeghs, and J. P. Kruth, *A priori Process Parameter Adjustment for SLM Process Optimization* (CRC Press-Taylor & Francis Group, Boca Raton, FL, 2012), 553.
18. W. Frazier, *J. Mater. Eng. Perform.* 23(6), 1917 (2014).
19. See http://www.realizer.com/rdesigner for information on a feed forward capability.
20. J. D. Williams and C. R. Deckard, *Rapid Prototyping J.* 4(2), 90 (1998).

21. S. Kolossov, E. Boillat, R. Glardon, P. Fischer, and M. Locher, *Int. J. Mach. Tools Manuf.* 44(2–3), 117 (2004).
22. K. Dai, X. X. Li, and L. L. Shaw, *Rapid Prototyping J.* 10(1), 24 (2004).
23. I. A. Roberts, C. J. Wang, R. Esterlein, M. Stanford, and D. J. Mynors, *Int. J. Mach. Tools Manuf.* 49(12–13), 916 (2009).
24. N. Contuzzi, S. Campanelli, and A. D. Ludovico, *Int. J. Simul. Model.* 10(3), 113 (2011).
25. M. Matsumoto, M. Shiomi, K. Osakada, and F. Abe, *Int. J. Mach. Tools Manuf.* 42(1), 61 (2002).
26. A. Hussein, L. Hao, C. Yan, and R. Everson, *Mater. Des.* 52, 638 (2013).
27. M. Zaeh and G. Branner, *Prod. Eng.* 4(1), 35 (2010).
28. P. Fischer, V. Romano, H. P. Weber, N. P. Karapatis, E. Boillat, and R. Glardon, *Acta Mater.* 51(6), 1651 (2003).
29. N. K. Tolochko, M. K. Arshinov, A. V. Gusarov, V. I. Titov, T. Laoui, and L. Froyen, *Rapid Prototyping J.* 9(5), 314 (2003).
30. A. V. Gusarov and I. Smurov, *Phys. Proc.* 5, 381 (2010).
31. W. E. King, H. D. Barth, V. M. Castillo, G. F. Gallegos, J. W. Gibbs, D. E. Hahn, C. Kamath, and A. M. Rubenchik, *J. Mater. Process. Technol.* 214(12), 2915 (2014).
32. N. K. Tolochko, T. Laoui, Y. V. Khlopkov, S. E. Mozzharov, V. I. Titov, and M. B. Ignatiev, *Rapid Prototyping J.* 6(3), 155 (2000).
33. A. V. Gusarov and J. P. Kruth, *Int. J. Heat Mass Transfer* 48(16), 3423 (2005).
34. X. C. Wang, T. Laoui, J. Bonse, J. P. Kruth, B. Lauwers, and L. Froyen, *Int. J. Adv. Manuf. Technol.* 19(5), 351 (2002).
35. C. Boley, S. Khairallah, and A. Rubenchik, *Appl. Opt.* 54(9), 2477 (2015).
36. FRED is distributed by *Photon Engineering*, LLC, Tucson, AZ.
37. E. D. Palik, *Handbook of Optical Constants of Solids* (Academic Press, Orlando, FL, 1985).
38. G. Friedman, *ParticlePack User's Manual* (Lawrence Livermore National Laboratory, Livermore, CA, 2011).
39. P. Meakin and R. Jullien, *J. Phys.* 48(10), 1651 (1987).
40. E. J. R. Parteli, in *Powders and Grains 2013*, edited by A. Yu, K. Dong, R. Yang et al. (American Institute of Physics, Melville, NY, 2013), 1542, 185.
41. I. Kovaleva, O. Kovalev, and I. Smurov, *Phys. Proc.* 56(0), 400 (2014).
42. E. J. R. Parteli and T. P€oschel, *Powder Technol.* 288, 96 (2016).
43. R. M. Kelkar, T. Anderson, P. Wang, and D. Bartosik, *DMLM: Effect of Bi-Modal Particle Size Distribution on Surface Finish* (Additive Manufacturing with Powder Metallurgy, 2014).
44. A. B. Hopkins, F. H. Stillinger, and S. Torquato, *Phys. Rev. E* 88(2), 022205 (2013).
45. A. M. Prokhorov, V. I. Konov, I. Ursu, and N. Mihailescu, *Laser Heating of Metals* (A. Hilger, Bristol, 1990).
46. A. M. Rubenchik, S. S. Q. Wu, V. K. Kanz, M. M. LeBlanc, W. H. Lowdermilk, M. D. Rotter, and J. R. Stanley, *Opt. Eng.* 53(12), 122506 (2014).
47. R. W. McVey, R. M. Melnychuk, J. A. Todd, and R. P. Martukanitz, *J. Laser Appl.* 19(4), 214 (2007).
48. R. P. Martukanitz, R. M. Melnychuk, M. S. Stefanski, and S. M. Copley, *J. Laser Appl.* 19(4), 214 (2007).
49. A. Rubenchik, S. Wu, S. Mitchell, I. Golosker, M. LeBlanc, and N. Peterson, *Appl. Opt.* 54(24), 7230 (2015).
50. V. Y. Bodryakov, *High Temp.* 51(2), 206 (2013).
51. See http://app.knovel.com/hotlink/toc/id:kpRVTPSCA1/recommendedvalues-offor recommended values of thermophysical properties for selected commercial alloys.
52. I. Yadroitsev, A. Gusarov, I. Yadroitsava, and I. Smurov, *J. Mater. Process. Technol.* 210(12), 1624 (2010).
53. A. V. Gusarov and E. P. Kovalev, *Phys. Rev. B* 80(2), 024202 (2009).
54. M. Rombouts, L. Froyen, A. V. Gusarov, E. H. Bentefour, and C. Glorieux, *J. Appl. Phys.* 97(2), 013533 (2005).
55. S. I. Anisimov, *High Temp.* 6(1), 110 (1968).
56. S. I. Anisimov, *Sov. Phys. JETP-USSR* 27(1), 182 (1968).
57. C. J. Knight, *AIAA J.* 17(5), 519 (1979).
58. A. Klassen, T. Scharowsky, and C. Korner, *J. Phys. D: Appl. Phys.* 47(27), 275303 (2014).

59. S. I. Anisimov and V. A. Khokhlov, *Instabilities in Laser-Matter Interaction* (CRC Press, Boca Raton, FL, 1995).

60. R. Rai, J. W. Elmer, T. A. Palmer, and T. DebRoy, *J. Phys. D: Appl. Phys.* 40(18), 5753 (2007).

61. G. G. Gladush and I. Smurov, *Physics of Laser Materials Processing: Theory and Experiment* (Springer, Berlin, Germany, 2011).

62. D. Steinberg, Lawrence Livermore National Laboratory, Report No. UCRL-MA-106439, 1996.

63. C. K€orner, E. Attar, and P. Heinl, *J. Mater. Process. Technol.* 211(6), 978 (2011).

64. C. K€orner, A. Bauereiß, and E. Attar, *Modell. Simul. Mater. Sci. Eng.* 21(8), 085011 (2013).

65. F. J. Gurtler, M. Karg, K. H. Leitz, and M. Schmidt, in *Lasers in Manufacturing*, edited by C. Emmelmann, M. F. Zaeh, T. Graf et al. (Elsevier Science BV, Amsterdam, the Netherlands, 2013), 41, 874.

66. R. Ganeriwala and T. I. Zohdi, *Proc. CIRP* 14, 299 (2014).

67. A. V. Gusarov, I. Yadroitsev, P. Bertrand, and I. Smurov, *J. Heat Transfer* 131(7), 072101 (2009).

68. M. F. Zaeh, G. Branner, and T. A. Krol, in *Innovative Developments in Design and Manufacturing: Advanced Research in Virtual and Rapid Prototyping*, edited by P. J. D. Bartolo, A. C. S. DeLemos, A. M. H. Pereira et al. (CRC Press-Taylor & Francis Group, Boca Raton, FL, 2010), p. 415.

69. N. E. Hodge, R. M. Ferencz, and J. M. Solberg, *Comput. Mech.* 54(1), 33 (2014).

70. J. A. Goldak and M. Akhlaghi, *Computational Welding Mechanics* (Springer, New York, 2005).

71. L. E. Lindgren, *Computational Welding Mechanics: Thermomechanical and Microstructural Simulations* (Woodhead and Maney Pub., Cambridge, UK; CRC Press, Boca Raton, FL, 2007).

72. P. Michaleris, *Finite Elem. Anal. Des.* 86, 51 (2014).

73. E. R. Denlinger, J. Irwin, and P. Michaleris, *J. Manuf. Sci. Eng.* 136(6), 061007 (2014).

74. J. M. Solberg, N. E. Hodge, R. M. Ferencz, I. D. Parsons, M. A. Puso, M. A. Havstad, R. A. Whitesides, and A. P. Wemhoff, Diablo User Manual, Livermore, CA, Report No. LLNL-SM-651163, 2014.

75. J. Schilp, C. Seidel, H. Krauss, and J. Weirather, *Adv. Mech. Eng.* 6, 217584 (2014).

76. A. S. Wu, D. W. Brown, M. Kumar, G. F. Gallegos, and W. E. King, *Metall. Mater. Trans.* A 45(13), 6260–6270 (2014).

77. C. Seidel, M. F. Zaeh, M. Wunderer, J. Weirather, T. A. Krol, and M. Ott, *Proc. CIRP* 25, 146 (2014).

78. D. Pal, N. Patil, K. Zeng, and B. Stucker, *J. Manuf. Sci. Eng.-Trans. ASME* 136(6), 061022 (2014).

79. D. Pal, N. Patil, and B. E. Stucker, paper presented at the *International Solid Freeform Fabrication Symposium–An Additive Manufacturing Conference* (Austin, TX, 2012).

80. National Research Council, *Assessing the Reliability of Complex Models: Mathematical and Statistical Foundations of Verification, Validation, and Uncertainty Quantification* (The National Academies Press, Washington, DC, 2012).

81. I. Yadroitsev and I. Smurov, *Phys. Proc. Part B* 5, 551 (2010).

82. J. P. Kruth, M. Badrossamay, E. Yasa, J. Deckers, L. Thijs, and J. Van Humbeeck, paper presented at the *16th International Symposium on Electromachining* (ISEM XVI, Shanghai, China, 2010).

83. K. Kempen, L. Thijs, E. Yasa, M. Badrossamay, W. Verheecke, and J.-P. Kruth, in *Twenty Third Annual International Solid Freeform Fabrication Symposium—An Additive Manufacturing Conference*, edited by D. Bourell (University of Texas at Austin, Austin, TX, 2011), p. 484.

84. G. W. Oehlert, *A first Course in Design and Analysis of Experiments* (W. H. Freeman, New York, 2000).

85. K. Fang, R.-z. Li, and A. Sudjianto, *Design and Modeling for Computer Experiments* (Chapman & Hall/CRC, Boca Raton, FL, 2006).

86. J. Delgado, L. Sereno, J. Ciurana, and L. Hernandez, *Methodology for Analyzing the Depth of Sintering in the Building Platform* (CRC Press-Taylor & Francis Group, Boca Raton, FL, 2012), p. 495.

87. C. Kamath, B. El-dasher, G. F. Gallegos, W. E. King, and A. Sisto, *Int. J. Adv. Manuf. Technol.* 74(1–4), 65 (2014).

88. T. W. Eagar and N. S. Tsai, *Weld. J.* 62(12), S346 (1983).

89. D. C. Montgomery, *Design and Analysis of Experiments* (John Wiley & Sons, Hoboken, NJ, 2004).

90. R. Bridson, in *ACM SIGGRAPH 2007 Sketches* (ACM, San Diego, CA, 2007), p. 22.

91. C. Kamath, *Scientific Data Mining: A Practical Perspective* (SIAM, Philadelphia, 2009).
92. M. A. Hall, *Correlation-Based Feature Selection for Discrete and Numeric Class Machine Learning* (Department of Computer Science, University of Waikato, Hamilton, New Zealand, 2000).
93. L. Breiman, *Classification and Regression Trees* (Wadsworth International Group, Belmont, CA, 1984).
94. C. E. Rasmussen and C. K. I. Williams, *Gaussian Processes for Machine Learning* (MIT Press, Cambridge, MA, 2006).
95. A. Okunkova, M. Volosova, P. Peretyagin, Y. Vladimirov, I. Zhirnov, and A. V. Gusarov, *Phys. Proc.* 56, 48 (2014).
96. A. Okunkova, P. Peretyagin, Y. Vladimirov, M. Volosova, R. Torrecillas, and S. V. Fedorov, *Proc. SPIE* 9135, 913524 (2014).
97. I. V. Zhirnov, P. A. Podrabinnik, A. A. Okunkova, and A. V. Gusarov, *Mech. Ind.* 16, 709 (2015).
98. A. Tarantola, *Inverse Problem Theory and Methods for Model Parameter Estimation* (SIAM, Philadelphia, PA, 2005).

chapter twenty seven

Calculation of laser absorption by metal powders in additive manufacturing

C. D. Boley, Saad A. Khairallah, and Alexander M. Rubenchik

Contents

We have calculated the absorption of laser light by a powder of metal spheres, typical of the powder employed in laser powder-bed fusion (LPBF) additive manufacturing (AM). Using ray-trace simulations, we show that the absorption is significantly larger than its value for normal incidence on a flat surface, because of multiple scattering. We investigate the dependence of absorption on powder content (material, size distribution, and geometry) and on beam size. OCIS codes: (1) (080.2710) Inhomogeneous optical media, (2) (080.5692) Ray trajectories in inhomogeneous media, and (3) (160.1245) artificially engineered materials.

27.1 Introduction

AM is a fast-growing technology for building the parts of a device [1]. In selective laser melting, the layers of a metal powder are melted in a controlled manner, forming successive slices of a part. This process is characterized by a number of parameters, including the powder material, the layer thickness and porosity, the laser beam size and profile, and the laser scan speed. Reliable process modeling is very useful in order to determine the optimal parameters and to anticipate possible problems in the build process.

An important component of modeling efforts is the description of the absorption of the metal powder and of the spatial distribution of the absorbed radiation. Direct measurements of the absorption are quite difficult [2]. Also, it is problematic to make use of measurements obtained without detailed specifications of the experiment, because the absorption depends on the parameters noted above, along with the distribution of particle sizes and the spatial distribution of the particles. Thus it is not sufficient to know the results for one particular powder of a given material and for a particular beam, as we will demonstrate below.

Similarly, the spatial distribution of absorbed radiation is difficult to obtain experimentally. These considerations reinforce the usefulness of absorption calculations.

A commonly used laser absorption model, proposed in [3], assumes diffusive radiation transport in the powder. This assumption, however, is not applicable for the thin, low-porosity metal powder layer used in the selective laser melting process, for which the thickness is a few powder particles. As we shall see, in this case the energy is typically absorbed in the top layer, and the absorption is highly nonuniform. These circumstances are inconsistent with a diffusion model.

Physically, the powder is an assembly of metal particles, taken here to be spheres, with sizes appreciably larger than the laser wavelength (taken as about 1 μm) and with a complex refractive index appropriate to the material and the wavelength. It is natural to use ray tracing to calculate the powder absorption. This has previously been considered, for example, in [4], but the angular and polarization dependence of the absorption of incident rays was neglected.

In the present chapter, we report the results of comprehensive absorption modeling, including all the effects mentioned above. A major challenge is the problem of tracing rays within an assembly of thousands of objects, while keeping track of the angle, polarization, power, and reflection/refraction of individual rays. However, this issue has long been the subject of study, and commercial software is available for handling it. Here we utilize the product FRED [5], a multipurpose optics code widely used in optical design and analysis. In our application, which differs from typical applications, we make extensive use of its ray-trace capability. In order to handle our problem, substantial scripting and postprocessing was required. Previously, we employed FRED in the similar problem of laser interactions with composite materials [6].

To begin the calculations, we consider a powder consisting of spheres of a single size, densely packed in a hexagonal structure. Six materials (Ag, Al, Au, Cu, stainless steel, and Ti) are considered. We first study the overall absorptivity of such a powder, by assuming a uniform beam of width large compared with the particle size, so that the absorption is nearly independent of the beam position. The calculations show that the resulting powder absorptivity is significantly higher than the absorptivity of a flat surface or of a single, isolated sphere, thus confirming the important role of multiple scattering. We demonstrate that most of the energy is absorbed in the top layer of the array.

A real powder has a distribution of sizes and is not densely packed. Therefore we use a particle-packing program [7] to set up a powder layer. The algorithm is similar to that of the rain-packing model [8]. Specific calculations are performed for two powders. The first powder is that used in the Concept Laser M2 metal powder bed fusion (PBF) AM machine [9]. Some experimental results obtained recently at Lawrence Livermore National Laboratory (LLNL, U.S. Department of Energy) with this device are described in [10]. For some specific calculations we use the parameters from these experiments. For the second powder, we consider a bimodal particle distribution that provides a higher powder packing density. We demonstrate that this can lead to a significant increase in the powder absorption, especially for highly reflective materials.

For a laser beam width comparable with the typical powder sphere size (a typical situation for the Concept Laser machine), the absorption is sensitive to the beam position. Calculating the absorption pattern along a track through the powder, we show that the absorption fluctuates noticeably along the path on a distance scale appreciably larger than the particle size.

Finally, we summarize the results and discuss their impact on the AM process.

27.2   *Absorptivity and ray-tracing calculations*

In practical applications, the typical particle radius (generally at least 10 μm) appreciably exceeds the laser wavelength, and ray tracing is applicable for the description of the interaction. For a beam striking the surface of a dielectric material at an angle θ to the normal, the absorptivity is given by the Fresnel formulas [11]:

$$\alpha_s(\theta) = 1 - \left| \frac{\cos\theta - \left(n^2 - \sin^2\theta\right)^{\frac{1}{2}}}{\cos\theta + \left(n^2 - \sin^2\theta\right)^{\frac{1}{2}}} \right|^2 \tag{27.1}$$

$$\alpha_p(\theta) = 1 - \left| \frac{n^2\cos\theta - \left(n^2 - \sin^2\theta\right)^{\frac{1}{2}}}{n^2\cos\theta + \left(n^2 - \sin^2\theta\right)^{\frac{1}{2}}} \right|^2 \tag{27.2}$$

in which the electric field is either perpendicular (S) or parallel (P) to the plane of incidence, and n is the complex index of refraction of the material (divided by the index of refraction of the external medium, which we take as unity). A general polarization can be expressed as a combination of S and P.

Figure 27.1 shows the absorptivity of stainless steel (SS), which has a refractive index $n = 3.27 + 4.48i$ at a wavelength of 1 μm [12], as a function of the incident angle. At perpendicular incidence, the absorptivity of each polarization is about 0.34. As the angle is increased, the absorptivity decreases smoothly for S polarization, whereas it increases to a maximum of about 0.75 at 80 degrees for P polarization. The absorptivity is greater in the latter case because the electric field has a component directed into the material.

For a large, uniform beam incident on an isolated sphere, the absorptivity can readily be calculated and has the pattern shown in Figure 27.2. In this case, the angle and the power split between S and P depend on the point of incidence on the sphere.

In our ray tracing, a ray is tracked from surface to surface. It has a particular power in each polarization state. After an interaction, the reflected ray either strikes a neighboring

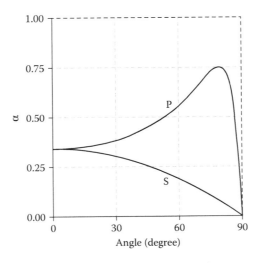

Figure 27.1 Absorptivity of 1 μm light incident on stainless steel.

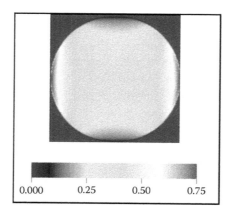

Figure 27.2 Absorptivity of a beam by a single sphere. The beam is polarized horizontally. The absorptivity falls to zero all along the edge, although this is not resolved in the graphic.

surface (sphere or substrate), or exits the system. In the former case, the refracted ray deposits power within the sphere or substrate, and this deposited power is not followed. After every reflection, therefore, the power of a ray decreases. We stop propagation of a ray when its power drops to less than 0.1% of its initial power. The number of rays varied from 50,000 to a few million, depending on the specific problem. This number was always chosen sufficiently high so that the results were insensitive to the specific choice.

27.3   Ideal powder array

We now turn to a powder of identical spheres, assuming ideal packaging, that is, hexagonal close-packing. We consider two layers of spheres, resting on a substrate of the same material. This is similar to the actual setup in LPBF AM. Calculations of the absorptivity were performed for several metals, illuminated perpendicularly from above, as shown in Figure 27.3. The refractive indices near a wavelength of 1 µm were taken from a data compilation [12].

First, calculations were performed for a uniform circular beam having a radius much larger than the radius of a sphere. The results, which clearly are independent of the particular sizes, are summarized in Table 27.1. Most important for each metal is the total

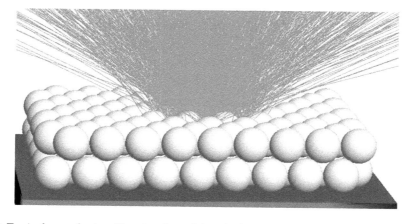

Figure 27.3 Typical rays during illumination of the ideal array.

Table 27.1 Absorption details (α denotes the absorptivity)

(1) Material	(2) Re(n)	(3) Im(n)	(4) α (flat surface)	(5) α (isolated sphere)	(6) α (top layer)	(7) α (bottom layer)	(8) α (substrate)	(9) α (spheres + substrate)	(10) α (spheres + substrate)/ α (flat surface)
Ag	0.23	7.09	0.018	0.020	0.072	0.047	0.010	0.13	7.2
Al	1.244	10.	0.047	0.056	0.15	0.063	0.011	0.22	4.7
Au	0.278	7.20	0.021	0.024	0.081	0.050	0.011	0.14	6.7
Cu	0.35	6.97	0.028	0.032	0.101	0.055	0.011	0.17	6.1
SS	3.27	4.48	0.34	0.36	0.53	0.062	0.013	0.60	1.7
Ti	3.45	4.	0.38	0.40	0.56	0.062	0.014	0.64	1.7

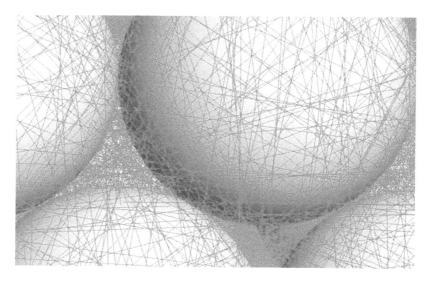

Figure 27.4 Detail of ray trajectories in Figure 27.3, showing multiple scattering from spheres.

absorptivity by the spheres and the substrate (column 9). This is to be compared with the absorptivity of the metal at normal incidence on a flat surface (column 4), and the average absorptivity of an isolated sphere illuminated by a uniform beam (column 5).

Note that the most of the power is absorbed in the top layer of the spheres (column 6). Little more than 1% of the power penetrates beneath the two layers to the substrate (column 8).

We see that the total absorptivity of the spheres is noticeably higher than either the normal-incidence value or the single-sphere value. This effect was observed in experiment [2]. The enhancement is due to multiple scattering, as illustrated in Figure 27.4. A ray can scatter repeatedly, leading to additional absorption relative to the case of a flat surface. Thus the relative increase in absorptivity is higher for highly reflective metals (Ag, Al, Au, Cu) than for moderately absorbing metals (SS and Ti). In the former case, this ratio (column 10) varies from 4.7 to 7.2, whereas in the latter case, the ratio decreases to 1.7.

More generally, one is interested in not only the total absorbed power but also the spatial distribution of the absorbed power. In some AM machines, the laser beam size is roughly comparable to the powder particle size. Here we consider a powder with spheres of radius 10 μm and a beam having a $1/e^2$ radius of 24 μm. From now on, the radius of a Gaussian beam always refers to the $1/e^2$ radius.

Figure 27.5 shows the distribution of absorbed irradiance along the top layer of a SS array, as the beam is rapidly scanned across the array. This distribution was obtained by calculating the absorbed irradiance patterns at a number of points along the path, and taking the average. It gives a qualitative picture of the absorbed irradiance on a timescale short compared to thermomechanical times, that is, for a sufficiently fast scanning speed. We see that the scattered light is well confined and that the typical absorption area is comparable to the beam area.

The absorptivity is sensitive to the beam size, and fluctuations of the absorptivity are smoothed with increasing beam size. As shown in Figure 27.6, the absorption fluctuates by about 20% for a Gaussian beam of radius 8 μm (80% of the particle radius), and by less than 0.1% when the radius is increased to 24 μm. The fluctuations become negligible for a beam

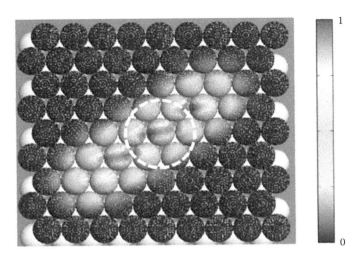

Figure 27.5 Irradiances (arbitrary scale) for 61 successive beam positions, from lower left to upper right, in steps of 2 μm. The irradiances pertain to the spherical surfaces. A sample beam spot (1/e^2 radius) is shown. The radius of a sphere is 10 mm.

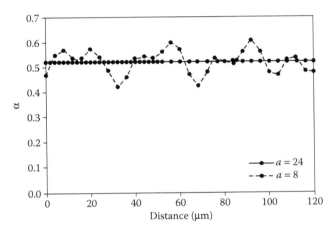

Figure 27.6 Spatial variations of the absorption along the beam path for beams of different radii.

radius of about twice the particle radius. Nevertheless, even in this case the distribution of absorbed power on a single sphere is very nonuniform.

27.4   Realistic powder array

A realistic powder has a distribution of sizes and a nonuniform geometrical arrangement, generally with a porosity greater than that of an ideal array.

To generate the powder geometry, we used a particle-packing program [7] with an algorithm similar to that of the rain model for random deposition [8]. The program randomly places powder particles, with a specified distribution of sizes, on a powder bed, up to the first contact with other particles or with the substrate. If the contact is with a

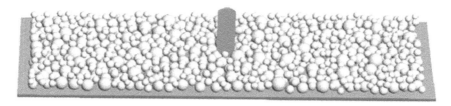

Figure 27.7 Powder with a Gaussian distribution of sizes. The length of the bed is about 1100 µm, and the beam path lengthwise.

particle, the particle is randomly perturbed, in an effort to achieve the minimum potential energy due to gravity. Finally, to simulate the removal of extra powder by a coater blade, the program inserts a plane at a specified distance from the substrate and removes all particles intersected by the plane or situated above it.

We consider two specific powder types: the first, shown in Figure 27.7, mimics the powder used in the Concept Laser machine [9]. The powder has a Gaussian distribution of radii, with an average radius of 13.5 µm, a full width at half maximum equal to 2.3 µm, radial cutoffs at 8.5 µm and 21.5 µm, and a powder thickness of 43 µm [13]. In the absorption calculations, the path of the beam extends along the length of the powder bed, as shown in the figure.

The SS absorption results encountered along the path are shown in Figure 27.8. We see that local variations in the powder structure give rise to sizeable fluctuations in the absorption. The fluctuations occur on a scale of about 100 µm, which is much larger than the typical sphere size. The mechanism for the fluctuations can be seen in the two inserts in Figure 27.8.

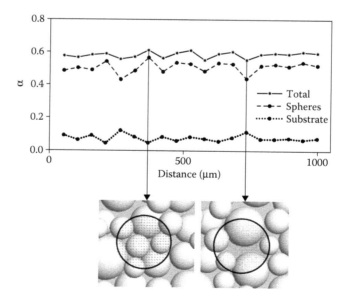

Figure 27.8 Absorption α as calculated along the beam path for the Gaussian powder of Figure 27.7. The material is stainless steel. The inserts show the powder and incident beam size ($1/e^2$) at locations with high absorption (left) and low absorption (right).

In the left insert, the incident beam has mainly struck small spheres, with larger spheres on the periphery. This results in multiple reflections and an increased absorption. In the right insert, on the other hand, much of the incident power has reached the substrate, producing fewer reflections and a decreased absorption. In detail, the absorption of the spheres alone has an average of about 0.50 with a standard deviation of about 0.07, whereas the total absorption (spheres plus substrate) has an average of 0.58 with a standard deviation of 0.03. The reason for the decrease in fluctuations of the total is that the contribution from the substrate tends to cancel that from the spheres (the spheres shield the substrate).

For the second example of a powder, shown in Figure 27.9, we consider a bimodal distribution characterized by a 7:1 ratio of radii and a volume fraction of small spheres equal to 20%, as discussed in [14]. This powder was chosen because of its high density, or low porosity. Following [14], we consider a large-sphere radius of 42 μm and a powder thickness of 50 μm.

Figure 27.10 shows the calculated absorption for SS along a 1 mm laser beam path. In this configuration, holes in the powder layer are practically absent. The absorption minima correspond to situations when the beam mainly strikes a large sphere, with much of the light directly reflected (left insert). The largest absorption occurs when the beam strikes a local assembly of small spheres, as seen in the right insert. The difference between these two cases lies in the ratio of the beam size to the size of the irradiated spheres, with a

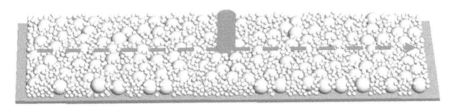

Figure 27.9 Powder with a bimodal distribution of sizes. The powder bed and the beam are as in Figure 27.7.

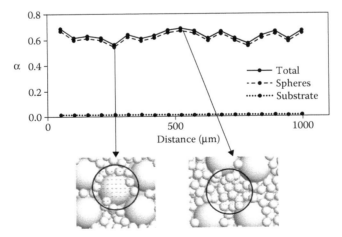

Figure 27.10 Absorption as calculated along the beam path for the bimodal powder of Figure 27.9 (stainless steel). The inserts show the powder and incident beam size at locations with low absorption (left) and high absorption (right).

Table 27.2 Total absorptivities for selected materials

Material	Ideal array (Table 27.1)	Gaussian array	Bimodal array
Ag	0.13	0.081	0.14
Au	0.14	0.093	0.16
SS	0.60	0.58	0.63

larger ratio offering more opportunity for multiple reflections. As in the previous case, the absorption fluctuates on a distance scale larger than a particle size, or about 100 μm.

Parenthetically, it should be noted that the problem of a powder structure producing a maximum density has been investigated in a number of studies, for example, in reference [15]. The structure is not completely disordered, because it includes both regular cells and long-distance correlations. It is not known whether such a structure can be reproduced with the packaging algorithm used here.

Returning to the Gaussian and bimodal powders, let us compare the overall results with those for the ideal powder of the previous section 27.4. The results are summarized in Table 27.2. We see that a change in the powder structure can noticeably affect the absorptivity. For a moderately absorbing metal such as SS, the difference is not large, about a few percent. On the other hand, for highly reflective metals such as silver and gold, the variation can be nearly a factor of 2. In these cases, multiple scattering is very important, and the powder geometry affects the total absorptivity.

27.5 Discussion and conclusions

We have developed a detailed ray-trace model that can be used to calculate the absorption and energy deposition in a metal powder, starting with the optical parameters of the constituents and the powder structure. We have found that the absorption is substantially increased relative to its flat-surface value because of multiple scattering. The effect is especially important for highly reflective metals, such as those used in the AM of jewelry. We demonstrated that, by optimization of the powder layer structure, one can increase the absorption by nearly a factor of 2.

Let us discuss the importance of the localized energy deposition. There are two general issues regarding absorption nonuniformity, one related to the nonuniformity of absorption within a single particle, and the other related to the nonuniformity of absorption on a larger scale given by the final beam size and the local structure of the actual powder.

Consider first the former issue. The time for homogenization of energy absorbed nonuniformly on a sphere with radius R, due to thermal conduction, is $\tau_c = R^2/D$, where D is the thermal diffusivity of the metal. Another typical time is the time needed to melt the material, or $\tau_m = RH_m/\alpha_o I$. Here H_m is the melting enthalpy per volume, I is the laser irradiance, and a_0 is the flat-surface absorptivity. For processing steel with a Concept Laser AM machine, one has $R \sim 10$ μm, $D \sim 0.04$ cm^2/s, ~8 kJ/cm^3, $\alpha_o \sim 0.3$, and $I \sim 10$ MW/cm^2. Therefore the thermal diffusion time is about 25 μs, and the melting time is shorter by nearly a factor of 10. This means that nonuniformity of absorption results in only partial melting of the particle. It can be shown that melt penetrates to the substrate more rapidly due to wetting and capillary forces [16]. Unmelted particle pieces can produce residual voids (e.g., between the particle and substrate) and defects.

This result differs from the conventional model [3] that assumes uniform, volumetric deposition. However, for lower intensities and materials with higher thermal conductivity, when the thermal diffusion rapidly homogenizes the temperature within the particle, our deposition model and the model [3] can produce similar results for the same deposited energy.

Regarding the issue of large-scale variability of absorption, we note that nonuniformity on the scale of 100 μm can produce fluctuations of the melt pool size and can explain the track modulation typically observed in experiments, for example in reference [10].

In conclusion, the fact that multiple scattering plays an important role means that the absorption value is strongly affected by the size distribution of the powder spheres and their geometrical arrangement. The nonuniformity of energy deposition affects the melt dynamics. Thus control of the powder structure can be an important tool for optimization of the laser PBF AM process, and ray-trace modeling is an effective method for achieving this control.

Acknowledgments

We thank W. E. King and W. A. Molander for helpful discussions. This work was performed under the auspices of the U.S. Department of Energy by Lawrence Livermore National Laboratory (LLNL) under Contract DE-AC52-07NA27344. This work was funded by the Laboratory Directed Research and Development Program at LLNL under project tracking code 13-SI-002.

References

1. T. Wohlers, Wohlers Report 2014. *3D Printing and Additive Manufacturing State of the Industry. Annual Worldwide Progress Report*, Wohlers Associates, Fort Collins, CO (2014).
2. N. Tolochko et al., Absorptance of powder materials suitable for laser sintering, *Rapid Prototyping J.* 6, 155 (2000).
3. A. V. Gusarov and J.-P. Kruth, Modelling of radiation transfer in metallic powders at laser treatment, *Int. J. Heat Mass Transfer* 48, 3423–3434 (2005).
4. X. C. Wang et al., Direct selective laser sintering of hard metal powders: Experimental study and simulation, *Int. J. Adv. Manuf. Technol.* 19, 351–357 (2002).
5. FRED is distributed by Photon Engineering, LLC, Tucson, AZ.
6. C. D. Boley and A. M. Rubenchik, Modeling of laser interactions with composite materials, *Appl. Opt.* 52, 3329–3337 (2013).
7. G. Friedman, ParticlePack user's manual, Lawrence Livermore National Laboratory, LLNL-SM-458031 (2011).
8. P. Meakin and R. Jullien, Restructuring effects in the rain model for random deposition, *J. Physique* 48, 1651–1662 (1987).
9. Concept Laser GmbH (www.concept-laser.de/en/home.html).
10. W. E. King, H. D. Barth, V. M. Castillo, G. F. Gallegos, J. W. Gibbs, D. E. Hahn, C. Kamath, and A. M. Rubenchik, Observation of keyhole-mode laser melting in laser powder-bed fusion additive manufacturing, *J. Mater. Process. Technol.* 214, 2915–2925 (2014).
11. L. D. Landau, E. M. Lifshitz, and L. P. Pitaevskii, *Electrodynamics of Continuous Media*, 2nd edition, Pergamon Press, London 1984.
12. E. D. Palik, ed., *Handbook of Optical Constants of Solids*, (Academic Press, Orlando, FL, 1985).
13. Powder CL 20ES, produced by Concept Laser GmbH (Ref. 9).
14. R. Kelkar, Effect of metal powder particle-size distribution on surface roughness for CoCrMo parts manufactured via direct metal laser melting (DMLM) process, *Additive Manufacturing with Powder Metallurgy*, Orlando, FL (2014).
15. A. B. Hopkins, F. H. Stillinger, and S. Torquato, Disordered strictly jammed binary sphere packings attain an anomalously large range of densities, *Phys. Rev. E* 88, 022205 (2013).
16. S. A. Khairallah and A. Anderson, Mesoscopic simulation model of selective laser melting of stainless steel powder, *J. Mater. Process. Technol.* 214, 2627–2636 (2014).

chapter twenty eight

The accuracy and surface roughness of spur gears processed by fused deposition modeling additive manufacturing

Junghsen Lieh, Bin Wang, and Omotunji Badiru

Contents

Abstract: The objective of this article is to study and develop an understanding of accuracy and roughness of spur gears created by additive manufacturing (AM or 3D printing) techniques. A number of ABS gears were printed and their corresponding steel gears were purchased. Geometries of these gears were measured and compared with those calculated with theoretical formulas. In order to observe the roughness of these gears, surfaces of these parts are photographed and the peak-to-valley heights of surface profile were measured.

28.1   Introduction

Since the merger of stereolithography (SL) process to produce three-dimensional (3D) parts in 1987 (SL is a laser-heated process capable of solidifying thin layers of UV-sensitive liquid polymer), AM has gradually become a popular approach for rapid prototyping. With the advancement in control development, fast material processing, heating methods, and cost reduction, more and more different AM technologies were evolved, typical examples are fused deposition modeling (FDM, a process that extrudes filament-form nonmetal materials to produce 3D objects), solid ground curing (SGC, a process that solidifies full layer of liquid polymer by ultraviolet light through a mask), digital light processing (DLP, an optical MEM technology similar to SL process but uses a digital micromirror projector), selective laser sintering (SLS, a process uses laser light to sinter material powders that bind them together to produce 3D parts), selective laser melting

(SLM, a process uses a laser beam to create 3D objects by fusing metal powders together), electron beam melting (EBM, a process uses an electron beam to heat and weld metal powders or wires together to create 3D parts), laminated object manufacturing (LOM, a process uses heat and pressure to fuse or laminate layers of material together and then cut into 3D shape with a PC-controlled laser or blade), and so on. With the cost of 3D printers being affordable, application of these AM processes to industrial, medical, aerospace, and personal use are widely accepted.

It was predicted that the global AM market (including hardware, software, materials, direct parts, and service and parts) would grow at 20%–30% annually, and by 2020 the annual revenue may reach as high as to U.S. \$11 billion. For inventors, the use of 3D printers to rapid prototyping can shorten the time between ideas and commercialization. For on-site engineers, the use of 3D printers to print replacement parts for malfunctional machinery can significantly reduce the time and cost. For home owners, the use of 3D printers is to fabricate replacement parts/decors and hobbyists to produce special parts for their toys and gadgets, 3D printers are regarded as the most idealized manufacturing process. As for the application to products with complex geometry (such as engine components), the use of 3D printers can be much more effective than conventional subtractive machining processes in terms of number of part count.

The advantages of AM include low energy consumption by simplifying fabrication steps, less waste than conventional machining processes, fewer part count for complex products, reduced time and cost (with less tooling and overhead), less design restrictions, shortened time between invention and commercialization, and rapid response to market change and service requirement. Business is therefore taking the advantages of the technology to produce parts and custom products with plastic, composite, or metal materials. AM is especially suitable for small-to-moderate batch production, such as defense, aerospace, and medical industries.

Although AM technology has many advantages, there are still a few obstacles yet to be overcome in order to broaden its application. Typical examples are tolerance (accuracy), surface roughness, strength, standards, and process control. Many applications require micronscale accuracy but so far not all AM processes can achieve this. The surface of products created by 3D printers could be very rough, therefore a postprocess is normally required for further refinement. Improved control systems are a key to improve the precision, quality, and reliability of the AM processes. The strength of 3D printed parts needs to be validated, and standards of the AM processes are yet to be established. For these to be resolved, there is a need for extensive testing, demonstration, data collection, and statistical analysis.

28.2   Fused deposition modeling

When using this technique, material filament is automatically supplied to an extrusion nozzle. As the nozzle is heated to a preset temperature, the material begins to melt and is squeezed out from the nozzle as shown in Figure 28.1. The nozzle can move in both horizontal and vertical directions based on the geometry and the support of part to be produced. The melted material is extruded out layer-by-layer from the nozzle as the platform moves from top to bottom. The platform will lower with one layer thickness on the vertical direction. The advantage of FDM-based 3D printers is that materials are easy to get, easy to operate, and low cost. The quality of created parts is based on printer itself and material used. Due to the above advantage, most personal users are considering this type of 3D printers.

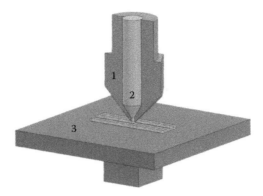

Figure 28.1 3D printing with FDM Process: (1) nozzle to extrude molten material, (2) deposited material film to form desired parts, and (3) controllable moving table.

28.2.1   Spur gears and theoretical involute generation

The spur gear is a simple and popular mechanical component for power and motion transmission. The gear consists of teeth and hub with appropriate bore for shaft and/or bearing mounting. To understand how the gear is formed and how it is engaged with the other gears, it is necessary to understand the nomenclature of the teeth as shown in Figure 28.2. For safe, quiet and long-term operations, the geometry of the teeth must be perfectly formed, smooth, and rigid.

The curve of gear tooth face is known as involute that may be constructed by using a string wrapping around the base circle and then tracing the end of the string by unwrapping the string. The commonly used method is to construct the involute by dividing the base

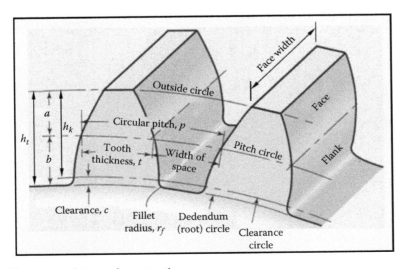

Figure 28.2 The nomenclature of gear teeth.

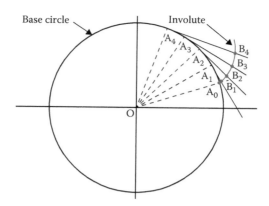

Figure 28.3 Discrete construction of involute curve.

circle into K equivalent sections and use the length of tangent lines to trace the involute curve as shown in Figure 28.3, where $OA_1 \perp A_1B_1$, $OA_2 \perp A_2B_2$, $OA_3 \perp A_3B_3$, $OA_4 \perp A_4B_4$, and $A_2B_2 = 2A_1B_1$, $A_3B_3 = 3A_1B_1$, $A_4B_4 = 4A_1B_1$, and so on.

Denote φ as the pressure angle and N the number of teeth, the major variables of the spur gear can be obtained from Table 28.1.

Table 28.1 Formulas for obtaining gear parameters

To obtain	Formula
Diametral Pitch (P_d)	$P_d = \dfrac{N}{D_p} = \dfrac{\pi}{p} = \dfrac{N+2}{D_o}$
Circular Pitch (p)	$p = \dfrac{\pi}{P_d}$
Tooth Thickness (t)	$t = D_p \sin\left(\dfrac{\pi}{2N}\right) \approx \dfrac{\pi D_p}{2N} = \dfrac{\pi}{2P_d}$
Outside Diameter (D_o)	$D_o = D_p + 2a = D_p + \dfrac{2}{P_d} = D_p\left(\dfrac{N+2}{N}\right)$
Addendum (a)	$a = \dfrac{1}{P_d} = \dfrac{D_p}{N} = \dfrac{2r_p}{N}$
Pitch Diameter (D_p)	$D_p = \dfrac{N}{P_d} = D_o - \dfrac{2}{P} = D_o - 2a = D_o\left(\dfrac{N}{N+2}\right)$
Base Diameter (D_b)	$D_b = D_p \cos\varphi$
Root Diameter (D_r)	$D_r = D_p - 2b$ (Dedendum diameter)
Dedendum (b)	$b = \dfrac{1.25}{P_d} = h_t - a$
Clearance (c)	$c = h_t - 2a$
Whole Depth (h_t)	$h_t = \dfrac{2.2}{P} + 0.002 \approx a + b$ (20 P_d and finer)
	$h_t = \dfrac{2.157}{P} \approx a + b$ (courser than 20 P_d)
Working Depth (h_k)	$h_k = 2a$

The formula representing the involute can be derived from Figure 28.4, where line s is of the same length as the string unwrapped from base circle s', and the radius of the base circle is r_b. Denote θ as the angle refers to x-axis and is expressed as

$$\theta = \frac{\pi}{2}t \text{ or } t = \frac{2\theta}{\pi} \tag{28.1}$$

where $t = 0 - 1$. As the string length s is the same as s', we get

$$s = s' = \frac{\pi r_b}{2}t = \frac{\pi r_b}{2}\frac{2\theta}{\pi} = r_b\theta \tag{28.2}$$

The coordinates of (x_c, y_c) and (x, y) are given below as

$$(x_c, y_c) = (r_b \cos\theta, \ r_b \sin\theta) \tag{28.3}$$

$$(x, y) = (x_c + s\sin\theta, \ y_c - s\cos\theta)$$

$$= r_b(\cos\theta + \theta\sin\theta, \ \sin\theta - \theta\cos\theta) \tag{28.4}$$

The radius r is obtained as follows:

$$r = \sqrt{x^2 + y^2} = r_b\sqrt{(\cos\theta + \theta\sin\theta)^2 + (\sin\theta - \theta\cos\theta)^2}$$

$$= r_b\sqrt{1 + \theta^2} = \sqrt{r_b^2 + (r_b\theta)^2} = \sqrt{r_b^2 + s^2} \tag{28.5}$$

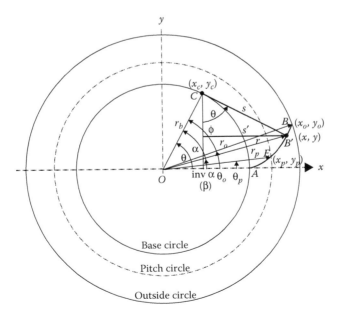

Figure 28.4 Continuous construction of involute curve.

This leads to

$$\theta = \sqrt{\left(\frac{r}{r_b}\right)^2 - 1} \tag{28.6}$$

At the intersections of pitch and outside circles, that is, points E and B, we have

$$s_p = r_b \theta_p \tag{28.7}$$

$$s_o = r_b \theta_o \tag{28.8}$$

Thus their corresponding angles and coordinates are

$$\theta_p = \sqrt{\left(\frac{r_p}{r_b}\right)^2 - 1} = \sqrt{\left(\frac{r_p}{r_p \cos\phi}\right)^2 - 1} = \tan\phi \tag{28.9}$$

$$(x_p, y_p) = r_b \left(\cos\theta_p + \theta_p \sin\theta_p, \; \sin\theta_p - \theta_p \cos\theta_p\right) \tag{28.10}$$

$$\theta_o = \sqrt{\left(\frac{r_o}{r_b}\right)^2 - 1} = \sqrt{\left(\frac{r_p + \dfrac{2r_p}{N}}{r_p \cos\phi}\right)^2 - 1} = \sqrt{\left(\frac{1 + \dfrac{2}{N}}{\cos\phi}\right)^2 - 1} \tag{28.11}$$

$$(x_o, y_o) = r_b \left(\cos\theta_o + \theta_o \sin\theta_o, \; \sin\theta_o - \theta_o \cos\theta_o\right) \tag{28.12}$$

The involute formula may also be derived by using the function of α or inv α, namely the involute function. From Figure 28.4, we get

$$\text{inv } \alpha = \beta = \theta - \alpha \tag{28.13}$$

$$\tan\alpha = \frac{B^*C}{OC} = \frac{s}{r_b} = \frac{r_b\theta}{r_b} = \theta \tag{28.14}$$

Substituting into Equation 28.13 yields

$$\text{inv } \alpha = \tan\alpha - \alpha \tag{28.15}$$

The coordinate (x, y) is

$$(x, y) = (r\cos\beta, r\sin\beta) = \left(r\cos(\text{inv } \alpha), r\sin(\text{inv}\alpha)\right) \tag{28.16}$$

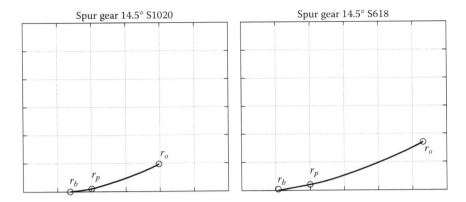

Figure 28.5 Computer generated tooth involute curves.

The coordinates at pitch and outside points (i.e., E and B) are

$$(x_p, y_p) = (r_p \cos \beta_p, r_p \sin \beta_p)$$ (28.17)

$$(x_o, y_o) = (r_o \cos \beta_o, r_o \sin \beta_o)$$ (28.18)

Where

$$\beta_p = \theta_p - \alpha_p = \theta_p - \tan^{-1} \theta_p = \tan \varphi - \varphi$$

$$\beta_o = \theta_o - \alpha_o = \theta_o - \tan^{-1} \theta_o$$

For demonstration purpose, the computer generated involute curves are shown in Figure 28.5.

28.2.2   Gear accuracy measurements

To evaluate the accuracy of FDM 3D printed gears, a number of gears were created and their corresponding off-the-shelf steel gears were purchased. The measurements were conducted with a gear tooth vernier caliper, as shown in Figure 28.6.

The following tables show the comparison between theoretical values and measured data: Table 28.2 is for S816 gear, Table 28.3 is for S1020 gear, and Table 28.4 is for S1626 gear. The ABS gears were produced by 3D printers and steel gears were purchased.

From these tables, it is observed that the tooth thickness of the gears created by the 3D printer is greater than that of gears purchased from market. When installing these gears into gearboxes (see Figure 28.7), the off-the-shelf gears can run smoothly, however, the 3D printed gears can hardly rotate.

Figure 28.6 Tooth thickness measurement with a gear tooth vernier caliper.

Table 28.2 Outside diameter and tooth thickness of S816 spur gear

Measurement	(Outside diameter)		Measurement	(Tooth thickness)	
	Steel	3D Printed		Steel	3D Printed
1	2.2485	2.2295	1	0.1945	0.2025
2	2.2485	2.2275	2	0.1950	0.2040
3	2.2485	2.2225	3	0.1945	0.2025
4	2.2480	2.2245	4	0.1950	0.2040
5	2.2485	2.2225	5	0.1950	0.2010
6	2.2485	2.2230	6	0.1950	0.1980
7	2.2485	2.2255	7	0.1950	0.1995
8	2.2485	2.2300	8	0.1945	0.2015
9	2.2485	2.2305	9	0.1945	0.2015
10	2.2480	2.2305	10	0.1940	0.2025
11			11	0.1940	0.2030
12			12	0.1950	0.2035
13			13	0.1945	0.2015
14			14	0.1940	0.1970
15			15	0.1940	0.2005
16			16	0.1940	0.2045
Average	2.2484	2.2266	Average	0.1945	0.2017
Theoretical		2.2500	Theoretical		0.1963

Table 28.3 Outside diameter and tooth thickness of S1020 spur gear

Measurement	(Outside diameter)		Measurement	(Tooth thickness)	
	Steel	3D Printed		Steel	3D Printed
1	2.1640	2.1800	1	0.1570	0.1630
2	2.1640	2.1785	2	0.1570	0.1620
3	2.1635	2.1830	3	0.1570	0.1625
4	2.1640	2.1865	4	0.1570	0.1630
5	2.1640	2.1860	5	0.1570	0.1665
6	2.1640	2.1795	6	0.1570	0.1615
7	2.1635	2.1740	7	0.1570	0.1580
8	2.1640	2.1760	8	0.1570	0.1605
9	2.1640	2.1760	9	0.1570	0.1620
10	2.1640	2.1775	10	0.1570	0.1640
11			11	0.1570	0.1650
12			12	0.1570	0.1615
13			13	0.1575	0.1630
14			14	0.1570	0.1650
15			15	0.1570	0.1640
16			16	0.1570	0.1665
17			17	0.1560	0.1600
18			18	0.1570	0.1600
19			19	0.1570	0.1615
20			20	0.1570	0.1635
Average	2.1639	2.1797	Average	0.1570	0.1627
Theoretical		2.2500	Theoretical		0.1571

Table 28.4 Outside diameter and tooth thickness of S1626 spur gear

Measurement	(Outside diameter)		Measurement	III (Tooth thickness)	
	Steel	3D Printed		Steel	3D Printed
1	1.7470	1.7545	1	0.0985	0.1090
2	1.7470	1.7615	2	0.0985	0.1065
3	1.7470	1.7635	3	0.0980	0.1060
4	1.7480	1.7585	4	0.0980	0.1050
5	1.7460	1.7510	5	0.0980	0.1085
6	1.7475	1.7565	6	0.0990	0.1115
7	1.7470	1.7555	7	0.0985	0.1105
8	1.7465	1.7575	8	0.0985	0.1085
9	1.7470	1.7510	9	0.0985	0.1085
10	1.7490	1.7620	10	0.0985	0.1065
11			11	0.0985	0.1075
12			12	0.0985	0.1095
13			13	0.0980	0.1095

(*Continued*)

Table 28.4 (Continued) Outside diameter and tooth thickness of S1626 spur gear

	(Outside diameter)			III (Tooth thickness)	
Measurement	Steel	3D Printed	Measurement	Steel	3D Printed
14			14	0.0985	0.1075
15			15	0.0985	0.1055
16			16	0.0985	0.1040
17			17	0.0985	0.1045
18			18	0.0985	0.1100
19			19	0.0995	0.1095
20			20	0.0990	0.1100
21			21	0.0980	0.1100
22			22	0.0985	0.1080
23			23	0.0980	0.1090
24			24	0.0985	0.1090
25			25	0.0985	0.1105
26			26	0.0980	0.1120
Average	1.7472	1.7572	Average	0.0984	0.1083
Theoretical		1.7500	Theoretical		0.0982

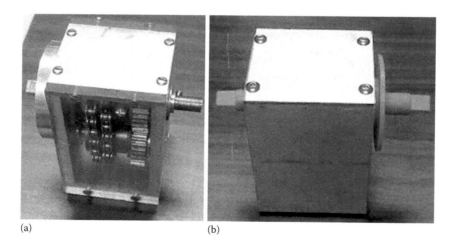

(a)   (b)

Figure 28.7 Gearboxes (a) with off-the-shelf gears and (b) with 3D printed gears.

28.3 Surface measurements

Two devices were used for the measurements: (1) digital microscope and (2) digital surface profile gage as shown in Figure 28.8. The surface photos are shown in Figure 28.9.

From the photos (a–d), 3D printers 1 and 2 provide very rough surfaces, the 3D printer 3 provides better surface finish (e), and the surface of purchased plate (f) has a very smooth surface. To show the depth of the surface profile, the surfaces of 3 gears were measured as shown in Table 28.5. Compared with purchased parts, the surface of the 3D printed gears is very rough.

(a)   (b)

Figure 28.8 **(See color insert.)** Surface measurement devices: (a) microscope and (b) surface profile gage.

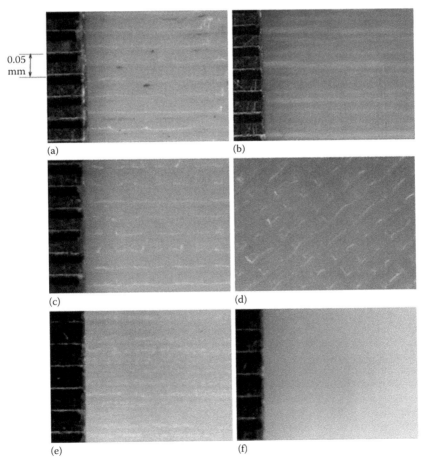

Figure 28.9 Surface photos, (a–b) printed by 3D printer 1, (c–d) printed by 3D printer 2, (e) printed by 3D printer 3, (f) purchased.

Table 28.5 The depth of surface profile

	Gear 1		Gear 2		Gear 3	
	mils	μm	mils	μm	mils	μm
1	6.1	154.9	4.9	124.5	4.6	116.8
2	4.6	116.8	5.0	127.0	5.1	129.5
3	6.2	157.5	5.4	137.2	5.5	139.7
4	4.6	116.8	4.4	111.8	4.6	116.8
5	5.4	137.2	5.8	147.3	4.1	104.1
6	4.2	106.7	6.1	154.9	4.8	121.9
7	5.8	147.3	5.0	127.0	4.4	111.8
8	6.6	167.6	5.6	142.2	4.6	116.8
9	4.7	119.4	5.0	127.0	3.8	96.5
10	6.0	152.4	4.9	124.5	4.8	121.9
11	6.4	162.6	5.6	142.2	5.7	144.8
12	6.5	165.1	5.4	137.2	5.7	144.8

28.4   Summary and future work

The purpose of this chapter is intended to understand the accuracy and surface roughness of gears produced by 3D printers. With the help of digital devices, the outside diameters, tooth thickness at the pitch circle, and roughness were measured. It was observed from the measured data that the accuracy and roughness of 3D printed gears need to be improved. The similar measurement approach may be extended to measure more complicated components fabricated by 3D printers, such as engines, medical devices, and so on. To have a comprehensive understanding of the performance of market available 3D printers, it will require more studies and tests of different components produced by other types of 3D printers.

Bibliography for further reading

1. Additive Manufacturing: Pursuing the Promise. *U.S. Department of Energy* (2012). https://www1.eere.energy.gov/manufacturing/pdfs/additive_manufacturing.pdf.
2. Buck, J., McMahan, T., and Huot, D. (2014). Space Station 3-D Printer Builds Ratchet Wrench to Complete First Phase of Operations, NASA. http://www.nasa.gov/mission_pages/station/research/news/3Dratchet_wrench.
3. Alec. (2015). China showcases large 3D printed metal frames for new generation of military aircraft. http://www.3ders.org/articles/20150717-china-showcases-large-3d-printed-metal-frames-for-new-generation-of-military-aircraft.html.
4. Dimitrov, D., Van Wijck, W., Schreve, K., and De Beer, N. (2006). Investigating the achievable accuracy of three dimensional printing. *Rapid Prototyping Journal*, 12(1), 42–52.
5. Lee, K. H., and Woo, H. (2000). Direct integration of reverse engineering and rapid prototyping. *Computers & Industrial Engineering*, 38(1), 21–38.
6. Lynn-Charney, C., and Rosen, D. W. (2000). Usage of accuracy models in stereolithography process planning. *Rapid Prototyping Journal*, 6(2), 77–87.
7. Kitson, P. J., Rosnes, M. H., Sans, V., Dragone, V., and Cronin, L. (2012). Configurable 3D-Printed millifluidic and microfluidic "lab on a chip" reactionware devices. *Lab on a Chip*, 12(18), 3267–3271.

8. Lanzetta, M., and Sachs, E. (2003). Improved surface finish in 3D printing using bimodal powder distribution. *Rapid Prototyping Journal*, 9(3), 157–166.

9. Campbell, T., Williams, C., Ivanova, O., and Garrett, B. (2011). Could 3D printing change the world. *Technologies, Potential, and Implications of Additive Manufacturing*, Atlantic Council, Washington, DC.

10. Duffy, D. C., McDonald, J. C., Schueller, O. J., and Whitesides, G. M. (1998). Rapid prototyping of microfluidic systems in poly (dimethylsiloxane). *Analytical chemistry*, 70(23), 4974–4984.

11. Islam, M. N., Boswell, B., and Pramanik, A. (2013, July). An investigation of dimensional accuracy of parts produced by three-dimensional printing. In *Proceedings of the World Congress on Engineering* (Vol. 1, pp. 3–5).

12. Bassoli, E., Gatto, A., Iuliano, L., and Grazia Violante, M. (2007). 3D printing technique applied to rapid casting. *Rapid Prototyping Journal*, 13(3), 148–155.

13. Ippolito, R., Iuliano, L., and Gatto, A. (1995). Benchmarking of rapid prototyping techniques in terms of dimensional accuracy and surface finish. *CIRP Annals-Manufacturing Technology*, 44(1), 157–160.

14. Vaezi, M., and Chua, C. K. (2011). Effects of layer thickness and binder saturation level parameters on 3D printing process. *The International Journal of Advanced Manufacturing Technology*, 53(1–4), 275–284.

15. Kruth, J. P., Mercelis, P., Van Vaerenbergh, J., Froyen, L., and Rombouts, M. (2005). Binding mechanisms in selective laser sintering and selective laser melting. *Rapid Prototyping Journal*, 11(1), 26–36.

16. Zein, I., Hutmacher, D. W., Tan, K. C., and Teoh, S. H. (2002). Fused deposition modeling of novel scaffold architectures for tissue engineering applications. *Biomaterials*, 23(4), 1169–1185.

17. Williams, J. M., Adewunmi, A., Schek, R. M., Flanagan, C. L., Krebsbach, P. H., Feinberg, S. E., Hollister, S. J., and Das, S. (2005). Bone tissue engineering using polycaprolactone scaffolds fabricated via selective laser sintering. *Biomaterials*, 26(23), 4817–4827.

18. Garcia, C. R., Rumpf, R. C., Tsang, H. H., and Barton, J. H. (2013). Effects of extreme surface roughness on 3D printed horn antenna. *Electronics Letters*, 49(12), 734–736.

19. Bose, S., Vahabzadeh, S., and Bandyopadhyay, A. (2013). Bone tissue engineering using 3D printing. *Materials Today*, 16(12), 496–504.

20. Calì, J., Calian, D. A., Amati, C., Kleinberger, R., Steed, A., Kautz, J., and Weyrich, T. (2012). 3D-printing of non-assembly, articulated models. *ACM Transactions on Graphics (TOG)*, 31(6), 130.

21. Nourghassemi, B. (2011). Surface Roughness Estimation for FDM Systems. A thesis presented to Ryerson University, Toronto, Canada.

22. Ramos, J. A., Murphy, J., Wood, K., Bourell, D. L., and Beaman, J. J. (2001, August). Surface roughness enhancement of indirect-SLS metal parts by laser surface polishing. In *Proceedings of the 12th Solid Freeform Fabrication Symposium* (pp. 28–38).

23. Budynas, R. G., and Nisbett, J. K. (2015). *Shigley's Mechanical Engineering Design*, 10th edition. New York: McGraw-Hill.

24. Juvinall, R. C., and Marshek, K. M. (2011). *Fundamentals of Machine Components Design*, 5th edition. New York: John Wiley & Sons.

25. Oberg, E. (2012). *Gears, Splines, and Cams, Machinery's Handbook*, 29th edition. New York: Industrial Press.

26. Boston Gear. Engineering Information—Spur Gears. *Gear Catalog*. https://www.bostongear.com/pdf/gear_theory.pdf.

27. Pro/ENGINEER. (2008). Involute Gears, Wildfire 3.0 Tips and Tricks, CADQUEST Inc.

28. Reyes, O., Rebolledo, A., and Sanchez, G. (2008). Algorithm to describe the ideal spur gear profile. In *Proceedings of the World Congress on Engineering*, Vol II, WCE 2008, London, UK.

29. SDP/SI. *Elements of Metric Gear Technology*. New York: Stock Drive Products/Sterling Instrument.

30. Wohlers Report 2014. *History of Additive Manufacturing*. Wohlers Associates.

31. Wile, R. (2013). Credit suisse: 3D printing is going to be way bigger than what the 3D printing companies are saying. *Business Insider*. http://www.businessinsider.com/the-3-d-printing-market-will-be-huge-2013-9

chapter twenty nine

Surface roughness of electron beam melting Ti-6Al-4V effect on ultrasonic testing

Evan Hanks, David Liu, and Anthony N. Palazotto

Contents

Experimental research is underway, focusing on the effect of surface roughness on defect detection using ultrasonic inspection. A nondestructive inspection technique for additively manufactured parts is necessary before such parts are utilized as production end items for use in aircraft battle damage repair applications such as the rapid, on demand, and 3D printing of aircraft replacement parts. This research will experimentally examine the effectiveness of ultrasonic testing on electron beam melting (EBM) additively manufactured samples made from Ti-6Al-4V. Experimental results will determine whether surface machining to remove roughness will effectively enhance detectability and resolution of ultrasonic inspection.

29.1 Introduction

Additive manufacturing (AM), commonly known as 3D printing, is quickly gaining popularity and increasing capability throughout multiple industries. In recent years, AM has moved from plastic models and polymer prototypes to complex geometries of various metals. The use of metals in AM is an exciting development, and opens many doors to the future of manufacturing. The basic principle of AM is to use a three-dimensional (3D) computer generated model to manufacture components using a layered approach. Each layer of material is a thin cross section of the part bonded to the previous layer.[1] EBM is a type of AM that uses powdered alloy melted layer-by-layer with an electron beam in vacuum. Data are digitally scanned or 3D modeled, and through software is sliced into individual layers. The thickness of these layers is controlled as a manufacturing constraint, which are typically 40–100 μm thick. This ability to manufacture customized end products has piqued the interest of many industries, including the biomedical field, and more recently, aerospace manufacturers.[2] The potential use of AM in the aviation industry includes the manufacture of low use and/or obsolete parts. The capability also exists to create items with complex geometries and to fill logistics shortfalls with a reduced customer wait.[3]

Titanium is a relatively lightweight structural material with high corrosion-resistant properties. Strengthening of this material can be achieved through alloying and heat treatment. For use in aircraft, it possesses a good strength-to-weight ratio, low density, high fracture toughness, and low heat treating temperatures.[4] Ti-6Al-4V is an alpha–beta alloy containing both alpha and beta phases at room temperature. Traditionally Ti-6Al-4V is available in all mill forms as well as castings and powder. In annealed or solution treated plus aged conditions, this product can suit a variety of applications. The useful temperature range for this alloy is from –195°C to 400°C (–320°F to 750°F). Strength properties of AM Ti-6Al-4V are shown close to those of cast Ti-6Al-4V. Compared to wrought products, they have a lower yield and ultimate strength and approximately 75% lower fatigue life.[5,6] As exciting as this technology is, many obstacles stand between the current state of use and qualification techniques for use on aircraft.[7]

Nondestructive inspection (NDI) techniques of additively manufactured products have begun on multiple fronts. NASA has performed initial testing which indicates, for an as-built condition component, the surface is too rough for reliable eddy current testing. Background noise produced by the rough surface masks any flaws or defects in the material. NASA has also experimented with the use of florescent penetrant inspection (FPI) and found that the porosity inherent to AM limits the effectivity of this technique. NASA's Johnson Space Center is currently exploring the use of ultrasonic testing on electron beam free form fabrication (EBFFF). However, no published report contains their findings.[8] The Air Force Research Laboratory (AFRL) at Wright–Patterson Air Force Base, Ohio, has also begun to explore NDI of AM and has found favorable results using computerized tomography (CT) scanning as shown in Figure 29.1. Although defects are clearly seen, there are still several downsides to CT scanning. The associated equipment is cost prohibitive to purchase and maintain for routine inspections. Additionally in its current use, inspections would necessitate removal of parts from service.

The purpose of this chapter is to present the results of research on ultrasonic inspection of EBM additively manufactured titanium–aluminum alloy. With the need to inspect

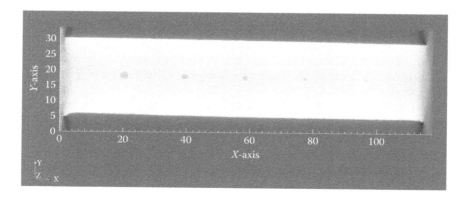

Figure 29.1 Computer Tomography (CT) Scan. 2D scan showing a single EBM sample in as manufactured condition with five imbedded spherical defects 0.100″, 0.080″, 0.060″, 0.040″, 0.020″ in diameter. Axis dimensions in mm.

255

0

Figure 29.2 Ultrasonic Scan. Sample B-41 using 5 MHz transducer through side surface, X–Z plane, with roughness of approximately 158 µm.

in-use and possibly field manufactured components, the primary focus of this research is on the use of existing field level ultrasonic inspection techniques. With the potential use for this technology in battle damage repair, the ability to inspect parts manufactured in austere locations is vital. Given the inherent surface roughness of EBM Ti-6Al-4V, a large scatter of ultrasonic waves is expected at the plane of entry. The intent of this research is to determine the impact of surface roughness on the ability to detect subsurface defects at common ultrasonic frequencies. As shown in Figure 29.2, the defects are masked by the scatter from the rough surface.

29.2   Experimental setup

The specimens used in this testing were designed by AFRL for the purpose of testing non-destructive and destructive inspection techniques. The final design of the samples were rectangular blocks 114.3 mm long (X), 25.4 mm wide (Y), and 25.4 mm tall (Z) illustrated in Figure 29.3. Two separate lots of Ti-6Al-4V powder from one manufacturer were used to build these samples. The first four samples were made from the first lot of powder and the remaining eight samples from the second. Specimens were designed as sets of two identical samples in each production run. These samples were given designation numbers of X-41 and X-42 where X indicates the production run. Each block was designed with five embedded spherical flaws ranging in size from 0.51 mm to 2.54 mm in diameter as listed in Table 29.1. These flaws were designed on the center line of the sample with the center of the spheres evenly spaced in ascending diameter as shown in Figure 29.3.

Following the standard sign convention in ASTM F2921-11, a part on the build plate is aligned to the X, Y, and Z axes, is described with a three-axis designation: (1) The first letter of the designation corresponds to the axis parallel to the longest dimension of the part,

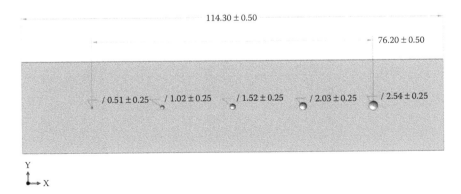

114.30 ± 0.50

76.20 ± 0.50

/ 0.51 ± 0.25   / 1.02 ± 0.25   / 1.52 ± 0.25   / 2.03 ± 0.25   / 2.54 ± 0.25

Y

X

Figure 29.3 CAD Design of Sample. 2-D Representation of center line of sample in the X–Y plane.

Table 29.1 Designed flaw sizes

Flaw	Diameter (mm)	Diameter (in)
1	0.51	0.02
2	1.02	0.04
3	1.52	0.06
4	2.03	0.08
5	2.54	0.10

(2) the second letter corresponds to the axis parallel to the second longest dimension, and (3) the third letter corresponds to the axis parallel to the shortest dimension. Experimental samples were designed using a 3D CAD model. The model was processed through a software package to create an STL file, which was then converted to an ABF file. This file contains all the data the system needs for each 2D layer comprising the entire manufactured part. These 2D layers were stacked in the +Z direction on the build bed of the machine to create the 3D samples. Samples for this research were manufactured at Oak Ridge National Laboratories in Oak Ridge, Tennessee, as part of a joint project with AFRL. A total of 12 samples were manufactured and used to collect the data in this research. These samples were produced as six sets of two. Each production run was given an alphabetic designation. Production runs A, B, D, and E were used to collect the data. The samples were manufactured using Arcam's A2 system and are listed in Table 29.2. The layered manufacturing process ends with the final melted layer on the upper surface, X–Y plane. Surface roughness on the top face is subsequently less than on any vertical surface. The profilometer used in measuring the roughness of the upper surface has a limit of 120 μm. When attempts were made to measure the side surface, the measurements were over this limit. To obtain a surface roughness for the side surfaces, photographs of the samples at 50X zoom were examined, and pixel size was used to determine the roughness. A representative sample produced surface roughness ranging from 138 μm to 178 μm. The roughness of the as-manufactured top surfaces of all samples are relatively close in magnitude with a mean of 18.2 μm and a standard deviation of 2.7 μm as shown in Table 29.2. For comparison, 220 grit sandpaper has a roughness of 18.5 μm and 36 grit is on the order of 150 μm.[9]

In order to determine the ability to detect anomalies in each specimen's as-manufactured condition, all samples were ultrasonically tested. Frequently ultrasonic information is displayed on an oscilloscope, with time on the horizontal axis and amplitude of ultrasonic energy received by the transducer on the vertical axis. As the transducer sends pulses of

Table 29.2 Surface roughness of upper surface (μm RMS)

Sample	Surface roughness
A-41	16.8
A-42	16.1
B-41	22
B-42	21.2
D-41	19.6
D-42	18.7
E-41	16.2
E-42	15.5

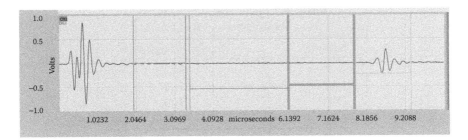

Figure 29.4 A-Scan. Echo returns displayed as energy amplitude through sample thickness. Sample B-41 using 2.2 MHz transducer.

energy, it receives echoes between pulses. These echoes are displayed on the oscilloscope as the amplitude of the return at the distance from the transducer, this produces what is referred to as an A-scan as shown in Figure 29.4.[10] Each set of samples were inspected at three frequencies and from two different sides to determine the effect of surface roughness. A three-axis controller was utilized to mount the transducer, allowing accurate step control to generate C-scans. An AFRL in-house data collection system was used to control the scanning unit and collect the ultrasonic return. The system was set up to take measurements at 0.06 mm intervals across the entire specimen. Scans were performed along the X-axis of the specimen at a rate of 600 mm/s. On completion of each scan the transducer was moved 0.06 mm in the Y-axis and another scan was completed. Sequential scans were taken until the entire upper surface was scanned. Data points collected through the thickness of each sample were taken at increments of 0.10 mm resulting in 254 points. At each step an A-scan was collected; all A-scans were compiled into a C-scan for each sample as shown in Figures 29.4 and 29.5. The C-scans represent the amplitude of the return at each gated section of the A-scan. Ultrasonic scans were completed using 2.2 MHz, 5 MHz, and 10 MHz transducers. Identical procedures were followed and scans accomplished for every sample set. C-scans were collected from each sample set through the upper surface, normal to the Z+ axis. The sample was then rotated about the X-axis. Scans were then taken of the Y- face of the block.

One sample of each set was sent for postprocess machining. For consistency, sample X-41 from each set was selected to receive postprocess machining. Of each of these six samples, the top and side face was machined using a shell cutter on a three-axis mill. Milled samples were measured for postmachining surface roughness, with results listed in Table 29.3. Postprocess machining produced a mean surface finish of 3.83 μm, a 79% reduction in roughness when compared to the as manufactured upper surface. Ultrasonic scans were once again performed on these samples at 2.2 MHz, 5 MHz, and 10 MHz on both machined surfaces.

Figure 29.5 C-Scan. Corresponding to Green Gate in Figure 29.4. A-Scan. Echo returns displayed as energy amplitude through sample thickness. Sample B-41 using 2.2 MHz transducer.

Table 29.3 Postmachining surface
roughness upper surface (µm)

Sample	Surface roughness
A-41	3.22
B-41	5.02
D-41	3.23
E-41	3.85

29.3 Data processing

Data collected on each sample through ultrasonic scans was loaded into MATLAB® for postprocessing. Raw data was read into an array where a Fourier Transform was performed to decompose the signal into its frequencies using the following equation:

$$Y_{p+1,q+1} = \sum_{j=0}^{m-1}\sum_{k=0}^{n-1} \omega_m^{jp}\omega_n^{kq}X_{j+1,k+1} \tag{19.1}$$

Using an input X Equation 19.1 uses ω_m and ω_n to represent complex roots of unity, $e^{-2\pi i/m}$ and $e^{-2\pi i/n}$ respectively. The notation i represents the imaginary unit, p and j are indices that range from 0 to $m-1$. Indices q and k run from 0 to $n-1$, whereas $p+1$ and $j+1$ run from 1 to m and the $q+1$ and $k+1$ run from 1 to n.[11] The resulting Y is a decomposed set of frequencies that comprised the original input signal. Decomposition allowed application of a filter to limit the range of frequencies in the data. For all samples, this range was set from 0.5 MHz to 12 MHz, reducing outside interference received by the transducer during data collection. Frequency filtered data was then inversely transformed into the time domain. In ultrasonic wave transmission, velocity in a medium is constant, therefore time corresponds to distance through the sample. As a result, the inversely transformed array contains filtered layers stacked either parallel or perpendicular to the build direction depending on scan orientation. To obtain the strongest ultrasonic returns from this array in a manageable form, the matrix of data corresponding to the center of the sample was selected. The center–plane matrix was combined with the five matrices above and five below the center, then normalized to form one representative matrix as shown in Figure 29.6.

The processed data compiled into a single grayscale image, as shown in Figure 29.6, allows for the use of image processing techniques. These techniques were used to measure the size and intensity of the flaws detected by the ultrasonic transducer. The single representative matrix was reduced to a grayscale image to facilitate image processing techniques. A single grayscale image is allowed for the use of an image erosion function. This was used to remove indications smaller than the transducer was

Figure 29.6 Combine Data from 11 Center Matrices. Sample B-41 5 MHz transducer.

physically capable of detecting. The erosion of an image uses a structuring element to compare an image pixel with its neighboring pixels. Generally, ultrasonic inspection techniques can only detect a flaw equivalent in size to one half the wavelength at the frequency used.[12] The size of the structuring element was set to one-half of the wavelength size for each frequency and the number was rounded down to the nearest whole number of pixels. An example, of this is shown in Figure 29.7. Image erosion ultimately reduced the intensity of small spikes in amplitude at lower levels while retaining pertinent data.

The final function used in this analysis was the Circular Hough Transform. The Hough Transform is the method behind circle detection used to measure and classify the flaws in the samples. The eroded images still contain a certain level of noise; the Hough Transform is an excellent tool because it is relatively unaffected by this noise.[13] The first step of this approach is the determination of the image pixels with the highest gradient. These pixels are identified and recorded. As groups of high gradient pixels are identified with a similar distance to a center point, those pixels are set as points on the circumference of the circle. User definable input arguments are:

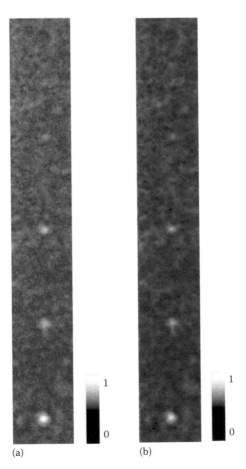

(a)   (b)

Figure 29.7 Image Erosion (a) Raw image prior to erosion and (b) eroded using disk shaped structure element. Sample A-41 10 MHz transducer side surface.

input image, radii range, object polarity, computation method, sensitivity, and edge threshold. Outputs include the following: coordinates of the center of any detected circles and the corresponding radii.[14] Radii range was set to 0.48 mm to 3 mm based on the design size of the spherical flaws. Image intensity corresponds to the amplitude of return of the ultrasonic wave during testing, therefore object polarity was set to detect bright objects in the image. Sensitivity was found using an iterative method, starting with a low value and incrementally increasing until the maximum number of known flaws were identified. Sensitivity was recorded for each sample set for use in later analysis.

29.4   Results

In order to compare the as manufactured to the machined surfaces, the representative images were joined along their long edge to form one image. The joined image contains the data for both the as manufactured and machined surfaces of sample A-41 along with the as manufactured surface of the corresponding A-42 as shown in Figure 29.8. The Hough Transform was then applied to this combined image at the optimal sensitivity for each sample and frequency. The resulting center points and radii were projected on the original image to show the result in Figure 29.9.

To best analyze the effect of surface area on the ultrasonic detection of flaws, the circular area of the flaws found using the Hough Transform was compared to the design area. For scans conducted on the side surface at 2.2 MHz and 5 MHz, no discernable flaws were detected, as shown in Figures 29.10a,c and 29.12a,c. The roughness of the surface masked the embedded flaws in the samples. Once this side surface was machined, the flaws became visible and were detectable using the Hough Transform as shown in Figure 29.10. Figure 29.10 also shows that the sensitivity required to detect the known flaws as the milled sample, produced false indications in the as manufactured samples. Figure 29.11 shows a set of samples, scanned at 5 MHz, at a sensitivity of 0.86 where all five design defects are seen in the milled sample, whereas only two were found in the exact same specimen prior to machining. At a frequency of 5 MHz and 2.2 MHz, no flaws were detected through the as manufactured side surface. As shown in Figure 29.12, all five design flaws were identified in sample A-41 on completion of postprocess machining, though not before. Scans performed at 10 MHz produced significantly different results from those at lower frequencies. Figure 29.13, shows that at 10 MHz, all flaws in sample A-41 are seen through the side surface, both before and after machining. Two of the five flaws in sample A-42 were also visible. With an increase in sensitivity of the Hough Transform of 0.02, four of the five flaws were detected in sample A-42 at 10 MHz from the side surface. Ultrasonic scans through the top surface at 10 MHz produce a scattered return of the flaws as shown in Figure 29.14. Data collected at 10 MHz through the side surface indicates that higher frequency inspections are less affected by the surface roughness of a sample. An inverse relation between flaw size and relative error was found at all frequencies. Figure 29.15 illustrates this tendency with consistent trend lines showing an inverse correlation between designed flaw size and relative error. Finally, with respect to surface roughness, Figure 29.16 shows data collected and processed to date. This Figure 29.16 indicates how relative error in flaw size increases with coarser surfaces.

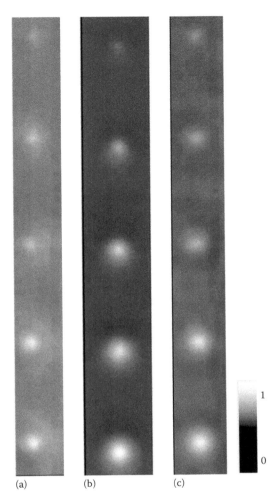

(a)   (b)   (c)

Figure 29.8 Joined Images (a) Sample A-41 in as manufactured condition, (b) Sample A-41 post-milling, and (c) Sample A-42 in as manufactured condition. 2.2 MHz transducer top surface.

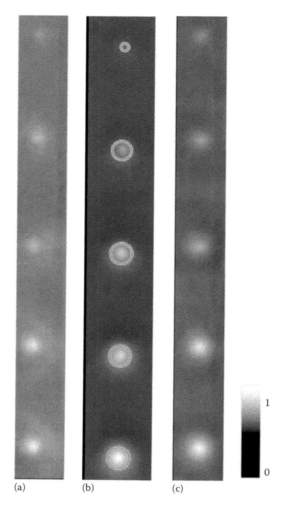

Figure 29.9 Joined Images with Hough Transform: (a) Sample A-41 in as manufactured condition; (b) Sample A-41 postmilling with Hough Transform, flaw sizes top to bottom 0.054 in., 0.136 in., 0.150 in., 0.148 in., 0.152 in.; and (c) Sample A-42 in as manufactured condition. 2.2 MHz transducer top surface.

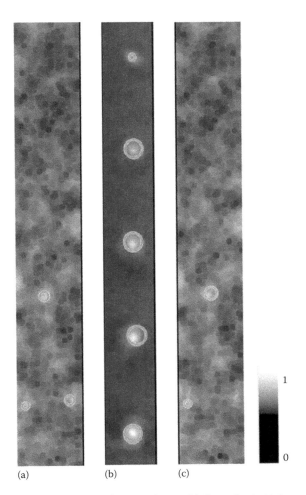

(a)   (b)   (c)

Figure 29.10 Joined Images with Hough Transform: (a) Sample A-41 in as manufactured condition; (b) Sample A-41 postmilling flaw diameter top to bottom: 0.050 in., 0.140 in., 0.143 in., 0.145 in., and 0.149 in.; and (c) Sample A-42 in as manufactured condition. 2.2 MHz transducer side surface.

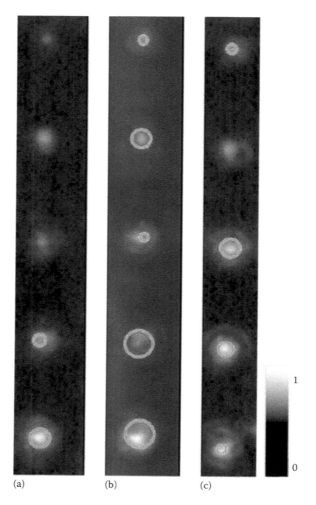

(a)   (b)   (c)

Figure 29.11 Joined Images with Hough Transform: (a) Sample A-41 in as manufactured condition; (b) Sample A-41 postmilling flaw diameter top to bottom: 0.063 in., 0.137 in., 0.059 in., 0.200 in., and 0.208 in.; and (c) Sample A-42 in as manufactured condition. 5 MHz transducer top surface.

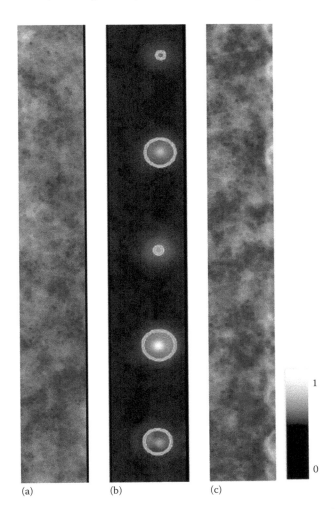

(a)   (b)   (c)

Figure 29.12 Joined Images with Hough Transform: (a) Sample A-41 in as manufactured condition; (b) Sample A-41 postmilling flaw diameter top to bottom: 0.058 in., 0.211 in., 0.060 in., 0.226 in., and 0.190 in.; and (c) Sample A-42 in as manufactured condition. 5 MHz transducer side surface.

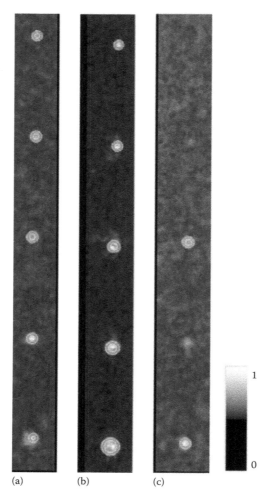

(a)   (b)   (c)

Figure 29.13 Joined Images with Hough Transform: (a) Sample A-41 in as manufactured condition; (b) Sample A-41 postmilling flaw diameter top to bottom: 0.049 in., 0.058 in., 0.070 in., 0.079 in., 0.081 in.; and (c) Sample A-42 in as manufactured condition. 10 MHz transducer side surface.

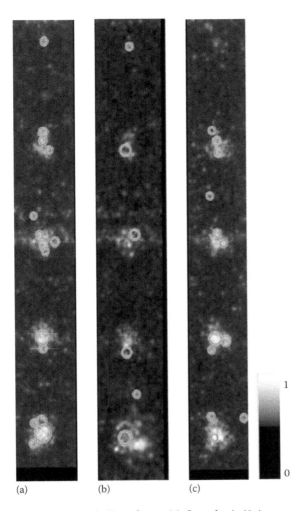

(a)   (b)   (c)

Figure 29.14 Joined Images with Hough Transform: (a) Sample A-41 in as manufactured condition; (b) Sample A-41 postmilling; and (c) Sample A-42 in as manufactured condition. 10 MHz transducer top surface.

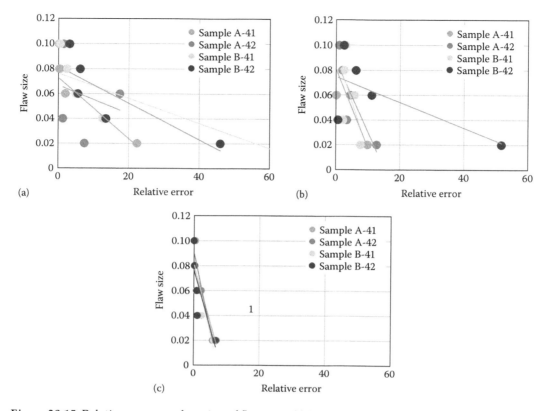

Figure 29.15 Relative error as a function of flaw size. (a) 2.2 MHz top surface, (b) 5 MHz top surface, and (c) 10 MHz top surface.

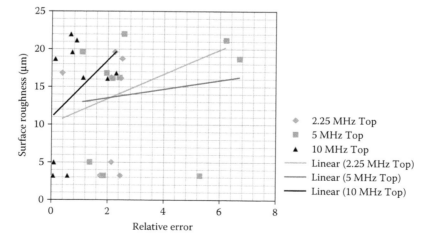

Figure 29.16 Relative error as a function of surface roughness.

29.5   Summary and Conclusions

During this preliminary work, the data suggests a relationship between frequency and surface roughness in ultrasonic inspections of additively manufactured Ti-6Al-4V. The data presented to this point suggests that at low frequency, 2.2 MHz and 5 MHz, ultrasonic scans produce no usable data when collected through an as manufactured surface perpendicular to the build direction of a sample. Milled surfaces and those parallel to the build direction produce usable data for the detection of flaws at low frequencies. These data suggest a relationship between the surface roughness of a sample and the frequency required to detect subsurface flaws. Sample sets B, D, and E are also in agreement, requiring postprocess machining for low frequency ultrasonic inspection through the side surface, but not the top surface. Postprocess machining capabilities are not always available or practical for an additively manufactured part. If NDI of a part is required, possibly in a deployed location, an ultrasonic scan at 10 MHz has shown the ability to detect flaws through an as manufactured side surface with low relative error. Current data also points to an increase in relative error as flaw size decreases. Data collection and processing from additional samples will provide further information on the effect of surface roughness on ultrasonic inspection. Future work in this area could expand the frequencies used for inspections or begin to test more complex geometries. Additional testing with field level, portable equipment could provide insight to the feasibility of on-aircraft inspections.

Acknowledgment

The authors would like to recognize and thank Mr. Dennis Lindell of the Joint Aircraft Survivability Program and Dr. Edwin Schwalbach of the Air Force Research Laboratories, Materials and Manufacturing Directorate, who graciously provided endless support and knowledge during this project.

References

1. I. Gibson, D. Rosen, and B. Stucker, *Additive Manufacturing Technologies: 3D Printing, Rapid Prototyping and Direct Digital Manufacturing*. New York: Springer, 2015.
2. A. Safdar, *Microstructures and Mechanical Properties of Electron Beam Rapid Manufactured Ti-6Al-4V*. Milmo: Media-Tryck, 2010.
3. H. Richards, *Topology Optimization of Additively Manufactured Penetrating Warhead*. Air Force Innstitute of Technology, Wright-Patterson AFB, Ohio, 2015.
4. Department of Defense, *Metallic Materials and Elements for Aerospace Vehicle Structures, Mil-Hdbk-5j*. Wright-Patterson: Department of Defense, 2003.
5. K. Rekedal, *Investigation of High-Cycle Fatigue Life of Selective Laser Melting and Hot Isostatically Pressed TI-6AL-4V*. Air Force Innstitute of Technology, Wright Patterson AFB, Ohio, 2015.
6. S. S. E. A. Shipp, *Emerging Global Trends in Advanced Manufacturing*. Alexandria, VA: Institute for Defense Analyses, 2012.
7. M. E. A. Koike, Evaluation of titanium alloys fabricated using rapid prototyping technologies-electron beam melting and laser beam melting. *Journal of Materials Processing Technology*, 4(10), 1776–1792, 2011.
8. J. E. A. Waller, *Nondestructive Evaluation of Additive Manufacturing: State-of-the-Discipline Report*. Hampton, VA: NASA Langley Research Center, 2014.
9. Nanovea, *Sandpaper Roughness Measurement Using 3D Profilometry*. Irvine, CA: Nanovea.

10. Department Of Defense, *Military Handbook Ultrasonic Testing Mil-Hdbk-728/6*. Department Of Defense, Watertown, MA, 1985.
11. Mathworks, *Matlab Documentation: Fast Fourier Transform*. Mathworks, Natick, MA, 2015.
12. AFLCMC/EZGTP, *Nondestructive Inspection Methods, Basic Theory, T.O. 33B-1-1*. Robins Air Force Base: United States Air Force, 2014.
13. R. Fisher, S. Perkins, A. Walker, and E. Wolfart. Hough Transform. *Image Transforms-*. Hypermedia Image Processing Reference, 2003. October 26, 2015.
14. Mathworks, *Matlab Documentation: Find Circles Using Hough Transform*. Mathworks, Natick, MA, 2015.

chapter thirty

Dynamic failure properties of additively manufactured stainless steel

Allison Dempsey, David Liu, Anthony N. Palazotto, and Rachel Abrahams

Contents

The Air Force Institute of Technology (AFIT), Ohio, United States is exploring how additive manufacturing (AM) might benefit aerospace structures. This method of manufacturing may reduce the waste of expensive materials, shorten logistics time, and enable optimized designs. However, before using finished AM parts in the United States Air Force, the weapon system program offices must understand any differences and uncertainty in material properties. This study correlates AM's effect on the microstructure, and consequently the dynamic properties of a stainless steel (SS) as compared to the conventional wrought material. Techniques such as energy dispersive X-ray spectroscopy (EDS) and electron backscatter diffraction (EBSD) reveal the composition and microstructure of the five different samples used in the study. Quasi-static tests and compression, indirection tension, and direct tension Split Hopkinson Pressure Bar (SHPB) tests are used to determine dynamic performance by subjecting materials to an *intermediate* (approximately $450 \ s^{-1}$) and *high* (approximately $900 \ s^{-1}$) strain rate.

30.1   Introduction

AM is generating excitement within the aerospace industry as a capability that can significantly impact how antiquated parts are manufactured and how novel parts are designed. The air logistic complexes (ALCs), responsible for management and certification of most of the parts on Air Force aircraft, identified the need to build this capability due to its speed and flexibility.[1] The Oklahoma City ALC is currently finalizing a strategic plan to integrate 3D printing in printing technology into nearly every aspect of its airpower sustainment mission, from making aircraft engine parts to printing electronic components.

30.2   Background

15–5PH SS is one of the most common steel alloys used in aerospace applications. 15–5PH has a desirable combination of high strength and corrosion resistance. Traditionally, aerospace parts made of this material are machined from wrought 15–5PH, but AM via direct metal laser sintering (DMLS) processes show great promise in producing near net-shape parts using 15–5PH powder. Although these materials have demonstrated similar static properties to the traditional wrought material, many AM materials are not yet fully characterized. Additionally, the potential for variability of the resultant product is still a concern due to a significant lack of standardization in the AM industry.[2] This research focuses on how postmanufacturing heat treatment changes the dynamic response of AM materials made with 15–5PH powder.

30.2.1   Additive manufacturing and direct metal laser sintering

AM is broadly described as a number of methods of producing parts by building up material instead of traditional subtractive processes. Due to the nature of formation, the AM material is subjected to different solidification protocols than traditionally cast or wrought material. This introduces a need to understand and manage the property ranges for materials considered for final part manufacturing.[2] DMLS is one of several techniques under the umbrella term selective laser melting (SLM). The SS specimens used in this study are made via SLM technique on an EOS GmbH M270 DMLS. DMLS processes builds up a part by sintering or melting layers of metal powder using a laser in an inert gas environment. The part is patterned off of a computer model design devolved into cross-sectional build layers. As each layer is sintered on the last, the build plate moves and another thin layer of powder deposits. As successive layers are created, a complete part is typically formed in one build. This is far different from the well investigated and standardized processes of forming and machining finished parts from the wrought stock material.[3]

30.2.2 Precipitation hardened stainless steel

SSs are based on an iron–chromium, iron–chromium–carbon, or iron–chromium–nickel combination, with additional alloys added to obtain desirable properties.[4] They are generally defined as greater than 10% Cr by weight. The focus material of this study is precipitation hardening (PH) SSs, which have added elements that strengthen the part when properly aged through the formation of precipitates. The PH stainless steels are classified by their room-temperature crystallographic structure such as austenitic, martensitic, or semiaustenitic. The increased strength caused by the precipitates means PH steels are often stronger and tougher than the non-PH variety with the same parent microstructure. The room-temperature crystallographic structure, dependent on the martensite-start (Ms) and martensite-finish (Mf) temperatures are shown in Figure 30.1. These define the temperature range beginning the transformation from an austenitic to martensitic structure on cooling from the solution-treatment temperature. When the Mf is just above room temperature, the material will transform completely to martensite on air cooling from the solution-treatment temperature. When the Ms is below room temperature, the resulting material is austenitic. Those with Ms just below room temperature are semiaustenitic.

Maximum strengthening of many types of martensitic and semiaustenitic PH SSs is obtained by aging at 850°F–950°F.[4] Higher temperatures within the precipitation range up to 1050°F increase in ductility and toughness but reduce both the yield and the ultimate tensile strength. This is known as *overaging*.[5] Austenitic PH SSs do not exhibit strengthening behavior when heat treated at these temperatures. This is because the structure of the austenite does not enter the martensite transformation range on cooling, and thus does not age on heating.[6] Therefore, because there are no transformation strains present in this structure to force the precipitation reaction, effective precipitate size and spacing distributions are not developed and they will not precipitate the hardened SS.

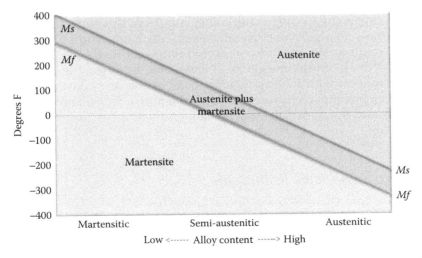

Figure 30.1 The effect of alloy content on the transformation temperature of PH stainless steels. (Reproduced from Hall, A.M. et al., *Thermal and Mechanical Treatment for Precipitation-Hardening Stainless Steels*, NASA Special Publication, Vol. 5089, 1967.)

30.2.3   15–5PH stainless steel

The intended material of study is 15–5PH, a martensitic precipitation hardenable SS. Its relative composition is approximately 75% Fe with 15% Cr, 5% Ni, and a 4% Cu precipitate. Nb, Mn, and Si are added for desirable additional impact on the material properties.[7] With no heat treatment, 15–5PH displays relatively good strength and ductility performance, but is not generally recommended for use without subsequent heat treatment.[8] It is often used when an application requires good tensile strength, creep and fatigue strength properties, in combination with moderate corrosion and heat resistance.[9] In the annealed condition it is essentially free of delta ferrite, which tends to lower the strength, ductility, and fabricability, and removes the ability to age harden by heat treatment.

The anticipated mechanical properties of AM metals are typically lower than their wrought counterparts.[1] The overriding goal of this project is to assess the dynamic characterics of a PH SS formed via an AM process, with and without subsequent heat treatment, to determine the variances of formation on the properties. This is done by both examination of the microstructure of representative samples of each test group and then dynamic testing at different strain rates. The test specimens, with the exception of the first compression test, are machined from cylinders produced by the EOS M270 machine in both a *vertical* orientation, built from a circular cross section, and a *horizontal* or varying rectangular cross section orientation. Sample sets are made from three heat treatment conditions: (1) no heat treatment, (2) H900 heat treatment, and (3) H1025 heat treatment.

A key difference of AM affecting the microstructure and consequently the material properties is the localized rapid heating and cooling of the sintered powder. This is likened to a series of small welds. The modes of solidification of welds are sometimes predicted by a composition diagram.[4] The composition diagrams for SSs use a calculated chromium equivalent (*Creq*) and a nickel equivalent (*Nieq*) to predict the microstructure of the material.[5] Generally, these diagrams suggest a composition with higher than 8% *Creq* and 4% *Nieq* is martensitic after undergoing fast and nonequilibrium cooling. Increasing values of *Creq* increase the chance of having a martensitic/ferritic mix, and increasing *Nieq* introduces a retained austenitic component into the martensite. The wrought 15–5PH material deviates from this expectation, slightly as it is expected to have an almost entirely martensitic structure with no delta ferrite even with a *Creq* of roughly 16%. It is also expected to have a very little retained austenite even though its *Nieq* is approximately 7%. It is possible; however, the trends of austenite and ferrite promotion with increasing alloy composition are consistent with the diagrams. *Creq* often includes chromium (Cr), molybdenum (Mo), silicon (Si), niobium (Nb), aluminum (Al), and titanium (Ti). *Nieq* is heavily influenced by carbon (C) and includes Manganese (Mn).

30.3   Methodology and techniques

30.3.1   Microscopy for microstructure examination

30.3.1.1   Optical microscopy

The first stage of this study is to examine the microstructure itself by optical microscopy and electron microscopy techniques such as energy dispersive X-ray spectroscopy (EDS) and electron backscatter diffraction (EBSD). Two different perspectives are obtained by cutting the AM cylinders and mounting them for examination as shown in Figure 30.2.

Representative samples

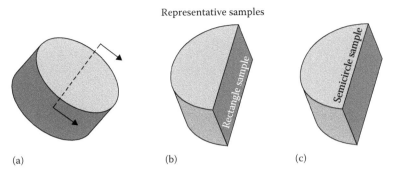

Figure 30.2 Example of specimens used in microscopic examination: (a) Three D sample, (b) rectangular sample, and (c) semicircular sample.

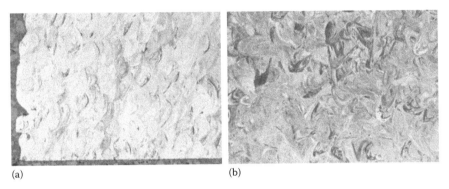

Figure 30.3 Optical microscopy examples (a) across the build direction and (b) across the build layers.

After mounting the samples in a phenolic compound and polishing as recommended, many of these were etched with a combination of phosphoric acid (H_3PO_4), water (H_2O), hydrofluoric acid (HF), and nitric acid (HNO_3) etchant. The revealed structure, shown in Figure 30.3 clearly illustrates the directional crescents formed from the individual sintering the melt pools.

30.3.1.2 Energy dispersive X-ray spectroscopy

EDS is a technique using energy detected from X-rays emitted from a sample during electron imaging to estimate the composition of a material.[10] It does this by using an electron beam to excite the atoms of a sample. As the atoms relax, the energy level of the resultant radiation uniquely indicates the element from which it came, and once converted and processed, provides information on the concentration of the elements present.[11] An example of the spectra for a sample of AM 15–5PH material is shown in Figure 30.4.

Although helpful in determining relative compositions, this method is semiquantitative at best. The breadth of each peak in Figure 30.4 indicates that the energy of an individual X-ray is not always measured exactly.[10] The amount of charge the X-ray generates in the detector is vulnerable to systemic and random error producing signal and background noise. Additionally, some elements are more easily detected at specific energy input ranges

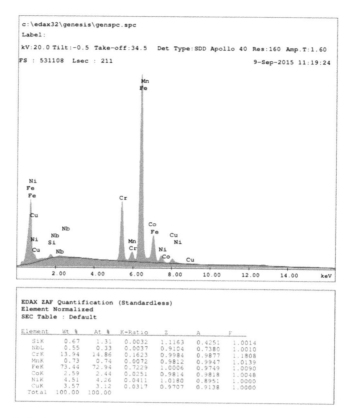

Figure 30.4 Example EDS spectrum for Build 3 material at 20 kV.

used in detection. Transition metals such as chromium, iron, copper, and nickel are typically detected even at extremely low concentrations. Conversely, the low energy X-rays produced by carbon, nitrogen, and oxygen atoms are much more difficult to detect and quantify. In this study, a setting of 20 kV was used to strongly distinguish the Ni, Fe, Cr, and Mn K-lines.

30.3.1.3   *Electron backscatter diffraction*

EBSD technique is used in this research to determine the relative composition of martensite, austenite, and ferrite present in each sample as well as the size and orientation of the grains. EBSD uses the beam of a scanning electron microscope (SEM) to collect crystallographic information about the microstructure of a material.[12] It uses a detector to reveal patterns diffracted from interaction of the beam with a point of interest on the sample. An algorithm is used for pattern recognition and indexing as shown in Figure 30.5. These patterns are then translated and turned into detailed maps showing the grain morphology, orientations, and boundaries of the sample region of interaction. This study conducts two scans of each material of interest, an area overview, and a small-scale focus.

Although powerful, this technique is not perfect. The method has difficulty in distinguishing between phases with similar crystal structures.[12] In 15–5PH, the martensite c to a ratio is close to 1, therefore it is difficult for the technique to differentiate between

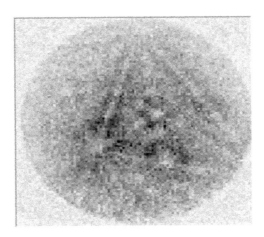

Figure 30.5 Image of a Kikuchi pattern found using EBSD.

the pattern produced by the body-centered tetragonal (BCT) martensite and the body-centered cubic (BCC) ferrite. For ease of comparison, in the EBSD scans used for this research, ferrite is selected to classify both the grain morphology and misorientation that is used to distinguish martensite from delta ferrite.

30.3.2 Split Hopkinson pressure bar

All test matrices in this study contain three groups according to heat treatment condition. The samples with no heat treatment are considered analogous to condition *A*, those with H900 heat treatment represent a *peak age*, and those subjected to H1025 heat treatment signify an *overage* condition. Additionally, specimens are designated by build orientation: horizontal or vertical. Compression SHPB, indirect tension, and direct tension SHPB tests were conducted within this study.

A SHPB, or Kolsky bar, is the most widely used characterization tool for the mechanical response of materials deformed at high strain rates (10^2–10^4 s^{-1}).[13] It measures the effect of a controlled impact by analyzing the stress wave propagation through the test apparatus and the material. With slight variations in setup, the SHPB can perform compression, direct tension, or indirect tension tests. Impact velocity, bar material, and specimen size are variables to achieve different strain rates.

The compression SHPB setup is composed of two elastic bars with a small 0.2 inch right cylinder specimen fixtured between the two bar ends. This setup is shown in Figure 30.6a.[13] When the incident (or input) bar is loaded by external impact, a compressive stress wave is generated and propagated toward the specimen. The strain conditions and reactions can be determined by measuring the reflected and transmitted waves through the bars.

Changing the setup enables the SHPB to apply a tension wave, either directly or indirectly. Both methods were used in this research. A round *dogbone* specimen with threaded ends, as shown in Figure 30.7, is screwed into the ends of each bar in the test section. The specimens used to conduct the tests of this material are virtually identical except for the threading specification. In the direct test, a tubular striker is driven by either a gas gun or a spring system. This tube slides on the incident bar until it impacts a flange or hard stop at the end of the incident bar. A tensile pulse is generated in the incident bar that propagates

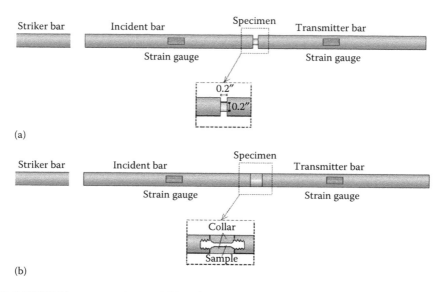

(a)

(b)

Figure 30.6 SHPB (a) compression and (b) indirect tension test setup.

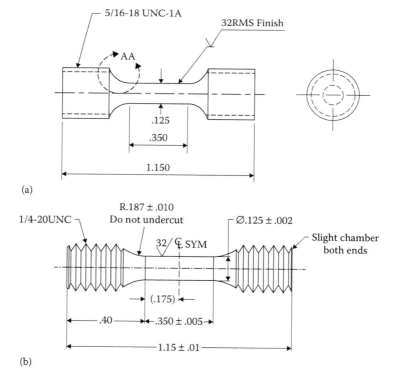

(a)

(b)

Figure 30.7 Examples of tension SHPB samples used in this research: (a) SHPB indirect tension sample and (b) SHPB direct tension sample.

to the specimen, subjecting it directly to tension. The indirect test uses a compression bar modified as illustrated in Figure 30.6b.[13] The main difference is the rigid collar placed over the specimen; on striker impact, the initial compression wave passes through the collar and transfers into the transmission bar, leaving the specimen virtually untouched. At the free end of the transmission bar, it is reflected back as a tensile wave. When this wave arrives at the specimen, the rigid collar cannot support the tensile wave and the specimen is subjected to a dynamic tensile pulse.[13]

30.4   Results

In this study, five AM *builds* were manufactured to obtain cylindrical specimens for examination. Only one build appears to match the constituency of the intended material, 15–5PH. It is unclear at this time why the resulting composition of the others varied from those parameters; the manufacturer is attempting to determine the root cause. However, this circumstance presented an opportunity to determine the effect of compositional variability on the microstructure and dynamic performance. Build 3, appearing compositionally closest to 15–5PH and utilized in all tests, provided a standard for comparison.

30.4.1   EDS results

EDS was used to gain a rough idea of the potential variation in the alloying composition between the builds. These results are not exact, particularly regarding the lower atomic mass and trace elements, but gives a good estimate of the relative amounts. The resultant percentage by weight found for each of the builds using EDS are summarized in Table 30.1.

30.4.2   EBSD results

EBSD scans conducted on the prepared samples of each build reveal the microstructure differences resulting from the build and composition variations. The software was

Table 30.1 Approximate percentage weight of alloy composition expected for 15–5PH and those of the 5 builds studied

Alloying elements	15–5 (AISI)	Build 1	Build 2	Build 3	Build 4	Build 5
C	0.07[a]	–	–	–	–	–
Mn	1	0.00	0.49	0.73	0	0
Si	1	0.47	0.48	0.67	0.70	0.79
Cr	14.00–15.50	9.40	9.06	13.94	12.73	19.06
Ni	3.50–5.50	9.50	9.01	4.51	6.14	8.275
Mo	0	1.79	1.93	0	0.82	0
Nb	0.15–0.45	0	0	0.55	0	0.37
Cu	2.50–4.50	2.68	2.50	3.57	2.91	4.51
Fe	71.9–77.7	72.18	70.44	73.44	76.70	66.99

[a] C cannot be semiquantitatively assessed using EDS.

set to find BCC-Fe (ferrite) and FCC-Fe (austenite) because EBSD does not distinguish well between the crystalline structures of the BCC ferrite and the slightly tetragonal martensite in this material. The nonheat treated samples reveal martensitic compositions of builds 1, 3, 4, and 5 of varying grain sizes and orientations. Build 2, however, shows an austenitic/ferritic structure.

30.4.2.1   EBSD Build 1

Build 1 consists of 15 AM specimens manufactured in the vertical orientation into 0.2 in. diameter by 0.2 in. length cylinder specimens intended directly for SHPB compression tests with no additional machining. The EDS compositional values show roughly equal amounts of Cr and Ni, approximately 9.5% by weight, respectively. As this build was vertically oriented and the EBSD scanned the top (circular section) of the sample, the resultant images are within the build plane. The EBSD scans conducted reveal a primarily martensitic structure with tiny indication of retained austenite. Therefore, it is expected that the samples of this build will age harden. Figures 30.8 and 30.9 show Build 1 and Build 2 of EBSD. Although the resolution of the figures appear to be the same, they are actually different in their original full-color plots.

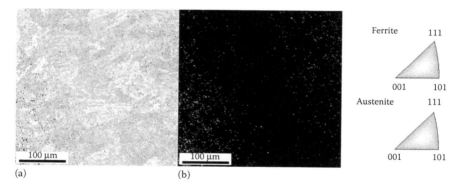

Figure 30.8 Build 1 EBSD, 100 µm scale: (a) Martensite and (b) Austenite.

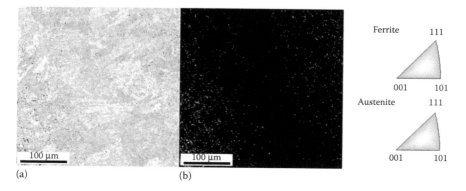

Figure 30.9 Build 2, EBSD, 100 µm scale: (a) Martensite and (b) Austenite.

30.4.2.2   EBSD Build 2

Build 2 is also a vertically oriented build of 30 samples, produced via AM into 0.5 in. diameter by 2 in. long cylinders. Tension SHPB test specimens of this build were machined into both indirect and direct tension SHPB test specimens. Even though the EDS approximates of the Cr and Ni compositions are similar, approximately 9% respectively, the scan of Build 2 shows a much larger percentage of retained austenite, approximately 21%. This significant amount of retained austenite is expected to lower the strength but increase the strain to failure of the resultant material. It will also decrease the age hardening response. The EBSD scans in Figure 30.9 are of the rectangular cross section, not within the build plane.

30.4.2.3   EBSD Build 3

Build 3 is the only build, by EDS evaluation, appearing to compositionally approach the 15–5PH SS. Of the 30 samples, 0.5 in. diameter by 2 in. long cylinders manufactured in the horizontal direction are machined into both compression and tension SHPB specimens. Before machining, the cylinders exhibited a much rougher surface than the vertically oriented build 2 and appeared significantly bowed. This bowing raises concerns that the test results were affected by potential residual stresses induced by the build. The EBSD scan of the rectangular cross section of this horizontally oriented material is shown in Figure 30.10: a finely grained martensite with a very small amount of retained austenite. This material is expected to age harden with heat treatment.

30.4.2.4   EBSD Build 4

According to the EDS results, Build 4 is compositionally more like build 3 than the other builds, although it has approximately 1% less Cr and 1.5% more Ni by weight. Following the guidelines of the composition diagrams, this difference in constituency may increase the amount of retained austenite, but otherwise this material is expected to exhibit very similar physical and material properties to 15–5PH.[5] Of the 30 samples, the 0.5 in. diameter by 2 in. long cylinders, 15 were made in the vertical orientation and 15 in the horizontal. They are also machined into tension SHPB specimens. The scan in Figure 30.11 of semicircular cross section (across the build layers) for the horizontal build reveals a slightly larger grained martensitic structure than Build 3. It has a very small amount of austenite detected in the nonheat treated condition, only about 3%.

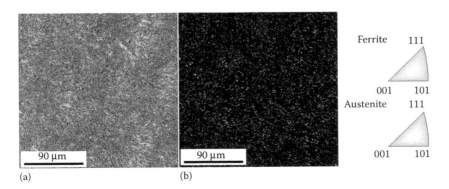

(a)   (b)

Figure 30.10 Build 3, 90 μm scale: (a) Martensite and (b) Austenite.

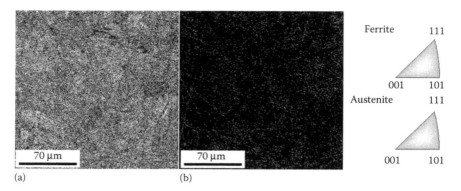

Figure 30.11 Build 4, 70 μm scale: (a) Martensite and (b) Austenite.

30.4.2.5 EBSD Build 5

Build 5 is quite compositionally different than the other builds. Where builds 1, 2, and 4 had too little Cr and too much Ni for 15–5PH, build 5 has more of both. It appears to approach the constituency of the 300 family, most closely 304. This variance is unexpected because it is made using fresh powder immediately after a thorough cleaning and system check of the EOS machine to avoid the compositional variance observed in the previous builds. According to the compositional diagrams, this high alloy material is likely to have martensite, austenite, and ferrite present.[4] However, according to the traditional Schaeffler diagram, the ferrite ratio expected is relatively low, and so the material is likely to be mostly martensitic.

As in build 4, build 5 consists of 15 AM 0.5 in. diameter by 2 in. long cylinders in the vertical orientation and 15 in the horizontal that were all machined into tension SHPB specimens. However, the microscopy samples are one end of a cylinder of each orientation. This may prove less representative of the center of the material. Figure 30.12 is a scan of the vertical build, in the build plane. These maps show an extremely fine grained material assessed as martensite, although it does not appear to look like classical martensite; it does not display the distinctive grain morphology of ferrite and has little to no austenite.

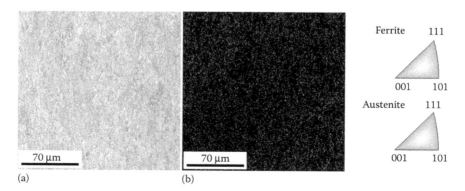

Figure 30.12 Build 5, 70 μm scale: (a) Martensite and (b) Austenite.

30.4.3   Quasi-static tests

Quasi-static testing is utilized to enable measurements of the mechanical parameters of the AM material in terms comparable to the reported values of the traditional 15–5PH material. Reported typical tensile strengths for 15–5PH H900 is 1380 MPa and for H1025 is 1170 MPa.[8] The percentage elongation is 10% longitudinal, 6% transverse for H900, and 12% and 7% respectively for H1025.[4] The results of testing builds 2 and 3 are in given in Table 30.2. These builds are selected because build 3 is most similar in composition to the wrought material and build 2 exhibited the most incongruous behavior. The results clearly illustrate the difference between the two. Table 30.2 mostly indicated that build 2 material exhibits no change with heat treatment except for a loss of ductility from the nonheat treated condition. In contrast, build 3 shows a distinct response to heat treatment by increasing from a tensile or max engineering strength of 1292–1572 MPa at H900 condition, approximately 200 MPa higher than predicted from the traditional values. The additional variable of build orientation is also present in addition to material composition, possibly affecting the offset yield, ultimate tensile strength, and percentage elongation. However, it is worth noting that the H900 in build 2 is approximately 20% lower than expected from the values of its traditional counterpart. Build 2 also demonstrates very little rise after yield before necking begins, meaning very little strain hardening. Build 3 has a very apparent strain hardening trend, albeit with an unexpected region of a concave rise after the linear elastic region. This is probably due to cracking observed in the material, likely resultant from residual stresses introduced during manufacturing. Both appear much more ductile than the minimum published values, although the percentage elongation in build 2 is higher than build 3.

30.4.4   Split Hopkinson pressure bar tests

The *Split Hopkinson pressure bar* (SHPB) tests, like the quasi-static tests, incorporate tests to compare the behavior of the three heat treatment conditions and the build orientation. If the tested material is martensitic, it should show an increase in strength and loss of ductility at H900 heat treatment and a lesser increase in strength and loss of ductility at H1025. The exact effect of the build orientation is not as easily predicted, although the influence of the sintering process in the different planes is expected to introduce some anisotropy. The results of SHPB compression tests conducted on build 1 and build 3 materials, indirect SHPB tension tests conducted on the build 2 and build 3 specimens, and direct SHPB tension tests on build 2, 3, 4, and 5 samples are shown in Tables 30.3 and 30.4.

Table 30.2 Selected quasi-static results

Build	Build 2			Build 3		
Orientation	Vertical			Horizontal		
Condition	No heat treatment	H900	H1025	No heat treatment	H900	H1025
Yield stress (MPa)	979.41	956.05	931.46	767.54	1119.65	1110.23
Max eng stress (MPa)	1076.95	1083.62	1092.45	1291.77	1571.79	1460.78
Elongation (%)	33.0	31.2	29.0	26.4	21.1	26.7

Table 30.3 Test results for builds 1, 2, and 3

Build		Method	Strain rate	Heat treatment	Approx max σ	Max E
1	Vertical	Compression	500 s^{-1}	No HT	1463.57	0.0976
				H900	1519.62	0.0951
2	Vertical	Tension, indirect	500 s^{-1}	No HT	937.24	0.2163
				H900	928.47	0.2253
				H1025	968.40	0.2207
			800 s^{-1}	No HT	1030.41	0.2233
				H900	1042.54	0.2163
				H1025	1025.71	0.2160
		Tension, direct	550 s^{-1}	No HT	1153.00	0.2800
				H900	1142.00	0.2730
				H1025	1144.00	0.2830
			875 s^{-1}	No HT	1182.00	0.2510
				H900	1158.00	0.2200
				H1025	1152.00	0.2490
3	Horizontal	Compression	400 s^{-1}	No HT	1560.26	0.2216
				H900	1884.62	0.1050
				H1025	1722.98	0.1576
		Tension, indirect	450 s^{-1}	No HT	1219.77	0.2237
				H900	1484.39	0.2023
				H1025	1457.06	0.2117
			750 s^{-1}	No HT	1180.19	0.2317
				H900	1506.77	0.2050
				H1025	1478.23	0.2060
		Tension, direct	500 s^{-1}	No HT	1326.00	0.2900
				H900	1644.00	0.2520
				H1025	1492.00	0.2590
			850 s^{-1}	No HT	1345.00	0.2780
				H900	1607.00	0.2290
				H1025	1464.00	0.2250

30.4.4.1 Compression tests

The results of compression tests conducted on Build 1 and Build 3 samples in Table 30.3 demonstrate a stark difference between these materials. Even though both appear stronger and less ductile on heat treating to H900, the 4% increase in strength of the build 1 material does not compare with the dramatic 20% jump as seen in the build 3 material. The ductility of the build 3 material also shows greater variation, the maximum strain drops from 0.220 to 0.105. Overall, build 1 has much less area under its curve, meaning lower toughness.

30.4.4.2 Indirect tension tests

The results of a full complement of SHPB indirect tension tests conducted on build 2 and build 3 materials are also recorded in Table 30.3. Although the results between the builds are noticeably different, the standard error between individual samples

Table 30.4 Test results for builds 4 and 5

Build		Method	Strain rate	Heat treatment	Approx max σ	Max E
4	Vertical	Tension, direct	500 s⁻¹	No HT	1360.00	0.2220
				H900	1730.00	0.1680
				H1025	1562.00	0.1830
			850 s⁻¹	No HT	1352.00	0.2010
				H900	1715.00	0.1580
				H1025	1526.00	0.1690
	Horizontal	Tension, direct	450 s⁻¹	No HT	1340.00	0.2450
				H900	1672.00	0.2010
				H1025	1557.00	0.1850
			800 s⁻¹	No HT	1377.00	0.2720
				H900	1677.00	0.1670
				H1025	1547.00	0.1850
5	Vertical	Tension, direct	475 s⁻¹	No HT	1367.00	0.2130
				H900	1651.00	0.1760
				H1025	1466.00	0.1720
			700 s⁻¹	No HT	1376.00	0.1980
				H900	1626.00	0.1520
				H1025	1466.00	0.1720
	Horizontal	Tension, direct	475 s⁻¹	No HT	1332.00	0.2790
				H900	1655.00	0.2300
				H1025	1466.00	0.2490
			850 s⁻¹	No HT	1345.00	0.3070
				H900	1650.00	0.2170
				H1025	1497.00	0.2100

in each group is less than 4% and is often less than 1%. This suggests that the tests are repeatable, the material is consistent within each build, and that the discernible trends between the test variables are likely due to the variables and not as a product of test inaccuracies. Build 2 overall shows a standard error of the entire population of 19 samples of no more than overall 2% and a variation of less than 2.2% between heat treatment conditions, whereas the build 3 displays clearly different behaviors, typically overall 6% standard error as well as a minimum of 5% up to 20% variation between the heat treatments.

30.4.4.3 Direct tension tests
The direct tension tests incorporate the results for one specimen of each orientation and heat treatment in most cases. These results display the influence of both test variables. The direct tension graphs were typically slightly higher and longer than their indirect counterparts but overall showed very similar trends. The graph of the high rate test outcomes is shown in Figure 30.13.

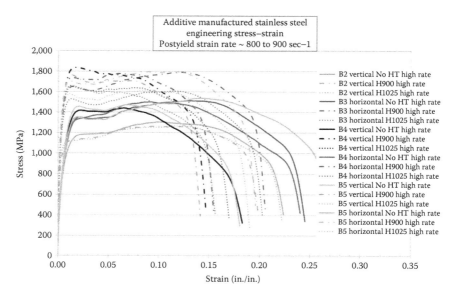

Figure 30.13 Direct tension high rate graph.

30.5 Discussion

Examination and testing of the five different builds showed some clear trends correlating with the observed microstructure. From the tests, and data presented in Figure 30.13, Tables 30.3, and 30.4, all of the materials except for build 2 exhibited their highest strengths after H900 heat treatment and *overaged* behavior at the H1025 condition. The orientation also makes a difference; the vertically oriented segments of builds 4 and 5 show distinctly different profiles than their horizontal counterparts. This anisotropy noticeably exchanges a slight strength advantage for the vertical builds with more strain hardening and significant higher ductility in the horizontal build material. The grain size is expected to play a part in strength values, but because builds 3, 4, and 5 are all fine grained, this is inconclusive at present.

For the considered applications, compositional variation is a potential issue. Figure 30.13 is a good illustration of the variation among the test results. This graph depicts the results of the *high* strain rate direct SHPB direct tension test. Of particular concern is the large discrepancy in performance from build 2 material. With a variation as much as 500 MPa (27%) in ultimate yield strength in the SHPB test from build 3. The result could negate a safety factor and prove catastrophic for a critical aircraft part. The tests show that the material in builds 1, 3, 4, and 5 are likely to be similar enough in heat treatment and performance under tension, to not immediately indicate that they are actually different materials. However, the graphs show that there is possible a reason for concern due to differing yield strengths, ductility, and toughness. The unexpected composition of build 1 and the strain hardening behavior and ultimate tensile strength of builds 4 and 5 will undoubtedly prove problematic for prediction of failure and repair. Although not tested, the fatigue, creep, and corrosion properties are potentially different than expected and will likely cause issues when installed on an aircraft.

30.6   Conclusions

Although each of these builds was made from the same powder type and process, the resultant material had significant compositional variations with different microstructure and dynamic properties. This adds additional design considerations when using AM materials for finished parts. Until the source of the compositional variability is found, or the potential impact is included into a safety factor consideration, certification of this process for AM manufacture is difficult. The wide disparities seen are an argument for the necessity to ensure that the composition of each AM part delivered is known. Additionally, considerations including build orientation, build support, and heat treatment change the resultant properties and will require full characterization of the material found to match the intended usage.

Acknowledgments

First printed in the 57th AIAA/ASCE/AHS/ASC Structures, Structural Dynamics, and Materials Conference, AIAA SciTech, Conference Proceedings in 2016 in San Diego, California.

References

1. Parker, J., Planning a larger role for 3-D printing. *Tinker Air Force Base Public Affairs*, October 19, 2015.
2. Gibson, I.W., D.W. Rosen, and B. Stucker, *Additive Manufacturing Technologies*. Springer, New York, 2014.
3. Kalpakjian, S., *Manufacturing Engineering and Technology*, 3rd ed. Addison-Wesley Publishing Company, Reading, MA, 1995.
4. Davis, J.R., *Stainless Steels. ASM Specialty Handbook*. ASM Internationals, Materials Park, OH, 1994.
5. Heat treating of stainless steels. In *Metals Handbook*, ASM International handbook Commette Vol. 2, 8th ed., pp. 243–253, 1964.
6. Hall, A.M., Hoenie, A.F., and Slunder, C.J., *Thermal and Mechanical Treatment for Precipitation-Hardening Stainless Steels*. NASA Special Publication, Vol. NASA-SP-5089, 1967.
7. i3DMFG, EOS Stainless Steel PH1 for EOSINT M270. Material Data Sheet, 2015.
8. Allegheny Technologies Incorporated, ATI15-5 Technical Data Sheet. March 15, 2012.
9. Hoenie, A.F., and Roach, D.B., New developments in high-strength stainless steels. Defense Metals Information Center Report 233, Fort Belvoir, VA, January 3, 1966.
10. Briggs, D., Brady, J., and Newton, B. *Scanning Electron Microscopy and X-Ray Microanalysis*, p15, Wiley, New York 2000.
11. EDAX. *Energy Dispersive Spectroscopy*. John Wiley & Sons, Chichester, UK, 2015.
12. Oxford Instruments. *EBSD Explained from Data Acquisition to Advanced Analysis*. Oxford Instruments, Oxfordshire, UK, 2015.
13. Chen, W.W., and Song, B., *Split Hopkinson (Kolsky) Bar Design, Testing and Applications*. Springer, New York, 2011.

chapter thirty one

Investigation of the high-cycle fatigue life of selective laser melted and hot isostatically pressed Ti-6Al-4V

Kevin D. Rekedal and David Liu

Contents

Experimental research was conducted on the fatigue life of selective laser melted Ti-6Al-4V. A thorough understanding of the fatigue life performance for additively manufactured parts is necessary before such parts are utilized as production end-items for real-world applications such as the rapid, on demand, and 3D printing of aircraft replacement parts. This research experimentally examines the fatigue life of Ti-6Al-4V material specimens built directly to net shape and then either stress-relieved or hot isostatic pressing (HIP). Experimental results will help determine whether HIP effectively reduces porosity and increases fatigue life when the specimen surface is not machined to remove surface roughness from the additive manufacturing (AM) process.

31.1   Introduction and background

Layer-based AM technology, commonly known as 3D printing, is widely utilized as a cost-effective method for rapid prototyping with polymer-based materials. More recently, AM technology has expanded to allow the processing of metals. Several metal-capable 3D-printing machines have been developed and marketed for commercial use.[1] A number of these commercial machines have the capability to produce parts with aerospace metals including high strength steels, nickel-based alloys, and titanium.[2] The availability of high-strength metals and machines capable of precision 3D manu-facturing provides opportunities for the rapid manufacturing of end-use parts for a wide array of applications. Potential aerospace applications include the fabrication of reduced weight, topology-optimized components typically impossible or impractical to manufacture by traditional means, and rapid, on-demand manufacturing of replacement parts when an existing spare is not immediately available.

Titanium-alloy aircraft parts, such as those made from Ti-6Al-4V (Ti-64), are widely used in both commercial and military aircraft systems. Due to their high material costs and relatively difficult machining characteristics, titanium parts were identified as a first likely application area for the AM of spare parts for Department of Defense (DoD) aircraft. However, there are many challenges and barriers that require attention before such additively manufactured components are qualified for use.[3] One possible approach to address some of these barriers is to demonstrate the mechanical properties of AM parts that can meet the same design requirements as their wrought-material equivalents.[4] For many of the AM powder-based metals available, and for selective laser melting (SLM) Ti-64 specifically, the strength and hardness properties meet and often exceed the typical values for the wrought material.[5,6]

Recent studies on fatigue performance have indicated that the high-cycle fatigue life of SLM Ti-64 is considerably lower than typical wrought material when the AM material is left in its as-built state, absent of postprocessing heat treatments.[4,7,8] Assessments of postprocessing heat treatment have shown a marginal benefit on increasing the fatigue life of SLM Ti-64 and furthermore, specimens that were hot isostatic pressing (HIP) had a fatigue life nearly equivalent to the expected life for typical wrought Ti-64.[8] Examination of the fracture surfaces indicates porosity voids within the material, an inherent characteristic of SLM-produced metal parts, is the primary driver for fatigue failure in SLM Ti-64 that has not been HIP-treated.[4,8] Recently published SLM Ti-64 fatigue testing results from Edwards and Ramulu found as-built specimens, where the surface was machined to net shape to remove the effects of surface roughness, and were not heat treated, had no discernible impact on fatigue life due to the initiation of cracks from internal pores within the material.[4]

The purpose of this work is to present the results of additional research on the fatigue life of SLM Ti-64. Due to the large number of processing variables involved with AM processes, and a high degree of data scatter in experimental results published to-date, additional fatigue life data are desired to gain a greater understanding of the fatigue life implications from various processing parameters and postprocessing treatments.[4] Existing research has assessed the SLM Ti-64 fatigue life impacts of various parameters including build orientation, surface machining, heat treatment, and HIP.[4,7,8] This new research focuses on the fatigue life impact of HIP when the surface has not been machined to remove the surface roughness. This condition is representative of the direct manufacture of an aircraft replacement part to net dimensions when surface machining during postprocessing is impractical or would negate the benefits of AM.

31.2   Methodology

To assess the impact of HIP on the high-cycle fatigue life of Ti-64, stress-life (S-N) plots were developed to compare the fatigue life of HIP-treated specimens with that of a stress-relieved baseline. At the completion of this study, a total of 65 fatigue specimens and 12 tensile specimens will be tested to provide stress-life data and static material properties. Fracture surfaces were examined using optical and scanning electron microscopes (SEMs) to gain insight into the fatigue fracture characteristics, crack initiation sites, and defects resulting from the SLM process.

31.2.1   Specimen manufacture

Ti-64 fatigue and tensile testing specimens were fabricated using an EOSINT M 280 machine that utilizes EOS GmbH-Electro Optical Systems' proprietary direct metal laser sintering (DLMS) process.[9] The M 280 has a 250 mm × 250 mm × 325 mm build volume and is optionally equipped with a 400Watt Ytterbium fiber laser.[10] Specimens were printed in the XZY orientation when described using the orthogonal orientation notation defined in ISO/ASTM Standard 52921 as shown in Figure 31.1.

The process parameters were established by the *Ti-64 Speed 1.0* EOS parameters that sets the layer thickness to 60 μm. Although a 30 μm layer thickness *performance* parameter set is available, per the EOSINT M 280 technical specification, the *speed* parameters have an optimal balance between production speed and 12 surface quality when using the 400 W laser.[12]

Tensile specimens were designed in accordance with the dimensions for a subsize rectangular tension test specimen specified by ASTM Standard E8/E8M.[13] Accordingly, the tensile specimens were designed to a gauge length of 25 mm, 3 mm thickness, and overall length of 100 mm as depicted in Figure 31.2. Fatigue specimens were designed

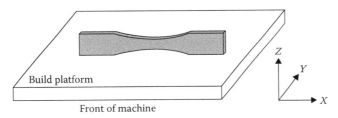

Figure 31.1 Build orientation of specimen using coordinate system defined by ISO/ASTM 52921. (From ISO/ASTM Standard 52921:2013(E), Standard Terminology for Additive Manufacturing Technologies-Coordinate Systems and Test Methodologies, 2013.)

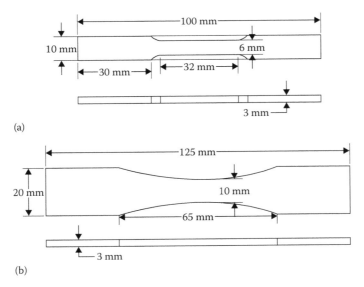

Figure 31.2 Test specimen design dimensions in accordance with ASTM E8/E8M and ASTM E466-07. (a) Tensile specimen and (b) Fatigue specimen.

in accordance with ASTM Standard E466-07.14.[13] Flat, dog-bone-shaped specimens were designed to match the gripping devices in available testing equipment. A continuous radius curvature was selected for the test section to drive failures toward the center of the specimen. Additionally, because the specimens were manufactured using a layer-by-layer additive process, the inclined surface of the continuous radius resulted in a *stair-stepping* effect along the edge of the specimen as illustrated in Figure 31.3 that may be representative of an AM-produced component with inclined surfaces fabricated directly to net dimensions. The fatigue specimens had a reduced width of 10 mm at the center of the test section, 3 mm thickness, and overall 125 mm length as depicted in Figure 31.2.

To fabricate the desired number of specimens, three separate production builds were required due to the number of specimens that could be built on a single substrate plate. After each build assembly cooled to room temperature, the specimens were cut from the substrate plate using a wire electrical discharge machining (EDM) process. Support material was required beneath the reduced area section of the test specimens to prevent the center of the bar from collapsing on itself during construction. The part can be designed to either utilize solid material for the support or allow the EOS software to add reduced density support material with a foam-like density. To remove the support material, the specified curvature for the lower half of the test specimen was traced by the wire EDM as the specimens were cut from the build plate as illustrated in Figure 31.4. This resulted in one edge of the reduced-area section being built directly by the SLM process with a stair-step characteristic and the opposing edge cut relatively smooth by the wire EDM. Optical microscope images are shown in Figure 31.5 to highlight the differences in surface appearance between the as-built edge and the wire EDM edge. Following removal from the build plate, the specimens were cleaned with an isopropyl alcohol solution and then either stress-relieved or HIP. For the stress-relieved configuration, specimens were processed in accordance with the parameters for heat-treated material identified in the material data sheet for EOS Ti-64 published by

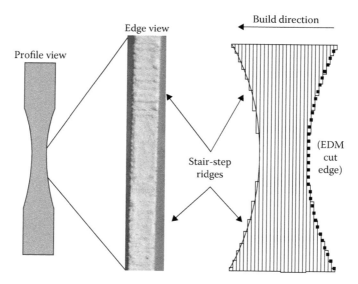

Figure 31.3 Depiction of the stair-step effect on inclined surfaces resulting from the layered build process.

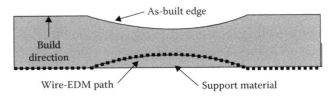

Figure 31.4 Cutting path of the wire EDM to remove specimens from the substrate plate.

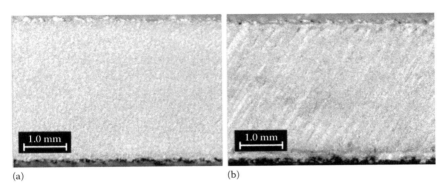

Figure 31.5 Comparison of the surface quality between the as-built edge and the edge cut by wire EDM. (a) Wire EDM surface and (b) As-built edge surface.

the manufacturer. This data sheet specifies a heat treatment at 800°C for 4 hours in an argon atmosphere in order to produce material properties meeting the minimum material requirements specified by ASTM F1472–08.15. For the HIP configuration, specimens were treated at 899°C for 2 hours at 101.7 MPa in accordance with the HIP parameters specified in ASTM F2924-14.16.

31.2.2 Static material properties

In order to verify the material test specimens manufactured for this study are consistent with the manufacture's published material property data for as-built and heat-treated material, static material properties were determined through tensile testing at room temperature performed in accordance with ASTM E8/E8M.[14] Testing was performed with an MTS systems landmark servo-hydraulic test system equipped with a 25 kN force-capacity load cell. Testing was accomplished on three different material configurations consisting of as-built (no heat treatment), stress-relieved (800°C for 4 hours), and HIP (899°C for 2 hours at 101.7 MPa). Three specimens for each configuration were tested. As per ASTM F2924, tensile properties should be determined following the ASTM E8/E8M test method using a strain rate of 0.003–0.007 mm/mm/min through the yield point.[15] To achieve a target strain rate of approximately 0.005 mm/mm/min through the yield point, a constant crosshead speed of 0.5 mm/min was utilized. Strain data was collected with a MTS model 632.53E-14 extensometer with a 12.7 mm gauge length. Force and strain data were recorded at a 60 Hz sampling rate. Test data was then plotted to obtain the ultimate tensile strength (UTS), modulus of elasticity (E), and 0.2% offset yield strength (YS) using the procedures in ASTM E8/E8M.[13] Elongation at break was verified by measurements of the final gauge length based on gauge-length markings placed on each specimen at a starting length of 25.4 mm.

31.2.3   Fatigue testing

High-cycle axial fatigue testing on the stress-relieved and HIP configurations was performed utilizing the same MTS landmark system used for tensile testing. Force-controlled, constant-amplitude testing was performed at room temperature following the procedures in ASTM 466–07 with the exception of surface preparation because testing of the as-built surface was desired for this study. Tests were conducted using a constant-amplitude sine wave at a frequency of 60 Hz and stress ratio of $R = 0.1$. Failure criteria was established as specimen separation or test run-out at 10 million (10^7) cycles. Various maximum stress levels were selected for the purpose of plotting a S-N curve for both the stress-relieved and HIP configurations. For the stress-relieved configuration, initial stress levels were chosen at various increments ranging from 220 to 600 MPa to provide a preliminary shape of the S-N curve and an initial approximation of where the mean fatigue strength lies. Based on these initial data points, the general shape of the S-N curve was modeled by the general expression in Equation 31.1 where N is the number of cycles, S_{max} is the maximum applied stress, A_1 and A_2 are the curve fitting parameters, and $\bar{\mu}$ is the mean fatigue strength.

$$\log N = A_1 + A_2(S_{max} - \bar{\mu}) \tag{31.1}$$

The mean fatigue strength at 10^7 cycles was determined using the up-and-down staircase method outlined in ASTM STP 588.[16] Using the staircase method, the first sample is run at the estimated fatigue limit. If the specimen fails, the next specimen is run at a lower stress level. If the specimen survives to 10^7 cycles, the next specimen is run at a higher stress level. Uniform incremental steps of 10 MPa for each stress level in the staircase were used to permit data analysis using the Dixon and Mood method described in Ref. 18. Using notation adapted from Pollack, the mean fatigue strength ($\bar{\mu}$) and standard deviation ($\bar{\sigma}$) are found using Equations 31.2 through 31.4.[17]

$$\bar{\mu} = S_0 + s \times \left(\frac{B}{A} \pm 0.5 \right) \tag{31.2}$$

$$\bar{\sigma} = \begin{cases} 1.62 \times s \times \left(\dfrac{A \times C - B^2}{A^2} + 0.029 \right), \text{ for } \dfrac{A \times C - B^2}{A^2} \geq 0.3 \\[2ex] 0.53 \times s, \text{ for } \dfrac{A \times C - B^2}{A^2} < 0.3 \end{cases} \tag{31.3}$$

$$A = \sum_{i=0}^{i\max} m_i, B = \sum_{i=0}^{i\max} im_i, C = \sum_{i=0}^{i\max} i^2 m_i \tag{31.4}$$

In this previous set of equations, i is an integer corresponding to the stress level where i_{max} represents the highest stress level in the staircase. Note in Equation 31.2, a positive or negative value is used inside the brackets. The addition operator is used when the majority of specimens in the staircase are failures, and the subtraction operator is used when the majority of specimens in the staircase are survivals after 10^7 cycles. For the case where the majority of specimens are survivals, $i=0$ corresponds to the minimum stress level at which a failure was observed, and m_i denotes the number of specimens failing at each stress level. Similarly, for the case where the majority of specimens are failures, $i=0$ corresponds to the minimum stress level at which a survival was observed, and m_i denotes the number of specimens that survived at each stress level. The maximum stress corresponding to $i=0$ is denoted S_0 and the symbol s denotes the size of the uniform stress increment.

31.3   Experimental results

The tensile testing results for the as-built, stress-relieved, and HIP configurations are shown below in Figure 31.6. From the data in Figure 31.6, the UTS, YS, and E for each specimen were determined. The average values are summarized in Table 31.1 along with the minimum and typical values as per the manufacture's material data sheet.[18] Based on the experimental tensile test results for the as-built configuration, the average UTS is 89.3 MPa below the typical value published by the manufacturer and the average YS is 121.8 MPa below the typical expected value. For the stress-relieved configuration, the UTS and YS are also below the typical values reported by the manufacturer, but exceed the minimum values to meet the requirements of ASTM F1472-08 by 6.9 MPa based on the average UTS and 2.4 MPa based on the average YS. Although the static material properties for the as-built and stress-relieved configurations are below the typical expected values reported by the manufacturer, meeting the minimum specified UTS and YS thresholds for the stress-relieved configuration provides some level of confidence for the SLM machine setup, processing parameters, and heat-treatment parameters used to manufacture specimens for this study resulted in samples with static material properties representative of the typical range expected from the EOS DMLS process.

Fatigue testing to-date includes 15 specimens in the stress-relieved configuration conducted at stress levels ranging from 200 to 600 MPa. The results of these tests are shown in Figure 31.7. When plotted in a log-linear scale, the data points in the finite-life region are approximately linear with a horizontal asymptote at the stress level corresponding to the mean fatigue strength. Therefore, the trends in the experimental data appear to agree with the theoretical model presented in Equation 31.1. The fitting parameters were determined to be 12.285 and −0.0111, respectively. To determine the mean fatigue strength, the staircase data shown in Figure 31.8 were analyzed using the Dixon–Mood equations as shown in Equations 31.2 through 31.4. At the present time, seven staircase data points were performed. The original Dixon–Mood theory is based on large sample theory requiring sample sizes on the order of 40–50. However, as reported by Pollack, and research done by

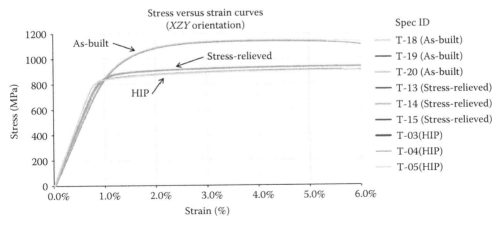

Figure 31.6 Experimental stress versus strain curves for the as-built, stress-relieved, and HIP configurations.

Table 31.1 Experimental static material properties of SLM Ti-64 in the XZY orientation compared to the manufacture's published material specifications

		As-built	Stress-relieved	HIP
UTS ± STD (MPa)	As tested	1140.7 ± 5.0	936.9 ± 3.6	910.1 ± 2.9
	EOS min	NA	930	NA
	EOS typical	1230 ± 50	1050 ± 20	NA
YS ± STD (MPa)	As tested	938.2 ± 7.7	862.4 ± 3.1	835.4 ± 3.8
	EOS min	NA	860	NA
	EOS typical	1060 ± 50	1000 ± 20	NA
E ± STD (GPa)	As tested	91.8 ± 0.5	98.0 ± 1.2	106.8 ± 1.3
	EOS min	NA	NA	NA
	EOS typical	110 ± 10	116 ± 10	NA

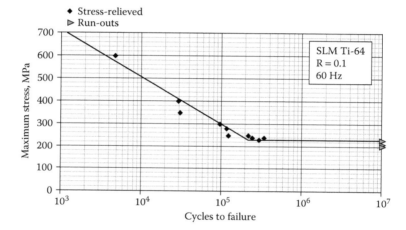

Figure 31.7 Experimental S-N curve for the stress-relieved configuration.

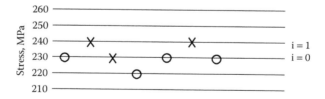

Figure 31.8 Staircase data for the stress-relieved configuration used to determine the mean fatigue strength.

Brownlee et al. has shown that the Dixon–Mood equations provide a reasonably reliable estimate of mean fatigue strength in sample sizes as small as 5–10.[17] From the staircase data in Figure 31.8, it was shown that the majority of the specimens were survivals. As such, the lowest stress level at which a failure was observed, 230 MPa, is denoted as $i=0$ and the subtraction operation in Equation 31.2 is used when calculating the mean fatigue strength. From this data, the mean fatigue strength for the stress-relieved configuration is 231.7 MPa with a standard deviation of 5.3 MPa.

All of the high-cycle fatigue failures in the stress-relieved configuration had cracks initiated from the as-built edge denoted in Figure 31.5. The majority of the stress-relieved

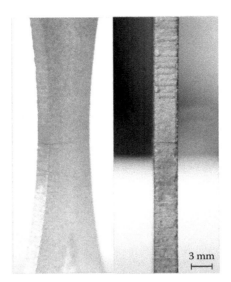

Figure 31.9 Typical fatigue crack location in a stress-relieved specimen initiating from the as-built edge of the specimen.

specimens failed slightly above or below the narrowest point of the specimen as shown in Figure 31.9. The crack initiation point most often occurred between the boundary of the small flat section at the center of the specimen and the first ridge of the stair-stepped region. The small flat section on the edge of the specimen results from the layered build process as illustrated in Figure 31.3. The fracture surfaces were examined under an optical microscope and SEM. Unlike previous studies by Edwards and Ramulu and Leuders et al. that noted a relatively high degree of internal porosity, relatively few pores, and defects that were visible on the fracture surfaces.[4,8] An example of an internal defect is shown in the optical microscope images in Figure 31.10. Although this image indicates the presence of a material void, matching material on the opposing fracture surface suggests this particular defect resulted from a lack of fusion. The rough surface of the fracture surfaces makes it difficult to discern precise fatigue-crack initiation sites. However, as shown in the SEM images in Figure 31.11, stress concentrations from surface defects and micronotches as a result of the unmachined surface appear to be likely contributors to

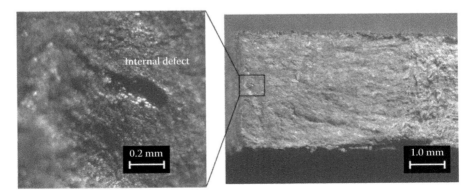

Figure 31.10 Optical microscope image of an internal defect found in a stress-relieved specimen.

mag	HV[kV]	WD[mm]	HFW[μm]
400x	15.00	18.4	640

mag	HV[kV]	WD[mm]	HFW[mm]
53x	15.00	18.6	4.8

Figure 31.11 SEM image of a surface defect in a stress-relieved specimen that is a suspected crack initiation site that resulted in premature failure of the specimen under high-cycle fatigue.

fatigue-crack initiation. As seen in the upper left-hand corner of the fracture surface in the right-hand image in Figure 31.11, the presence of a noticeable surface defect appears to be a likely crack initiation site. The specimen in this figure failed at a stress level of 230 MPa after only 297,000 cycles and was the only failure of the four specimens that were run at this stress level. The premature failure of this specimen is believed to be the result of the defect shown in Figure 31.11.

31.4   Conclusion

Based on axial tension and high-cycle fatigue testing performed to-date on stress-relieved samples of SLM Ti-64 produced by an EOSINT M 280 DLMS machine, the EOS machine is producing consistent material specimens with material properties in general agreement with manufacture's data sheet. Internal porosity observed on the fracture surfaces appears minimal indicating that the manufacturer-set processing parameters are well optimized for the material. However, additional metallographic analysis of polished samples is required to better characterize porosity. The preference of fatigue cracks to initiate from the as-built edge of the specimen rather than the smoother EDM-cut edge suggests high-cycle fatigue failure of the stress-relieved specimens is dominated by cracks initiating from the surface rather than from internal pores or defects. This observation is in contrast to a previous study by Edwards and Ramulu that concluded that the removal of surface defects by machining did not yield a significant increase in high-cycle fatigue life suggesting fatigue-crack initiation was dominated by subsurface initiation.[4] A possible explanation for this difference is the EOS machine used to produce the samples for the present study yielded less porosity than the machine and processing parameters used for the specimens in the study conducted by Edwards and Ramulu. It is also undetermined at this point what impact the microstructure plays in terms of resistance to crack initiation at stress concentration points. A previous study by Van Hooreweder et al. concluded that in machined samples free of surface defects, the inferior fatigue properties of SLM Ti-64 compared to conventionally manufactured parts are likely to be caused by anisotropy in the microstructure as opposed to the presence of pores and other internal defects.[19]

A comparison of the fatigue results obtained in this study to the results of similar past studies and the typical range for wrought material is shown in Figure 31.12.[19–21] Although

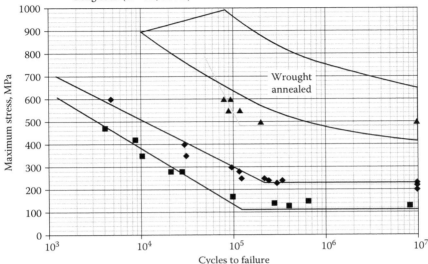

Figure 31.12 Stress-life comparisons of the experimental fatigue data with previous research on stress-relieved Ti-64 specimens and the typical expected range of wrought material.[19-21] The data-fit lines for the previous research were approximated by visual inspection for illustrative purposes.

the machines, surface quality, and test conditions vary between studies precluding a direct comparison of results, a more general comparison indicates that the stress-life data obtained for stress-relieved specimens during this study are within the range of previous studies despite the fact that the specimens for this study were built directly to net shape without any surface machining. Figure 31.12 also highlights high-cycle fatigue life for the unmachined stress-relieved parts fabricated for this study is lower than the expected fatigue life for typical wrought material according to Ti-64 fatigue data published by ASM International.[21]

Additional planned testing as part of this study will test as-built specimens that are HIP-treated to determine if HIP results in a measurable impact to high-cycle fatigue life. Due to the relatively low level of porosity observed in the specimens produced by the EOSINT M 280, it is not expected that HIP will result in the dramatic increase in high-cycle fatigue life observed in a previous study by Leuders et al. The Leuders et al. study reported an increase in mean fatigue life from 93,000 cycles to greater than 2 million cycles for HIP-treated specimens versus those of stress- relieved at 800°C when tested at a 600 MPa stress level.[8] Of central importance to this study is whether the HIP-treated specimens demonstrate improved high-cycle fatigue life when crack initiation is influenced by surface roughness and defects from an unmachined surface. Such data aims to assess whether HIP is a worthwhile endeavor for SLM Ti-64 parts built directly to net shape.

Acknowledgments

This research is sponsored by the Joint Aircraft Survivability Program. The authors would also like to express their gratitude to Dr. Alan Jennings and Dr. Timothy Radsick from the Air Force Institute of Technology for their support of this effort as research committee members and to Dr. Kathleen Shugart and Mr. Michael Velez from UES, Inc. for providing laboratory facilities, training, and support at the Air Force Research Laboratory Materials Characterization Facility.

First printed in the 56th AIAA/ASCE/AHS/ASC Structures, Structural Dynamics, and Materials Conference, AIAA SciTech, Conference Proceedings in 2015.

References

1. Gibson, I., Rosen, D.W., and Stucker, B., *Additive Manufacturing Technologies,* Springer, New York, 2010, Chaps. 1–2.
2. Ruan, J. et al., *A Review of Layer Based Manufacturing Processes for Metals, Solid Freeform Fabrication Symposium,* University of Texas Press, Austin, TX 2006, pp. 233–245.
3. Shipp, S.S. et al., *Emerging Global Trends in Advanced Manufacturing,* Institute for Defense Analyses, Alexandria, VA, 2012.
4. Edwards, P., and Ramulu, M., Fatigue performance evaluation of selective laser melted Ti–6Al–4V, *Materials Science and Engineering,* 598, 2014, 327–337.
5. Koike, M. et al., Evaluation of titanium alloys fabricated using rapid prototyping technologies—Electron beam melting and laser beam melting, *Materials,* 4(10), 2011, 1776–1792.
6. Yu, J., Rombouts, M., Maes, G., and Motmans, F., Material properties of Ti6Al4V parts produced by laser metal deposition, *Physics Procedia,* 39, 2012, 416–424.
7. Chan, K.S., Koike, M., Mason, R.L., and Okabe, T., Fatigue life of titanium alloys fabricated by additive layer manufacturing techniques for dental implants, *Metallurgical and Materials Transactions,* 44(2), 2013, 1010–1022.
8. Leuders, S. et al., On the mechanical behaviour of titanium alloy TiAl6V4 manufactured by selective laser melting: Fatigue resistance and crack growth performance, *International Journal of Fatigue,* 48, 2013, 300–307.
9. ASTM Standard F2792-10, Standard Terminology for Additive Manufacturing Technologies, ASTM International, West Conshohocken, PA 2012.
10. EOS GmbH Electro Optical Systems, System Data Sheet: EOSINT M 280, 2013.
11. ISO/ASTM Standard 52921:2013(E), Standard terminology for additive manufacturing technologies-coordinate systems and test methodologies, ASTM International, West Conshohocken, PA 2013.
12. EOS GmbH Electro Optical Systems, Technical Description: EOSINT M 280, 2010.
13. ASTM Standard E466-07, *Standard test methods for tension testing of metallic materials,* ASTM International, West Conshohocken, PA 2007.
14. ASTM Standard E8/E8M, *Standard test methods for tension testing of metallic materials,* ASTM International, West Conshohocken, PA 2013.
15. ASTM Standard F2924-14, Standard Specification for Additive Manufacturing Titanium-6 Aluminum-4 Vanadium with Powder Bed Fusion, ASTM International, West Conshohocken, PA 2014.
16. ASTM STP588, *Manual on statistical planning and analysis,* ASTM International, West Conshohocken, PA 1975.
17. Pollak, R.D., *Analysis of Methods for Determining High Cycle Fatigue Strength of a Material With Investigation of Ti- 6Al-4V Gigacycle Fatigue Behavior,* Ph.D. Dissertation, Air Force Institute of Technology, Wright-Patterson AFB, OH, 2005.
18. EOS GmbH Electro Optical Systems, Material Data Sheet: EOS Titanium Ti64, 2011.
19. Van Hooreweder, B., Boonen, R., Moens, D., Kruth, J., and Sas, P., *On the Determination of Fatigue Properties of Ti6Al4V Produced by Selective Laser Melting,* 53rd AIAA/ASME/ASCE/AHS/ASC Structures, Structural Dynamics and Materials Conference, Honolulu, Hawaii 2012, pp. 1–9.
20. Gong, H., Rafi, K., Gu, H., Starr, T., and Stucker, B., Analysis of defect generation in Ti–6Al–4V parts made using powder bed fusion additive manufacturing processes, *Additive Manufacturing,* 2014, Vol 1–4, 87–98.
21. Donachie, M.J. ed., *Titanium: A Technical Guide,* ASM International, Metals Park, OH, 1988, p. 189.

chapter thirty two

Impact response of titanium and titanium boride monolithic and functionally graded composite plates[*]

Reid A. Larson, Anthony N. Palazotto, and Hugh E. Gardenier

Contents

Functionally graded materials (FGMs) have gained significant interest within the research community in recent years. FGMs are advanced composites in which local material properties are tailored to suit application requirements by altering the volume fraction ratios of two or more constituents. In this article, the behavior of metal–ceramic FGM plates under low-velocity, medium-energy impact loading is considered using experimental and computational techniques. A series of impact tests were conducted on monolithic and functionally graded plates composed of titanium and titanium boride. The tests were performed using a vertical drop test apparatus in which highly controlled impacts of up to 108 J were delivered to the center of the top surface of each plate. The opposing bottom surface of each plate was instrumented with strain gauges wired into a high-speed data acquisition system to collect strain histories throughout the duration of the impact event.

[*] Reprinted from Larson, R., Palazotto, A., and Gardenier, H. Impact Response of Titanium and Titanium Boride Monolithic and Functionally Graded Composite Plates, *AIAA Journal*, 47(3), 676–691, 2009.

A sophisticated finite element method (FEM) of the test was constructed to simulate the conditions of the experiments. Two distinct material models were used in the finite element analyses (FEAs) to study the monolithic and graded plates. The first model used analytical expressions based on the local volume fractions of the constituents to generate homogenized-material properties for the mixtures of titanium and titanium boride. The second model randomly distributed cells containing titanium elements and titanium boride elements constrained to satisfy local volume fraction ratios in the monolithic and graded specimens. The strain histories from the experiments were compared with the analogous solutions from the FEM analyses to validate the computational models used in the study. Specifically, analyses with respect to historical trends, maximum strain magnitudes, and strain-rate effects were performed to gain insight into the impact response of the plate structures. The key contribution is validation of FGM models and a computational framework for studying the impact response of FGM plates as a foundation for investigations of more severe impact loads.

32.1   Introduction

FGMs are advanced composites with mechanical properties that vary continuously through a given dimension. The property variation can be accomplished by chemically or mechanically treating a single material locally to alter its characteristics or by varying the volume fraction ratio of two or more constituents along a given dimension. FGMs have generated a great deal of interest in recent years due to their flare advan for use in a wide variety of environments, including those structural applications in which extreme thermal and corrosion resistance are required [1]. In this article, the response of metal–ceramic functionally graded plates subject to impact loading is studied, both experimentally and computationally.

FGMs, in the general sense, have been available for centuries; in the sense of specially tailored engineering materials, the majority of research into these composites has occurred over the past two decades. Suresh and Mortensen [1] provided a comprehensive literature review of the state-of-the art of FGMs dated before 1998, and Birman and Byrd [2] compiled another extensive literature review covering FGM research from 1997 to 2007. Selected works pertinent to this investigation will be highlighted here. Lambros et al. [3,4] developed an inexpensive method for constructing polymer-based FGMs by treating a polyethylene derivative with ultraviolet light. Parameswaran and Shukla [5] developed another inexpensive technique for constructing FGMs by combining aluminum silicate spheres in a polyester resin matrix in which the volume fraction of the spheres was locally tailored to provide the property gradient. These methods can be desirable given the inherent cost and availability of FGM specimens. Reddy et al. [6–9], Loy et al. [10], and Pradhan et al. [11] have studied the behavior of a wide variety of FGM plate configurations under static and dynamic loading, as have others in the field [12–16]. To-date, only a few researchers have given consideration to studying impact response and wave propagation in functionally graded composites. Gong et al. [17] studied the low-velocity impact of FGM cylinders with various grading configurations. Bruck [18] developed a technique to manage stress waves in discrete and continuously graded FGMs in one dimension. Li et al. [19] first studied FGM circular plates under dynamic pressures simulating an impact load with a specific metal–ceramic system and using a rate-dependent constitutive relation they developed. Banks–Sills et al. [20] also studied an FGM system under dynamic pressures of various temporal applications. These works were all performed using analytical and computational techniques, but none of them were compared with physical or experimental data given the fact that very little test data of any kind associated with functionally graded composites can be found in the literature. This is due to (1) the difficulty of manufacturing FGMs, (2) the limited availability of such materials in industry and academia, and (3) the high cost associated with producing them.

The FGM system used exclusively in this research is a titanium–titanium–boride system developed by BAE Systems Advanced Ceramics in Vista, California. BAE Systems uses a proprietary *reaction sintering* process to produce Ti–TiB FGMs and monolithic composites. Commercially pure titanium (Ti) and titanium diboride (TiB_2) are combined in powder form in a graphite die according to prescribed volume fractions through the plate thickness. A catalyzing agent is applied to the construction, and the powders are subjected to extreme temperature (near the melting point of titanium) and pressure in a vacuum or inert gas environment. The catalyzing agent reacts with the T_i and T_iB_2 powders to form titanium boride (TiB) that crystallizes in a needle morphology. In the reaction process, almost no residual TiB_2 remains in the FGM. Through the sintering process, the powders adhere together and the Ti–TiB FGM or monolithic plate is the final product. This process can be used to construct monolithic composites of constant volume fraction or composites graded along the given dimensions. The change in composition of the constituents along a dimension is discrete and not truly continuous, although the distance over which a discrete change occurs can be very small and can closely approximate a continuous function over a larger distance. The FGM plates used in testing were graded over seven discrete layers of equal thickness with compositions ranging from 15% Ti–85% TiB to 100% Ti–0% TiB (see Table 32.2 for more precise details presented in section 32.3.1).

Ti–TiB composites are not new materials; in fact, the crystal structure of TiB was characterized by Decker and Kasper [21] as early as 1954. An extensive study of the microstructure and phases in Ti–TiB metal-matrix composites produced by reaction sintering was conducted by Sahay et al. [22] in 1999. The authors found that, at low volume fractions of TiB (up to Vf = 0.30), TiB whiskers are long, needle shaped, and randomly dispersed throughout the Ti matrix. At medium to high volume fractions of TiB (up to Vf = 0.86), colonies of densely packed TiB whiskers are formed. At very high volume fractions of TiB (Vf > 0:86), the TiB formed a very coarse, elongated structure with very few whiskers present. In general, small traces of residual TiB_2 were detected in samples as the volume fraction of TiB increased. The fact that TiB reinforcement is produced *in situ* by chemical reaction makes the direct measurement of the basic material properties of TiB difficult, although Atri et al. [23] and Panda and Ravichandran [24] have had some success using methods rooted in crystal physics and experimentation. Recent technological advances have made the construction of such composites easier to accomplish; thus, their availability to academia and industry has grown [25,26].

The key objectives of this study are to (1) design and conduct impact experiments on metal–ceramic FGM plates to collect strain histories from the plates over the duration of the impact event, (2) construct a finite element (FE) simulation of the impact experiment that can be easily replicated by scientists and engineers in practice, and (3) correlate the results from the experiments and FEMs and draw conclusions regarding the validity of analytical and computational techniques used to study the response of FGM plate structures.

This chapter is organized as follows: An experimental technique for obtaining strain histories in plates subjected to impact loading will be presented, and the results of a series of impact tests with monolithic and functionally graded Ti–TiB composites will be discussed. Next, the details associated with a FEM of the plate impact experiments developed to compare numerical simulations with the actual test data are presented. The FEM incorporates two classes of material models: (1) The first material model randomly distributes cells of Ti and TiB relative to local volume ratios and (2) The second material model homogenizes material properties locally according to an analytical function of the volume fractions of the constituents. Each impact test was simulated using the FEM, and the solutions from the computational model are compared with the experimental results: with respect to historical trends, maximum strain magnitudes, and strain-rate effects. The chapter concludes with a discussion of the impact response of the graded plates in testing and simulation, as well as of the effective modeling of FGMs in engineering practice. The key contribution of this work is the validation of FGM material models and a computational framework for studying the impact response of FGM plates as a foundation for investigations of more severe impact loads at higher velocities and energy levels.

32.2 Plant impact experiments

The first objective of this work was to design and conduct impact experiments on metal–ceramic FGM plates to collect strain histories from the plates over the duration of the impact event. A series of impact experiments were conducted using monolithic Ti and Ti–TiB plates along with seven-layer Ti–TiB FGM plates. The strain histories give insight to the dynamic behavior of the physical specimens under these conditions that can be later compared with the numerical simulation.

32.2.1 Test setup and hardware

The plate specimens used in the tests were 7.62×7.62 cm^2 and 1.27 cm thick; there were ten plates in all. Six of the plates were monolithic in composition: two plates consisted of the American Society for Testing and Materials grade 2 commercially pure titanium; two plates consisted of 85% Ti–15% TiB; two plates consisted of 15% Ti–85% TiB. The remaining four plates were seven-layer Ti–TiB FGM. Each plate was instrumented with three 350 Ω large deformation strain gauges as shown in Figure 32.1. The gauges were configured such that the bulk wires to the gauges were soldered to a terminal separate from the strain gauge and single-stranded jumper wires were then soldered to the actual gauge pads. This configuration is commonly used for dynamic tests in which inertial effects require minimizing the mass of the adhesive, strain gauge, and wiring assembly.

The Dynatup apparatus (developed by General Research Corp.), operated by the Air Vehicles Directorate (AFRL/RB) of the U.S. Air Force Research Laboratory at the Wright–Patterson Air Force Base, Ohio was chosen to deliver impacts to the specimens at various velocities and energy levels. The Dynatup is shown in Figure 32.2. The apparatus is designed to deliver impact energies up to 442 J to a specimen by converting a prescribed potential energy (PE) into kinetic energy (KE). The Dynatup can supply a gravity-driven vertical impact of variable energy to a specimen in which the energy is controlled by the height of the load cell above the specimen and the mass attached to the crosshead assembly. Pneumatic spring assists can be used to provide further PE to

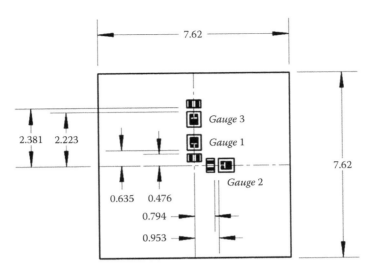

Figure 32.1 Specification for specimen plates and strain gauge locations. All dimensions are in centimeters unless otherwise specified. The gauges are mounted on the bottom surface of the plate (titanium surface on layer 7 on FGM). Each plate is 1.27 cm thick, and the strain gauges are 0.318 cm wide and 0.635 cm long.

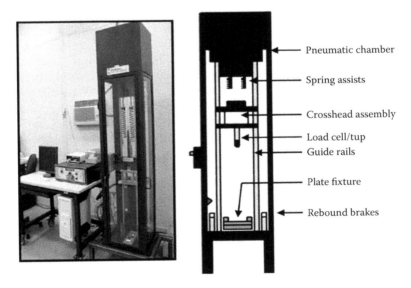

Figure 32.2 Dynatup apparatus and accompanying schematic.

the system to induce higher impact velocities and energy levels (this feature was not used in this study). A 2.54-cm-diameter tup was used to transfer the dynamic loads to each specimen. The tup is composed of hardened steel and is cylindrical in shape with a hemispherical tip. The speed of the tup at impact is measured by a velocity photodetector wired into the Dynatup data acquisition system. A set of pneumatic rebound brakes prevent a secondary impact from a rebound of the tup and crosshead.

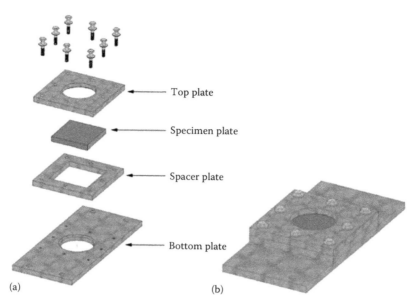

Figure 32.3 Two views of the plate fixture assembly: (a) exploded view and (b) assembled view.

A special fixture was constructed to hold the plate specimens for each test. Schematics of the fixture are shown in Figures 32.3 through 32.6. The fixture was specifically designed to configure the plate specimens to behave as close to a plate with a circular boundary condition as possible. The fixture consists of a bottom plate with a circular opening that rests on the base of the Dynatup, a spacer plate that serves to position the square plates properly in the fixture while additionally preventing crushing of the plate during installation and a top plate with a circular opening. The specimen plates are placed in the fixture such that the strain gauges lie on the (bottom) surface opposite the impact (top) surface. The components of the fixture were machined from 1.27-cm-thick 304 grade SS. The three components are fastened together with eight (American) 1/4–28 unified coarse threads (UNF) screws (these will be referred to as the *fixture screws* from this point forward). The holes for the screws were tapped on the bottom plate such that the screws could be threaded into this component. The fixture assembly then attaches to the Dynatup with four 3/8–16 unified fine threads (UNC) screws (these will be referred to as the *Dynatup screws* from this point forward).

The base of the Dynatup has tapped holes so that the screws can be threaded into. The fixture screws were each torqued to 20 N m, and the Dynatup screws were torqued to 35 N m. These values were determined to simultaneously (1) prevent crushing of the Ti–TiB composites, (2) prevent separation of the fixture assembly components during impact, and (3) ensure that the fixture assembly was secured in the Dynatup. The fixture and Dynatup screws are 18/8 grade SS, a broader category of SS that includes the 304 grade used for the fixture components. Figure 32.7 shows an FGM plate in the test fixture as installed in the Dynatup apparatus.

Once the plate and fixture assembly were installed in the Dynatup, the strain gauges were wired into a signal conditioning system using a standard 1/4 Wheatstone bridge arrangement with 5.0 V of excitation. The conditioner uses *dc* differential amplifiers with 80 kHz of bandwidth. The output from the signal conditioners/amplifiers was input to an

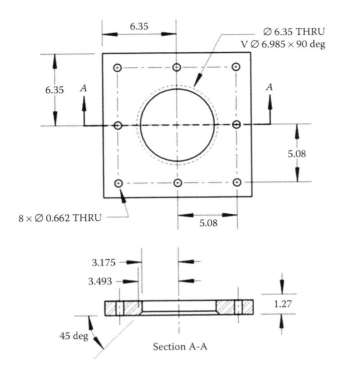

Figure 32.4 Specification for the top plate of the specimen fixture. All dimensions are in centimeters unless otherwise specified. The top plate is machined from 304 stainless steel.

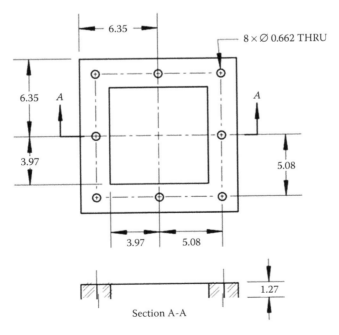

Figure 32.5 Specification for the spacer plate of the specimen fixture. All dimensions are in centimeters unless otherwise specified. The spacer plate is machined from 304 stainless steel.

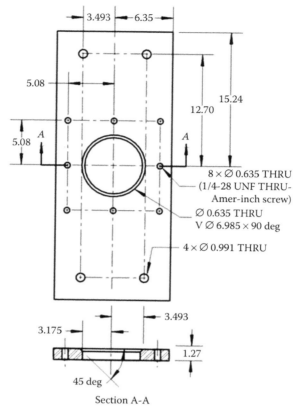

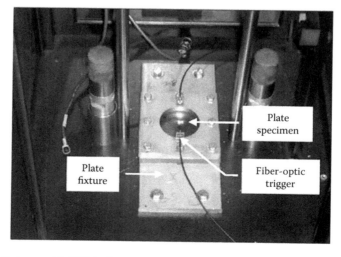

Figure 32.6 Specification for the bottom plate of the specimen fixture. All dimensions are in centimeters unless otherwise specified. The bottom plate is machined from 304 stainless steel.

Figure 32.7 Plate fixture with FGM plate installed in the Dynatup apparatus. Note the guide rails and rebound brakes from the Dynatup on either side of the fixture, as well as the fiber-optic sensor resting on the top plate.

oscilloscope programmed to collect 10,000 samples over a 2000 μs window. A fiber-optic sensor was used to automatically trigger data collection from the oscilloscope. The sensor, composed of a transmitter, receiver, and amplifier, emits a light beam that sends a voltage signal to the oscilloscope when the beam is interrupted. A single sensor was placed directly above the impact site on each specimen, and the beam was interrupted as the tup passed through the beam just before contact with the plate. By triggering data collection in this fashion, the strain histories from the impact event were wholly captured without interference or premature triggering of the system due to background noise. The next section 32.2.2 outlines the procedure by which the strain histories were collected.

32.2.2   Test procedure

The test procedure for collecting strain histories is summarized as follows:

1. The specimen is fitted with strain gauges as shown in Figure 32.1 on the bottom surface of the plate. The surface of the plate with the gauges will be the surface opposite to the impact surface.
2. The plate is installed and centered into the test fixture and each of the fixture screws is torqued to 20 Nm.
3. The specimen/fixture assembly is placed in the Dynatup and the Dynatup screws are torqued to 35 N m.
4. The Dynatup crosshead is raised above the impact surface of the plate to a prescribed height (see Table 32.1). The height is measured from the tip of the tup to the impact surface of the plate.
5. The fiber-optic sensor is armed; the oscilloscope and signal conditioner are prepared for data acquisition.
6. The Dynatup system is armed to release the crosshead for impact against the specimen plates.
7. The crosshead is released; the PE stored in crosshead assembly is converted into KE.
8. A velocity photodetector records the speed of the tup at impact; the tup impacts the top surface of the plate.
9. The local and global deformation of the plate are recorded through the strain histories collected by the three strain gauges attached to the bottom surface of the plate during the entire impact event.
10. The rebound brakes in the Dynatup engage and prevent the tup from multiple impacts due to rebound.
11. The strain history is recorded and the test is completed.

32.2.3   Test results

Ten specimens were tested using only gravity-driven impacts with the Dynatup. The speed of the tup just before impact was recorded, and the impact energy associated with the velocity was tabulated.

$$KE_1 + PE_1 + W_{1 \rightarrow 2} = KE_2 + PE_2 \tag{32.1}$$

$$\frac{1}{2}mv_1^2 + mgh_1 + W_{1 \rightarrow 2} = \frac{1}{2}mv_2^2 + mgh_2 \tag{32.2}$$

Table 32.1 List of plate impact experiments

Test number	Plate specimen	Crosshead mass, kg	Crosshead height, m	Ideal impact velocity, m/s	Actual impact velocity, m/s	Impact energy, J
1	100% Ti, 0% TiB monolithic	13.06	0.508	3.157	3.040	60.35
2	100% Ti, 0% TiB monolithic	13.06	0.635	3.530	3.476	78.90
3	85% Ti, 15% TiB monolithic	13.06	0.508	3.157	3.050	60.75
4	85% Ti, 15% TiB monolithic	13.06	0.635	3.530	3.479	79.04
5	15% Ti, 85% TiB monolithic	13.06	0.381	2.734	2.585	43.63
6	15% Ti, 85% TiB monolithic	13.06	0.508	3.157	3.050	60.75
7	Seven-layer Ti–TiB FGM	13.06	0.508	3.157	3.040	60.35
8	Seven-layer Ti–TiB FGM	13.06	0.635	3.530	3.412	76.02
9	Seven-layer Ti–TiB FGM	13.06	0.762	3.867	3.765	92.56
10	Seven-layer Ti–TiB FGM	13.06	0.889	4.176	4.078	108.6

One column of Table 32.1 shows the ideal impact velocity given the initial height above the plate specimen and the mass of the crosshead assembly. The ideal impact velocity assumes that, upon release of the crosshead from rest, all PE stored in the crosshead is converted to KE on impact with no work performed on the system or losses from external sources:where KE and PE are the kinetic and potential energy of the crosshead assembly in states 1 and 2, illustrated in Figure 32.8. $W_{1 \to 2}$ is the external work (or losses in the system) performed from state 1 to state 2. In the ideal scenario, $W_{1 \to 2}$ is zero; in the actual tests, losses due to friction or other external sources are included in this term and $W_{1 \to 2}$ is negative (indicating a loss). In Equation 32.2, m is the mass of the assembly, h is the height measured from the tip of the tup to the top surface of the specimen plate, v is the velocity of the crosshead assembly, and g is the gravitational constant. It is clear from the data that there are indeed losses in the system. These losses could be attributed to any or multiple factors including friction, vibration in the guide rails, drag, and uneven motion of the crosshead during a drop test.

Strain histories were successfully collected from eight of the ten tests. The tests involving the 15% Ti–85% TiB monolithic samples (tests 5 and 6) were the two unsuccessful tests. These specimens immediately after impact fractured severely and virtually no data was collected on these specimens as a result. The FGM specimen from test 10 also got fractured; however, a significant portion of the strain history was collected before failure and this history has been included.

Tests 5 and 6 were unsuccessful, as noted earlier, and these tests are not included. For brevity, only the strain histories from gauges 2 and 3 (see Figure 32.1) are shown.

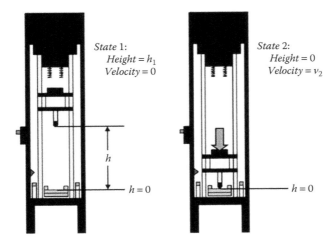

Figure 32.8 Illustration of kinetic, potential energy states in the Dynatup apparatus for the given impact test.

Note that gauges 1 and 2 were close in radial proximity and the data from these gauges are only slightly different in magnitude. The results from gauges 2 and 3, reading strains in perpendicular directions and themselves not as close in radial proximity, provide better results for discussion. Significant outliers were removed from the histories and these data were smoothed using the well-known robust locally weighted regression (or loess) technique developed by Cleveland [27] and Cleveland and Devlin [28] with a quadratic polynomial regression and weighted least squares over a range of 40 data points. The choice of weighted least-squares over 40 data points is large enough that significant oscillations in the plates are not lost while simultaneously ensuring that the noise present in the signals is eliminated. The reader will also note that in a few areas there are gaps in the strain histories. This is due to the removal of outliers from electrical shorts in the gauges that occurred during the impact event and history collection. The results from the tests will be discussed more extensively in Section 32.4 in which they are compared with the results from the finite element models.

32.3 Finite element models

The second objective of this work was to construct a finite element simulation of the impact experiment that could be easily replicated by scientists and engineers in practice. FEMs of the plate impact experiments were thus designed to numerically simulate the tests. In this section, information pertinent to the development of the models is presented. The commercial code ABAQUS [29] was used to simulate the tests. The simulations covered the 2000 μs window of the event. Explicit integration was used to solve the governing equations to take full advantage of the computational efficiency and the inherent effectiveness at solving dynamic and wave-oriented models. One thousand data points of strain were recovered from the solution database. Strain data were collected only from nodes directly under each strain gauge grid, and only strain outputs oriented along the principal direction of the strain gauges were used to compare with the experimental strain histories. This is an important point to remember because the state of strain is a complex three-dimensional state at virtually all points in the plate during the impact event, and the strain gauge measures only the component of strain (directly) along the principal direction of the gauge. The following paragraphs outline specific details associated with the material models, geometries, and meshes of the components, loads, constraints, boundary conditions, and contact interactions.

32.3.1 Material models

Two material models were used to simulate material properties in this work: (1) the two-phase material model and (2) the homogenized-layers material model. Before discussing the material models, the properties for Ti and TiBr will be assumed for the remainder of this chapter. The material properties for commercially pure titanium [30] are (1) elastic modulus, $E = 110$ GPa, (2) Poisson's ratio = 0:340, and (3) density, $\rho = 4510$ kg/m^3. The material properties for TiBr (provided via correspondence with BAE Systems) are (1) elastic modulus, $E = 370$ GPa, (2) Poisson's ratio = 0:140, and (3) density, $\rho = 4630$ kg/m^3.

The two-phase material model randomly distributes metal and ceramic-only cells (i.e., elements) constrained by the local volume fraction of the constituents. In the case of the Ti–TiB FGM system, each cell (or element) contains the material properties of only Ti or only TiBr. Suppose that a layer of FGM plate contains 100 elements and consists of 70% Ti and 30% TiB. Using the two-phase material model, 70 elements would be titanium and 30 elements would be TiBr in that layer. This model allows the random nature of the particulate distribution to be considered, which can be important to understanding how local distributions of constituents contribute to local effects such as wave propagation, plasticity, and damage (these effects are not studied in this chapter). The size and geometry of each cell is very important to the analysis and can ultimately affect the results. An illustration of the two-phase model is shown in Figure 32.9.

The homogenized-layers material model uses analytical equations relating the material properties of the constituents and the volume fractions of the constituents to the net material properties of the composite. There are many such analytical functions that have been developed over the years. These functions vary in complexity from relatively simple relations such as the classical rule of mixtures [31], to the self-consistent model [32–34] based on more rigorous physics and mathematics, and to those that account for statistical distributions of constituents and their geometries at the micromechanical level [35]. The advantage of using these functions is that definite quantitative properties can be obtained

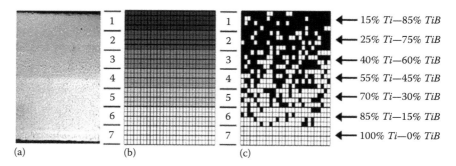

Figure 32.9 Cross sections of the seven-layer BAE Systems Ti–TiB FGM: (a) the actual FGM, (b) the two-phase model, and (c) the homogenized-layers model. Black cells represent TiB-only elements; white cells represent Ti-only elements. The homogenized layers are shaded based on the volume ratio of Ti to TiB.

for well-defined regions of constant volume fraction ratio, thus eliminating the need for extensive knowledge of the microstructure. These models are typically used when formulating the analytical behavior of FGM structures using principles of continuum mechanics.

The chosen analytical relations for homogenizing local material properties in the monolithic and functionally graded composites are the Mori–Tanaka estimates. Mori and Tanaka [36] demonstrated that, in two-phase composites, that is, a matrix with randomly distributed misfitting inclusions, the average internal stress in the matrix is uniform throughout the material and independent of the position of the domain in which the average is obtained. They also showed that the actual stress in the matrix is the average stress in the composite plus a locally varying stress, the average of which is zero in the matrix phase. Benveniste [37] used their analysis as the basis for developing equations that can be used to determine bulk and shear moduli for the composite material as a whole:

$$\frac{K-K_1}{K_2-K_1}=\frac{V_2^f C_1}{(1-V_2^f)+V_2^f C_1}; \frac{G-G_1}{G_2-G_1}=\frac{V_2^f C_2}{(1-V_2^f)+V_2^f C_2} \tag{32.3}$$

The subscripts 1 and 2 represent the individual constituents, whereas no subscript on K and G indicate the bulk and shear moduli for the composite. V^f is the volume fraction of a given constituent. The expressions in Equation 32.3 are explicit, and the variables C_1 and C_2 depend on the nature of the particle inclusions. Berryman [38,39] developed a framework for the two special cases of needle and spherical inclusions. The special case of needle inclusions will be the focus, as the TiB in the monolithic and graded composites is primarily in a whisker/needle morphology. The constants C_1 and C_2 are given as follows:

$$C_1 = \frac{K_1 + G_1 + (1/3)G_2}{K_2 + G_1 + (1/3)G_2} \tag{32.4}$$

$$C_2 = \frac{1}{5}\left(\frac{4G_1}{G_1+G_2}+2\frac{G_1+f_1'}{G_2+f_1'}+\frac{K_2+(4/3)G_1}{K_2+G_1+(1/3)G_2}\right) \tag{32.5}$$

$$f_1' = \frac{G_1(3K_1+G_1)}{(3K_1+7G_1)} \tag{32.6}$$

The elastic properties E and v for each composite layer of Ti–TiB can be solved for by taking the results from Equation 32.3 in each layer and relating those results to the definitions of bulk and shear moduli in terms of these properties:

$$K = \frac{E}{3(1-2v)}; G = \frac{E}{2(1+v)} \tag{32.7}$$

The Mori–Tanaka needle (MTN) estimates do not account for material density, a necessary material property in dynamic analyses. The density ρ of each composite layer is usually determined using the classical rule of mixtures, and Equation 32.8 shows the relation that was used to find this property in individual layers:

$$\rho = V_1^f \rho_1 + V_2^f \rho_2 \tag{32.8}$$

Table 32.2 Homogenized-material properties within FGM layers.

Layer	Volume Ti, %	Volume TiB, %	Mori–Tanaka needle estimate		
			Elastic modulus, GPa	Poisson ratio	Density, kg/m³
1	15	85	315.0	0.175	4612
2	25	75	282.7	0.196	4600
3	40	60	239.4	0.227	4582
4	55	45	201.4	0.256	4564
5	70	30	167.6	0.284	4546
6	85	15	137.4	0.312	4528
7	100	0	110.0	0.340	4510

Table 32.2 shows the effective material properties for each layer in the functionally graded plates using the MTN estimates and the rule of mixtures for density. An illustration of the homogenized-layers model compared with the two-phase model and an actual Ti–TiB FGM plate is shown in Figure 32.9.

32.3.2   Model components

The FEM of the test can be effectively divided into the following components: (1) the specimen plate fixture with its bottom, spacer, and top plates, (2) the eight fixture screws, (3) the four Dynatup screws, (4) the idealized tup, and (5) the specimen plate. The complete FEM mesh is shown in Figure 32.10. The entire model was meshed with linear, eight-noded, and three-dimensional solid continuum brick elements. Linear bricks of this type were chosen primarily for computational efficiency.

The top, bottom, and spacer plate components of the fixture were machined from 304 SS, as mentioned in the previous section 32.3.1. The parts were manually constructed and meshed in ABAQUS according to the specifications shown in Figures 32.4 through 32.6. This was done to maintain a consistent mesh and prevent the formation of irregularly

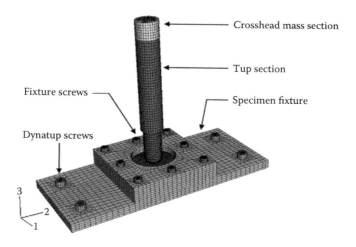

Figure 32.10 Finite element model of plate impact experiments.

shaped elements that could potentially cause numerical problems during solution. The bottom plate consists of 7935 nodes and 5656 elements, the spacer plate consists of 3800 nodes and 2560 elements, and the top plate consists of 4280 nodes and 2944 elements. The actual fixture components were visually inspected after testing and found to be virtually undeformed. The components are thus assumed to require only elastic material properties. The material properties used in the FEM for 304 grade SS [30] are (1) elastic modulus, $E = 193$ GPa; (2) Poisson's ratio = 0:290; and (3) density, $\rho = 8030$ kg/m³.

The eight fixture screws are American $1 = 4$–28 UNF composed of 18/8 grade SS (note that 18/8 SS is a broad category of SS alloys containing 18% chromium and 8% nickel; 304 SS alloy is a member of this category). The geometry of the screws was simplified in the FEM. Each hex screw's head is nominally 0.397 cm thick, and the width across the flats is 1.111 cm [30]. Round washers used in conjunction with the screws were also 18/8 SS and were nominally 0.198 cm thick with an outside diameter slightly larger than the width across the flm t of the head. The simplified FEM consists of a round cylinder for the head with a diameter equal to the width across the flats of the hex-head screw and a thickness equal to the hex head plus the washer thickness (thus effectively combining the washer and head into one geometry). The stud length is equal to the thickness of all three fixture plates sandwiched together (3.81 cm), and the diameter of the stud is 0.635 cm. Each of the eight fixture screws contains 1300 nodes and 1040 elements; the total number of nodes and elements contained in the fixture screws are, thus, 10,400 and 8,320, respectively. The four Dynatup attachment screws are American $3 = 8$–16 UNC and are also composed of 18/8 grade SS. Each hex screw's head is 0.595 cm thick, and the width across the flats is 1.429 cm [30]. The round washers used in conjunction with the screws were the same 18/8 SS as the screws and were 0.198 cm thick with an outside diameter slightly larger than the width across the flats of the hex head. The simplified FEM consists of a round cylinder for the head with a diameter equal to the width across the flats of the hex-head screw and a thickness equal to the hex head plus the washer thickness. The stud length is equal to the thickness of the bottom plate (1.27 cm), and the diameter of the stud is 0.953 cm. Each of the four Dynatup screws contains 641 nodes and 480 elements; the total number of nodes and elements for these components is thus 2,564 and 1,920, respectively. The same screws were used in all impact tests and were virtually undeformed after testing; therefore, only elastic properties were input into the finite element models. The same properties used for 304 SS were used for 18/8 SS.

The crosshead, load cell, and tup assembly in the Dynatup apparatus represent a very unique part of the FEM as a whole. The best scenario for modeling the Dynatup apparatus is to include nearly all the parts that make up the assembly, which is itself not a trivial matter. The primary function of the crosshead, load cell, and tup assembly is to transfer energy and momentum to the target specimen through contact between the tup and specimen. A highly simplified model of this assembly was developed that would serve both of these purposes (referred to as simply the *tup FEM* or *tup model* from this point forward). It consists of two unique sections: (1) a tup section and (2) a crosshead-mass section. The tup section is composed of two cylindrical pieces: (1) one piece is 3.175 cm in diameter and 13.97 cm long and (2) one piece is 2.54 cm in diameter and 4.445 cm long. At the end of the smaller cylinder is a hemispherical tip of radius 1.27 cm. The larger cylinder represents the load cell, and the smaller cylinder with hemispherical tip is the tup. The net length of this section is, thus, 19.69 cm. The crosshead-mass section is a cylinder 2.54 cm long and 3.175 cm in diameter. The net length of the entire tup model is, thus, 22.23 cm. The actual composition of the load cell and tup was not known and not provided for proprietary reasons. A set of material properties thus had to be assumed for the tup section, and 4340 hardened-alloy steel was chosen. Material properties for this steel alloy [30] are (1) elastic modulus, E = 200 GPa; (2) Poisson's ratio = 0:290; and (3) density, P = 7800 kg/m³. The total mass of the tup section, given the density of 4340 hardened-alloy steel, was 1.072 kg. The crosshead-mass section was designated as a pseudomaterial; the purpose of this section was to store the remainder of the entire mass of the crosshead-load cell-tup assembly without constructing the entire crosshead. From Table 32.1, the mass of the entire assembly during testing was 13.06 kg. Knowing the dimensions of the crosshead-mass section in the FEM, this remaining mass (11.99 kg) was distributed through this small volume and incorporated in the model by applying a calculated density to the elements of that section. Also noteworthy is the fact that the elastic modulus was set very high in comparison with the rest of the model components; essentially the crosshead-mass section behaves like an rigid mass attached to a deformable load cell tup. Material properties for this pseudomaterial are (1) elastic modulus, E = infinity GPa (i.e., a very large discrete number that ABAQUS can process); (2) Poisson's ratio = 0:300; and (3) density, ρ = 596123 kg/m³. The tup model contains total 12,190 nodes and 11,008 elements.

Finally, the plate specimens were constructed as 7:62 × 7:62 cm² and 1.27 cm thick. A computer script was programmed to generate the mesh and assign material properties to individual elements in the plate model. For the two-phase model, the individual Ti and TiB elements are randomly distributed according to their local volume fraction ratios by the computer script using a random number generator algorithm. For the homogenized-layers model, the material properties for the elements in each layer are calculated by the same equations discussed earlier and assigned as such (see Table 32.2). Eq 32.3-32.7

A study was undertaken to determine the mesh that would be most effective from both a computational and solution convergence standpoint. The plate mesh used here, based on the study, was 14 elements through the thickness of the plate and 42 element divisions along each side of the plate; thus, the total number of nodes and elements for the plates were 27,735 and 24,696, respectively. Increased mesh density for the plates showed virtually no improvement in the FEM solutions while significantly increasing computational expense. Material properties were assigned based on the use of the two-phase or homogenized-layers model discussed earlier in section 32.3.1 figure 32.11 shows a sample of the specimen plate using each model.

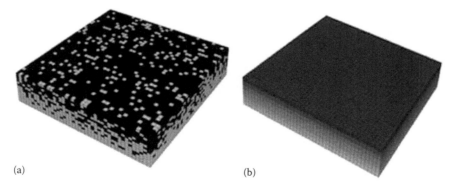

(a)   (b)

Figure 32.11 Comparison of the finite element representations of the FGM plates: (a) the two-phase material model and (b) the homogenized-layers model. Black cells represent TiB-only elements; white cells represent Ti-only elements. The homogenized layers are shaded based on the volume ratio of Ti to TiB.

32.3.3   *Boundary conditions, constraints, and loads*

The following boundary conditions and constraints were applied to the model. Note the orientation of the coordinate axes in Figure 32.10 when reading the following:

1. The bottom surface of the bottom plate was constrained from vertical displacement only (direction 3). This surface is in direct contact with the base of the Dynatup. The Dynatup base is made of 5.08 cm steel and is not easily deformable; thus, this assumption was deemed prudent.
2. An imaginary plane parallel to the 1–3 plane passing through the center of the fixture assembly was constrained such that all nodes and element faces contained directly in this plane were not allowed translations parallel to the two-axis direction. This included the bottom, spacer, and top plates of the fixture as well as the specimen plate; nodes or elements attached to the tup and plate fastener screws were not assigned this restriction.
3. An imaginary plane parallel to the 2–3 plane passing through the center of the fixture assembly was constrained such that all nodes and element faces contained directly in this plane were not allowed translations parallel to the one-axis direction. This included the bottom, spacer, and top plates of the fixture as well as the specimen plate; nodes or elements attached to the tup and plate fastener screws were not assigned this restriction.
4. The four Dynatup attachment screws were threaded into the Dynatup base. The FEM stud length of these screws was shortened to a length equal to the thickness of the bottom plate of the fixture (1.27 cm). Once threaded and torqued to the required level, it is assumed that the screws could not be pulled out of the Dynatup base. Thus, the bottom surface of each Dynatup screw (that is, the surface contained in the same plane as the Dynatup base-fixture bottom plate interface) was constrained from vertical deflection only.
5. The axes of the four Dynatup attachment screws were constrained from motion in the directions 1 and 2.
6. The axis of the tup was constrained from motion in the directions 1 and 2.
7. Recall that the fixture screws were threaded into the bottom plate only. The holes in the spacer and top plates were through-holes only. In the FEM, the holes in the bottom plate were made to be the same diameter as the stud diameter of the fixture screws; the holes in the other two plates were made slightly larger. Therefore, a small portion of the outside surfaces of the screws along the stud length will coincide with the surfaces inside the holes of the bottom plate throughout the thickness of the plate. To simulate a tight, rigid connection between the fixture screws and the threaded holes of the bottom plate, the nodes attached to the coincident surfaces of the bottom plate and the screws in each hole were constrained to have identical displacements in all three principal directions. This constraint is assumed to model a threaded connection without requiring an actual model of the threads themselves.

ABAQUS has the capability of modeling contact interactions between various components of the model assembly. Contact in the realm of finite element theory is a highly nonlinear analysis and requires special treatment [40,41]. The contact law used exclusively in this FEM was that of a rigid-hard contact. The main feature of this law is that it is essentially an on–off law in which the pressure applied from one object to another is zero when the objects are not in contact and positive when in contact; the magnitude of pressure is a function of the interpenetration of the two object surfaces in contact. This contact law was applied to the following surfaces: (1) the interface of the bottom and spacer plates of the fixture, (2) the interface of the spacer and top plates of the fixture, (3) the interface of the specimen plate and the fixture bottom plate, (4) the interface of the specimen plate and the fixture top plate, (5) the top surface of the specimen plate and the surface near the tip of the tup, (6) the top surface of the fixture top plate and each surface underneath the head of each fixture screw, and (7) the top surface of the fixture bottom plate and each surface underneath the head of each Dynatup attachment screw.

In the actual plate impact experiments, the primary load to the plates was the impact load delivered by the tup. In the assembly of the FEM, the tup was placed above the specimen plate with 0.1 mm of separation initially between the two objects. A velocity field was applied to the entire tup model in the FEM as an initial condition with a magnitude equal to the speed measured during the test (see Table 32.1). Incidentally, the velocity field has components in only the vertical direction 3; thus, the speed and the magnitude of that component of the velocity field are identical. The tup maintains this speed until contact is established between the tup and the specimen plate. Once this occurs, the dynamics of the system take over and the instantaneous velocity of the tup must be determined based on the solution to the FEM.

A note on the torque loads applied to the Dynatup and fixture screws is in order. Recall that each of the four Dynatup screws were torqued to a 35 N m load and the eight fixture screws were torqued to a 20 N m load. It was noticed that these torque loads did not appear to visually deform the fixture plates or the screws themselves during screw preloading or after impact tests occurred. A subsequent analysis of the force loading on the screws using analytical and finite element techniques verified the hypothesis that the deformation in the fixture plates and screws was negligible during preloading of the screws. For this reason, the torque loading on the screws was left out of the FEM. Further analysis showed that the deformation of the specimens during impact induced noticeable lifting forces and potential separation of the fixture plates from the specimen leveraging against the fixture. This action produced stresses in the fixture screws as they resist the leveraging and fixture plate separation. With the screws in direct contact with the fixture component plates and threading accounted for in the constraints of the system, it was deemed that this was sufficient to emulate the behavior of the fixture plate fastening.

32.4   *Experiment versus FEM strain histories*

The third and final objective of this work was to correlate the results from the experiments and finite element models and draw conclusions regarding the validity of analytical and computational techniques used to study the response of FGM plate structures. In this section, the results from the experiments and FEAs are presented for direct comparison. The primary goals are to assess the validity of the material models (two phase vs. homogenized layers) and determine whether FGMs can be effectively modeled and studied using finite element methods. Comparisons and analysis will be performed in three specific areas: (1) a qualitative comparison of the strain history plots from the experiments and FEM, (2) a quantitative assessment of the predicted maximum strains in the plates during the impact events, and (3) an investigation of the strain rates applied to the plates during each impact load.

32.4.1   Experiment versus FEM strain histories

The strain histories from strain gauges 2 and 3 are plotted along with the results from the two-phase material model and the Mori–Tanaka homogenized-material model with needle inclusions. Strain data were collected only from nodes directly under each strain gauge grid, and only strain components oriented along the principal direction of the strain gauges were used to compare with the experimental strain histories. Figures 32.12 through 32.19 show the results from the FEMs next to the experimental data.

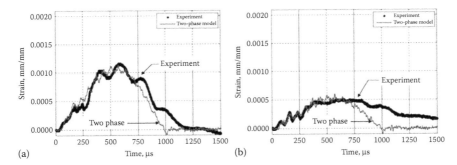

Figure 32.12 Test 1 strain histories versus FEM results: (a) strain gauge 2 and (b) strain gauge 3. Specimen was 100% Ti–0% TiB monolithic with an impact velocity of 3.040 m/s.

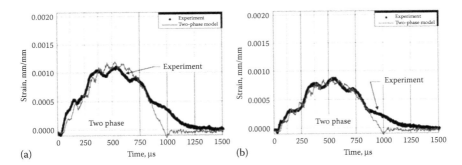

Figure 32.13 Test 2 strain histories versus FEM results: (a) strain gauge 2 and (b) strain gauge 3. Specimen was 100% Ti–0% TiB monolithic with an impact velocity of 3.476 m/s.

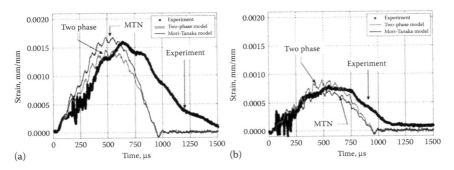

Figure 32.14 Test 3 strain histories versus FEM results: (a) strain gauge 2 and (b) strain gauge 3. Specimen was 85% Ti–15% TiB mono/lithic composite with an impact velocity of 3.050 m/s.

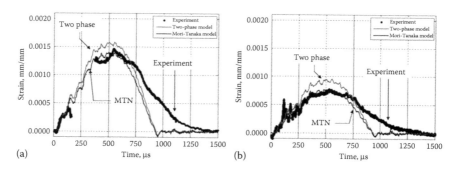

Figure 32.15 Test 4 strain histories versus FEM results: (a) strain gauge 2 and (b) strain gauge 3. Specimen was 85% Ti–15% TiB monolithic composite with an impact velocity of 3.479 m/s.

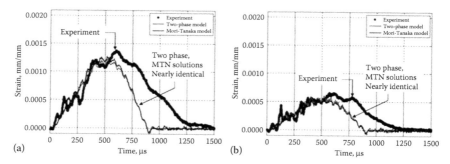

Figure 32.16 Test 7 strain histories versus FEM results: (a) strain gauge 2 and (b) strain gauge 3. Specimen was seven-layer FGM with an impact velocity of 3.040 m/s.

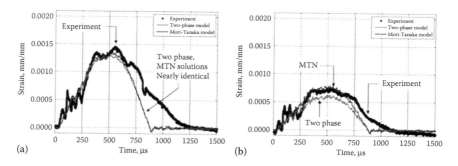

Figure 32.17 Test 8 strain histories versus FEM results: (a) strain gauge 2 and (b) strain gauge 3. Specimen was seven-layer FGM with an impact velocity of 3.412 m/s.

The Ti plates in tests 1 and 2 (Figures 32.12 and 32.13) were modeled using only the two-phase material model. Note that the MTN estimates for the case of 100% Ti would simply return material properties and material distribution precisely the same as the two-phase material model; therefore, the two analyses are redundant. The histories predicted by the FEM match up extremely well with the test results. This is very important in that it validates the modeling of the FEM given the geometry, loading, and constraints on the model.

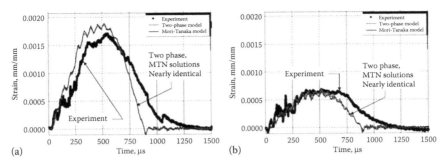

Figure 32.18 Test 9 strain histories versus FEM results: (a) strain gauge 2, and (b) strain gauge 3. Specimen was seven-layer FGM with an impact velocity of 3.765 m/s.

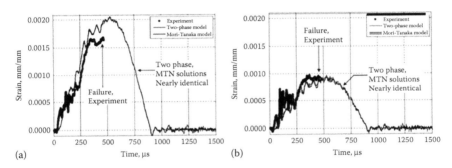

Figure 32.19 Test 10 strain histories versus FEM results: (a) strain gauge 2, and (b) strain gauge 3. Specimen was seven-layer FGM with an impact velocity of 4.078 m/s.

The histories from the 85% Ti–15% TiB monolithic plate tests (tests 3 and 4, Figures 32.14 and 32.15) match well with the exception of the histories in test 3, gauge 2. The response in gauge 2 indicates that the impact event is occurring over a larger period than reflected in gauge 3, lending credence to the theory that the unloading of the gauges may be a source of the discrepancy between test and simulation. Here, the two-phase and homogenized-layers models were used in these FEMs and are plotted against the experimental histories. This was the next step in the FEM validation of the impact tests. These monolithic plates have a constant volume fraction ratio of Ti to TiB. The homogenized-layers model of this specimen is a plate composed of a single layer of constant material properties. The two-phase model is composed of a random distribution of Ti to TiB in which 85% of the elements in the plate are Ti and 15% of the elements in the plate are TiB. These tests and FEMs thus present an added level of complexity above the plate specimens composed of pure Ti and demonstrate that the material models and FEM are suitable for extension to modeling the more complex FGM plates.

Figures 32.16 through 32.19 show the histories for the FGM plates. Again, the FEM simulations match the strain histories reasonably well from a qualitative standpoint. Recall that the FGM plate in test 10 failed midway through the impact event. Given the degree of correlation to this point, it is reasonable to assume that the FEM simulates the data well and the predicted histories from the FEM indicate how the plate would behave had failure not occurred.

Whether or not it is valid to assume that failure will occur in the FGM under the conditions of test 10 in all cases would require more testing. No histories, experimental or simulated, are included for tests 5 and 6 due to the catastrophic failure of the plates.

As a reasonable correlation between the test results and the FEM strain histories was demonstrated, more detailed analysis can be presented. By simple inspection, it can be seen that there are many strong correlations between the FEMs and the test data, and there are also areas of discrepancy that need to be addressed. First, the FEMs can predict the peak maximum values of the strains for both gauges at both locations very effectively. In most cases, both the two-phase model and the homogenized-layer models are within a reasonable error of the test results (see section 32.4.2). In some cases the two-phase and homogenized-layer models are nearly identical in response, and in other cases the maximum strain predicted is slightly greater or less than the counterpart. The two-phase model is generated by randomly distributing cells of Ti and TiB, and so it is not unexpected that this can and will occur.

Some of the test data show highly distinct and sharp oscillations very early in the response that disappear as the deflections become larger through the course of the impact event. These oscillations are not reflected in the FEM and are undoubtedly related to noise in the data or electrical shorting of the strain gauges (the flin the F of the single-stranded jumper wires during impact loading were found to account for this). The FEM clearly shows several vibration modes being excited on impact from the tup. Some of these modes are not clearly discernible in the test results. On the other hand, it appears that there are some lower frequency modes being excited in the test data that are not reflected in the FEM, especially near the peaks of the response (especially gauge 2). This could be attributed to many possibilities, including error in applied material properties and the nature of the boundary conditions and constraints applied in the FEM. Addressing the former, Hill and Lin [42] reported material properties while testing Ti–TiB FGM specimens that were significantly lower than the properties predicted by Mori–Tanaka estimates or other similar models. This is partially due to the difficulty in controlling the exact volume fractions of Ti to TiB given that a chemical reaction is required to produce the TiB *in situ*, but may also be related to the sintering process itself. Additionally, the strain gauge adhesive may inherently damp out some of the oscillations that occur in the plate and are thus not registered in the strain histories from the tests. The histories from the finite element models are obtained from strain recorded directly from nodes in the vicinity of the strain gauge grids on the surface of the plate. Note that damping was not included in the FEM simulations presented here. Values for damping coefficients were not available for the Ti–TiB system at the time of this study. An informal study of the matter showed that the addition of artificial values of damping to the FEM significantly increased the computational cost of running the simulations with virtually no effect on the strain histories.

32.4.2 *Maximum strains during impact*

Tables 32.3 through 32.5 show the maximum value of strain from each experiment versus that predicted by the finite element models. In most cases, both the two-phase model and the homogenized-layer models are within a reasonable error of the test results. The majority of the tabulated strains from the FEM simulations are within approximately 10% of the experimental results. A few gauges have differences that are higher; the largest difference is shown to be 22% (found in test 4, gauge 3; see Table 32.5). Tests in which the plates and gauges failed are not tabulated.

Table 32.3 Maximum strain comparison between experiment and FEM results, strain gauge 1

Test number	Plate specimen	Impact energy, J	Experiment maximum strain	Two-phase FEM maximum strain	MTN FEM maximum strain	Maximum percent difference
1	100% Ti, 0% TiB monolithic	60.35	Gauge failed	0.001943	0.001943	
2	100% Ti, 0% TiB monolithic	78.90	0.002602	0.002845	0.002845	9.33
3	85% Ti, 15% TiB monolithic	60.75	0.001711	0.001843	0.001684	7.70
4	85% Ti, 15% TiB monolithic	79.04	0.001764	0.001998	0.001833	13.21
5	15% Ti, 85% TiB monolithic	43.63	Plate failed			
6	15% Ti, 85% TiB monolithic	60.75	Plate failed			
7	Seven-layer Ti–TiB FGM	60.35	0.001459	0.001647	0.001621	12.82
8	Seven-layer Ti–TiB FGM	76.02	0.001783	0.001763	0.001737	−2.56
9	Seven-layer Ti–TiB FGM	92.56	0.001889	0.001894	0.001885	0.25
10	Seven-layer Ti–TiB FGM	108.6	Plate failed	0.002071	0.002057	

Figure 32.20 shows a plot of the radial strain distribution in the FGM plate in test 8 at the point of maximum transverse deflection along the axes on the bottom surface of the plate where the strain gauges were installed. One plot shows the strain distributions along the two axes for the two-phase model, whereas the second plot shows the same strain distributions along the same axes using the homogenized-layers MTN model. Also shown (on separate plots for purposes of clarity) are the areas, or *windows*, covered by each gauge based on the schematic in Figure 32.1. It is very clear, simply from these FEM-based distributions, that the maximum radial strain varies significantly over the small area occupied by each gauge. For this reason, the nodes directly under each strain gauge grid were averaged to give the results plotted in Figures 32.12 through 32.19. However, it is easy to see that small changes to the gauge grid location can significantly affect a strain history's maximum recorded value. This can account for some error between the strain histories given in Tables 32.3 through 32.5.

32.4.3 Strain rates in loading and unloading

Strain rates are important to the discussion of nearly all impact events. Here the average strain rates in the plate from both the experiments and the FEMs will be compared using the homogenized-layers FGM model (the strain rates predicted by the two-phase FEM are virtually identical to the homogenized-layers FEM). The impact interaction between each plate and tup occurs over two significant periods of time: (1) a *loading* period and (2) an *unloading* period. Figure 32.21 shows how the strain rates in loading and unloading in the experiments and FEMs are determined. A linear, least-squares curve fit to the data in the

Table 32.4 Maximum strain comparison between experiment and FEM results, strain gauge 2

Test number	Plate specimen	Impact energy, J	Experiment maximum strain	Two-phase FEM maximum strain	MTN FEM maximum strain	Maximum percent difference
1	100% Ti, 0% TiB monolithic	60.35	0.001173	0.001131	0.001131	−3.54
2	100% Ti, 0% TiB monolithic	78.90	0.001161	0.001239	0.001239	6.69
3	85% Ti, 15% TiB monolithic	60.75	0.001603	0.001461	0.001679	−8.86
4	85% Ti, 15% TiB monolithic	79.04	0.001469	0.001581	0.001401	7.60
5	15% Ti, 85% TiB monolithic	43.63	Plate failed			
6	15% Ti, 85% TiB monolithic	60.75	Plate failed			
7	Seven-layer Ti–TiB FGM	60.35	0.001391	0.001227	0.001281	−11.78
8	Seven-layer Ti–TiB FGM	76.02	0.00468	0.001307	0.001358	−10.97
9	Seven-layer Ti–TiB FGM	92.56	0.001720	0.001867	0.001890	9.86
10	Seven-layer Ti–TiB FGM	108.6	Plate failed	0.002045	0.002061	

loading and unloading portions of the strain histories was applied. The linear data fit was used to determine a general slope to the line through the data. The slope of that line is the strain rate for that part of the curve. The strain rate is not, in general, a constant through the duration of the impact events. For each oscillation in the history, the strain rate is changing dynamically. However, the linear fit to the data in the loading and unloading portions of the curve allows general trends to be assessed. The line fit was taken so that the peak at maximum strain from each gauge was not included in the strain-rate calculation. The duration of the experimental histories was slightly longer than that of the FEMs and so the times over which the strain rates were assessed in the loading and unloading portions of the curves were adjusted accordingly.

Tables 32.6 through 32.9 show tabulated strain rates for the Dynatup and FEM impact tests for both loading and unloading at the three gauge locations. Also shown are the tup impact velocities for each of the tests for reference. In general, the strain rates show a trend in which increasing tup velocity results in an increasing magnitude of strain rate. This trend is not followed in all instances, but this can be attributed to the positioning of the gauges in the experiments, which can have a profound effect on the strain histories. Further, the FEM was used to match the experimental strain histories as close as possible and accounted for this potential variation in gauge placement. Note that the magnitudes of strain rates are lower, for the most part, the farther away the gauges are from the center. The strain histories at the gauge locations all reach maximum values at the same time, and the strain magnitude is dictated by the location of the gauge with respect to the center of the plate. Thus, gauges farther away from the center reach a lower magnitude of maximum strain at the same time instant as gauges closer to the center reach a larger

Table 32.5 Maximum strain comparison between experiment and FEM results, strain gauge 3

Test number	Plate specimen	Impact energy, J	Experiment maximum strain	Two-phase FEM maximum strain	MTN FEM maximum strain	Maximum percent difference
1	100% Ti, 0% TiB monolithic	60.35	0.005157	0.006026	0.0006026	16.86
2	100% Ti, 0% TiB monolithic	78.90	0.009269	0.009241	0.009241	−0.31
3	85% Ti, 15% TiB monolithic	60.75	0.008058	0.009139	0.007308	13.41
4	85% Ti, 15% TiB monolithic	79.04	0.007834	0.009579	0.007774	22.27
5	15% Ti, 85% TiB monolithic	43.63	Plate failed			
6	15% Ti, 85% TiB monolithic	60.75	Plate failed			
7	Seven-layer Ti–TiB FGM	60.35	0.006638	0.006103	0.005626	−15.24
8	Seven-layer Ti–TiB FGM	76.02	0.007573	0.006394	0.008032	−15.56
9	Seven-layer Ti–TiB FGM	92.56	0.007016	0.006498	0.001885	−7.39
10	Seven-layer Ti–TiB FGM	108.6	Plate-Failed	0.009787	0.009522	

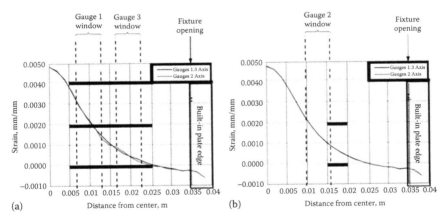

Figure 32.20 Comparison of radial strain distribution along plate aces containing strain gauges on bottom of surface of plate after impact at instant of maximum transverse deflection: (a) two-plane model and (b) Mori–Tanaka homogenized-layers model. Specimen was seven-layer FGM with an impact velocity of 3.412 m/s (test 8).

magnitude of maximum strain. As the strain rate is more or less the difference in strain divided by the time period the difference is measured, it is easy to see how the rates near the circular opening of the fixture would be less than those near the center of the plates.

The maximum magnitude of strain rate for any of the impact tests is 8:847 s^{-1} in the unloading of the pure Ti plate in test 2, predicted by the FEM. These rates of strain,

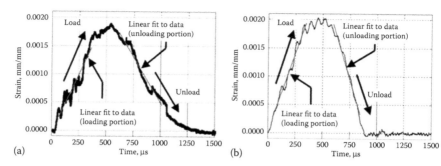

Figure 32.21 A straight line was fit to the data in the loading and unloading portions of the strain history curves to determine the strain rates in each respective part of the curve. Shown are the data from FGM plate test 9, strain gauge 2: (a) experimental and (b) FEM with homogenized-layers MTN model.

Table 32.6 Strain rate applied to plate during loading portion of impact event; experiment

Test number	Plate specimen	Actual tup velocity, m/s	Gauge 1	Gauge 2	Gauge 3
1	100% Ti, 0% TiB monolithic	3.040	Gauge failed	2.184	0.967
2	100% Ti, 0% TiB monolithic	3.476	6.175	2.754	2.161
3	85% Ti, 15% TiB monolithic	3.050	3.979	3.302	1.800
4	85% Ti, 15% TiB monolithic	3.479	3.759	3.367	1.781
5	15% Ti, 85% TiB monolithic	2.585	Plate failed	Plate failed	Plate failed
6	15% Ti, 85% TiB monolithic	3.050	Plate failed	Plate failed	Plate failed
7	Seven-layer Ti–TiB FGM	3.040	2.509	2.467	1.110
8	Seven-layer Ti–TiB FGM	3.412	3.927	3.389	1.643
9	Seven-layer Ti–TiB FGM	3.765	4.135	4.091	1.379
10	Seven-layer Ti–TiB FGM	4.078	5.167	4.678	2.140
			Strain rates, loading (1 = s)		

Table 32.7 Strain rate applied to plate during loading portion of impact event; FEM (MTN)

Test number	Plate specimen	Actual tup velocity, m/s	Strain rates, loading (1 = s)		
			Gauge 1	Gauge 2	Gauge 3
1	100% Ti, 0% TiB monolithic	3.040	4.295	2.374	1.162
2	100% Ti, 0% TiB monolithic	3.476	6.738	2.821	2.032
3	85% Ti, 15% TiB monolithic	3.050	4.035	4.055	1.640
4	85% Ti, 15% TiB monolithic	3.479	4.964	3.756	2.038
5	15% Ti, 85% TiB monolithic	2.585	Plate failed	Plate failed	Plate failed
6	15% Ti, 85% TiB monolithic	3.050	Plate failed	Plate failed	Plate failed
7	Seven-layer Ti–TiB FGM	3.040	4.099	3.152	1.272
8	Seven-layer Ti–TiB FGM	3.412	4.850	3.726	2.077
9	Seven-layer Ti–TiB FGM	3.765	5.527	4.261	1.724
10	Seven-layer Ti–TiB FGM	4.078	5.852	4.507	2.508

Table 32.8 Strain rate applied to plate during unloading portion of impact event; experiment

Test number	Plate specimen	Actual tup velocity, m/s	Gauge 1	Gauge 2	Gauge 3
1	100% Ti, 0% TiB monolithic	3.040	Gauge failed	−2.111	−0.363
2	100% Ti, 0% TiB monolithic	3.476	−4.463	−1.746	−1.589
3	85% Ti, 15% TiB monolithic	3.050	−2.966	−2.393	−1.331
4	85% Ti, 15% TiB monolithic	3.479	−3.207	−2.580	−1.350
5	15% Ti, 85% TiB monolithic	2.585	Plate failed	Plate failed	Plate failed
6	15% Ti, 85% TiB monolithic	3.050	Plate failed	Plate failed	Plate failed
7	Seven-layer Ti–TiB FGM	3.040	−2.449	−2.074	−1.096
8	Seven-layer Ti–TiB FGM	3.412	−3.522	−2.432	−1.723
9	Seven-layer Ti–TiB FGM	3.765	−3.245	−3.599	−1.477
10	Seven-layer Ti–TiB FGM	4.078	Plate-failed	Plate failed	Plate failed

Table 32.9 Strain rate applied to plate during unloading portion of impact event; FEM (MTN)

Test number	Plate specimen	Actual tup velocity, m/s	Gauge 1	Gauge 2	Gauge 3
1	100% Ti, 0% TiB monolithic	3.040	−5.862	−3.283	−1.604
2	100% Ti, 0% TiB monolithic	3.476	−8.847	−3.764	−2.682
3	85% Ti, 15% TiB monolithic	3.050	−4.879	−4.904	−1.975
4	85% Ti, 15% TiB monolithic	3.479	−5.498	−4.164	−2.231
5	15% Ti, 85% TiB monolithic	2.585	Plate failed	Plate failed	Plate failed
6	15% Ti, 85% TiB monolithic	3.050	Plate failed	Plate failed	Plate failed
7	Seven-layer Ti–TiB FGM	3.040	−5.368	−4.119	−1.608
8	Seven-layer Ti–TiB FGM	3.412	−6.006	−4.620	−2.548
9	Seven-layer Ti–TiB FGM	3.765	−6.397	−4.906	−1.898
10	Seven-layer Ti–TiB FGM	4.078	Plate failed	Plate failed	Plate failed

as tabulated for these tests, are very low compared with what would be experienced from a high-speed impact, such as would occur from a projectile in space [43]. In applications such as these, the rates of strain can be in excess of 103–105. Given rates as high as these, localized wave response and the effects of rate-sensitive constitutive models need to be included [43].

The rates experienced by the FGM plates here in these tests are low enough that the global effects dominate the solution, and the local wave effects are so dominated by the global response of the structure that they are virtually indistinguishable. The key conclusion based on this analysis is that elastic, rate-independent material properties are sufficient for studying the Ti–TiB FGMs under these impact loading conditions.

An interesting trend is that the magnitude of strain rate in the experiments is greater in the loading of the plate and lower in the unloading of the plate. Just the opposite is true with the FEMs. In the FEMs, the magnitude of strain rate is greater in the unloading than in the loading. The strain rates with respect to the loading of the plates experimentally and through the FEM show good correlation. The FEMs predict loading slopes (that is, strain rates) comparable to the experimental data for the gauges up to the point of maximum strain. As mentioned, the strain rates associated with each plate's unloading history do not correlate well. In fact, it appears that the rate the plate returns from maximum strain to *zero* strain is significantly less than that of the FEM. It is not likely that this behavior is associated with the monolithic or FGM composites themselves, indicating an invalid material model or FEM. One possibility is that it is an effect of the response of the adhesive used for the strain gauges unloading in a different manner than when it loads in tension to maximum strain. This effect could cause a measured strain different than the actual strain associated with the plate. This would additionally explain why gauges from the same specimen indicate larger or smaller windows for the total time over which the impact event occurs. Another explanation for the discrepancy is that the constraints and boundary conditions applied to the FEM are too restrictive when compared with the actual tests. The constraints and boundary conditions in the FEM could cause the simulated plates to rebound more quickly than the actual plates while simultaneously neglecting friction and leveraging effects that could slow the unloading response of the plate.

32.5   Conclusion

In this chapter, the response of FGMs subject to impact loading has been considered. A titanium–titanium–boride metal–ceramic composite system was chosen for a series of plate impact experiments. Strain histories were successfully collected from the impact experiments. The test results were compared with finite element simulations using two-phased and homogenized-layer material models to emulate the material properties of the Ti–TiB monolithic and graded plates. The FEM simulations compared well with the experimental data, and some inferences about differences between the test results and simulations were made. The key conclusion of this work is that the two-phase and homogenized-layer material models appear to be adequate for studying elastic FGM plate dynamics and work well within the more general finite element framework as demonstrated by the correlation between experiment and simulation. The major contribution of this work is the validation of FGM material models and a computational framework for studying the impact response of FGM plates as a foundation for investigations of more severe impact loads at higher velocities and energy levels.

The greatest challenge in working with FGMs is determining accurate and consistent material properties for the mixture of materials. This is especially true with a metal–ceramic combination. The finite element methods used to analyze the impact responses of FGM specimens worked very well overall but undoubtedly could have improved with more knowledge of the basic properties of the layers in the FGMs. The powder sintering process used to construct the materials also adds complexity as the microstructure of sintered materials is of an inherently different nature than wrought materials or even many other metal-matrix composites. Until more accurate material characterizations are available, the general two-phase and homogenized-layer material models are adequate for studying dynamic loading in the elastic regime.

Acknowledgments

The views expressed in this article are those of the authors and do not reflect the official policy or position of the U.S. Air Force, Department of Defence, or the U.S. Government. This work was funded by the U.S. Air Force Research Laboratory, Air Vehicles Directorate, Structural Science Center (AFRL/RBS), and the Dayton Area Graduate Studies Institute (DAGSI) under the AFRL/DAGSI Ohio Student–Faculty Research Fellowship Program.

The authors would like to further acknowledge Brian Smyers, Richard Wiggins, and Brett Hauber (AFRL/RBS); Daniel Ryan (U.S. Air Force Institute of Technology); Kevin Poormon (University of Dayton Research Institute); and Rachael DeRoche (Pennsylvania State University) and Joe Sabat (U.S. Air Force Academy) for their various contributions to this research.

References

1. Suresh, S., and Mortensen, A., *Fundamentals of Functionally Graded Materials: Processing and Thermomechanical Behaviour of Graded Metals and Metal-Ceramic Composites*, IOM Communications, Cambridge, UK, 1998.

2. Birman, V., and Byrd, L. W., Modeling and analysis of functionally graded materials and structures, *Applied Mechanics Reviews*, 60, 2007, 195–216. doi:10.1115/1.2777164.

3. Lambros, J., Santare, M. H., Li, H., and Sapna, G. H., III, A novel technique for the fabrication of laboratory scale functionally graded materials, *Experimental Mechanics*, 39(3), 1999, 184–190. doi:10.1007/BF02323551.

4. Lambros, J., Narayanaswamy, A., Santare, M. H., and Anlas, G., Manufacture and testing of a functionally graded material, *Journal of Engineering Materials and Technology*, 121, 1999, 488–493. doi:10.1115/1.2812406.

5. Parameswaran, V., and Shukla, A., Processing and characterization of a model functionally gradient material, *Journal of Materials Science*, 35, 2000, 21–29. doi:10.1023/A:1004767910762.

6. Reddy, J. N., Wang, C. M., and Kitipornchai, S., Axisymmetric bending of functionally graded circular and annular plates, *European Journal of Mechanics-A, Solids*, 18(2), 1999, 185–199.

7. Reddy, J. N., Analysis of functionally graded plates, *International Journal for Numerical Methods in Engineering*, 47, 2000, 663–684. doi:10.1002/(SICI)1097-0207(20000110/30)47:1/3<663::AID-NME787>3.0.CO;2-8.

8. Reddy, J. N., and Cheng, Z. Q., Three-dimensional thermomechanical deformations of functionally graded rectangular plates, *European Journal of Mechanics. A, Solids*, 20(5), 2001, 841–860.

9. Reddy, J. N., and Cheng, Z. Q., Frequency correspondence between membranes and functionally graded spherical shallow shells of polygonal planform, *International Journal of Mechanical Sciences*, 44(5), 2002, 967–985. doi:10.1016/S0020-7403(02)00023-1.

10. Loy, C. T., Lam, K. Y., and Reddy, J. N., Vibration of functionally graded cylindrical shells, *International Journal of Mechanical Sciences*, 41(3), 1999, 309–324. doi:10.1016/S0020-7403(98)00054-X.

11. Pradhan, S. C., Loy, C. T., Lam, K. Y., and Reddy, J. N., Vibration characteristics of functionally graded cylindrical shells under various boundary conditions, *Applied Acoustics*, 61, 2000, 111–129. doi:10.1016/S0003-682X(99)00063-8.

12. Woo, J., and Meguid, S. A., Nonlinear analysis of functionally graded plates and shallow shells, *International Journal of Solids and Structures*, 38, 2001, 7409–7421. doi:10.1016/S0020-7683(01)00048-8.

13. Yang, J., and Shen, H.-S., Dynamic response of initially stressed functionally graded rectangular thin plates, *Composite Structures*, 54, 2001, 497–508. doi:10.1016/S0263-8223(01)00122-2.

14. Yang, J., and Shen, H.-S., Vibration characteristics and transient response of shear deformable functionally graded plates in thermal environments, *Journal of Sound and Vibration*, 255(3), 2002, 579–602. doi:10.1006/jsvi.2001.4161.

15. Vel, S. S., and Batra, R. C., Exact solution for thermoelastic deformations of functionally graded thick rectangular plates, *AIAA Journal*, 40(7), 2002, 1421–1433. doi:10.2514/2.1805.

16. Prakesh, T., and Ganapathi, M., Axisymmetric flexural vibration and thermoelastic stability of fgm circular plates using finite element method, *Composites. Part B, Engineering*, 37, 2006, 642–649. doi:10.1016/j.compositesb.2006.03.005.

17. Gong, S. W., Lam, K. Y., and Reddy, J. N., The elastic response of functionally graded cylindrical shells to low-velocity impact, *International Journal of Impact Engineering*, 22(4), 1999, 397–417.

18. Bruck, H. A., A one-dimensional model for designing functionally graded materials to manage stress waves, *International Journal of Solids and Structures*, 37, 2000, 6383–6395. doi:10.1016/S0020-7683(99)00236-X.

19. Li, Y., Ramesh, K. T., and Chin, E. S. C., Dynamic characterization of layered and graded structures under impulsive loading, *International Journal of Solids and Structures*, 38, 2001, 6045–6061. doi:10.1016/S0020-7683(00)00364-4.

20. Banks-Sills, L., Eliasi, R., and Berlin, Y., Modeling of functionally graded materials in dynamic analyses, *Composites. Part B, Engineering*, 33, 2002, 7–15. doi:10.1016/S1359-8368(01)00057-9.

21. Decker, B. F., and Kasper, J. S., The crystal structure of TiB, *Acta Crystallographica*, 7, 1954, 77–80. doi:10.1107/S0365110X5400014X.

22. Sahay, S. S., Ravichandran, K. S., and Atri, R., Evolution of microstructure and phases in situ processed Ti–TiB composites containing high volume fractions of TiB whiskers, *Journal of Materials Research*, 14(11), 1999, 4214–4223. doi:10.1557/JMR.1999.0571.

23. Atri, R., Ravichandran, K. S., and Jha, S. K., Elastic properties of in-situ processed Ti–TiB composites measured by impulse excitation of vibration, *Materials Science and Engineering A*, 271(1), 1999, 150–159.

24. Panda, K. B., and Ravichandran, K. S., First principles determination of elastic constants and chemical bonding of titanium boride (TiB) on the basis of density functional theory, *Acta Materiala*, 54, 2006, 1641–1657.

25. Panda, K. B., and Ravichandran, K. S., Synthesis of ductile titanium–titanium boride (Ti–TiB) composites with beta-titanium matrix: The nature of TiB formation and composite properties, *Metallurgical and Materials Transactions A: Physical Metallurgy and Materials Science*, 34, 2003, 1371–1385. doi:10.1007/s11661-003-0249-z.

26. Ravichandran, K. S., Panda, K. B., and Sahay, S. S., A TiBw-Reinforced Ti composite: processing, properties, application prospects, and research needs, *JOM The Journal of the Minerals, Metals and Materials Society*, 56(5), 2004, 42–48. doi:10.1007/s11837-004-0127-1.

27. Cleveland, W. S., Robust locally weighted regression and smoothing scatterplots, *Journal of the American Statistical Association*, 74(368), 1979, 829–836. doi:10.2307/2286407.

28. Cleveland, W. S., and Devlin, S. J., Locally weighted regression: An approach to regression analysis by local fitting, *Journal of the American Statistical Association*, 83(403), 1988, 596–610. doi:10.2307/2289282.

29. ABAQUS, Software Package, Ver. 6.6-1, Dassault Systèmes SIMULIA, Providence, RI, 2008.

30. Oberg, E., Jones, F. D., Horton, H. L., and Ryfell, H. H. (eds.), *Machinery's Handbook*, 26th ed., Industrial Press, New York, 2000.

31. Daniel, I. M., and Ishai, O., *Engineering Mechanics of Composite Materials*, 2nd ed., Oxford University Press, New York, 2006.

32. Hill, R., A self-consistent mechanics of composite materials, *Mechanics of Composite Materials*, 13, 1965, 213–222.

33. Hashin, Z., and Shtrikman, S., A variational approach to the theory of the elastic behaviour of multiphase materials, *Journal of the Mechanics and Physics of Solids*, 11, 1963, 127–140. doi:10.1016/0022-5096(63)90060-7.

34. Hashin, Z., Assessment of the self-consistent scheme approximation: conductivity of particulate composites, *Journal of Composite Materials*, 2(3), 1968, 284–300. doi:10.1177/002199836800200302.

35. Yin, H. M., Paulino, G. H., Buttlar, W. G., and Sun, L. Z., Micromechanics-based thermoelastic model for functionally graded particulate materials with particle interactions, *Journal of the Mechanics and Physics of Solids*, 55, 2007, 132–160. doi:10.1016/j.jmps.2006.05.002.

36. Mori, T., and Tanaka, K., Average stress in matrix and average elastic energy of materials with misfitting inclusions, *Acta Metallurgica*, 21, 1973, 571–574. doi:10.1016/0001-6160(73)90064-3.

37. Benveniste, Y., A new approach to the application of mori–tanaka's theory in composite materials, *Mechanics of Materials*, 6, 1987, 147–157. doi:10.1016/0167-6636(87)90005-6.

38. Berryman, J. G., Long-wavelength propagation in composite elastic media I: Spherical inclusions, *Journal of the Acoustical Society of America*, 68(6), 1980, 1809–1819. doi:10.1121/1.385171.

39. Berryman, J. G., Long-wavelength propagation in composite elastic media II: Ellipsoidal inclusions, *Journal of the Acoustical Society of America*, 68(6), 1980, 1820–1831. doi:10.1121/1.385172.

40. Belytschko, T., Liu, W. K., and Moran, B., *Nonlinear Finite Elements for Continua and Structures*, Edward Arnold, London, 1960.

41. ABAQUS Theory Manual, ABAQUS Software Package, Ver. 6.6, Dassault Systèmes SIMULIA, Providence, RI, 2008.
42. Hill, M. R., and Lin, W., Residual stress measurement in a ceramic-metallic graded material, *Journal of Engineering Materials and Technology*, 124, 2002, 185–191. doi:10.1115/1.1446073.
43. Zukas, J. A., Nicholas, T., Swift, H. F., Greszczuk, L. B., and Curran, D. R., *Impact Dynamics*, Wiley, New York, 1982.

chapter thirty three

Laser powder-bed fusion additive manufacturing

Physics of complex melt flow and formation mechanisms of pores, spatter, and denudation zones

Saad A. Khairallah, Andrew T. Anderson,
Alexander M. Rubenchik, and Wayne E. King

Contents

Abstract: This study demonstrates the significant effect of the recoil pressure and Marangoni convection in laser powder-bed fusion (LPBF) of 316L stainless steel (SS). A three-dimensional (3D) high fidelity powder-scale model reveals how the strong dynamical melt flow generates pore defects, material spattering (sparking), and denudation zones. The melt track is divided into three sections: (1)

a topological depression, (2) a transition, and (3) a tail region, each being the location of specific physical effects. The inclusion of laser ray-tracing energy deposition in the powder-scale model improves over traditional volumetric energy deposition. It enables partial particle melting, which impacts pore defects in the denudation zone. Different pore formation mechanisms are observed at the edge of a scan track, at the melt-pool bottom (during collapse of the pool depression), and at the end of the melt track (during laser power ramp down). Remedies to these undesirable pores are discussed. The results are validated against the experiments and the sensitivity to laser absorptivity is also discussed.

33.1   Introduction

Additive manufacturing (AM) is paving the way toward the next industrial revolution [1]. The essence of this advancement is a part that is produced from a digital model by depositing material layer-by-layer, in other words, 3D printing the model. This technique is in contrast with the traditional subtractive and formative manufacturing approaches. It also eliminates most of the constraints that hinder optimal design, creativity and ease of manufacturing of complex parts [2,3].

A promising future is in store for LPBF AM. However, widespread adoption of LPBF with metallic parts hinges on solving a main challenge: the requirement that the final product should meet engineering quality standards [4]. This includes reducing porosity, because pore defects have one of the most adverse effect on mechanical properties. Experimental advances on this front rely on trial and error methods, which are costly and time inefficient. An attractive alternative to answering this challenge is through modeling and predictive simulation.

The finite element method (FEM) is the most popular numerical method for simulation of metal powder bed AM processes. Critical reviews by Schoinochoritis et al. [5] and King et al. [6] discuss different FEM models, assumptions, and results. The emphasis is how to get the most out of FEM simulations while avoiding computational expense. Some simplifications include: (1) treating the powder as a homogeneous continuum body with effective thermomechanical properties, (2) treating the laser heat source as a homogeneous model that deposits laser energy volumetrically like with De-Beer–Lambert's law or one derived for deep powder bed [7], and (3) ignoring melt-pool dynamics and therefore assuming a steady state. Take for example the work of Gu et al. [8] who employed a commercial code based on the finite volume method (FVM) to highlight the significant effect of Marangoni convection on heat and mass transfer in a continuum 3D model. In that model, the discrete nature of the powder is not accounted for; hence, the melt flow is symmetric along the melt track and does not exhibit fluctuations that may be introduced by a randomly packed powder bed.

This current chapter falls outside the FEM body of work. Our approach is to study the LPBF problem with a fine-scale model that treats the powder bed as randomly distributed particles. There are few studies that follow this mesoscopic approach.

In [9], Gutler et al. employ a volume of fluid method (VOF) and were the first to show more realism with a 3D mesoscopic model of melting and solidification. However, a single size powder arranged uniformly was represented at a coarse resolution that does not resolve the point contacts between the particles. This chapter makes qualitative correlations with experiments.

Körner et al. [10] used the lattice Boltzmann method (LBM) under the assumption that the electron beam melting process can be represented in 2D. One big hurdle in this method is the severe numerical instabilities occurring when accounting for the temperature. Körner uses the multidistribution function approach to reduce these limitations under the assumption that the fluid density is not strongly dependent on temperature. The method has been applied in 2D to study single layer [11] and layer-upon-layer consolidation [12], and shows the importance that the powder packing has on the melt characteristics. Their observation of the undesirable balling effect was attributed to the local powder arrangement [11]. Recently, a 2D vapor recoil pressure model was added in [13] to improve the melt-depth predictions. The Marangoni effect is neglected. In [14], a 3D model that does not include recoil, Marangoni, or evaporation effects was used to establish process strategies suitable to reduce build time and cost while enabling high-power electron beam applications.

Khairallah et al. [15] reported on a highly resolved model in 3D that considers a powder bed of 316L SS with a size distribution taken from experimental measurements. Khairallah et al. emphasized the importance of resolving the particle point contacts to capture the correct reduced effective thermal conductivity of the powder and the role of surface tension in breaking up the melt track into undesirable ball defects at higher laser scan speeds due to a variant of Plateau–Rayleigh instability theory [16].

A recent mesoscopic study by Lee and Zhang [17] introduces the powder into the model using the discrete element method. Their VOF study emphasizes the importance of particle size distribution and discusses the smoothing effect of small particles on the melt. They agree with Khairallah et al. [15] that balling is a manifestation of Plateau–Rayleigh instability and add that higher packing density can decrease the effect. Recoil and evaporation effects are neglected.

Recently, Qiu et al. [18] performed an experimental parameter study, whereby the surface roughness and area fraction of porosity were measured as a function of laser scan speed. They noted that the unstable melt flow, especially at high laser scan speed, increases porosity and surface defects. Based on a computational fluid dynamics (CFD) study of regularly packed powder of a single large size of 50 μm, they believe that the Marangoni and recoil forces are among the main driving forces for the instability of melt flow.

This chapter describes a new high-fidelity mesoscopic simulation capability developed to study the physical mechanisms of AM processes by eliminating certain physical assumptions that are prevalent in the literature due to modeling expense. The model uses a laser ray-tracing energy source and is in 3D to account for the fluid flow effects due to the recoil pressure, the Marangoni effect, and evaporative and radiative surface cooling. The new findings point out the importance of the recoil pressure physics under the laser and its dominant effect on creating a topological depression (similar to a keyhole) with strong complex hydrodynamic fluid flow coupled to a Marangoni surface flow. A vortex flow results in a cooling effect over the depression, which coupled to evaporative and radiation cooling over an expanded recoiled surface, regulates the peak surface temperatures. This finding should benefit part scale and reduced order modeling efforts, among others, that limit heat transfer to just conduction and therefore suffer from uncontrolled peak surface temperatures and may have to resort to model calibration to capture the effect.

This study, other than detailing the dominant physics in LPBF, reveals the formation mechanisms for pore defects, spatter, and the so-called denudation zone where powder particles are cleared in the vicinity of the laser track. Several authors report experimentally observing these effects, however, they formulate assumptions for formation mechanisms because, experimentally, it is challenging to dynamically monitor the LPBF process at the microsecond and micrometer scales. For example, Thijs et al. assume that some particles located in the denudation zone melt incompletely and create pore defects [19] and that other pores form due to the collapse of a keyhole [20]. Qiu et al. [21] observe open pores and assume that the incomplete remelting of the previous layer generates spherical pores.

The present study explains how three kinds of pore defects (depression collapse, lateral pores, open and trapped pores) are generated and discusses strategies to avoid them. This study, thanks to the laser ray-tracing energy source and the inclusion of recoil pressure, is also able to describe the physical mechanisms behind sparking [22], spattering, and denudation [23,24].

Experimental validation with sensitivity to the choice of laser absorptivity is also presented. The model makes use of the ALE3D [25] massively-parallel multi-physics code. Code details and SS material properties can be found in [15,26].

33.2 Model: Underlying physics and validation

33.2.1 Volumetric versus ray-tracing laser heat source

LPBF is a heat-driven process, which needs to be modeled accurately. This study uses a ray-tracing laser source (200 Watts) that consists of vertical rays with a Gaussian energy distribution ($D4\sigma = 54$ µm) scanning at 1.5 m/s. The laser energy is deposited at the points of powder-ray intersections. To reduce the computational complexity, the rays are not followed on reflection. The direct laser deposition is an improvement over volumetric energy deposition (energy as a function of fixed Z-axis reference) used commonly in the literature. First, in reality the heat is generated where the laser rays hit the surface of the powder particles and diffuses inward, whereas homogeneous deposition heats the inner volume of the particle uniformly. Second, the rays track the surface and can reproduce shadowing. In Figure 33.1a, a 150 W Gaussian laser beam is initially centered above a 27 µm particle sitting on a substrate and moved to the right at 1 m/s. For volumetric energy deposition,

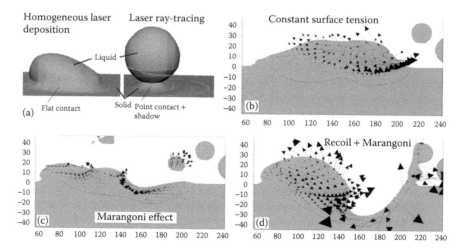

Figure 33.1 Incremental physics fidelity, significantly alters the heat transfer, melt-pool depth, and flow. The red pseudocolor corresponds to temperature scale capped at 4000 K, blue pseudocolor is 293 K. The red contour line is the melt line. The powder particle is illuminated by a laser (power 150 W) moving to the right (speed 1 m/s) for 10 µs. The melt tracks are 2D slices of 3D simulations (laser power is 200 W and scan speed is 1.5 m/s) demonstrating the effect of improved physics modeling on the melt pool (see Sections 33.2.2 and 33.3).

melting happens simultaneously everywhere inside the particle. The wetting contact with the substrate increases rapidly, which artificially increases heat dissipation. On the other hand, with realistic laser ray tracing, melting is nonuniform as it occurs first at the powder particle surface. More heat accumulates inside the powder particles compared with the homogeneous laser deposition because it releases to the substrate slowly through a narrow point contact. If insufficient heat is deposited, the particles are partially melted and contribute to surface and pore defects as discussed in Section 33.3.2.5. The laser ray-tracing heat source helps to better couple the physics behind surface heat delivery and melt hydrodynamics.

33.2.2   *Temperature-driven 3D flow effects: Surface tension, Marangoni convection, and recoil pressure*

Figure 33.1b–d illustrate the significant change of melt-pool characteristics as more temperature dependent physics is included. If surface tension (177 N/m) is assumed to be temperature independent, unphysical effects are observed. The melt pool is the shallowest with a constant surface tension in Figure 33.1b and shows a balling effect due to surface tension tendency to minimize surfaces by creating liquid spheres. The melt flow is also driven by buoyancy.

In Figure 33.1c, the strong temperature gradients below the laser necessitate enabling temperature dependent surface tension $\sigma(T) = 3.282 - 8.9e^{-4}\,T$, where T is temperature in Kelvin. This creates Marangoni effects. It drives the melt flow from the hot laser spot toward the cold rear. This serves to increase the melt depth, recirculate the melt flow (hence, cool the location of the laser spot) and create spattering as liquid metal with low viscosity ejects away from the surface.

The next increment in physics fidelity in Figure 33.1d comes from recognizing that the surface temperatures below the laser spot can easily reach boiling values. The vapor recoil pressure adds extra forces to the surface of the liquid that create a melt-pool surface depression below the laser. As the applied heating in LPBF does not cause extreme vaporization (ablation), the model does not resolve the vapor flow discontinuities and expansion from the liquid phase to ambient gas [27,28], nor does it include the mass lost to vaporization. In this study, a simplified model due to Anisimov [29] is employed, which has been used previously [26,30,31]. The recoil pressure P depends exponentially on temperature,

$$P(T) = 0.54 P_a exp^{-\frac{\lambda}{K_B}\left(\frac{1}{T} - \frac{1}{T_b}\right)}$$

where:
  $P_a = 1$ bar is the ambient pressure
  $\lambda = 4.3$ ev/atom is the evaporation energy per particle
  $K_B = 8.617 \times 10^{-5}$ ev/K is Boltzmann constant
  T is the surface temperature
  $T_b = 3086$ K is the boiling temperature of 316L SS

By combining the Marangoni effect with recoil pressure, the melt depth significantly increases, which also increases the surface area of the melt pool (by creating a depression; see Section 3.1) and helps further with cooling due to additional evaporative and radiative surface cooling. In fact, among the three 2D melt pool slices, the last shows the least amount of stored heat (shown in red pseudocolor).

33.2.3   Surface cooling: Evaporative and radiative cooling

As it is essential to calculate the surface temperature accurately, extra care is given to account for thermal losses. An evaporative cooling term is calculated at the surface interface and has the big role of limiting the maximum surface temperature under the laser, because the flux of evaporated metal vapor increases exponentially with T. According to Anisimov's theory [29], around 18% of the vapor condenses back to the surface due to large scattering angle collisions in the vicinity of the liquid and hence reduces the cooling effect. The net material evaporation flux is $J_v = 0.82AP(T)/\sqrt{2\pi MRT}$ and is consistent with the recoil pressure $P(T)$ derivation, where A is a sticking coefficient, which is close to unity for metals, M is the molar mass, R is the gas constant, and T is the surface temperature.

The model neglects evaporative mass loss, because the amount is negligible. As a conservative mass loss estimate, consider an area of 1 mm × 54 μm fixed at 3000 K for 0.67 ms. The mass loss amounts to ~0.1 μg, which is much less than the mass of an average SS particle with a radius of 27 μm.

In addition to evaporative cooling, radiative cooling that follows the Stefan–Boltzmann law, $R = \sigma\varepsilon\left(T^4 - T_0^4\right)$, assuming black body radiation, is included. Note that compared to the total deposited laser energy, the radiation heat losses are quite small. Here the Stephan's constant is $\sigma = 5.669 \times 10^{-8}\,\text{W/m}^2\text{K}^4$. The emissivity, ε, varies with temperature and surface chemistry and therefore is hard to represent [32]. For simplicity, an average value for the emissivity is taken to be 0.4 for the solid SS and 0.1 for the liquid state. T_0 is the ambient temperature. The model assumes that the lateral sides of the problem domain are insulated, whereas the bottom surface uses a boundary condition that approximates the response of a semi-infinite slab.

33.2.4   Experimental model validation and sensitivity to material absorptivity

The highest temperature gradients exist soon after the laser is turned on. For a laser power of 200 W and laser scan speed of 1.5 m/s, the surface melt-pool shape settles into quasi-steady-state about ~225 μs after the laser is turned on as shown in Figure 33.2. The width of the melt pool is observed to fluctuate along the solidified track. On the other hand, the melt depth increases until it stabilizes earlier at ~100 μs.

Table 33.1 shows that the melt depth, for a constant absorptivity of 0.35, yields very good quantitative agreement with the experiment. The second row shows a sensitivity study of melt-pool depth and width on absorptivity. The melt-pool depth is sensitive to laser absorptivity, whereas the width does not vary much as it depends mostly on beam size. Taking a constant absorptivity (which is a common approach [33]), is a main approximation in the model. A depression forms below the laser that could absorb more heat due to multiple reflections as shown in Figure 33.2. Experimentally, a plasma/metal vapor plume can change the absorptivity along the pool depth. However, incorporating a variable absorptivity is quite complex and not necessary for this model because the depression is not as deep as a keyhole [30].

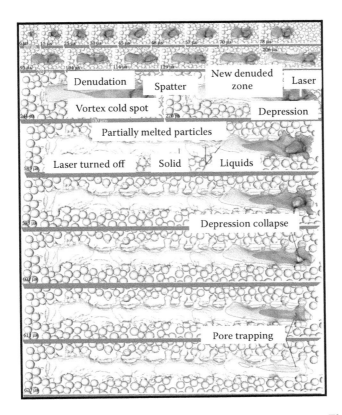

Figure 33.2 Time snapshots showing the evolution of the surface temperature. The laser scan speed is 1.5 m/s and moving to the right with a power of 200 W. The liquid melt pool is confined within the colored regions ($T > 1700$ K). The surface melt reaches a steady state late in time around 229 µs. The laser creates a topological depression, which is the site of forward and sideways spatter, and also contributes to the denudation process. The laser is turned off at 585 µs. Later in time, the depression collapse creates a trapped pore beneath the surface.

Table 33.1 Simulation and experiment data (separated by /) comparison of depth D [µm] and width W [µm] at different laser scan speeds S [mm/s] and powers P [Watts]. The power density is over a diameter given by $D4\sigma = 54$ µm. The material absorptivity (*abs.*) is held constant in the first row. The experimental uncertainty is 5 µm and the simulation's melt depth is on the order of zone size, which is 3 µm. The width fluctuates more than depth. The second row tests the sensitivity of the results to the absorptivity. An absorptivity of 0.35 shows the best agreement with the experiment

P300S1800	D68/65	P200S1200	D70/68	P150S800	D69/67
abs. 0.35	W96 ± 8/94	abs. 0.35	W94 ± 12/104	abs. 0.35	W89 ± 4/109
P200S1500	D45/57	P200S1500	D54/57	P200S1500	D60/57
abs. 0.3	W80 ± 5/84	abs. 0.35	W80 ± 9/84	abs. 0.34	W80 ± 4/84

33.3   Results and discussions

33.3.1   Anatomy of a melt track

It is possible to subdivide the melt track into three differentiable regions: (1) a *depression* region located at the laser spot, (2) a *tail end* region of the melt track located near the end, and (3) a *transition* region in between (see Figure 33.3 at 241 μs). This choice of subdivision is based on the exponential dominance of the recoil force at the depression and the dominance of surface tension in the cooler transition and tail regions.

The depression may be viewed as a source of fluid. Although the flow at the depression is complex, the flow in the transition zone has a net surface velocity component (V_x) in the negative direction (to the rear). The velocity snapshots from 215–270 μs in Figure 33.3 show a

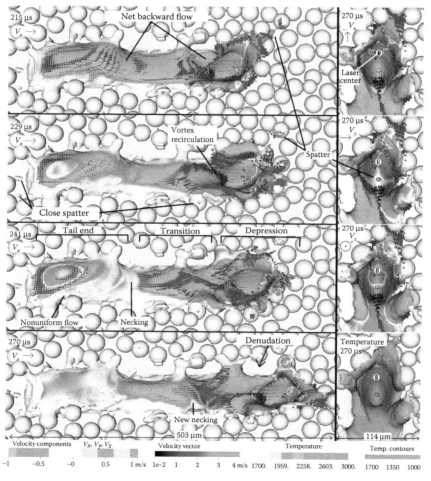

Figure 33.3 Time snapshots of the melt flow in Figure 33.1 showing spattering and denudation. The melt has a large backward flow (blue color; $V_x < 0$) due to Marangoni effect and recoil, compared to forward flow ($V_x > 0$; red color). The backward net flow breaks up later in time at the necking. The velocity scale is capped at +−1 m/s for better visualization. The right panel magnified view at 270 μs (flow rotated by +90°) shows the velocity components (V_x, V_y, and V_z) and the temperature (with contour lines) at the depression. The white letter *O* shows that the laser center is not at the bottom of the depression.

dominant blue region ($V_x < 0$) behind the depression region. At 225 μs, the surface melt-pool shape achieves a steady state. The backward flow starts to break up at the tail end of the track. Later (at 241 μs and 270 μs), it becomes easy to distinguish the three regions: the depression, the transition, and the tail. When placed in the laser reference frame, this flow breakup is reminiscent of the Plateau–Rayleigh instability in a cylindrical fluid jet that breaks into droplets, which has been observed in LPBF experiments [24,34]. This is a manifestation of nature's way of minimizing surface energy using surface tension. The melt track achieves a lower surface energy by transitioning from the segmented cylinder [24] observed in the transition region to the segmented hemispherical-like tail-end region [16]. The necking locations where the melt track dips or even disappears correspond to the necking of a narrow cylindrical fluid jet prior to break up into droplets. These dips cool down quickly. It is possible to control the magnitude of the fluctuations in the tail-end regions by adjusting the laser speed for a given power, and hence averts major balling, by controlling the heat content over time in the melt track. Less heat content gives the surface tension less time to completely break the flow [15]. For the current simulation parameters (scan speed lower than in [15]), the balling instability is mild.

33.3.2   Effects of a strong dynamical melt flow

33.3.2.1   Depression formation

Figure 33.4 shows a time series of track cross sections for a fixed position with the laser moving out of the plane. They highlight the formation of the depression region, which is marked by the highest temperatures achieved on the track (See Figure 33.2). In this region, which is directly under the laser, the recoil effect is dominant due to its exponential dependence on temperature and creates a noticeable topological depression. At 45 μs, the momentum imparted by hot spatter falling ahead of the depression moves the particles lying ahead of the laser. After 58 μs, the particles melt within 20 μs ahead of the Gaussian laser center. The particle sizes follow a normal distribution centered at 27 μm, with a full width at half maximum of 1.17, and with tail cutoffs at 42 μm and 17 μm. The smaller one melts completely before the larger one and hence increases the particle thermal contact area (see discussion on the laser source in Section 2.1 and Figure 33.1). The ensuing liquid has a large speed lateral flow component ~4–6 m/s directed away from the center of the hot spot, which is marked by a narrow black temperature contour line (3500 K). The center of the laser reaches the slice ~30 μs after first signs of powder melting. With surface temperatures approaching the boiling temperature, the recoil pressure applies an exponentially increasing force normal to the surface, which accelerates the liquid away from the center as the velocity vectors show at 76 μs. The result is a depression with a thin liquid boundary layer at the bottom. It is mostly thin at the bottom of the depression, where the temperature is the highest. The vertical velocity component of the liquid is negative at the bottom of the depression where the recoil force is digging the hole, and is positive along the sidewalls and the rim where the liquid escapes vertically at relatively high speed (~1 m/s) and contributes to spattering as shown in Figure 33.3 at 270 μs.

This depression is closely related to the keyhole cavity observed in welding [33]. Also, King et al. [35], experimentally observed keyhole-mode melting in laser powder bed fusion (PBF) and ascribe this to a surface threshold temperature close to boiling. The recoil force is the main driving force for the keyhole-mode melting. Many numerical models for keyhole-mode laser welding involve simplifying assumptions. They typically balance the recoil force, the surface tension pressure, and the hydrostatic liquid pressure. Furthermore, the models can be 2D and often consider heat transfer by conduction only, without accounting for the influence of convection on heat dissipation. As similar underlying physics processes also occur in LPBF, these simplifying approaches have also been adopted when developing LPBF

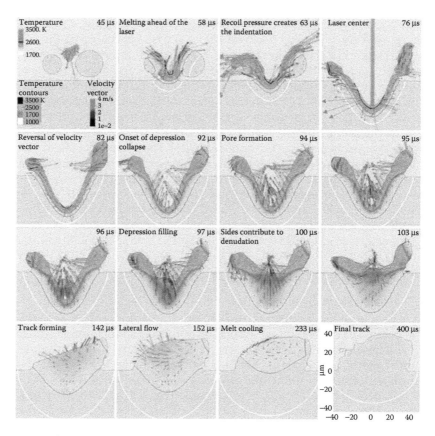

Figure 33.4 Lateral 2D slices of the track from Figure 33.1 showing the temperature and velocity field of the melt as the laser scans (direction out of page) by a fixed location. They show the events before the arrival of the laser center (45–76 µs), the indentation formation (76–82 µs), the indentation collapse and formation of a pore (92–103 µs), and the asymmetrical flow pattern due to an asymmetrical cooling as the melt solidifies (142–400 µs).

models [5]. However, missing the effects, such as convective cooling, of the strong dynamical flow shown in Figures 33.3 and 33.4 may limit the range of predictability of these models.

33.3.2.2   *Depression collapse and pore formation mechanism*

At 82–92 µs (Figure 33.4), the laser's hottest spot has just passed through the plane of the figure. The temperature at the back of the depression decreases, which is indicated by the recession of the black temperature contour line (~3500 K). Behind the hottest spot, a decrease in temperature is accompanied by an exponential decrease in recoil force; however, the surface tension increases at lower temperatures and overcomes the recoil force effect, which was keeping the depression open. As a result, the melt-flow velocity-vector field reverses direction toward the center in Figure 33.4 starting at 82 µs. This reversal is abrupt and causes the sidewall to collapse within 5 µs. Gravity is included in the model but has negligible effect on this timescale. This fast flow increases the chance of trapping gas bubbles and therefore forming pores at the bottom of the track. The sequences at 94–97 µs show this pore formation mechanism.

Figure 33.5a shows another possible mechanism for pore formation due to a vortex, represented by a velocity vector field circulating counterclockwise that follows the

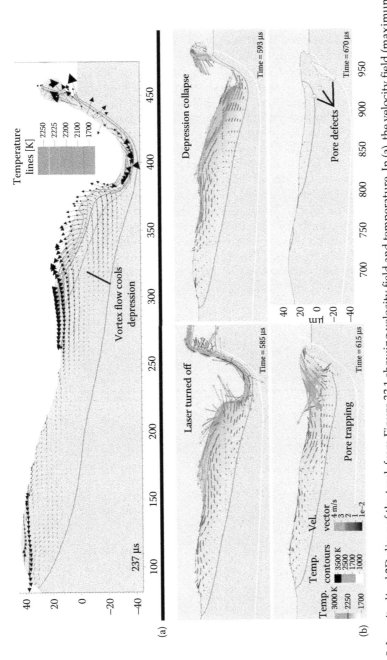

Figure 33.5 Longitudinal 2D slices of the track from Figure 33.1 showing velocity field and temperature. In (a), the velocity field (maximum magnitude 9 m/s) shows a vortex pushing cold temperature contour lines into hotter regions behind the depression zone. In (b), the figures at 585–670 μs show another process where pores are formed. On turning the laser off, the depression collapse creates three pores.

depression closely from the rear. We speculate that it could trap bubbles and/or seed a bigger pore by pore coalescence, meanwhile the solid front advancing from the bottom would catch the bubble and freeze it into a permanent pore.

The vortex has another effect. It helps with cooling as it brings colder liquid back to the depression. The vortex is visible in Figure 33.3 (270 μs) as a small red patch ($V_x > 0$), at the back wall of the depression, surrounded by a blue region ($V_x < 0$). Figure 33.5a shows cold temperature contour lines pushing hotter ones toward the depression. Figure 33.2 (241 μs) shows this cooling effect as a cold blue patch ($T < 2258$ K) mixing with hotter yellow region. The vortex only ceases to exist after the laser is turned off at 585 μs.

King et al. [35] observed pores in keyhole-mode laser melting in laser PBF experiments on 316L SS. King et al. followed a similar scaling law as Hann et al. [36] to analyze their findings. Hann et al. derived a scaling law to classify a variety of materials with different welding process parameters. The general welding data seem to collapse to one curve under the assumption that the melt depth divided by the beam size is a function of $\Delta H/hs$, which is the deposited energy density, divided by the enthalpy at melting. King et al. showed that similar scaling applied well to laser-bed fusion and found that the threshold to transition from conduction to keyhole-mode laser melting is $\Delta H/hs \approx (30 \pm 4)$. They concluded that "going too far below the threshold results in insufficient melting and going too far above results in an increase in voids due to keyhole-mode melting." With a ratio of $\Delta H/hs = 33$, the simulation model in this study is at the threshold, and indeed it shows a relatively small keyhole like depression and small number of pores, as is evident from the 3D view as shown below in Figure 33.6.

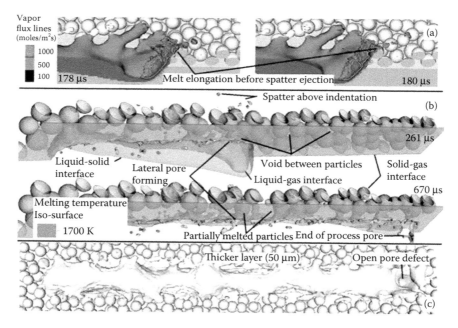

Figure 33.6 Formation of defects and spatter (laser power 200 W, speed 1.5 m/s). In (a), the 3D selection clips show an elongated fluid column breaking into spatter due to high vapor flux, that is, evaporations. In (b), the snapshots reveal a lateral pore forming out of the voids that exist between the particles. After turning the laser off at 585 μs, an end of process pore forms. It is capped. In (c), another end of process pore is shown. But it is open. These pores can seed in more defects in the subsequent layers.

33.3.2.3 Denudation mechanism

At 100–400 μs in Figure 33.4, the liquid fills the depression and grows in height. A lateral liquid flow is noted due to asymmetrical cooling in the transition region. This is due to partially melted particles (see Figure 33.2 at 585 μs) that remain in touch with the melt track and dissipate heat laterally. The surface tension will then pull surface fluid toward the cold spot (Marangoni effect) and hence bias any lateral circulation. This is undesirable because these can possibly create bridges with gaps underneath and seed further defects in the next deposited layer.

Most often, the side particles melt completely and are trapped in the flow in the transition region. The cause is liquid that circulates around the rim of the depression and resembles a teardrop. This pattern is observed in traditional welding. It is visible in Figure 33.3 (270 μs, V_y) where the flow alternates between red ($V_y < 0$) and blue ($V_y > 0$) two times around the depression rim: Once ahead of the depression, to indicate motion away from the laser spot, and one last time to indicate fluid coming from the sides and joining to form the transition region.

This circular motion has a wider diameter than the melt-track width. This can be seen in Figure 33.4 at 100 μs where the melt-temperature contour line in the substrate does not extend far enough to contain the melt above. The liquid that spills over to the sides catches the neighboring particles and drags them into the transition zone, behind the depression, hence creating what is known as the denudation zone along the sides of the track [23,24]. The velocity vectors in the snapshot series from 241 μs to 270 μs in Figure 33.3 show the denudation from top view: The flow at 241 μs overlaps with particles that disappear later at 270 μs. The mechanism for the denudation is enhanced by the high velocity circular flow (1–6 m/s). Yadroitsev et al. [23,24] observed the denudation zones experimentally and attributed it partially to particles in the immediate vicinity of the track as seen in this study.

33.3.2.4 Spatter formation mechanism

Figures 33.3 and 33.4 at 45 μs show the build up of liquid that develops ahead of the depression and the laser spot. This build up is similar in nature to the *bow wave* that develops as a boat moves through the water or to the motion of snow rolling over in front of a snowplow. The liquid colored in red in Figure 33.3 moves up the front wall of the depression and spills over onto the powder particles ahead of the laser beam. This is an important feature as liquid can be pinched off in this process and be deposited as spatter particles in the powder bed.

Figure 33.6a details how this liquid build-up (or *bow wave*) leads to spattering, which is experimentally observed in [22]. The high vapor surface flux (referred to as gas plume in [22]) exerts a pressure force that ejects liquid metal. When the liquid metal elongates, it thins out and breaks up into small droplets due to surface tension tendency to minimize surface energy. Figures 33.3 and 33.6a show that the elongation is in the radial direction to the laser spot and pointing away from the melt pool.

33.3.2.5 Lateral shallow pores and trapped incompletely melted particles

Another pore formation mechanism takes place in the transition region. The strong high-speed flow along the depression rim that brings in the particles and hence creates the denudation zone also mixes in voids that originally existed between the particles. One realistic effect of the ray-tracing laser source is that it allows for partial melting of particles. If a particle does not melt completely and merge with the melt pool, the voids present between the particles may contribute to pore defects. The snapshots in Figure 33.6b show a partially melted particle below which, a shallow lateral pore on the order of 5 μm is generated. These trapped particles are also defects that increase surface roughness and "deteriorate the wetting behavior of the next layer and act as the origin of continued layer instability" according to D. Wang et al. [37].

Thijs et al. [19] observed lateral pores when a laser scan was performed with hatch spacing equal to the melt-pool width. This means that the neighboring scanning vectors do not overlap each other. They observed the pores between the tracks running parallel to the scan directions. When viewed from the front side of the part, in the direction of the scan, these pores were vertically aligned, and the line repeated in a periodic way along the edges of the melt-track width. Although these pores are observed at the part level, the defects are seeded at the single layer level [38] and are most likely related to trapped partially melted particles. These pores are certainly undesirable but fortunately, it is possible to eliminate them by appropriately overlapping the neighboring scan tracks. The remedy is to adjust the hatch spacing process parameter to create a 25% scan overlap suggested by Thijs et al. [19].

33.3.2.6   *End of process pores*

Thijs et al. [20] report on keyhole pores at the end of the scan track. The current model also shows that an opportunity for pores to arise occurs on switching off the laser. The snapshots taken, after the laser is turned off at 585 μs, in Figures 33.5 (585–670 μs) and 33.6b show a large ellipsoidal pore getting trapped beneath the surface due to a fast laser ramp-down (1 μs). Two other small spherical pores form this way. Figure 33.6c offers a different scenario whereby different random powder packing (thicker layer) randomly leaves an uncapped narrow depression.

The remedy for these kinds of pores is to allow the surface tension ample time to smooth the surface. So the laser should be ramped down slowly, on the order of few $t_\sigma = 27$ μs, given by a characteristic timescale for surface tension ($t_\sigma = \sqrt{\rho L^3/\sigma}$, where ρ is density, σ is surface tension, and L is a characteristic length scale).

33.4   *Conclusion*

In conclusion, this study demonstrates the importance of recoil pressure and Marangoni convection in shaping the melt-pool flow and how denudation, spattering, and pore defects emerge and become part of a laser bed-fusion process. The physics processes involved are intimately coupled to each other because they all have a strong dependence on the temperature.

Although radiation cooling scales as T^4, the evaporative cooling is more efficient at limiting the peak surface temperature because of its exponential dependence on T. This has a strong effect on the magnitude of the recoil pressure because the latter also grows exponentially with the temperature. The recoil force overcomes the surface tension, which opposes the compressive effect of the recoil force, and therefore creates the depression and material spatter. Upon cooling below the boiling point, the surface tension takes over and causes pores to form on depression wall collapse. The surface tension effects dominate in the transition region where a strong flow (Marangoni effect) takes place. This flow helps with cooling of the depression, creating the denudation zone, pulling in adjacent particles and creating side pores close to partially melted particles. Eventually the transition zone thins out due to the melt flow breaking up and forming the tail-end region. The latter is subject to irregular flow that is short lived due to the drop in temperatures and solidification.

Deep and narrow depressions should be avoided in order to decrease pore formation due to depression collapse. One should also note that, upon changing direction along a scan track, the laser intensity should be decreased; otherwise, extra heat deposited could lead to a deep and narrow depression, which collapses and forms pores. An appropriate scan vector overlap can increase the densification by eliminating partially melted and trapped particles and any associated shallow lateral pores. Also, a gentle ramping down of the laser power, on the order of few t_σ, can prevent end of track pores and side surface roughness.

Acknowledgment

We acknowledge valuable input from Wayne King. This work was performed under the auspices of the U.S. Department of Energy by Lawrence Livermore National Laboratory under Contract DE-AC52 - 07NA27344. This work was funded by the Laboratory Directed Research and Development Program under project tracking code 13-SI-002. The LLNL document review and release number is LLNL-JRNL-676495.

References

1. B. Berman, 3-D printing: The new industrial revolution, *Business Horizons*, 55, 155–162, 2012.
2. M. C. Leu, D. W. Rosen, and D. L. Bourell, *Roadmap for Additive Manufacturing Identifying the Future of Freeform Processing*, The University of Texas at Austin, Austin TX, 2009.
3. S. Srivatsa, *Additive Manufacturing (AM) Design and Simulation Tools Study*, Air Force Research Laboratory, Wright-Patterson AFB OH 45433, 2014.
4. Anon, *3D printing and the new shape of industrial manufacturing*. Delaware: Price water house Coopers LLP, 2014.
5. B. Schoinochoritis, D. Chantzis, and K. Salonitis, Simulation of metallic powder bed additive manufacturing processes with the finite element metho: A critical review, *Institution of Mechanical Engineers*, Vol 0, 1–22, 2014.
6. W. E. King, A. T. Anderson, R. M. Ferencz, N. E. Hodge, C. Kamath, S. A. Khairallah, A. M. Rubenchik, Laser powder bed fusion additive manufacturing of metals, *Applied Physics Review 2*, 2, 041304, 2015.
7. A. F. Gusarov, and J. P. Kruth, Modelling of radiation transfer in metallic powders at laser treatment, *International Journal of Heat and Mass Transfer*, 48(16), 3423–3434, 2005.
8. P. Yuan, and G. Gu, Molten pool behaviour and its physical mechanism during selective laser melting of TiC/AlSi10Mg nanocomposites: Simulation and experiments, *Journal of Physics D: Applied Physics*, 48, 16, 2015.
9. F. J. Gutler, M. Karg, K. H. Leitz, and M. Schmidt, Simulation of laser beam melting of steel powders using the three-dimensional volume of fluid method, *Physics Procedia*, 41, 874–879, 2013.
10. E. Attar, and C. Korner, Lattice Boltzman model for thermal free surface flows with liquid-solid phase transition, *International Journal of Heat and Fluid Flow*, 32, 156–163, 2011.
11. C. Korner, E. Attar, and P. Heinl, Mesoscopic simulation of selective beam melting processes, *Journal of Materials Processing Technology*, 211, 978–987, 2011.
12. C. Korner, A. Bauerei, and E. Attar, Fundamental consolidation mechanisms during selective beam melting of powders, *Modelling and Simulation in Materials Science and Engineering*, 21, 18, 2013.
13. A. Klassen, A. Bauerei, and C. Korner, Modelling of electron beam absorption in complex geometries, *Journal of Physics D: Applied Physics*, 47, 11, 2014.
14. R. Ammer, U. Rude, C. Korner, M. Markl, Numerical investigation on hatching process strategies for powder-bed-based additive manufacturing using an electron beam, *The International Journal of Advanced Manufacturing Technology*, 78, 239–247, 2015.
15. S. A. Khairallah, and A. Anderson, Mesoscopic simulation model of selective laser melting of stainless steel powder, *Journal of Materials Processing Technology*, 214, 2627–2636, 2014.

16. A. V. Gusarov, and I. Smurov, Modeling the interaction of laser radiation with powder bed at selective laser melting, *Physics Procedia*, 5, 381–394, 2010.

17. Y. S. Lee, and W. Zhang, Mesoscopic simulation of heat transfer and fluid flow in laser powder bed additive manufacturing, in *International Solid Free Form Fabrication Symposium*, Austin, TX, 2015, pp. 1154–1165.

18. C. Qiu et al., On the role of melt flow into the surface structure and porosity development during selective laser melting, *Acta Materialia*, 96, 72–79, 2015.

19. L. Thijs, F. Verhaeghe, T. Craeghs, J. V. Humbeeck, and J. P. Kruth, A study of the microstructural evolution during selective laser melting of Ti-6Al-4V, *Acta Materialia*, 58, 3303–3312, 2010.

20. K. Kempen, J. P. Kruth, J. Van Humbeeck L. Thijs, Fine-structured aluminium products with controllable texture by selective laser melting of pre-alloyed AlSi10Mg powder, *Acta Materialia*, 61, 1809–1819, 2013.

21. N. J. E. Adkins, M.M. Attallah, C.L. Qiu, Microstructure and tensile properties of selectively laser-melted and of HIPed laser-melted Ti–6Al–4V, *Materials Science and Engineering. A*, 578, 230–239, 2013.

22. Y. Kawahito, K. Nishimoto, and S. Katayama Hiroshi Nakamura, Elucidation of melt flows and spatter formation mechanisms during high power laser welding of pure titanium, *Journal of Laser Applications*, 27(3), 32012–32022, 2015.

23. P. Bertrand, G. Antonenkova, S. Grigoriev, and I. Smurov I. Yadroitsev, Use of track/layer morphology to develop functional parts by selective laser melting, *Journal of Laser Application*, 25, 5, 2013.

24. A. Gusarov, I. Yadroitsava, and I. Smurov I. Yadroitsev, Single track formation in selective laser melting of metal powders, *Journal of Materials Processing Technology*, 210, (12), 1624–1631, 2010.

25. C. R. McCallen, ALE3D: Arbitrary Lagrange Eulerian three- and two dimensional modeling and simulaiton capability, July 18, 2012.

26. A. Anderson, A. M. Rubenchik, J. Florando, S. Wu, and H. Lowdermilk S. A. Khairallah, Simulation of the main physical processes in remote laser penetration with large laser spot size, *AIP Advances*, 5, 47120, 2015.

27. Z. Zhaoyan, and G. George, Theory of shock wave propagation during laser ablation, *Physical Review B*, 69, 235403–235403, 2004.

28. M. Aden, E. Beyer, and G. Herziger, Laser-induced vaporization of metal as a Riemann problem, *Journal of Physics D–Applied Physics*, 23, 655–661, 1990.

29. S. I. Anisimov, and V. A. Khokhlov, *Instabilities in Laser-Matter Interaction*. Boca Raton, FL: CRC Press, 1995.

30. V. V. Semak, W. D. Bragg, B. Damkroker, and S. Kempka, Transient model for the keyhole during laser welding, *Journal of Physics D–Applied Physics*, 32, 61–64, 1999.

31. V. Semak, and A. Matsunawa, The role of recoil pressure in energy balance during laser materials processing, *Journal of Physics D–Applied Physics*, 30, 2541–2552, 1997.

32. H. Schopp et al., Temperature and emissivity determination of liquid steel S235, *Journal of Physics D–Applied Physics*, 45, 235203, 2012.

33. R. Rai, J. W. Elmer, T. A. Palmer, and T. DebRoy, Heat transfer and fluid flow during keyhole mode laser welding of tantalum, Ti–6Al–4V, 304L stainless steel and vanadium, *Journal of Physics D–Applied Physics*, 40, 5753–5766, 2007.

34. J. -P. Levy, G. Klocke, F. Childs, T.H.C. Kruth, Consolidation phenomena in laser and powder-bed based layered manufacturing. *CIRP Annals—Manufacturing technology*, 56, 730–759, 2007.

35. W. E. King et al., Observation of keyhole-mode laser melting in laser powder-bed fusion additive manufacturing, *Journal of Materials Processing Technology*, 214, 2915–2925, 2014.

36. D. B. Hann, J. Iammi, and J. Folkes, A simple methodology for predicting laser-weld properties from material and laser parameters, *Journal of Physics D–Applied Physics*, 44, 445401, 2011.

37. D. Wang, X. Liu, D. Zhang, S. Qu, J. Ma, G. London, Z. Shen, W. Liu X. Zhou, 3D-imaging of selective laser melting defects in a Co–Cr–Mo alloy by synchrotron radiation micro-CT, *Acta Materialia*, 98, 1–16, 2015.

38. J. Schwerdtfeger, R. E. Singer, and C. Koerner, In Situ flaw detection by IR-imaging during electron beam melting, *Rapid Prototyping Journal*, 18, 259–263, 2012.

chapter thirty four

Measurement science needs for real-time control of additive manufacturing powder-bed fusion processes

Mahesh Mani, Shaw Feng, Brandon Lane, Alkan Donmez, Shawn Moylan, and Ronnie Fesperman

Contents

Abstract: Additive manufacturing (AM) is increasingly used in the development of new products: from conceptual design to functional parts and tooling. However, today, variability in part quality due to inadequate dimensional tolerances, surface roughness, and defects, limits its broader acceptance for high-value or mission-critical applications. Although process control in general can limit this variability, it is impeded by a lack of adequate process measurement methods. Process control today is based on heuristics and experimental data, yielding limited improvement in part quality. The overall goal is to develop the measurement science* necessary to make in-process measurement and real-time control possible in AM. Traceable dimensional and thermal metrology methods must be developed for real-time closed-loop control of AM processes. As a precursor, this report presents a review on the AM control schemes, process measurements, and modeling and simulation methods as it applies to the powder bed fusion (PBF) process, though results from other processes are reviewed where applicable. The aim of the review is to identify and summarize the measurement science needs that are critical to real-time process control. We organize our research findings to identify the correlations between process parameters, process signatures, and product quality. The intention of this report is to serve as a background reference and a go-to place for our work to identify the most suitable measurement methods and corresponding measurands for real-time control.

Keywords: additive manufacturing powder bed fusion real-time controlmeasurement science correlations process parameters process signatures product quality

34.1 Introduction

AM is increasingly used in the development of new products: from prototypes to functional parts and tooling. AM [1] is also referred to as rapid prototyping, additive fabrication, freeform fabrication, three-dimensional (3D) printing, and rapid manufacturing, and uses advanced technologies to fabricate parts by joining and building up material layer-by-layer. According to [2] "the expected long-term impact is in highly customized manufacturing, where AM can be more cost-effective than traditional methods." According to an industry report by Wohlers Associates [3], by 2015, the sale of AM products and services could reach $3.7 billion worldwide, and by 2019, exceed $6.5 billion. However, research is still required to fully realize the potential of AM, particularly for complex metal components (e.g., aerospace parts or automotive parts).

The widespread adoption of AM is challenged by part quality issues, such as dimensional and form errors, undesired porosity, delamination of layers, as well as poor or undefined material properties. Once the input material is established, part quality issues may be attributed to the AM process parameter settings, typically chosen today by a trial-and-error method. This approach is time consuming, inaccurate, and expensive. It is important to establish correlations between the AM process parameters and the process/part characteristics, to ensure desirable part quality and promote widespread adoption of AM technology. Once the correlations are established, in-process sensing and real-time control of AM process parameters can be done to minimize variations during the AM build process to ensure resulting product quality and production throughput.

According to a roadmap workshop on the measurement science needs for metal-based AM [4,5] hosted by the National Institute of Standards and Technology (NIST), closed-loop control systems for AM was identified as an important technology and measurement

---

* Measurement science broadly includes: development of performance metrics, measurement and testing methods, predictive modeling and simulation tools, knowledge modeling, protocols, technical data, and reference materials and artifacts; conduct of intercomparison studies and calibrations; evaluation of technologies, systems, and practices, including uncertainty analysis; development of the technical basis for standards, codes, and practices in many instances via test-beds, consortia, standards and codes development organizations, and/or other partnerships with industry and academia.

challenge vital for: (1) monitoring of process and equipment performance, (2) assurance of part adherence to specifications, and (3) the ability to qualify and certify parts and processes. Part quality in AM, defined by geometry, mechanical properties, and physical properties, is highly variable thereby limiting AM's broad acceptance. This variability can be reduced through robust process control.

Based on a literature review, the scope of this report is to identify the measurement science needs for real-time monitoring and control of PBF processes. The report is subsequently organized as follows: Section 34.2 first presents an overview of the PBF process. Section 34.3 presents a literature review according to the review strategy to potentially identify the correlations between process parameters, process signatures, and product quality. Section 34.4 then presents the implications for real-time process control followed by a summary on the potential research opportunities. Section 34.5 concludes the report.

34.2 Overview of PBF process

PBF is one of the seven categories of AM processes defined in ASTM F2792 [1]. PBF processes use thermal energy to selectively fuse areas of a layer of powder using laser or an electron beam as the energy source [1]. When the energy source traces the geometry of an individual layer onto the top surface of the powder bed, the energy from the beam spot is absorbed by the exposed powder causing that powder to melt. This small molten area is often described as the melt-pool. Individual powder particles are fused together when the melt-pool resolidifies. After one layer is completed, the build platform is lowered by the prescribed layer thickness, and a new layer of powder from the dispenser platform is swept over the build platform, filling the resulting gap and allowing a new layer to be built. Figure 34.1 depicts one such process that uses a laser beam as the energy source. When a part build is completed, it is fully buried within the powder in the build platform.

There are several different types of PBF commercial systems that can produce either polymer or metal parts. Today, most of the commercially available metal-based AM systems are PBF processes [3]. Some varieties/variations of PBF processes use low power lasers to bind powder particles by only melting the surface of the powder particles (called selective laser sintering or SLS) or a binder coating the powder particles. These processes produce green parts that require further postprocessing to infiltrate and sinter the parts to

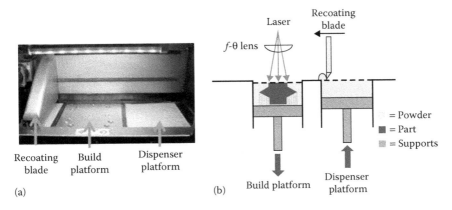

Figure 34.1 Components of the build chamber: (a) photograph showing the positions of the build platform, powder dispenser platform, and recoating blade and (b) schematic depicting the process of recoating and spreading a new layer of powder over the previously fused layers of the part.

make them fully dense. Another class of PBF processes uses high power energy beams to fully melt the powder particles, which then fuse together to the previous layer(s) when the molten material cools, for example, SLM, direct metal laser sintering (DMLS), or electron beam melting (EBM). Repeating this process, layer-by-layer, directly results in a part with near 100% density, even in metals. These processes are of primary interest to this study. General specifications for metal-based PBF systems can be seen in the Appendix.

34.3   Literature review

The central idea to the review strategy followed in this report is to identify the correlations between process parameters, process signatures, and product qualities to exploit these relationships in the monitoring and control solutions. AM *process parameters* are the *inputs* and primarily determine the rate of energy delivered to the surface of the powder and how that energy interacts with material. We categorize process parameters into either controllable (i.e., possible to continuously modify), such as laser power and scan speed, or predefined (i.e., set at the beginning of each build) material properties, such as powder size and distribution. The *process signatures* are dynamic characteristics of the powder heating, melting, and solidification processes as they occur during the build. These are categorized into either observable (i.e., can be seen or measured), such as melt-pool shape and temperature, or derived (i.e., determined through analytical modeling or simulation), such as melt-pool depth and residual stress. Process signatures significantly influence the final product qualities. Those *product qualities* are categorized into geometrical, mechanical, and physical qualities. Identifying the correlations between process parameters, process signatures, and product qualities, as shown in Figure 34.2, should facilitate the development of the in-process sensing and real-time control of AM process parameters to characterize and control the AM PBF process.

   We group the review into three categories: *control schemes, process measurements,* and *modeling and simulation* efforts as applicable to real-time process control.

34.3.1   Current control schemes in AM

This section reviews previous research efforts that are directly or potentially applicable to a closed-loop adaptive control system that utilizes melt-pool temperature and size, layer-by-layer part geometry, or defect characteristics as feedback.

34.3.1.1   PBF-related process control

In the reported studies, the melt-pool temperature and size are most often assumed to be the critical control factor influencing the outcome of the process.

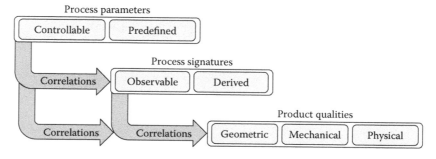

Figure 34.2 Correlations between process parameters, process signature, and product qualities.

The group at the Catholic University of Leuven developed a control system for a laser-based PBF system based on real-time monitoring of the melt-pool [6]. The melt-pool was monitored using a complementary metal-oxide-semiconductor (CMOS) camera and a photodiode placed coaxially with the laser. The image from the camera was used to determine the melt-pool geometry. Based on their observation they found the photodiode signal correlated well with the melt-pool area. They used this area-based signature as feedback to control the laser power and showed improved surface roughness. Later, they extended their process control efforts by introducing an online control methodology using two complementary measurement systems: (1) visual inspection of powder deposition and (2) real-time monitoring of melt-pool, that is, measuring both melt-pool geometry and infrared (IR) radiation intensity signal [7]. They state that the melting process is influenced by more than 50 parameters, which are classified as input parameters (such as scanning, deposition, and atmosphere) and boundary conditions (such as material properties, geometric parameters, and machine parameters), and concede that monitoring or controlling all parameters is a significant challenge. This work extended their measurement system to include a visible- light camera overlooking the entire build platform, which detected defects due to recoating blade wear and local damage of the blade. The same melt-pool monitoring system calculated melt-pool geometry (characterized as length-to-width ratio) in real-time. Results showed increasing photodiode intensities apparently due to defective layer-size control. This was attributed to overheating of the melt-pool during acute corners of the laser scan path. The optical system was further developed to detect process failures in each build layer by mapping the melt-pool temperature signatures as a function of the X–Y laser beam position on each layer [8]. Using such maps in real-time, the group was able to detect deformation due to thermal stresses and overheating zones due to overhangs.

Mumtaz and Hopkinson studied the effect of heat delivered to the melt-pool, that is, the laser–material interaction zone, to determine the roughness of the surface generated by the solidified melt-pool [9]. Heat-affected zone (HAZ) is the area near and including the melt-pool that is directly affected by high local temperatures. Using a pulsed laser system, they experimented with various pulse shapes to distribute energy within a single laser pulse. It was proposed that the use of pulse shaping would offer precise and tailored control over the heat input and would allow refining and improvement over the use of standard rectangular pulses. The height of the laser-induced plasma plume was measured using a video camera to identify the correlation between the pulse shapes and the amount of spatter generated during processing. The added degree of control through pulse shaping resulted in a combined lower surface roughness on the top and side of the part.

Ning et al. studied the accuracy of a PBF system by investigating the percentage shrinkage due to different geometric shapes. They experimentally studied the effect of 2D geometric shape factors on dimensional accuracy and later used that information to analyze the effect of different geometric shapes on the dimensional accuracy of the part. They regarded a change in the dimensional accuracy of the 2D layer as a composite effect of the voxels. Each hatch vector (identified as a dexel) on a 2D layer was used to denote a corresponding voxel. Based on this model, different geometric shapes can be regarded as different combinations of dexels. Analyzing the accuracy due to the effect of geometric shapes can be considered similar to analyzing the effect of the dexels and their interaction. Based on an empirical relationship, they developed a speed compensation method. The method involved controlling the scan speed and laser power separately or together for individual dexels to improve the accuracy of the fabricated parts [10]. Simchi, Petzoldt, and Pohl reported on improving the accuracy of the sintered parts by using an integrated beam compensation technique, where the laser beam diameter is offset to compensate for the observed dimensional error as a result of the shrinkage. The process was strongly affected by shape, size, and distribution of the particles, and the chemical constituents of the powder. It was evident that the final part density strongly depends on the duration time of the laser beam on the surface of the powder particles. The study purported that by using optimized process parameters, such as scanning speed and scanning pattern accompanied by predefined powder characteristics such as particle size and distribution, high-density functional prototypes with superior mechanical properties can be produced. Further sintering behavior, mechanical properties, and microstructural features of the multicomponent iron-based powder were studied and presented [11]. Similar works based on laser beam offset were also reported in Refs [12,13].

34.3.1.2   NonPBF related process control

Although the application of control systems specific for PBF processes in the literature is sparse, research on controlling other AM processes, notably in directed energy deposition (DED) processes, has been reported in the past two decades. DED processes use thermal energy to fuse materials by melting as they are being deposited [1].

Doumanidis and Kwak describe an optimized closed-loop control system (based on lumped parameter multiinput multioutput) for DED processes [14]. The control scheme is based on measuring bead profile geometry using a laser optical scanner and infrared (IR) pyrometry. The control involves modulating process input parameters, such as thermal source power, source velocity, material transfer rate, and direction of material transfer with respect to source velocity. Using analytical models based on mass, momentum, and energy balance of melt-pool, as well as solid conduction in the substrate, they generated relationships between input parameters and the bead profile. A simplified proportional-integral-derivative (PID) control system was implemented using the cross-sectional area of the bead as the scalar error (actual versus expected area) and the thermal source velocity as the input parameter. Due to practical limitations, the bead profile measurements are time delayed compared to process parameter inputs, which are handled by using a *Smith-predictor* scheme in the controller.

The control of melt-pool size under steady-state conditions over the full range of process variables was reported for a particular DED process (defined in this case as laser engineered net shaping, LENS) [14]. The control later extended to consider melt-pool size under transient conditions and as a function of process size scale [15–18]. Numerically determined melt-pool temperature response times were used to establish a lower bound on the response times for thermal feedback control systems. Similar works have been reported in Ref. [19].

Cohen developed a control system for droplet-based DED processes using the part geometry to determine the locations of subsequent droplets to compensate for geometric inaccuracies [20]. Using geometric measurements and a model of the target object, the system chooses appropriate locations for subsequent droplets such that the fabricated part ultimately matches the target geometry. The system chooses these deposition locations from a set of candidate locations by selecting *best* candidates with the highest scores, as defined by a user-selected scoring algorithm.

Bi et al. investigated a closed-loop control of a DED process, based on the IR-temperature signal, for deposition of thin walls [21]. A PID controller was built between a photodiode and laser in the control system. The IR-radiation from the melt-pool was detected by the photodiode and converted to a temperature signal. The actual value of the temperature signal was compared with a set-value. The PID-controller created a control variable out of the deviation to regulate the laser power, so that the melt-pool temperature was controlled. The results showed that the process control with a path-dependent set-value could notably improve the homogeneity of the microstructure and hardness as well as the dimensional accuracy of the deposited samples.

Hu and Kovacevic studied real-time sensing and control to achieve a controllable powder delivery for the fabrication of functionally-graded material using DED processes [22]. An optoelectronic sensor was developed for sensing the powder delivery rate in real-time at a high sampling frequency. To achieve consistent processing quality, a closed-loop control system was developed for heat input control in the DED process based on the observed IR image of the HAZ. The experimental results of closed-loop controlled DED showed improvement in the geometrical accuracy of the part being built. A 3D finite element method (FEM) was developed to explore the thermal behavior of the melt-pool. The results from the finite element thermal analysis were intended to provide guidance for the process parameter selection and an information base for further residual stress analysis [22,23].

Process maps have often been used as a method to optimize AM processes. For the DED processes, Birnbaum et al. considered the transient behavior of melt-pool size, due to a step change in laser power or velocity, for dynamic feedback control of melt-pool size using IR imaging techniques. They modeled the relationship between the process variables (laser power and velocity) and the desired melt-pool size [17]. They proposed a process map approach to condense results from a large number of simulations over the full range of process variables into plots that process engineers could readily use. Bontha et al. addressed the ability of thermal process maps for predicting and controlling the microstructure in DED materials [24]. The focus of the work was the development of thermal process maps relating solidification cooling rate and thermal gradient (key parameters controlling microstructure) to DED process variables (laser power and velocity).

A closed-loop DED system with image feedback control was patented in 2002 [25]. The feedback controls material deposition using real-time analysis of IR radiation images. From the imaging data intrinsic parameters such as temperature distribution, size and shape of the molten pool, maximum degree of pool superheating, the trailing thermal gradient, and thickness of the deposition are extracted. A feedback-based control system then compares the current intrinsic parameters with the target intrinsic parameters to generate new control values (laser power and traverse velocity) based on the feedback-driven adjustments and the predetermined operating schedule. The resulting system can fabricate components with a several-fold improvement in dimensional tolerances and surface finish.

The issue of residual stress control for laser-based AM processes has also been addressed using the process map approach [26,27]. The thermal gradient behind the melt-pool was used to predict changes in residual stress based on thermal simulation results. A method of stress reduction by localized part preheating via a dual-beam laser or electron beam system was also proposed [28].

Table 34.1 in Appendix A summarizes the research efforts applicable to AM PBF control schemes.

34.3.2   *Process measurements*

As mentioned in Section 34.3.1.1, quality of the parts resulting from PBF processes varies significantly and depends on many interrelated influencing factors such as powder characteristics, process parameters, geometry, and other surrounding conditions. To clarify these relationships, researchers use a variety of measurement techniques. This section focuses on the preprocess, in-process, and postprocess measurements described in literature to identify correlations (discussed in Section 34.4) between the key process parameters, process signatures, and product qualities.

Table 34.1 Summary of the research efforts applicable to AM PBF and related
nonPBF control schemes

Control parameter	Setup	Correlations	Control	Reference
		PBF related		
Reference	CMOS camera and planar photodiode coaxial with the laser	Photodiode signal intensity and melt-pool area. Melt-pool dimensions as a function of C, Y and positions of laser beam on the X-Y plane	Area-based signature as feedback to control the laser power	[6–8]
Surface roughness of solidified melt-pool	Pulsed laser system, video camera	Heat intensity and surface roughness. Pulse shapes and material spatter	Investigative	[9]
Part geometry	CMM, beam compensation	Shrinkage due to different geometric shapes	Laser beam, laser power, and scanning speed	[10,12,13]
		Non-PBF related		
Bead profile geometry	Laser optical scanner, IR pyrometer	Input parameters and bead profile	Control bead cross sectional area and with a single process input parameter along with the inverse source velocity	[14]
Part geometry	FDM and compensation algorithm	Geometric measurements and a model of the target object	Compensation droplets to match the target geometry	[20]
IR-temperature signal	PID-controller was built between a Gephotodiode and laser	Laser path versus homogeneity of the microstructure, hardness, and dimensional accuracy	Laser path versus homogeneity of the microstructure, hardness, and dimensional accuracy	[21,25]
Delivered powder volume	Optoelectronic sensor for powder delivery	Thermal variation and processing quality	Controllable powder delivery and heat input	[22,23]
Melt-pool size	Thermal imaging, process map	Transient behavior of melt-pool size and laser power or velocity	Dynamic feedback for desired melt-pool size	[15–19,24,26,28]

34.3.2.1 Preprocess measurements

Preprocess measurements are generally not directly applicable to *in situ* feedback control. However, they can potentially be used to define appropriate system input parameters, or supplement a process model for use in feed-forward control. They are also crucial to establishing relationships between input process parameters and process and part characteristics. These measurements often relate to material properties (density, thermal conductivity, etc.) and intrinsic properties of the system (laser power, powder absorptivity, etc.). Kruth et al. provided a list, based on a literature review, of additional material related properties that significantly affect melt-pool signatures: surface tension, viscosity, wetting, thermo-capillary effects, evaporation, and oxidation [29].

Researchers at the National Institute of Standards and Technology (NIST) summarized metal powder characterization methods, in particular those that measure and describe powder size and distribution [30]. Another NIST study measured size distribution, particle morphology, chemistry, and density of powders, and compared sample-to-sample consistency and variability from recycling of used metal powders [31]. Amado et al. also reviewed and demonstrated multiple methods of flowability characterization for polymer PBF powders for SLS applications [32]. Although these works thoroughly described powder characterization techniques, they did not investigate the relationships between variations in these characteristics and resulting process signatures or final part quality.

The role of powder size and size distribution in sintering kinetics is well understood, that is, it affects the relative density of the powder, which in turn affects the activation energy required for heated particles to coalesce [33,34]. Smaller powder sizes with higher relative powder densities require less energy to sinter. It is known that a wider distribution of particles sizes can allow for higher powder density, because smaller particles can fit in the gaps between larger particles. McGeary demonstrated that specific ratios of bimodally distributed powder sizes can achieve an optimal packing density of 84% with a 1:7 size ratio and a 30% weight fraction consisting of the smaller size [35]. Multimodal distributions could achieve even higher densities.

Higher relative density in powders improves the process by reducing internal stresses, part distortion, and final part porosity [29]. High relative densities increase the relative thermal conductivity of the powder bed [36,37] (which is further discussed in Section 3.3). However, this decreases the absorptivity of the laser energy in AM systems, counteracting the benefits of a lowered energy barrier [38]. In some instances, these effects may negate each other. For example, Karlsson et al. measured little difference in hardness, elastic modulus, surface roughness, and macro and microstructure in laser beam melting of Ti-6Al-4V builds when comparing two powder size distributions of 25–45 μm and 45–100 μm [39]. Liu et al. also tested two powder distributions (narrow and wide with similar mean values) in the PBF process under varying scan speeds and laser power levels. They found that the wider particle size distribution, that is, with a higher relative powder density, resulted in higher part density requiring less laser energy intensity [40]. Spierings et al. showed that unless a certain relative powder density is achieved, a lower scan speed (e.g., higher energy density) is required to produce fully dense parts [41,42]. Differences in the relation of the powders to the densities, the layer thicknesses, and laser scan speeds indicate that powder grain size distribution should be taken into account for optimal results.

Further, local thermal conductivity has an effect on melt-pool signatures and thus part quality (see Section 3.3). Although metal powder thermal conductivity has been measured in multiple instances [43], conductivity of the fully dense material is generally better known and easier to measure. This measurement can be supplemented to models to derive the effective powder conductivity. Gusarov et al. demonstrated a method to calculate effective thermal conductivity of powders in which the relative density, the sphere packing coordination number (i.e., the mean number of the nearest neighbors to each particle), and the interparticle contact size were shown to have the greatest effect [37].

Finally, there are certain preprocess measurements not involving input materials. For example, some part quality issues may stem from machine errors. These may include motion and positioning errors (with well-established measurement guidelines that may be taken from machine tool standards, e.g., ISO 230–1), or errors in the laser optics and scanning system. These error sources and solutions for increased precision through better design or feedback control are not unique to AM, but relevant also to other manufacturing processes.

34.3.2.2   In-process measurements

The primary focus of research in-process monitoring has been associated with determining the geometry and the temperature profile of the HAZ. IR thermography and pyrometry are two well-developed nonintrusive techniques for the measurement of surface temperatures. There is also some reported work on the in-process monitoring of the dimensional accuracy, errors, and defects during the build process. A few reports also discuss the in-process measurement of strain-stress.

34.3.2.2.1   Surface temperature measurement: Thermographic imaging of AM processes can be grouped based on the optical path used by the imaging system. In coaxial systems, the imager field of view aligns with the laser beam through the beam scanning optics [8,44–48]. In these systems, the field of view follows the melt-pool throughout its scan trajectory. Alternatively, the imager may be set externally to the build chamber to view the build through a window [49–53]. An improvised method was developed by Craeghs et al. [8]. Using the coaxial system, they mapped the charge-coupled device (CCD) camera and photo detector signals stemming from the melt-pool in the build plane using the XY laser scan coordinates. This created mapped images of the entire build area, with more local and detailed signatures of the melt-pool. Through this method, they could detect part deformation and overheating near overhanging structures through measured changes in the photo-detector signal. A lower signal resulted from the laser defocusing on distorted surfaces. A higher signal resulted on overhang surfaces that had less heat sinking support structure, and thereby poorer surface quality.

There are several known difficulties with thermography of additive processes. First and foremost, the imaged object's emissivity must be known in order to determine a true thermodynamic temperature from radiation-based measurements. Emissivity is likely different for the melt-pool, unconsolidated powder, and solidified surface, so a thermal image composed of all three components could give deceptive temperature predictions. For example, Rodriguez et al. noted that the powder areas surrounding the solidified part surfaces glowed brighter than the part in thermal images even though the powder was likely of lower temperature [52]. This was attributed to the lower emissivity of the part surface, which reduced the imaged radiant intensity in these areas. Several techniques have been used to determine emissivity of different build components in AM systems: (1) assume a certain imaged area is at the liquidus or solidus temperature of the melt and use this as a reference emissivity [50,51,54], (2) create an emissivity reference by building and imaging a blackbody cavity [52,55], or (3) only provide temperature without correction for emissivity (e.g., apparent or brightness temperature) or provide raw sensor signal values [56]. Another challenge, in particular with coaxial systems, is that f-theta lenses used in scanning systems induce chromatic or spectral aberrations. This requires that the only radiation sensor systems with narrow bandwidth near that designed for the f-theta lens may be used accurately [8,45,57]. Finally, metallic debris from the HAZ can coat a window or viewport used in an AM imaging system, and disturb temperature measurements by changing the radiation transmission through the window [49,51,58]. This is particularly troublesome in EBM systems, and prompted Dinwiddie et al. to create a system to continuously roll new kapton film over the viewport in order to provide new, unsullied transmission [49].

Several studies using thermography are of particular interest in relating process signatures to either input parameters or product qualities. Krauss et. al described the radiance (not temperature) images of the HAZ, captured by a microbolometer, in terms of area, circularity, and aspect ratio [56]. They compared these measurands versus scan speed, laser power, hatch distance, scan vector length, layer thickness, and changes when the melt-pool passes over an artificial flaw. Despite the relatively slow exposure time and limited resolution, they showed that size of the HAZ area was the most suitable measurand to detect deviations in scan velocity or laser power.

Yadroitsev et al. noted how melt-pool temperature, width, and depth in single track scans in selective laser melting (SLM) of Ti-6Al-4V increased with laser power and *irradiance time*, defined as the ratio of laser spot diameter to scanning speed [48][165]. Peak melt temperature increased with both power and irradiance time, but was more sensitive to power over the ranges measured. Melt-pool width and depth were measured from cross sections cut from the melted tracks. They thoroughly characterized the microstructure of the SLM material for two scan strategies, and multiple postbuild heat treatments. However, no definitive comparison of microstructure to the SLM process parameters or the thermal measurements was highlighted.

Hofmeister et al. empirically correlated cooling rate behind the melt-pool to the melt-pool size and noted how these changed depending on proximity to the build substrate and thus local average thermal conductivity in a LENS process [54]. They also noted that calculating cooling rate is more difficult in a real-time monitoring system, and measuring melt-pool length as a corollary signature is more feasible. Similar to Yadroitsev et al., however, distinct correlations between thermographic process signatures to microstructure were not exemplified.

Santosprito et al. describe a thermography-based system to record the movement of heat movement through the laser track [59]. As defects (cracks, porosity, etc.) create lower conductivity regions and affect heat flow, they can be detected using thermography. However, because the changes due to these defects are small, they created new algorithms such as asymmetrical spatial derivative analysis, asymmetrical time derivative analysis, and asymmetrical line profile analysis (using multiple image frames and image subtraction) to improve the effectiveness of the defect detection. It was reported that a minimum defect size around 400 μm is detectable with this system.

Dinwiddie et al. developed a high speed IR thermographic imaging system with an integration time of 1.0 ms, retrofitted to a commercial electron beam machine, to monitor beam-powder interaction, quantify beam focus size, and detect porosity [60]. To overcome the contamination of the optics due to free metal ions released during the process, they designed a shutterless viewing system allowing continuous IR imaging of the beam-powder interaction. The chapter describes the design of the system as well as examples of how to use this system in e-beam focus measurement (that requires spatial calibration), detection of over-melting during preheat, and porosity detection. However, because there was no temperature calibration, the images could not be converted to true temperatures. In another study, Dinwiddie et al integrated an extended range IR camera into a fused deposition modeling (FDM) machine for imaging of the parts through the front window of the machine [61]. Another IR camera was integrated to the liquefier head to obtain higher resolution images of the extrusion process.

Price et al. described another implementation of near-IR (wavelengths in the range of 780 nm to 1080 nm) thermography (with 60 Hz frame rate) for an EBM process [62]. They mounted the IR camera in front of the observation window of the machine and monitored the process as it goes through various stages, such as platform heating, powder preheating, contour melting, and hatch melting. They were able to measure the melt-pool size and the temperature profile across the melt-pool. However, they stated that the assumptions about the emissivity values are sources of uncertainty. The spatial resolution of the imaging system was reported as 12 μm when using a close-up lens.

Pavlov et al. described pyrometric measurements taken coaxial with the laser to monitor the temperature of the laser impact zone to detect deviations of process signatures that correlate to deviations of process parameters from their set values [63]. This approach relies on the sensitivity of the temperature of HAZ with respect to process parameters. The laser impact zone surface temperature was measured using a bicolor pyrometer (1.26 μm and 1.4 μm wavelengths with 100 nm bandwidth) covering a circular area of 560 μm diameter with 50 ms sampling time. A laser spot size of 70 μm diameter results in about a 100 μm remelted powder track. A 400 μm diameter optical fiber was used to collect temperature information. Temperature was represented as digital signal levels. Using this system, they investigated three strategies: (1) time variance of pyrometer signal during laser scanning of multiple tracks, (2) changes in pyrometer signal as a function of hatch spacing (with thin and thick powder layers), and (3) pyrometer signal changes as a function of layer thickness. The authors used this measurement method to differentiate the three process strategies proposed. They found that the pyrometer signal from the laser impact zone is sensitive to the variation of the main operational parameters (powder layer thickness, hatch distance between consecutive laser beam passes, scanning velocity, etc.), and could be used for online control of manufacturing quality [63]. Similar work was reported in Ref. [45].

34.3.2.2.2 Residual Stress: There are a number of techniques to measure strains and residual stresses in metal components. However, the relative part sizes and other physical attributes associated with the scanned region make it extremely difficult to apply direct methods of measurement. There are a number of reported indirect measurement techniques applicable. These indirect methods monitor physical attributes that are representative of the strains and residual stresses. Indirect techniques are based on strain or displacement measurement relating to the rebalancing of internal stresses that are released when material is removed or allowed to deform [64,65].

Several researchers have reported on surface distortion measurement methods while investigating residual stresses [66–68]. Robert described a method that involves capturing the topography of the upper surface laser using a scanning confocal microscopy and deriving the platform's surface displacement by mapping the surface positions before and after the direct laser melting process [69]. Shiomi et al. discussed the use of strain gages mounted to the build platform to measure residual stress *in situ* [70]. They were able to measure the strain changes in a build platform when SLS-induced layers were successively milled off. They found that the residual stresses decreased (i.e., stress relief) as more layers were removed from the built part.

More recently Van Belle et al. investigated residual stresses induced during a PBF process [71]. A strain gauge rosette was mounted under a support platform. By monitoring the variation of the strain gauge data, residual stress corresponding to elastic bending is calculated in the support and the part, using force balance principles.

34.3.2.2.3 Geometric measurements: There is not much work that focuses on the in-process geometric measurements. Cooke and Moylan showed that process intermittent measurements can be viable for both process improvement and characterization of internal part geometries. Process intermittent measurements were compared to contact and noncontact measurements of the finished parts to characterize deviations in printed layer positions and changes in part dimensions resulting from postprocess treatments [72].

Pedersen et al. [73] discussed a vision system for enhancing build-quality and as a means of geometrical verification. Given the very nature of layered manufacturing, a generic geometry reconstruction method was suggested, where each layer is inspected prior to addition of the successive layers. The hypothesis was that, although most AM processes have a tendency to accumulate stresses and suffer from elastic deformations, the nondeformed layers characterized by such systems will yield sufficient data to assess whether defects of internal geometries are present. This includes visually present defects from the inspected layers.

Kleszczynki et al. used a high resolution CCD camera with a tilt and shift lens to correct the image mounted on the observation window of a commercial PBF machine [74]. The camera has a field of view of 130 mm × 114 mm with a pixel size of 5.5 × 5.5 μm. They categorized potential error sources during the build process and collected images representing these errors.

Table 34.2 in Appendix A summarizes the research efforts on in-process measurement.

34.3.2.3 Postprocess measurements

The postprocess measurements have in general focused on the part quality and are based on the following categories: dimensional accuracy, surface roughness, porosity, mechanical properties, residual stress, and fatigue. Parts, in the context of this review, consist of standard material testing specimens, process/design-specific specimens, and functional parts. This section captures relevant findings and correlations that have come from the postprocess measurements.

34.3.2.3.1 Dimensional accuracy: Several chapters discuss dimensional accuracy with examples. Yasa et al. investigated the elevated edges of parts, using a contact surface profilometer and optical microscope, built using different laser power levels, speeds, and scan strategies [75]. The chapter identified that certain process parameters and scanning strategies could improve flatness of elevated surface. Abd-Elghany evaluated PBF processed parts with low-cost powders by measuring dimensions before and after finishing by shot-peening process. Using a 3D scanner it was observed that the part was 2%–4% larger than designed before shot peening, and 1.5% after shot peening. It was also noted that the tolerances were not uniform and varied in the z direction [76]. Mahesh et al. investigated the controllable and uncontrollable parameters in a PBF process [13]. They identified correlation between the controllable process parameters such as scanning speed, laser power, and scanning direction on the geometrical profiles of the geometric benchmark part. They reported that the preferred settings of control parameters based on the analysis of the mean dimensional errors for the specific geometric features on the benchmark part. Paul and Anand developed a mathematical analysis of the laser energy required for manufacturing a simple part based on laser energy expenditure (minimum total area for sintering) of SLS process and its correlation to the geometry [77]. Khaing et al. studied the design of metal parts fabricated by PBF [78]. A coordinate measuring machine (CMM) was used to measure the dimensional accuracy of the parts. They observed deviations along the X- and the Y-axis. The values along the Y-axis were the most accurate. They concluded that the optimization of the process parameters and the accuracy of the laser scanning units were crucial to improve the dimensional accuracy.

Table 34.2 Research on in-process measurement

Purpose of in-process measurement	Measurement setup	Reference
Surface temperature measurement	IR thermography and pyrometry, emissivity reference	[8,44–63]
Correlate deviations of process signatures to input parameters	sBi-color pyrometer	[63]
Determine temperature and time history of temperature distribution in melt-pool area	Co-axial measurement system uses a bi-color pyrometer	[45]
Determine melt-pool size and temperature	Photodiode and CMOS	[57]
Use temperature maps to detect deformation due to thermal stresses and overheating zone due to overhangs	Co-axial near-IR (780 nm to 950 nm) temperature measurement system consisting of a planar (?) photodiode and high-speed CMOS camera.	[8]
Monitor beam-powder interaction, quantify beam focus size, and detect porosity	IR-thermography imaging system	[60]
Monitor melt-pool dynamics by introducing additional illumination source for high resolution imaging at high scanning velocities	Co-axial optical system	[47]
Measure the melt-pool size as well as the temperature profile across the melt-pool	Near-IR (780 to 1080 nm) thermography (with 60 Hz frame rate)	[62]
Track movement of heat through the laser track	Thermography-based system	[59]
Strain measurement	Surface distortion measurement, strain gages mounted to the build platform	[64–68, 70,71]
Geometric measurements	Vision system	[72–74]

Krol et al. studied the prioritization of process parameters for an efficient optimization of AM by means of a FE method. They stated that the scanning speed, the support geometry, the preheating temperature of the substrate, and the scanning pattern were the most influential parameters for dimensional accuracy [79]. Similarly Delgado et al. [80] and Wang et al. [81] also reported on the influence of process parameters on part quality. Table 34.3 in Appendix A summarizes the related research on dimensional accuracy as it applies to part quality.

34.3.2.3.2 Surface quality: Abd-Elghany and Bourell evaluated the surface finish of the PBF processed part with layer thickness of 30 μm, 50 μm, and 70 μm. The roughness of the top and side surfaces was measured using a scanning electron microscopy (SEM), equipped with an energy dispersive X-ray (EDX) analyzer. The results of this study indicated that large particles inside thick layers could increase surface roughness because the volume of particles has a tendency to form voids when they are removed in finishing processes. It was also noted that the side surface was smoother at the bottom than at the top [76]. Mumtaz and Hopkinson investigated the laser pulse shaping on thin walls of parts built by PBF by relating pulse shape, thin-wall width, and plasma plume height to surface roughness using a profilometer, digital calipers, and digital video camera. The results of

Table 34.3 Dimensional accuracy research summary

Purpose	Variable	Instruments	Correlations	Reference
Evaluate SLM of low cost powders	Layer thickness. Laser scanning speed	Reinshaw Cyclone II 3D scanner (scan probe)	Measured dimensions before finishing were 2–4% larger than designed, after finishing dimensions were 1.5% larger, tolerances were not uniform and varied in the z-direction, no shrinkage	[76]
Investigate elevated edges	Laser power, speed, scan strategy, edge height	Contact surface profilometer, optical microscope	Not possible to eliminate the built up edge, however, appropriate process parameters and scanning strategies can improve flatness	[75]
Influence of process parameters on dimensional accuracy	Laser power, speed, scan strategy, layer thickness	Profilometer, CMM	Dimensional errors and control can be specific geometric profiles	[13]
Analysis of the laser energy required for manufacturing	Part geometry, slice thickness and the build orientation	Mathematical analysis	Laser energy expenditure of SLS process and its correlation to the geometry	[77]
Design of metal parts fabricated by PBF	Laser power, speed, scan strategy, layer thickness	CMM	Process parameters and the accuracy of the laser scanning units were crucial to improve the dimensional accuracy	[78]
Investigate deformations and deviations of geometry of thin walls in SLM	Size and position	CMM	Deviations ranged from 0.002 mm to 0.202 mm for position and size, respectively	[79]
Influence of process parameters on part quality	Laser power, speed, scan strategy	X-ray spectroscopy, scanning electron microscope, energy-dispersive X-ray spectroscopy, surface profilometer, universal testing machine, hardness tester	Build direction has a significant effect on part quality, in terms of dimensional error and surface roughness	[80]
Quality optimization of overhanging surfaces	Inclined angle (part), scan speed, and laser power	Camera, CMM	Better controlling part orientation and energy input will improve overhanging surface quality	[81]

this study indicated that the wall width varied with the pulse shape, which in turn influenced the melt-pool width. A suppressed pulse shape that consisted of a high peak power, low energy, and short time duration proved to be the most effective pulse shape for PBF [9]. Meier and Haberland investigated various process parameters to evaluate their influence on part density and surface quality for parts fabricated by PBF [82]. Approaches to improve density, surface quality, and mechanical properties were also presented. Related research was also reported in Refs [42,75,80]. Table 34.4 in Appendix A summarizes the related research on surface quality.

Table 34.4 Surface quality research summary

Purpose	Variable	Instruments	Correlations	Reference
Evaluate SLM of low cost powders	Roughness	High sensitivity digital scale, Reinshaw Cyclone II 3D scanner, SEM, JOEL JSM5200, EDX analyzer	Large particles inside thick layers increased surface roughness. Side surface was smoother at the bottom than at the top	[76]
Investigate pulse shaping on SLM of thin walled parts	Pulse shape, roughness, width, degree of plasma plume	Profilometer, digital calipers, digital video camera	Pulse shaping was shown to reduce spatter ejection. Improve top surface toughness, and minimize melt-pool width	[9]
Investigate failures	Layer thickness, scanning speed, orientation, energy density, part density and roughness	SEM	A narrow processing window exists that produces 100% part density and the best surface quality	[82]
Investigate elevated edges	Laser power, speed, and scan strategy, edge height	Contact surface profilometer, optical microscope	Edge height ranged from 10 μm to 160 μm, not possible to eliminate the built up edge, however, appropriate process parameters and scanning strategies can improve flatness.	[75]
Influence of particle size distribution on surface quality and properties	Particle size, layer thickness	Mechanical testing	Optimized powder granulations generally lead to improved mechanical properties	[42]
Influence of process parameters on part quality	Scanning speed, layer thickness, and building direction	X-ray spectroscopy, scanning electron microscope, energy-dispersive X-ray spectroscopy, surface profilometer, universal testing machine, hardness tester	Mechanical properties and surface finish sensitive to the build direction and layer thickness	[80]

34.3.2.3.3   Mechanical properties: Meier and Haberland investigated failures in tensile tests of stainless steel (SS) and cobalt-chromium parts. The findings showed that the density measurements do not identify deficient connections of consecutive layers, and vertically fabricated specimens have lower tensile strengths and elongations [82]. Abd-Elghany and Bourell also characterized hardness and strength as a function of layer thickness and scan speed using hardness, tensile, and compression tests for SLM process. The findings conclude that hardness is not much affected within the range of process parameters studied; however, variations in hardness due to surface porosity were observed. Strength was good at low scanning speeds and thin layers. The parts became brittle with higher layer thickness due to porosity and microcracking.

Compression testing resulted in shapes identical to the buckling of solid parts, that is, layers were very coherent and did not separate or slip due to secondary shear forces [76]. Sehrt and Witt investigated a dynamic strength and fracture toughness on a cylindrical beam and disk by the rotating bending fatigue tests. Specimens were investigated at defined oscillating stresses and the resulting number of cycles that led to the failure of the specimen was determined. The findings showed that fatigue strength was comparable to conventionally manufactured parts [83]. Storch et al. [84] analyzed material properties of sintered metals to qualify metal-based powder systems in comparison to conventional materials used in automotive engines and power trains. Key observations included material properties being sensitive to the build direction and that material strength increases with the chamber atmospheric temperature.

By studying the material properties and the process parameters, Gibson and Shi concluded that the powder properties directly affect the process, which in turn affect the mechanical properties of the resultant component [85]. The research concluded that the knowledge of the effects of sintering and postprocesses must be incorporated into design and postprocessing.

Wegner and Witt developed a statistical analysis to correlate part properties with main influencing factors. According to their study, PBF shows nonlinear correlations among multiple parameter interactions. The four main influences on mechanical properties (i.e., tensile strength, Young's modulus, elongation) were scan spacing, scan speed, layer thickness, and interaction of scan spacing and layer thickness [86].

Manfredi et al. reported on the characterization of aluminum alloy in terms of size, morphology, and chemical composition, through the measurement and evaluation of mechanical and microstructural properties of specimens built along different orientations parallel and perpendicular to the powder deposition plane [87].

Yadroitsev and Smurov studied the effects of the processing parameters such as scanning speed and laser power on single laser-melted track formation. Experiments were carried out at different laser power densities (0.3 parameter[6] W/cm^2) by continuous wave Yb-fiber laser. Optimal ratio between laser power and scanning speed (process map) for 50 µf layer thickness was determined for various SS grade material powders. A considerable negative correlation is found between the thermal conductivity of bulk material and the range of optimal scanning speed for the continuous single track sintering [88]. Related research was also reported in Refs [42,80,89,90].

Table 34.5 in Appendix A summarizes the related research on mechanical properties.

34.3.2.3.4   Residual stress: With rapid heating and cooling inherent in any PBF process, especially in a process that fully melts metal powder, thermal stress and residual stress certainly affect the resulting parts. These residual stresses are most apparent when they cause warping of the part, features, or build platform. As such, residual stress has

Table 34.5 Mechanical properties research summary

Purpose	Variable	Instruments	Correlations	Reference
Investigate failures	Layer thickness, Scanning speed, orientation, energy decay, part density and roughness	SEM	Density measurements do not identify efficient connections of consecutive layers, vertically fabricated specimens have lower tensile strengths and elongations	[82]
Evaluate PBF of low cost powders	Layer thickness, scan speed	Micro–Vickers, stress–strain	Hardness not as affected by the parameters, however, variations due to surface porosity were observed. Strength was best t low speeds and thickness. Part became brittle with higher layer thickness due to porosity and micro-cracking	[76]
Investigate dynamic strength and fracture toughness on a cylindrical beam and disk	Standard exposure strategies	Rotating bending fatigue tests	Fatigue strength was comparable to conventionally manufactured parts	[83]
Qualifying metal based powder systems for automotive	Build orientation, surface finish and temperature	Material analysis, tensile test, compression test	Material properties are sensitive to the build direction. Surface treatment potential method to increase material properties. Materials strength decreases with high temperatures	[84]
Study on material properties and process parameters	Material properties	Material analysis, tensile test	Powder properties directly affect the process in turn affect the mechanical properties	[85]
Correlation of process parameters and part properties in laser sintering	Laser power, scan spacing, scan speed, powder bed temperature layer thickness, energy density	Tensile test	Four main influences on mechanical properties were scan spacing, can speed, layer thickness, and interaction of scan spacing and layer thickness	[86]

(Continued)

*Table 34.5 (**Continued**)* Mechanical properties research summary

Purpose	Variable	Instruments	Correlations	Reference
Characterization of Commercial AlSiMg Alloy Processed through Direct Metal Laser Sintering	Build orientations	Light microscopy; electron microscopy	Difference in mechanical and microstructural properties of specimens built along different operations	[87]
Investigate single layer track stability	Powder input, scanning speed, laser power	Optical granulomophometer, real-time optical sieving system, image and analysis software	Negative correlation is found between the thermal conductivity of bulk material and the range of optimal scanning speed for the continuous single track sintering	[88]
PBF of dies	Later offset	Single/dual lasers, Vickers hardness testing machine	Vickers hardness decreases as beam offset increases. Reheating increases bending strength	[91]
Analyze the influence of the manufacturing strategy on the internal structure and mechanical properties of the components	Hatch distance, build orientation	Granulomophometer, INSTRON	Two-zone method created the lowest porosity <1% yield ultimate tensile strength was consistent with both vertical and horizontal build directions, Young's modules is 1.5 times higher for horizontal builds	[89]
Influence of particle size distribution on surface quality and properties	Particle size, layer thickness	Mechanical testing	Optimized powder granulation generally lead to improved mechaicaly properties	[42]
Effect of PBF layer on quality	Gas flow direction	Porosity measurements, mechanical testing	Gas temperature/flow effects part quality	[92]

(Continued)

Table 34.5 (Continued) Mechanical properties research summary

Purpose	Variable	Instruments	Correlations	Reference
Influence of process parameters on part quality	Scanning speed, layer thickness, and building direction	X-ray spectroscopy, Scanning Electron Microscope, Energy-dispersive X-ray spectroscopy, surface profilometer, universal testing machine, hardness tester	For SLM process, the build direction has no influence on mechanical properties	[80]
Designing material properties locally, PBFs	Energy density, modulus, yield strength	Brinell test	Hardness is influenced by the pore structure	[93]
Effect of geometry on shear modules	Pitch of the spring as geometric factor	Compression test	Geometry has a major effect on the produced mechanical properties	[94]
Investigate the effects of preheating on the distortion of Al parts	Preheat temperature	3D Optical measurement system	Hardness decreases with preheat temperature	[90]

been widely studied by AM researchers [69–71,90–105]. Mercelis and Kruth described the two mechanisms causing the residual stress: the large thermal gradients that result around the laser spot and the restricted contraction during the cooling that occurs when the laser spot leaves the area [99]. Withers and Bhadeshia discussed the techniques used to measure residual stress, and most of these methods are performed postprocess and often require some sort of specimen destruction [103]. The methods include hole drilling (distortion caused by stress relaxation), curvature (distortion as stresses rise or relax), X-ray diffraction (atomic strain gauge), neutrons (atomic strain gauge), ultrasonics (stress related changes in elastic wave velocity), magnetic (variation in magnetic domains with stress), and Raman spectroscopy. Shiomi discussed the use of strain gauges mounted to the build platform to measure residual stress *in situ* [70]. Van Belle et al. expanded on this method, using a table support mounted to the bottom of the build platform [71]. The table support was designed to amplify strain and was instrumented with strain gages to measure that strain. A thermocouple was also mounted close to the strain gauge to record the temperature evolution for the thermal strain. The removed layer method was used and modified to determine the residual stress in the part and the support during the layer addition with the measured strains.

It was observed that many researchers linked process parameters to the residual stress present in the resulting parts and investigated strategies to reduce the residual stresses. The most commonly discussed method of reducing residual stress was through postprocess heat treatment [70,97,99,102], although these results have little impact on process control. Residual stresses were also significantly reduced by heating the build platform [70,99], that is, higher heating temperatures resulting in lower residual stresses. The path the laser beam follows to trace and fill the geometry (i.e., scan strategy) of each layer has also been shown to influence the residual stress present [99], and the layer thickness used to build the part [71,98]. Table 34.6 in Appendix A summarizes the related research on residual stress as it applies to part quality.

34.3.2.3.5 Porosity/Density: The effects of various process parameters on part density for many materials have been investigated and the contributors causing porosity have been identified. Laser power, scan speed, scan spacing, and layer thickness can be directly related to energy density and thus to part density. Several researchers have studied the effects of energy density parameters on different materials like 316L SS [41,82,106,107], 17–4 precipitation hardening (PH) steel [93], Ti-6Al-4V [107], and American Iron and Steel Institute (AISI)-630 steel [93]. Their efforts suggest a correlation between the energy density and the part density. Parthasarathy evaluated the effects of powder particle size, shape, and their distribution on the porosity of 316L SS [108]. Porosity/density has a direct effect on the mechanical properties of components fabricated by PBF [109]. Internal and external pores, voids, and microcracks introduced during fabrication act as stress concentrators that cause premature failure and thus compromising part quality. Fully dense parts (100% relative density), however, have shown to have mechanical properties equal to or better than the properties of wrought materials.

Morgan et al. investigated the effects of remelting on the density of the part [106]. The density increased with decreasing scan speed. Density decreases with decreasing scan spacing but not significantly. The plasma recoil compression forces can modify melt-pool shape and affect density. There appeared to be a maximum energy density associated with part density. Gu et al. studied the influences of energy density on porosity and microstructure of PBF 17–4PH SS parts [110]. They showed that coupons fabricated using the same energy density level using different laser powers and scan speeds showed

Table 34.6 Residual stress research summary

Purpose	Variable	Correlations	Reference
Measurement residual stress	Laser scanning, heating	Base plate heating, re-scanning, and heat treatment reduced residual stress	[70]
Residual stresses in PBF	Material properties, sample and substrate height, the laser scanning strategy and heating conditions	Heat treating re-scanning, and heating of the base plate helps relieve residual stress	[99]
Effects of positioning powders and thickness on residual stresses	Position and thickness	Stress magnitude decreased moving towards inner layers.	[98]
Investigate heat treatment of PBF components	Temperature and time	The most promising heat treatment consisted of a moderate cooling rate after solution treatment at 1,055°C	[102]
Investigate fatigue and crack growth of TiAl6V4 PBF in the z-direction	Temperature, atmosphere	Micron sized pores mainly affect fatigue strength, residual stresses have a strong impact on fatigue crack growth	[97]
Investigate residual stress and density	Laser power, heating	Observed deformation was due to residual stress. Stressed were found to be very high and approached and exceeded the yield strength	[104]
Effect of PBF layout on quality	Gas flow direction	Gas temperature/flow effects part quality	[92]
Investigate heat treatment on residual stress, tensile strength, and fatigue of SLM components	Temperature, time, gas, and hot isostatic pressing	Heat treating reduced residual and tensile stress and increased fatigue life	[105]
Investigate the influence of material properties on residual stress	Density, micro hardness, curl-up angle	Micro-cracking and the formation of oxides effect residual stress, material properties influence was obscured	[100]
Measure residual stress to validate numerical model	Strain, temperature, cooling time	Residual stresses are largest for large layer thickness (mm) and long cooling time	[71,98]
Investigate the effects of preheating on the distortion of Al parts	Preheat temperature	Reduction in distortion begins at a preheat temperature of 150°C, distortion is no longer observed at a preheat temperature 250°C and above, additionally, hardness decreases with preheat temperature	[90]

significantly different levels of porosity. Two types of porosity formation mechanisms were identified and discussed. Balling phenomena and high thermal stress cracking were mainly responsible for the porosity that occurs at very high laser power and scan speed, whereas insufficient melting is the primary reason for crevices filled with many unmelted powders at very low laser power and scan speed. Also, pores in coupons manufactured using both high laser power and scan speed exhibit smaller size and more circular shape in comparison with pores in coupons manufactured using both low laser power and scan speed.

Chatterjee et al. investigated the effects of the variation of sintering parameters: layer thickness and hatching distance on the density, hardness, and porosity of the sintered products [111]. Applying statistical design of experiments and regression analysis, they observed that the increasing layer thickness and hatching distance results in an increase in porosity that diminishes the hardness and density.

Related research was also reported in [42,76,80,89,112,113].

Table 34.7 in Appendix A summarizes the related research on porosity and density as it applies to part quality.

34.3.2.3.6 Fatigue: Fatigue performance is crucial if AM parts are to be used as functional components in dynamic environments, for example, aircraft engines. Under dynamic conditions, AM parts have shown to have a high sensitivity to surface quality and internal pores that act as stress risers. Researchers have recently reported on studies to characterize fatigue performance, endurance limit, and fracture behavior of AM components for various materials that include 15–5 PH, 17–4 PH, 316L SS, AlSi10Mg, Ti-6Al-4V, and CPG2Ti [42,83,97,110,116–120]. Sehrt and Witt [83] investigated the dynamic strength and fracture toughness of 17–4 PH SS components using Woehler fatigue tests (i.e., rotating bending test) and compact tension tests [ASTM E399, DIN EN ISO 1237]. They found that the fatigue strength and the critical stress intensity factor for additively manufactured 17–4 PH components are comparable to conventionally-manufactured components. Other researchers performed high cycle fatigue (HCF) tests described by ASTM E466 [97,116,117,119]. Leuders et al. studied the effects of heat treatment and hot isostatic pressing (HIP) for vertically built specimens and found that fatigue life increased with increasing temperature [97]. By closing near surface pores, HIP was found to increase the fatigue life of Ti-6Al-4V to a level above two million cycles. In addition to evaluating the fatigue of vertically built Ti-6Al-4V specimens, Rafi et al. also evaluated 15–5PH specimens [121]. Titanium alloy Ti-6Al-4V and 15–5PH specimens were heat treated at 650°C for four hours and at 482°C for precipitation hardening, respectively. Their results suggested that the fatigue life of PBF Ti-6Al-4V specimens is better than cast and annealed specimens. However, the endurance limit of 15–5 PH was reduced by 20% when compared to conventionally-manufactured components. Spierings et al. compared the endurance limit for as-built, machined, and polished specimens [110]. Like Rafi et al., they also reported that the endurance limit for 15–5 PH was reduced by 20%. Similarly, the endurance limit for 316L was reduced by 25% when compared to conventionally manufactured components. Spierings et al. also reported that as-built specimens were weakest and polished specimens were only slightly better than machined [119]. Brandl et al. studied the effects of heat treatment and vertical build orientation on the HCF performance of AlSi10Mg samples. The authors concluded that a combination of heat treatment (300°C) and peak-hardness increases fatigue resistance and neutralizes the effects of build orientation. Additionally, the fatigue resistance of PBF AlSi10Mg samples was very high when compared to standard cast samples [116]. To further investigate the practicality of using SLM

Table 34.7 Porosity/density research summary

Purpose	Variable	Correlations	Reference
Comparison of density of 316L	Layer thickness, particle size, distribution, Mettler balance	Basic powder requirements identified	[41]
Study PBF	Layer thickness, scanning speed, orientation, energy density, part density and roughness	Density measurements do not identify deficient connections of consecutive layers, a narrow processing window (energy density) exists that produces 100 % part density and the best surface quality	[82]
Investigate the effects of remelting on density	Scan speed, scan spacing, pulse frequency	Increase density with decreasing scan speed, density decreases with decreasing scan spacing (although not significant)	[106]
Investigate the influence of laser remelting on density	Scan spacing, scan speed, number of remelting scans, laser power	Higher remelting scan speed with low laser power exhibits very-low porosity, additional remelting did not significantly change porosity. Increased energy by decreasing the scan spacing and increasing the number of scans increased porosity, but not as bad as not remelting	[107]
Designing material properties locally in PBF process	Build orientation, layer thickness, scan speed, laser power, heat treat, and energy density	Generated a curve for density as a function of specific energy input, Boccaccini equation can be used to predict modulus as a function of porosity, hardness is influenced by pore structure	[93]
Investigate the density of PBF powders: gas atomized and water atomized	Particle size, shape, and distribution	Lower laser power, higher scan speed, and thicker layer yields worsened wetting characteristic characterized by fluctuant surface, gas atomized powder produces denser structures, pore size increased with increaser hatch spacing	[108]
Influences of energy density on porosity and microstructure of PBF 17-4PH	Layer thickness, scanning speed, orientation, energy density, part density and roughness	Energy density may not be a good indicator for porosity level of SLM manufactured parts. Balling phenomena and high thermal stress cracking are mainly responsible for the porosity	[109,110]

(Continued)

Table 34.7 (Continued) Porosity/density research summary

Purpose	Variable	Correlations	Reference
Effects of the variation of sintering parameters	Layer thickness and hatching distance	Increasing layer thickness and hatching distance results in an increase in porosity that diminishes the hardness and density.	[111]
Study the influence of the hatch distance on internal structure and porosity	Hatch distance, build orientation	Porosity increased as hatch distance increased, to-zone method created the lowest porosity <1% yield ultimate tensile strength was consistent with both vertical and horizontal build directions. Young's modulus is 1.5 times higher for horizontal builds	[89]
Investigate residual stress in PBF	Specimen thickness	Produced parts with 1.4% porosity	[114]
Investigate increased production with increased laser power	Laser power, scan speed	With 1 KW lasers scan speed, scan spacing, and build rate can be significantly increased	[115]
Reduce required laser power and increase scan rate by investigating the effects on porosity	Hatch distance, scan speed	Lows can speeds generate roughness set layer thickness, high scan speeds led to lower relative densities due to insufficient powder melting	[114]
Comparison of density measurement techniques	Particle size, layer thickness	Porosity is less controllable at high scan speeds, Archimedes method has lower uncertainty and greater repeatability	[42]
To investigate the influence of volume energy density on porosity	Laser power, scan speed, hatching distance, layer thickness	The volume energy density, including all four investigated parameters, shows a strong influence on the overall porosity	[113]
Evaluate PBF of low cost of powders	Part geometry, dimensional tolerance, surface quality, density, mechanical properties and microstructure	The volume energy density, including all four investigated parameters, shows a strong influence on the overall porosity	[76]
Investigate density and residual stress within PBF specimens	Laser power, heating	Archimedes-method yielded an average density of 99.75% and pixel analysis yielded an avg. of 99.7% Sharp-edged defects and near circular voids existed SLM can produce near full dense parts	[104]

Table 34.8 Fatigue related research summary

Purpose	Variables	Correlation	References
Investigate dynamic strength and fracture toughness	Standard exposure strategies	SLM fatigue strength comparable to conventional manufactured parts	[83]
Investigate fatigue and crack growth of TiAl6V4 PBF in the z-direction	Temperature, atmosphere	Micron sized pores mainly affect fatigue strength, residual stresses have a strong impact on fatigue crack growth	[97]
Functional parts for formula race car	Static and dynamic stress	Parts can be manufactured with SLM, brackets survived a year of racing	[42]
Investigate microstructure, high cycle fatigue, and fracture behavior of PBF samples	Build platform temperature, vertical build orientation, and heat treat	Post heat treatment has the most considerable effect and the building direction has the least on fatigue. Fatigue of samples is higher than standard DINEN 1706	[116]
Investigate and compare fatigue performance PBF stainless steel parts to conventionally processed materials	Static and dynamic stress	As fabricated were the weakest, polished was slightly better than machined	[119]
Fatigue performance of Ti-6Al-4V	Roughness	Drastic decrement of fatigue limit due to poor surface quality	[120,121]
Fatigue performance of PBF parts	Temperature, vertical/horizontal build orientation, and heat treat	Horizontally built samples showed relatively better tensile properties as compared with the vertically built samples	[117]

components as functional parts, Spierings et al. 2011, successfully designed, fabricated, and tested brackets used for supporting the suspension of a formula race car [42]. Table 34.8 in Appendix A summarizes the related research on fatigue as it applies to part quality.

34.3.3 *Modeling and simulation*

Science-based predictive models are crucial to predict the material behavior that accounts for the changes in material properties. Detailed understanding of material changes during melting (microstructural changes, phase transformations) would enable optimization and control of the processes improving overall product quality. Such capabilities integrated into the current control schemes can potentially cater to much desired feed forward and feedback capabilities. Many models have been developed for simulating highly dynamic and complex heating, melting, and solidification of materials during PBF processes. Dynamics imply heating, melting, wetting, shrinking, balling, solidification, cracking, warping, and so on in a very short period of time. Complexity implies highly coupled heat and metallurgical interactions in the AM process. This section provides a literature review of available modeling and simulation research works with the following objectives: (1) evaluate currently available physics-based, numerical models that describe the PBF processes and (2) investigate observable and derived process signatures that are necessary for closed-loop control.

Zeng et al. [122] thoroughly reviewed the development and methodology in modeling and simulation research for PBF processes. Therefore, construction of the numerical models is only briefly reviewed here with select examples highlighted. Though much focus of AM modeling papers is on development and model verification, many offer insight into process parameter relationships. The use of modeling to guide process control development is not limited by the models, but by the focus of the modeling efforts. Here, we attempt to extract what information from modeling and simulations may be utilized in control schemes, and identify those derived process signatures that *require* modeling and simulation if they hope to be controlled.

34.3.3.1   Modeling and simulation methods

Nearly all models of the PBF and DED processes include the following input parameters in one form or another: (1) a heat source representing the laser with associated power and profile shape and (2) a body of powder with associated geometry, boundary conditions (typically radiation and convective top surface with either adiabatic or isothermal bottom surface), and thermomechanical material properties. These are modeled either numerically (e.g., through multiphysical finite element analysis [FEA]) or analytically with varying degrees of dimension, geometry, scale, and with varying modeled phenomena or subprocesses. In 3D FEM, laser heat sources are typically modeled as a Gaussian-shaped surface flux with variable power or radius, or as an internal heat generation [123]. Many use a laser *absorptance* factor relating to the fraction of laser energy converted to thermal energy, and/or an *extinction coefficient* or *penetration depth* of the laser energy into the powder. Gusarov et al. developed an analytical model for absorptance, extinction coefficient, and reflected radiation based on multiple laser reflections and scattering through the open pores of a powder bed [38,124]. Various other empirical or analytical submodels are also used for temperature, phase, or powder density-dependent thermal conductivity and specific heat [123,125–129].

Analytical models mostly use the 3D *Rosenthal* solution for a moving point heat source [130]. However, its limited complexity allows it only to verify more complex results from numerical methods (e.g., finite element [FE] results from [131]). Other, more complicated analytical models typically use numerical methods such as finite difference to solve for laser radiation interactions [132].

Some analytical models use nondimensional parameters, which aid in comparison of models and experiments across varying scales and conditions. Vasinonta et al. developed nondimensional parameters that relate input parameters and results of DED process simulations to material parameters based on the Rosenthal solution [26,133]. Others who develop nondimensional parameters include Chen and Zhang [134,135] for the SLS process, and Gusarov et al. described results using traditional heat-transfer nondimensional parameters such as Peclet number using the laser beam width as a characteristic length [136]. For a more thorough analysis of potential nondimensionalized parameters for the PBF process, see [137].

A relatively new method for modeling hydrodynamic effects in the melt-pool is the lattice Boltzmann method (LBM). This method uses particle collision instead of Navier–Stokes equations in fluid dynamics problems. The LBM can model physical phenomena that challenge continuum methods, for example, influence of the relative powder density, the stochastic effect of a randomly packed powder bed, capillary and wetting phenomena, and other hydrodynamic phenomena [138]. For example, Korner et al. demonstrated multiple melt-pool morphologies could result from the stochastically varying local powder density near the scanned region, or effect of changing the bulk powder density. They also developed a process map for scan morphology as a function of laser speed and power for one specified powder packing density. LBM is very computationally intensive, because multiple simulations are needed (by varying input parameters) to extract parameter-signature relationships. For further reference on LBM methods in AM, see [138–141].

34.3.3.2 Parameter-signature-quality relationships

In general, for single scan tracks in powder-bed type processes, the melt-pool and high temperature zone form a comet-like shape, with a high temperature gradient in the leading edge of the melt-pool, and lower temperature on the trailing edge [36,126,142], similar to results from Hussein et al. in Figure 34.3.

As mentioned in Section 34.3.1.1, melt-pool size and temperature are already being used as feedback parameters in closed-loop control schemes. Melt-pool size as a single-valued measurand is not always defined explicitly in reported simulation results. This is likely due to the fact that full characterization of the melt-pool throughout its volume is possible, and single-value measurands are found to be too simplistic. However, length (in the scan direction), depth, width, and area values are sometimes used to relate to process parameters. Often in AM modeling literature, a plot of the melt-pool temperature versus some cross section distance is given [123,142–144]. Melt-pool size may be inferred and related to input parameters, though it is not often expressed as a single-value measurand (e.g., the melt-pool is x mm). Soylemez et al. mentioned that while melt-pool cross-sectional area is a key descriptor, melt-pool length was known to affect deposited bead shape, so they proposed using length-to-depth ratio (L/d) as a descriptor in their process mapping efforts [145]. Childs et al. also mentioned that L/d ratio determined the boundary between continuous and balled tracks when scanning on powder beds without a solid substrate [146].

Typically, the melt-pool size and temperature increase with laser power; however, the relationship with scan speed is more complicated. For stationary pulsed laser tests (e.g., [147]), the effects of longer pulse durations are related to lower scan speeds and resulting higher temperature. Multiple simulation efforts have addressed the trends in temperature and size of the melt-pool with process parameters, which are organized in Table 34.9. It was shown in [142] that the width and depth decreased slightly with scan

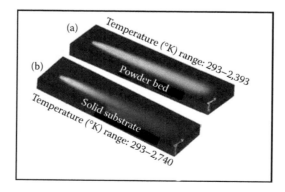

Figure 34.3 FE simulation surface temperature results showing comet-like shape, and the temperature distribution's relation to proximity to high conductive zones (e.g., solid substrate, image [b]) or low conductive zones (e.g., powder bed, image [a]). (From Hussein, A. et al., *Mater. Des.*, 52, 638–647, 2013.)

Table 34.9 Commonly observed melt-pool signatures and related process parameters as evidenced in AM models and simulations

Melt-pool signature	Relationship	Measurand	References
Temperature (peak)	Increases	Laser power	[123,143,144,147–149]
	Decreases	Scan speed	[142,143,148–150]
	Decreases	Thermal Conductivity	[142,146]
Size**	Length, width, and depth increase	Laser power	[26,36,123,134,143,144, 147–149]
	Width - decrease	Scan speed	[136,142,148]
	Length - increase	Scan speed	[136,142]
	Depth - decrease	Scan speed	[36,134,136,142,149]
	Length, width, and depth increase	Thermal conductivity*	[142,146,149]

*Used as a general term assuming higher conductivity in proximity to previously solidified regions, the build plate, or build up of solidified layers
**Measurement of the fused or solidified material mass or size may be used as an indicator of melt-pool size.

speed (from 100 to 300 mm/s), whereas the length of the melt-pool in the scan direction increased, contributing more to the overall melt-pool size. This was for the single-layer model geometry shown in Figure 34.3. Chen and Zhang also showed depth decreasing with speed, but change in length was less pronounced [134]. Chen and Zhang also created simulations where melt-pool depth was kept constant, which required more input power at the higher speeds. The thin-wall geometry modeled in [26,27] (not PBF) showed that melt-pool length decreased with increasing scan speed, though at much lower speeds (<10 mm/s). One interesting approach by Birnbaum et al. used a FEM to look at transient changes to melt-pool geometry given a step change in laser power with the specified intent to apply in thermal imaging feedback control [17].

Modeling offers a comprehensive analysis of the melt-pool, to deduce the irregular shape and temperature contours in the interior and not just the surface. Surface level measurements of melt-pool signatures are leading efforts *in situ* process control. Modeling and simulation can relate these melt-pool signatures to the complex and dynamic characteristics internal to the melt-pool, powder bed, or the solid part itself, such as residual stresses, porosity, or metallic phase structure.

One promising application of AM simulation to closed-loop control is the ability to study the effect of variable thermal conductivity on melt-pool signatures, and thus the part quality. The fully solidified part exhibits higher thermal conductivity than the surrounding powder, thereby conducting more heat from the laser source, reducing the melt-pool temperature but increasing its size. Multiple AM models have shown this phenomenon or studied it in detail [134,142,146]. Hussein et al. showed how the melt-pool and trailing hot zone changed temperature and shape depending on whether the laser is scanned over powder bed (low thermal conductivity) or solid substrate (high conductivity) [142]. Scanning over the powder bed produced lower peak temperatures in the melt-pool but higher temperatures in the trailing region for the first scan. However, this trend changed such that subsequent scans over the solid substrate always resulted in lower temperatures. Chen and Zhang simulated multiple layers while keeping melt-pool depth constant [134]. They showed that more power was necessary as build layers increased to maintain the processing depth, indicating that more heat was conducting into the solid layers.

Wang came to the same conclusion, but for multiple layers in a thin-wall geometry [127]. The relationships between melt-pool signatures and changes in thermal conductivity have guided the use of feedback controlled melt-pool size. However, there are other critical phenomena that are less understood, but may be addressed through intelligent melt-pool monitoring guided by results from modeling and simulations.

The time history of temperature plays a crucial role in residual stresses and build-direction variability in density and material phase structure. Although extremely important to final part quality, these phenomena are difficult to measure *in situ* during a build. In the future, successful models may be able to predict these phenomena to be exploited in feed-forward control schemes. In a series of papers, Wang et al. [127,128] looked at time history of temperature in each layer as the build progresses in a DED system. Subsequent scans on new layers reheated the base layers, which turned originally hard martensitic layers to softer, tempered martensite, whereas new layers stayed consistently hard. By increasing scan speed and laser power (keeping melt-pool size constant), the number and consistency of hard, martensitic layers could be increased because the lower layers were subjected to shorter heating from upper layer builds. Others have studied this lower layer reheating phenomena [126,143] and its effect on residual stresses [142,151–153].

Others [142,144,148] also studied preheating and postheating of a surface point before and after the laser scan had passed on one layer (rather than subsequent layers). Under certain conditions, locations on previously scanned tracks were remelted. This number of remelting cycles increases for narrower hatch spacing. For constant hatch spacing, Yin et al. showed that lower scan speeds promoted remelting primarily due to the resulting higher temperatures [144]. However, one can assume that under different conditions, a slower scan speed would allow points on adjacent tracks to cool enough not to be remelted. This remelting effect has been shown experimentally to relate to part quality (e.g., surface roughness, mechanical properties, porosity) [154].

Hussein et al. also studied thermal stresses in powder-bed geometry for multiple layers [142]. Their results showed that regions in the build experience thermal expansion and contraction is based on the local temperature history and build geometry. It was also demonstrated that the relationships between the melt-pool signatures and residual stresses are very complex; therefore melt-pool monitoring may not provide enough information to predict residual stress formation. Nickel et al. specifically investigated effects of scanning pattern on residual stress and part deformation [155]. Though this forms an excellent guide to optimal scanning patterns developed before the build takes place, it is unlikely that scan patterns can be effectively changed *in situ* to control stress without affecting other part qualities such as porosity, homogeneity, or strength. Vasinonta et al. mapped residual stress in thin wall formation, and proposed that build plate and part preheating is much more effective in reducing residual stresses than varying scan speed or laser power [26,27]. Though Vasinonta et al. did not include reheating of lower layers or adjacent scan tracks, this may indicate that control schemes that target minimization of residual stress may focus on monitoring build plate and chamber temperature, rather than monitoring melt-pool signatures. As mentioned, scan pattern has been shown to relate to residual stress formation, though this may be more difficult to adaptively control than build plate or chamber temperature.

The reheating phenomenon also has an effect on metallic phase structure, (not to be confused with the more often modeled powder-liquid-solid phases). Wang and Fenicelli et al. [127,131] looked at metal phase change based on temperature cycle history and volume fraction of three possible phases (in 410 SS) using commercial welding simulation software. In the simulation results in [127], they observed that the high temperatures

caused by the initial pass by the DED system laser would create a high-strength, martensitic microstructure. Key to these phase changes was the high rate of cooling observed in
their model, a consequence of the material thermal properties, boundary conditions, and
overall geometry. In [156], they extended the model to predict thermally and mechanically
induced residual strain versus laser power, scan speed, and powder flow rate (in a DED
system), then compared to neutron-diffraction strain measurement results from [157] with
good agreement for the range of parameters studied. Though results were complex and
cannot all be detailed here, one interesting result showed that residual stress in the laser
scan direction changed from compressive to tensile when scan speed doubled from
4.2 mm/s to 8.5 mm/s, while maintaining the steady melt-pool size by adjusting laser
power (increasing with scan speed, but decreasing with pass number).

Modeling and simulation can link measurable melt-pool or process signatures to
immeasurable but critical phenomena like instantaneous material phase and microstructure. However these complex relationships require an organized and simplified methodology to implement *in situ* control. Perhaps the best method is through development of
process maps, which several research groups have developed using modeling and simulations for the DED process for process control. Vasinonta et al. used a FE method to
develop process maps for the DED manufacturing of thin walls, and put results in term
of nondimensional parameters based on the Rosenthal moving point source solution
[26,130,133]. Bontha et al. used a 2D analytical (Rosenthal) and FEMs to calculate cooling
rates in DED processing of Ti-6Al-4V as a function of laser power, traverse speed, and
increasing build depth [158]. These are overlaid onto previously developed process maps
that detail expected microstructure forms for different ranges of thermal gradients versus
solidification rates (*G–R plot* or *solidification map* [159]). Soylemez et al. formed process maps
that linked melt-pool signatures to laser power versus scan velocity (called a *P–V map*)
using a 3D FE simulation of single bead deposition [145], then later Gockel and Beuth combined the maps to show how specific combinations of laser power and speed can achieve
constant grain size and tailored morphology in an electron beam wire feed process as
shown in Figure 34.4 [160]. They proposed use of this hybrid microstructure map, which

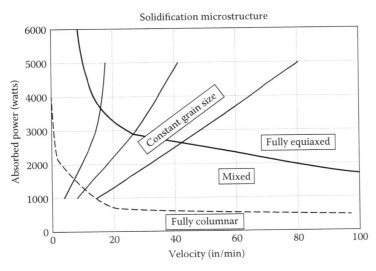

Figure 34.4 Microstructure P–V map for wire-fed E-beam Ti-6Al-4V. (From Gockel, J. and Beuth, J. L.,
Solid Freeform Fabrication Proceedings, Austin, TX, pp. 666–674, 2013.)

depends on simulation data to develop, for *real-time indirect microstructure control through melt-pool dimension control*. Though microstructure control is the primary focus in [160], it may be possible to extend this methodology to develop process maps for residual stress [26,27]. Much of this reviewed process mapping work was centered at Carnegie Mellon University, Pittsburgh, Pennsylvania and Wright State University, Fairborn, Ohio, and a thorough review of these efforts is given by Beuth et al., including a list of patent applications submitted by the authors [161].

34.4 Implications for process control

Based on the review presented in Section 34.3, this section first identifies and categorizes the process parameters, process signatures, and product qualities as reported in the literature to systematically analyze the needed correlations among them. Next, the section presents the research opportunities specifically for the real-time control of AM PBF processes.

34.4.1 Parameters-signatures-qualities categorization

As summarized in Section 34.3, the influence of AM process parameters on the resultant part quality in general has been widely studied and reported. To establish foundations for process control, we subcategorize the process parameters and process signatures and product quality according to the abilities to be measured and/or controlled. *Process parameters* are input to the PBF process and they are either potentially controllable or predefined. Controllable parameters (e.g., laser and scanning parameters, layer thickness, and temperature) are used to control the heating, melting, and solidification process and thus control the part quality. Predefined parameters, for example, include part geometry, material, and build plate parameters. Controllable process parameters generally correlate to the observable and derived *process signatures* (e.g., melt-pool size, temperature, porosity, or residual stress). Derivable parameters cannot be directly measured but can be calculated with a numerical model, such as the maximum depth of a melt-pool. For purposes of correlations we further subdivide the process signatures into three categories: melt-pool, track, and layer. Process signatures determine the final *product qualities* (geometric, mechanical, and physical). Developing correlations between the controllable process parameters and process signatures should support feed forward and feedback control, with the goal of embedding process knowledge into future control schemes. Figure 34.5 categorizes and lists the process parameters, process signatures, and product qualities to derive the needed correlations.

The main process controllable parameters include the following:

1. *Laser beam velocity*: quantifies the scanning speed and direction of the laser beam.
2. *Laser power*: quantifies the power of the laser beam.
3. *Laser beam diameter*: quantifies the diameter of the laser beam scanning the powder bed.
4. *Layer thickness variation*: quantifies the variation to the preset powder layer thickness for refilling the previously fabricated sublayer.
5. *Inert gas flow*: quantifies the inert gas flowing above the powder bed for cooling using two subparameters: *flow rate* and the *flow pattern*, such as laminar flow, turbulent flow, or transient flow, of the inert gas.
6. *Scanning pattern*: quantifies the order of the scanning directions of the laser beam.

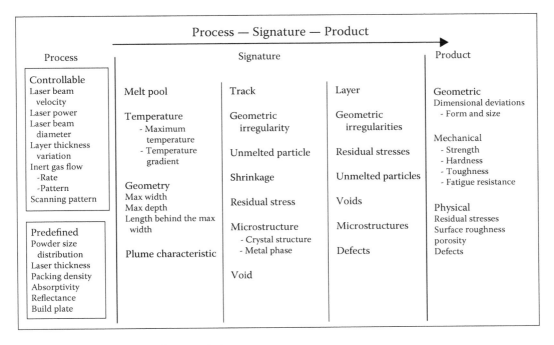

Figure 34.5 Parameters in the correlations.

Predefined process parameters are those process-related parameters that are defined prior to laser scanning and cannot be changed during scanning. The following are the pre-defined parameters:

1. *Powder size distribution*: quantifies the particle size distribution of the metal powder.
2. *Layer thickness*: quantifies the predetermined thickness of powder layer for each layer of scanning.
3. *Packing density*: quantifies the density of powder in the powder chamber after packing.
4. *Absorptivity*: quantifies the coefficient of the heat absorbed per unit mass of powder.
5. *Reflectance*: quantifies the ratio of the heat reflected by the powder bed to the heat delivered by the laser beam.
6. *Build plate*: indicates the type of plate that is used to fabricate a product.

Melt-pool, a subcategory of process signature, has the following parameters:

1. *Temperature*: includes two subparameters namely the *maximum temperature* of the melt-pool, and the *temperature gradient* of the melt-pool.
2. *Geometry*: includes three subparameters namely *maximum width* of the melt-pool, *maximum depth* of the melt-pool, and *length* of the melt-pool behind the maximum width.
3. *Plume characteristic*: characterizes the plume.

Track, another subcategory of process signature, has the following parameters:

1. *Geometric irregularity*: indicates irregularities in the track (e.g., balling, voids, disconti-nuity, and delamination) causing the fabricated track to deviate from the desired track.
2. *Unmelted particle*: indicates the location of an unmelted particle in the track.
3. *Shrinkage*: indicates the size reduction due to cooling and solidification of the track.
4. *Residual stress*: quantifies residual stress in the track due to shrinkage or deformation, such as bending and twisting.
5. *Microstructure*: indicates microstructure of the track denoted using two subparameters: *Crystal structure* (including grain size and grain growth direction) and *Metal phase*.
6. *Void*: indicates the location and shape of an empty space, such as pore, crack, and delamination, in the track.

Layer, the other subcategory of process signature, has the following parameters:

1. *Geometric irregularities*: indicates irregularities in the layer. Combined shape irregu-larities from all the tracks in a layer can make the entire fabricated layer to deviate in shape.
2. *Residual stresses*: indicates the residual stresses and stress distribution in the layer.
3. *Unmelted particles*: indicates particles, which are not melted by the laser beam, in the layer.
4. *Voids*: quantifies empty spaces, such as pores, cracks, and delamination, in the layer.
5. *Microstructure*: indicates the crystal structures and metal phase in the layer.
6. *Defects*: quantifies imperfections (e.g., delamination, discontinuity, and severe defor-mation) in the layer such that the product can be disqualified if the defect cannot be remedied in fabricating the succeeding layers.

The category of product includes the following:

1. *Dimensional deviation*: quantifies the deviation of the measured dimension from the nominal dimension due to form and size errors.
2. *Mechanical property*: quantifies mechanical performance of the product, such as strength, hardness, toughness, and fatigue resistance.
3. *Surface roughness*: quantifies the roughness of a surface of the product.
4. *Porosity*: quantifies the amount of voids in the product.
5. *Defects*: quantifies imperfections in the product that makes the product fail to per-form by design.
6. *Residual stress*: quantifies unintended residual stress in the product.

34.4.2 *Correlations*

With the parameters individually defined in the previous section, this section describes qualitative correlations to describe the cause-and-effect relationship between process con-trol parameters, process signatures, and product quality. The correlations are synthesized according to literature review in Section 34.1, particularly, Section 34.3. Most reviewed papers discussed the correlations between process parameters and product quality (e.g., increasing laser power can improve product mechanical strength due to deeper and wider melting). Those papers that discussed process signatures mostly focused on melt-pool

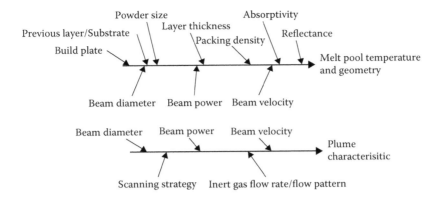

Figure 34.6 Correlations between process and melt-pool signature.

temperature and area. Process parameters along with signatures in general have not yet been directly related to product quality.

From the literature, process parameters are driving factors that determine a melt-pool formation. Figure 34.6 shows the correlations between controllable process parameters and melt-pool signature parameters. *Melt-pool temperature* and *melt-pool geometry* depend on the controllable (beam diameter, beam power, and beam velocity) and predefined parameters (reflectance, absorptivity, packing density, layer thickness, powder-size distribution, previous layer/substrate, and build plate). *Plume characteristic* generally depends on the beam diameter, beam power, beam velocity, scanning strategy, and inert gas flow (including flow rate flow pattern).

Note that in the paragraph text, the *causes* of cause-and-effect relationships are capitalized and the *effects* are capitalized and italicized for reading convenience. The *effects* are bolded in the figures that follow.

After the melt-pool cools, the metal solidifies and forms a track. From Figure 34.7, *shrinkage* depends on the controllable process parameters namely the layer thickness variation and powder packing density. The thicker the layer, the more the metal shrinks. The higher the powder packing density, the less the metal shrinks. The *geometric irregularity* depends on melt-pool temperature, melt-pool geometry, shrinkage, beam velocity, and layer thickness. If the *melt-pool temperature* is too high, the shape of the track will be wider due to extreme melting. If the *melt-pool geometry* is larger than the desired geometry, the track shape will become too large. *shrinkage* deforms the shape of the track from the shape of the powder layer. If the beam velocity is too fast, balling occurs and causes *geometric irregularity* in the track.

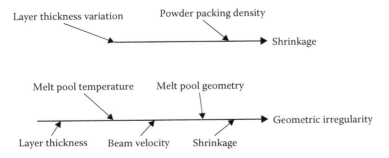

Figure 34.7 Correlations between melt-pool and track (1/3).

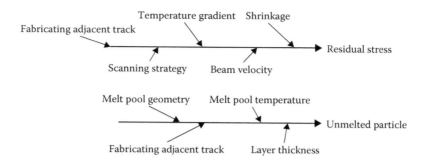

Figure 34.8 Correlations between melt-pool and track (2/3).

From Figure 34.8, *residual stress* is the maximum residual stress in the track and depends on shrinkage, temperature gradient, fabricating adjacent track, beam velocity, and scanning strategy. The more the melt-pool shrinks during solidification, the higher the residual stress is. Similarly, the steeper the temperature gradient, the higher is the residual stress. *Unmelted particles* depends on melt-pool geometry, melt-pool temperature, layer thickness, and fabricating adjacent track (Figure 34.8). If the melt-pool temperature is lower than the ideal temperature, unmelted particles can occur because of incomplete melting. If the melt-pool geometry is irregular, some particles cannot have sufficient heat to melt and become unmelted particles. The thicker the layer, more particles in the bottom of the melt-pool tend to exist. Fabricating an adjacent track can remelt the unmelted particles.

From Figure 34.9, *voids* depend on melt-pool geometry, melt-pool temperature, and fabricating adjacent track. Similar to unmelted particles, if the melt-pool geometry is irregular, some particles will not have the sufficient heat to melt, and pores will be in the track. Similarly, if the melt-pool temperature is lower than the ideal temperature, unmelted particles can occur because of incomplete melting, and pores will be in the track. Fabricating an adjacent track can remelt the unmelted particles and, thus, remove voids. *Microstructure* includes grain size, grain growing direction, and metal phase and depends on the following melt-pool parameters: Melt-pool temperature, temperature gradient, beam velocity, and fabricating adjacent track. The three parameters that is, melt-pool temperature, temperature gradient, and beam velocity affect grain sizes, grain growing directions, and metal phases of the track. Fabricating adjacent track remelts a portion of the previous track as a heat treatment and thus affects the microstructure of the track.

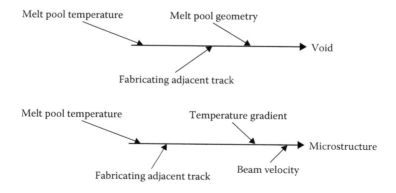

Figure 34.9 Correlations between melt-pool and track (3/3).

After tracks are fabricated, a layer of metal is formed. Figure 34.10 shows the layer related signatures: Geometric irregularities, residual stresses, and unmelted particles. *Geometric irregularities* of the layer depends on the combined track geometric irregularities. *Residual stresses* of the layer depends on the combined track residual stresses and fabricating other layers. Fabricating other layers can release or worsen the residual stress in the layer. The *unmelted particles* parameter is derived from the combined track unmelted particles.

Figure 34.11 shows the other layer related signatures: Voids, microstructures, and defects. *Voids* are derived from both the voids in tracks and between tracks parameter and the geometric irregularity parameter. *Microstructures* depend on the combined track microstructures parameter. *Defects* depend on the shape irregularities, combined track microstructures, residual stresses, and unmelted particles. *Defects* indicate the locations, and the types of defects in a layer. If the defects can be remedied in the succeeding layer fabrication, the defects will not be the reason to stop the fabrication process; otherwise, the fabrication process should be stopped to avoid making a product with defects.

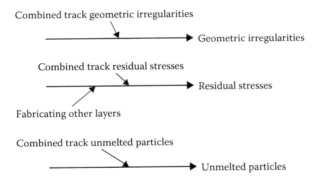

Figure 34.10 Correlations between track and layer (1/2).

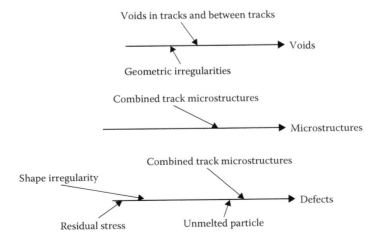

Figure 34.11 Correlations between track and layer (2/2).

Product quality directly depends on dimensional deviations, surface roughness, mechanical properties, residual stresses, porosity, and defects. *Dimensional deviations* includes form and size deviations from the desired form and dimensions. From Figure 34.12, *dimensional deviations* depend on combined layer dimensions and combined layer geometric irregularities. *Surface roughness* depends on voids (voids on the product surface) and geometric irregularities (geometric irregularities on the product surface). *Mechanical properties* (including part mechanical strength, hardness, toughness, and fatigue performance) depends on the combined layer microstructures, the geometric irregularities, voids, unmelted particle, and combined layer residual stress.

From Figure 34.13, *residual stresses* in the product depends on geometric irregularities, combined microstructures, voids, unmelted particles, and combined layer residual stress. The combined layer residual stress is main contributor to the residual stress in the product. *Porosity* depends on voids in all the layers. Finally, *defects* (includes delamination, substandard mechanical properties, and out of tolerances) depends on combined voids in layers, unmelted particles, geometric irregularities, and residual stresses.

From the above discussions, various correlations have been qualitatively connected through cause-and-effect diagrams from process parameters to process signatures and to

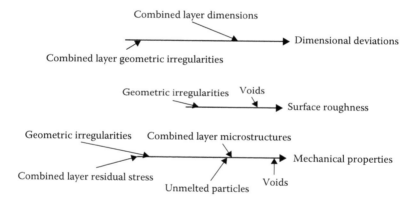

Figure 34.12 Correlations between layer and product (1/2).

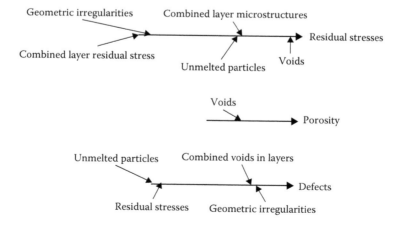

Figure 34.13 Correlations between layer and product (2/2).

part qualities. Change in one process parameter can affect multiple signatures and multiple part qualities. Part quality generally depends on multiple process parameters. Process and product usually follow a multiple input and multiple output relationship.

There are potentially other missing parameters. One possible missing process signature is the heat absorption, before the actual melt-pool formation. The heat absorption signature can include the heat absorption rate and the temperature raising profile. More research in this subject is needed.

34.4.3 Research opportunities

For design of AM PBF process control there must be further development of parameter-signature-quality relationships and relative sensitivities of those relationships through experiments and simulations. Existing control design for the DMD process focuses on measuring and controlling melt-pool signatures (size and temperature) by varying laser parameters (power and scan speed), and there is reason to believe that PBF process control will follow similar trends. Therefore, for controller development, research results ought to focus on the parameter-signature-quality relationships and sensitivities, with particular focus on measureable melt-pool signatures, and controllable process parameters.

In addition to further defining these process relationships, new traceable measurement methods and identification of new measurable process signatures are necessary. Two issues: residual stress and varying metallic phase structure, are particularly problematic in PBF processes yet there are few or no *in situ*, nonintrusive measurement methods available to detect these phenomena as they vary during a build. Melt-pool signatures (e.g., size and temperature) are the most often considered measurands for *in situ* feedback control. However, there is potential for other, less considered signatures that may offer greater sensitivity to process variations or simplified measurement, for example, measurements of the laser ablation plume size, or the spectral measurements of the ablation zone [162–164]. Methods for controlling porosity, surface finish, and residual stress will be necessary for increasing the endurance limit.

Most of the reviewed literature has limited analysis of measurement error and traceability, and there is a need for better measurement uncertainty evaluations and reporting. First, simulations require *accurate* and *repeatable* measurements for validation. For example, there are simulations that correlate temperature to melt-pool size. In such cases, a large uncertainty in a temperature evaluation will result in an uncertainty of the melt-pool size, and therefore inadequate comparison of measurement data with the model output. Better understanding of measurement uncertainty assists system controller design by identifying the necessary level of precision required to attain the goals of the control system.

It is well known that the relationships between parameters in the PBF process are complex. Process maps, such as those in [24,26,27,133,158,161], will be a key tool to organize and communicate the complex, multidimensional parameter relationship topology. These maps will be essential for multiinput, multioutput (MIMO) control algorithm design, and model-based predictive controller design.

The AM process control design landscape is so far limited in variety, with most examples using melt-pool temperature and/or size to control laser power or speed. This method could very well be the most effective; however, there is wider potential for different levels of control loops. For example, control loops may occur discretely between completion of each build layer rather than continuously (e.g., the powder-bed temperature mapping by Craeghs et al. [8]). However, it is yet unclear which signatures are best modeled or measured, and which input parameters are best controlled for which timescale (either continuously or discrete inter-layer). It is a worthwhile endeavor to create an AM control loop architecture that identifies the multiple potential control loops, and provides a basis for identifying which loops are optimal for controlling which parameter-signature-quality relationship.

34.5   Conclusions

This chapter presented a review on the AM process control schemes, process measurements, and modeling and simulation methods as applied to the PBF process, though related work from other processes were also reviewed. This background study is aimed to identify and summarize the measurement science needs that are critical to real-time AM process control. The report was organized to present the correlations between process parameters, process signatures, and product quality. Based on the review, we presented the implications for process control highlighting the research opportunities and future directions. For example, we found reported correlations between the laser power (process parameter) and the melt-pool surface geometry and surface temperature (process signatures) on the resulting relative density of the part (part quality). Melt-pool size and temperature have already been used as feedback parameters in closed-loop control schemes. Considering residual stresses as another example, researchers have identified that an increase in the build platform temperature correlates to lower residual stresses. There were also reported correlations on the residual stress to the scan strategy and layer thickness used to build. In the future work, newer process signatures and corresponding correlations will have to be investigated for newer control schemes.

Future work at NIST will also involve the development of a benchtop open architecture AM research platform to test and demonstrate the in-process measurement and control methods. Such a benchtop platform will enable us to directly observe melting and solidification of metal powders, integrate process metrology tools, and implement software interfaces and data acquisition for process measurements, as well as test the control algorithms. The AM community can benefit from such a test platform to implement, test, and validate a real-time and closed-loop control of AM processes.

Disclaimer

Certain products or services are identified in the chapter to foster understanding. Such identification does not imply recommendation or endorsement by the NIST, nor does it imply that the products or services identified are necessarily the best available for the purpose.

Appendix A: AM PBF machine specifications

Typically, metal PBF machines have build volumes on the order of 250 mm × 250 mm × 200 mm. The metals that are available for production are stainless steels, tool steels, titanium alloys, nickel alloys, aluminum alloys, cobalt chrome alloys, and bronze alloys. Layer thicknesses are typically between 0.02 mm and 0.10 mm. The process builds in an inert environment of nitrogen or argon (though some processes, especially electron-beam based processes, build in a vacuum). Laser-based systems typically deflect the laser beam off two mirrors and through some optics (often an f-theta lens) to focus the beam to a 0.05 mm to 0.5 mm beam width on the top surface of the powder bed. The beam is scanned by a galvanometer system that rotates the deflecting mirrors. Laser scan speeds can be as fast as 7 m/s. Parts are typically built by first tracing the laser spot over the perimeter of the layer's geometry, then filling the area with a raster or hatch pattern.

References

1. ASTM Standard 2792, *Standard Terminology for Additive Manufacturing Technologies*, ASTM International, West Conchocken, PA, 2012.
2. Department of Commerce, *Fact Sheet: Additive Manufacturing*, 15-Aug-2012. [Online]. Available: http://www.commerce.gov/news/fact-sheets/2012/08/15/fact-sheet-additive-manufacturing (Accessed December 15, 2014).
3. T. Wohlers, U.S. Manufacturing Competitiveness Initiative Dialogue, presented at the *Council on Competitiveness*, Oak Ridge, TN, April 18, 2013.
4. Energetics Inc. for National Institute of Standards and Technology, Measurement science roadmap for metal-based additive manufacturing, May-2013. [Online]. Available: http://www.nist.gov/el/isd/upload/NISTAdd_Mfg_Report_FINAL-2.pdf (Accessed December 15, 2014).
5. D. L. Bourell, M. C. Leu, and D. W. Rosen, *Roadmap for Additive Manufacturing: Identifying the Future of Freeform Processing*, University of Texas at Austin, Austin, TX, 2009.
6. T. Craeghs, F. Bechmann, S. Berumen, and J.-P. Kruth, Feedback control of layerwise laser melting using optical sensors, *Phys. Procedia*, 5, 505–514, 2010.
7. T. Craeghs, S. Clijsters, E. Yasa, and J.-P. Kruth, Online quality control of selective laser melting, in *Solid Freeform Fabrication Proceedings*, Austin, TX, 2011, pp. 212–226.
8. T. Craeghs, S. Clijsters, J.-P. Kruth, F. Bechmann, and M.-C. Ebert, Detection of process failures in layerwise laser melting with optical process monitoring, *Laser Assist. Net Shape Eng. 7 LANE 2012*, 39, 753–759, 2012.
9. K. A. Mumtaz and N. Hopkinson, Selective laser melting of thin wall parts using pulse shaping, *J. Mater. Process Technol.*, 210(2), 279–287, January 2010.
10. Y. Ning, Y. Wong, J. Y. Fuh, and H. T. Loh, An approach to minimize build errors in direct metal laser sintering, *Autom. Sci. Eng. IEEE Trans. On*, 3, 73–80, 2006.
11. A. Simchi, F. Petzoldt, and H. Pohl, On the development of direct metal laser sintering for rapid tooling, *J. Mater. Process. Technol.*, 141, 319–328, 2003.
12. X. Wang, Calibration of shrinkage and beam offset in SLS process, *Rapid Prototyp. J.*, 5(3), 129–133, 1999.
13. M. Mahesh, Y. Wong, J. Fuh, and H. Loh, A Six-sigma approach for benchmarking of RP&M processes, *Int. J. Adv. Manuf. Technol.*, 31(3–4), 374–387, 2006.
14. C. Doumanidis and Y.-M. Kwak, Geometry modeling and control by infrared and laser sensing in thermal manufacturing with material deposition, *J. Manuf. Sci. Eng.*, 123(1),45–52, 2001.
15. P. Aggarangsi, J. L. Beuth, and M. L. Griffith, Melt pool size and stress control for laser-based deposition near a free edge, in *Solid Freeform Fabrication Proceedings*, Austin, TX, 2003, pp. 196–207.
16. P. Aggarangsi, J. L. Beuth, and D. D. Gill, Transient changes in melt pool size in laser additive manufacturing processes, in *Solid Freeform Fabrication Proceedings*, Austin, TX, 2004, pp. 163–1747.
17. A. Birnbaum, P. Aggarangsi, and J. L. Beuth, Process scaling and transient melt pool size control in laser-based additive manufacturing processes, in *Solid Freeform Fabrication Proceedings*, Austin, TX, 2003, pp. 328–339.
18. A. Birnbaum, J. L. Beuth, and J. W. Sears, Scaling effects in laser-based additive manufacturing processes, in *Solid Freeform Fabrication Proceedings*, Austin, TX, 2004, pp. 151–162.
19. M. R. Boddu, R. G. Landers, and F. W. Liou, Control of laser cladding for rapid prototyping—A review, in *Solid Freeform Fabrication Proceedings*, Austin, TX, 2001, pp. 6–8.
20. D. L. Cohen, *Additive Manufacturing of Functional Constructs Under Process Uncertainty*, PhD Dissertation, Cornell University, Ithaca, NY, 2010.

21. G. Bi, A. Gasser, K. Wissenbach, A. Drenker, and R. Poprawe, Characterization of the process control for the direct laser metallic powder deposition, *Surf. Coat. Technol.*, 201(6), 2676–2683, 2006.

22. D. Hu and R. Kovacevic, Sensing, modeling and control for laser-based additive manufacturing, *Int. J. Mach. Tools Manuf.*, 43(1), 51–60, 2003.

23. D. Hu, H. Mei, and R. Kovacevic, Improving solid freeform fabrication bylaser-based additive manufacturing, *Proc. Inst. Mech. Eng. Part B J. Eng. Manuf.*, 216, 1253–1264, 2002.

24. S. Bontha and N. Klingbeil, Thermal process maps for controlling microstructure in laser-based solid freeform fabrication, presented at the *Solid Freeform Fabrication Proceedings*, 2003, pp. 219–226.

25. M. L. Griffith, W. H. Hofmeister, G. A. Knorovsky, D. O. MacCallum, M. E. Schlienger, and J. E. Smugeresky, *Direct laser additive fabrication system with image feedback control*, U. S. Patent No. 6,459,951, 2002.

26. A. Vasinonta, J. L. Beuth, and M. L. Griffith, Process maps for controlling residual stress and melt pool size in laser-based SFF processes, in *Solid Freeform Fabrication Proceedings*, Austin, TX, 2000, pp. 200–208.

27. A. Vasinonta, M. L. Griffith, and J. L. Beuth, Process maps for predicting residual stress and melt pool size in the laser-based fabrication of thin-walled structures, *J. Manuf. Sci. Eng.*, 129(1), 101–109, 2007.

28. P. Aggarangsi and J. L. Beuth, Localized preheating approaches for reducing residual stress in additive manufacturing, in *Solid Freeform Fabrication Proceedings*, Austin, TX, 2006, pp. 709–720.

29. J.-P. Kruth, G. Levy, F. Klocke, and T. H. C. Childs, Consolidation phenomena in laser and powder-bed based layered manufacturing, *CIRP Ann.Manuf. Technol.*, 56(2), 730–759, 2007.

30. A. Cooke and J. A. Slotwinski, *Properties of metal powders for additive manufacturing: a review of the state of the art of metal powder property testing*, US Department of Commerce, National Institute of Standards and Technology, NISTIR 7873, 2012.

31. J. A. Slotwinski, E. J. Garboczi, P. E. Stutzman, C. F. Ferraris, S. S. Watson, and M. A. Peltz, Characterization of metal powders used for additive manufacturing, *J. Res. Natl. Inst. Stand. Technol.*, 19, 2014.

32. A. Amado, M. Schmid, G. Levy, and K. Wegener, Advances in SLS powder characterization, in *Solid Freeform Fabrication Proceedings*, Austin, TX, 2011, 7, 12.

33. I. Robertson and G. Schaffer, Some effects of particle size on the sintering of titanium and a master sintering curve model, *Metall. Mater. Trans. A*, 40(8), 1968–1979, 2009.

34. H. Su and D. L. Johnson, Master sintering curve: A practical approach to sintering, *J. Am. Ceram. Soc.*, 79(12), 3211–3217, 1996.

35. R. K. McGeary, Mechanical packing of spherical particles, *J. Am. Ceram. Soc.*, 44(10), 513–522, 1961.

36. G. B. M. Cervera and G. Lombera, Numerical prediction of temperature and density distributions in selective laser sintering processes, *Rapid Prototyp. J.*, 5, 21–26, 1999.

37. A. V. Gusarov, T. Laoui, L. Froyen, and V. Titov, Contact thermal conductivity of a powder bed in selective laser sintering, *Int. J. Heat Mass Transf.*, 46, 1103–1109, 2003.

38. A. V. Gusarov and J.-P. Kruth, Modelling of radiation transfer in metallic powders at laser treatment, *Int. J. Heat Mass Transf.*, 48, 3423–3434, 2005.

39. J. Karlsson, A. Snis, H. Engqvist, and J. Lausmaa, Characterization and comparison of materials produced by Electron Beam Melting (EBM) of two different Ti–6Al–4V powder fractions, *J. Mater. Process Technol.*, 213(12), 2109–2118, Dec. 2013.

40. B. Liu, R. Wildman, C. Tuck, I. Ashcroft, and R. Hague, Investigation the effect of particle size distribution on processing parameters optimisation in Selective Laser Melting process, in *Solid Freeform Fabrication Proceedings*, Austin, TX, 2011.

41. A. B. Spierings and G. Levy, Comparison of density of stainless steel 316L parts produced with selective laser melting using different powder grades, in *Solid Freeform Fabrication Proceedings*, Austin, TX, 2009, pp. 342–353.

42. A. B. Spierings, N. Herres, and G. Levy, Influence of the particle size distribution on surface quality and mechanical properties in AM steel parts, *Rapid Prototyp. J.*, 17(3), 195–202, 2011.

43. E. Tsotsas and H. Martin, Thermal conductivity of packed beds: A review, *Chem. Eng. Process. Process Intensif.*, 22(1), 19–37, 1987.
44. S. Berumen, F. Bechmann, S. Lindner, J.-P. Kruth, and T. Craeghs, Quality control of laser- and powder bed-based Additive Manufacturing (AM) technologies, *Phys. Procedia*, 5, Part B, 617–622, 2010.
45. Y. Chivel and I. Smurov, On-line temperature monitoring in selective laser sintering/melting, *Phys. Procedia*, 5, 515–521, 2010.
46. T. Craeghs, F. Bechmann, S. Berumen, and J.-P. Kruth, Feedback control of layerwise laser melting using optical sensors, *Phys. Procedia*, 5, 505–514, 2010.
47. P. Lott, H. Schleifenbaum, W. Meiners, K. Wissenbach, C. Hinke, and J. Bültmann, Design of an optical system for the in situ process monitoring of Selective Laser Melting (SLM), *Phys. Procedia*, 12, Part A, 683–690, 2011.
48. I. Yadroitsev, P. Krakhmalev, and I. Yadroitsava, Selective laser melting of Ti6Al4V alloy for biomedical applications: Temperature monitoring and microstructural evolution, *J. Alloys Compd.*, 583, 404–409, 2014.
49. R. B. Dinwiddie, R. R. Dehoff, P. D. Lloyd, L. E. Lowe, and J. B. Ulrich, Thermographic in-situ process monitoring of the electron-beam melting technology used in additive manufacturing, in *Proceedings of the SPIE*, 2013, 8705, 87050K–87050K–9.
50. S. Price, J. Lydon, K. Cooper, and K. Chou, Experimental temperature analysis of powder-based electron beam additive manufacturing, in *Proceedings of the 24th Solid Freeform Fabrication Symposium*, Austin, TX, 2013, pp. 162–173.
51. S. Price, J. Lydon, K. Cooper, and K. Chou, Temperature Measurements in Powder-Bed Electron Beam Additive Manufacturing, in *Proceedings of the ASME 2014 International Mechanical Engineering Congress & Exposition*, Montreal, Canada, 2014.
52. E. Rodriguez, F. Medina, D. Espalin, C. Terrazas, D. Muse, C. Henry, and R. Wicker, Integration of a Thermal Imaging Feedback Control System in Electron Beam Melting, *WM Keck Cent. 3D Innov.* University of Texas at El Paso, El Paso, TX 2012.
53. A. Wegner and G. Witt, *Process monitoring in laser sintering using thermal imaging, presented at the SFF Symposium*, Austin, TX, 2011, pp. 8–10.
54. W. Hofmeister and M. Griffith, Solidification in direct metal deposition by LENS processing, *JOM*, 53(9), 30–34, 2001.
55. R. B. Dinwiddie, V. Kunc, J. M. Lindal, B. Post, R. J. Smith, L. Love, and C. E. Duty, Infrared imaging of the polymer 3D-printing process, presented at the *SPIE Sensing Technology+ Applications*, 2014, 910502–910502.
56. H. Krauss, C. Eschey, and M. Zaeh, Thermography for monitoring the selective laser melting process, in *Proceedings of the 23rd Annual International Solid Freeform Fabrication Symposium*, Austin, TX, 2012, pp. 999–1014.
57. J.-P. Kruth, J. Duflou, P. Mercelis, J. Van Vaerenbergh, T. Craeghs, and J. De Keuster, On-line monitoring and process control in selective laser melting and laser cutting, in *Proceedings of the Laser Assisted Net Shape Engineering 5 (LANE)*, 2007, 1, 23–37.
58. G. Bi, B. Schürmann, A. Gasser, K. Wissenbach, and R. Poprawe, Development and qualification of a novel laser-cladding head with integrated sensors, *Int. J. Mach. Tools Manuf.*, 47(3), 555–561, 2007.
59. S. P. Santospirito, K. Słyk, B. Luo, R. Łopatka, O. Gilmour, and J. Rudlin, Detection of defects in laser powder deposition (LPD) components by pulsed laser transient thermography, *SPIE Defense, Security, and Sensing*, 2013, 87050X–87050X–11.
60. R. B. Dinwiddie, R. R. Dehoff, P. D. Lloyd, L. E. Lowe, and J. B. Ulrich, Thermographic in-situ process monitoring of the electron-beam melting technology used in additive manufacturing, in *Proceedings of the SPIE*, 2013, 8705, 87050K–87050K–9.
61. R. B. Dinwiddie, L. J. Love, and J. C. Rowe, Real-time process monitoring and temperature mapping of a 3D polymer printing process, in *Proceedings of the SPIE*, 2013, 8705.
62. S. Price, K. Cooper, and K. Chou, Evaluations of temperature measurements by near-infrared thermography in powder-based electron-beam additive manufacturing, in *Solid Freeform Fabrication Symposium*, 2012, pp. 761–773.

63. M. Pavlov, M. Doubenskaia, and I. Smurov, Pyrometric analysis of thermal processes in SLM technology, *Phys. Procedia*, 5, 523–531, 2010.
64. K. Masubuchi, *Analysis of welded structures: Residual stresses, distortion, and their consequences*, Pergamon Press, New York, 1980.
65. B. Ekmekçi, N. Ekmekçi, A. Tekkaya, and A. Erden, Residual stress measurement with layer removal method, *Meas. Tech.*, 1, 3, 2004.
66. N. W. Klingbeil, J. L. Beuth, R. K. Chin, and C. H. Amon, Residual stress-induced warping in direct metal solid freeform fabrication, *Int. J. Mech. Sci.*, 44, 1, 57–77, 2002.
67. Z. Gan, H. W. Ng, and A. Devasenapathi, Deposition-induced residual stresses in plasma-sprayed coatings, *Surf. Coat. Technol.*, 187(2), 307–319, 2004.
68. G. Branner, M. Zaeh, and C. Groth, Coupled-field simulation in additive layer manufacturing, presented at the *Proceedings of the 3rd International Conference on Polymers and Moulds Innovations*, 2008, pp. 184–193.
69. I. A. Roberts, Investigation of residual stresses in the laser melting of metal powders in additive layer manufacturing, 2012.
70. M. Shiomi, Residual stress within metallic model made by selective laser melting process, *CIRP Ann.*, 53, 195–198, 2004.
71. L. Van Belle, G. Vansteenkiste, and J. C. Boyer, Investigation of residual stresses induced during the selective laser melting process, *Key Eng. Mater.*, 554–557, 1828–1834, 2013.
72. A. L. Cooke and S. P. Moylan, Process intermittent measurement for powder-bed based additive manufacturing, *Proceedings of the 22nd International SFF Symposium-An Additive Manufacturing Conference*, August 2011, pp. 8–11.
73. D. B. Pedersen, L. De Chiffre, and H. N. Hansen, *Additive Manufacturing: Multi Material Processing and Part Quality Control*, Technical University of Denmark, 2013.
74. S. Kleszczynski, J. zur Jacobsmühlen, J. Sehrt, and G. Witt, Error detection in laser beam melting systems by high resolution imaging, Presented at the *Solid Freeform Fabrication Symposium*, Austin, TX, 2012.
75. E. Yasa, J. Deckers, T. Craeghs, M. Badrossamay, and J.-P. Kruth, Investigation on occurrence of elevated edges in selective laser melting, Presented at the *International Solid Freeform Fabrication Symposium*, Austin, TX, 2009, 673–685.
76. K. Abd-Elghany and D. L. Bourell, Property evaluation of 304L stainless steel fabricated by selective laser melting, *Rapid Prototyp. J.*, 18(5), 420–428, 2012.
77. R. Paul and S. Anand, Process energy analysis and optimization in selective laser sintering, *J. Manuf. Syst.*, 31(4), 429–437, 2012.
78. M. Khaing, J. Y. Fuh, and L. Lu, Direct metal laser sintering for rapid tooling: processing and characterisation of EOS parts, *5th Asia Pac. Conf. Mater. Process*, 113(1–3), 269–272, 2001.
79. T. A. Krol, C. Seidel, and M. F. Zaeh, Prioritization of process parameters for an efficient optimisation of additive manufacturing by means of a finite element method, *Eighth CIRP Conf. Intell. Comput. Manuf. Eng.*, 12, 169–174, 2013.
80. J. Delgado, J. Ciurana, and C. A. Rodríguez, Influence of process parameters on part quality and mechanical properties for DMLS and SLM with iron-based materials, *Int. J. Adv. Manuf. Technol.*, 60(5–8), 601–610, 2012.
81. D. Wang, Y. Yang, Z. Yi, and X. Su, Research on the fabricating quality optimization of the overhanging surface in SLM process, *Int. J. Adv. Manuf. Technol.*, 65(9–12), 1471–1484, 2013.
82. H. Meier and C. Haberland, Experimental studies on selective laser melting of metallic parts, *Mater. Werkst.*, 39(9), 665–670, 2008.
83. J. Sehrt and G. Witt, Dynamic strength and fracture toughness analysis of beam melted parts, in *Proceedings of the 36th International MATADOR Conference*, Manchester, UK, 2010, pp. 385–388.
84. S. Storch, D. Nellessen, G. Schaefer, and R. Reiter, Selective laser sintering: Qualifying analysis of metal based powder systems for automotive applications, *Rapid Prototyp. J.*, 9(4), 240–251, 2003.
85. I. Gibson and D. Shi, Material properties and fabrication parameters in selective laser sintering process, *Rapid Prototyp. J.*, 3, 129–136, 1997.
86. A. Wegner and G. Witt, Correlation of process parameters and part properties in laser sintering using response surface modeling, in *Proceedings of the Laser Assisted Net Shape Engineering 7 (LANE)*, 2012, 39, 480–490.

87. D. Manfredi, F. Calignano, M. Krishnan, R. Canali, E. P. Ambrosio, and E. Atzeni, From powders to dense metal parts: Characterization of a commercial AlSiMg alloy processed through direct metal laser sintering, *Materials*, 6(3), 856–869, 2013.
88. I. Yadroitsev and I. Smurov, Selective laser melting technology: From the single laser melted track stability to 3D parts of complex shape, *Phys. Procedia*, 5, Part B, 551–560, 2010.
89. I. Yadroitsev, L. Thivillon, P. Bertrand, and I. Smurov, Strategy of manufacturing components with designed internal structure by selective laser melting of metallic powder, *Appl. Surf. Sci.*, 254, 980–983, 2007.
90. D. Buchbinder, W. Meiners, N. Pirch, K. Wissenbach, and J. Schrage, Investigation on reducing distortion by preheating during manufacture of aluminum components using selective laser melting, *J. Laser Appl.*, 26(1), 012004, 2013.
91. F. Abe, K. Osakada, M. Shiomi, K. Uematsu, and M. Matsumoto, The manufacturing of hard tools from metallic powders by selective laser melting, *J. Mater. Process. Technol.*, 111(1–3), 210–213, 2001.
92. S. Dadbakhsh, L. Hao, and N. Sewell, Effect of selective laser melting layout on the quality of stainless steel parts, *Rapid Prototyp. J.*, 18(3), 241–249, 2012.
93. A. B. Spierings, K. Wegener, and G. Levy, Designing material properties locally with additive manufacturing technology SLM, in *Solid Freeform Fabrication Proceedings*, Austin, TX, 2012, pp. 447–455.
94. M. A. Saleh and A. E. Ragab, Ti-6Al-4V Helical spring manufacturing via SLM: Effect of geometry on shear modulus, in *Proceedings of the International MultiConference of Engineers and Computer Scientists*, Hong Kong, 2013, 2.
95. C. Casavola, S. L. Campanelli, and C. Pappalettere, Preliminary investigation on distribution of residual stress generated by the selective laser melting process, *J. Strain Anal. Eng. Des.*, 44(1), 93–104, 2009.
96. J.-P. Kruth, J. Deckers, E. Yasa, and R. Wauthlé, Assessing and comparing influencing factors of residual stresses in selective laser melting using a novel analysis method, *Proceedings of the Institution of Mechanical Engineers, Part B: Journal of Engineering Manufacture*, 226(6), 980–991, 2012.
97. S. Leuders, M. Thöne, A. Riemer, T. Niendorf, T. Tröster, H. A. Richard, and H. J. Maier, On the mechanical behaviour of titanium alloy TiAl6V4 manufactured by selective laser melting: Fatigue resistance and crack growth performance, *Int. J. Fatigue*, 48, 300–307, 2013.
98. C. Casavola, S. Campanelli, and C. Pappalettere, Experimental analysis of residual stresses in the selective laser melting process, presented at the *2008 SEM International Conference and Exposition on Experimental and Applied Mechanics*, Orlando, FL, 2008.
99. P. Mercelis and J.-P. Kruth, Residual stresses in selective laser sintering and selective laser melting, *Rapid Prototyp. J.*, 12(5), 254–265, 2006.
100. B. Vrancken, R. Wauthlé, J.-P. Kruth, and J. Van Humbeeck, Study of the influence of material properties on residual stress in selective laser melting, *Proceedings of the Solid Freeform Fabrication Symposium*, 2013.
101. M. F. Zaeh and G. Branner, Investigations on residual stresses and deformations in selective laser melting, *Prod. Eng.*, 4(1), 35–45, 2010.
102. T. Sercombe, N. Jones, R. Day, and A. Kop, Heat treatment of Ti-6Al-7Nb components produced by selective laser melting, *Rapid Prototyp. J.*, 14(5), 300–304, 2008.
103. P. J. Withers and H. K. D. H. Bhadeshia, Residual stress. Part 2—Nature and origins, *Mater. Sci. Technol.*, 17(4), 366–375, 2001.
104. C. R. Knowles, T. H. Becker, and R. B. Tait, Residual stress measurements and structural integrity implications for selective laser melted Ti-6AL-4V, *South Afr. J. Ind. Eng.*, 23, 119–129, 2012.
105. M. Thone, S. Leuders, A. Riemer, T. Tröster, and H. Richard, Influence of heat-treatment on selective laser melting products–eg Ti6Al4V, presented at the *Solid Freeform Fabrication Symposium SFF*, Austin, TX, 2012.
106. R. Morgan, C. Sutcliffe, and W. O'neill, Density analysis of direct metal laser re-melted 316L stainless steel cubic primitives, *J. Mater. Sci.*, 39, 1195–1205, 2004.

107. E. Yasa, J. Deckers, and J.-P. Kruth, The investigation of the influence of laser re-melting on density, surface quality and microstructure of selective laser melting parts, *Rapid Prototyp. J.*, 17(5), 312–327, 2011.

108. J. Parthasarathy, B. Starly, S. Raman, and A. Christensen, Mechanical evaluation of porous titanium (Ti6Al4V) structures with electron beam melting (EBM), *J. Mech. Behav. Biomed. Mater.*, 3(3), 249–259, 2010.

109. J.-P. Kruth, M. Badrossamay, E. Yasa, J. Deckers, L. Thijs, and J. Van Humbeeck, Part and material properties in selective laser melting of metals, in *Proceedings of the 16th International Symposium on Electromachining*, Shanghai, China, 2010.

110. H. Gu, H. Gong, D. Pal, K. Rafi, T. Starr, and B. Stucker, Influences of energy density on porosity and microstructure of selective laser melted 17–4 PH stainless steel, in *Solid Freeform Fabrication Proceedings*, Austin, TX, 2013, 37.

111. A. Chatterjee, S. Kumar, P. Saha, P. Mishra, and A. R. Choudhury, An experimental design approach to selective laser sintering of low carbon steel, *J. Mater. Process Technol.*, 136, 151–157, 2003.

112. K. A. Ghany and S. F. Moustafa, Comparison between the products of four RPM systems for metals, *Rapid Prototyp. J.*, 12(2), 86–94, 2006.

113. H. Stoffregen, J. Fischer, C. Siedelhofer, and E. Abele, Selective laser melting of porous structures, in *Solid Freeform Fabrication Proceedings*, 2011, 680–695.

114. C. Casavola, S. Campanelli, and C. Pappalettere, Experimental analysis of residual stresses in the selective laser melting process, *Proceedings of the 10th International Congress and Exposition*, Orlando, FL, 2008.

115. D. Buchbinder, H. Schleifenbaum, S. Heidrich, W. Meiners, and J. Bültmann, High power selective laser melting (HP SLM) of aluminum parts, *Phys. Procedia*, 12, Part A, 271–278, 2011.

116. E. Brandl, U. Heckenberger, V. Holzinger, and D. Buchbinder, Additive manufactured AlSi10Mg samples using Selective Laser Melting (SLM): Microstructure, high cycle fatigue, and fracture behavior, *Mater. Des.*, 34, 159–169, 2012.

117. R. Khalid, N. V. Karthik, T. L. Starr, and B. E. Stucker, Mechanical property evaluation of Ti-6Al-4V parts made using electron beam melting, in Solid *Freeform Fabrication Proceedings*, Austin, TX, 2012, pp. 526–535.

118. P. Lipinski, A. Barbas, and A.-S. Bonnet, Fatigue behavior of thin-walled grade 2 titanium samples processed by selective laser melting. Application to life prediction of porous titanium implants, *J. Mech. Behav. Biomed. Mater.*, 28, 274–290, 2013.

119. A. B. Spierings, T. L. Starr, and K. Wegener, Fatigue performance of additive manufactured metallic parts, *Rapid Prototyp. J.*, 19(2), 88–94, 2013.

120. E. Wycisk, C. Emmelmann, S. Siddique, and F. Walther, High cycle fatigue (HCF) performance of Ti-6Al-4V alloy processed by selective laser melting, *Adv. Mater. Res.*, 816, 134–139, 2013.

121. H. K. Rafi, T. L. Starr, and B. E. Stucker, A comparison of the tensile, fatigue, and fracture behavior of Ti-6Al-4V and 15-5 PH stainless steel parts made by selective laser melting, *Int. J. Adv. Manuf. Technol.*, 69(5–8), 1299–1309, 2013.

122. K. Zeng, D. Pal, and B. E. Stucker, A review of thermal analysis methods in laser sintering and selective laser melting, Presented at the *Solid Freeform Fabrication Symposium*, Austin, TX, 2012.

123. R. B. Patil and V. Yadava, Finite element analysis of temperature distribution in single metallic powder layer during metal laser sintering, *Int. J. Mach. Tools Manuf.*, 47, 1069–1080, 6.

124. A. V. Gusarov, Homogenization of radiation transfer in two-phase media with irregular phase boundaries, *Phys. Rev. B*, 77(14), 144201, 2008.

125. S. Kolossov, E. Boillat, R. Glardon, P. Fischer, and M. Locher, 3D FE simulation for temperature evolution in the selective laser sintering process, *Int. J. Mach. Tools Manuf.*, 44(2–3), 117–123, 2004.

126. I. A. Roberts, C. J. Wang, R. Esterlein, M. Stanford, and D. J. Mynors, A three-dimensional finite element analysis of the temperature field during laser melting of metal powders in additive layer manufacturing, *Int. J. Mach. Tools Manuf.*, 49(12–13), 916–923, 2009.

127. L., Felicelli, S., Gooroochurn, Y., Wang, P. T., Horstemeyer, M. F. Wang, Optimization of the LENS® process for steady molten pool size, *Mater. Sci. Eng. A*, 474, 148–156, 2008.

128. L. Wang and S. Felicelli, Process modeling in laser deposition of multilayer SS410 steel, *J. Manuf. Sci. Eng.-Trans. ASME*, 129, 1028–1034, 2007.
129. J. D. Williams and C. R. Deckard, Advances in modeling the effects of selected parameters on the SLS process, *Rapid Prototyp. J.*, 4(2), 90–100, 1998.
130. D. Rosenthal, *The theory of moving sources of heat and its application to metal treatments*, ASME, Cambridge, MA, 1946.
131. L. Wang and S. Felicelli, Analysis of thermal phenomena in LENS (TM) deposition, *Mater. Sci. Eng.*, 435, 625–631, 2006.
132. A. V. Gusarov and I. Smurov, Modeling the interaction of laser radiation with powder bed at selective laser melting, *Phys. Procedia*, 5, 381–394, 2010.
133. A. Vasinonta, M. L. Griffith, and J. L. Beuth, A process map for consistent build conditions in the solid freeform fabrication of thin-walled structures, *J. Manuf. Sci. Eng.*, 123(4), 615–622, 2000.
134. T. Chen and Y. Zhang, Numerical simulation of two-dimensional melting and resolidification of a two-component metal powder layer in selective laser sintering process, Numer. *Heat Transf. Part Appl.*, 46, 633–649, 2004.
135. T. Chen and Y. Zhang, Thermal modeling of laser sintering of two-component metal powder on top of sintered layers via multi-line scanning, *Appl. Phys. A*, 86(2), 213–220, 2007.
136. A. V. Gusarov, I. Yadroitsev, P. Bertrand, and I. Smurov, Heat transfer modelling and stability analysis of selective laser melting, *Appl. Surf. Sci.*, 254(4), 975–979, 2007.
137. M. Van Elsen, F. Al-Bender, and J.-P. Kruth, Application of dimensional analysis to selective laser melting, *Rapid Prototyp. J.*, 14, 15–22, 2008.
138. C. Körner, E. Attar, and P. Heinl, Mesoscopic simulation of selective beam melting processes, *J. Mater. Process. Technol.*, 211(6), 978–987, 2011.
139. R. Ammer, M. Markl, U. Ljungblad, C. Körner, and U. Rüde, Simulating fast electron beam melting with a parallel thermal free surface lattice Boltzmann method, in *Proceedings of ICMMES-International Conference for Mesoscopic Methods for Engineering and Science*, Taipei, Taiwan, 2012, 67, 318–330.
140. E. Attar, Simulation of Selective Electron Beam Melting Processes, Dr.-Ing., University of Erlangen, Nuremberg, Germany, 2011.
141. W. Zhou, D. Loney, A. G. Federov, F. Degertekin, and D. Rosen, Lattice Boltzmann simulations of multiple droplet interactions during impingement on the substrate, in *Solid Freeform Fabrication Proceedings*, Austin, TX, 2013, pp. 606–621.
142. A. Hussein, L. Hao, C. Yan, and R. Everson, Finite element simulation of the temperature and stress fields in single layers built without-support in selective laser melting, *Mater. Des.*, 52, 638–647, 2013.
143. L. Dong, A. Makradi, S. Ahzi, and Y. Remond, Three-dimensional transient finite element analysis of the selective laser sintering process, *J. Mater. Process. Technol.*, 209, 700–706, 2009.
144. J. Yin, H. Zhu, L. Ke, W. Lei, C. Dai, and D. Zuo, Simulation of temperature distribution in single metallic powder layer for laser micro-sintering, Comput. *Mater. Sci.*, 53(1), 333–339, 2012.
145. E. Soylemez, J. L. Beuth, and K. Taminger, Controlling melt pool dimensions over a wide range of material deposition rates in electron beam additive manufacturing, in *Solid Freeform Fabrication Proceedings*, Austin, TX, 2010, 571–582.
146. T. H. C. Childs, C. Hauser, and M. Badrossamay, Selective laser sintering (melting) of stainless and tool steel powders: Experiments and modelling, *J. Eng. Manuf.*, 219(4), 339–357, 2005.
147. M. Shiomi, A. Yoshidome, F. Abe, and K. Osakada, Finite element analysis of melting and solidifying processes in laser rapid prototyping of metallic powders, *Int. J. Mach. Tools Manuf.*, 39(2), 237–252, 1999.
148. R. Li, Y. Shi, J. Liu, H. Yao, and W. Zhang, Effects of processing parameters on the temperature field of selective laser melting metal powder, *Powder Metall. Met. Ceram.*, 48(3–4), 186–195, 2009.
149. L. Wang, S. D. Felicelli, and J. Craig, Thermal modeling and experimental validation in the LENS process, in *Solid Freeform Fabrication Proceedings*, Austin, TX, 2007, 100–111.
150. D. Gu and Y. Shen, Effects of processing parameters on consolidation and microstructure of W– Cu components by DMLS, *J. Alloys Compd.*, 473, 107–115, 2009.
151. L. Van Belle, G. Vansteenkiste, and J. C. Boyer, Comparisons of numerical modelling of the selective laser melting, *Key Eng. Mater.*, 504–506, 1067–1072, 2012.

152. M. Labudovic, D. Hu, and R. Kovacevic, A three dimensional model for direct laser metal powder deposition and rapid prototyping, *J. Mater. Sci.*, 38(1), 35–49, 2003.
153. M. Matsumoto, M. Shiomi, K. Osakada, and F. Abe, Finite element analysis of single layer forming on metallic powder bed in rapid prototyping by selective laser processing, *Int. J. Mach. Tools Manuf.*, 42, 61–67, 2002.
154. K. Guan, Z. Wang, M. Gao, X. Li, and X. Zeng, Effects of processing parameters on tensile properties of selective laser melted 304 stainless steel, *Mater. Des.*, 50, 581–586, 2013.
155. A. H. Nickel, D. M. Barnett, and F. B. Prinz, Thermal stresses and deposition patterns in layered manufacturing, *Mater. Sci. Eng. A*, 317(1–2), 59–64, 2001.
156. L., Felicelli, S., Pratt, P. Wang, Residual stresses in LENS-deposited AISI 410 stainless steel plates, *Mater. Sci. Eng. A*, 496, 234–241, 2008.
157. P. Pratt, S. Felicelli, L. Wang, and C. Hubbard, Residual stress measurement of laser-engineered net shaping AISI 410 thin plates using neutron diffraction, *Metall. Mater. Trans. A*, 39, 3155–3163, 2008.
158. S. Bontha, N. W. Klingbeil, P. A. Kobryn, and H. L. Fraser, Thermal process maps for predicting solidification microstructure in laser fabrication of thin-wall structures, *J. Mater. Process. Technol.*, 178(1), 135–142, 2006.
159. P. A. Kobryn and S. L. Semiatin, Microstructure and texture evolution during solidification processing of Ti–6Al–4V, *J. Mater. Process. Technol.*, 135(2–3), 330–339, 2003.
160. J. Gockel and J. L. Beuth, Understanding Ti-6Al-4V microstructure control in additive manufacturing via process maps, in *Solid Freeform Fabrication Proceedings*, Austin, TX, 2013, 666–674.
161. J. L. Beuth, J. Fox, J. Gockel, C. Montgomery, R. Yang, H. Qiao, E. Soylemez, P. Reeseewatt, A. Anvari, and S. Narra, Process mapping for qualification across multiple direct metal additive manufacturing processes, in *Solid Freeform Fabrication Proceedings*, Austin, TX, 2013, 655–665.
162. L. Song and J. Mazumder, Real time Cr measurement using optical emission spectroscopy during direct metal deposition process, *Sens. J. IEEE*, 12(5), 958–964, 2012.
163. L. Song, V. Bagavath-Singh, B. Dutta, and J. Mazumder, Control of melt pool temperature and deposition height during direct metal deposition process, *Int. J. Adv. Manuf. Technol.*, 58(1–4), 247–256, 2012.
164. K. Bartkowiak, Direct laser deposition process within spectrographic analysis in situ, in *Proceedings of the Laser Assisted Net Shape Engineering* 6 (LANE), 2010, 5, Part B, 623–629.
165. E. Louvis, P. Fox, and C. J. Sutcliffe, Selective laser melting of aluminium components, *J. Mater. Process. Technol.*, 211(2), 275–284, 2011.

chapter thirty five

Denudation of metal powder layers in laser powder-bed fusion processes

*Manyalibo J. Matthews, Gabe Guss, Saad A. Khairallah,
Alexander M. Rubenchik, Philip J. Depond, and Wayne E. King*

Contents

Abstract: Understanding laser interaction with metal powder beds is critical in predicting optimum processing regimes in laser powder-bed fusion (PBF) additive manufacturing (AM) of metals. In this work, we study the denudation of metal powders that is observed near the laser scan path as a function of laser parameters and ambient gas pressure. We show that the observed depletion of metal powder particles in the zone immediately surrounding the solidified track is due to a competition between outward metal vapor flux directed away from the laser spot and entrainment of powder particles in a shear flow of gas driven by a metal vapor jet at the melt track. Between atmospheric pressure and ~10 Torr of Ar gas, the denuded zone width increases with decreasing ambient gas pressure and is dominated by entrainment from inward gas flow. The denuded zone then decreases from 10 to 2.2 Torr reaching a minimum before increasing again from 2.2 to 0.5 Torr where metal vapor flux and expansion from the melt pool dominates. The dynamics of the denudation process were captured using high-speed imaging, revealing that the particle movement is a complex interplay among melt-pool geometry, metal vapor flow, and ambient gas pressure. The experimental results are rationalized through FE simulations of the melt-track formation and resulting vapor flow patterns. The results presented here represent new insights to denudation and melt-track formation that can be important for the prediction and minimization of void defects and surface roughness in additively manufactured metal components.

35.1 Introduction

Laser powder-bed fusion (LPBF) of metal powders is currently the dominant method for producing 3D-printed metal structures. Although the new design freedoms afforded by AM of components directly from digital files has impacted multiple areas of industry, the resulting properties of the material in the printed part generally do not match those of wrought or cast metal [1,2]. On the one hand, grain refinement hardening or internal dislocation effects through the rapid thermal cycling typical of LPBF can lead to stronger materials compared with more traditional methods [3]. On the other hand, voids associated with keyhole-mode melting (due to strong vaporization) or the incomplete melting of powder (i.e., lack of fusion defects) can have significant and negative effects on mechanical properties such as fatigue [4]. Residual stress, nonequilibrium material phase, and high surface roughness are also known to degrade ultimate part performance. It is well known that the scan strategy used (laser power, beam size, scan speed, and hatch spacing) can have a strong effect on porosity and void generation. In particular, careful choice of the hatch spacing is important in order to avoid linear void structures associated with powder denudation effects [5,6]. Denudation, or the apparent clearing of powder around a single-track bead, has been observed in the literature, but to our knowledge, the detailed physics that produces denudation has not been reported [5,7–9]. Moreover, current modeling and interpretation of the experimental data is focused on only heating and melting of the powder, without regard to complete two-phase flow behavior that includes the ambient gas. As we will show here, the ambient gas and the induced powder motion is in fact important for both the denudation process and for the incorporation of powder into the melt track that forms the building block of a LPBF manufactured part.

In this work, we study the denudation of Ti alloy and steel alloy powders under varying laser conditions and ambient gas pressures using a pressure-controlled single-track test chamber and high-speed imaging. Our principal finding is that, for a typical LPBF environment (Ar gas at 760 Torr), the dominant driving force for denuding powder near a melt track is the entrainment of particles by surrounding gas flow. The flow is induced by the intensive evaporation that occurs within the laser spot and pressure drop inside the associated vapor jet due to the Bernoulli effect. To a lesser extent, particles within a few particle widths of the melt track can also be consumed through direct contact with the liquid metal and capillary forces. The vapor-driven entrainment causes particles not only to be incorporated into the melt track, an effect intimately related to final bead size of the track, but also to eject vertically and rearward, relative to laser scan direction and redistribute elsewhere on the powder bed. Inbound particles, under certain conditions, can impact the melt pool and yet remain semi-solid, resulting in track roughness. When ambient gas mass flux is reduced through a reduction of pressure below ~50 Torr, the induced convective flow first increases then vanishes below ~10 Torr, revealing an outward expulsion of nearby particles away from the laser spot due to direct vapor momentum transfer from the melt pool. Interestingly, a minimum in denudation width occurs are ~4 Torr at which the effects balance and the denuded zone is very sharply defined in comparison to higher pressures. Our experimental results are supported by powder-scale FE modeling of the laser-powder-melt interaction. The findings presented here shed new light on denudation effects that can lead to void defects and layer nonuniformity, and help explain some unresolved observations in the literature related to process-property correlation and highlight the deficiency of existing single phase flow models.

35.2 Experimental details

A 600 W fiber laser (JK lasers, model JK600FL) is directed through a three-axis galvanometer scanner (Nutfield technologies, Hudson, United States) and into a 15 × 15 × 15 cm³ vacuum chamber through a high purity fused silica window. The ~f/20 optical system results in a focused $D4\sigma$ diameter of ~50 μm at the sample. The vacuum chamber is evacuated using a turbomolecular pump and purged with argon. Residual oxygen content is measured using a photoluminescence quenching meter (Ocean Optics Neo Fox) and the concentration of oxygen was below 0.01%, the lower limit of the sensor's measurement range. Pressures between 0.5 and 500 Torr were controlled using a combination of purge and pumping rates. A separate setup was used outside the chamber at ~760 Torr of flowing Ar gas to image the melt-pool formation using an off-axis (~45°) high-speed camera (Shimadzu HPV-2), microscope optics (Mititoyo 10x/0.28NA, Infinity K2), and a 10 nm band pass filter centered at 638 nm to reject incandescent emission from the melt pool. Imaging was performed at 500,000 frames per second, with an optical resolution of ~5 μm. A ~1 W, 638 nm diode laser was used to illuminate the surface. In both configurations, a 25.4 mm diameter, 3.2 mm thick build plate of the same composition as the powder with a bead-blasted and ultrasonic cleaned surface was used. For the pressure-dependent studies, gas atomized ~30 μm Ti-6Al-4V powders (*Ti64*) were manually applied using a stainless steel (SS) razor edge to simulate the spreading process in production machine. Nominal thicknesses for the metal powder layers for all measurements were ~60 μm. In addition to the Ti64 samples, samples of 316L SS and pure Aluminum powders were also prepared in the same way for ambient pressure and high-speed imaging measurements.

35.3 Model for simulating melt-pool dynamics and vapor flow patterns

The multiphysics simulation code (ALE3D, Lawrence Livermore National Laboratory) was used to simulate melt-pool formation [10]. Complete details of the method are given in [11]. ALE3D uses an operator splitting approach to advance the simulation. The Lagrangian-motion component moves the material adiabatically in response to forces, using single point Gauss quadrature for strain mapping and nodal force calculations based on face normal. The thermal component moves heat within the materials without any material motion, using nodal temperature-based integration. The advection component remaps the mesh back to its original configuration. All three components occur sequentially at every time step. All boundaries of the computational domain are fixed for the hydrodynamics component. In the thermal component, all boundaries except the bottom are treated as insulated. The bottom face used a custom thermal boundary condition that mimics the response of a semi-infinite body at this interface. The laser energy deposition model uses the simplified version of ray tracing [12]. The code takes into account the evaporation process including the recoil momentum produced by metal vapor ejection and evaporation cooling. The code also describes the melt motion induced by the recoil pressure and by the surface tension including melt spattering effects. Not included in the code at present is the effect of the ambient gas explicitly (complete two-phase flow) and as a result, as shown below, some physics is missed in the modeling and our experimental results stimulates further code development.

35.4   Results

Figure 35.1a shows a wide view optical image of displaced powder surrounding melted tracks created by laser scanning at 2 m/s, incident powers ranging from 10 to 350 W and at 0.2 Torr of Ar gas. The lighter contrast in the image represents the resolidified melt, whereas the background surface is comprised of powder particles. The powder thickness in the image shown in Figure 35.1a is ~60 μm. As a function of increasing power, the width of both the track and the denuded zone (DZ) increases. The width of the DZ was analyzed using a particle detection method described in the supplemental material. The width of the resolidified melt track was measured using confocal height microscopy. Figure 35.1b

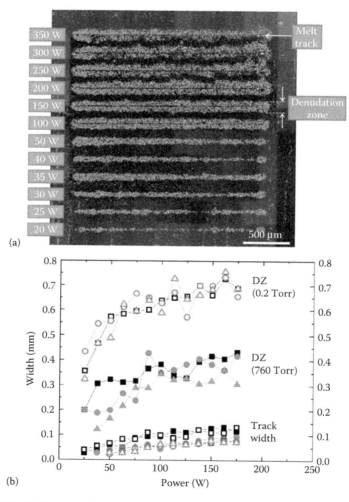

Figure 35.1 (a) Wide field image of denuded zones around melt tracks created by LPBF as a function of laser power and at a scan rate of 2 m/s. The melted track appears as a shiny semi-continuous line. The denuded zone surrounds each track and appears dark in contrast above the track and light in contrast below the track. (b) Measured denuded zone (DZ) and resolidified track widths as a function of laser power, scan rate and ambient Ar pressure. Open symbols represent 0.2 Torr, while the solid symbols represent 760 Torr. Scan rates: square ¼ 0.5 m/s, circle ¼ 1.5 m/s, triangle ¼ 2 m/s.

shows the laser power and scan rate dependence of both the DZ and the resolidified track, under different ambient pressures (0.2 and 760 Torr). A factor of almost 2x increase in denudation width is observed with a decrease of pressure from 760 to 10 Torr, whereas the melt-track width remains roughly constant as a function of pressure. Apparent in the low-pressure case is a change in DZ width slope with laser power. Interestingly, the laser scan rate appears to have little effect on the DZ width, whereas the effect on melt-track width is more noticeable, particularly in going from 0.5 to 1.5 m/s.

To better resolve the DZ as a function of ambient Ar pressure, high-resolution imaging and laser confocal scanning profilometry were performed on a set of single tracks at 1.4 m/s and 225 W laser scan rate and power respectively. The optical micrographs shown in the top portion of Figure 35.2 display the progression of the DZ and local powder morphology, as pressure is decreased from 220 to 0.5 Torr. The data shown in Figure 35.2 was derived from the same sample, spread from a single dose of powder. Powder morphologies observed between 220 and 760 Torr were virtually identical and therefore not included in the figure. At 220 Torr, the DZ is visible and roughly 500 μm in width but the resolidified melt track is barely visible due to an overlay of powder. As pressure decreases from 220 to 10 Torr, the DZ increases and the overlay of powder over the track becomes lighter. The DZ then *decreases* with decreasing pressure from 10 to 2.2 Torr, with an abrupt change between

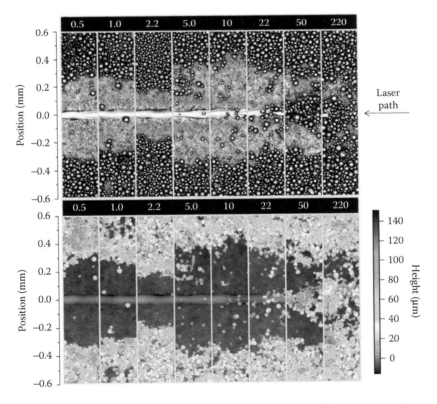

Figure 35.2 Montage of 1.2–0.25 mm optical micrographs (top) and height maps (bottom) of the solidified melt track within a powder layer following scanning laser exposure at 225 W and 1.4 m/s as a function of ambient Ar pressure (shown above image slices in Torr). Three distinct regions can be identified near the laser path center, namely track accumulation zone, the denuded zone (DZ), and the background powder zone.

5 and 2.2 Torr. We note that for the 2.2 Torr case, the DZ zone edge is very clearly defined and relatively powder-free. The DZ then increases again for pressures below 2.2 Torr, and powder particles again appear distributed between the melt track and DZ edge. We note that the size distribution of particles that remain in the DZ varies with pressure as well, with smaller particle content increasing with decreasing pressure between 5 and 220 Torr. Conversely, below 2.2 Torr, larger particles appear to be favored in the DZ. The bottom portion of Figure 35.2 shows the corresponding height map for the optical micrograph and quantifies the pileup of powder particles on the melt track at the higher pressures (22–220 Torr). Note that between 5 and 220 Torr, the edge of the powder layer appears near the nominal thickness of ~60 μm, whereas from 2.2 to 0.5 Torr the powder thickness at the DZ edge is ~2x this value, suggesting that material has been pushed away from the track center. We note that the particle size distribution in the powder surrounding the DZ appears qualitatively the same for all cases except the 2.2 Torr case, where the powder particles appear somewhat smaller. Assuming that smaller particles are more easily displaced by the forces that create the DZ than larger particles, their apparent prevalence at 2.2 Torr suggests that these forces are at a relative minimum, which is consistent with the minimum width of the DZ at this pressure.

To probe in more detail the dramatic change in DZ width at low pressures, we varied the pressure in steps as small as 0.1 Torr near the transition point at ~5 Torr. We used a particle detection algorithm to locate particles within the DZ, and used the 50% transition point from the low density in the DZ to the surrounding powder layer to define the DZ width (see Supplemental Material for additional details). Figure 35.3 shows plotted along the right axis the (log) pressure dependence of the fractional change in particle density (i.e., fraction of particles displaced relative to the original powder density) within the DZ over the central 1.5 mm of a 3 mm track. The left axis of Figure 35.3 shows the change in DZ width as a function of (log) ambient Ar pressure. Between 500 and ~35 Torr the DZ

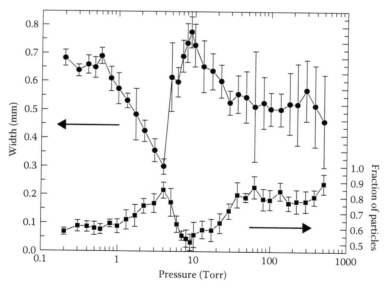

Figure 35.3 Denuded zone (DZ) width as a function of ambient Ar pressure, displayed on left axis. The right axis shows the estimated fraction of detected particles within the DZ. In both cases, the error bars displayed correspond to the standard deviation of the sampled data. See Supplemental Material for further details.

is measured to be ~500 μm, but as observed in the images of Figure 35.2, the DZ changes substantially below 35 Torr. Specifically, we find that the DZ increases with decreasing pressure to ~800 μm at 10 Torr, then to ~300 μm at 4 Torr before increasing again to ~650 μm for pressures between 0.2 and 0.6 Torr. We note that the decrease in DZ width is particularly large (~50%) as pressure is decreased from 5 to 4 Torr, whereas the decrease in DZ from 3 to 4 Torr is more modest (~14%). Interestingly, the increase in DZ width below 4 Torr is found to follow to scale with pressure P as ~$P^{-0.05}$. In both the DZ width and density fraction data, the error bars shown reflect the standard deviation in particle density along the 1 mm length sampled at 50 μm intervals (30 samples).

A second configuration was used in order to resolve powder dynamics at atmospheric (760 Torr) pressure. A laminar Ar flow was established over a 60 μm thick powder layer that was placed under a microscope and imaged with a high-speed camera. Complete video recordings of several measurements can be found in the online supplemental material (see *high-speed video* files). Images were recorded at 500,000 frames per second with and illuminated using a continuous wave laser diode. Figure 35.4 displays a series of images taken at the beginning and end of a 200 μs capture showing a melt pool forming and *traveling* left to right as the laser spot is scanned at 2 m/s with an incident power of 105 W through a Ti64 powder bed. The orange arrow indicates the location of the laser path. One can observe that as the melt pool makes contact with the powder in the immediate vicinity of the laser path, particles are melted and incorporated into the melt pool. However, a second process of powder incorporation into the melt pool comes by way of particles moving toward the melt pool, as we discuss in detail below.

Highlighted in the background of the +0 and +200 μs frames in Figure 35.4 are selected individual powder particles (red dots) and their trajectory over 200 μs as the laser beam

Figure 35.4 High speed imaging of melt track progression and powder movement under the influence of the hot vapor Bernoulli effect. The laser scan path over 200 ms is shown as the horizontal arrow (in orange) at þ0 ms in the upper left, and initial selected particle positions shown by red dots. After þ200 ms (top right image) the particles highlighted at þ0 ms have been displaced, with a fraction of them colliding with the melt pool (shown as blue arrow trajectories) and being incorporated into the melt and the remaining fraction being swept away from the melt pool (shown as yellow arrow trajectories) in a rearward direction relative to the laser scan direction. The series of four images at the bottom of the montage display intermediate times, with an orange arrow indicating the location of the laser progressing left to right through the frames. The images shown were captured at 500 k fps with an illumination diode bandpass filter to block melt pool incandescence, with a laser scanning speed and incident power of 2 m/s and 105 W respectively.

passes nearby creating a melt pool. A clear initial movement of particles toward the approximate location of the laser spot (and near the maximum in surface temperature) is observed. As shown at +200 µs, a fraction of the inbound particles collide with the melt pool and are thus incorporated into the melt (shown by cyan paths), whereas the remaining fraction of particles are swept above and away from the melt pool in a rearward direction relative to the laser scan direction. This removal of nearby powder particles and motion of powder particles farther away to partially fill the zone of swept away particles is consistent with the appearance of *piled up* particles at high pressures observed in Figure 35.2. Incidentally, as is more clearly captured in the supplemental material, one can also observe *spatter* from the melt pool as molten droplets are ejected at speeds of up to ~10 m/s (an additional high-speed video data set in the supplementary material *SS316L montage* from SS particles in which the laser band pass filter is removed most clearly highlights the faster moving spatter that appear white due to incandescence). In contrast, the cold particle motion captured in Figure 35.4 occurs at speeds closer to ~2 m/s. Although most of the hot droplets ejected from the melt pool were ejected rearward and were similar in size to the original powder, liquid droplets could be observed ejected forward and were much larger (similar in size to the track width, 60–100 µm). A detailed discussion of this spatter behavior is beyond the scope of this chapter. However, it is noted that material spatter, particularly large droplets, can have negative effects on the material and mechanical properties of final LPBF parts [13,14].

From the high-speed imaging data, we observe that inward particle motion leads to (1) addition of powder to the melt track not immediately in the laser beam path and (2) dispersal of powder upward and rearward thus contributing further to a denuded zone. Although the high-speed imaging was only performed at atmospheric pressures, the lack of pile up of material around the fusion zone below ~50 Torr in the images of Figure 35.2 implies that less material is incorporated into the melt track and, as a consequence, we expect melt heights and volumes to increase with increasing ambient pressure. Figure 35.5 shows the average height of the resolidified tracks as a function of pressure corresponding to the data in Figures 35.2 and 35.3 (1.4 m/s, 225 W), after the ~60 µm powder layer was brushed off of the sample. The error bars shown in the Figure 35.5

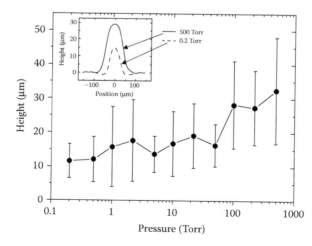

Figure 35.5 Average height of the resolidified melt tracks as a function of pressure for u ¼ 1.4 m/s and P ¼ 225 W. The error bars correspond to the standard deviation of the measured height along 3 mm of track at a 1 mm sampling interval. The inset shows the average height lineout for 0.2 and 500 Torr.

correspond to the standard deviation of the measurements along the 3 mm track at 1 μm sampling intervals. At near-atmospheric pressure (500 Torr), the track height is ~30 μm, decreasing to ~15 μm near 100 Torr, and gradually decreasing nonmonotonically to ~11 μm at 0.2 Torr. We note that both the height and the variation in the standard deviation are somewhat consistent with the morphologies observed in Figure 35.2: in cases where the DZ is relatively powder free, both the height and the standard deviation are at a local minimum pointing to the stochastic nature of powder incorporation through apparent flow-driven entrainment. Although not shown, no significant difference was observed between the 1.4 m/s, 225 W data shown in Figure 35.5 and the other two cases studied (0.6 m/s, 150 W and 2 m/s, 350 W).

The effect of denudation on the deposition pattern produced by multiple, overlapping tracks has been studied in the past [7,15] and is relevant when considering a complete AM process. For example, Thijs et al. noted the presence of elongated pores (as shown in Figure 35.5b of [7]) that they attributed to powder denudation and accumulation of surface roughness between layers. In Figure 35.6, we present the multiple track patterns for different ambient pressures and track overlap of ~30%, laser power of 225 W, and scan speed of 1.4 m/s. Figure 35.6a displays the optical micrograph, whereas Figure 35.6b shows the height maps corresponding to the images in Figure 35.6a. We can see that at high pressures, the melting and spreading of the additional powder supplied by the vapor flow allow for deposition layers roughly 30 to 50 μm thick, whereas at low pressures only the first 1–2 scans produce deposition tracks with any appreciable height. At the highest pressure shown, 500 Torr, a gradient in height is observed with the height decreasing from 60 μm at the first track to less than 20 μm for the last track. This behavior is consistent with the pile up of powder near a resolidified track following a single scan, because material pulled away in the DZ to build early tracks leaves later tracks relatively powderless and unable to produce thick tracks. It is worth noting that this effect has been observed for both single track experiments on bare plates [5], as well as multiple track scans performed in a commercial PBF system [16]. In particular, Yadroitsev et al. [16] showed that denudation leads

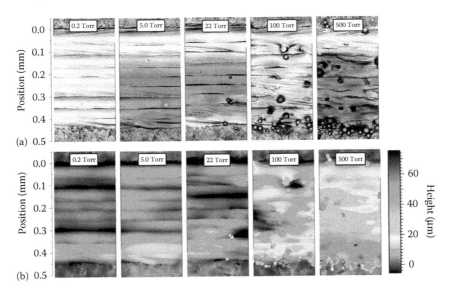

Figure 35.6 Montage of micrographs (a) and confocal height maps (b) of 9 overlapping laser tracks as a function of ambient pressure.

to a decrease in track height with successive scans which are strongly affected by hatch spacing. Thus, when interpreting the single track data presented here (and elsewhere), it is important to consider how denudation plays a role in determining optimal process parameters and scan strategies.

Turning now to the powder-scale model simulation results, Figure 35.7a displays a cut-through view of a simulated melt track within a full layer of metal powder for power of 300 W, $D4\sigma$ beam diameter of 54 μm, and scan speed of 1.8 m/s where the laser beam is traveling left to right. The times displayed in Figure 35.7a correspond to time since the initiation of the track at the left of the simulated domain. As mentioned above, an explicit two-phase flow with gas is not simulated. However, in terms of vapor pressure calculations, an ambient pressure of 760 Torr is used. One can see a consolidated melt track roughly 96 μm in width, preceded by a 68 μm deep depression created through vapor recoil pressure. Under these conditions, the DZ is simulated to be on the order of 2 particle widths (~60 μm) and much smaller than observed experimentally at any ambient pressure. Nonetheless, capillary action of nearby melted particles leads to a small DZ on the order of the size of an individual particle. The color scale in Figure 35.7a corresponds to the log of the mass flux density of the vapor, peaking at ~1000 mole/m²s. Notably, the peak in the vapor flux does not occur at the bottom of the depression but at the leading

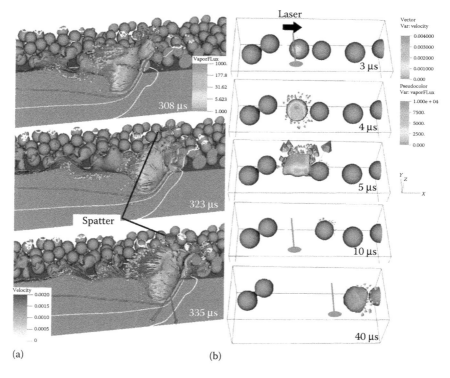

(a)   (b)

Figure 35.7 Powder-scale finite element model simulations of laser powder bed fusion of metal powder. (a) Log of vapor flux [moles/m²s] overlaid on simulated powder bed morphology showing the extent and directionality of the metal vapor at three snapshots in time after the beginning of the track. The velocity [m/s] vectors indicate surface motion. (b) Simulations of an edge illuminated particle near the melt track being ejected through self-recoil vapor pressure and traveling outward towards nearby powder particles.

edge wall (right side of depression in Figure 35.7a), thus directing vapor motion upward and backward. Also indicated in Figure 35.7a by arrows is the surface motion induced by both recoil pressure and Marangoni convection.

Figure 35.7b consists of simulation snapshots of an array of powder particles interacting with a laser beam of power of 300 W, $D4\sigma$ beam diameter of 54 µm, and scan speed of 2 m/s with the laser moving from left to right. The particles are initially slightly displaced from the centerline, so as to put in evidence the particle motion as a result of a lateral force. The times shown in Figure 35.7b indicate elapsed time since the initiation of the track. Overlaid on the Figure 35.7b in pseudocolor is the log of the vapor flux magnitude and directional arrow vectors associated with the motion of each grid point in the simulation domain. As the beam approaches a particle, it heats the particles nonuniformly. As the beam radius is on the order of a particle diameter, this nonuniform heating leads to a strong lateral (y direction) thermal gradient. Under typical processing conditions, the deposited energy is enough to heat one side of a particle to the boiling temperature. Due to the exponential dependence of vapor flux on temperature, a highly localized recoil force is created on the side of the particle at time 5 µs (center frame of Figure 35.7b). This force is strong enough to propel the particle sideways at speeds close to 20 m/s and through momentum transfer with neighboring particles might explain the denudation and pile up effects occurring at low pressures.

35.5   Discussion

Prior to this study, a clear picture of the causes for denudation was not available. Moreover, the effect of pressure on the track consolidation process was not explored whatsoever. We now discuss the underlying physics driving the phenomena observed in the high-speed video and in *ex situ* morphology characterization as a function of pressure. The principal processes to address are laser absorption into the powder, melt-pool formation, and evaporation and motion of the two-phase fluid system. The laser parameters used in this study (50–300 W, 0.5–2 m/s, 50 µm $D4\sigma$ diameter) correspond to irradiance levels near ~10 MW/cm^2 and effective dwell times (beam diameter divided by scan speed) of 10's of µs. It can be shown that the time required to melt a 30 µm diameter Ti64 metal sphere in loose contact with surrounding material is approximately 2 µs. As a result, particles melt mainly on the front edge of the laser spot and most of the beam interacts with the melt (see modeling Figure 35.6 and experimental Figure 35.4).

Several complex processes influence the interaction of the rapidly formed melt pool and the surrounding particles. The melted metal wets the substrate and the surface tension will spread the melt pool around the beam path. This capillary-driven motion and thermal transport into the substrate leads to a melt-track width that is typically 2 to 3 times that of the incident laser beam. Additionally, the melt surface is easily heated over the boiling point leading to vapor recoil momentum [17] that can further drive the melt outward. Marangoni convection, that will tend to drive an outward, circulating flow in the melt pool, is also known to be important [18]. When the melt encounters a particle, wetting and particle melting rapidly occurs. Surface tension then acts to *pull* melted particles into the melt pool, and further adds to the dynamic motion of the liquid around the laser spot. Correspondingly, a small denudation zone is created through the immediate incorporation of particles into the melt pool. However, this effect involves only the particles directly interacting with the melt over a distance of approximately the particle diameter, and cannot explain the much wider denudation zone observed experimentally.

We now consider the action of the metal vapor flux on the melt pool and solid metal particles. As the laser spot size is comparable to the size of the powder particles, fluid instabilities driven by vapor recoil and spatially varying absorptivity [19] lead to highly dynamic motion of the melt pool that can be seen in the supplemental video material. The temperature-dependent recoil pressure exerted on the melt pool due to evaporated metal atoms is given by [20]

$$P(T) = P_0 e^{\left[\chi\left(\frac{1}{T_b} - \frac{1}{T}\right)\right]}$$ (35.1)

where P_0 is the ambient gas pressure, χ is the evaporation energy per atom, and T_b is the boiling point of the liquid metal. For 316L, $\chi \sim 4.3$ eV and $T_b = 3560$ K. We then employ our powder-scale simulations to calculate the peak surface temperature for a 300 W laser power and 1.8 m/s scan speed to arrive at $T \sim 4000$ K and recoil pressures around $4.7P_0$. The pressures exerted on the liquid surface drive fluid motion at velocities given approximately by [21]

$$u \sim \sqrt{\frac{2P}{\rho}}$$ (35.2)

where ρ is the liquid metal density. Using Equation 35.2, we find that melt-pool velocities are approximately 10 m/s, agreeing well with the observed motion captured in the high-speed video. We should emphasize however, that these are only rough approximations to the general behavior and that the dynamics and spatial nonuniformity of the laser heating neglected here could play an important role. Nonetheless, the vapor pressure and associated flux generated are sufficient to create localized fluid flow that can interact with the surrounding powder layers.

Both the experimental data and modeling presented thus far emphasize that the denudation is a complex phenomena driven by a few competing effects. The vapor flow induces an inward ambient gas flow that entrains particles and results in the denudation zone. On the other hand, a few effects can act to move the particles away from the track also leading to denudation. We already discussed the particle side ejection (Figure 35.7b) due to the nonuniform illumination as a possible mechanism. It was also demonstrated by modeling (see Figure 35.7a) that the vapors from a fully developed melt pool are ejected partially back but also sideward due to the curved melt-pool surface. Vapor flow from the curved melt-pool surface can thus move particles outside the track in multiple directions. In addition, metal vapor flux emitted from the melt pool at low pressures may sufficiently expand into a wide plume as compared to a more narrow jet at higher pressures. At high ambient pressure the experimental results demonstrate that the inward particle motion results from a metal vapor jet and Bernoulli effect-driven gas flow, as depicted in Figure 35.8; at low pressure this mechanism is arrested and denudation could be explained by outward gas expansion and the ejection of particles from the track region leading to particle pile up on the edge of denudation zone. Despite the unavailability of high-speed video data at low pressures, the pressure-dependent experiments help to clarify the relative roles of the competing effects that are at play during the LPBF process.

Our experiments demonstrate the important effect of the surrounding gas on the powder particle movement for pressures above ~5 Torr. As we mentioned above, at typical experimental parameters we have intensive evaporation of metal within the melt pool. Metal vapor is expelled normal to the surface with a velocity close to the thermal velocity associated with the surface temperature. Our simulations give a maximum surface temperature estimate for Ti64 of about 3000 K at 225 W, 1.4 m/s and a maximum vapor velocity

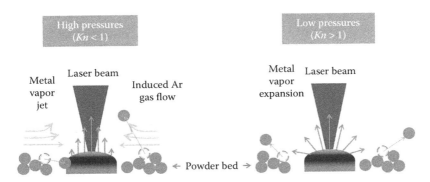

Figure 35.8 Schematic depicting the action of evaporated metal flux on the flow pattern of the surrounding Ar gas and displacement of particles in the powder bed. At high pressures the Knudsen number (Kn) is small as shown on the left, and the low pressure zone near the melt pool caused by high speed metal vapor flux induces Ar gas flow through the Bernoulli effect which in turn results in powder particle entrainment. As shown in the left diagram for $Kn < 1$, particles are either drawn into the melt pool, adding to melt pool material consolidation, or are ejected upward (and rearward) as shown in the high speed video. Conversely, at low pressures where the surrounding Ar gas transport is defined by a molecular flow, Kn is large and the metal vapor flux can expand laterally which will tend to push powder particles outward with minimal material consolidation into the melt pool. Particle collisions from glancing irradiation conditions as discussed in the text can also act to expel powder outward.

of about 700 m/s. The localized directed ejection of vapor is an example of a well-known *submerged jet* problem discussed in detail by Landau [21] and numerically modeled for laser generated plumes by Ho et al. [22]. However, using a velocity of 700 m/s and estimating a vapor stream size similar to the laser spot size we derive a Reynolds number Re~3500 implying a highly turbulent flow, thus deviating from the simple case of a laminar submerged jet [19,21]. Balance of momentum and mass indicates that the jet pulls in the surrounding gas [19]. In the submerged jet case, the jet propagates within a narrow cone with an angle of ~25 degrees thus defining the region over which the surrounding gas is drawn in. The total flux in the jet, Q, grows as $Q\mu x$, where x is the distance from the surface and the average jet velocity drops as $1/x$. The velocity of the incoming Ar gas flow over the adjacent powder is 10's of times smaller than the metal vapor jet velocity [21], about 10 m/s in our case, but can be high enough to entrain the particles and explain the observations for pressures greater than ~10 Torr. The entrainment of particles in the convective flow that was observed in our experiments can be expected from the arguments given above and will lead to particles far (several particle diameters) from the laser spot moving toward and becoming consolidated with the melt pool. As depicted in Figure 35.8 for a simplified flow pattern, those particles that can attain significant vertical momentum from particle-particle collisions or are entrained by the vapor stream will tend to follow the vapor flow and are expelled, whereas those scattered downward or with negligible vertical momentum will collide and are available to merge with the melt pool. Thus, in terms of source material for the final weld bead produced by LPBF, the convective flow due to evaporation and subsequent particle entrainment are shown to be important factors. It is interesting to note that the induced gas flow extends beyond the width of the melt pool and continues to supply particles to the laser-scanned region even when the melt-pool temperature cools down. These particles therefore may not melt completely and can be a source of surface roughness and porosity. To the author's knowledge, this physical phenomenon has not been brought to light prior to this study and till now has not been addressed through modeling.

From the preceding explanation of the cause of the DZ via particle entrainment by convective flow at high pressure, one might expect that the reduction of the ambient pressure would suppress the denudation. Surprisingly, the experiments demonstrate that with decreasing pressure the denudation is even more pronounced, and displays widely varying behavior for pressures below 10 Torr. This observation is consistent with the notion of competing vapor flow phenomena as discussed earlier in section 35.4. For pressures above 10 Torr, the increased range of the convective flow can be understood through the reduction in ambient pressure resistance, which leads to an increase in vapor velocity emanating from the melt pool. The increase of vapor flux with decreasing ambient pressure was studied by Ho et al. for microsecond vapor plume evolution from nanosecond pulsed ablation of Au substrates [22], and is consistent with the present interpretation of increased particle entrainment. Below ~10 Torr, a dramatic decrease in DZ is observed, and is indicative of the onset of rarified or molecular flow. The Knudsen number (Kn) for particles in a gas flow is given by $Kn = \lambda/d$ where d is the particle diameter and λ is the gas mean free path given by

$$\lambda = \frac{\mu}{P_0} \sqrt{\frac{\pi k_B T}{2m}} \tag{35.3}$$

where $\mu = 22.3$ μPa·s is the dynamic viscosity of Ar, $P_0 = 760$ Torr is ambient pressure, $m = 6.6 \times 10^{-26}$ kg is the mass of an Ar atom. Using these values, we arrive at a mean free path of $\lambda = 69$ nm at 760 Torr, 300 K that increases to 10 μm at 5 Torr, 300 K. However, because μ increases with temperature for Ar gas and the gas next to the DZ can get to temperatures near 2000 K, we estimate that μ~75 μPa·s yielding a mean free path near 80 μm at 5 Torr and a Kn value of approximately unity. At this pressure, the momentum transfer from the transported gas is negligible and the entrainment that leads to the DZ ceases to be effective.

From the preceding arguments, the DZ is determined to be influenced by both the presence of the ambient gas that is capable of transferring momentum to the powder, the change in resistive gas pressure from ambient Ar gas, and the vaporization of the metal melt pool and subsequent convective flow. With the first two of these drivers removed for $Kn > 1$, we are left with only metal gas flux from the evaporating metal, to explain the increase of the DZ for pressures below 5 Torr. One possible explanation could be that particles adjacent to the melt pool may receive *glancing* laser heating such that a vapor flux is generated that propels the particle away from the laser spot center. As can be seen, frames from the modeling results in the top of Figure 35.7. The recoil momentum of evaporated material pushes the particle out of the DZ, and collisions with other particles transfers momentum such that all adjacent particles start to move. The observation of the powder pile up on the edge of the denudation zone visible in Figure 35.3 for $p < 5$ Torr supports this explanation. However, careful inspection of the DZ at low (<5 Torr) and high (>10 Torr) shows that in the low pressure case, mostly larger particles remain, whereas at higher pressures the opposite is true. If edge vaporized particles knocking adjacent particles away from the laser path were the cause of the DZ at low pressures, one would expect both small and large particles to be knocked away. An additional consideration is the expansion of the metal vapor at low pressures. Although a relatively narrow vapor jet is expected for the case of a *submerged jet* as argued above, the angle of vapor expansion will increase with decreasing ambient pressure. This effect of ambient gas pressure on vapor jet expansion can similarly be observed in the case of high altitude rocket exhaust plumes [22] and is a consideration in rocket engine design: At low altitudes and high pressures, rocket exhaust is confined to a narrow jet, whereas at high altitudes and low pressures a much wider plume can be observed. Additionally, up until now it has been assumed that the melt pool presents a vaporizing surface that directs flow normal to the build plane; however, the dynamic motion of the melt pool leads to transient shapes that are nearly tangentially directed (see high-speed video) and could enhance outward flow along the surface, pushing particles away and causing the DZ to appear. This outward flow would tend to push out smaller particles in the same way that they are preferentially pulled in at higher pressures.

In our analysis of denudation, we considered other possible effects, such as rapid thermal expansion of the particles and/or substrate, expansion of gas trapped within void spaces, and thermal buoyancy effects, which we briefly discuss here. Very rapid heating of the powder results in thermal expansion of the powder with high velocity and can produce lateral particle motion. Substrate heating can result in displacement of the substrate in the vertical direction and subsequent powder ejection similar to that of the laser cleaning process [23]. The analysis of our model shows that for the parameters of interest, the vertical velocities in denudation zone are too small to produce the denudation effect. It was suggested (Y. Chivel in [24]) that the heated gas trapped in the porous structures can move the particles aside. Although this mechanism may contribute to the low pressure and outward movement of particles, it cannot explain the inward movement at higher pressures. Finally, convection due to buoyancy of the heated gas above the melt pool can compete with vapor flux from the evaporating metal. For an Ar gas density differential, $\Delta\rho$ ~0.4 kg/m^3, for gas heated between 300 and 2000 K the velocity due to convected gas can be approximated through balance with viscosity as $u \approx d^2 g \Delta\rho/\mu$ that yields velocities near 10 μm/s, far too slow to explain the current results.

35.6   Conclusion

We have presented a detailed study of the denudation of Ti and steel alloy powders under varying laser conditions and ambient gas pressures. Our key finding is that, for a typical LPBF environment (Ar gas at 760 Torr), the dominant driving force for denuding powder near a melt track is the entrainment of particles by surrounding gas flow, due to the Bernoulli effect induced by the vaporizing melt-track center. We have demonstrated that the gas flow affects the height of the track and the extent of denudation thus will have significant influence on ultimate processing quality of LPBF parts. Our FE modeling results indicate a melt pool that directs vapor normal and rearward relative to laser scan direction, consistent with the observed motion of entrained particles. Moreover, through high-speed video, a portion of the entrained particles are observed to be a source of material that ultimately resides in the resolidified track after being consumed by the melt pool that can be directly related to the final surface roughness The drag force on the particles due to the shear flow slows down with decreasing gas density and at some pressure reverses the particle motion. At this pressure the width of the denudation zone is observed to be at a minimum. For the parameters of experiments presented here, the pressure for this minimum is about 4 Torr. We note that at this pressure the mean free path for Ar atoms ranges from ~10 μm at 300 K to ~85 μm at 2000 K and can become larger than the particle size ($Kn > 1$), thus driving the nature of the gas transitions from a fluid to a rarified flow. At low pressures, the sources of denudation is postulated to be that of either tangentially vaporizing particles that collide with nearby particles as observed in our simulations, or from a dynamically undulating melt pool that transiently directs metal vapor flow along the surface entraining particles and sending them outward.

Acknowledgments

The authors acknowledge M. Wang and A. Anderson of LLNL, I. Yadroitsev of the Central University of Technology, South Africa and D. Novikov of IPG Photonics for stimulating and enlightening conversations. This work was funded through a Laboratory Directed Research and Development grant 15-ERD-037 and performed under the auspices of the U.S. Department of Energy by Lawrence Livermore National Laboratory under contract DE-AC52-07NA27344.

References

1. D.D. Gu, W. Meiners, K. Wissenbach, R. Poprawe. Laser additive manufacturing of metallic components: Materials, processes and mechanisms, *Int. Mater. Rev.* 57 (2012) 133–164.
2. L.E. Murr, S.M. Gaytan, D.A. Ramirez, E. Martinez, J. Hernandez, K.N. Amato, P.W. Shindo, F.R. Medina, R.B. Wicker. Metal fabrication by additive manufacturing using laser and electron beam melting technologies, *J. Mater. Sci. Technol.* 28 (2012) 1–14.
3. B. Song, S.J. Dong, S.H. Deng, H.L. Liao, C. Coddet. Microstructure and tensile properties of iron parts fabricated by selective laser melting, *Opt. Laser Technol.* 56 (2014) 451–460.
4. R. Fadida, D. Rittel, A. Shirizly. Dynamic mechanical behavior of additively manufactured Ti6Al4V with controlled voids, *J. Appl. Mech. Trans. ASME* 82 (2015) 041004.
5. I. Yadroitsev, P. Bertrand, I. Smurov. Parametric analysis of the selective laser melting process, *Appl. Surf. Sci.* 253 (2007) 8064–8069.
6. A.V. Gusarov, I. Yadroitsev, P. Bertrand, I. Smurov. Heat transfer modelling and stability analysis of selective laser melting, *Appl. Surf. Sci.* 254 (2007) 975–979.
7. L. Thijs, F. Verhaeghe, T. Craeghs, J. Van Humbeeck, J.P. Kruth. A study of the micro structural evolution during selective laser melting of Ti-6Al-4V, *Acta Mater.* 58 (2010) 3303–3312.
8. X.B. Su, Y.Q. Yang. Research on track overlapping during selective laser melting of powders, *J. Mater. Process. Technol.* 212 (2012) 2074–2079.
9. N.T. Aboulkhair, I. Maskery, C. Tuck, I. Ashcroft, N.M. Everitt. On the formation of AlSi10Mg single tracks and layers in selective laser melting: Microstructure and nano-mechanical properties, *J. Mater. Process. Technol.* 230 (2016) 88–98.
10. C. McCallen. *ALE3D: Arbitrary Lagrange Eulerian Three- and Two Dimensional Modeling and Simulation Capability.* Livermore, CA: Lawrence Livermore National Laboratory, 2012.
11. S.A. Khairallah, A. Anderson. Mesoscopic simulation model of selective laser melting of stainless steel powder, *J. Mater. Process. Technol.* 214 (2014) 2627–2636.
12. C.D. Boley, A.M. Rubenchik. Modeling of laser interactions with composite materials, *Appl. Optics* 52 (2013) 3329–3337.
13. Y. Liu, Y. Yang, S. Mai, D. Wang, C. Song. Investigation into spatter behavior during selective laser melting of AISI 316L stainless steel powder, *Mater. Des.* 87 (2015) 797–806.
14. M. Simonelli, C. Tuck, N.T. Aboulkhair, I. Maskery, I. Ashcroft, R.D. Wildman, R. Hague. A study on the laser spatter and the oxidation reactions during selective laser melting of 316L stainless steel, Al-Si10-Mg, and Ti-6Al-4V, *Metall. Mater. Trans A* 46A (2015) 3842–3851.
15. I. Yadroitsev, L. Thivillon, P. Bertrand, I. Smurov. Strategy of manufacturing components with designed internal structure by selective laser melting of metallic powder, *Appl. Surf. Sci.* 254 (2007) 980–983.
16. I. Yadroitsev, P. Bertrand, G. Antonenkova, S. Grigoriev, I. Smurov. Use of track/layer morphology to develop functional parts by selective laser melting, *J. Laser Appl.* 25 (2013) 052003.
17. G.G. Gladush, I. Smurov. *Physics of Laser Materials Processing: Theory and Experiment.* Berlin, Germany: Springer, 2011.
18. T.R. Anthony, H.E. Cline. Surface rippling induced by surface-tension gradients during laser surface melting and alloying, *J. Appl. Phys.* 48 (1977) 3888–3894.

19. A.F.H. Kaplan. Local absorptivity modulation of a 1 mu m-laser beam through surface waviness, *Appl. Surf. Sci.* 258 (2012) 9732–9736.

20. D. Bauerle. *Laser Processing and Chemistry.* Berlin, Germany: Springer, 2011.

21. L.D. Landau, E.M. Lifshitz. *Fluid Mechanics.* 2nd ed. Butterworth-Heinemann, Langford Lane, Kidlington, Oxford OX5 1GB, United Kingdom, 1987.

22. J.R. Ho, C.P. Grigoropoulos, J.A.C. Humphrey. Computational study of heat-transfer and gas-dynamics in the pulsed-laser evaporation of metals, *J. Appl. Phys.* 78 (1995) 4696–4709.

23. B. Luk'yanchuk. *Laser Cleaning.* Singapore: World Scientific, 2002.

24. Y. Chivel. Ablation phenomena and instabilities under laser melting of powder layers. *8th International Conference on Photonic Technologies LANE 2014: Bayerisches Laserzentrum GmbH,* Fürth, Germany 2014. pp. 1–7.

chapter thirty six

Tension-compression fatigue of an oxide/oxide ceramic composite at elevated temperature

Marina B. Ruggles-Wrenn and R. L. Lanser

Contents

Abstract: Tension-compression fatigue behavior of an oxide-oxide ceramic-matrix composite (CMC) was investigated at 1200°C in air and in steam. The composite is comprised of an alumina matrix reinforced with Nextel™720 alumina–mullite fibers woven in an eight harness satin weave (8HSW). The composite has no interface between the fiber and matrix, and relies on the porous matrix for flaw tolerance. Tension-compression fatigue behavior was studied for fatigue stresses ranging from 60 to 120 MPa at a frequency of 1.0 Hz. The R ratio (minimum stress to maximum stress) was −1.0. Fatigue run-out was defined as 10^5 cycles and was achieved at 80 MPa in air and at 70 MPa in steam. Steam-reduced fatigue lives by an order of magnitude. Specimens that achieved fatigue run-out were subjected to tensile tests to failure to characterize the retained tensile properties. Specimens subjected to prior fatigue in air retained 100% of their tensile strength. The steam environment severely degraded tensile properties. Tension-compression fatigue was considerably more damaging than tension-tension fatigue. Composite microstructure, as well as damage and failure mechanisms were investigated.

Keywords: Ceramic-matrix composites (CMCs) Oxides Fatigue High-temperature properties Mechanical properties Fractography

36.1   Introduction

Advanced applications such as aircraft turbine engine components, land-based turbines, hypersonic missiles, and flight vehicles and, most recently, spacecraft reentry thermal protection systems have raised the demand for structural materials that exhibit superior long-term mechanical properties and retained properties under high temperature, high pressure, and varying environmental factors. CMCs, capable of maintaining excellent strength and fracture toughness at high temperatures are prime candidate materials for such applications. As these applications require exposure to oxidizing environments, the thermodynamic stability and oxidation resistance of CMCs are vital issues. The need for environmentally stable composites motivated the development of CMCs based on environmentally stable oxide constituents [1–4].

Oxide/oxide CMCs exhibit damage tolerance combined with inherent oxidation resistance [3,5]. Moreover, oxide-oxide CMCs have displayed excellent high-temperature mechanical properties [4,6–9]. However, recent studies revealed dramatic degradation of mechanical performance of oxide-oxide CMCs and their constituents at elevated temperature in steam [10–21]. When a CMC is subjected to mechanical loading at elevated temperature in steam, multiple degradation and failure mechanisms may operate simultaneously. These may include environmentally assisted subcritical crack growth, grain growth, and matrix densification, and loss of SiO_2 as $Si(OH)_4$.

Numerous recent studies investigated mechanical behavior of oxide-oxide CMCs at elevated temperature [6–18,22–24]. Porous-matrix oxide/oxide CMCs exhibit several behavior trends that are distinctly different from those exhibited by traditional dense-matrix CMCs with a fiber-matrix interface. Most SiC-fiber-containing CMCs exhibit longer life under static loading and shorter life under cyclic loading [25]. For these materials, fatigue is significantly more damaging than creep. Conversely, in the case of porous-matrix CMCs creep loading was found to be considerably more damaging than fatigue [9,10]. Furthermore, both creep resistance and fatigue performance of Nextel™720/alumina composite were significantly degraded in the presence of steam [10–18].

Efforts to assess the life-limiting behavior of oxide-oxide CMCs under cyclic loading focused mainly on tension-tension fatigue. Yet, in many potential applications, porous-matrix oxide/oxide CMCs may be subjected to fatigue loading with negative ratios of minimum to maximum stress. Therefore a thorough understanding of tension-compression fatigue performance of oxide-oxide CMCs in service environments is critical to their acceptance for high-temperature structural applications. This study investigates the tension-compression fatigue behavior of an oxide-oxide CMC consisting of a porous alumina matrix reinforced with Nextel™720 fibers. Tension-compression fatigue tests were conducted at 1200°C in air and in steam environments. The composite microstructure, as well as damage and failure mechanisms are discussed.

36.2   Material and experimental arrangements

The material studied was Nextel™720/alumina (N720/A), an oxide-oxide CMC (manufactured by ATK-COIC, San Diego, CA) consisting of a porous alumina matrix reinforced with Nextel™720 fibers woven in an eight harness satin weave (8HSW). There is no fiber coating. The damage tolerance of the N720/A CMC is enabled by the porous matrix. The composite was supplied in a form of 5.76-mm thick panels comprised of 24 0°/90° woven layers, with a density of ~2.84 g/cm³, a fiber volume of ~44.2%, and matrix porosity of ~22.3%. The fiber fabric was infiltrated with the matrix in a sol–gel process. The laminate was dried with a

(a)   (b)

Figure 36.1 As-received material: (a) overview and (b) porous nature of the matrix is evident.

vacuum bag technique under low pressure and low temperature, and then pressureless sintered [26]. The overall microstructure of the CMC is presented in Figure 36.1.

All tests were performed at 1200°C using the experimental setup detailed elsewhere [10,18,27]. Prior to testing, extreme care was taken to align the mechanical testing system using the MTS alignment fixture and the alignment specimen instrumented with eight strain gages. In all tests, the misalignment was limited to 0.015% of bending. Note that the N720/A composite exhibits no loss of stiffness with increasing temperature in the 23°C–1200°C range [11,28]. Hence the possibility of macroscopic bending during tests due to loss of stiffness with increasing temperature is unlikely. As compressive loading, and thus the potential for buckling failure modes, was involved in the cycle type, specimens with hourglass-shaped gage section (Figure 36.2) were used in all tests. The stress concentration inherent in an hourglass specimen was assessed. Finite element analysis (FEA) of the specimen shows that the axial stress at the edges in the middle of the hourglass section is only 3.5% higher than the average axial stress.

Deionized water was used to generate steam for testing in steam. Chemical analysis of water entering the steam generator revealed trace amounts (below 10 ppb) of Al, B, Fe, and Zn. Chemical analysis of condensed water exiting the steam generator revealed trace amounts (10–30 ppb) of Al, B, and Fe, and slightly higher but still negligible amounts (55–80 ppb) of Zn. We believe that these levels of impurities are too low to cause contamination of the test specimens and to influence the mechanical performance of the N720/A composite. In all tests, a specimen was heated to test temperature at 1°C/s, and held at temperature for additional 45 min prior to testing. The same procedures were used for testing in air and in steam.

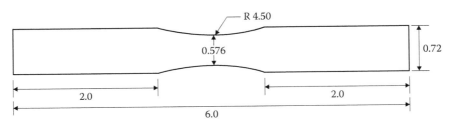

Figure 36.2 Test specimen. All dimensions in inches.

Tension-compression fatigue tests were performed in load control with an *R* ratio (minimum to maximum stress) of −1.0 at 1.0 Hz. Fatigue run-out was defined as 10^5 cycles. This cycle count represents the number of loading cycles expected in aerospace applications at that temperature. Cyclic stress–strain data were recorded throughout each test, so that modulus change as well as variations in maximum and minimum strains with fatigue cycles and/or time could be examined. All specimens that achieved run-out were tested in tension to failure at 1200°C in air to determine the retained tensile properties. Fracture surfaces of failed specimens were examined using an optical microscope (Zeiss Discovery V12) and a scanning electron microscope (SEM, Quanta 450).

36.3   Results and discussion

36.3.1   Tension-compression fatigue

Results of the tension-compression fatigue tests are shown in Figure 36.3 as maximum stress versus cycles to failure (S-N) curves, where results of the tension-tension fatigue tests from prior work [18] are also included. It is noteworthy that all fatigue failures occurred during the compressive portion of the fatigue cycle.

At 1200°C in air, the fatigue run-out was achieved at 80 MPa (40%UTS), suggesting that the fatigue limit is between 80 and 90 MPa. The tension-compression cycling is considerably more damaging than tension-tension fatigue. Including compression in the load cycle caused dramatic reductions in fatigue life of N720/A composite. For a given stress level, the cyclic lives obtained in tension-compression fatigue can be three orders of magnitude lower than those produced under tension-tension fatigue [18]. The run-out stress in tension-tension fatigue was a high 170 MPa, more than twice the run-out stress of 80 MPa obtained in tension-compression fatigue. Furthermore, while in tension-tension fatigue, a run-out of 10^5 cycles was achieved at 125 MPa, tension-compression cyclic life at 120 MPa was a very poor 199 cycles. Including compression in the fatigue cycle reduced fatigue life by 99% for σ_{max} of 120 MPa.

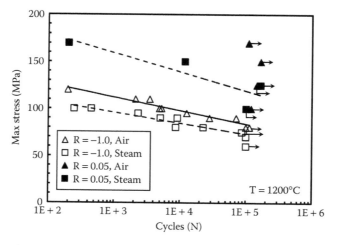

Figure 36.3 Fatigue S-N curves for N720/A at 1200°C in air and in steam. Arrow indicates that failure of specimen did not occur when the test was terminated. Tension-tension fatigue data from Ruggles-Wrenn, M. B. et al., *Int. J. Fatig.*, **30**, 502–516, 2008.

Presence of steam causes noticeable degradation in fatigue performance of the N720/A composite. In steam the tension-compression fatigue run-out was reached only at 70 MPa (35%UTS). The reduction in cyclic life due to steam was 80%–92% for σ_{max} of 80 MPa. Reductions in cyclic lifetimes due to steam were observed under tension-tension fatigue in prior work [10,18]. However, detrimental effect of steam was considerably more pronounced in the case of tension-tension fatigue than under tension-compression fatigue. The cyclic lifetimes produced in tension-compression fatigue in steam were nearly an order of magnitude lower than those produced in air. Contrastingly, in the case of tension-tension fatigue steam reduced the cyclic lifetimes by two to three orders of magnitude for the max stress ≥ 150 MPa [10,18]. Nevertheless, in steam the tension-compression fatigue is still more damaging than tension-tension fatigue. Including compression in the load cycle causes a nearly 100-fold reduction in cyclic life in steam.

Evolution of hysteresis stress–strain response of N720/A composite with cycles at 1200°C in air and in steam is typified in Figure 36.4. It is seen that the hysteresis response produced in tension-compression tests in steam is qualitatively similar to that in air. The hysteresis loops are nearly symmetric about the origin. Such symmetry is maintained for the duration of the test. For each cycle the tensile (compression) modulus was calculated as the slope of the tensile (compressive) portion of the hysteresis loop within the linear region. In all tests, the tensile modulus was approximately the same as the compression modulus for a given cycle. In all tests, the tensile and compressive moduli decrease with fatigue cycling. Progressive decrease in tensile (compressive) modulus is accompanied by an increase in cyclic tensile (compressive) strain. The apparent stiffening observed during compression in Figures 36.4a and b is attributed to mechanical impediment of crack closure by matrix debris [29,30].

Figure 36.5 shows maximum and minimum strains versus fatigue cycles for tests conducted at 1200°C in air and in steam. In all tests performed in this work the evolution of minimum strain with cycles, mirrors the evolution of maximum strain. Notably, lower maximum strains were accumulated in tests performed with higher levels of maximum stress. Generally, lower strain accumulation with cycling indicates that less damage has occurred, and that it is mostly limited to some additional matrix cracking. However, in this case, low accumulated strains are more likely due to early bundle failures leading to specimen failure. Similar conclusion was reached in the study of tension-tension fatigue of N720/A at 1200°C [10,18]. It is noteworthy that the maximum strains measured during tension-compression fatigue are lower than the strains accumulated during tension-tension cycling. For example, the tensile strains attained in tension-compression fatigue tests with σ_{max} of 95 and 110 MPa do not exceed 0.36%, whereas tensile strains accumulated in tension-tension fatigue tests with σ_{max} of 100 MPa performed in prior work [10,18] reached 0.6%. Results in Figure 36.5 also reveal that strain accumulation is accelerated in the presence of steam.

Of importance in cyclic fatigue is the reduction in stiffness (hysteresis modulus determined from the maximum and minimum stress–strain data points during a load cycle), reflecting the damage development during fatigue cycling. The change in normalized modulus (i.e., modulus normalized by the modulus obtained in the first cycle) with fatigue cycles at 1200°C is shown in Figure 36.6. The rate of modulus decay and thus the rate of damage accumulation accelerate slightly with increasing maximum stress. It is noteworthy that although some specimens tested in air achieved fatigue run-out of 10^5 cycles, a decrease in normalized modulus with cycling was still observed. Decay in normalized modulus is accelerated in the presence of steam, suggesting an increase in the rate of damage accumulation in steam. The degrading effect of steam on the evolution of the

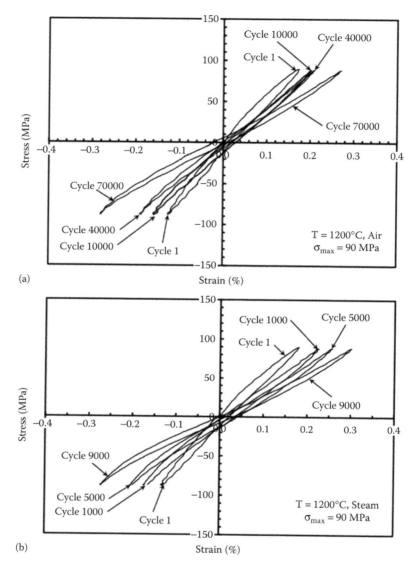

Figure 36.4 Typical evolution of stress–strain hysteresis response of N720/A composite with fatigue cycles at 1200°C (a) in air and (b) in steam. σ_{max} = 90 MPa.

normalized modulus is observed for all maximum stress levels. This result is consistent with the decreased number of cycles to failure produced in steam.

Retained strength and stiffness of the specimens that achieved fatigue run-out were evaluated in tensile tests performed at 1200°C (Figure 36.7). The specimens subjected to 10^5 cycles of prior tension-compression fatigue with σ_{max} = 80 MPa at 1200°C in air exhibited with no loss of tensile strength. However, a modulus loss of 45% was observed. In contrast, prior tension-compression fatigue with σ_{max} of 60 and 70 MPa in steam caused significant degradation of tensile strength. Specimens subjected to 10^5 fatigue cycles in steam retained only 62%–83% of their tensile strength and less than 50% of their modulus.

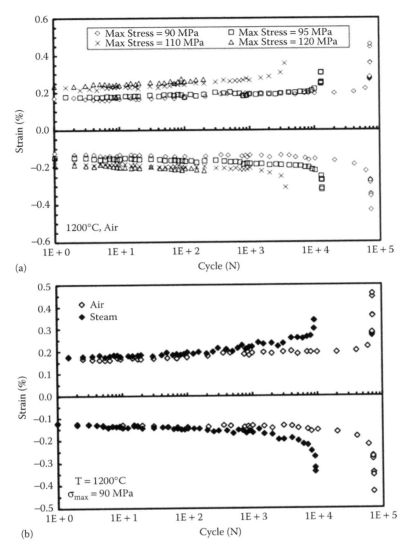

Figure 36.5 Maximum and minimum strains versus fatigue cycles at 1200°C: (a) in air and (b) in air and in steam.

Prior tension-tension fatigue in steam also causes degradation of tensile strength and stiffness [10,18]. However, the strength and modulus loss were greater in the case of prior tension-compression fatigue.

The considerable loss of tensile strength suggests that prior fatigue in steam has caused significant degradation of the Nextel™720 fibers. It is recognized that the superior high-temperature creep performance of the Nextel™720 fibers is due to the high content of mullite, which has a much better creep resistance than alumina [31,32]. Recently Wannaparhun et al. [33] reported that SiO_2 could be leached from Nextel™720 fiber at 1100°C in water-vapor environment. Results of prior work [10,18] suggest that depletion of the mullite phase in the Nextel™720 fiber may also be responsible for the deterioration

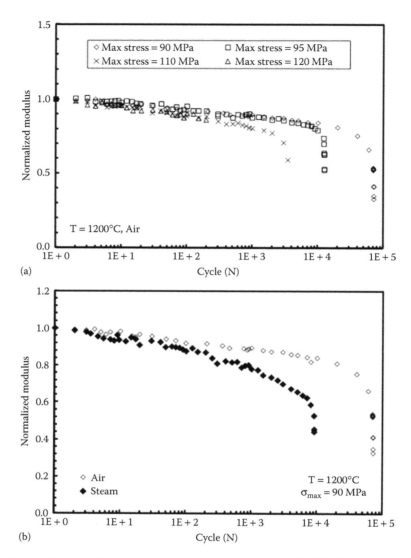

Figure 36.6 Normalized tensile modulus versus fatigue cycles at 1200°C: (a) in air and (b) in air and in steam.

of the N720/A fatigue performance at 1200°C in steam. These findings were supported by Armani et al. [19], who tested Nextel™720 fiber tows in creep at 1100 and 1200°C in air and in steam. Armani and coworkers reported that in steam the mullite in the Nextel™ 720 fibers decomposed to porous alumina. A layer of porous alumina with thick, plate-like grains formed on the surfaces of Nextel™ 720 fibers tested in steam. This porous alumina layer was up to 2.2 μm thick with grain size of 100–200 nm for fibers tested at 1200°C, and ~0.5 μm thick with grain size of 50–100 nm for fibers tested at 1100°C. The formation of porous alumina layers over 2 μm thick significantly decreases the load-bearing capacity of the 10–12 μm diameter fibers. Hence we believe that the decomposition of mullite and formation of porous alumina layers are behind the loss of the composite tensile strength after 10^5 fatigue cycles (nearly 28 h) at 1200°C in steam.

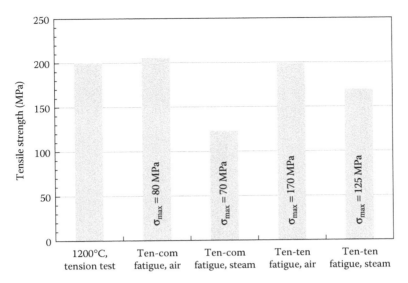

Figure 36.7 Retained tensile strength of the N720/A specimens subjected to prior fatigue in laboratory air and in steam environment at 1200°C. Tension-tension fatigue data from Ruggles-Wrenn, M. B. et al., *Int. J. Fatig.*, **30**, 502–516, 2008.

36.3.2 Composite microstructure

Optical micrographs of fracture surfaces obtained in tension-compression fatigue tests conducted at 1200°C in air and in steam are shown in Figure 36.8. Brushy fracture surfaces and long (~22 mm) damage zones indicative of fibrous fracture are produced in fatigue tests performed with σ_{max} of 100 MPa in air (Figure 36.8a) and in steam (Figure 36.8b). Not surprisingly, the effects of steam on fracture surface appearance are minimal. These specimens produced the shortest fatigue lives in their respective test environments; hence, the 100 MPa fatigue test in steam was of a fairly short duration (<8 min). In contrast, steam has a pronounced effect on the fracture surfaces obtained in tests of longer duration (>24 h). The fracture surface obtained in air with σ_{max} of 80 MPa (Figure 36.8c) is similar to that obtained in the fatigue test of a shorter duration performed in air with σ_{max} of 100 MPa (Figure 36.8a). Uncorrelated fiber fracture and a fairly long damage zone are still observed. Conversely, the fracture surface of the specimen tested in fatigue with σ_{max} of 75 MPa in steam (Figure 36.8d) is dominated by coordinated fiber failure and has a significantly shorter damage zone, suggesting that alumina matrix has densified. The loss of matrix porosity and matrix densification are likely to decrease damage tolerance and degrade composite performance under tensile loading. However, these observations do not explain why tension-compression fatigue is so much more damaging than tension-tension cycling in air as well as in steam.

To gain insight into the mechanisms responsible for drastic reductions in fatigue life seen when compression is included in the load cycle, we examine the fracture surfaces with an SEM. The key feature of the fracture surfaces produced in tension-compression fatigue tests in this study is the proliferation of compression curl fiber fractures (Figure 36.9). Compression (or cantilever) curl is a telltale feature of flexural fiber fracture. The existence of a compression curl is a sign that the fiber was loaded primarily in bending. The crack initiates on the tension side of the fiber, propagates from the tension side into the

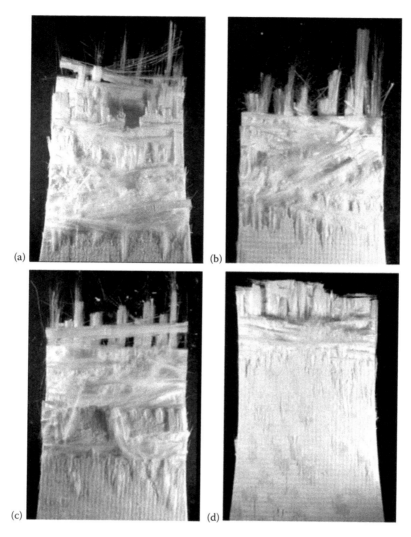

Figure 36.8 Fracture surfaces of N720/A specimens tested at 1200°C in tension-compression fatigue: (a) in air, σ_{max} = 100 MPa, N_f = 4902 cycles, (b) in steam, σ_{max} = 100 MPa, N_f = 450 cycles, (c) in air, σ_{max} = 80 MPa, N_f = 113382 cycles, and (d) in steam, σ_{max} = 75 MPa, N_f = 86548 cycles.

compression side, then slows down and changes direction resulting in a compression curl fiber fracture. The origin of fracture on a fracture surface is located opposite the compression curl. It is recognized that the compressive failure in fiber-reinforced composites is generally associated with micro-buckling or kinking of the fibers [34–37]. Figure 36.10 shows an example of fiber microbuckling seen in this study. In a 0/90 cross ply CMC, compressive failure begins with nucleation of axial cracks between adjacent fibers in the 90° plies [17,38–41]. These cracks grow subcritically to gradually form shear zones, which induce 0° ply flexure and cause buckling of the 0° fibers. Flexural stresses in fibers produced by in-phase buckling lead to the formation of kink zones and subsequent fracture of brittle fibers [42,43]. The fracture surfaces obtained in tension-compression fatigue tests performed in this study exhibit abundance of compression curl fiber fractures caused by

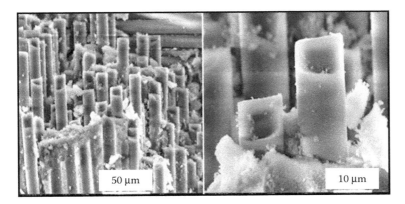

Figure 36.9 Fracture surfaces of N720/A specimen tested in tension-compression fatigue at 1200°C. Compression curl fiber fractures.

Figure 36.10 Fracture surface of N720/A specimen tested in tension-compression fatigue at 1200°C. Fiber microbuckling.

fiber microbuckling. We conclude that shearing and bending fracture of 0° fiber bundles occurs during compression portion of every fatigue cycle. In the case of the N720/A composite, the porous matrix is exceptionally weak and the fibers bear most of the load. Buckling and fracture of the 0° bundles leads to the loss of the composite's load-bearing capacity. As a result, tension-compression cycling becomes much more damaging than the tension-tension fatigue for N720/A composite.

Results of prior work [10–16,18,19] suggest that at 1200°C in steam, the loss of matrix porosity and the degradation of N720/A fibers due to mullite decomposition work together to reduce the tension-tension fatigue performance. In the case of tension-compression fatigue, degradation of N720/A fibers in steam facilitates widespread breakage of fibers due to microbuckling at lower compressive loads. As a result, tension-compression fatigue performance is degraded in the presence of steam.

36.4 Concluding remarks

Tension-compression fatigue behavior of the N20/A composite was studied at 1.0 Hz at 1200°C in air and in steam. Fatigue stress levels ranged from 60 to 120 MPa. The fatigue run-out was achieved at 80 MPa (40%UTS) in air and at 70 MPa (35%UTS) in steam. Presence of steam noticeably degrades tension-compression fatigue performance of N720/A. Steam decreases tension-compression fatigue lives by nearly an order of magnitude. Prior fatigue in air causes no reduction in tensile strength, suggesting that no damage occurred to the fibers. In contrast, prior fatigue in steam can reduce tensile strength by nearly 40%. Tension-compression cycling is considerably more damaging than tension-tension fatigue. Including compression in the load cycle severely degraded fatigue lifetimes in both air and steam.

The damage and failure of the composite in tension-compression fatigue at 1200°C in air are due to extensive fiber breakage due to fiber microbuckling during compression portion of the cycle. The presence of steam causes decomposition of mullite and formation of porous alumina layers on the fiber surfaces, thus decreasing the load-bearing capacity of the N720 fibers. The decomposition of mullite and formation of porous alumina layers are behind the reduced tension-compression fatigue performance of the N720/A composite in steam.

Acknowledgment

The authors thank L. Zawada and C. Przybyla (Air Force Research Laboratory) for providing the test material.

References

1. A. Szweda, M. L. Millard, and M. G. Harrison, Fiber Reinforced Ceramic Composite Member and Method for Making. U. S. Pat. No. 5 601–674, 1997.
2. F. W. Zok, Developments in Oxide Fiber Composites. *J. Am. Ceram. Soc.*, **89**, 3309–3324, 2006.
3. R. J. Kerans, R. S. Hay, T. A. Parthasarathy, and M. K. Cinibulk, Interface design for oxidation resistant ceramic composites. *J. Am. Ceram. Soc.*, **85**, 2599–2632, 2002.
4. B. Kanka and H. Schneider, Aluminosilicate fiber/mullite matrix composites with favorable high-temperature properties. *J. Eur. Ceram. Soc.*, **20**, 619–623, 2000.
5. M. A. Mattoni, J. Y. Yang, C. G. Levi, F. W. Zok, and L. P. Zawada, Effects of combustor rig exposure on a porous-matrix oxide composite. *Int. J. Appl. Ceram. Technol.*, **2**, 133–140, 2005.
6. E. A. V. Carelli, H. Fujita, J. Y. Yang, and F. W. Zok, Effects of thermal aging on the mechanical properties of a porous-matrix composite. *J. Am. Ceram. Soc.*, **85**, 595–602, 2002.
7. K. A. Keller, T. Mah, T. A. Parthasarathy, E. E. Boakye, P. Mogilevsky, and M. K. Cinibulk, Effectiveness of monazite coatings in oxide/oxide composites after long term exposure at high temperature. *J. Am. Ceram. Soc.*, **86**, 325–332, 2003.
8. S. Hackemann, F. Flucht, and W. Braue, Creep investigations of alumina-based all-oxide ceramic matrix composites. *Compos. A*, **41**, 1768–1776, 2010.
9. L. P. Zawada, R. S. Hay, J. Staehler, and S. S. Lee, Characterization and high temperature mechanical behavior of an oxide/oxide composite. *J. Am. Ceram. Soc.*, **86**, 981–990, 2003.
10. J. M. Mehrman, M. B. Ruggles-Wrenn, and S. S. Baek, Influence of hold times on the elevated-temperature fatigue behavior of an oxide-oxide ceramic composite in air and in steam. *Compos. Sci. Technol.*, **67**, 1425–1438, 2007.

11. M. Ruggles-Wrenn and J. Braun, Effects of steam environment on creep behavior of Nextel™720 alumina ceramic composite at elevated temperature. *Mater. Sci. Eng. A*, **497**, 101–110, 2008.

12. M. B. Ruggles-Wrenn and C. L. Genelin, Creep of Nextel™720/alumina-mullite ceramic composite at 1200°C in air, argon, and steam. *Compos. Sci. Technol.*, **69**, 663–669, 2009.

13. M. B. Ruggles-Wrenn, P. Koutsoukos, and S. S. Baek, Effects of environment on creep behavior of two oxide/oxide ceramic-matrix composites at 1200°C. *J. Mater. Sci.*, **43**, 6734–6746, 2008.

14. M. B. Ruggles-Wrenn and T. Kutsal, Effects of steam environment on creep behavior of Nextel™720/alumina-mullite ceramic composite at elevated temperature. *Compos. A*, **41**, 1807–1816, 2010.

15. M. B. Ruggles-Wrenn and M. Ozer, Creep behavior of Nextel™720/alumina-mullite ceramic composite with +/−45 fiber orientation at 1200°C. *Mater. Sci. Eng. A*, **527**, 5326–5334, 2010.

16. M. B. Ruggles-Wrenn, G. T. Siegert, and S. S. Back, Creep behavior of Nextel™720/alumina ceramic composite with +/−45 fiber orientation at 1200°C. *Compos. Sci. Technol.*, **68**, 1588–1595, 2008.

17. M. B. Ruggles-Wrenn and N. R. Szymczak, Effects of steam environment on compressive creep behavior of Nextel™720/alumina ceramic composite at 1200°C. *Compos. A*, **39**, 1829–1837, 2008.

18. M. B. Ruggles-Wrenn, G. Hetrick, and S. S. Baek, Effects of frequency and environment on fatigue behavior of an oxide-oxide ceramic composite at 1200°C. *Int. J. Fatig.*, **30**, 502–516, 2008.

19. C. J. Armani, M. B. Ruggles-Wrenn, R. S. Hay, and G. E. Fair, Creep and microstructure of Nextel™ 720 fiber at elevated temperature in air and in steam. *Acta Mater.*, **61**, 6114–6124, 2013.

20. C. J. Armani, M. B. Ruggles-Wrenn, G. E. Fair, and R. S. Hay, Creep of Nextel™ 610 fiber at 1100°C in air and steam. *Int. J. Appl. Ceram. Technol.*, **10**, 276–284, 2013.

21. R. S. Hay, C. J. Armani, M. B. Ruggles-Wrenn, and G. E. Fair, Creep and microstructure evolution of Nextel™ 610 fiber in air and steam. *J. Eur. Ceram. Soc.*, **34**, 2413–2426, 2014.

22. E. Volkmann, K. Tushtev, D. Koch, C. Wilhelmi, G. Grathwohl, and K. Rezwan, Influence of fiber orientation and matrix processing on the tensile and creep performance of Nextel 610 reinforced polymer derived ceramic matrix composites. *Mater. Sci. Eng. A*, **614**, 171–179, 2014.

23. E. Volkmann, A. Dentel, K. Tushtev, C. Wilhelmi, and K. Rezwan, Influence of heat treatment and fiber orientation on the damage threshold and the fracture behavior of Nextel fiber-reinforced mullite-SiOC matrix composites analyzed by acoustic emission monitoring. *J. Mater. Sci.*, **49**, 7890–7899, 2014.

24. E. Volkmann, L. Lima Evangelista, K. Tushtev, D. Koch, C. Wilhelmi, and K. Rezwan, Oxidation-induced microstructural changes of a polymer derived Nextel™ 610 ceramic composite and impact on the mechanical performance. *J. Mater. Sci.*, **49**, 710–719, 2014.

25. S. Lee, L. Zawada, J. Staehler, and C. Folsom, Mechanical behavior and high-temperature performance of a Woven Nicalon™/Si-N-C ceramic-matrix composite. *J. Am. Ceram. Soc.*, **81**, 1797–1811, 1998.

26. R. A. Jurf and S. C. Butner, Advances in oxide-oxide CMC. *J. Eng. Gas Turbines Power Trans ASME*, **122**, 202–205, 2000.

27. M. B. Ruggles-Wrenn and T. Jones, Tension-compression fatigue behavior of a SiC/SiC ceramic matrix composite at 1200°C in air and in steam. *Int. J. Fatig.*, **47**, 154–160, 2013.

28. M. B. Ruggles-Wrenn, S. Mall, C. A. Eber, and L. B. Harlan, Effects of steam environment on high-temperature mechanical behavior of Nextel™720/alumina (N720/A) continuous fiber ceramic composite. *Compos. A*, **37**, 2029–2040, 2006.

29. M. Bouquet, J. M. Birbis, and J. M. Quenisset, toughness assessment of ceramic matrix composites. *Compos. Sci. Technol.*, **37**, 223–248, 1990.

30. G. Camus, L. Guillaumat, and S. Baste, Development of damage in a 2D Woven C/SiC composite under mechanical loading: I. Mechanical characterization. *Compos. Sci. Technol.*, **56**, 1363–1372, 1996.

31. D. M. Wilson and L. R. Visser, High Performance oxide fibers for metal and ceramic composites. *Compos. A*, **32**, 1143–1153, 2001.

32. D. M. Wilson, S. L. Lieder, and D. C. Luenegurg, microstructure and high-temperature properties of Nextel 720 fibers. *Ceram. Eng. Sci. Proc.*, **16(5)**, 1005–1014, 1995.

33. S. Wannaparhun and S. Seal, A combined spectroscopic and thermodynamic investigation of Nextel-720/alumina ceramic matrix composite in air and water vapor at 1100°C. *J. Am. Ceram. Soc.*, **86**, 1628–1630, 2003.

34. B. W. Rosen, Mechanics of Composite Strengthening. Ch 3. In: *Fiber Composite Materials*, Metals Park, OH: ASM, 1965.

35. L. B. Greszczuk, Microbuckling failure of circular fiber-reinforced composites. *AIAA J.*, **13**, 1311–1318, 1975.

36. A .S. D. Wang, A Non-Linear Microbuckling Model Predicting the Compressive Strength of Unidirectional Composites. ASME Paper 78-WA/Aero 1, 1978.

37. H. T. Hahn and J. G. Williams, Compressive Failure Mechanisms in Unidirectional Composites. In: Whitney JM, editor, *Composite Materials: Testing and Design (Seventh Conference)*, *ASTM STP 893*, American Society for Testing and Materials, West Conshohocken, Pennsylvania, 115–139, 1986.

38. J. Lankford, The effect of hydrostatic pressure and loading rate on compressive failure of fiber-reinforced ceramic-matrix composites. *Compos. Sci. Tech.*, **51**, 537–543, 1994.

39. J. Lankford, Compressive failure of fiber-reinforced composites: Buckling, kinking, and the role of the interphase. *J. Mat. Sci.*, **30**, 4343–4348, 1995.

40. J. Lankford, Shear versus dilatational damage mechanisms in the compressive failure of fiber-reinforced composites. *Compos. A*, **26**, 213–222, 1997.

41. M. Wang and C. Laird, Damage and fracture of a cross Woven C/SiC composite subject to compression loading. *J. Mat. Sci.*, **31**, 2065–2069, 1996.

42. R. Arrowood and J. Lankford, compressive fracture processes in an alumina/glass composite. *J. Mat. Sci.*, **22**, 3737–3747, 1987.

43. J. Lankford, Strength of monolithic and fiber-reinforced glass ceramics at high rates of loading and elevated temperature. *Ceram. Eng. Sci. Proc.*, **9**, 843–849, 1988.

chapter thirty seven

Effects of steam environment on fatigue behavior of two SiC/[SiC+Si$_3$N$_4$] ceramic composites at 1300°C

Marina B. Ruggles-Wrenn and Vipul Sharma

Contents

Abstract: The fatigue behaviors of two SiC/[SiC+Si$_3$N$_4$] ceramic matrix composites (CMC) were investigated at 1300°C in laboratory air and in steam. Composites consisted of a crystalline [SiC+Si$_3$N$_4$] matrix reinforced with either Sylramic™ or Sylramic-iBN fibers (treated Sylramic™ fibers that possess an *in situ* BN coating) woven in a five-harness satin weave fabric and coated with a proprietary boron-containing dual-layer interphase. The tensile stress–strain behaviors were investigated and the tensile properties measured at 1300°C. Tension-tension fatigue behaviors of both CMCs were studied for fatigue stresses ranging from 100 to 180 MPa. The fatigue limit (based on a run-out condition of 2 × 10^5 cycles) in both air and steam was 100 MPa for the CMC containing Sylramic™ fibers and 140 MPa for the CMC reinforced with Sylramic-iBN fibers. At higher fatigue stresses, the presence of steam caused noticeable degradation in fatigue performance of both composites. The retained strength and modulus of all run-out specimens were characterized. The materials tested in air retained 100% of their tensile strength, whereas the materials tested in steam retained only about 90% of their tensile strength.

Keywords: Ceramic-matrix composites (CMCs), Fatigue, High-temperature properties, Fractography

37.1   Introduction

Advances in power generation systems for aircraft engines, land-based turbines, rockets, and, most recently, hypersonic missiles and flight vehicles have raised the demand for structural materials that have superior long-term mechanical properties under high temperature, high pressure, and varying environmental factors, such as moisture. On account of their low density, high strength, and fracture toughness at high temperatures, silicon carbide fiber-reinforced silicon-carbide matrix composites are currently being evaluated for aircraft engine hot-section components [1–4]. In these applications, the composites will be subjected to sustained and cyclic loadings at elevated temperatures in oxidizing environments. Therefore a thorough understanding of high-temperature mechanical behavior and performance of SiC/SiC composites in service environments is critical to design with, and life prediction for these materials.

The main advantage of CMCs over monolithic ceramics is their superior toughness, tolerance to the presence of cracks and defects, and noncatastrophic mode of failure. This key advantage is achieved through a proper design of a fiber/matrix interphase, which serves to deflect matrix cracks and to prevent early failure of the fibrous reinforcement [5–9]. The most significant problem hindering SiC-fiber-containing CMCs is oxidation embrittlement [10]. Typically the embrittlement occurs once oxygen enters through the matrix cracks and reacts with the fibers and the fiber coatings [11–13]. The degradation of fibers and fiber coatings is typically accelerated by the presence of moisture [14–16]. Composite degradation may be further accelerated by cyclic loading, as the reaction gases are expelled from matrix cracks during unloading, and oxidizing environment is drawn into the composite through the matrix cracks during reloading [10].

Several recent studies evaluated mechanical behavior of high-performance SiC/SiC CMCs at elevated temperature. Morscher et al. [17] studied creep at 1315°C in air of the composites consisting of high modulus SiC fibers (Hi-Nicalon S) and a melt-infiltrated SiC matrix. Ojard et al. [18,19] and Morscher et al. [4] reported on the elevated-temperature mechanical performance of a ceramic composite, consisting of a melt-infiltrated SiC matrix reinforced with Sylramic-iBN SiC fibers, a CMC system developed at NASA Glenn Research Center [20]. The present chapter aims to evaluate the fatigue behavior of two high-performance polycrystalline SiC fiber-reinforced ceramic composites, which are made by polymer infiltration and pyrolysis (PIP) method. Polymer infiltration and pyrolysis is an attractive processing approach because of its relatively low cost. In addition, this processing method allows near-net-shape molding and fabrication, resulting in nearly fully dense composites [21–23]. The materials studied in this effort are reinforced with Sylramic™ and Sylramic-iBN fibers. Fatigue tests were conducted at 1300°C in air and steam environments for stress levels ranging from 100 to 180 MPa. Resulting fatigue performance imposes limitations on the use of these materials in high-temperature applications.

37.2   Material and experimental arrangements

The materials studied were two SiC fiber-reinforced composites with a PIP-derived crystalline [SiC+Si_3N_4] matrix manufactured by COI Ceramics, Inc. (San Diego, CA). The first composite (Syl/[SiC+Si_3N_4]) was reinforced with Sylramic™ fibers, whereas the second composite (Syl-iBN/[SiC+Si_3N_4]) contained Sylramic-iBN, that is, treated Sylramic™ fibers that possess an *in situ* BN coating. In processing of both composites, the woven five-harness satin weave (5HSW) fiber cloth was coated with a proprietary boron-containing dual-layer interphase, stacked and infiltrated with a mixture of polymer, filler particles, and solvent.

In the case of the Syl-iBN/[SiC+Si$_3$N$_4$] composite, additional filler designed to inhibit oxidation was also introduced. During pyrolysis at temperatures >1000°C, the polymer was pyrolyzed to a crystalline [SiC+Si$_3$N$_4$] matrix. The infiltration and pyrolysis procedure was repeated several times to increase the density of the matrix. Both composites were composed of 12 plies of woven fabric in a 0°/90° layup, with a finished fiber volume of ~42% and an open porosity of ~3%.

A servocontrolled MTS mechanical testing machine equipped with hydraulic water-cooled wedge grips, a compact two-zone resistance-heated furnace, and two temperature controllers were used in all tests. An MTS TestStar II digital controller was employed for input signal generation and data acquisition. Strain measurement was accomplished with an MTS high-temperature air-cooled uniaxial extensometer of 12.5 mm gage length. Tests in steam environment employed an alumina susceptor (tube with end caps), which fits inside the furnace. The specimen gage section is located inside the susceptor, with the ends of the specimen passing through slots in the susceptor. Steam is introduced into the susceptor (through a feeding tube) in a continuous stream with a slightly positive pressure, expelling the dry air and creating a near 100% steam environment inside the susceptor. For elevated temperature testing, thermocouples were bonded to the specimen using alumina cement (Zircar) to calibrate the furnace on a periodic basis. The furnace controllers (using noncontacting thermocouples exposed to the ambient environment near the test specimen) were adjusted to determine the settings needed to achieve the desired temperature of the test specimen. The determined settings were then used in actual tests. The power settings for testing in steam were determined by placing the specimen instrumented with thermocouples in steam environment and repeating the furnace calibration procedure. Fracture surfaces of failed specimens were examined using scanning electron microscope (SEM) (FEI Quanta 200 HV) as well as an optical microscope (Zeiss Discovery V12).

All tests were performed at 1300°C. Dog bone shaped specimens of total length 152 mm with an 8-mm-wide gage section were used in all tests. In all tests, a specimen was heated to test temperature at 1°C/min, and held at temperature for additional 25 min prior to testing. In air, tensile tests were performed in stroke control with a constant displacement rate of 0.05 mm/s. Tension-tension fatigue tests were conducted in load control with an R ratio (minimum to maximum stress) of 0.05 at a frequency of 1.0 Hz. Fatigue run-out was set to 20×10^5 cycles. The 2×10^5 cycle count represents the number of loading cycles expected in aerospace applications at that temperature. Fatigue run-out limits were defined as the highest stress level, for which run-out was achieved. Cyclic stress–strain data were recorded throughout each test. Thus stiffness degradation and strain accumulation with fatigue cycles and/or time could be examined. All specimens that achieved run-out were subjected to tensile test to failure at 1300°C in laboratory air to determine the retained strength and stiffness. It is noteworthy that in all tests reported below, the failure occurred within the gage section of the extensometer.

37.3   Results and discussion

37.3.1   Monotonic tension

Tensile stress–strain behavior of the two composites at 1300°C is typified in Figure 37.1. For Syl/[SiC+Si$_3$N$_4$], the average ultimate tensile strength (UTS) was 188 MPa, elastic modulus, 129 GPa, and failure strain, 0.16%. For Syl-iBN/[SiC+Si$_3$N$_4$], the average UTS was 241 MPa, elastic modulus, 147 GPa, and failure strain 0.25%. The Syl-iBN/[SiC+Si$_3$N$_4$] composite displayed an ultimate strength advantage of ~53 MPa over the Syl-iBN/[SiC+Si$_3$N$_4$] CMC.

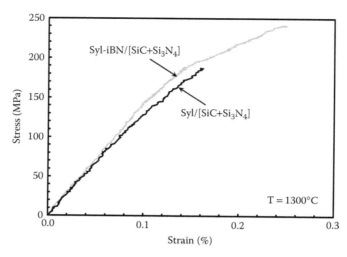

Figure 37.1 Tensile stress–strain curves obtained for Syl/[SiC+Si$_3$N$_4$] and Syl-iBN/[SiC+Si$_3$N$_4$] ceramic composites at 1300°C in laboratory air.

This is attributed primarily to the *in situ* BN coating that protects the fibers from detrimental environmental effects introduced during processing [24]. The average proportional limit for Syl-iBN/[SiC+Si$_3$N$_4$] was determined to be a high 144 MPa. The stress–strain curve obtained for Syl-iBN/[SiC+Si$_3$N$_4$] exhibits a nearly bilinear behavior typical of a brittle ceramic composite. The stress–strain behavior is linear up to the proportional limit, where nonlinear behavior caused by matrix cracking occurs. Afterward, the stress–strain curve continues with a decreased slope. Conversely, the stress–strain curve obtained for Syl/[SiC+ Si$_3$N$_4$] at 1300°C is nearly linear to failure and does not exhibit a clear proportional limit.

37.3.2 Tension-tension fatigue

Results of the tension-tension fatigue tests are summarized in Table 37.1. Figure 37.2 shows stress versus cycles to failure (S–N) curves for both composites. For Syl/[SiC+Si$_3$N$_4$] the fatigue limit was 100 MPa (53%UTS) at 1300°C in air and in steam. For Syl-iBN/[SiC+Si$_3$N$_4$] the fatigue limit was 160 MPa (66%UTS) in air and 140 MPa (58%UTS) in steam. Presence of steam noticeably degrades fatigue performance of both composites at higher fatigue stress levels. For Syl/[SiC+Si$_3$N$_4$], the reduction in fatigue life due to steam was 48% at the fatigue stress of 120 MPa and 77% at the fatigue stress of 140 MPa. For Syl-iBN/[SiC+Si$_3$N$_4$], the loss of fatigue life due to steam was 23% at the fatigue stress of 160 MPa and 75% at the fatigue stress of 180 MPa.

Maximum and minimum strains as functions of cycle number obtained at 1300°C in air and in steam are presented in Figure 37.3a and b for Syl/[SiC+Si$_3$N$_4$] and Syl-iBN/ [SiC+Si$_3$N$_4$], respectively. It is seen that ratcheting (progressive strain accumulation with cycles) takes place in all fatigue tests conducted in this study. For both composites, ratcheting develops gradually, and is more noticeable in the latter part of the fatigue tests. Earlier onset of ratcheting is observed in tests with higher fatigue stress levels. Results in Figure 37.3 reveal that the Syl-iBN/[SiC+Si$_3$N$_4$] produced larger maximum strains during fatigue cycling than the Syl/[SiC+Si$_3$N$_4$]. At 1300°C in air, maximum strains produced by Syl/[SiC+Si$_3$N$_4$] ranged from 0.14% to 0.16%, whereas those produced by Syl-iBN/[SiC+Si$_3$N$_4$] ranged from 0.19% to 0.21%. As shown in Figure 37.3, both composites generated somewhat

Table 37.1 Summary of fatigue results for Syl/[SiC+Si$_3$N$_4$] and Syl-iBN/[SiC+Si$_3$N$_4$] ceramic composites at 1300°C in laboratory air and steam environments

Test environment	Max stress (MPa)	Cycles to failure	Failure strain (%)
$J!J![SiC+SisN4]			
Laboratory air	100	200,000*	0.055
Laboratory air	120	28,515	0.027
Laboratory air	140	10,104	0.036
Steam	100	200,000*	0.088
Steam	120	14,688	0.054
Steam	140	2,329	0.060
Syl-iBNl [SiC+SisN4]			
Laboratory air	140	200,000*	0.073
Laboratory air	160	200,000*	0.080
Laboratory air	180	22,808	0.072
Steam	140	200,000*	0.101
Steam	160	153,143	0.106
Steam	180	5,765	0.098

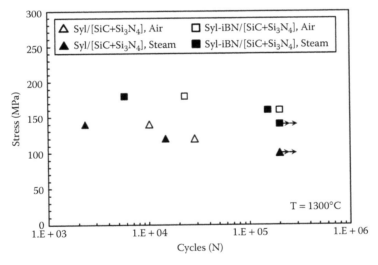

Figure 37.2 Fatigue S-N curves for Syl/[SiC+Si$_3$N$_4$] and Syl-iBN/[SiC+Si$_3$N$_4$] ceramic composites at 1300°C in air and in steam. Arrow indicates that failure of specimen did not occur when the test was terminated.

larger strains in steam than in air. At 1300°C in steam, maximum strains reached 0.17% for Syl/[SiC+Si$_3$N$_4$] and 0.22% for Syl-iBN/[SiC+Si$_3$N$_4$].

Of interest is the reduction in modulus (hysteresis modulus determined from the maximum and minimum stress–strain data points during a load cycle), reflecting the damage development during fatigue cycling. Figure 37.4 shows change in normalized modulus (i. e., modulus normalized by the modulus obtained on the first cycle) with fatigue cycles. It is noteworthy that although some tests achieved run-out of a small decrease in normalized modulus with cycling was still observed. This drop in modulus is hypothesized to be due to a low level of microcracking. As seen in Figure 37.4, modulus loss increased

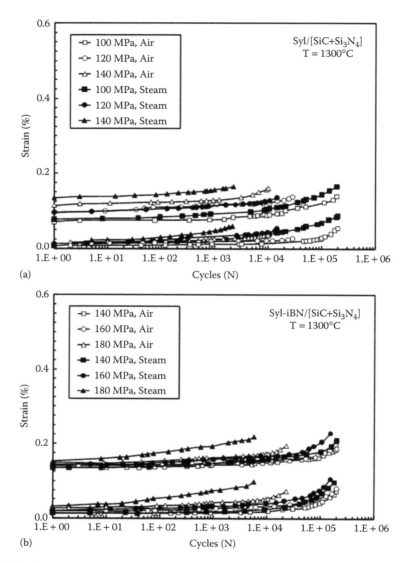

Figure 37.3 Maximum and minimum strains as functions of cycle number at 1300°C in air and in steam for: (a) Syl/[SiC+Si$_3$N$_4$] and (b) Syl-iBN/[SiC+Si$_3$N$_4$] ceramic composites.

with increasing fatigue stress for both composites. Continuous drop in modulus observed at higher fatigue stress levels suggests progressive damage accumulation with continued cycling. Furthermore, decrease in normalized modulus becomes more pronounced in steam environment for both CMCs. In the case of Syl/[SiC+Si$_3$N$_4$], the normalized modulus loss was limited to 15% in air and to 22% in steam. In the case of Syl-iBN/[SiC+Si$_3$N$_4$], the decrease in normalized modulus reached 25% in air and 33% in steam. This suggests accelerated damage growth in steam. The damage development in these composites is likely to proceed by the oxidation-induced growth of matrix cracks. Progressive matrix cracking exposes fibers to oxidizing environment and thus accelerates intrinsic fiber degradation. Moreover, once oxygen enters the matrix cracks, BN and SiC react to form gaseous species

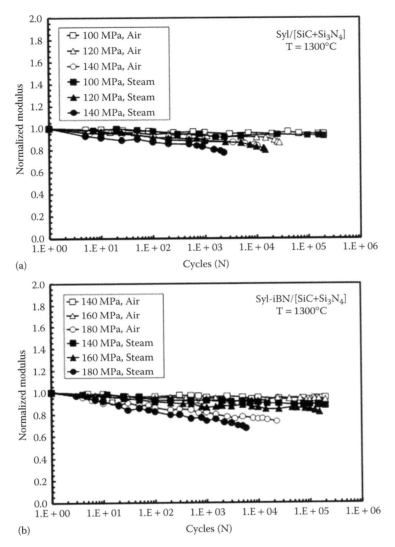

Figure 37.4 Normalized modulus versus fatigue cycles at 1300°C in air and in steam for: (a) Syl/[SiC+Si$_3$N$_4$] and (b) Syl-iBN/[SiC+Si$_3$N$_4$] ceramic composites.

and solid borosilicate reaction products that bond exposed fibers together [4], causing fiber failures due to local stress concentration from load sharing created by strongly fused fibers.

Retained strength and stiffness of the fatigue specimens that achieved run-out are summarized in Table 37.2. Evaluation of retained properties is useful in assessing the damage state of the composite subjected to prior loading. It is seen that specimens of both composites subjected to prior fatigue in air exhibited no loss of tensile strength, irrespective of the fatigue stress level. However, modulus loss of 5–6% was observed for both CMCs. In contrast, prior fatigue in steam caused reduction in both strength and stiffness of the two composites. Strength loss in steam was ~10% for both materials. Modulus loss in steam was 7% for Syl/[SiC+Si$_3$N$_4$] and 12% for Syl-iBN/[SiC+Si$_3$N$_4$].

Table 37.2 Retained properties of the Syl/[SiC+Si$_3$N$_4$] and Syl-iBN/[SiC+Si$_3$N$_4$] specimens subjected to prior fatigue in laboratory air and in steam environment at 1300°C; retained properties measured at 1300°C in laboratory air

Fatigue stress (MPa)	Fatigue environment	Retained strength (MPa)	Strength retention (%)	Retained modulus (GPa)	Modulus retention (%)	Strain at failure (%)
§JJJ![SiC+SisN4]						
100	Air	206	100	122	95	0.20
100	Steam	171	91	120	93	0.15
Syl-iBNl [SiC+SisN4]						
140	Air	246	100	140	95	0.27
160	Air	247	100	137	94	0.29
140	Steam	215	89	129	88	0.23

37.3.3   Composite microstructure

All specimens tested in this study were examined with optical and SEMs to elucidate failure and damage mechanisms. Optical micrographs of fractured Syl/[SiC+Si$_3$N$_4$] and Syl-iBN/[SiC+Si$_3$N$_4$] specimens are shown in Figure 37.5a and b, respectively. The fracture surfaces in Figure 37.5 are similar in appearance, both are relatively flat and perpendicular to the loading direction. Note that the fracture surfaces in Figure 37.5 are typical and representative of all optical micrographs obtained in this study. Furthermore, no distinctive features attributable to test type (monotonic tension vs. fatigue) or test environment (air vs. steam) could be discerned at the low magnification such as that of Figure 37.5. Therefore for the sake of brevity, optical micrographs of other fractured specimens are not shown.

The SEM micrographs of the fracture surfaces of the Syl/[SiC+Si$_3$N$_4$] specimen tested in fatigue at 1300°C in air and in steam are shown in Figures 37.6 and 37.7, respectively. All images obtained at lower magnification in air (Figure 37.6a and d) and in steam (Figure 37.7a and d) show very similar nearly planar fracture surfaces. Yet the specimen in Figure 37.6a achieved fatigue run-out in air and failed in a subsequent tension test, whereas the specimen in Figure 37.7d failed after only 2,329 fatigue cycles in steam. The influence of test environment becomes noticeable in images obtained at intermediate magnification.

(a)   (b)

Figure 37.5 Fracture surfaces obtained in fatigue tests conducted at 1300°C in air: (a) Syl/[SiC+Si$_3$N$_4$], σ_{max} = 140 MPa and (b) Syl-iBN/[SiC+Si$_3$N$_4$], σ_{max} = 140 MPa.

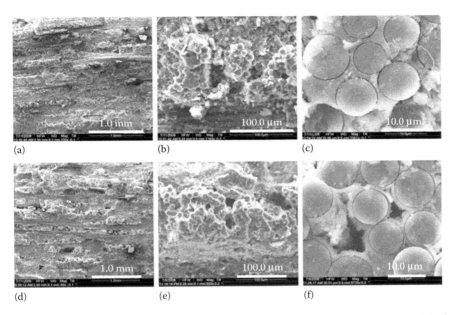

Figure 37.6 SEM micrographs of fracture surfaces of Syl/[SiC+Si$_3$N$_4$] specimens tested in fatigue at 1300°C in air: (a)–(c) with σ_{max} = 100 MPa and (d)–(f) with σ_{max} = 140 MPa.

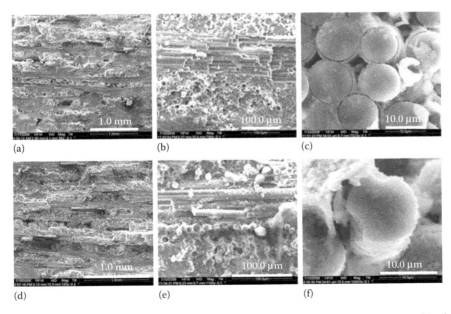

Figure 37.7 SEM micrographs of fracture surfaces of Syl/[SiC+Si$_3$N$_4$] specimens tested in fatigue at 1300°C in steam: (a)–(c) with σ_{max} = 100 MPa and (d)–(f) with σ_{max} = 140 MPa.

The SEM micrographs of specimens tested in air (Figure 37.6b and e) show fracture along different planes, fiber debonding, and pullout of fiber bundles. In contrast, the fracture surfaces of the specimens tested in steam (Figure 37.7b and e) show fracture surfaces that are almost completely planar. The degrading effects of steam environment as well as those of higher fatigue stress level are clearly revealed in higher magnification images. Consider the specimen in Figure 37.6c that achieved run-out with $\sigma_{max} = 100$ MPa in air. Its fibers, fiber coating, and matrix show little or no physical degradation. In contrast, the matrix of the specimen subjected to fatigue in air with $\sigma_{max} = 140$ MPa (Figure 37.6f) appears to be severely damaged, although its fibers and fiber coating remain relatively intact. Similar observations can be made regarding the specimen that achieved run-out with $\sigma_{max} = 100$ MPa in steam (Figure 37.7c). Although its fibers and fiber coating appear undamaged, the matrix between the fibers has disintegrated in places. Severest degradation is seen in the case of specimen tested in steam with $\sigma_{max} = 140$ MPa (Figure 37.7e), which also produced the shortest fatigue life of 2,329 cycles. The fibers are severely damaged, the fiber coating is absent, and matrix appears to have disintegrated.

Figures 37.8 and 37.9 show the fracture surfaces of the Syl-iBN/[SiC+Si$_3$N$_4$] specimens tested in fatigue at 1300°C in air and in steam, respectively. Notably the SEM micrographs obtained at low magnification in air (Figure 37.8a and d) and in steam (Figure 37.9a and d) are alike. All show the nearly planar fracture surfaces. Furthermore the fracture surfaces in Figures 37.8a and d and Figures 37.9a and d differ little from the low magnification images obtained for the Syl/[SiC+Si$_3$N$_4$] composite. The same can be said about the images obtained at intermediate magnification (Figure 37.8b and e) and (Figure 37.9b and e). At this magnification the characteristics observed in the Syl-iBN/[SiC+Si$_3$N$_4$] fracture surfaces are akin to those seen in the Syl/[SiC+Si$_3$N$_4$] fracture surfaces in Figure 37.6b and e) and Figures 37.7b and e. Higher magnification images in Figures 37.8c and f and Figures 37.9c and f are of more interest. Fracture surface of the Syl-iBN/[SiC+Si$_3$N$_4$] specimen that achieved fatigue run-out at 140 MPa in air (Figure 37.8c) reveals some damage to

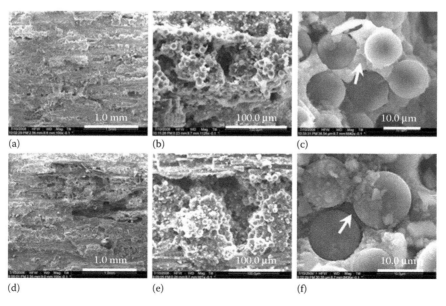

Figure 37.8 SEM micrographs of fracture surfaces of Syl-iBN/[SiC+Si$_3$N$_4$] specimens tested in fatigue at 1300°C in air: (a)–(c) with $\sigma_{max} = 140$ MPa and (d)–(f) with $\sigma_{max} = 180$ MPa.

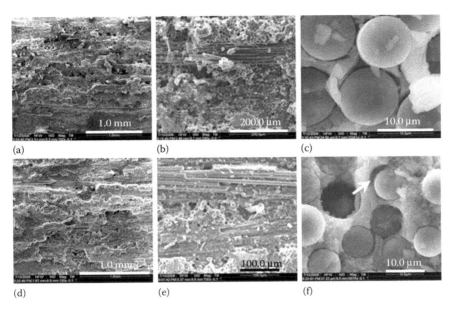

Figure 37.9 SEM micrographs of fracture surfaces of Syl-iBN/[SiC+Si$_3$N$_4$] specimens tested in fatigue at 1300°C in steam: (a)–(c) with σ_{max} = 140 MPa and (d)–(f) with σ_{max} = 180 MPa.

the matrix and degradation of the interphase (arrow). However, fibers appear to be undamaged. The fracture surface in Figure 37.8f of the specimen that failed after 22,808 cycles with σ_{max} = 140 MPa in air also shows damage to the interphase (arrow), whereas the fibers remain intact. The damage to the interphase becomes more pronounced in steam, especially in tests of longer duration. Figure 37.9c shows the fracture surface of the specimen that achieved run-out with σ_{max} = 140 MPa in steam. Interphase material is severely degraded, matrix oxidation is also evident. Yet individual fibers still appear to be physically intact. The fracture surface of the specimen fatigued with σ_{max} = 180 MPa in steam (Figure 37.9e) also shows degradation of the interphase (arrow) and matrix oxidation. Once again, the fibers do not appear to have suffered significant damage. It is possible that the *in situ* grown BN layer provided an oxidation resistant physical barrier that delayed fiber damage whenever the fiber tows were exposed to oxidizing environment during matrix cracking.

37.4   Concluding remarks

The tensile stress–strain behaviors of the Syl/[SiC+Si$_3$N$_4$] and Syl-iBN/[SiC+Si$_3$N$_4$] composites were investigated and the tensile properties measured at 1300°C. The UTS of the Syl-iBN/[SiC+Si$_3$N$_4$] composite was ~28% higher than that of the Syl/[SiC+Si$_3$N$_4$] composite. This improvement in ultimate tensile strength is attributed to the *in situ* BN coating that protects fibers from environmental degradation during processing.

Tension-tension fatigue behavior of both composites was studied at 1300°C in air and in steam. Fatigue stress levels ranged from 100 to 140 MPa for the Syl/[SiC+Si$_3$N$_4$] composite and from 140 to 180 MPa for the Syl-iBN/[SiC+Si$_3$N$_4$] composite. Of the two materials studied, the Syl-iBN/[SiC+Si$_3$N$_4$] composite exhibits a considerably better fatigue performance at 1300°C in both air and steam environments. The fatigue limit of the Syl/[SiC+Si$_3$N$_4$] composite is 100 MPa (53%UTS) in both air and steam. The fatigue limit of the Syl-iBN/[SiC+Si$_3$N$_4$] CMC is 160 MPa (66%UTS) in air and 140 MPa (58%UTS) in steam. The presence of steam degrades fatigue performance of both materials at higher fatigue stress levels. The detrimental effect of steam on fatigue life of the Syl/[SiC+Si$_3$N$_4$] composite become noticeable as the fatigue stresses exceed 64%UTS. For Syl-iBN/[SiC+ Si$_3$N$_4$], the damaging effects of steam become significant when the fatigue stresses exceeds 67%UTS.

Both composites retain 100% of their tensile strength after achieving fatigue run-out at 1300°C in air. However, modulus loss of 5–6% is observed for both materials. Following fatigue run-out in steam both materials retain only ~90% of their tensile strength. Modulus loss is limited to 7% for Syl/[SiC+Si$_3$N$_4$] and to 12% for Syl-iBN/[SiC+Si$_3$N$_4$].

Acknowledgment

The authors thank Dr. G. Fair for many valuable discussions. The financial support of the Air Force Research laboratory Materials and Manufacturing Directorate (Dr. M. Cinibulk) is highly appreciated.

References

1. Brewer, D.: HSR/EPM combustor materials development program. *Mater Sci Eng A.* A261, 284–91 (1999).
2. Brewer, D., Ojard, F., Gibler, M.: Ceramic matrix composite combustor liner rig test. ASME Turbo Expo 2000, Munich Germany, May 8–11, 2000, ASME Paper 2000-GT-0670.
3. Corman, G.S., Luthra, K.: Silicon melt infiltrated ceramic composites (HiPerComp). In: Bansal, N. (ed.) *Hand book of ceramic composites*, 99–115. Kluwer Academic, NY (2005).
4. Morscher, G.N., Ojard, G., Miller, R., Gowayed, Y., Santhosh, U., Ahmad, J., John, R.: Tensile creep and fatigue of Sylramic-iBN melt-infiltrated SiC matrix composites: Retained properties, damage development, and failure mechanisms. *Comp Sci Tech.* 68, 3305–13 (2008).
5. Evans, A.G., Zok, F.W.: Review: The physics and mechanics of fiber-reinforced brittle matrix composites. *J. Mater. Sci.* 29, 3857–3896 (1994).
6. Kerans, R.J., Parthasarathy, T.A.: Crack deflection in ceramic composites and fiber coating design criteria. *Composites A* 30, 521–524 (1999).
7. Kerans, R.J., Hay, R.S., Parthasarathy, T.A., Cinibulk, M.K.: Interface design for oxidation-resistant ceramic composites. *J. Am. Ceram. Soc.* 85(11), 2599–2632 (2002).
8. Marshall, D., Evans, A.G.: *Acta Metall.* 37, 2567–83 (1989).
9. Naslain, R.: Design, preparation and properties of non-oxide CMCs for application in engines and nuclear reactors: an overview. *Comp Sci Tech.* 64, 155–170 (2004).
10. McNulty, J.C., He, M.Y., Zok, F.W.: Notch sensitivity of fatigue life in a Sylramic™/SiC composite at elevated temperature. *Comp Sci Tech.* 61, 1331–38 (2001).
11. Prewo, K.M., Batt, J.A.: The oxidative stability of carbon fibre reinforced glass-matrix composites. *J. Mater. Sci.* 23, 523–527 (1988).
12. Mah, T., Hecht, N.L., McCullum, D.E., Hoenigman, J.R., Kim, H.M., Katz, A.P., Lipsitt, H.A.: Thermal stability of SiC fibres (Nicalon). *J. Mater. Sci.* 19, 1191–1201 (1984).
13. Heredia, F.E., McNulty, J.C., Zok, F.W., Evans, A.G.: An oxidation embrittlement probe for ceramic matrix composites. *J. Am. Ceram. Soc.* 78, 2097–100 (1995).
14. More, K.L., Tortorelli, P.F., Ferber, M.K., Keiser, J.R.: Observations of accelerated silicon carbide recession by oxidation at high water-vapor pressures. *J. Am. Ceram. Soc.* 83(1), 211–213 (2000).
15. More, K.L., Tortorelli, P.F., Ferber, M.K., Walker, L.R., Keiser, J.R., Brentnall, W.D., Miralya, N., Price, J.B.: Exposure of ceramic and ceramic-matrix composites in simulated and actual combustor environments. In: *Proceedings of international gas turbine and aerospace congress.* Paper No. 99-GT-292 (1999).

16. Ferber, M.K., Lin, H.T., Keiser, J.R.: Oxidation behavior of non-oxide ceramics in a high-pressure, high-temperature steam environment. In: Jenkins, M.G., Lara-Curzio, E., Gonczy, S.T. (eds) Mechanical, Thermal, and Environmental Testing and Performance of Ceramic Composites and Components. 210–215. ASTM STP 1392, American Society for Testing and Materials, OH (2000).
17. Morscher, G.N., Pujar, V.V.: Creep and stress-strain behavior after creep for SiC fiber reinforced, melt-infiltrated SiC matrix composites. *J. Am. Ceram. Soc.* 89(5), 1652–1658 (2006).
18. Ojard, G., Gowayed, Y., Chen, J., Santhosh, U., Ahmad, J., Miller, R., John, R.: Time-dependent response of MI SiC/SiC composites part I: standard samples. *Ceram. Eng. Sci. Proc.* 28(2), 145–154 (2007).
19. Ojard, G., Calomino, A., Morscher, G., Gowayed, Y., Santhosh, U., Ahmad, J., Miller, R., John, R.: Post creep/dwell fatigue testing of MI SiC/SiC composites. *Ceram. Eng. Sci. Proc.* 28(2), 135–143 (2007).
20. DiCarlo, J.A., Yun, H.-M., Morscher, G.N., Bhatt, R.T.: SiC/SiC composites for 1200°C and above. In: Bansal, N. (ed.) *Hand book of ceramic composites,* 77–98. Kluwer Academic, NY (2005).
21. Lundberg, R., Pompe, R., Carlsson, R., Goursat, P.: Fibre reinforced silicon nitride composites. *Comp. Sci. Tech.* 37, 165–176 (1990).
22. Sirieix, F., Goursat, P., Lemcote, A., Dauger, A.: Pyrolysis of polysilazanes: relationship between precursor architecture and ceramic microstructure. *Comp. Sci. Tech.* 37, 7–19 (1990).
23. Gonon, M.F., Fantozzi, G., Murat, M., Disson, J.P.: Association of the CVI process and of the use of polysilazane precursor for the elaboration of ceramic matrix composites reinforced by continuous fibres. *J. Eur. Ceram. Soc.* 15, 185–190 (1995).
24. Yun, H.M., Gyekenyesi, J.Z., Chen, Y.L., Wheeler, D.R., DiCarlo, J.A.: Tensile behavior of SiC/SiC composites reinforced by treated Sylramic SiC fibers. *Ceram. Eng. Sci. Proc.* 22(3), 521–531 (2001).

section three

Application section

chapter thirty eight

3D product design, evaluation, justification, and integration

Adedeji B. Badiru

Contents

Abstract: The emerging proliferation of 3D printing has made it imperative that careful and structural assessment be instituted for 3D-printing products. The conventional product development environment is vastly different from what 3D printing will require. Hitherto, individuals and organizations have been jumping on the 3D-printing bandwagon without strategic consideration of downstream and upstream aspects of 3D printing of products. This chapter introduces the application of the existing DEJI model for 3D-product design, evaluation, justification, and integration. The model is recommended as a complementary approach to executing new 3D-printing products. The approach will facilitate a better alignment of product technology with future development and needs. Some of the benefits that can be derived from using a systems model to guide 3D-printing product development include operational effectiveness, raw material efficiency, higher return on investment curve, rapid product deployment, growth potentials, flexibility, and anywhere-anytime production agility.

38.1 Introduction to DEJI model

Making things matter matters. Making things is essential for economic development. But, how do we make things from a systems perspective to meet all requirements? This chapter advocates using a structured approach to link all aspects of making things. Systems integration is crucial in highly technical products, not only for the current operational need,

D	E	J	I
Design	Evaluation	Justification	Integration
Design product for execution in a 3D-production environment.	Evaluate proposed 3D-printing product for operational requirements.	Justify 3D printing on account of cost, time, and performance.	Integrate 3D-printing product with conventional products and operations.

Figure 38.1 Stages of implementation of *DEJI* model for product development.

but also for future operations in a dynamic environment. Due to its dynamism and multifaceted operations, the aerospace industry can use a new perspective for aerospace technology capitalizing on the emergence of 3D printing. Applying a technique such as the DEJI (Design, Evaluate, Justify, and Integrate) model introduced by Badiru (2012) will call early attention to integration needs of conventional products and 3D-printing products. Figure 38.1 illustrates the DEJI model for a product life cycle.

The technique (Badiru, 2010) is unique among product development tools and techniques because it explicitly calls for a rejustification of the product within the product development life cycle. This is important for the purpose of determining when a program should be terminated even after going into production and what realignment of resources may be needed to keep the product current with new technological developments. If the program is justified, it must then be integrated and *accepted* within the ongoing business of the enterprise (Giachetti, 2010). Department of Defense (DoD) has expressed the desire to have an integrated design and redesign of a product as it goes through its lifecycle. The *DEJI* model facilitates such a recursive design-evaluate-justify-integrate process for product evolution feedback looping. The biggest challenge for any program management endeavor is coordinating and integrating the multiple facets that affect the final outputs of a program, where a specific output may be a physical product, a service, or a desired result. Addressing the challenges of program execution from a systems perspective increases the likelihood of success. The *DEJI* model can facilitate program success through structural implementation of design, evaluation, justification, and integration. Although originally developed for product development projects, the model is generally applicable to all types of programs because every program goes through the stages of process design, evaluation of parameters, justification of the product, and integration of the product into the core business of the organization. The model can be applied across the spectrum of the following elements of an organization:

1. People
2. Process
3. Technology

Table 38.1 presents the implementation elements of the *DEJI* model. The model is complemented by existing tools and techniques of process improvement, such as DMAIC (Define, Measure, Analyze, Improve, Control), SIPOC (Suppliers, Inputs, Process, Outputs, Customers), DRIVE (Define, Review, Identify, Verify, Execute), PDCA (Plan, Do, Check, Act), 6S (Sort, Stabilize, Shine, Standardize, Sustain, Safety), CEDAC (Cause and Effect

Table 38.1 Taxonomy of implementation elements of DEJI model

DEJI model	Functional characteristics	Tools, techniques, and models
Design	• Define goals • Set functional derivatives • Set performance metrics • Identify milestones • Assess credibility of design • Assess agility of design • Assess stability of design • Assess integrity of design • Assess flexibility of design • Assess eligibility of design • Assess sustainability of design • Benchmark design references • Assess environmental impact	1. Parametric assessment 2. Project state transition 3. Value stream analysis 4. Functional variants analysis 5. State-space modeling 6. Design of experiments
Evaluate	• Assess practicality • Assess acceptability • Assess amenability • Assess desirability • Assess technology state • Measure parameters • Assess attributes • Benchmark results	1. Pareto distribution 2. Life cycle analysis 3. Risk assessment 4. Test and evaluation 5. Proprietary analysis
Justify	• Assess lifecycle cost • Do benefit-cost analysis • Assess economics • Assess technical output • Assess affordability • Assess technical feasibility • Justify payback period	1. Benefit-cost ratio 2. Payback period 3. Present value 4. Utility models 5. Accuracy metrics
Integrate	• Identify continuum • Align with prevailing goals • Embed in normal operation • Verify symbiosis • Leverage synergy • Identify overlaps • Match for continuity • Watch for duplication • Focus on implementation • Assess sustainability of system	1. Specific, Measurable, Aligned, Realistic, and Timed (SMART) linkages 2. Process improvement 3. Quality control 4. Integrated learning curves 5. Calculus of variables

Diagram with the Addition of Cards), and OODA loop (for Observe, Orient, Decide, and Act). Thus, DEJI not only addresses the product development, but also considers process improvement requirements for developing the product. The benefit of using the model provides is that it explicitly calls for using existing analytical tools and techniques for implementing product design, evaluation, justification, and integration. The justification for the *DEJI* model as a systems tool for manufacturing can be found in several reports of failed integration in large and complex defense industry products. One 2014 case example involved the Air Force Expeditionary Combat Support System (ECSS), which reportedly cost more than $1 billion and lost over 600 jobs, and yet was never fielded. At the time of its cancellation in 2014, after eight years of futile efforts, the program was criticized for unsuccessfully using commercial off-the-shelf software, with some modifications, to replace more than 200 computer logistics systems. Apparently, the program management did not follow the sequence of designing, evaluating, justifying, and integrating dissimilar products.

38.2   Design: First stage of DEJI model

Product design in a 3D-printing environment is more than product architecture. There are new product evolutionary nuances that must be taken into account. Direct digital manufacturing, additive manufacturing (AM) or 3D printing, without the benefit of intermediate prototyping, implies that more care must be exercised at the product design stage. The design process must be more deliberate with regard to the product attributes expected at each stage of the product. The technique of state-space modeling is helpful in achieving the stage-to-stage expectations of a product during manufacturing. Product design should be structured to follow point-to-point transformations. A good technique to accomplish this is the use of state-space transformation, with which we can track the evolution of a product from concept stage to a final product stage. For this purpose, the following definitions are applicable:

> *Product state*: A state is a set of conditions that describe the product at a specified point in time. The *state* of a product refers to a performance characteristic of the product that relates input to output such that a knowledge of the input function over time and the state of the product at time $t = t_0$ determines the expected output for $t \geq t_0$. This is particularly important for assessing where the product stands in the context of new technological developments and the prevailing operating environment.
>
> *Product state-space*: A product *state-space* is the set of all possible states of the product lifecycle. State-space representation can solve product design problems by moving from an initial state to another state, and eventually to the desired end-goal state. The movement from state-to-state is achieved by means of actions. A goal is a description of an intended state that has not yet been achieved. The process of solving a product problem involves finding a sequence of actions that represents a solution path from the initial state to the goal state. A state-space model consists of state variables that describe the prevailing condition of the product. The state variables are related to inputs by mathematical relationships. Examples of potential product state variables include the following: schedule, output quality, cost, due date, resource, resource utilization, operational efficiency, productivity throughput, and technology alignment. For a product described by a system of components, the state-space representation can follow the quantitative metric below:

$$Z = f(z, x)$$

$$Y = g(z, x)$$

where f and g are vector-valued functions. The variable Y is the output vector, whereas the variable x denotes the inputs. The state vector Z is an intermediate vector relating x to y.

In generic terms, a product is transformed from one state to another by a driving function that produces a transitional relationship given by

$$S_s = f\left(x \mid S_p\right) + e$$

where:
     S_s is the subsequent state
     x is the state variable
     S_p is the preceding state
     e is the error component (equivalent to a contingency buffer)

The function f is composed of a given action (or a set of actions) applied to the product. Each intermediate state may represent a significant milestone in the project. Thus, a descriptive state-space model facilitates an analysis of what actions to apply in order to achieve the next desired product state. Putting the above equation in simple operational terms, gives us the type of statement below:

> Action *A* applied to Initial State "*I*" will result in Output "*j*," which will require so many units of resource type "*k*."

If the above reasoning is applied iteratively throughout the product lifecycle, all the players involved at each stage of the product will have a better handling of what is required at each stage and how to move to the next desired stage. In a dynamic DoD type of product development, having a current view of the product will facilitate better control of the product development process. Thus, applying a quantitative assessment of the DEJI model to the FIST (Fast, Inexpensive, Simple, and Tiny) concept developed by Dan Ward (Ward, 2012) can pave the way for realizing the much-sought-after product development reform in a real sense.

38.3   *Product transformation due to technology changes*

For a conventional manufacturing environment, Table 38.2 shows an example of the representation of the transformation of a product from one state to another through the application of human or machine actions. This simple representation can be expanded to cover several components within the product information framework. Hierarchical linking of product elements provides an expanded transformation structure. The product state can be expanded in accordance with implicit requirements. These requirements might include grouping of design elements, linking precedence requirements (both technical and procedural), adapting to new technology developments, following required communication links, and accomplishing reporting requirements. The actions to be taken at each state depend on the prevailing product conditions. The nature of subsequent alternate states depends on what actions are implemented. Sometimes there are multiple paths that can lead to the desired end result. At other times, there exists only one unique path to the desired objective. In conventional practice, the characteristics of the future states can only be recognized after the fact, thus, making it impossible to develop adaptive plans. By comparison, in a 3D-printing environment, the state-to-state transformation of the product is much more constrained by virtue of the direct digital manufacturing process. This means that more deliberate stages of the 3D-printing product must be embedded into

Table 38.2 Design transformation due to technology changes

State	Inputs	State transformations	Outputs
S_0	Initial state (Raw material)	—	—
T_1	Planning	$S_1 = T_1(S_0)$	Product specs
T_2	Defining	$S_2 = T_2(S_1)$	Problem statement
T_3	Formulating	$S_3 = T_3(S_2)$	Overall function
T_4	Synthesizing	$S_4 = T_3(S_3)$	Subfunction
T_5	Abstracting	$S_5 = T_5(S_4)$	Basic operation
T_6	Varying effects	$S_6 = T_6(S_5)$	Effect variants
T_7	Varying effectors	$S_7 = T_7(S_6)$	Effector variants
T_8	Representing principles	$S_8 = T_8(S_7)$	Solution principles
T_9	Combining	$S_9 = T_9(S_8)$	Assembly variants
T_{10}	Combining	$S_{10} = T_{10}(S_9)$	System variants
T_{11}	Varying forms	$S_{11} = T_{11}(S_{10})$	Varying forms
T_{12}	Laying out	$S_{12} = T_{12}(S_{11})$	Qualitative layout
T_{13}	Dimensioning	$S_{13} = T_{13}(S_{12})$	Scale layout
T_{14}	Analyzing	$S_{14} = T_{14}(S_{13})$	Preliminary layout
T_{15}	Elaborating	$S_{15} = T_{15}(S_{14})$	Final layout
T_{16}	Detailing	$S_{16} = T_{16}(S_{15})$	Detailed drawing
T_{17}	Production preparation	$S_{17} = T_{17}(S_{16})$	Production docs
T_{18}	Producing	$S_{18} = T_{18}(S_{17})$	Final product
T_{19}	Marketing	$S_{19} = T_{19}(S_{18})$	Product delivery
T_{20}	User satisfaction	$S_{20} = T_{20}(S_{19})$	Customer feedback

the product design upfront. In the implementation of the *DEJI* model for conventional manufacturing, adaptive plans can be achieved because the events occurring within and outside the product state boundaries can be taken into account. However, in a 3D-printing case, once printing starts, the boxed manufacturing environment is mostly impervious to external factors.

If we describe a product by P state variables s_i, then the composite state of the product at any given time can be represented by a vector **S** containing P elements. That is,

$$\mathbf{S} = \{s_1, s_2, ..., s_P\}$$

The components of the state vector could represent either quantitative or qualitative variables (e.g., cost, energy, color, time). We can visualize every state vector as a point in the state space of the product. The representation is unique because every state vector corresponds to one and only one point in the state-space. Suppose we have a set of actions (transformation agents) that we can apply to the product information so as to change it from one state to another within the project state-space. The transformation will change a state vector into another state vector. A transformation may be a change in raw material or a change in design approach. The number of transformations (or actions) available for a product may be finite or countably infinite. We can construct trajectories that describe the potential states of a product evolution as we apply successive transformations with respect to technology forecasts. Each transformation may be repeated as many times as needed. Given an initial state $\mathbf{S}_0$, the sequence of state vectors is represented by the following:

$$\mathbf{S}_1 = T_1(\mathbf{S}_0)$$

$$\mathbf{S}_2 = T_2(\mathbf{S}_1)$$

$$\mathbf{S}_3 = T_3(\mathbf{S}_2)$$

$$....$$

$$\mathbf{S}_n = T_n(\mathbf{S}_{n-1})$$

The final State, $\mathbf{S}_n$, depends on the initial state $\mathbf{S}$ and the effects of the actions applied.

38.4   Evaluation: Second stage of DEJI model

A product can be evaluated on the basis of cost, quality, schedule, and meeting requirements. There are many quantitative metrics that can be used in evaluating a product at this stage. Learning curve productivity is one relevant technique that can be used because it offers an evaluation basis of a product with respect to the concept of growth and decay. The half-life extension (Badiru, 2010, 2012) of the basic learning is directly applicable because the half-life of the technologies going into a product can be considered. In today's technology-based operations, retention of learning may be threatened by fast-paced shifts in operating requirements. Thus, it is of interest to evaluate the half-life properties of learning curves. Information about the half-life can tell us something about the sustainability of learning-induced technology performance. This is particularly useful for designing products whose life cycles stretch into the future in a high-tech environment, such as the F-22A Raptor. Figure 38.2 shows a graphical representation of performance as a function of time under the influence of performance decay. Technology performance degrades as time progresses. Our interest is to determine when performance

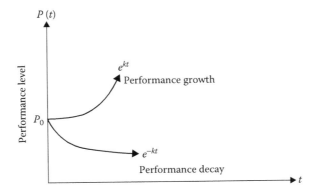

Figure 38.2 Concept of learning curve growth and decay.

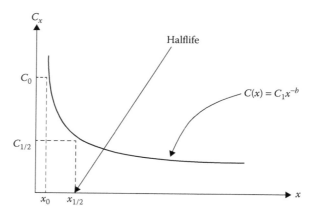

Figure 38.3 Profile of a learning curve with half-life point.

has decayed to half of its original level. Figure 38.3 shows an example of a learning curve with the half-life indicated.

38.5   *Justification: Third stage of DEJI model*

We need to justify a program on the basis of quantitative value assessment. The systems value model (SVM) is a good quantitative technique that can be used here for project justification on the basis of value. The model provides a heuristic decision aid for comparing project alternatives. It is presented here again for the present context. Value is represented as a deterministic vector function that indicates the value of tangible and intangible attributes that characterize the project. It is represented as:

$$V = f\left(A_1, A_2, \dots, A_p\right)$$

where V is the assessed value and the A values are quantitative measures or attributes. Examples of product attributes are quality, throughput, manufacturability, capability, modularity, reliability, interchangeability, efficiency, and cost performance. Attributes are considered to be a combined function of factors. Examples of product factors are market share, flexibility, user acceptance, capacity utilization, safety, and design functionality. Factors are themselves considered to be composed of indicators. Examples of indicators are debt ratio, acquisition volume, product responsiveness, substitutability, lead time, learning curve, and scrap volume. By combining the above definitions, a composite measure of the operational value of a product can be quantitatively assessed. In addition to the quantifiable factors, attributes, and indicators that impinge on overall project value, the human-based subtle factors should also be included in assessing overall project value.

38.6   *Contemporary earned value technique*

A companion analytical technique to use for the justification stage is the conventional earned value technique (EVT), which can be used for cost, quality, and schedule elements of product development with respect to value creation. The technique involves developing

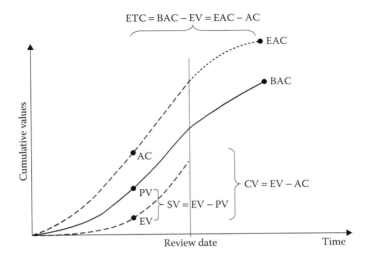

Figure 38.4 Earned value in product lifecycle.

important diagnostic values for each schedule activity, work package, or control element as shown in Figure 38.4. The definitions of the variables in the figure are summarized below:

Planned Value (PV): This is the budgeted cost for the work scheduled to be completed on an activity up to a given point in time.

Earned Value (EV): This is the budgeted amount for the work actually completed on the schedule activity during a given time period.

Actual Cost (AC): This is the total cost incurred in accomplishing work on the schedule activity during a given time period. AC must correspond in definition, scale, units, and coverage to whatever was budgeted for PV and EV. For example, direct hours only, direct costs only, or all costs including indirect costs. The PV, EV, and AC values are used jointly to provide value performance measures of whether or not work is being accomplished as planned at any given point in time. The common measures of project assessment are cost variance (CV) and schedule variance (SV).

Cost Variance (CV): This equals earned value minus actual cost. The cost variance at the end of the project will be the difference between the budget at completion (BAC) and the actual amount expended.

$$CV = EV - AC$$

Schedule Variance (SV): This equals earned value minus planned value. Schedule variance will eventually become zero when the project is completed because all of the planned values will have been earned.

$$SV = EV - PV$$

Cost Performance Index (CPI): This is an efficiency indicator relating earned value to actual cost. It is the most commonly used cost-efficiency indicator. CPI value less than 1.0 indicates a cost overrun of the estimates. CPI value greater than 1.0 indicates a cost advantage of the estimates.

$$CPI = \frac{EV}{AC}$$

Cumulative CPI (CPI^C): This is a measure that is widely used to forecast project costs at completion. It equals the sum of the periodic earned values (Cum. EV) divided by the sum of the individual actual costs (Cum. AC).

$$CPI^C = \frac{EV^C}{AC^C}$$

Schedule Performance Index (SPI): This is a measure that is used to predict the completion date of a project. It is used in conjunction with CPI to forecast project completion estimates.

$$SPI = \frac{EV}{PV}$$

Estimate to Complete (ETC) based on new estimate: Estimate to complete equals the revised estimate for the work remaining as determined by the performing organization. This is an independent noncalculated estimate to complete for all the work remaining. It considers the performance or production of the resources to date. The calculation of ETC uses two alternate formulae based on earned value data.

ETC based on atypical variances: This calculation approach is used when current variances are seen as **atypical** and the expectations of the project team are that the similar variances will not occur in the future.

$$ETC = BAC - EV^C$$

where BAC = Budget at Completion.

ETC based on typical variances: This calculation approach is used when current variances are seen as **typical** of what to expect in the future.

$$ETC = \frac{BAC - EV^C}{CPI^C}$$

Estimate at Completion (EAC): This is a forecast of the most likely total value based on project performance. EAC is the projected or anticipated total final value for a schedule activity when the defined work of the project is completed. One EAC forecasting technique is based on the performing organization providing an estimate at completion. Two other techniques are based on earned value data. The three calculation techniques are presented below. Each of the three approaches can be effective for any given project because it can provide valuable information and signal if the EAC forecasts are not within acceptable limits.

EAC using a new estimate: This approach calculates the actual costs to date plus a new ETC that is provided by the performing organization. This is most often used when past performance shows that the original estimating assumptions were fundamentally flawed or that they are no longer relevant due to a change in project operating conditions.

$$\mathbf{EAC = AC^C + ETC}$$

EAC using remaining budget: In this approach, EAC is calculated as cumulative actual cost plus the budget that is required to complete the remaining work; where the remaining work is the budget at completion minus the earned value. This approach is most often used when current variances are seen as *atypical* and the project management team expectations are that similar variances will not occur in the future.

$$\mathbf{EAC = AC^C + (BAC - EV)}$$

where (BAC – EV) = remaining project work = remaining PV.

EAC using cumulative CPI: In this approach, EAC is calculated as actual costs to date plus the budget that is required to complete the remaining project work, modified by a performance factor. The performance factor of choice is usually the cumulative CPI. This approach is most often used when current variances are seen as *typical* of what to expect in the future.

$$\mathbf{EAC = AC^C + \frac{(BAC - EV)}{CPI^C}}$$

Other important definitions and computational relationships are summarized below:

Earned   → Budgeted cost of work actually performed
Planned   → Budgeted cost of work scheduled
Actual   → Cost of actual work performed
Ending CV   = Budget at completion – Actual amount spent at the end
                     = BAC – EAC = VAC (Variance at Completion)
EAC   = ETC + AC = (BAC – EV) + AC = AC + (BAC – EV)
ETC   = EAC – AC = BAC – EV

38.7   Integration: Fourth stage of DEJI model

Without being integrated, a system will be in isolation and it may be worthless. We must integrate all the elements of a system on the basis of alignment of functional goals. The overlap of systems for integration purposes can conceptually be viewed as projection integrals by considering areas bounded by the common elements of subsystems as shown in Figure 38.5. Quantitative metrics can be applied at this stage for effective assessment of the product state.

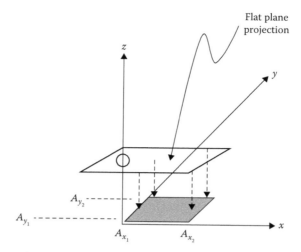

Figure 38.5 Technology alignment surface for product integration.

Following are the guidelines and important questions relevant for product integration:

- What are the unique characteristics of each component in the integrated system?
- How do the characteristics complement one another?
- What physical interfaces exist among the components?
- What data/information interfaces exist among the components?
- What ideological differences exist among the components?
- What are the data flow requirements for the components?
- What internal and external factors are expected to influence the integrated system?
- What are the relative priorities assigned to each component of the integrated system?
- What are the strengths and weaknesses of the integrated system?
- What resources are needed to keep the integrated system operating satisfactorily?
- Which organizational unit has primary responsibility for the integrated system?

In complex system of systems, integration from one stage to the other is a critical requirement for product success. 3D printing is still an emerging technology. The underlying chemistry and material science are well understood. How these interplay in a human environment is the topic desiring additional research and development efforts. Systems-based models, such as the *DEJI* model can open new insights into product design, evaluation, justification, and implementation integration. Figure 38.6 presents a comprehensive view of how several factors and functional forms revolve around the model. The design of a 3D printed product requires new consideration to accommodate a new digital manufacturing scenario. New power requirements, new lighting, and clean-room type of environment are some of the unique aspects that may come into play for 3D-printing operations. Unlike a conventional production system, the design of a 3D-printing product suggests the need to get it right the first time.

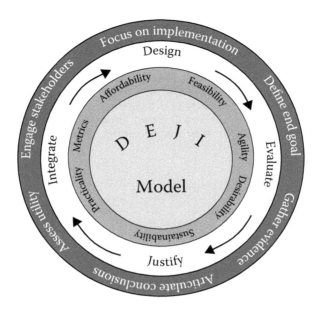

Figure 38.6 3D printed product lifecycle elements in DEJI model.

38.8   *Conclusion*

Recent cost over-run and technology misalignment reports on some DoD product development programs have necessitated the search for alternate or additional techniques for assessing technical products. This has presented an additional robust technique that can complement existing efforts. The chapter presents an application of the *DEJI* product development model as a complementary approach to executing new technology-based product acquisition. The proposed approach will facilitate a better alignment of product technology with future development and needs. The stages of the model involve design, evaluation, justification, and integration. Existing analytical tools and techniques are presented for implementing the stages. The methodology of the chapter adds to the repertoire of tools and techniques available to aerospace manufacturers planning to take advantage of the 3D-printing revolution. In its embryonic stage of the technology, 3D printing should complement rather than compete with conventional manufacturing techniques.

References

Badiru, A. B. (2012), Application of the DEJI model for aerospace product integration, *Journal of Aviation and Aerospace Perspectives*, 2(2), 20–34, 2012.

Badiru, Adedeji B. (2012), Half-life learning curves in the defense acquisition life cycle, *Defense Acquisition Research Journal (DARJ)*, 19(3), 283–308.

Badiru, A. B. (1995), Multivariate analysis of the effect of learning and forgetting on product quality, *International Journal of Production Research*, 33(3), 777–794.

Giachetti, Ronald E. (2010), *Design of Enterprise Systems: Theory, Architecture, and Methods*, Taylor and Francis/CRC Press, Boca Raton, FL.

Ibidapo-Obe, O., O. S. Asaolu, and A. B. Badiru (1998) Optimal and multicriteria probabilistic design of homogeneous beams, *Engineering Design & Automation*, 4(3), 335–341.

Sieger, David, A. B. Badiru, and M. Milatovic (2000), A Metric for agility measurement in product development, *IIE Transactions*, 32, 637–645.

Thal, Alfred E., A. B. Badiru, and R. Sawhney (2007), Distributed project management for new product development, *International Journal of Electronic Business Management*, 5(2), 93–104.

Ward, D. (2012), Faster, Better, Cheaper: Why not pick all three, *National Defense Magazine*, 1–2.

chapter thirty nine

3D printing rises to the occasion

ORNL group shows how it is
done, one layer at a time

Leo Williams

Contents

Things have come a long way since the mid-1980s when 3D Systems cofounder Chuck Hull worked out the technology to print objects in three dimensions, one very thin layer at a time (Williams, 2013).

Hull called his new technology *stereolithography*. In it, a guided beam of ultraviolet light focused on a vat of liquid polymer, solidifying areas where it hits. When one layer is complete, the platform holding the object lowers a bit, and the process is repeated.

The technology was impressive but limited, with the printed objects serving as prototypes but not much else. In the intervening decades, and especially in the past few years, 3D printing has made it to the big time, taking off both in capability and application.

Consider the following:

Electron beam melting systems create intricate, high-quality components by sweeping a precise layer of metal power over an object, and selectively melting it to the object. Swedish manufacturer Arcam AB has used this process to produce more than 30,000 acetabular cups, the components in a hip replacement that attach to the hip socket and hold the ball joint. These printed components are literally walking all over Europe.

Boeing uses 3D printing—also called additive manufacturing (AM)—to produce more than 20,000 military aircraft parts, and GE Aviation has announced that it will produce more than 100,000 additive-manufactured components for its LEAP and GE9X jet engines by 2020.

39.1 3D printing at ORNL

Oak Ridge National Laboratory's (ORNL's) focus on printing is led by the Deposition Science & Technology (DST) Group within the Manufacturing Demonstration Facility. The DST is young, created in 2013. According to group leader Chad Duty, it works with a variety of advanced manufacturing technologies such as carbon fiber, magnetic field processing, and printed electronics. Figures 39.1 through 39.3 illustrate some of the 3D printed products at ORNL.

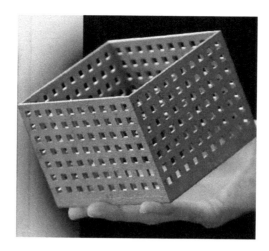

Figure 39.1 **(See color insert.)** A Perforated metal box produced by an Arcam 3D printer. This detailed *calibration* part illustrates some of the versatility of 3D printing. (Courtesy of Jason Richards.)

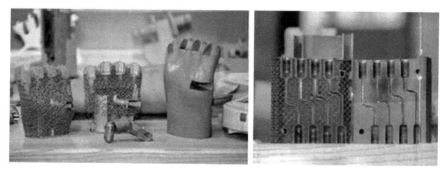

Figure 39.2 **(See color insert.)** 3D printing can build products from a variety of materials for products ranging from heavy equipment to biomedical implants. (Courtesy of Jason Richards.)

Figure 39.3 **(See color insert.)** ORNL scientist Chad Duty removes a finished part from a production-grade 3D printer. (Courtesy of Jason Richards.)

In addition, of course, the group works on AM. In this realm, its role is a combination of research and education. On the research side the group is making use of ORNL's unique strengths, including materials science, neutron imaging, and supercomputing.

"There are several areas where it makes sense for a national lab to be doing this," Duty says. "One is that we can leverage all the historical strengths of a national lab and bring it to bear on this new technology front."

For instance, warping is a big issue for systems that build parts one layer at a time. The component gets very hot at this point of melting, but surrounding areas may stay cool, depending on the technology. Electron beam systems keep the whole assembly at about 700 degree Celsius (approximately 1,300 degree Fahrenheit), which helps to minimize warping. Laser beam systems, on the other hand, do not heat the surround material, so warping is a greater issue.

Duty's group is working with Ralph Dinwiddie or ORNL's Scattering and Thermophysics Group to measure temperatures across the printing surface as the component is being produced. The thermal imaging techniques pioneered by this collaboration will allow for a better understanding of temperature differences and, ultimately, the development of ways to reduce warping.

39.2   3D and neutrons

Quality control is also an issue, one that is especially important in areas such as medical implants or aerospace manufacturing. In response, Duty's group is working with ORNL neutron scientists at both the Spallation Neutron Source and the High Flux Isotope Reactor, using the unique ability of neutrons to look inside materials without damaging them.

"The neutron source can bring a new way of inspecting these materials, optimizing them and reducing residual stress in these components," Duty says. "There's no other way to nondestructively evaluate those kinds of systems."

Supercomputing is also helpful when Duty and his colleagues look at structural issues associated with AM, he has been working in this area with Sreekanth Pannala of ORNL's Computer Science and Mathematics Division.

"When you're trying to model a weld beam or weld line, it's very complex," Duty says. "You've got thermal issues. You've got microstructural issues. You've got moving interface. You've got the phase change of materials from liquid to solid."

"It's a pretty complicated system, and that's just for one weld beam line across a material. In a cubic inch in the Arcam electron beam technology, we've got five miles of weld line," Duty says.

39.3   Bigger, Faster, Cheaper

Duty and his colleagues are also working in a wide range of other areas designed to push forward both the technology and its application in American manufacturing. They are working to improve the materials being used, contributing to the development of high-performance metal alloys and stronger polymers that incorporate carbon fibers. They are working to improve the manufacturing process in an initiative Duty identifies with the slogan "Bigger, Faster, Cheaper."

And they're working to educate manufacturers about what the technology can and can't do. This involves helping manufacturers learn new things and unlearn old notes.

"We're in a kind of second birth for additive manufacturing," Duty explains. "It went through a phase (in the 1980s) called 'rapid prototyping,' and some people were saying, 'It can do everything; we'll do away with all other types of manufacturing.' And then it cooled off.

"A lot of folks when through that cool-off period. They tried it, it didn't work, and they've written it off. And they think this is the second verse of the same song. We help them take another look at it."

One challenge that requires both the technical expertise found across the laboratory and Duty's personal role as a technology evangelist is something he refers to as "the valley of death"—the collection of practical limitations that prevents a wonderful idea in the laboratory from making it into production.

"That's kind of why my group exists; to help companies transition things that are really cool to things that are commercially relevant," Duty says. "One of the things we do in a manufacturing demonstration facility is demonstrate the technology, show people what can be done with that technology, start their wheels turning, and help them work through problems in their industry where it can be useful."

In fact, Duty says, additive manufacturing is not appropriate in every situation. If you're producing 10,000 simple, inexpensive brackets a day for the automotive industry, chances are pretty good that making that bracket on a 3D printer is not a good idea. On the other hand, if you're making a low-volume component that is expensive, complex and specialized, Duty and his colleagues would like to talk with you.

"The areas where it's really getting initial traction are where you should expect," Duty says. "It will be in those areas that have really high margins, like biomedical, aerospace, defense and nuclear. In general, these are areas where the parts are really complex, highly customized and produced in low volumes."

Complexity is not a bad thing in AM; it is not even that much of a challenge. As these systems build a structure one later at a time, they do not care how complex it is. In fact, a complex mesh is an easier job for AM than a chunk of metal. The Arcam system, for instance, can lay down about 5 cubic inches of material in an hour, whether that material is spread out over a fine mesh or plopped down as a cube.

This new reality takes some time for getting used to.

"It's completely non-intuitive," Duty noted, "which is why it's so paradigm-shifting. People think, 'I can really make this into something useful, bit it would really complicate the design.' And we say, 'That's good; we can make it faster and cheaper for you if you do that."

For example, his group was working with a company that produced impeller bladders, rotors within a pipe that have the job of increasing or decreasing fluid pressure. The company asked them to duplicate the part exactly.

The problem—or more accurately, the opportunity—was that the piece had been designed for casting, with the angles, thickness, and other compromises that are necessary when you pour molten metal into a mold.

Duty and his colleagues certainly could make an exact duplicate of the impeller, but they are convinced they could do better. So they asked the company to redesign the piece for AM.

"They took two days to do the redesign," he says, "and we made the other version. We tested it out; it met all the performance metrics, and it was 56 percent lighter because they got the wall thickness down. For a rotating piece of machinery, weight reduction is huge."

The story illustrates both the potential of AM and the challenge it presents to existing ideas of what manufacturing involves.

"If you're trying to use additive manufacturing to make the exact same thing that you're already making, you're using it wrong," Duty explains. "You're not thinking about it right. The real potential and opportunity here is to do things that you can't do today."—Leo Williams

Reference

Williams, Leo, "3D Printing rises to the occasion," ORNL Review, Oak Ridge National Lab, Oak Ridge, Tennessee, Vol. 46, Nos. 2 & 3, 2013, pp. 6–9.

chapter forty

3D printing implications for STEM education

John L. Irwin

Contents

40.1 Introduction

The driving force for the advancement in the popularity of 3D printing in the education market is that 3D printers have become more affordable to purchase commercially and even more so when constructed utilizing open-source hardware and when operated using open-source software. As predicted by Campbell et al. (2012), the future of additive manufacturing (AM) from the perspective of three key elements: applications, materials, and design is that as the primary patents expire medium cost, AM hardware developed will foster an increased market demand which in turn will accelerate the rate of entry-level suppliers into the market with new and improved materials. Commercially, a high risk aspect of the business is hardware improvements plus the software maintenance and development, which in an open-source environment is passed on to the open-source community that collectively strive to improve upon previous versions.

In an education setting, it is important that the equipment used by students is safe, robust, and can perform to meet the requirements of the lesson or project with the end goal to increase student's learning of subject matter. With that criteria in mind and with the overabundance of 3D-printing systems on the market, making the decision for which printer to purchase is challenging. This chapter is not intended to single out one specific AM system as the best for Science, Technology, Engineering and Mathematics (STEM) education but is intended to provide examples of how 3D printing can be implemented in the classroom for attainment of curricular goals and objectives. For instance according to Campbell et al. (2012), designers can be educated in the future to pay less attention to design for manufacturing limitations, and will be able to develop design concepts inspired from art and nature that can also perform efficiently and ergonomically while being aesthetically pleasing. Also, during the early education of future engineers and scientists, having access to AM technology in K-12 classrooms allows the opportunity for increased design, build, and test experiences.

40.2 Background

Traditionally, the availability of AM technology in the education system has been reserved for higher education and utilized mainly for creating models for design visualization purposes. Institutions having an engineering or design curriculum are encouraged to include curricular elements that are used in industry to allow students to experience technology that they will encounter later in industry. For instance, stated in the accreditation body for engineering and technology program (ABET) (2013), in the guidelines for facilities, there are requirements that programs include modern tools to enable students to attain the student outcomes and to support program needs, and that students must be provided appropriate guidance regarding their use. The AM technology in this sense can be viewed as a piece of equipment, such as a lathe or mill that has a certain level of complexity to operate and maintain, and is necessary to master in order to manufacture a design.

More important, is the question whether learning how to operate the material removal process equipment and/or utilizing AM equipment improve(s) engineering skills as a result of becoming familiar with their capabilities and/or limitations. Since AM equipment is becoming increasingly available to students from K-12 through higher education, it is a topic of increased research. As stated by Malicky et al. (2010), while integrating a machine shop class into an engineering curriculum, it adds depth, ownership, and integration of the entire learning experience and active learning design–build experiences provide students for constructing more theoretical knowledge structures.

Several initiatives are underway to develop support for STEM educators in the use of AM technologies. In an AM based study described by Tseng et al. (2014), a National Science Foundation funded project to support STEM learning environments is designed to prepare students for the needs of industry and promoting AM technologies. The project plans to develop an online 3D-virtual facility and cyber tutor system to develop AM relevant curriculum to cultivate technical success for engineering students through seminars, workshops, and internship. Other organizations, such as The Square One Education Network, support integrating STEM experiences for K-12 teachers and students who creatively use best practice instruction through unique project designs such as the innovative vehicle design underwater robotics program (Square One, 2014). In order to support these initiatives, professional development workshops are provided for teachers where they build their own 3D printers to immediately implement in their classrooms.

In this chapter, the implications of 3D printing in STEM education will be discussed in relation to: selection, operation, and maintenance of common commercial and open-source AM software and hardware; implementation of AM technologies in design, build, and test course projects; and the use of AM technologies to enhance K-12 engineering outreach and STEM-related curriculum. Personal experiences will be shared throughout the chapter describing first-hand the use of AM technologies while teaching STEM-related topics in various education settings.

40.3 3D printer selection

The budget is the bottom line for most education-related projects that will determine what type of AM technology to implement. The most popular technology used by education is the fused deposition modeling (FDM) commercialized by Stratasys Corporation and developed by S. Scott Crump (Stratasys, 2014). The open-source replicating rapid prototyper (RepRap) project community uses a different terminology, fused filament fabrication, to describe the AM technology commonly used for modeling, prototyping, and production applications. In this AM process, the material is laid down in layers from a spool of filament while being extruded through a heated nozzle. This AM technology allows the user to see the part being produced layer-by-layer, which is analogous to watching a honeycomb being constructed by a bee colony in time-lapse photography. The print head, building platform, and extruder motor moving in perfect timing is like an amazing computer-controlled symphony where each instrument is in harmony. From an education standpoint, the observation of the 3D printer extruding layers of a part being produced gives the opportunity for the observer to learn principles of material properties, motor control, signal processing, temperature measurement, and sensing among others.

In addition to the categories of commercial and open-source 3D-printing systems, there is also the consideration of size of the build envelope to be considered. Most open-source 3D printers are considered to be of the desktop variety being that they are small enough to fit on an office desk like the desktop document printers. These printers generally have a build envelop similar to the RepRap Prusa Mendel size of around 200 mm wide × 200 mm deep and 140 mm high or 8″ × 8″ × 5.5″ United States Customary System (USCS) units, and the size of the 3D printer is just 500 mm (20″) × 400 mm (16″) not taking up much space on a desk. The other category of 3D printers generally of the commercial variety are stand-alone systems similar to the Stratasys Fortus 900C that has a build envelop of 914 × 610 × 914 mm (36″ × 24″ × 36″), physical dimensions of 2772 × 1683 × 2027 mm (109.1″ × 66.3″ × 79.8″), and weighs over 6000 pounds.

To classify the 3D printers even further would be to divide the commercial 3D printers into those of industrial- and consumer-based markets. In some education settings, an industrial-based model is necessary for the strength, accuracy, and/or fine resolution if the parts are to be used for material science and product development research. Although for the use of the 3D printer in education for product realization, fit and finish, assembly/disassembly, or ergonomic design, a consumer-based 3D printer is more applicable. Also in the open-source or Do it yourself (DIY) category of 3D printers, there are kits available to buy that are preassembled ready to operate, or 3D printers that can be built from a material bill of material and 3D-printed parts produced on another 3D printer. For instance, the cost to build a Prusa Mendel version of a self-RepRap is approximately $550 for the components including the necessary 3D-printed parts, which need to be created on an existing machine. This 3D printer can be assembled in as little as 2 days (Kentzer et al., 2011; Wittbrodt et al., 2013). The RepRap has already demonstrated utility in a wide range of educational environments (Gonzalez-Gomez et al., 2012; Pearce, 2012, 2014). The RepRap Project has an open-source hardware design using open-source software with the intent to improve on the design and software with each future generation (RepRap, 2013). The 3D printer extrudes either polylactic acid (PLA), in which most of the parts for the printer are made from, or acrylonitrile butadiene styrene (ABS), which is used for parts requiring high heat resistance (Figure 40.1).

A commercially available desktop 3D printer developed for the consumer and education market similar to the open-source RepRap is the XYZPrinting Da Vinci 1.0 3D printer featuring a 5.9″ W × 7.8″ D × 7.8″ H (150 × 200 × 200 mm) build area for under $500 and the Da Vinci 2.0 with two extruders (the second extruder to be used for soluble support material) for under $650. The XYZPrinting desktop 3D printers also are supplied with commercially supported proprietary software, but unlike most commercial 3D printers are

Figure 40.1 **(See color insert.)** MOST Prusa Mendel RepRap.

able to use standard 1.75 diameter ABS filament similar in cost as purchased by the spool. The Stratasys Corporation has a desktop 3D printer aimed at the desktop market called the Mojo, which has a $127 \times 127 \times 127$ mm ($5'' \times 5'' \times 5''$) build platform, has build and support material, commercial software, but the material has to be purchased from Stratasys. The price of a Statasys Mojo is a significantly more expensive system at just under $11,000 with supplies.

A huge advancement in open-source 3D printing for education use in the K-12 system is the MOST Delta RepRap design, which costs about $400 and can be built in approximately 8 hours once the full bill-of-materials has been collected (Figure 40.2).

Figure 40.2 **(See color insert.)** Customized MOST Delta RepRap.

Table 40.1 3D printer selection

Model	Market	Price	Features
Stratasys Fortus 400MC	Commercial: industry/ university research education	$200,000	11 material options Build and support 406 × 355 × 406 mm (16″ × 14″ × 16″)
Stratasys Mojo	Commercial: industry/ consumer/education	$11,000	ABS*plus* Sr-30 support 127 × 127 × 127 mm (5″ × 5″ × 5″)
XYZPrinting Da Vinci 1.0 and 2.0	Commercial: consumer/ education	$500 $650	ABS cartridges in 12 colors 150 × 200 × 200 mm (5.9″ × 7.8″ × 7.8″)
MOST Mendel Prusa RepRap	Open source: consumer/ education	$550	ABS or PLA 200 × 200 × 140 mm (8″ × 8″ × 5.5″)
MOST Delta RepRap	Open source: consumer/ education	$400	PLA 250 × 250 mm dia (9.8″ × 9.8″ dia)

The value of high-school students becoming active participants in the ongoing evolution of the RepRap 3D-printer design can be more valuable than simply incorporating 3D printing into the classroom alone. Having access to a machine that they can take apart, fix, upgrade and try design experiments on provides a much richer learning environment than *press print* experiences with proprietary desktop 3D printers, which are normally sold at roughly 400% of the cost of a RepRap and frequently cannot be user-repaired. For example, the frame structure of the Delta design is made with fairly simple parts of metal or wood, which means that these parts can actually be made by students in any school with access to a basic wood or metal shop. So with the capacity for students to design and create their own customized parts for these machines, the educational value and *student buy-in* are increased even further (Irwin et al., 2014).

Reviewing the selection for a 3D printer cannot be just based on price alone, since there are many factors such as size or space restrictions, end use of the 3D print for product realization or research purposes, and the ability to depend on a commercially developed and maintained equipment and software or rely on the open-source community to continue to develop and advance the product. There are pros and cons to these alternative options, and in sections 40.5 & 40.6 they will become more apparent to the reader in which system is optimal for each circumstance. The information, shown in Table 40.1, describes just a few of the systems on the market that could be considered as education style 3D printers.

40.4   Operation and maintenance

Utilizing a 3D printer in an education setting also depends on the style of printer that is being used and for what purpose. One scenario is that a commercially available industry style 3D printer is normally operated and maintained by one or two dedicated staff or faculty who have attended the vendor training and have been certified as competent in loading material, changing print heads, removing jams, running prints, and taking care of general maintenance issues. These machines usually are accompanied by a service contract that will cover a major component failure and travel of a service technician to perform any out of the ordinary maintenance. In other words, the student will not normally be involved in the process of 3D printing until after the part has been removed from the machine. Most times, the staff or faculty person will also perform the software manipulation of the file in preparation for 3D printing. This process is somewhat different depending on which commercial software being used, but in general consists of loading a stereolithography file that has been exported from the computer-aided design (CAD) software package or downloaded from a 3D printing website, manipulating the file in the most appropriate orientation, scaling if necessary, slicing the file, and if necessary editing the slice file layer-by-layer to repair layers or modify raster thicknesses.

Another scenario for 3D printing in the educational environment is that a desktop 3D printer, either commercial or open source, can be operated by the STEM student. Some schools and public buildings have incorporated 3D printing stations around campus. Even some Home Depot hardware stores have set up MakerBot 3D printer kiosks (Williams, 2014), and there are plans to have 3D-printing vending machines such as the DreamVendor 2.0 promising to have machines at street corner gas stations, drug stores and malls (Orr, 2014), while other visionaries are planning to implement 3D printing kiosks in toy stores for customized pendants, chess pieces, rings, or other novelty items (Oravecz, 2014). Most of these situations involve the customer selecting or developing a custom design to be 3D printed and a trained store technician operating the machine to produce the 3D-printed part.

In an education setting, there is training necessary to understand how to operate and maintain a commercial or open-source 3D printer. This is an important aspect, since for full educational value, the AM experience should involve a level of engagement in not just the design of the part to be 3D printed, but also the software manipulation and operation of the 3D printer. This is analogous to the traditional manufacturing methods where it helps to be a better designer of sheet metal and structural steel products if the designer has experienced the manufacturing process of welding.

Designing a product to solve an engineering problem incorporates elements of the engineering process as outlined by Dieter (2012), as the embodiment stage of design where concepts are developed to take physical form to represent the arrangement of physical elements of a product to carry out its function. After the model has undergone initial analysis and parametric aspects of the product have been considered to create the best performance and manufacturing methods, it is usually appropriate to create a prototype. The prototype model for designing purpose can be used to perform further tests to determine its function, for instance an innovative design for a wind turbine blade can be 3D printed at a small scale for testing its speed at various wind speed conditions.

The operation of the software generally involves two applications which are the *slicing* software used to create the layers that generate the G-code used to create the 3D printer movement, and the software that controls the 3D printer to set the offset values for the nozzle height in relation to the printer build platform, the manual manipulation of the nozzle temperature, extrusion of material, and setting the home position prior to starting the 3D print. The commercial software *Insight* is used for the Stratasys Fortus series 3D printers, and *Cura* is used by many of the RepRap open-source community users. Each software uses a stereolithography file format (.stl) that is exported from the CAD software or can be downloaded from internet services such as Thingiverse.com, which has .Stl files available of premade designs. In this software, it takes some experience to become familiar with the most advantageous settings for variables such as extruder speed, nozzle temperature, and layer thickness to produce accurate and high-resolution parts. The commercial software offers less adjustment of variables manually, since these are preset depending on layer height and material type. Also, the placement of the part is determined to provide either strength attributes due to the direction of layers, time necessary to print, or amount of support material. For example, a tall and narrow part may be rotated on the build platform to lie on its side to provide a greater surface area to adhere to the plate and also have fewer layers in the Z-direction of height usually requiring less time to produce.

The 3D printer control software is then required to send the G-code to the 3D printer and manipulate the 3D printer movements. Before starting a new 3D print in a commercial printer, a calibration routine is used when a new material is placed in the machine. Also, each material requires a different print nozzle; therefore, the exact height to the build platform is calculated through the calibration routine. A similar routine is required in open-source 3D printers if the printer becomes out of adjustment due to wear. The height calibration is accomplished through repeated adjustment of end stop limit switches and manual motion switches in the software for the Z-motor movement. When the G-code file is loaded, it can be adjusted as far as location on the build platform, or multiple G-code files can be arranged on the build platform. Finally, the 3D print can be started, paused, and stopped from the control software and monitored for progress.

Operation of the 3D printer hardware is similar to that of a standard paper printer. The machine needs to be powered on, material must be loaded, and common maintenance issues are when filament jams in the nozzle or in the extruder. The hardware needs general attention to moving parts that need lubrication and ensuring that cooling fans are operating efficiently. Before sending the G-code file, the build platform must be clean, leveled, and must be free of any existing 3D prints. Operation of the larger commercial 3D printers is much the same as the desktop printers, but they are usually covered under warranty or maintenance agreements for larger issues dealing with items other than general maintenance.

40.5 AM technologies in design, build, and test scenarios

Typically in industry, the traditional use of AM technologies is implemented in design projects toward the end of the embodiment stage after analysis has taken place. The use of less expensive desktop printers can allow the prototype to be used during the conceptual stage in the design process to evaluate alternative part design solutions or assembly configurations. With less invested in the prototype process in terms of time and cost, it allows the designer the freedom to investigate alternative solutions in more depth. For instance, according to a report by Higginbotham (2012), Ford Motor Company Engineers have been all provided with Makerbot desktop 3D printers to implement the iterative design process.

Engineering and engineering technology curricula are required to incorporate projects where students experience real-world examples of design–build–test that can be completed in one or two semesters inexpensively and with tangible results. Typically students participate in a two semester capstone project sequence to meet the requirements for graduation and to comply with ABET standards (ABET, 2013). An example of industry expressing the need for engineers to be practiced in design, build, and test is clear in one of the recommendations from the American Society of Mechanical Engineers (ASME) Vision 2030 report where the current weaknesses of graduates expressed by their employers, as well as the early career engineers themselves are that mechanical engineering technology (MET) programs should strive toward creating curricula that inspire innovation, creativity, and entrepreneurship and that mechanical engineering (ME) programs should include an increased emphasis on practical applications of how devices are made and work (Perry and Kirkpatrick, 2012). The ASME Vision 2030 report states: "To address these weaknesses, an increase in and enrichment of applied engineering design–build experience throughout degree programs is urged."

There are studies to show the effect of rapid prototyping on the engineering design process for practicing engineers and industrial designers (Evans, 2002; Hallgrimsson, 2012; Jones and Richey, 2000), but the literature is lacking regarding the effectiveness of rapid prototyping in teaching and learning the design process. Since 3D Printing has become accessible using desktop 3D printers, a lecture on 3D Printers also can be accompanied by a live demonstration. For example, the instructor can simply roll a cart with the 3D Printer and laptop into the classroom and lead the students through the steps to create a plastic part from a CAD model. Students then follow-up the lecture with a lab activity such as, creating a CAD model that will integrate with an electronic communication device: Individualized cellular phone cases, speaker mechanisms, or charging docks are modeled, 3D printed, checked for fits and tolerances as well as function (Irwin et al., 2014).

An open-source RepRap style 3D printer can be thought of as a design, build, and test project in itself when the kit of parts is provided to a student group to build. The kit can be assembled in as little as 8 hours and then calibrated and tested in another 4 hours. While building the 3D printer, engineering students will first hand see the challenges with the design of parts in having to attach pieces together with a set of tools provided. One design for assembly principle of uniformity in fastener sizes is immediately apparent while continuously searching for the correct size socket head tool to use for attaching the components. Also, the design for ergonomics has plenty of room for improvement in most of the RepRap designs, because the parts and fasteners are very small allowing very little clearance for tools and hands to reach into the confined areas. The open-source community has embraced the idea of improving RepRap 3D printer designs and allowing the public to take advantage of these design improvements.

For example, an engineering student employee working as a machinist's assistant operated the prototype machine partially for performing tasks related to his job function, but also to assist other students with their senior projects. While using the 3D printer modifications were made to the 3D printer, for example, a hinged accessible extruder was added to facilitate changing filament and clearing of jams. The student employee simply downloaded the part designed by another user to solve this filament accessibility issue, printed the part, and installed it on the machine. Additional designs have been uploaded by the RepRap open-source community, such as a Double Whammy Prusa that uses a *Push-me-pull-you extruder drive* (http://www.thingiverse.com/thing:46800) along with a pair of *Acarius10's Minimalistic J Head X Carriages* allowing the 3D printer to make two copies of the same print at once (http://www.thingiverse.com/thing:45883) (see Figure 40.3).

On the RepRap 3D printer, the extruder hot end can be a problem spot with plugging of the 0.50 mm nozzle the most significant issue. The small orifice is difficult to clear if foreign, high glass temperature plastic or non-plastic material plugs it. Clearing protocols have been developed, but this issue is the bane of 3D printer operators and great care is taken by filament manufacturers and printer operators to avoid it. A second issue is related to the design of hot ends. The goal of the hot end is to heat filament to its glass temperature immediately before deposition while keeping the filament before the nozzle as cool and stiff as possible since that filament serves as the piston driving the heated plastic through the orifice. Incoming filament keeps the temperature before the hot end at a sufficient low temperature, but interruption of feeding for long periods of time can lead to the filament heating to its glass temperature and deforming. This leads to kinks within the hot end that can be difficult to extract.

This is a perfect instance for a group of engineering students to solve a real engineering problem. Several extruder hot-end variations offer the opportunity for design, build, and test scenarios in order to optimize the 3D printer performance. One design solution is to machine the nozzle parts from more durable brass material and Polyether Ether

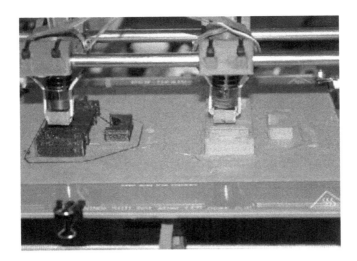

Figure 40.3 **(See color insert.)** Double Whammy Prusa modification.

Ketone (PEEK) stock to replace the purchased nozzle to create a possible solution to one of the limitations of the RepRap 3D printer.

University research is also possible using AM technologies both at the graduate and undergraduate level. An application of research in 3D printing at the graduate level using open-source 3D printers is that unlike commercial printers the plastic filament used for extruding is purchased in a spool that is open to environmental conditions of temperature and humidity versus a sealed cartridge of material. The printing envelop in most open source 3d printers is open to the environment, where in most commercial 3D printers the area where the layers are being extruded onto the build platform is sealed and controlled using fans, dryers, and sensors. For engineering students with plastics manufacturing background, this offers an opportunity to study the effects of temperature and humidity on the quality of the 3D-printing plastic parts.

Temperature and humidity effects were investigated in a research study by Irwin and Garg (2014), where samples were 3D printed at various conditions that could be experienced in the environment where a 3D printer may be used. The technique used for sample creation utilized an open-source MOST RepRap Prusa Mendel 3D printer. The samples produced at various temperature and humidity conditions were measured for dimensional accuracy. The analysis of the data of the project is done to study the variation in parameters of volume and cross-sectional area of the sample cube. During the first phase, the results were analyzed for constant humidity with varied temperature. Since the most samples were produced for 30% RH, the analysis was restricted to these samples. During second phase, two further results were analyzed for variation in area and volume at the humidity levels of 30%, 40%, 50%, and 60% at increasing temperatures from 15°C to 40°C. The samples in Phase 2 provided a more comprehensive set of data to analyze. Results show that the error in sample volume decreases as humidity increases. Based on the results, recommendations such as strategic cooling of parts could result in a greater dimensional accuracy. Practical implications offer 3D printer users suggestions for the optimal temperature and humidity for 3D printing, which can be controlled using a potable humidifier with humidity level indicator and a thermostat adjustment in the 3D-printing room.

Finally, to consider, the economic viability for a typical household consumer to utilize a 3D printer, Wittbrodt et al. (2014) report on the life-cycle economic analysis (LCEA) of

RepRap technology for an average U.S. household. The economic costs of a selection of 20 open-source printable designs, which are the products that a typical household might purchase, are quantified for print time, energy, and filament consumption and compared to low and high Internet market prices for similar products without shipping costs. The results show that the avoided purchase cost savings would range from $300 to $2000/year providing a simple payback time for the RepRap in 4 months to 2 years and provide an return on investment (ROI) between 20% and 40%. The conclusion from this study is that the RepRap is an economically attractive investment for the average U.S. household.

40.6   AM technologies in K-12 STEM outreach activities and curriculum

STEM is an acronym for science, technology, engineering, and mathematics fields of study meant to improve the U.S. competitiveness by guiding curriculum and influencing education policy. STEM education begins with K-12 educators, who are struggling with how to implement the Next Generation Science Standards (NGSS) that now place explicit emphasis on the relationship of engineering to science. The NGSS guidelines suggest that science curriculum should have activities with an iterative process involving: defining the problem, developing possible solutions, and optimizing design solutions. The NGSS guidelines include a framework with eight practices including number six, *constructing explanations and designing solutions*, which is where one major distinction is made between science and engineering practice. The goal of science is to construct theories about the natural world where the goal of engineering design is to find solutions to problems that can be manifested in a physical product, plan, or mechanical device (NGSS, 2013). Even though just a handful of states have officially adopted the NGSS standards, the National Science Teachers Association (NSTA) called for states to adopt NGSS in the November 2013 publication "Position Statement" that outlines the steps needed to ensure that all students have the skills and knowledge required for STEM careers and includes teacher professional development (NSTA, 2013).

Workshops and grant opportunities are available for teachers to receive professional development to assist in implementation of the NGSS, and some of these involve teaching about AM technologies. According to a report by Smith (2014), a NASA educator, Todd Ensign, taught a one-day workshop on 3D printing that utilized the resources of a local industry to provide teachers lessons on aviation. He trained 30 teachers from the region on how 3D-printing models could be designed and used for K-12 lessons. Ensign is quoted in the article as saying "It's such a compelling tool, I feel like it could achieve our goal to generate more student enrollment in STEM careers, we are facing a huge shortage in those careers as the baby boomers are retiring."

The commercial 3D printer company MakerBot has classroom-based 3D CAD models available to download for free to utilize in the classroom including a frog dissection kit and the Great Pyramid of Giza. In addition, there is a movement called MakerBot Academy to encourage creativity and design for teachers, parents, and kids by offering 3D models to be downloaded from Thingiverse. The Makerbot Academy highlights teachers using 3D printing in their classrooms and encourages teachers to use fund-raising methods such as Donors Choose.org to collect donations for purchasing the Makerbot Academy 3D Printing Bundle at a price of $2,000.00 to provide a 3D printer to a classroom. An example of a school using a Makerbot 3D Printer purchased via the fund-raising technique is Madison Middle School in Tampa, Florida, where students learn the basics of engineering and then move on to aerospace engineering (Donors Choose, 2014). The STEM curriculum starts them on the pathway toward taking AP courses in high school. In the applied science and math course, students design various airfoil shapes to test in a small wind tunnel using smoke, so the students can see the airflow. The 3D printer offers the ability for students to change the airfoil design and then test the design to observe the effects of turbulence.

In 2013, an *innovative additive manufacturing* (IAM) workshop was the first of its kind funded by the Square One Education Network and Michigan Tech Open Sustainability Technology (MOST). K-12 teachers applied to attend the workshop as teams of two from the same school or school district. Twelve teams were selected to attend the workshop where each team built and commissioned a MOST Prusa Mendal RepRap 3D printer to take back to their school. The workshop was so successful that it was repeated in 2014 offering a similar format for teams of two, but this time around they built a MOST Delta RepRap 3D printer, which can be completed in less time allowing each teacher a 3D printer for their own classroom. The workshop used a wiki-based, fully illustrated, self-paced program with four experienced facilitators available to help the 24 participants as they worked. In addition to building the printers, teachers were introduced to a completely free and open-source software tool chain (3D modeling, tool path generation, and printer interface) used to design and print models.

Feedback from an online postworkshop survey was positive including one survey respondent who commented:

> This conference was an amazing revitalization on my own excitement for teaching and working with kids. I couldn't have taken more away in 4 days than I did and I haven't been this excited about getting back to school in decades!!!

The workshops have also spurred the development of a virtual community to support K-12 educational 3D printing where teachers can share their course projects, lesson plans, or student's work in solving engineering design problems using 3D CAD and 3D printing. An example of a student-developed design is a bilge pump motor mount and propeller guard for use as a thruster on an underwater remote-operated vehicle (ROV) (see Figure 40.4).

Two professional development courses for K-12 science teachers have been offered as part of a Master's degree in Applied Science Education at Michigan Technological University. They were partially funded by two different Title II Improving Teacher Quality grants awarded during the 2009–2010 and 2013–2014 cycles. *The engineering process* course

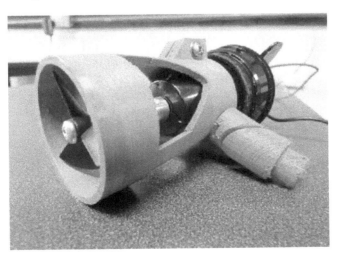

Figure 40.4 **(See color insert.)** High school student ROV project design.

and *engineering applications in the physical sciences* course have each provided practicing science teachers exposure to open-source 3D-printing technology and techniques to implement NGSS framework practices in their classrooms. Then to facilitate the implementation of the techniques learned in these two courses, the grant funded a 3D-printing workshop in the same format as the IAM workshop for teachers to ring a 3D printer back into their classrooms.

In the requirements of the grant-sponsored 3D workshop, unlike the IAM workshops, participants earned Master's level credit, took pre- and postevaluations of their understanding of 3D printing, and were required to submit lesson plans and examples of student learning after implementing the 3D printers in their classrooms. The 18 grant participants had an increase of 32% from pre- to postevaluation on a 10-question test. Also, the student data collected from the participants is valuable to assess the impact of the 3D printers on students' understanding of engineering design process concepts. Some examples of lesson plans implemented include: designing parts for mouse trap vehicles, personalized name plates, parts that are made from a plant source, parts for ROVs, product finding inspiration from a living organism, and device to solve a classroom problem, (a doorstop for example), all of which have students solve engineering problems by designing, building, and testing their solutions.

One example of student reflection after completing the 3D-printing lesson to have students create parts for mouse trap vehicles exhibits an understanding of the advantages of making prototypes for testing purposes in the answer to the question: Do you feel your final project was a success? Explain.

> I'm mad that it crashed every time we tested it. It was too light for the power we had from the snapping mouse trap. After talking to Mr. H, we think we will do better if we add more weight with washers. We will try that next week. I think it was cool anyway, and want to build a better one next time. I know that Thingiverse has more mouse trap cars; I want to print those and see if they work better. We learned a lot from what went wrong, I think we will be successful with the car soon.

40.7 Summary

The use of AM technology in STEM education offers a vehicle to teach the NGSS requirements for constructing explanations and designing solutions. The K-12 teachers as well as higher education can implement AM technologies into their classrooms to improve engineering education. The open-source desktop RepRap 3D printers offer a less-expensive alternative with the advantage of allowing students to *tinker* with the mechanics of the 3D printer and enjoy near-immediate gratification from experiencing the engineering process first hand. The commercial desktop and large-scale 3D printers allow engineering students to research AM technology to produce larger more refined parts from a variety of materials to be used in a wider variety of applications.

References

ABET (2013). Criteria for accrediting engineering programs; 2014–2015, Baltimore, MD. Retrieved from http://abet.org/eac-criteria-2014-2015/.

Campbell, I., Bourell, D., and Gibson, I. (2012). Additive manufacturing: Rapid prototyping comes of age. *Rapid Prototyping Journal,* 18 (4): 255–258.

Dieter, G.E. and Schmidt, L.C. (2012). *Engineering Design*, 5th ed. McGraw-Hill Book Co. New York, NY.

Donors Choose (2014). 3D printing airfoils to lift up success. Retrieved from http://www.donorschoose. org/donors/proposal.html?id=1225078&utm_source=dc&utm_campaign=typ_cover&utm_ medium=email&utm_content=Project&pma=true.

Evans, M. (2002). The integration of rapid prototyping within industrial design practice. PhD diss., Loughborough University.

Gonzalez-Gomez, J., Valero-Gomez, A., Prieto-Moreno, A., and Abderrahim, M. (2012). A new open source 3d-printable mobile robotic platform for education. *Advances in Autonomous Mini Robots* (pp. 49–62). Berlin, Germany: Springer.

Hallgrimsson, B. (2012). A model for every purpose: A study on traditional versus digital model-making methods for industrial designers. Retrieved from www.idsa.org/sites/default/files/ A%20Model%20for%20Every%20Purpose.pdf.

Higginbotham, S. (2012). Ford's gift to engineers: MakerBot 3D printers. Retrieved from http:// www.businessweek.com/articles/2012-12-21/fords-gift-to-engineers-makerbot-3d-printers.

Irwin, J., Pearce, J., Anzalone, G., and Oppliger, D. (2014). The RepRap 3-D printer revolution in STEM education. *ASEE 2014 Conference Proceedings* (p. 13). Indianapolis, IN: ASEE.

Irwin, J. and Garg, A. (2014). A study of temperature and humidity effects on desktop 3D printers. *ATMAE 2014 Conference Proceedings* (p. 32). St. Louis, MO: ATMAE.

Jones, T.S. and Richey, R.C. (2000). Rapid prototyping methodology in action: A developmental study. *Educational Technology Research and Development*, 48 (2): 63–80.

Kentzer, J., Koch, B., Thiim, M., Jones, R.W., and Villumsen, E. (2011). An open source hardware-based mechatronics project: The replicating rapid 3-D printer. *Mechatronics (ICOM), 2011 4th International Conference*. doi:10.1109/ICOM.2011.5937174.

Malicky, D., Kohl, J., and Huang, M. (2010). Integrating a machine shop class into the mechanical engineering curriculum: Experiential and inductive learning. *International Journal of Mechanical Engineering Education*, 38 (2): 135.

NGSS (2013). Next generation science standards. Retrieved from www.nextgenscience.org/.

NSTA (2013). Position statement: The next generation science standards. Retrieved from http:// www.nsta.org/about/positions/ngss.aspx.

Oravecz, J. (2014). Young visionaries at PieceMaker Technologies Inc. see future in 3-D. Retrieved from http://triblive.com/business/headlines/5953166-74/printing-piecemaker-kiosks#axzz3FE8S1q7p.

Orr, T. (2014). DreamVendor 2.0: 3D printing kiosks coming to a corner near you. Retrieved from http://3dprint.com/4621/dream-vendor-3d-printing-kiosks/.

Pearce, J.M. (2012). Building research equipment with free, open-source hardware. *Science*, 337 (6100): 1303–1304.

Pearce, J.M. (2014). *Open-Source Lab: How to Build Your Own Hardware and Reduce Research Costs*. Elsevier, Cambridge, MA.

Perry, T. and Kirkpatrick, A.T. (2012). AC 2012-4832: ASME's Vision 2030's import for mechanical. *ASEE 2012 Conference Proceedings* (p. 10). San Antonio, CA: ASEE.

RepRap (2013). Welcome to RepRap. Retrieved from http://reprap.org/wiki/Main_Page.

Smith, F. (2014). NASA educator sees bright STEM future in 3D printing. Retrieved from http://www. edtechmagazine.com/k12/article/2014/05/nasa-educator-sees-bright-stem-future-3d-printing.

Square One (2014). Square one education network. Retrieved from http://www.squareonenetwork. org/.

Stratasys (2014). About fused deposition modeling. Retrieved from http://www.stratasys. com/3d-printers/technology/fdm-technology.

Tseng, T., Chiou, R., and Belu, R. (2014). Fusing rapid manufacturing with 3D-virtual facility and cyber tutor system into engineering education. *ASEE 2014 Conference Proceedings* (p. 12). Indianapolis, IN: ASEE.

Williams, R. (2014). Home depot sets up MakerBot 3D printer kiosks in 12 stores. Retrieved from http:// hothardware.com/News/Home-Depot-Sets-Up-MakerBot-3D-Printer-Kiosks-In-12-Stores/.

Wittbrodt, B. T., Glover, A. G., Laureto, J., Anzalone, G. C., Oppliger, D., Irwin, J. L., & Pearce, J. M. (2013). Life-cycle economic analysis of distributed manufacturing with open-source 3-D printers. *Mechatronics*, 23(6), 713–726.

chapter forty one

Additive manufacturing applicability for United States Air Force Civil Engineer contingency operations

Seth N. Poulsen and Vhance V. Valencia

Contents

41.1   Introduction

Additive manufacturing (AM) is a relatively new technique that is gaining popularity in many applications. One of these developing applications is the use of AM machines for the production of supplies in remote, austere, or deployed locations. United States Air Force (USAF) Civil Engineers (CEs) is one of the many organizations that often labor in such contingency environments. Recently, a research project was undertaken to determine how

AM techniques can be beneficially applied in Air Force Civil Engineer contingency operations and to predict the appropriate timeframe for this novel application.

41.1.1   *Proposed air force additive manufacturing application*

USAF CEs are responsible for the construction, operation, maintenance, repair, and disposal of USAF civil infrastructure systems on Air Force bases throughout the United States and abroad. These CEs manage a diverse portfolio of infrastructure that includes the following: facilities, roads, runways, water distribution, and other systems. CEs are responsible for the maintenance of these systems not only on large primary bases, but also in contingency locations that are often remote, isolated, and austere.

Maintaining infrastructure at contingency locations poses unique and significant challenges. One of these challenges is the supply of tools and parts required for infrastructure maintenance activities. Due to their remote or isolated location, contingency locations often prove to be challenging to supply. As a result, initial CE teams typically deploy with a toolkit that provides them with an initial capability to maintain and repair the location's infrastructure. These deployable CE toolkits are known as unit type code toolkits, or equipment only UTCs.

To determine if AM machines would be beneficial if included in CE equipment UTCs and to predict the appropriate timeframe for including AM machines in UTCs, this research project was conducted from August through December of 2014. This research proposed that an appropriate machine be chosen and added to an existing CE UTC. This addition would provide the method for requesting and conveying an AM machine to a contingency location based on an existing Air Force program and construct the UTC management system. Specifically, this research was designed to answer four investigative questions:

1. *What categories of AM machine are currently well-suited for utilization in CE equipment UTCs?*
   Many types, makes, and models of AM machines are on the market today. This question seeks to understand which of these various machines would be suitable for CE applications. This question does not look at companies or brands, but instead analyzes the various raw materials and build processes currently available in the AM industry.
2. *What attributes make an AM machine well-suited for use in a CE equipment UTC?*
   This question focuses on the specific attributes necessary in an AM machine for CE contingency applications. It seeks to understand the desired qualities of an ideal AM machine and will focus on machine *-ilities* such as reliability, usability, quality, maintainability, and others. These properties are not necessarily fundamental requirements of an AM machine, but knowing which of these attributes are most important can assist in selecting the best machine for contingency engineering (de Weck, Roos, & Magee, 2011, p. 66).
3. *Has the AM industry currently reached a point at which the selected categories of AM machines embody these beneficial attributes?*
   This question seeks to understand the status of current AM practices and future possibilities. AM is not a new technology; in fact, similar methods for creating objects layer-by-layer have been in use since the 1890s (Bourell, Beaman Jr., Leu, & Rosen, 2009, p. 5). Since that time, AM technology has been continually progressing. This question seeks to determine if AM technology has progressed far enough

today, or if the technology needs to further mature, to be suitable for CE contingency applications.

4. *What are the most promising benefits of including an AM machine in a CE UTC?*
   The reasons for using an AM machine in a contingency environment differ significantly from the reasons for using one in a lab in the United States. Therefore, this question is designed to provide an understanding of why it would be beneficial to deploy AM machine downrange.

This research utilized a Delphi study to answer these questions. This study was designed to elicit opinions and predictions from a panel of experts who are knowledgeable about AM and/or CE UTCs. This study combined and refined the cumulative knowledge of these experts through multiple rounds of questions that address this topic. Therefore, the ultimate results of this research are a compilation of the predictions of these panel members. These predictions are used to determine if AM machines would be beneficial if included in CE equipment UTCs and to predict the appropriate timeframe for including AM machines in UTCs.

41.1.2   Military additive manufacturing applications

The benefits and possibilities of AM have not been overlooked by the U.S. Military services. Significant research is being conducted by the services and other military-sponsored organizations. A few of the ongoing Army, Marine, Navy, and Air Force AM research efforts and applications are reviewed in this section.

The Army began researching AM in the 1990s, looking at stereolithography (Zimmerman & Allen, 2013, p. 13). One of the most interesting and recent applications that the Army has employed is the mobile expeditionary labs (Ex Labs), which were delivered to the Rapid Equipping Force in 2012. These Ex Labs contain an AM machine along with traditional manufacturing equipment and are rapidly deployable to forward operating locations to provide custom engineering and prototyping (United States Army Rapid Equipping Force, 2014).

The U.S. Marine Corps has also been actively pursuing AM technology. A 2014 report outlined several AM applications for the Marines, including inventory reduction capabilities, reduction in transportation costs, and reduction in manufacturing costs (Robert W. Appleton & Company, Inc., 2014, p. 25). These are the same benefits that appeal to many military individuals and organizations.

The U.S. Navy has taken the lead in AM research and has various projects that include AM machines. The recent *Print the Fleet* workshop the Navy held at Dam Neck, Virginia, illustrates the importance the Navy is placing on AM. This workshop was designed to "introduce 3D printing and additive manufacturing to Sailors and other [Navy] stakeholders" (Stinson, 2014). Additionally, the Navy is now utilizing AM machines in all four of its shipyards for rapid prototyping and custom part fabrication (Cullom, 2014). The Navy is also experimenting with AM at sea and has installed AM machines on the USS Essex (Cullom, 2014) and the USS Enterprise (Campbell & Ivanova, 2013, p. 74).

Finally, the Air Force is researching possibilities for the application of AM. Currently, the Air Force employs AM for "design iteration, prototyping, tooling, and fixtures, and for some noncritical [aircraft] parts" (Mack, 2013). Recently, the Air Force awarded several multimillion dollar contracts that use AM for both research and production, including one contract for F-35 parts (3DSystems, 2012) and another for rocket engine parts (Leopold, 2014).

41.1.3   Contingency additive manufacturing applications

AM provides a highly customizable and self-contained manufacturing process. As such, it has been considered for application in remote, isolated, and austere contingency environments as a means of producing necessary items while minimizing warehousing requirements. Contingency applications are currently being researched by the Department of Homeland Security, the National Defense University, and the U.S. Army and Navy.

The Department of Homeland Security is assessing the possible applications for AM machines in disaster response scenarios. They are currently evaluating the possibilities for deploying AM machines to a disaster location and providing a central library of digital 3D models that can be physically produced anywhere as needed (Lacaze, Murphy, Mottern, Corley, & Chu, 2014). Additionally, The Center for Technology and National Security Policy at National Defense University has recognized the potential for AM application in contingency environments. As a result, the center recently issued a challenge to "examine the uses of additive manufacturing for humanitarian assistance and disaster relief operations" (McNulty, Arnas, & Campbell, 2012, p. 11).

Furthermore, the U.S. Army has recognized the benefits of using AM in deployed locations. As previously noted, the Army forward deployed AM machines in their Ex Labs in 2012 (United States Army Rapid Equipping Force, 2014). Finally, the Navy is researching the use of AM machines in contingencies on the open seas. They are currently testing the benefits of AM machines deployed on the USS Essex (Cullom, 2014) and the USS Enterprise (Campbell & Ivanova, 2013, p. 74), as previously discussed in section 41.1.2. These applications show that testing and researching AM application in contingency environments is moving forward and many organizations already recognize the benefits that AM machines can provide in unique situations.

41.2   Air force unit type codes

The Air Force defines a UTC as "a potential capability focused upon accomplishment of a specific mission that the military service provides" (United States Department of the Air Force, 2006, p. 87). Therefore, each UTC is not just a toolkit: it is an enabler used to accomplish a certain mission or task. A UTC may include the following: tools, equipment, and supplies and it may also include AF personnel. Some UTCs consist only of equipment, some contain only personnel, and some are a combination of both personnel and equipment (United States Department of the Air Force, 2006, p. 87).

Every UTC is identified by several pieces of information: a unique number, a mission capabilities statement (MISCAP), a personnel number, and a material weight. The number that defines each UTC is a five-digit alphanumeric code, which uniquely identifies a UTC and indicates the functional area responsible for the UTC. The MISCAP is a brief "statement of the capabilities of the force identified by each UTC" (United States Department of the Air Force, 2012, p. 66). The personnel number associated with each UTC is known as the Authorized Personnel (AUTH) number. This number indicates the quantity of personnel assigned to a specific UTC; it is zero if no personnel are assigned to a UTC. The weight for a UTC indicates the weight of all material contained in the kit in total short tons (ST). This value is crucial for determining the options for deploying a UTC. This number is zero if the UTC consists of personnel only (United States Department of the Air Force, 2012, p. 66).

41.2.1   UTC utilization

When planning for military or contingency situations, war planners use UTCs to under-stand and anticipate the total manpower and logistics chain required to support an opera-tion (United States Department of the Air Force, 2012, p. 66). A war planner anticipating a military requirement will turn to a list of UTCs to find a predefined capability that will meet the need. Thus, a UTC is the basic building block utilized to meet peacekeeping, humanitarian relief, and rotational operation needs in contingencies from small to large (United States Department of the Air Force, 2006, p. 88).

Equipment UTCs are warehoused and maintained at a primary base in the continental United States. When needed, an equipment-only UTC will be picked up from its storage location and delivered via air cargo to the requisite deployed location. This system allows for the UTC to be continually maintained and ready for rapid deployment at any time (United States Department of the Air Force, 2006, p. 88).

41.2.2   Civil engineer UTCs

The USAF maintains thousands of UTCs and of these, 96 are specific to CEs (Grissett, 2014). CE-specific UTCs are designated as *4F9XX*, where *XX* indicates the designation for a specific UTC. These UTCs meet a variety of engineering needs and each is specifically tailored to pro-vide a capability that may be needed in a wide range of contingency environments. Two gen-eral engineering kits will be examined in further detail: the 4F9ET Engineer Force Equipment Kit and the 4F9RY Rapid Engineer Deployable Heavy Operational Repair Squadron Eng-ineer (RED HORSE) Equipment Kit. The MISCAPS for these UTCs are given in Table 41.1.

The 4F9ET is a general engineer force equipment kit used for light construction. It is an equipment-only UTC and is designed to be paired with two personnel-only UTCs. When these three UTCs are deployed together, the capability to establish, operate, and sustain a contingency location is delivered. This kit contains basic tools and equipment

Table 41.1 4F9ET and 4F9RY mission capability statements

UTC	MISCAP
4F9ET	Engineer force equipment set to support two 4FPET UTCs. Supports missions (including recovery) to establish, operate, and sustain contingency operating locations, aerial ports, enroute bases, natural disaster recovery operations, and joint-base support. Provides equipment for initial beddown of bare base and/or forward operating locations. May be augmented with one or more 4F9EF UTCs based on mission requirements.
4F9RY	REDHORSE (RH) equipment UTC to support lead C2 element (hub) of a deployed RH squadron responsible for managing RH construction projects in a theatre of operations. It must be combined with a 4FPRY UTC to support RH beddown. Vehicle maintenance, services, design, and engineering support surveying, drafting, and material testing capabilities. Requires a 4F9GP UTC for precision survey requirements using global positioning system equipment. Horizontal/vertical construction capability is obtained when combined with one or more of the following RH UTC combinations. Horizontal construction teams 4F9RU/4FPRU or 4F9RV/4FPRV UTCs and/or vertical construction teams 4F9RS/4FPRS or 4F9RT/4FPRT UTCs. When combined with a 4FPRY, this UTC contains enhanced logistics and communication capability.

for electricians, structural craftsmen, pavements craftsmen, heating, ventilation and air conditioning (HVAC) technicians, and others. Some of the items included are hammers, saws, tape measures, pliers, rakes, crowbars, concrete floats, drills, chisels, screwdrivers, levels, helmets, padlocks, ladders, and other tools needed to establish and maintain an air base (Air Force Civil Engineer Center, 2014).

The 4F9RY is a basic UTC for heavy construction. This kit, tailored for use by a RED HORSE unit, has more robust capability for construction, paving, and logistics. This kit is an equipment-only kit designed to be used by the RED HORSE personnel in the 4FPRY UTC. The kit includes many items that are in the 4F9ET and adds larger items like power distribution panels, latrines, heaters, water purification systems, a tactical radio kit, a welding kit, fuel tanks and pumps, a skid steer loader, and trucks and tents (Air Force Civil Engineer Center, 2014).

41.3   The Delphi study technique

The Delphi method was created out of necessity in the early 1950s. The need for the methodology arose from the RAND Corporation's work on a U.S. military project (Linstone and Turoff, 2011). During this project, significant amounts of forecasting for previously unstudied topics were being undertaken. To ascertain the most accurate predictions, the RAND Corporation turned to leading experts in the field in an effort to gain valuable insight. RAND solicited input from these individuals in several, anonymous rounds and consolidated the varied insights in their report. This was the first research to utilize what would come to be known as a Delphi study.

The Delphi approach was named after the Oracle at Delphi, a prominent figure in ancient Greek mythology who *was able to predict the future with infallible authority* (Clayton, 1997, p. 374). The name is fitting as a Delphi study is often used to predict the future or to *address what could/should be* (Miller, 2006, p. 1). This stands in contrast to a traditional survey that is designed to understand or represent *what is* (Miller, 2006). Although the Delphi technique is not a statistical method for creating new knowledge, it is nonetheless a powerful tool for making the best use of available information (Powell, 2003, p. 380). The Delphi technique is well-suited to determining or developing possible program alternatives and to collecting informed judgments on a topic that spans a range of disciplines (Delbecq, Ven, & Gustafson, 1975, p. 11).

The Delphi is a good tool for use when planning for the future and looking at program alternatives. This technique is specifically designed to "predict or forecast future events and relationships in order to make appropriate and reasonable plans or changes" (Ludwig, 1997). This is often the case in emerging industry or when applying a new technique in a novel application. In such a situation, the Delphi method excels at predicting the future possibilities as it provides a *flexible and adaptable tool to gather and analyze the needed data* (Hsu & Sandford, 2007, p. 5).

The Delphi technique is also beneficial for garnering expert judgment in a multidisciplinary topic. The technique is specifically designed to "gather information from those who are immersed and imbedded in the topic of interest and can provide real-time and real-world knowledge" (Hsu & Sandford, 2007, p. 5). Again, this is particularly useful in emerging technologies and their novel application. As the Delphi method relies upon targeted experts rather than random individuals, a Delphi is designed to combine the knowledge and opinions of the participants and to structure and organize their communications (Keeney, Hasson, & McKenna, 2006, p. 206) in an area of uncertainty or where empirical evidence is lacking or yet to be created (Powell, 2003, pp. 376–377).

The Delphi technique is a unique tool, suited to unique research applications. In particular, it is a powerful method for forecasting future alternatives and possibilities (Miller, 2006) and for gathering cutting edge, real-time, and real-world expert opinions (Hsu & Sandford, 2007, p. 5). Although this method differs from more traditional survey- or statistical-based methodologies, it is a powerful tool when appropriately applied to predictive research.

41.3.1   Delphi application

The unique nature and benefits of the Delphi technique make it well-suited for determining the possibilities for future application of AM technology in CE contingency operations. Therefore, a Delphi study was designed to elicit opinions and predictions from a panel of experts who are knowledgeable about AM and/or CE UTCs to further explore this novel AM application. To conduct this study, four rounds of questionnaires were distributed to the panel participants via electronic mail. The questionnaires in each round were tailored to generate panel discussion about possible AM applications for CE UTCs. Further, each round built on answers from the previous rounds.

41.3.2   Delphi study participants

In order to conduct this Delphi study, the panel of expert participants first needed to be selected. For this study, the panel consisted of 20 individuals. Each panel member was then hand selected for their past experience and specialized knowledge. Ten individuals were selected for inclusion in the panel as AM experts and ten were chosen as CE UTC experts.

The first group of members selected consists of ten individuals who are AM experts. These panel members were chosen from members of the America Makes organization. America Makes is an organization based in Youngstown, Ohio that aims to "accelerate the adoption of additive manufacturing technologies in the U.S. manufacturing sector and to increase domestic manufacturing competitiveness" and consists of individuals who "are at the forefront of new 3D-printing materials, technologies, and education" (America Makes, 2014). Delphi participants for this research were chosen from America Makes members based on their experience in academia or industry. Each of the selected participants had a minimum five years of experience in AM. Further, these members had a working knowledge of various types of AM processes and the respective capabilities and limitations of each type. The demographics of these members are presented in Table 41.2.

The second group of Delphi panel members was selected for their experience in CE UTC use and management. These members were selected from members of the Air Force Civil Engineer Center (AFCEC). AFCEC is the Air Force organization responsible for the planning and policy for all CE UTCs. The individuals selected for this Delphi are those

Table 41.2 Delphi panel demographics

	AM experts	CE UTC experts
Number of panelists	10	10
Gender	100% Male	88% Male/12% Female
Age	38–73	35–45
Experience (Years)	5–23	10–23
Education	Associates Degree—17%	Associates Degree—75%
	Doctor of Philosophy—83%	Bachelor of Science—25%

who are currently responsible for those plans and policies. Panelists had a minimum of three years of experience managing or creating policy for CE UTCs. Further, they had a working knowledge of CE UTC contents and requirements. The demographics of these members are presented in Table 41.2.

41.4 Results

The four investigative questions for this research were used as a basis and a guide for the Delphi study. Each of these questions was tailored to discern if AM machines would be beneficial if included in an Air Force CE equipment UTC and to predict the appropriate timeframe for this inclusion. Each of these four questions is presented here and answers to the questions will be drawn from the results of the Delphi study.

41.4.1 What categories of AM machine are currently well-suited for utilization in CE equipment UTCs?

At the beginning of the Delphi study, the panel members were asked to identify which categories of AM machines were most likely to endure into the future. The categories from *ASTM F-42 Classification of Additive Manufacturing Processes* (ASTM International, 2012) were used as a basis for the responses to this question. The panel members were divided in their opinions on this question, and no process stood out as being most enduring. However, the most common responses were—in order of frequency—powder bed fusion (PBF), directed-energy deposition, material extrusion, and binder jetting as shown in Table 41.3. These processes were identified by five or more respondents, which constitutes one-third of the panel members. Therefore, these four processes were identified for further discussion.

To further understand this question, in the second round of the study, panel members were asked if a plastic- or polymer-based AM machine would be most useful for this application or if a metal AM machine would be preferable. Panel responses were varied on this topic as shown in Table 41.4. It appears that either type of machine would be beneficial, with metal being slightly favored. However, the results obtained were not strong enough to conclude that either type of machine would be better for this application.

In the final round of Delphi questions, two final statements were presented to panel members in an effort to determine which categories of AM machine are currently well-suited for utilization in CE equipment UTCs. These statements were generated by panel discussion, and were presented in an effort to establish a final consensus among panel members for which category of AM machine is best suited for CE application.

Table 41.3 AM machine category potential

Which categories of AM machines show potential to endure many years into the future?	
AM machine category	Percentage of 12 respondents
Powder-bed fusion	67%
Directed energy deposition	58%
Material extrusion	50%
Binder jetting	42%
Material jetting	25%
Vat photopolymerization	17%

Table 41.4 Beneficial nature of AM machine types

How beneficial do you believe inclusion of each of these types of AM machine would be in a CE UTC?

Not at all beneficial	1	2	3	4	Very beneficial		
	Mean	**Median**	**Mode**				
Plastic/Polymer	2.87	3	3				

Box plot (y-axis 0 to 4.5): UTC Experts, AM Experts

| Metal | 3.13 | 3 | 3 | | | | |

Box plot (y-axis 0 to 4.5): UTC Experts, AM Experts

Panel members were asked to rate their level of agreement with the statements on a 1–5 *Likert* scale as shown in Table 41.5.

Some dissension appeared among panel members in responding to these statements that address the type of AM machine that should be used for Air Force CE contingency operations as shown in Table 41.6. The panel members were not decisive in selecting either the type or material that should be used for this application. Part of this lack of agreement may have arisen due to the dual nature of the wording of these questions, which may have created confusion or biased the results of the first part of the question, if a panel member did not agree with the second part or vice versa. Regardless, throughout the study there was lack of consensus on these points. It is clear that further research is needed in this area to determine which categories of AM machine are best suited for use in CE UTCs.

Table 41.5 Delphi study Likert response scale

Response	Likert value
AGREE with the main point	1
SOMEWHAT AGREE with the main point	2
NEITHER agree nor disagree with the main point	3
SOMEWHAT DISAGREE with the main point	4
DISAGREE with the main point	5

Table 41.6 Best AM machine category and type

The types of AM machine best suited for use in deployed CE operations are powder-bed fusion or directed energy deposition machines. Two other promising options are material extrusion and binder jetting.

Mean	Median	Mode
2.5	3.0	3.0

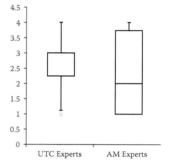

An AM machine that uses metal raw material is most likely to be the best option for CE applications but plastic/polymer machines should not be completely ruled out.

Mean	Median	Mode
2.1	1.5	1.0

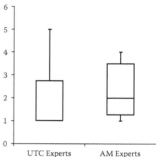

Ultimately, the panel members were unable to predict categories of AM machine that are currently well-suited for utilization in CE equipment UTCs. This uncertainty likely arose for two main reasons: (1) The panel was made of two disparate groups of experts, neither of which is well-versed in the other's field and (2) The information provided in this Delphi study was likely insufficient to make an appropriate decision on the topic.

First, the Delphi study conducted for this research pooled together two very different groups of experts. One group of individuals is very knowledgeable about AM technology and research, but is not familiar with CE UTCs or their contents and purpose. The second group is well-versed in CE UTC employment, but is largely unaware of the current state of the AM industry. At the beginning of the study, a document containing background information on each of these two diverse subjects was provided to the panel members in order to give each group some knowledge about the other's area of expertise. However, it appears that this information was insufficient.

Ideally, this research could be repeated with individuals who are experts in both AM technology and CE UTCs. However, there are few, if any, individuals who meet this requirement. As a powerful alternative, panel members could be brought together for an educational seminar to learn about each topic and share ideas before this Delphi is repeated. This would produce panel members who would be better able to make educated guesses and predictions about the confluence of these two fields.

Second, further background research should be conducted to understand the contents of CE UTCs. This research should focus on the parts, tools, and supplies in UTCs that can be produced by AM. These candidate items should also be divided into two groups: metal and polymer/plastic. Had this information been provided in this Delphi study, it is likely that the panel would have been better suited to determine the categories of AM machine that are currently well-suited for utilization in CE equipment UTCs.

41.4.2 *What attributes make an AM machine well-suited for use in a CE equipment UTC?*

In the first round of the Delphi study, the panel was asked about what qualities of an AM machine would be desirable for CE UTC applications. The results from this study show that the Delphi panel members consider usability to be the most important quality for a deployed AM machine. This quality was identified by 75% of the Delphi participants as being important. Additional desirable qualities that were cited by more than half of the panel members are reliability, flexibility, and adaptability. Finally, quality and safety were also identified by at least one-third of the respondents as an important AM machine attributes. The results of this first round of questions are presented in Table 41.7.

The responses from the first round question were used to create an overall panel opinion. This opinion stated that "The quality of an AM machine most important to consider for deployed CE applications is usability. Additionally, reliability, flexibility, and adaptability are qualities of secondary importance." This statement was presented to the panel and the members were asked how strongly they agreed or disagreed with the statement based again on the 1–5 Likert scale. Once responses were analyzed, it was found that there was strong agreement among panel members on this topic as shown in Table 41.8.

Table 41.7 AM machine qualities

What qualities of an AM machine would make it well-suited for use in a CE UTC or in a deployed or field operating environment?	
AM machine quality	Percentage of 16 respondents
Usability	75%
Reliability	63%
Adaptability	63%
Flexibility	56%
Quality	50%
Safety	31%
Interoperability	13%
Resilience	6%
Other	0%

Table 41.8 Most important AM machine qualities

The quality of an AM machine most important to consider for deployed CE applications is usability. Additionally, reliability, flexibility, and adaptability are qualities of secondary importance.	
Mean	**Median**	**Mode**
1.6	1.0	1.0

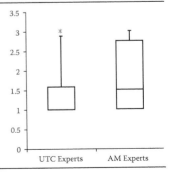

From these responses, it is apparent that considering usability of an AM machine is of paramount importance when determining what makes an AM machine well-suited for use in a CE equipment UTC. It is also important to consider reliability, flexibility, and adaptability when making this determination. It is interesting to note that the UTC experts were more strongly in agreement on this point than the AM experts were, as shown in the box and whisker plots in Table 41.8. This division occurred because the UTC experts are most concerned with actual use of the machine and the operation of the machine in a contingency environment. However, the AM experts are less interested in the qualities of the machine itself, and more concerned with the items it produces. Regardless, both groups strongly agreed that usability is the most important attribute that makes an AM machine well-suited for use in a CE equipment UTC.

41.4.3   Has the AM industry currently reached a point at which the selected categories of AM machines embody these beneficial attributes?

The first round of this Delphi study attempted to determine the time frame in which including an AM machine in CE UTCs is expected to be beneficial. Of panelists who responded, 81% agree that including an AM machine would be beneficial within the next 10 years. Further, 37% of respondents agree that doing so would be beneficial today. These statistics are presented in Table 41.9.

In addition to determining when AM machine application would be beneficial, the Delphi panel also noted that a pilot study implementation would be a good starting point for this application. The panel agreed more strongly that AM industry has reached a point today where a pilot study would be beneficial and that the industry will have progressed further allowing full-scale implementation after conclusion of the study. This is illustrated in Table 41.10, with answers being rated on the same Likert scale presented in Table 41.5.

Based on these responses, panel members were hesitant to agree that the AM industry has currently reached a point at which AM machines embody the beneficial attributes of usability, reliability, flexibility, and adaptability. However, the panel does agree that at this time, a pilot study would be beneficial to begin testing for these AM machine attributes in a deployed location. Further, once this study has been completed, the industry will be nearing a point where full-scale deployment of AM machine for CE contingency operations will be beneficial.

Table 41.9 AM application time frame

Do you think AM technology has currently reached a point where including an AM machine in a UTC would be beneficial? If not, when do you think technology will progress far enough that inclusion would be beneficial?

Timeframe	Percentage of 16 respondents
Today	37%
1–5 years	25%
5–10 years	19%
More than 10 years	19%

Table 41.10 AM pilot study

Creating a pilot study or case study for AM-machine deployment would be very beneficial and technology has progressed far enough that this initial study can be performed today. Within five years it is expected that this technology will have progressed enough that full-scale deployment of an AM machine will be beneficial.

Mean	Median	Mode
2.0	1.0	1.0

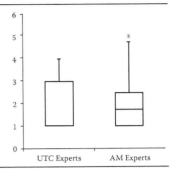

41.4.4   What are the most promising benefits of including an AM machine in a CE UTC?

In addition to predicting the suitability of AM machines for CE contingency operations, this Delphi study also endeavored to determine what benefits civil engineers could expect to realize from including an AM machine in their UTC toolkits. At the beginning of the study, panel members were asked "What are some potential benefits of including an AM machine in a CE UTC?" An open response space was provided and the ideas generated by the panel were compiled based on theme as shown in Table 41.11.

The panel members agreed (63% of respondents) that the most important benefit of including an AM machine is the ability to create various necessary parts. Additionally, the panel members identified that the ability for rapid, on-demand production of these parts would be beneficial (31%). Other benefits mentioned multiple times included prototyping and model building, tool production, reduced inventory, and better supply chain options. In the next round of the study, panelists were asked to decide which of these six benefits were the most promising. The results of this question are presented in Table 41.12.

Table 41.11 Potential AM benefits

What are some potential benefits of including an AM machine in a CE UTC?	
Potential benefit	Percentage of 16 respondents
Part production	63%
Rapid/on-demand production	31%
Prototyping/models	25%
Tool production	25%
Reduced inventory	19%
Local production	13%
Better supply chain options	13%
Inexpensive	6%
Construction	6%
Increased UTC capability	6%
Easy to transport	6%

Table 41.12 Most promising AM benefits

After compiling the possible BENEFITS of including an AM machine in a CE UTC, six common themes were discovered among Round 1 respondents. Of these six possible benefits, which do you believe are the most promising for the future of deployed civil engineer operations?

Most promising	1	2	3	4	5	6	Least promising
Possible benefit							Box plot

AM machines can be used to produce necessary and specialized tools on-site

AM machines can be used to produce spare parts when needed

AM capabilities can enable a reduction of inventory of parts, tools, and so on.

(*Continued*)

Table 41.12 (Continued) Most promising AM benefits

After compiling the possible BENEFITS of including an AM machine in a CE UTC, six common themes were discovered among Round 1 respondents. Of these six possible benefits, which do you believe are the most promising for the future of deployed civil engineer operations?

Most promising	1	2	3	4	5	6	Least promising
Possible benefit							Box plot

AM machines allow on-demand and rapid production		

AM machines allow production of prototypes and models on-site and in real-time	3.9	

AM machines allow independence from some aspects of a traditional supply chain	4.5	

From the responses received to this question, it was determined that the most promising benefit of deploying an AM machine for CE contingency operations is the ability to produce specialized tools rapidly, and on demand. The second most promising benefit is the ability to produce spare parts rapidly, and on demand. However, it should be noted that the two groups of panelists were divided on these answers. The UTC experts were more interested in producing spare parts that would allow CE personnel to create a wide range of parts with minimal warehousing requirements.

Table 41.13 The most promising AM benefit

The most promising benefit of a deployed AM machine for the CE community is the ability to create specialized tools and parts on-site and on demand.		
Mean	**Median**	**Mode**
1.6	1.0	1.0

This clearly reflects the operational desires of the CE community. However, the AM experts felt this was one of the least promising benefits. Rather, the AM experts strongly agree that the most promising use for AM machines is the creation of specialized part on-site and on demand. This more accurately reflects the AM industry as these machines currently do not rival mass production manufacturing techniques for large runs of simple items.

Finally, the insight gained from the first two questions was compiled into a statement of the panel opinion. The panel members were again asked to rate agreement with the statement on a Likert scale of 1–5. The results are presented in Table 41.13.

Once again, the difference in opinion of the two groups of experts is evident in these responses. However, after this round, the panel appeared to be largely in agreement. The most promising benefit of a deployed AM machine for the CE community is the ability to create specialized tools and parts on-site and on demand.

41.5   *Conclusions of research*

These four investigative questions provide context and background to determine if AM machines would be beneficial if included in an Air Force CE equipment UTC and to predict the appropriate time frame for this inclusion. Based on the answers to these three investigative questions, the members pooled in this Delphi study believe that

1. Including an AM machine in a new UTC would be beneficial in meeting deployed CE requirements.
2. AM technology has currently reached a point at which a pilot study would be beneficial to validate the benefits of including an AM machine in a CE equipment UTC.
3. Within the next five years, AM technology will have progressed far enough that a full-scale deployment of AM machines in CE UTCs will be beneficial.

These statements, drawn from experts who participated in the Delphi study conducted for this research, identify that AM machines can in fact be beneficially used in a CE contingency environment. This novel application of AM technology is currently untested but the panel of experts assembled for this research believes that now is the time to begin planning for this integration.

Acknowledgments

This work was made possible through a research funding from the Air Force Civil Engineer Center, Requirements and Acquisition Division (AFCEC/CXA). A special thanks to Mr. Craig Mellerski and Dr. Joseph Wander for their continued support, leadership, and guidance in this and other research efforts.

References

3DSystems. (2012, October 23). *3DSystems Press Releases*. Retrieved from 3D Systems Receives U.S. Air Force Rapid Innovation Fund Award: http://www.3dsystems.com/press-releases/3d-systems-receives-us-air-force-rapid-innovation-fund-award

Air Force Civil Engineer Center. (2014). *4F9ET engineer force equipment set*. United States Air Force, Tyndall Air Force Base, Florida.

America Makes. (2014). *Mission Statement*. Retrieved from America Makes: https://americamakes.us/about/mission

Bourell, D. L., Beaman Jr., J. J., Leu, M. C., & Rosen, D. W. (2009). A Brief History of Additive Manufacturing and the 2009 Roadmap for Additive Manufacturing: *Looking Back and Looking Ahead*. US – TURKEY Workshop On Rapid Technologies, (pp. 5-11).

Anderson, D. R., Sweeney, D. J., & Williams, T. A. (1999). *Statistics for business and economics*. Cincinnati, OH: South-Western College Publishing.

Campbell, T. A., & Ivanova, O. S. (2013). Additive manufacturing as a disruptive technology: Implications of three-dimensional printing. *Technology and Innovation, 15*, 67–79.

Clayton, M. J. (1997). Delphi: A technique to harness expert opinion for critical decision-making tasks in education. *Educational Psychology, 17*(4), 373–386.

Cullom, P. (2014, July 15). *Navy Live: The Official Blog of the United States Navy*. Retrieved October 28, 2014, from 5 Things to Know About Navy 3D Printing: http://navylive.dodlive.mil/2014/07/15/5-things-to-know-about-navy-3d-printing/

De Weck, O. L., Roos, D., & Magee, C. L. (2011). Engineering Systems. Cambridge: The MIT Press.

Delbecq, A. L., Ven, A. H., & Gustafson, D. H. (1975). *Group techniques for program planning: A guide to nominal group and Delphi processes*. Middleton, WI: Green Briar Press.

Grissett, D. A. (2014, June 30). Delphi Description Document—UTC Experts. (S. N. Poulsen, Interviewer).

Hsu, C.-C., & Sandford, B. A. (2007). The Delphi technique: Making sense of consensus. *Practical Assessment, Research & Evaluation, 12*(10), 1–8.

Keeney, S., Hasson, F., & McKenna, H. (2006). Consulting the Oracle: Ten lessons from using the Delphi technique in nursing research. *Journal of Advanced Jursing, 53*(2), 205–212.

Lacaze, A., Murphy, K., Mottern, E., Corley, K., & Chu, K. D. (2014, May). 3D printed rapid disaster response. *In SPIE* Sensing Technology+ Applications (pp. 91180B–91180B). International Society for Optics and Photonics.

Leopold, G. (2014, August 21). *Defense Systems*. Retrieved December 2014, from Air Force enlists 3D printing for rocket engines: http://defensesystems.com/articles/2014/08/21/air-force-aerojet-3d-printing-rocket-engines.aspx

Linstone, H. A., & Turoff, M. (2011). Delphi: A brief look backward and forward. *Technological Forecasting & Social Change, 78*, 1712–1719.

Ludwig, B. (1997). Predicting the future: Have you considered using the Delphi Methodology? *Journal of Extension, 35*(5), 1–4.

Mack, R. (2013, August 2). *AFRL gives Barn Gang a 3-D look at Air Force's future*. Retrieved December 2014, from http://www.wpafb.af.mil/news/story.asp?id=123358364

McNulty, C. M., Arnas, N., & Campbell, T. A. (2012). *Toward the printed world: Additive manufacturing and implications for national security*. Washington, DC: Institute for National Strategic Studies, National Defense University.

Miller, L. E. (2006). Determining what could/should be: The Delphi technique and its application. *Annual meeting of the Mid-Western Educational Research Association.*

Powell, C. (2003). The Delphi technique: Myths and realities. *Journal of Advanced Nursing, 41*(4), 376–382.

Robert W. Appleton & Company, Inc. (2014). *Additive manufacturing overview for the United States marine corps.* Sterling Heights, MI: Robert W. Appleton & Company, Inc.

Stinson, T. N. (2014, June 27). *www. Navy.mil.* Retrieved October 28, 2014, from Dam Neck Explores Future of 3D Printing for Navy: http://www.navy.mil/submit/display.asp?story_id=81936

United States Army Rapid Equipping Force. (2014). *REF Rapid Equipping Force - United States Army.* Retrieved March 2015, from Expeditionary Labs: http://www.ref.army.mil/exlab.html

United States Department of the Air Force. (2006, March 13). *Air force operations planning and execution* (AFI 10-401). Washington D.C.

United States Department of the Air Force. (2012, September 20). *Deployment planning and execution* (AFI 10-403). Washington D.C.

Zimmerman, B. A., & Allen, E. E. (2013). *Analysis of the potential impact of additive manufacturing on Army logistics.* Monterey, CA: Naval Postgraduate School.

chapter forty two

Additive manufacturing applications for explosive ordnance disposal using the systems engineering spiral process model

Maria T. Meeks, Bradford L. Shields,
Eric S. Holm, and Vhance V. Valencia

Contents

42.1   Introduction

A strength of additive manufacturing (AM), or 3D printing, is the ability to produce unique objects from imagination to reality, relatively quick. This strength lends itself to military applications, which often utilize systems made from one-of-a-kind components. The purpose of this research is to demonstrate the applicability of AM technology to military applications, focusing specifically on two needs of the explosive ordnance disposal (EOD) unit at Wright–Patterson Air Force Base (AFB), Ohio. These needs were the following:

1. Attach environmental sensors to a remote-controlled robot.
2. Have a ready source of replica munitions for unit training.

42.2   Military applications of additive manufacturing

Revolutionizing the military supply chain, AM can create needed components or tools in austere areas that are either far removed from supply lines or on the frontlines of the battlefield. Designs can be made anywhere in the world and sent electronically to a strategically placed AM center on the battlefield.

Also, the military will continue to maintain legacy systems for life spans longer than originally intended. A challenge of the United States military is maintaining a supply inventory of spare parts for multiple weapon systems (Brown, Davis, Dobson, & Mallicoat, 2014). As legacy weapon systems continue to age, repair parts needed to maintain the systems become increasingly difficult to obtain. AM can create replacement parts for legacy systems that may not have the availability of repair parts compared to newer systems. Instead of going through a lengthy acquisition process to acquire a critical replacement part that has since gone out of production, AM printers could print the part on demand (Brown et al., 2014). On-demand production could eliminate the need of maintaining costly supply warehouses.

The Navy has introduced 3D printers on some ships in a program called *Print the Fleet* (Tadjdeh, Navy Beefs Up 3D Printing Efforts, 2014). The goal of the program is to introduce sailors to AM and investigate the applicability for the Navy. The Navy is currently using 3D printing for tooling, modeling, and prototyping (Tadjdeh, Navy Beefs Up 3D Printing Efforts, 2014). Vice Admiral Phil Cullom, deputy chief of naval operations for fleet readiness and logistics believes that "3D printing and advance manufacturing are breakthrough technologies for our maintenance and logistics functions in the future" (Tadjdeh, Navy Beefs Up 3D Printing Efforts, 2014). Cullom states the advantages of 3D printing to the Navy are "rapid repairs, print tools, and the immediate availability of parts, and the reduction of inventory of spares."

The United States Army has deployed the Army Rapid Equipping Force's Expeditionary Lab Mobile (ELM) for short (Parsons, 2013). The Rapid Equipping Force is an Army organization whose purpose is to quickly provide deployed Army units with advanced government and commercially available solutions that meet urgent requirements. The outside of the ELM resembles a metal shipping container. Inside, the ELM contains 3D printers, computers, and milling machines. Each ELM is manned by two engineers. The engineers can use the ELM 3D printers and milling machines to create parts from plastic, steel, or aluminum. Satellite communications allow the ELM engineers to communicate with colleagues anywhere in the world. Westley Brin, a civilian for the Army's Rapid Engineering Force, said "the technology has allowed troops to modify systems with proprietary designs to better fit their needs or make them more efficient in the field" (Parsons, 2013). One example, is the modification of a flashlight used by soldiers in Afghanistan that would accidently go off, which could give away the patrols position at night. Using the traditional defense acquisition process to field a new flashlight would have taken months, if not years. AM allowed a solution to be tested and fielded quickly on the battlefield (Parsons, 2013).

The Air Force has yet to deploy enterprise-wide AM efforts like that of the Army and the Navy. Small pockets of research into this capability are currently on-going, and this chapter outlines the efforts of one of those areas. Due to the nature of their mission, the Air Force EOD community is a natural fit for the unique capabilities of AM, specifically polymer printing. This chapter outlines the mission of Air Force EOD and then discusses recent research successes employing AM for EOD.

42.3   EOD mission needs

EOD duties are embodied within the Civil Engineer Squadron at nearly every Air Force installation. The official U.S. Air Force website documents the mission statement for the EOD career field as

"Explosive ordnance disposal (EOD) work begins in dangerous situations and ends in safe solutions. EOD members apply classified techniques and special procedures to lessen or totally remove the hazards created by the presence of unexploded ordnance. This includes conventional military ordnance, criminal and terrorist homemade items, and chemical, biological, and nuclear weapons. In addition to manufactured munitions, EOD Technicians also deal with improvised explosive devices. They are also experts in chemical, biological, incendiary, radiological, and nuclear materials. EOD personnel provide support to VIPs, help civilian authorities with bomb problems, teach troops about bomb safety, and aid local law enforcement. Some duties are dangerous, but EOD members are fully trained and equipped to safely deal with any situation. EOD personnel are part of an elite group of highly trained technicians that have a proud heritage of protecting personnel and property from the effects of hazardous unexploded ordnance" (U.S. Air Force, 2006).

With respect to the AM applications developed within this chapter, focus was placed on two components of the EOD mission: (1) the adequate training of EOD technicians and (2) the efficient use of equipment and tools already within the typical EOD shop inventory.

42.4   Robot specifications

The EOD technicians at Wright–Patterson AFB utilize the Northrup–Grumman Remotec® unmanned ground vehicle (UGV) for hazardous duty operations, including field inspection and detonation of explosive devices (Figure 42.1). The specifications of this UGV are sensitive in nature, and therefore details are not included in this chapter. The primary focus of the AM applications discussed here applies only to the robot arm assembly (Figure 42.2), which will be specified in detail using measurements, drawings, and photographs.

The UGV performs a critical function for technicians by accessing dangerous areas with unconfirmed threats, assisting with the identification of ordnance using optical cameras and video feed, and even disarming the ordnance. As each of these capabilities is critical to the efficient neutralization of unexploded ordnance, the research team ensured that all of these capabilities were maintained when finding AM solutions for the UGV. Full range of motion of the arm assembly and visibility of the sensor display were additional vital design drivers for the prototype developed in this chapter.

Figure 42.1 Northrop Grumman Remotec® unmanned ground vehicle. (Cooper, J., Langley EOD: Serving at home, overseas, *Peninsula Warrior*, August 5, 2011.)

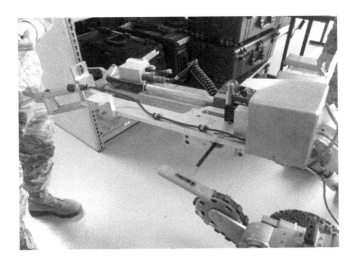

Figure 42.2 Arm assembly on the Northrup–Grumman Remotec® UGV.

42.5   *Sensors used in field applications*

The EOD technicians at Wright–Patterson AFB typically utilize four different sensors for environmental sampling and ordnance testing:

1. Victoreen® Fluke® Biomedical 451P Pressurized μR Ion Chamber Radiation Survey Meter (Figure 42.3)
2. MultiRAE® PGM 6248 Wireless Portable Multithreat Monitor for Radiation and Chemical Detection (Figure 42.4)
3. IndentiFINDER® R400 Handheld Radiation Detector (Figure 42.5)
4. Smiths Detection® LCD 3.2E Handheld CWA and TIC Detector (Figure 42.6)

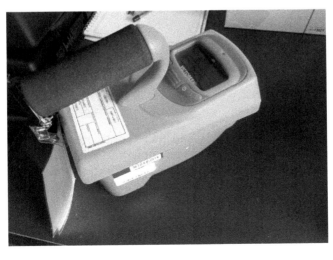

Figure 42.3 Victoreen® Fluke® Biomedical 451P Pressurized μR ion chamber radiation survey meter.

Figure 42.4 MultiRAE® PGM 6248.

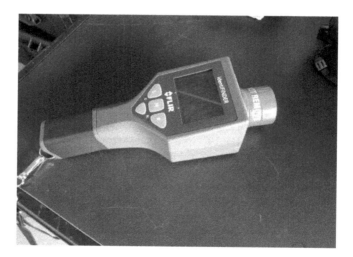

Figure 42.5 identiFINDER® R400 handheld radiation detector.

All sensors operate independently, and there is not a recurring need at this time to attach more than one sensor to the UGV at any one time. Table 42.1 was developed to summarize key information regarding each sensor as all of the sensors have different dimensions, weights, and functions. The Victoreen® Fluke® Biomedical 451P sensor was used for the prototype design due to its significant size, specifically its depth. Weight was not a concern during prototype design.

Currently, EOD technicians at Wright–Patterson AFB use adhesive tape to secure environmental sensors to the UGV and spend valuable time removing each sensor from the robot. Although this is an effective low-cost solution, the process does take a significant

Figure 42.6 Smiths Detection® LCD 3.2E Handheld CWA and TIC Detector.

Table 42.1 Summary of sensors

Sensor	Function	Dimensions (inches)	Weight (lbs)	Source
Victoreen® Fluke® Biomedical 451P Pressurized µR ion chamber radiation survey meter	Radiation survey of environment	4 (w) × 8 (d) × 6 (h)	2.4	(Fluke Biomedical, retrieved February 23, 2015)
MultiRAE® PGM® 6248	Multithreat monitor for radiation and chemical detection	3.8 (w) × 2.6 (d) × 7.6 (h)	1.9	(RAE Systems, retrieved February 25, 2015)
identiFINDER® R400	Detects, locates, measures, and identifies radionuclides and isotopes	9.8 (w) × 3.7 (d) × 2.9 (h)	3.2	(FLIR, retrieved February 26, 2015)
Smiths Detection® LCD 3.2E (aka JCAD M40)	Real-time detection of chemical and toxic substances	4.3 (w) × 7.1 (d) × 2.0 (h)	1.15	(Smiths Detection, retrieved February 26, 2015)

amount of time when the field is replacing sensors and removing adhesive residue from the sensors and robot once an operation is complete.

42.6 Additive manufacturing lab equipment and processes

This section outlines the materials, equipment, and design and production processes used in this research. With the exception of digital scanning, all work accomplished for this project was accomplished at the Air Force Institute of Technology (AFIT) 3D-printing laboratory.

42.6.1 Printer and material type

The laboratory equipment used for the design, production, and postprocessing of the EOD prototypes included the following:

- *Printer*: 3D Systems® ProJet™ 1500, film transfer photopolymer machine, which uses photopolymer plastic material with ultraviolet (UV) curing inside printer (Figure 42.7).
- *Solvent wash*: Polycarbonate solvent washer and water rinse.
- *Curing*: UV lamp cabinet (Figure 42.8).

Figure 42.7 **(See color insert.)** ProJet™ 1500 printer, opened to view printing bed (right).

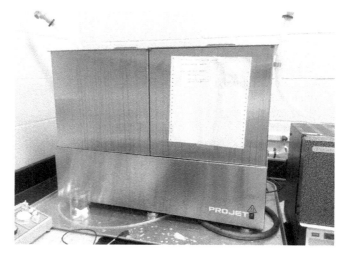

Figure 42.8 ProJet™ solvent washing basin.

42.6.2   Lab processes

The production of the EOD sensor bracket started with the designs being created in Solidworks® computer software program. This modeling software allowed the design team to have a firm grasp on the exact shape and dimensions of the bracket prior to actually creating the prototype. A variety of other modeling software programs have the same capability as Solidworks®, so the chosen software just depends on the user's abilities. As long as the file can be converted to an STL format, then the printer is nondiscriminatory toward any program.

Following the prototype design in Solidworks®, the 3D printer's software imported the file and created support structures for the model. Support structures are thin rods printed in order to connect the printed model to the printed mat, and hold any hanging structures within the print. Once the printer has run its initialization process the prototype is ready to print. The driving factor in print time is the total height of the print in the Z direction. Knowing that driving factor, the rule of thumb for printing is approximately 4 hours per inch printed in the Z direction. The actual time can definitely change, however this rule allowed for consistent estimates within 30 minutes. Once the print is completed, the prototype moves on to postprocessing.

Postprocessing is often an overlooked part of the AM process; however it was of relative importance to optimize postprocessing in this research effort because the entire design and production phase was accomplished in approximately nine weeks. The postprocessing of prototypes followed three basic steps:

1. Solvent wash removes uncured material from prototype surface and reduces *tackiness*.
2. UV lamp cabinet cures material and increases strength of prototype.
3. Remove prototype from printing mat, break part away from supports and smooth surface with tools.

Several challenges were encountered in the postprocessing of prototypes. An imperfect method of detaching the printed part from its supports often left uneven surfaces that required additional tooling. Also, thinner dimensions on the printed part were at risk for breakage. Interior supports are not easily accessible to remove, and sometimes required much effort to remove completely. Finally, the printing mat on which the part is produced is extremely difficult to remove from printer plate and required rigorous cleaning between prints.

42.7   Research question and goals

The goal of this research is to design and produce a solution using AM technology. The overarching question for this research was: How would additive manufacturing capability improve EOD operations at Wright–Patterson AFB? To adequately address the goal of this research, three investigative goals supporting this research were identified:

1. Design and manufacture a universal sensor bracket for the Northrup–Grumman Remotec® UGV using AM.
2. Adequately replicate EOD training aids using AM to meet EOD training objectives.
3. Calculate the cost benefit of using AM for a universal sensor bracket and training aid replication in the EOD unit at Wright–Patterson AFB.

42.8   Selection and application of a systems engineering design process

Of paramount importance in this research was to implement a systematic and iterative process that (1) ensures the research team adequately covered all necessary factors of design and (2) followed a process by which future AM applications for EOD could follow. The project team held several project meetings with the EOD unit to discuss their end-user needs and desires. Internal deliberations with the project team were further held to discuss the most applicable systems engineering design processes. The processes were distilled to three: the *Vee* process model, the spiral process model, and the waterfall process model. Before further work on design factors for the final product, the project team selected the most appropriate design process. This section outlines the three different systems engineering processes and the final selection, the spiral process model, used for this project.

42.8.1   "Vee" process model

The *Vee* process model, shown in Figure 42.9, provides a framework for more clarification and focus on the user needs throughout the design and production process. By starting the process with the user needs and ending with a user-validated system, the model helps capture an understanding of the user's desires for the system being designed.

   The process diagram, made of a V-shape, shows the flow of the design process moving from left to right along the V. The left side of the shape consists of the *Project Definition*, or decomposition, and definition activities. In this part of the process, a systems engineer must understand how to define the requirements, allocate the system functions, and have a detailed need of the components within the system. The right side of the model is where project integration occurs and verification of a design takes place. The Vee model is designed to be iterative and constant testing and verification takes place that may result in rethinking initial concepts to ensure that the system is designed according to the needs of the user (Blanchard & Fabrycky, 2011). It is crucial for the entire system to meet all the specifications laid out in the planning steps of the process.

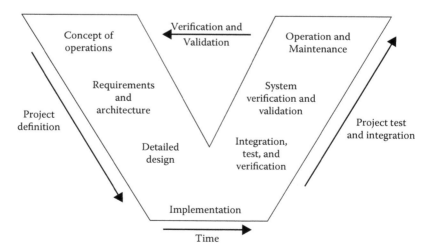

Figure 42.9 *Vee* process model.

42.8.2 Waterfall process model

Also commonly used within systems engineering, and shown in Figure 42.10, is the waterfall process model. Initially used for software development, the design varies between five and eight steps depending on the size and complexity of the project. The five main steps used most often includes requirement specifications, system design, implementation, testing, and maintenance (Blanchard & Fabrycky, 2011). Where this method differs from other processes is the continuous feedback it provides throughout the course, up and down the chain of command.

42.8.3 Spiral process model

The final model evaluated in this project is the spiral process model, presented in Figure 42.11. The model is "intended to introduce a risk-driven approach for the development of products or systems" (Blanchard & Fabrycky, 2011). As in the other two approaches, constant feedback through verification and validation is required in the spiral process. However, this model makes the process of requirements, design, and conception cyclical and explicitly calls for a risk analysis before moving onto the next spiral (i.e., iteration). While developing different prototypes, a design team using the spiral process model continually walks through each step in the design chain to ensure that it meets all the desired specifications.

After the design team had weighed their three options, they decided the spiral process model was the correct fit for creating the EOD robot bracket. The model's purpose of calculating risk suited the intent of cost savings with the bracket; as well, the risk analysis helped to identify the different end-products intended to use in holding extremely sensitive equipment (for the EOD bracket) and low-cost and replaceable items (for training aids). This process explicitly stepped the design team through the steps of understanding design drivers, constraints, functionality needs, and different prototype designs multiple times. The framework the spiral process model provided helped the design team address unforeseen project challenges and provided a path for which to work through these issues. Section 42.9 discusses the implementation of the spiral process model to what the team termed *design drivers*. The following section discusses design constraints, functionality requirements, and prototype development of an EOD bracket.

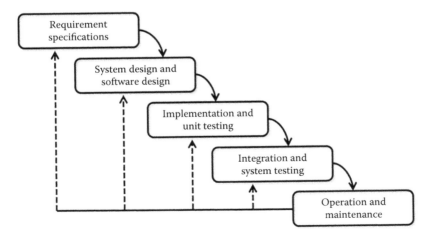

Figure 42.10 Waterfall process model.

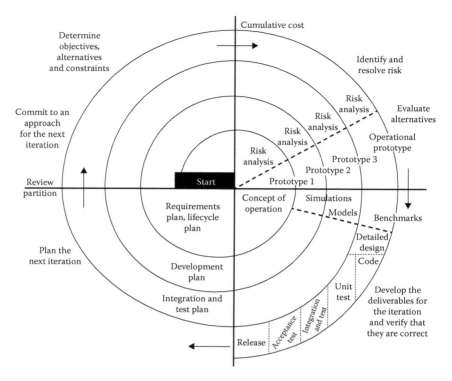

Figure 42.11 Spiral process model.

42.9 Addressing design drivers for EOD bracket and training aids

Designing any part or product in the manufacturing industry will always involve multiple iterations and design changes due to certain constraints and requirements being levied on the design process. In this project, the drivers consisted of requirements tasked by the EOD shop. In addition to the requirements, the tolerances and limitations of the actual 3D printer placed considerable constraints on the design as well.

Although these drivers and restrictions drove the team to update and make small modifications in the designs, the complexity of design changes for AM is almost like that of erasing a misspelled word with a pencil eraser. The ease with which the design group was able to build, print, test, and then make the necessary small changes to the model is clearly apparent when compared to traditional methods of manufacturing and prototyping. Hod Lipson describes the advantages of prototyping along with the cost and time of getting a prototype in his text *Fabricated: The New World of 3D Printing*:

> "Prototypes give both the marketing team and designers a sense of the product design's ergonomics and the spatial relationships of its parts. In the old days, when prototyping was still a slow and expensive process, it was risky for a company to cut a corner and just trust that a design would work out in real life" (Lipson & Kurman, 2013, p. 31).

The ability to rapidly make changes and print new iterations allowed a quick turnaround for testing and feedback of the model. This section will walk through the biggest drivers, the development of different iterations, the myriad of design considerations, and the notable changes made throughout the entire process. Each of these drivers will be discussed within the context of the two distinct products developed: sensor bracket for EOD and training aids for EOD.

42.10   Design constraints

Constraints for design of the sensor bracket and training aids are discussed here.

42.10.1   Design constraints: Sensor bracket for EOD robot

As mentioned in the original design requirement, the task was to develop and create a mount for the Northrup–Grumman Remotec® UGV, Figure 42.1, utilized by the Wright–Patterson AFB EOD technicians. This mount is to serve as the bracket for where to attach four separate operational Hazardous Materials (HAZMAT) sensors, which currently are connected in the field using only duct tape. During both real world and training responses, switching between the different sensors expends critical time for the EOD Airman. Before even diving into the measurements and in-depth considerations for the mount, there were several constraints and requirements that the design group had already brainstormed as needed for the design. Those design considerations included the following:

- The mount must quickly attach to the robot without any outside materials being used.
- The mount must be durable and reliable enough for any EOD training or real-world situation.
- The mount must make attaching a sensor more convenient and worthwhile for the EOD Airman.
- The mount must be entirely printed and assembled while minimizing any additional material needed from outside sources.
- The mount must be durable and strong enough to secure the different sensors due to their expensive cost and the rough terrain where the utilization of EOD robots is critical.
- The mount must not impede any other functions of the robot, even with an attached sensor.

After sharing these initial considerations, the EOD Airman shared what they needed most out of the mount and how it could solve a real need for them during training and real world missions. The design considerations and EOD requirements, and, the constraints of the 3D printer, will be discussed in greater detail below.

42.10.1.1   Shape/size of sensor

As seen in Table 42.1, the four different sensors the EOD shop asked the design team to consider varied in both size and weight. Just like in any engineering design, when multiple things are being designed for, the largest or heaviest becomes the driving factor in conception. For example, in the designing of bridges and roads, the axle weight of a tractor trailer is designed for instead of designing for the weight of just a Ford Focus or other typical vehicle. In the consideration of the EOD robot, it was the size of Victoreen® Fluke® Biomedical 451P Pressurized μR Ion Chamber Radiation Survey Meter, shown in Figure 42.3, which drove the overall dimensions of the bracket. With an approximate width of 4″, depth of 8″, height of 6″, and handle diameter of 1.5″, this became the largest sensor the bracket would have to hold. The other sensors had similar widths, within approximately 1″; however they had significant variations in length and height. Also, none of the other sensors consisted of a possible design consideration for a sensor handle.

42.10.1.2   *Anchor points to robot arm assembly*

Another consideration in the design was that the EOD robot had two locations for which to attach the mount, as shown in Figure 42.12. One location for attaching was on the right side of the robot arm, whereas the other was a little further down the arm on the left side (orientation is looking down the arm from the main body of the robot). EOD airman identified a multitude of additional training and operational applications for the anchor points on the right side of the arm (same orientation as before); therefore, the team decided to focus solely on designing the mount to attach to the anchor points on the left side (same orientation).

The bracket has to be designed to fit three specific anchor locations on the actual arm of the robot. The first two points are symmetrical and consist of 3/8″ diameter holes in the arm for which the bracket must align. Figure 42.12 show the holes on the arm of the robot where the cylinders extruding from the bottom side of the designed mount must align. Once aligned with the two holes, the principal method for attaching the mount is a hook. The hook swings up and out of the robot arm and clamps down on a 3/8″ diameter rod just inside the right wall of the mount. As shown in Figure 42.12, the attachment hook is in the locked position. With the requirements and constraints identified, the design team began developing prototypes for which to test by EOD.

Figure 42.12 Anchor points for bracket on the EOD robot arm assembly.

42.10.2   Design constraints: Training aids

During the literature review, it was found that AM had been used at least once to create training aids for EOD. The design constraints that Tan (2014) listed in *Advanced Ordnance Teaching Materials* include the following:

1. Mechanisms must be easily resettable to the unarmed position.
2. Mechanisms must be true to ordnance.
3. Mechanisms must be clearly visible.
4. Models must be physically robust and able to withstand repeated disassembly, table-top drops, and general classroom rigors.

The constraints faced by the project team were those due to the quality of 3D printers and methods of AM available to the team. Print quality and the ability to accurately reproduce replicas is largely a factor of printer type and skill in using 3D modeling software. Additionally, any training aids produced would be limited to the size of the printing bed. Producing training by AM must also be less than the cost needed to buy the aids through commercial sources. The material of the training aid is limited to the plastic polymer of the printer.

42.11   Functionality requirements

It is the intent of any designer to have their product benefit the user in some fashion. For the EOD robot and the part being designed, it is essential that the sensor mount decrease the time and effort it takes an Airman to attach a sensor to the robot. For the training aids, they must serve to enhance the learning and training experience of the EOD Airmen.

42.11.1   Functional requirements: Sensor bracket for EOD robot

Several functional items held significant importance during the actual design of the mount due to specific training and operational requirements necessary for EOD operations to be successful. Those elements included the following:

- The ability of the EOD Airman operating the robot to view the screen on the sensor using only the tilt of the robot arm and an attached camera.
- The size of the sensor mount could not impede any motion of the robot arm.

These two functional requirements are discussed in more detail below and determined the success in helping EOD become more efficient in their mission.

42.11.1.1   Visibility of sensor during operation

When the EOD robot is being used in a training or real-world environment, the robot operator must have a visual on the sensor screen attached to the robot. That visibility comes from a camera located on the top of the robot. The EOD airman informed the design team that covering the sensor screen would impede their ability to read data when in the field. Along with the other requirements and constraints, the design team took this need for visibility and incorporated it into each of their design iterations.

42.11.1.2   Maximum bracket size (to avoid robot arm range of motion)

The EOD robot has multiple uses when in a training or combat zone; so, the last thing the bracket needs to do is impede the robot's ability to complete its mission. Based on the movements of the arm, the design team concluded that the bracket could not be more than 8″ wide and 10″ long or it would hinder the motion. Also, if the bracket is too high, then it would begin to impede the vision of the user. Based on these functionality requirements, the design team began developing what they felt was the best prototype to fulfill the needs of EOD.

42.11.2   Functional requirements: Training aids

The main type of training aids for EOD is ordnance replicas. The primary purpose of any training aid for EOD is to teach ordnance identification. The secondary purpose is to teach fuse mechanisms (Tan, 2014). To be highly functional, a training aid must be nearly identical in appearance to the real ordnance. Shape is the most important factor used in identification. Other factors include the following: color, special identification markings, and materials.

42.12   Design factors for printing

The final design driver was consideration of the design itself when created through AM. The printing process presents opportunities to develop unique, one-of-a-kind designs, but the process must also consider these factors as constraints. This section discusses printing orientation, the importance of material minimization, and the use of fasteners and joints in AM. As these areas apply across the spectrum of possibilities in production through AM, the following is applicable to both the EOD robot bracket and the EOD training aids.

42.12.1   Printing orientation

Most 3D printers are limited based on the orientation of the part being printed. It is common to experience weaker and fewer satisfactory prints the more a design causes the printer to print in the Z direction (Smyth, 2013). The design team read about these limitations during their literature review and also experienced the same constraints when printing their products, so it is very common. The way around the limitations is to understand how the printer works.

If a design is broken down and the parts placed in a more X–Y plane, then the print has a greater chance of succeeding and not having layer failures. The design team took this advice into consideration, and it is the reason that all the plans include some assembly process.

42.12.2   Minimize material

As discussed during the failures of the first few iterations, minimizing the print material was crucial to the design of the EOD bracket and the training aids. The weight of the bracket and training aids in the first few iterations caused problems with the print. Also, the amount of printing material used increased the cost to print these prototypes. Taking the cost of the print into account during the risk analysis was critical due to the end goal being significant AFIT and EOD cost savings.

42.12.3   Fasteners and joints

With the fasteners and joints, certain printers work better than others due to the overall tolerances. The design team developed several iterations that included a variety of assemblies and joints. Based on their literature review and experience, it was safe to assume that a 0.2″ was the minimum design tolerance needed for any fastener or joint (Smyth, 2013). When the team printed the design, that design tolerance was successful in printing pieces that fit together approximately half the time.

42.13   EOD sensor bracket and training aid prototypes

This section discusses the prototyping iterations for both the EOD sensor bracket and the EOD training aids. Through applying the systems engineering spiral process model, several iterations of prototypes were necessary before arriving at the final design. This section provides details and figures for each of the prototypes for both products illustrating the improvements mode through each iteration of the spiral process model.

42.13.1   Sensor bracket for EOD robot

42.13.1.1   Iteration 1

For the first prototype iteration, the primary consideration was to design for the actual base shape for each sensor. Due to the form and size of the sensors, it was easiest to create a rectangular base on which the sensor would rest. The base had to be completely flat and designed to fit the largest sensor. The largest sensor had dimensions of approximately 4.5″ × 6″, as shown in Figure 42.3. Although the design team developed the base, it was also imperative for the design team to create method for securing the sensor from the sides.

With the height of 6″ for the sensor driving the layout, the mount had to ensure that the sensor could not tip over, while the robot was in motion. The probability of the sensor tipping over when the robot was in use drove the first iteration of the design. The first design consisted of tall side walls to keep the sensor from tipping. Figure 42.13 shows the basic model with side walls of the first iteration. Consistent throughout the different iterations are alignment pegs seen on the bottom of the model in Figure 42.14.

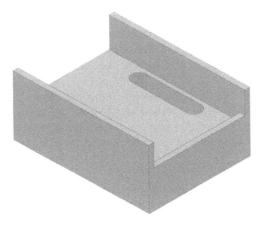

Figure 42.13 Iteration 1 isometric view.

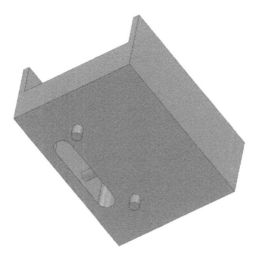

Figure 42.14 Iteration 1 bottom view.

Due to the tolerances of the 3D printer and materials used, the threaded holes did not adequately secure the sensors in place. The bracket also required a significant amount of printer polymer when printing and was relatively heavy when completed. With the inability to secure the sensor and the weight of the bracket itself, the design team scrapped the threaded holes and developed a second iteration of the model layout.

42.13.1.2   Iteration 2

The second iteration assembly, as shown in Figure 42.15, was designed to include actual holes in the top of the side walls of the bracket in order for EOD personnel to use small bungee cords or rubber restraints to secure the sensor. Figure 42.15 shows these holes located on the second iteration of the assembled mount. Although this method of using bungee cords or rubber restraints went against the fourth initial consideration listed above, the design team felt it was the simplest and most inexpensive method for securing the sensor.

The walls themselves were developed separately from the base and could move in or out due to 1/4¼″ holes for the walls to fit inside. The tolerances of the printer also caused problems with the wall pegs fitting inside of the holes on the base of the bracket. Using sandpaper to decrease the thickness of the pegs, the walls finally fit into the base. This method worked well, but the shear strength of the wall pegs did not hold up even during the simplest of tests.

As for the base of the bracket, the design team cut down on the overall volume by approximately 15% due to the design of the holes to hold the walls in place. Still, the bracket was heavy and used about a third of a printer cartridge when printing. Overall, the design team liked the method of adjustable walls and moved forward with creating a lighter base.

42.13.1.3   Iteration 3

The third iteration began with the idea of reinforcing the shear strength of the wall connections, but still allowing the walls to be adjustable based on the width of the sensor being used. With the need for the reduction of the base in volume and weight, the design team hollowed out the sides of the base and developed a method for allowing the walls to

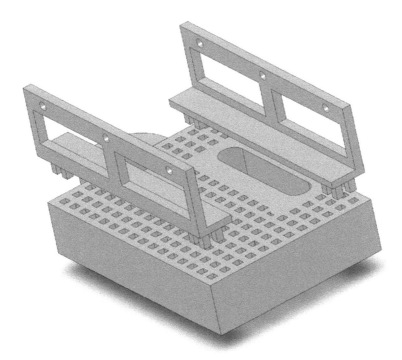

Figure 42.15 Iteration 2 isometric exploded view.

move in or out. A specially designed tab connected to the walls would allow them to move in, but would not slide back out without the user pushing in that particular tab. As shown in Figure 42.16, Iteration 3 design assembly and breakout exhibits how the walls were to move in and out based on the user's needs. The base was also reduced by almost 40% in volume compared to the Iteration 1 bracket.

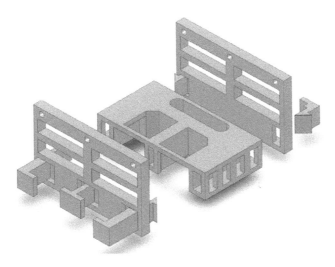

Figure 42.16 Iteration 3 isometric exploded view.

Unfortunately, the print material used resulted in the thin tabs being extremely brittle. Instead of the tabs bending and coming back to their original shape, they broke only when slightly bent. Printer tolerances also did not accurately print the holes or guides correctly. As a result, the design team realized that ultimately relying on precise measurements and accurate prints was the reason for the downfall of the designs up to that point. They again went back to the drawing board with hopes of developing a simpler and more usable design.

42.13.1.4   Iteration 4

The design of Iteration 4 came about from the small successes and failures during the past three iterations. The base of the bracket stayed the same overall shape; however, it was hollowed out to make it lighter in weight. Scrapping the walls, the design team put in their place, three tie points on either side of the bracket. On one side, the tie points stick out less from the bracket. Those tie points will have small rope or rubber cord secured to them. On the other side, the tie points stick out further allowing for the cord wrapping and tying. This method of quickly securing the cord will allow the sensor to be swiftly and securely tightened to the bracket. Figure 42.17 shows the design of Iteration 4.

Just as described in the spiral model process, the design team has to continuously evaluate their design, evaluate the risk, and then often times return back to design/redesign stages to repeat the processes in a new iteration. This process of evaluation and iteration is the reason behind why the systems engineering spiral process model has proven successful in helping designers grasp necessary changes and improvements for their designed products and systems (Figure 42.18).

42.13.2   Training aids

For the training aids, the design team failed to leave the first iteration of the spiral process model. This section outlines the work started, mostly scanning and digitizing, of the provided training aids. Unfortunately, severe limitations in the lab equipment hampered the efforts for creating valid training aids for use. Available in the labs for use were two scanners: the Makerbot Digitizer and the FARO Edge ScanArm. The two scanners can be seen in Figures 42.19 through 42.23.

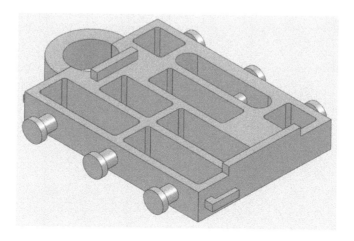

Figure 42.17 Iteration 4 isometric view.

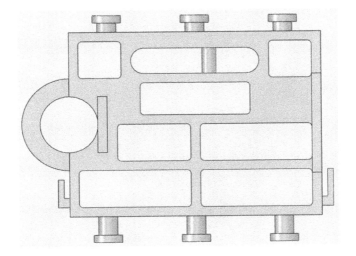

Figure 42.18 Iteration 4 top view.

Figure 42.19 An image of the Makerbot Digitizer (photo courtesy by author).

The most important design consideration in producing an EOD training aid is ensuring that the aid is identical to the actual explosive by both shape and color. It was hoped that 3D scanning would be the quickest method to create a printed object identical to the original training aid. Unfortunately, there were several limitations with using the scanner. It was difficult to capture complex geometries with the Makerbot. As shown in Figure 42.23 the fins at the base of the mortar were not captured by the Makerbot. The FARO scan did a better job at digitizing intricate details of the model, but both 3D digital models from the Makerbot and the FARO required heavy alterations in a modeling program prior to printing the part.

Figure 42.20 An inert EOD training aid being scanned by the MakerBot Digitizer (photo courtesy by author).

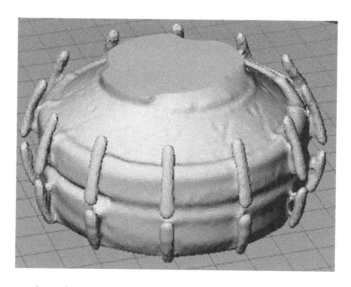

Figure 42.21 A screenshot of a 3D image of a mine scanned in the MakerBot Digitizer (photo courtesy by author).

The Makerbot's scanning ability declined when intense light shined on the scanner. The scanning improved when the scanner was moved to a room with moderate indoor light only. Knowledge of the scanners was also a limitation. It was the project team's first attempt at using the scanners, and several rounds of trial and error were required to get a scan at an acceptable level of detail (Figures 42.24 and 42.25).

One conclusion drawn from scanning the training aids was that knowledge of computer-based 3D modeling software can greatly increase success in 3D printing. It will

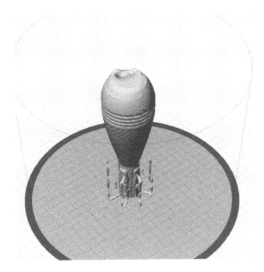

Figure 42.22 A screenshot of a 3D image of an inert mortar made by the Makerbot Digitizer (photo courtesy by author).

Figure 42.23 An image of the FARO® Edge ScanArm®. The mortar that was scanned by the FARO® can be seen at the right of the scanner on the blue FARO® desk (photo courtesy by author).

be difficult, if not impossible, to go directly from a digital scan to a print without some adjustments in a 3D modeling software. An alternative approach to scanning could have been to reverse engineer the training aids. 3D scanning could have been used to get exact measurements of the shape geometries for each of the training aids. With exact measures, the training aids could be modeled in a 3D modeling program from scratch for later printing.

Figure 42.24 A screenshot of the 3D image produced of the mine by the FARO® Edge ScanArm®.

Figure 42.25 A screenshot of the 3D image produced of the mortar by the FARO® Edge ScanArm®.

42.14   *Conclusions and significance of research*

This final section outlines the conclusions gained from the research and the significance of the work. Here, we present a cost savings estimate for EOD operations if both AM and the spiral process model are adopted across the Air Force for all EOD operations. The following cost analysis provided in this research suggests that significant operational cost savings can be realized through leveraging AM and the spiral process model for training aids. Cost savings are presented in both monetary costs and labor costs for EOD shops across the Air Force. Following the cost analysis is the significance of applying the systems engineering spiral process model to EOD needs across the Air Force.

42.14.1   Cost savings analysis

Although actual prints of EOD training aids did not occur, this research conducted a cost analysis to see if printing training aids would be more economical versus buying these aids through commercial vendors. The initial capital investment and costs for each individual training aid were estimated separately. Table 42.2 shows the initial capital investment required for the ProJet™ 1500 Printer. Table 42.3 shows the estimated cost of several different types of training aids. These estimates were based on calculated volumes from the digitization step of the training aids project.

A list of assumptions that were used in the cost analysis includes the following:

1. The 2 kg AM print cartridge prints 2 kg of material.
2. Each printer cartridge costs $562.00.
3. Density of the ProJet printing material at liquid (30 C) is 1.08 g/cm³.
4. There is no material shrinkage during photopolymerization.
5. Each training aid would be made with one part.
6. 2 man hours are required to initialize the printer and complete postprocessing of the final part.
7. The labor rate is based on the month basic pay of an E-5 with 8 years of service ($2,951) and that 240 hours are worked each month.
8. A learning-curve approach in estimating the labor hours was not used. For example, the time required for someone who is new to 3D printing to make the first training aid would probably be longer than someone who is highly experienced. This additional learning time to make the first series of training aids was not considered.
9. The cost to periodically inspect and maintain the printer was not included in the individual cost of the training aid.
10. Design costs were not included.

The first step in estimating the cost of an AM training aid was to determine the material cost of a training aid. The material cost is equal to the unit cost of the material multiplied by the total volume of the training aid (Equation 42.1).

$$\text{Material Cost} = \text{Material Unit Cost} \times \text{Volume of Training Aid} \qquad (42.1)$$

The material unit cost was calculated by dividing the cost of the printer cartridge by its volume as shown below (Equation 42.2).

Table 42.2 Initial capital investments for the ProJet™ 1500 printer

Initial capital investments	Unit cost ($)
Projet 1500 printer	14,500
Projet 1500 finisher	750
Projet 1500 wash-basin	400
Clean-a-part cleaning solution 4 gallons	165
Material/Supplies	
Supplies/hand tools for postprocessing	100
Total	15,915

Table 42.3 Estimated costs for several different training aids

EOD training aid	Material cost to be reproduced by AM ($)	Labor cost ($)	Postprocessing cleaning fluid ($)	Total material and labor cost by AM ($)	Commercial cost ($)	Most cost effective method
D-06	18.21	36.89	11.00	66.10	50.00	Commercial
D-48.1	22.76	36.89	11.00	70.65	250.00	AM
40mm Grenade	39.46	36.89	11.00	87.35	190.00	AM
Mine	51.60	36.89	11.00	99.49	226.00	AM
Mortar	69.81	36.89	11.00	117.70	100.00	Commercial

$$\text{Unit Cost} = \frac{\text{Total Cartridge Cost}}{\text{Cartridge Volume}} = \frac{\$562}{1851.85\,\text{cm}^3} = \$0.303/\text{cm}^3 \qquad (42.2)$$

The volumes of several different types of commercially available training aids were determined by measuring the amount of water that was displaced when the training aids were placed in water. 3D scanning could also have been an alternative method of determining the volume of the training aids. The volume of each print cartridge was calculated by dividing its mass by its density. The calculation to determine the volume of the print cartridge is given below (Equations 42.3 and 42.4).

$$\text{Volume} = \frac{\text{Mass}}{\text{Density}} \qquad (42.3)$$

$$1851.85\ \text{cm}^3 = \frac{2\ \text{kg}}{1.08\ \text{g cm}^{-3}} \qquad (42.4)$$

Labor costs were limited to only the time required to operate the 3D-printing machine and postprocess the training aid after printing. It was assumed that this took two hours. The total printing time would be longer than two hours, but most likely the 3D-printing machine would not be manned during the entire printing process. The hourly labor rate was based on the monthly basic pay of an E-5 with 8 years of service ($2,951) and 240 working hours each month.

When calculating the material cost for each of the training aids, it was assumed that each aid would be one part. 3D printing allows the production of innovative shape geometries that can reduce the amount of material needed to make a part. A skilled designer could create a part that is hollow inside to reduce the material volume of the object. By reducing the volume, the total cost of the part could be substantially reduced. Reducing the volume may also require the training aid to be made as an assembly of multiple parts. This would increase the time to run and operate the 3D-printing machine and the time needed to postprocess the parts, which would in-turn increase the labor costs per training aid.

AM is economical when the cost to produce the training aid through AM is less than commercial sources. As the commercial cost for the training aid increases, it becomes more worthwhile to make the aid through AM. One downside to producing training aids through AM is that the ProJet printer only produces parts out of plastic, and the more expensive training aids were made out of metal.

42.14.2 Significance of research

In addition to potential cost savings, this chapter concludes with a brief discussion of the significance of this research, that is, the application of the systems engineering spiral process model was a significant contributor to the success of developing a working prototype of an EOD sensors bracket through AM. Coupling the new technology of AM with a matured systems engineering design process proved successful for this proof-of-concept project. The new technology allowed a project team with little-to-no experience in EOD to identify an issue, develop a plan of action, and proceed with the product development in an iterative manner to produce a working prototype in a matter of weeks.

As shown in the cost analysis, acquiring the equipment necessary for AM is insignificant for large and governmental organizations such as the Air Force. In addition, the skills necessary for AM are more easily acquired than that of traditional manufacturing. Therefore, this research suggests that AM through the spiral process model can be easily scaled up in a large organization such as the United States Air Force and result in significant cost savings and improvements in operations and training.

Summary

Given that the military will continue to maintain legacy systems for life spans longer than originally intended, this research addressed a method to meet the challenge of maintaining a supply inventory of spare parts. AM offers this solution. Rather than undergo a lengthy acquisition process to acquire a critical replacement parts, AM printers could print the part on demand and on-demand production could eliminate the need of maintaining costly supply warehouses and logistics supply chains.

The purpose of this research is to demonstrate the applicability of AM technology to military applications. Through working with the EOD unit at the Wright–Patterson AFB, OH, this research explored the following mission needs:

1. Attach environmental sensors to a remote-controlled robot.
2. Have a ready source of replica munitions for unit training.

This research demonstrated that the systems engineering spiral process model and the AM, when coupled together, can result in working prototypes at significant cost savings. Further, this research suggests that this type of production can be scaled up to meet Air Force enterprise-wide needs for the EOD community.

Acknowledgments

This work was made possible through a research funding from the Air Force Civil Engineer Center, Requirements and Acquisition Division (AFCEC/CXA). A special thanks to Mr. Craig Mellerski and Dr. Joseph Wander for their continued support, leadership, and guidance in this and other research efforts.

Bibliography

Blanchard, B. S., & Fabrycky, W. J. (2011). *Systems engineering and analysis.* Boston, MA: Prentice Hall.

Brown, R., Davis, J., Dobson, M., & Mallicoat, D. (2014, May–June). 3D Printing, How Much Will It Improve the DoD Supply Chain of the Future? *Defense Acquisition, Technology and Logistics*, pp. 6–10.

Cooper, J. (2011, August 5). Langley EOD: Serving at home, overseas. *Peninsula Warrior.*

FLIR. (n.d.). *identiFINDER Radiation Detection.* Retrieved February 26, 2015, from FLIR.com: http://www.flir.com/threatdetection/display/?id=63333.

Fluke Biomedical. (n.d.). *451P Pressurized µR Ion Chamber Radiation Survey Meter.* Retrieved February 23, 2015, from Fluke Biomedical: http://www.flukebiomedical.com/biomedical/usen/radiation-safety/survey-meters/451p-pressurized-ion-chamber-radiation-detector-survey-meter.htm?PID=54793.

Lipson, H., & Kurman, M. (2013). *Fabricated: The new world of 3D printing.* Indianapolis, IN: John Wiley & Sons.

MakerBot® Industries LLC. (2015). Retrieved from www.makerbot.com.

Parsons, D. (2013, May). 3D Printing Provides Fast, Practical Fixes. *National Defense.*

RAE Systems. (n.d.). *MultiRAE PRO - Wireless Portable Multi-Threat Monitor for Radiation and Chemical Detection (PGM-6248) by RAE Systems.* Retrieved February 25, 2015, from RAE Gas Detection: http://www.raegasdetection.com/index.cfm/product/244/multirae-pro.cfm.

Smiths Detection. (n.d.). *LCD 3.2E Handheld CWA & TIC Detector.* Retrieved February 26, 2015, from Smiths Detection: http://www.smithsdetection.com/index.php/products-solutions/chemical-agents-detection/59-chemical-agents-detection/lcd-3-2e.html?lang=en#.VO-X4co5AkI.

Smyth, C. (2013). *Functional design for 3D printing: Designing 3D printed things for everyday use.* CreateSpace Independent Publishing Platform.

Tadjdeh, Y. (2014, October). Navy beefs up 3D printing efforts. *National Defense*, pp. 24–26.

Tan, A. (2014, Summer). Advanced ordnance teaching materials. *The Journal of ERW and Mine Action*, 39–42.

U.S. Air Force. (2006, September 1). *Explosive Ordnance Disposal - 3E8 × 1.* Retrieved February 25, 2015, from U.S. Air Force Homepage: http://www.af.mil/AboutUs/FactSheets/Display/tabid/224/Article/104600/explosive-ordnance-disposal-3e8×1.aspx.

chapter forty three

Proof-of-concept applications of additive manufacturing in air force explosive ordnance disposal training and operations

Abdulrahman Sulaiman Alwabel, Nathan Greiner, Sean Murphy, William Page, Shane Veitenheimer, and Vhance V. Valencia

Contents

43.1 Introduction

Explosive ordnance disposal (EOD) is a critical and high-risk mission within the military. There are many unique challenges to training and operations that involve low-production, high-cost equipment and high weight requirements during dismounted operations. This leads to a prime application for additive manufacturing (AM), commonly called 3D printing. 3D printing allows for the production of highly complex and varied shaped parts, with little lead time or manufacturing skill needed for effective low-production runs. This research focuses on the following AM applications in EOD: munition training aids, telescopic manipulator pole attachments, shape charges, and water disruptors used in EOD training and operations. Through the systems engineering *Vee* design model, designs for each category were created, printed, and tested in conjunction with the Wright–Patterson Air Force Base EOD unit. The resulting prints were compared with currently available equipment and training items of the same type in order to determine the effectiveness of AM for United States Air Force EOD applications.

43.2 Background

EOD is a critical and high-risk mission set of the United States' Department of Defense (DoD). Some of the traditional missions of EOD technicians include the following: (1) clearing unexploded ordnance (UXO) from training ranges, (2) providing defense support to civil authorities, and (3) assisting with the protection of high-ranking government officials (GAO, 2013). The Global War on Terrorism in Iraq and Afghanistan introduced a burgeoning use of improvised explosive devices (IEDs) that greatly increased the demand for, and strain on, EOD forces. From June 2003 to May 2010, EOD troops were involved in more than 86,000 IED incidents in the two countries (CSIS, 2010). As a result of this growing need, the number of EOD forces increased from 3,600 to 6,200 between 2002 and 2012 to meet the growing demands of wartime operations and the developing challenge of dealing with IEDs (GAO, 2013).

IEDs present a unique challenge to military operations because the *builder has had to improvise with the materials at hand*, which can be almost anything (GS, 2011). With this dynamic and ever changing threat, there have been many iterations of strategy changes to counter IED attacks. The deputy director of the Joint IED Defeat Organization in 2008 warned of the potential of wasting resources developing solutions that are too late or redundant: "They would be wasting their intellectual capital on the wrong problem or on a misunderstanding of a problem" (AFCEA, 2008).

As the number of EOD technicians increases and the threat of evolving IED strategies continue to be a wartime concern, access to shared information, adaptable tools, and realistic training becomes paramount. With the threat of wasting resources on solutions that are slow to counter the IED threat, additive manufacturing (AM) looks to be a natural ally to EOD forces.

As a relatively new technology, AM, more popularly known as 3D printing, is quickly gaining attention across the DoD. Resources are being allocated to further AM capabilities within and around all components of the DoD. Differently active duty services have been steadily increasing their uses of AM over the past three years, and research labs and contractors are even further along in using this adaptive technology.

The Navy first introduced 3D printers in March of 2013, with their Print the Fleet program. The initiative placed 3D printers on ships and trained six sailors per ship in the appropriate maintenance and repair for the machines, along with basic CAD skills. The sailors also have the ability to send more complicated design requests to skilled design engineers stateside (INSINNA, 2014). The ability to coordinate designs from around the world, followed by immediate on-site manufacturing is an exciting opportunity. A problem can be understood, a solution designed, manufactured, tested, modified, and remanufactured in a fraction of the time compared to traditional engineering and manufacturing. Through this program, the Navy is realizing the greatly positive impact of AM.

In the private sector, defense contractors across the United States also realize the potential of AM in providing flexibility and advancement in their fields. The Air Force Research Laboratory has implemented the project *America Makes* to look further into the capabilities of 3D printing (Lonardo, Conner, & Gorham, 2015). This is a public–private project that demonstrates how leveraging this new technology can help American companies and defense partners.

For these reasons, and at the request of Air Force EOD personnel, research and development of solutions using AM has been accomplished in three areas: (1) EOD training aids, (2) linear shaped charges and projected water disruptor, and (3) attachments for telescopic manipulators. These areas were chosen for the cost associated with the current products and the variety required in their design.

43.3   Vee model design method

In order to accomplish all three of these projects in just 10 weeks, an agile and versatile methodology was required. The research team chose to use the systems engineering Vee model design method, a model well-suited for the close coupling of the development and progress of the design with the customer's and/or stakeholder's needs (Frosber, Mooz, & Cotterman, 2000). By using this design method, the research team was able to not only meet their short timelines, but also established a lasting relationship with their EOD customers (Figure 43.1).

The basic Vee model is split into three sections: (1) project definition, (2) implementation, and (3) project test and integration. However, in order to implement the Vee model correctly, the project team needs to continually accomplish the *verification and validation* portion of the model. Verification is accomplished by including the customer in the design conversation from the beginning of the design. This allows for the proper amount of verification to take place before and during the design phase of the project. Without verifying designs with the customer's needs, the team may be wasting many hours of effort designing a product that will not be used. The immediate customer for this project was the 788th Explosive Ordnance Disposal Flight, located at Wright–Patterson Air Force Base, Ohio. The research team kept in constant contact with this unit throughout the process of defining the scope of the three projects. Multiple redirections were accomplished within the first month of design, allowing for more time detailing the final designs.

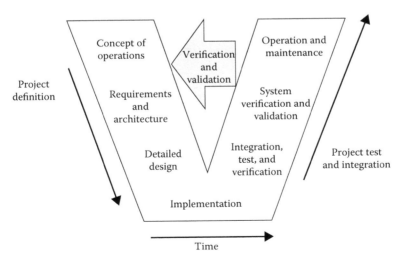

Figure 43.1 Vee model method. (Adapted from Forsberg, K. et al., *Visualizing Project Management: A Model for Business and Technical Success,* 2nd ed., New York, John Wiley and Sons Inc., 2000.)

Validation, the next vital component of the Vee design model, is a coordination step that usually takes place further along during the design process. Validation is crucial as it tests the functionality of the designed product. Although a design team accomplishes the verification of user needs, validation must also occur that provides evidence that the final product meets the user's expectations. Without validation, the project may completely fail. The 788th EOD Flight continually provided assistance with these validation tests for the three projects. Feedback from the EOD unit included the following: (1) adjusting the weight of training aids, (2) the hooks requiring more strength, and (3) necessary weight reductions for printed shape charges. These examples and many other feedback inputs were received through the constant coordination with the EOD technicians. The Vee model provided an excellent structure for creating products from scratch, especially as both the design team and customer were new users of the technology.

Sections 43.4–43.7 detail the development and testing of the three projects accomplished through this research. Training aids, linear shaped charges and projected water disruptors, and attachments for telescopic manipulators are discussed here. An overview for each project is provided and the results of the additive manufactured design are described.

43.4 Training aids

Dating back as far as the U.S. Civil War, EOD personnel have been called upon to disable a wide array of UXO (Perkins, 1996). This wide array of UXOs sometimes proves difficult to train for as many munitions that might still be encountered are no longer manufactured and, therefore, are nearly impossible to obtain for training. On top of this, almost every country that has produced explosive ordnance have used different designs and design details, making a large variety of shapes and forms to identify (denix.osd.mil, n.d.).

UXOs are found in areas where munitions have been disposed of or deployed, such as military bases, war zones, and military training ranges. In the United States alone, there are "over 11,000,000 acres of land potentially contaminated with UXO (not including Air Force sites)" (Mackenzie, Dugan, Division, & Kolodny, 1994). The dangerous nature of UXOs has prompted many governmental safety sites to educate their citizens on staying away and reporting the item to proper authorities (defense.gov.au, n.d.; denix.osd.mil, n.d.; nra.gov.la, n.d.).

For personnel training, recognition of munitions is the first critical step to ensure that proper safe disposal actions are taken. As shown by Defense Shield student UXO guide, there are slight variations in identifying UXOs based on many minute details, such as size, fin type, weld type, location of markings, and so on (Defense Shield, n.d.). The preferred method to train new EOD personnel on different munition types is to show a trainee how to identify and disable an unexploded munition with inert training aids. For rare items, they are specially manufactured or replicas purchased from specialized vendors. These purchases have proven costly because of the limited and sensitive market for the products. Items such as this, rare and specialty manufacturing, can easily be created using 3D modeling and AM. Once each munition is digitized using 3D software, the files can be shared across EOD forces for virtual visualization. Further, each EOD squadron with access to a 3D printer would be able to create physical training aids that can be manipulated and *safed* (i.e., disposed of) during training. AM has the potential to expose many more personnel to rare munitions than current procedures allow.

Using AM to develop and manufacture training aids goes beyond the U.S. military. For example, the Advanced Ordnance Teaching Materials kit has recently been sold to the United Nations, PeaceTrees Vietnam, and Switzerland's International Committee for the Red Cross. These detailed kits cost U.S. $7,000 but can be carried on a plane (unlike traditional training aids in this field) and contain many interworking parts to help teach how land mines work. The creators, a MIT professor and a retired Army EOD technician, develop and manufacture all the training aids using 3D printing. Their goal is to help rid the world of its 110 million active landmines, which account for more than 800 deaths monthly. These hands-on training tools prove to be widely more affective teaching measures when compared to traditional books and pamphlets.

43.5   Results

The Vee design model was used on training aids recreating a VS-50 antipersonnel mine and a token example of a mortar round. Both designs were scanned using the MakerBot Digitizer 3D Scanner, recreated in SolidWorks Education Edition 2013–2014, then printed on a ProJet 3500 HDMax. The mortar round was considered a token sample due to unclear scan and build space size limitations of the ProJet 3500. Given the poor scanning and the size limitation, a slightly shortened facsimile mortar resulted and many key mortar recognition details were missing. Scans and the resulting stereolithography (ST) files were obtained from previous researchers to begin the design process. The images from the scans are shown in Figure 43.2.

From these STL files, dimensions were unable to be obtained, leading the research team to search for appropriate sizes for input into the modeling software. Solidworks Part Files (SLDPRT) were then created with dimensioning, a hollow design to reduce material usage, and detail finishing.

The first iteration of designed parts was rough due to the lack of scan clarity but were suitable for an initial print to allow collaboration with trained EOD personnel. Based on feedback from the first prints, an additional training item of an attachable mortar fuse was designed and each item was adapted over several designs and print iterations as shown in Figures 43.3 and 43.4.

Figure 43.2 Scans of antipersonnel mine and indeterminate mortar.

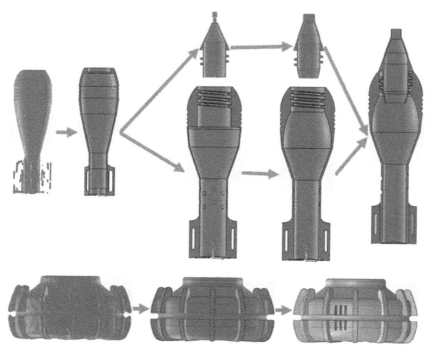

Figure 43.3 Evolution of design iterations.

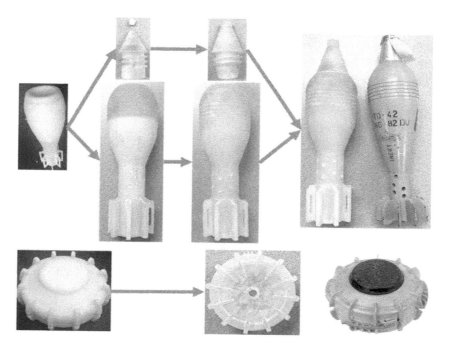

Figure 43.4 Evolution of print iterations compared to actual inert munitions.

43.6   *Linear shaped charges and projected water disruptors*

43.6.1   *Linear shaped charges*

Shaped charges, or explosive charges with a lined cavity, date back to as early as 1792. At that time, Franz von Baader, a mining engineer, noticed that when a hollow, or cavity, is created in the charge, the energy released by the explosive blast can be focused on a small area (Walters & Zukas, 1989). After that initial discovery, extensive research from around the world developed applications for various militaries. Several warheads, grenades, and other munitions were developed to make use of these effects (Walters & Zukas, 1989). Further research by Charles E. Munroe concluded that by applying explosives into a cavity placed against a steel plate shaped the explosive blast into an exact replica of the cavity. Munroe, in his experiments, placed explosives into a raised shape above a steel plate and successfully cut *U.S. NAVY* into the plate using the explosives (Walters & Zukas, 1989). Current applications employed by EOD units of this type of charge include cutting open car doors, truck beds, doors, and to disable IEDs.

Although linear shaped charges are effective, they are also expensive to manufacture and are ordered and delivered in large quantities. Due to these limitations, they are not ideally suited for training exercises and make prime candidates for manufacturing with AM. The 788th EOD Flight asked the research team to model an Mk 7 linear shaped charge, and to print and test against the original. In initial design discussions with the EOD technicians, they injected one further unique requirement: Make the AM printed shaped charge to be *EOD-proof*. That is, they asked for a more robust design in the model given the propensity for technicians to roughly handle their equipment and tools. In order to satisfy this requirement, the research team increased the thickness of each part of the linear charge to achieve robustness making the walls of the final print approximately 1/3" thick, see Figures 43.5 and 43.6.

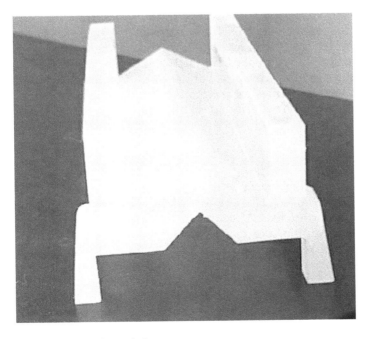

Figure 43.5 Printed Mk 7 linear shaped charge.

Figure 43.6 Original Mk 7 (left), printed Mk 7 with 1-sheet Al (center left); printed Mk 7 with 2-sheets Al (right).

Table 43.1 Results of Mk 7 versus printed Mk 7 and printed Mk 7s with Al

Shaped charge used	Average width	Average depth
Mk 7	0.2972″	0.5821″
Printed Mk 7	0.7785″	0.2641″
Printed Mk 7 w/1 Al	0.7587″	0.2556″
Printed Mk 7 w/2 Al	0.7932″	0.2645″

Verification and validation occurred during an explosives test. Working with the 788 EOD personnel, the research team was able to test an original Mk 7 against three different AM designs: a printed Mk 7, a printed Mk 7 with one layer of 0.025″ thick aluminum, and a printed Mk 7 with two layers of 0.025″ thick aluminum. The sheets of aluminum were added in an attempt to mimic the effect of the steel in the original Mk 7. For the test, the Mk 7 and its AM replicas were placed on steel *truth plates* and the resulting cavity from the detonation was measured. The resulting widths and depths of these cavities are shown in Table 43.1.

43.6.2   Results

From this test, the research team found significant differences in the cavities formed between the original Mk 7 and the printed Mk 7s. As shown in Table 43.1, the cuts in the truth plates from the printed Mk 7s were twice as wide and half as deep as the cut from the original, steel Mk 7. Additionally, feedback from the EOD technicians indicated that they used nearly three times the volume of C4 explosives for the printed Mk 7s. Given the larger volume of explosives, it was expected that the 3D printed shaped charges would have had better performance (Figure 43.7).

Upon further investigation, the research team found that the poor performance was due to a mistake in the design of the charge. While the lower angle of the charge, the portion closest to the truth plate, was the same as the original, the upper angle, that which was holding the explosive, was twice as wide. This discrepancy occurred when the digital

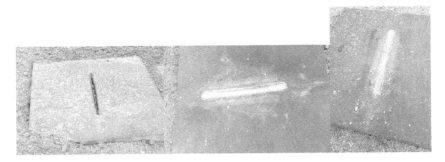

Figure 43.7 Truth plate of original Mk 7 (left), truth plate of printed Mk 7 (center), and truth plate of printed Mk 7 w/1-sheet Al (right).

model thickened the parts; the research team failed to check every angle and the resulting volume of the charge. Given this finding, the resulting discrepancy in the cuts can be explained and followed the Munroe effect.

Finally, the added aluminum had no effect on the cut width or depth. With this result, we conclude that the volume of the metal sheets is critical to plasma forming to assist the cut. With the selected metal sheets, their thicknesses were not enough to form a significant amount of plasma to vary the dimensions of the cuts.

Although not an unqualified success, this first round of testing has resulted in additional research questions. The 788 EOD Flight has asked the researchers to experiment with different plastics to see if some of these unintended effects can be mitigated. Additionally, determining the correct amount of metal and if the type of metal liner to assist in cuts is a potential area for further research.

43.6.3   *Projected water disruptors*

Projected water disruptors are a form of linear shaped charges that utilize water as the cutting jet medium. These are preferred in IED operations for disabling initiators without causing a spark that may set off the explosive component. By using projected water disruptors, technicians have better forensic analysis and intelligence gathering capabilities necessary for IED source identification. Currently available disruptors are expensive and are accountable items, limiting the availability to use them in training.

The research effort in this area focused on designing and testing an additively manufactured projected water disruptor shaped charge for use in training. Personnel from the 788 EOD Flight showed the research team multiple examples of projected water disruptors used in the field, and provide several suggestions for methods of improvement. The research team then took these suggestions and came up with an initial design through Google SketchUp, Figure 43.8.

As part of the design verification process of the Vee design model, the research team brought the print to the 788 EOD Flight and received valuable feedback to improve and refine the design. Such improvements included correcting errors with the top cover and caps, increasing the diameter of the necks to allow easier water filling, adding threads to the necks and caps to ensure a tight connection, increasing the diameter of the hole for the detonating chord, and decreasing the gap between the interior pyramidal-shape to precisely fit the depth of a shaped charge. These fixes resulted in the second prototype shown in Figure 43.9.

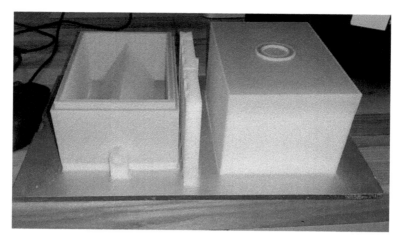

Figure 43.8 Printed projected water disruptor prior to postprocessing.

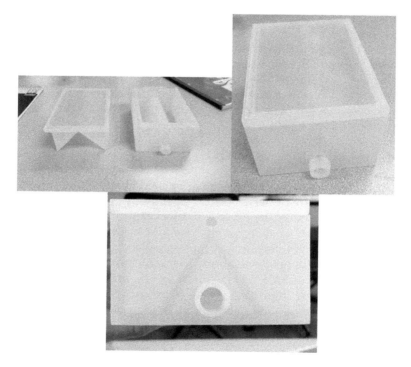

Figure 43.9 Postprocessed projected water disruptor.

Again, the research team encountered issues with the top cover. After consideration, it was discovered that the error was due to how the top cover was modeled in Google SketchUp. 788 EOD personnel also added further points of refinement such as increasing the tolerance of the hole for the detonating cord, and increasing the tolerance of the space between the interior sides of the top and bottom pieces.

Unfortunately, problems with the design using SketchUp hampered the research team's efforts and were unable to test at the time of publishing this book. The learning curve using SketchUp led to several misprints (early prototypes), and the research team is currently debugging the design. The flaw came from not understanding how SketchUp models solids: When a surface is deleted in SketchUp, the interior of the shape appears to be hollow. However, the software still models the shape as a solid although it appears to be a void. This caused multiple prints of the top cover to be printed as a solid.

The research team employed Google SketchUp for this portion of the research as a proof-of-concept for any EOD personnel to download the software and model a design. Open source tools such as SketchUp allow for wide proliferation of a modeling tool with a simple user interface that could, potentially and drastically change how operations are conducted throughout the EOD career field.

43.6.4 Omnidirectional water disruptor

Within the project water disruptor portion of this research, a second item was created for this project. An omnidirectional water disruptor is a tool used in separating simple IED device triggers from their attached explosive charge. In particular, this particular tool is used when the precise location of the trigger is difficult to determine or access. The device works by transferring the force of a cylindrically shaped explosive in a 360° dispersal pattern by surrounding the explosive with water on all sides.

The basic design for this need stems from the operational environment. Off-the-shelf water bottles are commonly accessible in today's military operations throughout the world. EOD technicians for the past decade have been adapting water bottles into omnidirectional water disruptors with detonation cord and water. From initial design discussions, the research team determined that technicians desired a method to control the amount of explosives used and the exact shape (i.e., straight) for this type of disruptor. Figure 43.10 shows the first iteration of this disruptor. This version integrated the bottle cap with the tube holding the charge (Figure 43.11).

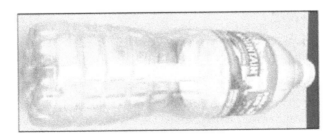

Figure 43.10 Omnidirectional water disruptor inserted into commercial water bottle.

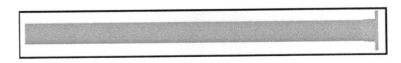

Figure 43.11 Omnidirectional water disruptor SolidWorks design.

Figure 43.12 Omnidirectional water disruptor application test.

After verification, the design for this project transitioned from the integrated cap-and-tube to only the tube. The integrated cap-and-tube would break at the cap threads when attached to a bottle. Therefore, the design was simplified to the design shown in Figure 43.9. This new design allows the user to heat a standard bottle cap near its melting point, then push one end through the plastic, thereby allowing the use of the bottle's manufacturing thread pattern. The disruptor is gradually widened near the top platform allowing the device to be firmly connected to the cap. Finally, a flat surface for the tube was designed to allow for a flush connection to the cap. This simplified design was tested with the 788 EOD flight, and performed the desired task of removing an initiator from a mock device. In general, the technicians were pleased with the results, and the results for the test are shown in Figure 43.12.

43.7   *Telescopic manipulators*

This third and final project of the research concerned creating attachments for telescopic manipulators. Telescopic manipulators are carbon fiber telescoping poles that allow EOD personnel to manipulate objects from a distance of 12.5'. The poles come with a number of attachments that can accomplish different tasks. The tasks can range from cutting, carrying, and dragging objects. When these attachments break, EOD personnel have to wait weeks for replacements.

When the telescopic manipulators are not operational, an EOD technician is required to be closer to a munition in order to disarm it. The closer the technician is to the munition; the greater the risk of a serious or fatal accident. Injuries from explosives can be put into the following four categories: primary, secondary, tertiary, and quaternary (CDC, n.d.). Primary injuries are caused by over pressurization wave with body surfaces. Secondary injuries are caused by fragmentation. Tertiary injuries are from an individual being thrown and landing on the ground or object. Quaternary injuries are those that cannot be contributed to other categories. Examples of quaternary injuries are burns and building collapses. The best way to limit blast effects and associated injuries is maximizing the distance between an individual and the possible source of an explosion. The telescopic manipulators enable an EOD technician to position themselves during disarmament in a way that limits the possible four types of blast injuries.

43.7.1   Methods

As with the previous two efforts, the requirements for this task were determined by meeting with the local EOD shop. Technicians from the shop requested three attachments: (1) a wall attachment, (2) standard hook, and (3) a double hook (Figure 43.13). The wall attachment has an adhesive pad applied to the bottom and is designed to stick to surfaces. This attachment is connected to the telescopic manipulator by a rope and provides a leverage point to disarm explosives. The standard hook is designed to expose buried wires. This allows the technician the ability to get a better assessment of the UXO and its immediate surroundings. Finally, the double hook is used to dig in the ground for wires, open doors to suspected vehicle borne IEDs, and to reposition shape charges. Figures 43.14 through 43.16 show the file CAD drawings for each attachment.

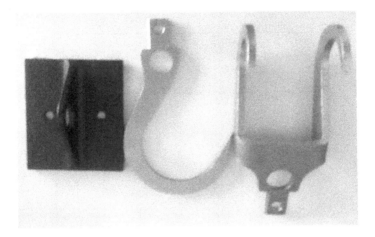

Figure 43.13 Metal attachments.

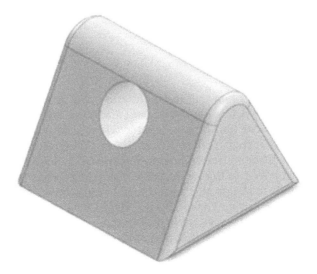

Figure 43.14 AM printed wall attachment.

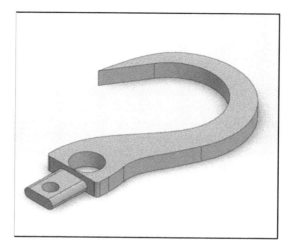

Figure 43.15 AM printed standard hook.

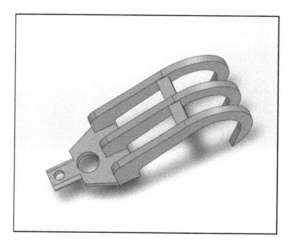

Figure 43.16 AM printed triple hook.

After three rounds of verification and validation, the research team settled on the designs found in Figures 43.13, the standard hook, and 43.14, the triple hook. The standard hook and the triple hook were found to be reliable when used in a back and forth motion, but are noticeably weaker if they are used in a side to side motion. To provide additional strength for side to side motion, the AM printed standard hook had to be thickened. The thickness is double compared to the metal hook. The AM printed triple hook had a third prong and support between the prongs added to provide additional strength.

After testing the standard and triple hooks, the EOD shop requested a J-hook knife attachment to be created for the telescopic manipulator. Current J-knives are considered hand tool and do not exist as an attachment for the manipulator. This attachment is used to cut detonation wires. The attachment is designed to have a ceramic blade slide into the top and a blade into the bottom as shown in Figure 43.17. Ceramic blades are used because the blades do not create a static electricity. This is the key as static electricity is capable of

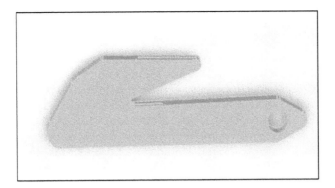

Figure 43.17 AM printed J-hook knife attachment.

triggering a detonation (SDMS, n.d.). The top and bottom blades create a V-shape to cut the wires. The blades will be held in place with friction, and once the blades are in the attachment they are considered permanent. The attachment could be designed where the blades can be replaced, but the attachment will have to be made thicker to accommodate an opening clip.

Finally, given the success of the previous prototypes, an S-hook was created in addition to the EOD shop requests, as shown in Figure 43.18. This attachment was created due to simplicity and it is included in the attachments kits. The attachment was not well-received because it is not used as much as the other attachments. This highlighted the importance of having the customers to be part of the design process. During this research, the team found that the EOD shop did not want the kits completely recreated; they merely wanted those attachments and tools the technicians find the most useful.

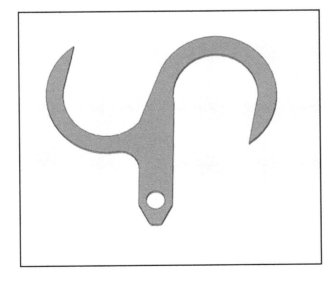

Figure 43.18 AM printed S-hook attachment.

43.7.2   Risk analysis

Tensile and compressive strengths of the attachments were not conducted as part of this research. As a proof-of-concept, this project showed that creating useful attachments using AM was possible. The EOD shop found the J-hook and wall attachment to be the most useful. The standard hook and double hook have potential, but technicians preferred the metal attachments due to their robustness and overall strength. They found that the AM printed hooks are significantly weaker and do not want to risk the attachment breaking while disarming an item. However, the technicians did agree that the AM printed parts would be useful while waiting for replacement metal attachments to arrive. The metal attachments can take up to 16 weeks to get to their final location. Being able to have an AM printed part will be better than having no useful attachments while waiting for replacements. A more in depth analysis should be carried out to determine optimized designs for these attachments.

43.7.3   Weight analysis

A benefit to AM is the ability to reduce the weight of a part. Reducing weight is important for a technician travelling by his own feet as it requires less energy to carry and operate. A U.S. Army study showed that an increase in weight for a mission results in an increase in the amount of time the soldier requires completing the mission and increases the chance of injury for the soldier (Polcyn, Bensel, Harman, Obusek, Pandorf, & Frykman, (2002)). On average, the AM parts are less than half the weight of the metal parts, as shown in Table 43.2. Attachment Weight Comparison. The total weight reduction for all the attachments amounts to less than a half pound. Although this amount of weight reduction is insignificant, if it is possible to reduce the weight of *everything* a technician carries, then there is potential for significant weight reduction over the entire attachments kit. All attachments were weighed on a standard food scale for accuracy to the nearest gram. The weight recorded for the AM parts is the weight after postprocessing.

43.7.4   Cost analysis

The metal attachments cost on average of $25–$40, depending on the complexity of the attachment. This cost includes tax and shipping worldwide. By comparison, the 3D printed attachments cost 32 cents a gram and support material cost 20 cents a gram. The amount of support material used depends on the part orientation during printing and part geometry. The double hook has the highest cost due to the amount of support material required for a successful printing. The wall attachment has a higher weight but lower

Table 43.2 Attachment weight comparison

	Metal (grams)	3D printed (grams)
Standard hook	75	25
Double hook	175	46
Hook seat belt cutter	N/A	32
Wall attachment	80	65
S-hook	90	29

Table 43.3 Attachment cost comparison

	Metal ($)	3D printed ($)
Standard hook	23	10
Double hook	31	24
Hook seat belt cutter	39	12
Wall anchor	15	15
S-hook	27	12

cost than the double hook because the weight includes support material inside the structure. Table 43.3 shows the cost comparison of the metal attachments to the 3D printed attachments.

The costs of the attachments can be reduced by optimizing the designs and determining the optimal orientation on the print bed. Further study will be required to lower the costs. 3D printed attachments are less durable as compared to the metal attachments. Using a 3D printer allows an EOD technician to replace an attachment in a day. Delivery for a metal attachment replacement can take 12–16 weeks to arrive to the technician. During this time the technician will have to make do with other attachments. This adds time to a mission, exposing the disarming team for a longer period of time. Another benefit of using 3D printers is that it enables an EOD unit to adapt to the enemy but allowing for design changes to the attachments.

43.7.5   Results

The attachments can be recreated using the AM machines at a lower cost compared to the metal attachments. These parts can be printed within hours, which is considerably faster than the weeks required for a part to show up in a deployment environment. The drawback is that the parts are made of a thermoplastic instead of metal to be economical. These parts are considerably less durable compared to the metal parts. There is a higher chance of an attachment breaking with the plastic during a mission. Spending time to replace the attachment will increase the time the technicians are vulnerable to the enemy. These printed parts are a good stop gap between a metal attachment breaking and a new attachment getting to the technician downrange.

43.8   Conclusion

This research explored the design, verification, and validation procedure for applying AM technology for specific Air Force EOD needs. In particular, the research team focused in three areas: (1) EOD training aids, (2) linear shaped charges and projected water disruptor, and (3) attachments for telescopic manipulators. These areas were chosen for the cost associated with the current products and the variety required in their design.

Through the systems engineering *Vee* design model, designs for each of the three areas were created, printed, and tested in conjunction with input and feedback from technicians at the Wright–Patterson Air Force Base EOD unit. EOD is a critical and high-risk mission within the military. With the many unique challenges to training and operations AM promises to produce these low-volume, high demand components. In a matter of 10 weeks, novice researchers were able to manufacture highly complex and varied shape parts, with little lead time and manufacturing skill. With further study, testing, verification, and validation, AM promises to meet the needs of EOD.

Acknowledgments

This work was made possible through a research funding from the Air Force Civil Engineer Center, Requirements and Acquisition Division (AFCEC/CXA). A special thanks to Mr. Craig Mellerski and Dr. Joseph Wander for their continued support, leadership, and guidance in this and other research efforts.

References

AFCEA. (2008). Retrieved from http://www.afcea.org/content/?q=improvised-explosive-devices-multifaceted-threat.

CDC. Retrieved from http://www.cdc.gov/masstrauma/preparedness/primer.pdf.

Center for Disease Control. (n.d.). *Explosions and Blasts a Primer for Clinicians*. Retrieved from cdc.gov: http://www.cdc.gov/masstrauma/preparedness/primer.pdf.

Center for Strategic & International Studies. (2010). Retrieved from http://csis.org/files/publication/101110_ied_metrics_combined.pdf.

Defense Shield. (n.d.). UXO Student Handout.

defense.gov.au. (n.d.). Australian Government, Department of Defence Home Page. Retrieved March 2, 2016, from http://www.defence.gov.au/uxo/what_is_uxo.asp.

denix.osd.mil. (n.d.). Unexploded Ordnance (UXO), FAQs. Retrieved March 2, 2016, from http://www.denix.osd.mil/uxo/UXO411/FAQs.cfm.

Forsberg, K., Mooz, H., & Cotterman, H. (2000). *Visualizing project management: A model for business and technical success* (2nd ed.). New York: John Wiley and Sons Inc.

Government Accountability Office. (2013). Retrieved from http://gao.gov/products/GAO-13-385.

Globalsecurity. (2011). Retrieved from http://www.globalsecurity.org/military/intro/ied.htm.

Insinna, V. (2014). Military Scientists Developing New 3-D Printing Applications. *National Defense*, 98(727), 22–23. Retrieved from http://search.ebscohost.com/login.aspx?direct=true&db=aph&AN=96262135&lang=pt-br&site=ehost-live

Lonardo, R., & Conner, B. (2015, October). Additive Manufacturing Provides Agility for Defense Contractors. *NDIA's Business and Technology Magazine*. Retrieved from http://www.nationaldefensemagazine.org/archive/2015/October/Pages/AdditiveManufacturingProvidesAgilityforDefenseContractors.aspx

Mackenzie, C. M., Dugan, R. E., Division, T., & Kolodny, M. A. (1994). Detecting UXO: putting it all into perspective, 2496, 94–99. Retrieved from http://proceedings.spiedigitallibrary.org/proceeding.aspx?articleid=1001256

nra.gov.la. (n.d.). UXO contamination in Lao PDR. Retrieved March 2, 2016, from http://www.nra.gov.la/uxo.html.

Perkins, B. E. (1996). Live Civil War Ammo Found - Archaeology Magazine Archive. Retrieved March 2, 2016, from http://archive.archaeology.org/9611/newsbriefs/civilwar.html.

Polcyn, A. F., Bensel, C. K., Harman, E. A., Obusek, J. P., Pandorf, C., & Frykman, P. (2002). *Effects of Weight Carried By Soldiers : Combined Analysis of four Studies on Maximal Performance, Physiology, and Biomechanics*. Retrieved from http://handle.dtic.mil/100.2/ADA400722

SDMS. (n.d.). *Ceramic Knife*. Retrieved from sdms.co.uk, http://www.sdms.co.uk/products/ceramic-bomb-disarming-tool.

Walters, W. P., & Zukas, J. A. (1989). *Fundamentals of shaped charges*. New York: John Wiley.

chapter forty four

Wing design utilizing topology optimization and additive manufacturing

David Walker, David Liu, and Alan Jennings

Contents

This research is a follow-on effort to a topology optimization (TO) study evaluating the ability to computationally reduce mass of a wing structure while maintaining structural performance. The static loading conditions were obtained through computational fluid dynamics (CFD) analysis of the wing and the results were applied to calculate the baseline displacement, von-Mises stresses, and buckling conditions. TO was then performed both locally on the ribs of wing and on a global scale where the entire internal structure of the wing acted as the design space. Additionally, a skin-thickness optimization was performed, and the integration of a fuel-tank as a functional component and as a load bearing structure. The local rib design was then manufactured in both plastic and aluminum to show the capabilities of additive manufacturing (AM) with aircraft components.

44.1   Introduction

This research is a continuation effort of a previous study on the feasibility of utilizing TO and AM to design and fabricate an aircraft structure. Through traditional manufacturing means, TO designs are difficult to manufacture. However, AM allows for more radical design features due to its build-up nature. With the previous research from Walker et al., the selected TO process was verified through examination via Altair's Optistruct software.[1] Additionally, the wing from a Van's RV-4 homebuilt aircraft was used as a baseline structure for the optimization. Finally, a plastic three-dimensional (3D) wing section was designed and built through AM. In this case, a thin skin was applied to a design space with two structural constraints running laterally through the wing.

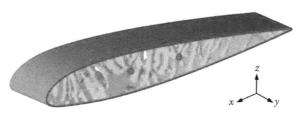

Figure 44.1 Postprocess topology optimized RV-4 wing segment with density fraction set to 38%.

Figure 44.2 3D printed wing section for optimized RV-4 wing segment.

The skin was subjected to two-dimensional (2D) forces in the x and z direction based on the 212 mph (94.8 m/s) maximum cruise velocity at a cruise altitude of 10,000 ft (air density = 0.958 kg/m³). The optimization was based on minimizing compliance constrained to a maximum volume fraction of 38%.[1] The resulting design and fabrication are seen in Figures 44.1 and 44.2, respectively.

This continuation effort enhances the intent of the previous research by establishing objectives to meet the overall intent of this study; develop a procedure that is used to reduce the weight of an aircraft while maintaining structural integrity of the wing. As the intent of this study is to utilize AM, radical design attributes often seen from TO are realistically manufacturable. Through TO, it is important to formulate a means to apply realistic loading conditions on a model, determine what attributes are important for an optimization, and find a means for verification. Additionally, AM enables further design considerations when not constrained by traditional manufacturing means. In this case, a dual-purpose structure that serves as a load bearing component was considered as a way to fully utilized AM capabilities. Therefore, the following five objectives were established:

1. Determine pressure loading on the wing for critical phases of flight and apply the values toward the analysis and optimization processes
2. Perform a computational analysis on the baseline aircraft wing to determine localized stress and displacement values
3. Generate a computational analysis on the optimized designs used for comparative purposes
4. Integrate a traditionally independent component into the optimized wing as a dual-purpose structure
5. Interpret full-scale design to meet AM constraints and produce the model

As discussed in Chapter 19, TO is the mathematical approach of finding the optimal material distribution over a given design space. For nearly all of this research, the designs were constrained to a set volume fraction of the design space with the objective of minimizing compliance. The TO method used is the power-law approach, otherwise referred to as the simple isotropic material with penalization (SIMP) method. The SIMP method considers all material properties of a discretized design space to be constant. However, the elemental density factor, ρ, varies as a value between 0 and 1. Elements of $\rho = 0$ have no density and elements of $\rho = 1$ are 100% dense. Intuitively, this is not manufacturable because most of the material density is constant. Therefore, final designs are often interpretations of the TO results; a threshold is set in which values below are considered a void and values above are considered solid material.[2] The SIMP method uses a penalization power, p, on the density fraction to steer the value toward a solid or a void. Applying a power to ρ reduces ambiguous moderate density areas near the chosen threshold value by steering the density fraction lower.[3] The procedure is deemed acceptable for any material with a Poisson's ratio of 1/3 as long as the penalization power is ≥ 3.[2] Mathematically, the SIMP method for minimum compliance problem is shown in Equation 44.1. In this case, U is the global displacement vector, F is the force vector, K is the global stiffness matrix, u_e is the local displacement vector, k_0 is the local stiffness matrix, and ρ_{min} is minimum relative density.

$$\text{min: } c(\rho) = U^T K U = \sum \rho_e^p u_e^T k_0 u_e$$

$$\text{Subject to: } = \frac{V(\rho)}{V_0} = \phi$$

$$: KU = F$$

$$: 0 < \rho_{min} \leq \rho \leq 1$$

(44.1)

The Altair Engineering suite of computational software was the primary tool used for this research. Particularly, the Optistruct optimization software was used for all topology and sizing optimizations. Optistruct utilizes the SIMP with a penalization power of 1.0 for 2D elements and 2.0 for 3D. Under certain settings, the penalty is increased to 3.0. If k_0 is the real stiffness matrix of the element, then the penalized stiffness is $\bar{k}$, as seen in Equation 44.2.[4]

$$\bar{k}(\rho) = \rho^p k_0$$

(44.2)

For this research, all optimizations were subjected to an objective of minimum compliance constrained by a set volume fraction. The selected volume fraction value was chosen through an iterative process and was dependent on the specific problem setup. Multiple loading conditions were applied and the solver simultaneously calculated each incremental optimization condition independently during each iteration. This was done as a means to ensure the solution satisfied each condition independently. This was done through the weighted compliance formulation shown in Equation 44.3. If C_w is the global compliance, w_i is the incremental weight factor, C_i is the incremental compliance, u_i is the incremental displacement matrix, and f_i is the applied incremental force; then C_w is shown in Equation 44.3. Even though it is possible to vary the weighting for each load condition; all were weighted evenly for this research.[4]

$$C_w = \sum w_i C_i = \frac{1}{2} \sum w_i u_i^T f_i$$

(44.3)

44.2   Methodology

The first step in analysis and optimization was to calculate realistic dynamic pressure values for the surface of the wing. These values were then set as the loading conditions for the TO. In order to do so, computational fluid dynamics (CFD) was utilized. A computer-aided design (CAD) model of the wing was developed and inserted into Altair's virtual wind tunnel (VWT) CFD program. VWT allows a user to set inlet and environmental conditions that are applied to the model. In this case, a generic fuselage was used to simulate the variation in lateral surface pressure experienced on an aircraft wing. An image of the model in VWT is shown in Figure 44.3.

The desire of this research was to develop a wing that would satisfy the loading conditions for all phases of flight. In order to develop an optimization suitable for all flight profiles, the aerodynamic conditions for the most severe cases were used.[5] For an aircraft, this occurs at the limits of its structural performance.[5] In the case of the RV-4, the envelope is at –3G, +6G, and +6G with full aileron deflection for both a left and right roll.[5] Using the RV-4 gross weight of 1,500 lbs, the –3G and +6G angle of attack was calculated at –10.80 and +15.56 , respectively. The CFD model was then modified for an aileron deflection of ±25 to account for maximum roll. VWT results output localized surface pressure at all points along the wing. These values were then easily interpreted into Optistruct as loading conditions for the wing model. Loading was applied on each element on the discretized surface of the wing as a pressure value. The next step was to develop a proper baseline model for comparative analysis with the final optimized wing. Therefore, a simplified RV-4 wing was created and analyzed using finite element analysis (FEA). The wing design was adapted from schematic drawings provided by Van's Aircraft.[6] The baseline wing design consists of a series of ribs with a main and trailing spar, surrounded by a thin sheet-metal skin. Each rib of the baseline wing has the same design and consists of a series of holes along the centerline to save mass. The sheet-metal skin surrounding the ribs provide the aerodynamic shape. This practice is extremely common for aircraft wing design, especially among low-cost general aviation aircraft. All components are composed of 2024-T4 aluminum ranging between 0.813 mm (0.025 in) to 1.016 mm (0.032 in) thick. For this study, flanges and connectors were disregarded. Initially, structural performance was calculated without considering the fuel tank. The FEA model was built using 2D surface

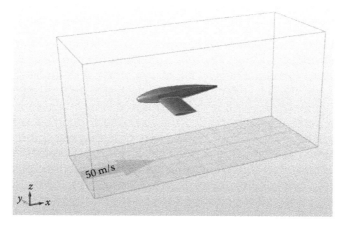

Figure 44.3 Van's RV-4 CFD wing model attached to generic fuselage for surface pressure analysis.

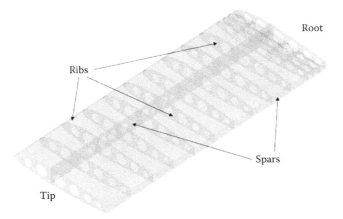

Figure 44.4 Baseline RV-4 wing without fuel tank for FEA analysis.

Table 44.1 Baseline RV-4 wing surface area, volume, and mass properties

	Each rib	Total rib	Front spar	Rear spar	Skin	Total
Surf Area (m^2)	0.128	1.795	0.506	0.284	6.479	9.064
Volume (m^3)	0.10×10^{-3}	1.46×10^{-3}	0.51×10^{-3}	0.29×10^{-3}	5.23×10^{-3}	7.49×10^{-3}
Mass (kg)	0.289	4.059	1.429	0.803	14.539	20.831

Table 44.2 Displacement and stress of baseline wing

Condition	Total displacement	Tip displacement	von-Mises stress
+6g	0.0425 m (1.67 in.)	0.0338 m (1.33 in.)	187.1 MPa
−3g	0.0211 m (0.83 in.)	0.0167 m (0.66 in.)	120.5 MPa
+6g, Roll Right	0.0492 m (1.93 in.)	0.0390 m (1.54 in.)	207.9 MPa
+6g, Roll Left	0.0313 m (1.23 in.)	0.0225 m (0.89 in.)	138.9 MPa

structures with an applied thickness. Figure 44.4 is the baseline wing model. The wings were analyzed for structural stress, displacement, and buckling. In addition, the mass and volume properties for the wing are shown in Table 44.1.

The resulting tip displacement, total displacement, and peak von-Mises stresses of the baseline wing analysis are summarized in Table 44.2. Total displacement is the maximum displacement seen at any point on the wing, whereas tip displacement is the maximum displacement seen at the tip of the wing. This was seen with skin material between the ribs that were displaced more from the starting point than the tip of the wing. An example displacement profile is seen in Figure 44.5a for the +6G, roll right loading condition, and a stress profile in Figure 44.5b. Displacement is in meters (m) and von-Mises stress is in Pascals (Pa).

In addition to displacement and stress, the first three modes of buckling were analyzed. In this case, all buckling occurred near the root of the wing on the skin surface. Figure 44.6 is an image of the buckling that occurred in the wing for the +6G, roll right flight maneuver. Figure 44.6 is an image of the buckling load factor (BLF) and Table 44.3 is a summary of BLF for all tested flight conditions. BLF is the ratio of the localized applied loading to the load at which buckling would occur. In other words, higher BLF values show a higher resistance to buckling. BLF more than 1.0 means a structure will not buckle.

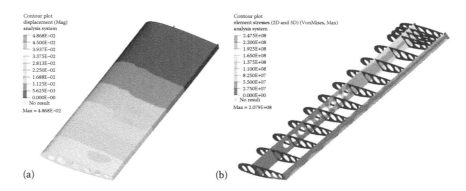

(a)   (b)

Figure 44.5 Wing stress contours for the +6G, roll-right flight profile of the baseline wing: (a) displacement profile and (b) von-Mises stress profile.

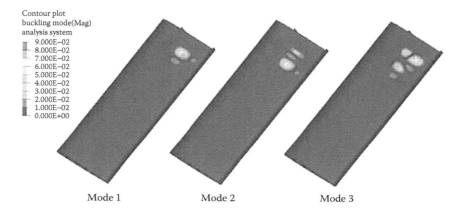

Mode 1   Mode 2   Mode 3

Figure 44.6 Buckling analysis for +6g, rolling right for baseline wing.

Table 44.3 Buckling load factor (BLF) values of baseline wing

	Buckling load factor		
Condition	Mode 1	Mode 2	Mode 3
+6g	0.095	0.102	0.106
−3g	0.0104	0.129	0.153
+6g, Roll Right	0.081	0.086	0.089
+6g, Roll Left	0.143	0.155	0.165

Once the baseline wing was analyzed, the TO was setup and calculated. Optimizations were performed using a variety of methods. The first optimization was performed as a local design, only concentrating on individual ribs. In this case, all other structures remained constant. Figure 44.7 is the design space of the local TO. Once again all optimizations were performed with an objective of minimizing compliance while constrained to a maximum volume fraction. The specific volume fraction chosen was an iterative process specific to each optimization type. This allowed the design to be fine-tuned using engineering judgment to provide the best output. In addition, there was a focus on optimizing with and without a pattern repetition. Pattern repetition ensures that each rib is of similar design.

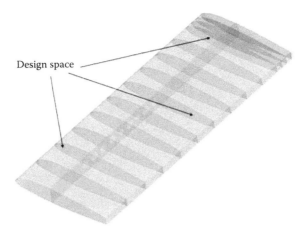

Figure 44.7 Local to 2D rib problem setup.

This was done to provide a direct comparison with the baseline rib. When pattern repetition is not used, each rib is uniquely designed. The local design is also not necessarily limited by traditional manufacturing considerations, but is simply analyzed for comparative purposes. As with all optimizations, the pressure loading obtained from the CFD analysis was applied only to the skin of the wing. The wings were structurally constrained at the spars and along the entire skin surface on the wing root.

A global optimization considers the entire internal structure as the design space. Doing so allows the optimizer to calculate the entire internal structure of the wing. This procedure was initially done without any structural material along the edges of the wings, as shown in Figure 44.8. It was additionally performed, with a structural support rib and rear spar to maintain the aerodynamic shape of the wing. As TO utilizes variable density material, areas of low density are disregarded in the final design when using the threshold method. Therefore, the edges of the wings were left without support material. Figure 44.9 shows the design space of the global optimization with the tip rib and rear spar. Note in both cases, the spars at the root of the wing were maintained in the same location as the local optimization.

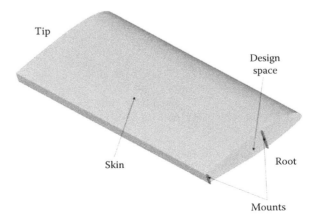

Figure 44.8 3D global optimization problem setup.

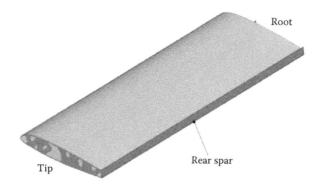

Figure 44.9 3D global optimization problem with tip ribs and rear spar.

Another design consideration as a means to reduce mass was to implement a dual-purpose structure that serves as both a functional component and a load-bearing structure. In this case, the fuel tank was used. For comparative analysis, the baseline wing performance was calculated with the fuel tank. For the baseline FEA, the fuel tank was built as an independent component consisting of an additional spar running the length of the tank and a slightly different rib design. The baseline wing with a fuel tank is shown in Figure 44.10. Once again, this wing was analyzed for stress, displacement, and buckling. Table 44.4 outlines the surface area, volume, and mass of the wing.

Similar to the baseline wing without an integrated fuel tank, structural properties were calculated through FEA. Table 44.5 shows the displacement and stress of the baseline fuel tank. Table 44.6 summarizes the BLF for this wing. Stress, displacement, and buckling occurred in a very similar manner to the baseline wing. Therefore, these figures are not

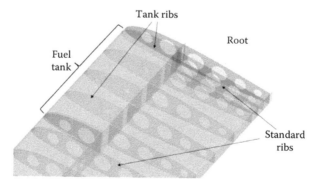

Figure 44.10 Baseline RV-4 wing with fuel tank for FEA analysis.

Table 44.4 Baseline RV-4 wing with fuel tank surface area, volume, and mass properties

	Total rib	Spars	Fuel tank	Skin	Total
Surface Area (m^2)	1.545	0.792	0.624	6.479	9.440
Volume (m^3)	1.21×10^{-3}	0.80×10^{-3}	0.56×10^{-3}	5.23×10^{-3}	7.85×10^{-3}
Mass (kg)	3.482	2.236	1.565	14.539	21.822

Table 44.5 Displacement and stress of baseline wing with fuel tank

Condition	Total displacement	Tip displacement	von-Mises stress
+6g	0.0416 m (1.64 in.)	0.0323 m (1.27 in.)	221.4 *MPa*
−3g	0.0201 m (0.79 in.)	0.0156 m (1.49 in.)	130.6 *MPa*
+6g, Roll Right	0.0487 m (1.92 in.)	0.0379 m (1.49 in.)	254.5 *MPa*
−6g, Roll Left	0.0306 m (1.20 in.)	0.0204 m (0.80 in.)	164.4 *MPa*

Table 44.6 Buckling load factor values of baseline wing with fuel tank

	Buckling load factor		
Condition	Mode 1	Mode 2	Mode 3
+6g	0.094	0.102	0.111
−3g	0.109	0.134	0.157
+6g, Roll Right	0.079	0.086	0.093
+6g, Roll Left	0.142	0.153	0.167

shown. Note the slightly higher levels of von-Mises stress in the wing. These peak levels occurred at the root primarily along the main spar. However, the von-Mises stress and BLF values remained relatively similar when compared to the baseline wing.

Finally, a skin thickness optimization was performed on both the local and global designs. With a skin optimization, material is distributed in a manner that satisfies the objective most appropriately. In this case, the elemental thickness at each point on the skin and spars were used as a variable. Even though this is not a TO, it was used as another means to improve final structural performance. Once again, the optimization was performed with an objective to minimize compliance while constraining the maximum volume. The volume for all skin optimizations was constrained to 0.00523 m³ which is the volume of material for the baseline wing. Therefore, minimal mass savings are seen using this method. However, structural performance did change. Minimum and maximum skin thickness at any point on the skin was set to 0.635 mm to 2.500 mm, respectively. The minimum thickness value is the thinnest found at any location on the RV-4 aircraft and the maximum limit is approximately four times that value. This optimization for the skin and spars performed concurrently to streamline the optimization process. The ribs were optimized prior and completely independent of the thickness optimization.

44.3   Results

The first optimization result is shown in Figure 44.11. The variable density values between 0 and 1 are displayed. Note the darker shaded areas are approaching a density fraction of 0, and the lighter areas are approaching 1. Prior to FEA analysis, this design was post-process to provide a manufacturable part. In this case, a volume fraction of 0.15 was used. Overall, there was a total mass savings of 18.5% among the ribs, which equates to a 3.7% total mass savings in the wing. Table 44.7 outlines the surface area, volume, and mass of this optimized design compared to the baseline wing. The design that utilized pattern repetition is shown in Figure 44.12 and the surface area, volume, and mass details are outlined in Table 44.8. The images of displacement, von-Mises stress, and buckling are not shown here, but are summarized later in this section.

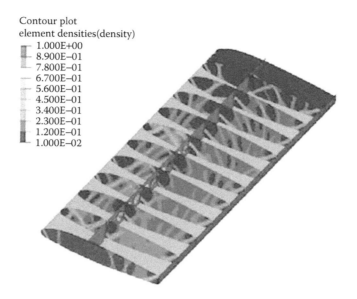

Figure 44.11 Element density for 2D rib optimization without pattern repetition.

Table 44.7 TO without pattern repetition surface area, volume, and mass compared to baseline wing

	Total rib	Spars	Skin	Total
Surface Area (m^2)	1.379	0.790	6.479	8.468
Volume (m^3)	1.10×10^{-3}	0.80×10^{-3}	5.23×10^{-3}	7.13×10^{-3}
Mass (kg)	3.306	2.231	14.539	20.076
Baseline wing mass (kg)	4.059	2.231	14.539	20.831
Savings from baseline (%)	18.5	0	0	3.7

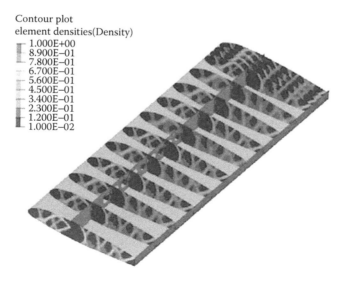

Figure 44.12 Element density for 2D rib optimization with pattern repetition.

Table 44.8 TO with pattern repetition surface area, volume, and mass compared to baseline wing

	Total Rib	Spars	Skin	Total
Surface Area (m^2)	1.582	0.790	6.479	8.730
Volume (m^3)	1.29×10^{-3}	0.80×10^{-3}	5.23×10^{-3}	7.36×10^{-3}
Mass (kg)	3.575	2.231	14.640	20.446
Baseline wing mass (kg)	4.059	2.231	14.539	20.831
Savings from baseline (%)	12.0	0	−0.1	1.9

An individual redesigned rib was also designed as a means of comparing structural performance for a TO design compared to the baseline rib. The practice of using large holes to reduce weight is common in most traditionally manufactured aircraft due to its simplicity. This is a heuristical process that aircraft designers use as a fast and efficient way to design the ribs. However, this is not necessarily the case for more radical wing shapes. An established TO process provides an opportunity to quickly develop a better design that can be manufactured through traditional means. Figure 44.13 is a comparison between the individual redesigned rib for this study and the baseline rib. This rib is 9.5% lighter than the baseline rib and does not exceed stress limits from the baseline FEA.

The skin thickness optimization resulted in a thick region of material near the root of the ring, as shown in Figure 44.14. In this design, the optimized rib was used for the problem setup and the skin and spars were optimized concurrently. Lighter regions of this result approach the thickness maximum and the darker regions approach the minimum. This material distribution is expected because a stronger root intuitively resists bending on the wing is in-line with the minimizing compliance objective. Overall, this wing

To rib
■ Original baseline rib

Figure 44.13 Shape comparison between baseline rib and TO rib.

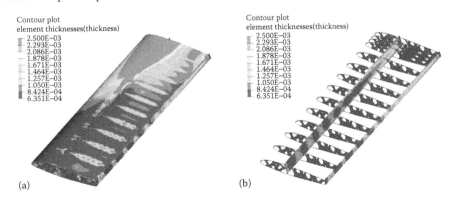

Figure 44.14 Local TO results for spar and skin free-sizing optimization of local wing design: (a) optimized skin and (b) optimized spars.

experienced similar peak stress levels compared to the baseline wing, but with much less displacement. Once again, this design is extremely difficult to manufacture without AM. However, it used the same amount of material as the baseline.

As buckling of the baseline wing occurred on the skin at the root of the wing, the skin thickness optimization had a significant impact on the buckling location and the BLF. As shown in Figure 44.15, buckling shifted laterally toward the wingtip. In addition, the BLF values decreased significantly. This has a significant result because it shows buckling location is controlled through the optimization process. It is important to consider this during problem setup and analysis.

For the global optimization, the same objective and constraints were used as for the local designs. The difference in these designs was the entire internal structure that was considered a design space. Figure 44.16a–d is a series of views of the TO for the global optimization with a tip rib and rear spar for skin support. Even though other global TO designs were calculated, they yielded similar results. For this design, a threshold of 0.20 was used to easily view the internal structure. For all global results, an I-beam like structure was formed laterally on the wing. This makes sense because I-beam structures oppose bending. These designs also show a need for very little support near the tips of the wing. Intuitively, some support is required. This is most likely a function of variable density TO. Unfortunately, a software fault prevented an FEA or a postprocessed design.

In order to integrate a dual-purpose component into the wing, the fuel tank design was influenced by previous optimizations. Areas where the TO created high-density elements, notably near the root of the wing along the main spar, were considered a template for the fuel tank. Loading conditions on the wing surface remained the same. Pressure from the fuel was not considered for this optimization. Figure 44.17 is the resulting TO for the baseline wing with the integrated fuel tank. In this optimization, only the ribs were considered a design space. After this design was postprocessed, a skin and spar thickness optimization was also conducted. Surface area, volume, and mass is compared to the baseline wing in Table 44.9.

Finally, the fuel tank was integrated into the global design. The fuel tank for this design is similar in shape to the local TO, but with slight shaping to mimic the I-beam like structure. The results in Figure 44.18 show high concentrations of material near the tank. Once again, there is obviously a need for support material near the tip of the wing.

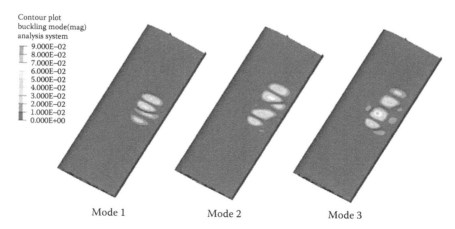

Mode 1   Mode 2   Mode 3

Figure 44.15 Buckling analysis for +6g, rolling right for spar and skin free-sizing optimization of local wing design.

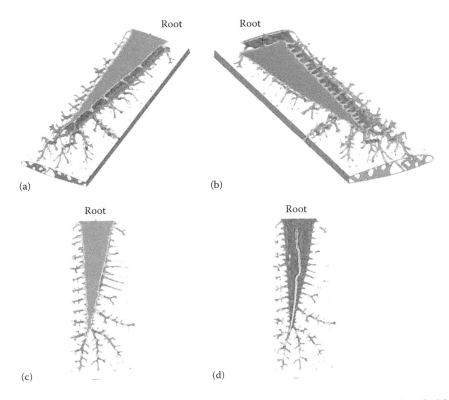

Figure 44.16 Various view orientations for global TO with density fraction threshold of 0.20: (a) upper trailing edge Iso, (b) lower leading edge Iso, (c) top, and (d) top, sliced.

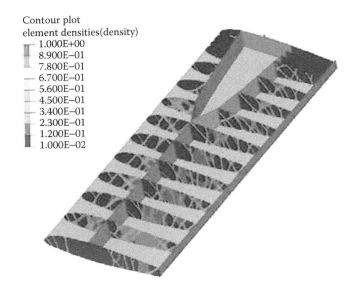

Figure 44.17 Results for fuel tank TO with rib only design space.

Table 44.9 TO of ribs with integrated fuel tank surface area, volume, and mass compared to baseline wing

	Total rib	Spars	Fuel Tank	Skin	Total
Surface Area (m^2)	0.812	0.616	0.543	6.479	8.45
Volume (m^3)	0.66×10^{-3}	0.63×10^{-3}	0.55×10^{-3}	5.23×10^{-3}	7.10×10^{-3}
Mass (kg)	1.835	1.740	1.534	14.640	19.479
Baseline wing with tank mass (kg)	3.482	2.236	1.565	14.539	21.822
Savings from baseline (%)	47.4	22.1	1.9	−0.1	10.8

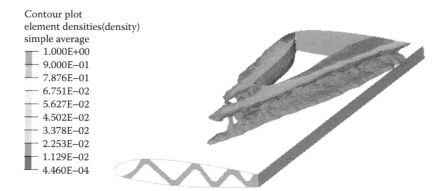

Figure 44.18 Results for fuel tank global TO with density fraction threshold of 0.20.

However, the low density fraction indicates that only minimal support is needed for these loading conditions. As with the earlier global design, these optimizations could not be exported into a working CAD file to be analyzed or manufactured.

A relatively recent concept for TO is the use of lattice structures. Lattice structures replace a partial dense element with a series of lattices to represent the comparable density fraction of the element. Figure 44.19 is the global TO with a fuel tank processed to consider lattice structures. This capability is new in the Optistruct optimization software and did not support CAD export at the time of this research. FEA analysis was conducted on this design.

Table 44.10 outlines the performance of this design along with the other optimized wings and the baseline wing. All of the localized optimizations at least partially reduced mass compared to the baseline wing. The best improvement was seen with optimization

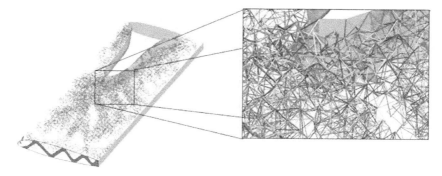

Figure 44.19 Lattice structure design for fuel tank Global TO.

Table 44.10 Performance summary of all TO results

Design iteration	Mass (*kg*)	Peak Disp (m)	Peak stress (Pa)	Min BLF (n = 1)	Max BLF (n = 1)
Baseline	20.831 (0%)	0.0492 (0%)	207.9 (0%)	0.081 (0%)	0.143 (0%)
Baseline w/Tank	21.822 (0%)	0.0487 (0%)	254.5 (0%)	0.079 (0%)	0.142 (0%)
Rib TO w/o Pat Rep	20.076 (−3.7%)	0.0510 (+3.6%)	203.4 (−2.2%)	0.079 (−2.5%)	0.141 (−1.4%)
Rib TO w/Pat Rep	20.446 (−1.9%)	0.0549 (−11.6%)	255.7 (−23.0%)	0.082 (+1.2%)	0.155 (+8.4%)
Redesigned rib	20.548 (−1.4%)	0.0499 (+1.4%)	222.9 (+7.2%)	0.073 (−9.9%)	0.134 (−6.3%)
Reint. Rib w/Sizing Opt	20.078 (−3.7%)	0.0214 (−56.5%)	207.1 (−0.4%)	0.207 (+155.6%)	0.271 (+89.5%)
Local Tank w/Sizing Opt	19.479 (−10.8%)	0.0194 (−60.0%)	254.4 (0%)	0.277 (+250.6%)	0.317 (+122.7%)
Lattice design	27.910 (+27.8%)	0.0271 (−44.4%)	254.4 (0%)	0.193 (+144.3%)	0.224 (+57.7%)

using the skin thickness optimizations. This indicates a significant advantage in manipulating thickness to provide support in critical areas. Peak von-Mises stresses remained similar, but the displacement was significantly decreased and the BLF was significantly increased. Without the thickness optimization, there was improvement in performance, but to a lesser degree. The only global design examined through FEA was the lattice structure design. Even though the mass of this design was significantly increased, performance was greatly enhanced. The inability to modify this design limited the engineering to compare mass and structural performance.

The final step of this research was to verify the feasibility of utilizing AM for production. Even though AM is often regarded as an ability to manufacture any design, this is not necessarily the case. All forms of AM have their limitations. The first AM model, shown in Figure 44.20, is a photopolymer print of the TO rib design with integrated fuel tank. Note the top skin of the wing was removed to show the internal structure. Detail in this part is relatively high, with a build layer of 16 microns. All models for this research were printed

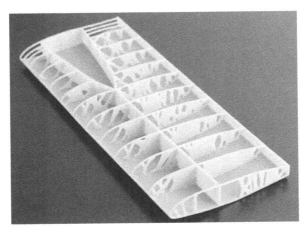

Figure 44.20 Plastic 3D printed model of local to integrated with fuel tank with top skin removed.

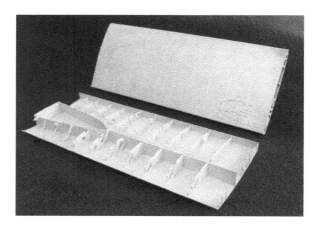

Figure 44.21 Aluminum AM model of local TO integrated with fuel tank with top skin removed.

at 1/12th scale. In order to build the part without structural failure, the thickness was significantly increased relative to the actual design. Even so, great care was taken during postprocessing to avoid damaging the part. Overall, postprocessing consisted of simply removing the soft support material with a water-jet.

An aluminum model of the same wing was printed using direct metal laser sintering (DMLS). In this case, the build layer is 40 microns. A significant disadvantage to DMLS is the inability to have sharp overhangs in the material. This made applying the top skin difficult because using removable support material was not feasible for the internal structure. Postprocessing of the metallic AM parts is also more involved. The aluminum AM parts are shown in Figure 44.21.

As previously discussed in this section, a lattice structure model was not exportable with the available software. However, a lattice structure model was designed by Within Engineering based on the given loading conditions. Based on the initial conditions provided, Within Engineering was able to generate a global TO solution and then incorporate their lattice structure. This model was also built out of aluminum using AM. The AM model is shown in Figure 44.22. Notice the lattice structure has varying lengths

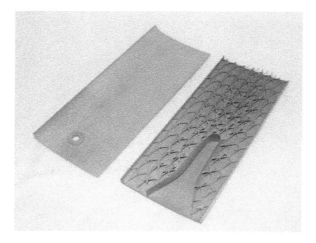

Figure 44.22 Aluminum AM model of lattice structure design.

and diameters based on the structural requirement in the local region. Unfortunately, an FEA analysis on this design was not completed at the time of this research.

44.4   Recommendations and future analysis

Through this research, it was quickly realized the significance of mesh size for any TO. The chosen mesh size for all 3D optimizations was 1.5 cm on each side of the cube. With this size, computational run time was on the order of 36 hours per optimization. A true, high-fidelity optimization would require much finer mesh, but the computation power required would increase exponentially. It is recommended to establish a solid strategy in setting up the model, and then decreasing the mesh size to allow for more accurate results. This is especially important when determining the placement of thin structures, such as the ribs in this research. It is possible that a finer mesh would result in significantly different results for the global optimization indicated by the somewhat sporadic material placement near the tip of the wing. In addition, it is recommended to consider a broader spectrum of loading conditions to best represent the stresses experienced by the aircraft. The wings in this research only considered static surface pressure conditions. Dynamic loading and forces created from control surface deflection will most likely impact the final results.

It is recommended that future work should focus on lattice structures as a means to interpret variable density TO. Lattices are an excellent way to represent the true intent of a variable density element. AM allows for manufacturing of these designs that were not previously feasible. Finally, future work should consider a broader variety of objectives and constraints. Even though the designs in this research showed improvement in the regions tested, greater focus on the real objective of the optimization may yield superior results.

Acknowledgments

The authors thank Mr. Dennis Lindell at the Joint Aircraft Survivability Program Office (JASP) for supporting this research.

Results presented were first printed in the 57th AIAA/ASCE/AHS/ASC Structures, Structural Dynamics, and Materials Conference, AIAA SciTech, Conference Proceedings in 2016. Discussion has been adapted to a chapter presentation.

Nomenclature[*]

C_i	Incremental compliance
C_w	Global compliance
F	Force vector
f_i	Applied incremental force
K	Global stiffness matrix
k_0	Local stiffness matrix
$\bar{k}$	Penalized stiffness
p	Penalization power
U	Global displacement vector
u_e	Local displacement vector
u_i	Incremental displacement matrix

---

[*] This material is declared a work of the U.S. Government and is not subject to copyright protection in the United States.

V	Volume
V_e	Baseline volume
w_i	Incremental weight factor
ρ	Density fraction
ρ_e	Local density
ρ_{min}	Minimum relative density
ϕ	Desired volume fraction

References

1. Walker, D., Liu, D., Jennings, A., Topology optimization of an aircraft wing, in *56th AIAA/ASCE/AHS/ASC Structures, Structural Dynamics, and Materials Conference,* 2015.
2. Sigmund, O., Design of multiphysics actuators using topology optimization; One material structures, in *Computer methods in Applied Mechanics and Engineering,* 190(49), 6577, 2001.
3. Sigmund, O., A 99 line topology optimization code written in Matlab, *Structural and Multidisciplinary Optimization,* Springer-Verlag, 21, 120–127, 2001.
4. Altair Engineering Inc., Optistruct 12.0 User's Guide, 2013.
5. Raymer, D. P., Aircraft design: A conceptual approach, *American Institute of Aeronautics and Astronautics,* Reston, VA, 2012.
6. Van's Aircraft, Inc., RV-4 Construction Manual, *RV-4 Preview Plans,* Aurora, Oregon, Jan. 2005.

chapter forty five

Topology optimization of a penetrating warhead

William T. Graves, Jr., David Liu, and Anthony N. Palazotto

Contents

Experimental and analytical research is underway to determine the optimum topology of a hard target penetrating warhead. By removing some of the warhead's exterior case mass and replacing it with an optimized interior support structure, a design with comparable stiffness and increased lethality is achieved. This research shows the potential for the design and production of warheads tailored uniquely to their intended targets, which are additively manufactured as needed by operational military forces.

45.1   Introduction

As long as humans have built fortified structures to protect valuable assets, they have also designed projectiles to defeat those structures. On the modern battlefield, precision-guided munitions have vastly increased the lethality of air-delivered weapons because their widespread proliferation began during the Gulf War I. Despite increases in guidance capability,

Figure 45.1 An internal view of a recently tested penetrating warhead design. (From Richards, H. K. and Liu, D., Topology Optimization of Additively-Manufactured, Lattice-Reinforced Penetrative Warheads, *56th AIAA/ASCE/AHS/ASC Structures, Structural Dynamics, and Materials Conference*, 2015.)

however, the design of penetrating warheads themselves has remained unchanged over the same time period. Generally speaking, the currently fielded engineering solution to creating a target penetrating warhead is simply to increase case thickness, without adding any internal structure.[1] Although this approach is effective in getting an explosive charge through thick barriers, it severely hinders the lethality of the weapon because most of the explosive energy released on detonation is expended in simply breaking up the outer case. The relationship between case mass, explosive mass, and initial fragment velocity (as a measure of lethality) is given by the Gurney model, presented in Equation 45.1.[2]

$$V_0 = \sqrt{2E}\left(\frac{M}{C} + \frac{1}{2}\right)^{-1/2}$$
(45.1)

In this model, V_0 is the initial velocity of the ejected fragments, M is the mass of the outer shell of the warhead, C is the mass of high explosive within the warhead, and $\sqrt{2E}$ is an empirically derived constant that is unique to the explosive used. It is therefore the goal of this research to create a warhead design that increases V_0 by decreasing M, without sacrificing penetrative performance of the weapon.

To improve on this thick-walled warhead design, topology optimization via finite element analysis (FEA) is used to determine the appropriate structural layout that would enable a thin-walled warhead to maintain comparable stiffness to a thick-walled design. Loading conditions for optimization are determined from live-fire test data[3] by creating dynamic, nonlinear FEA simulations that are designed to match known test results. Additive manufacturing (AM) is utilized in the production of the topology optimized penetrator design, as the internal structural layout of such designs is prohibitively difficult to manufacture by traditional (subtractive) manufacturing methods. As an example of the complexity of structures recently researched in this field, a penetrator model designed by Richards and Liu is included as shown in Figure 45.1.[4]

45.2   Impact simulation

Previous work in the field of optimized warhead design has produced live-fire test data of both standard and optimized penetrating warheads.[3] Additionally, other research conducted by Teng et al.[5] and Tai and Tang[6] has explored the numerical simulation of concrete penetration events by steel projectiles using live-fire test data as a baseline for accuracy evaluation. For the simulations conducted in this study, the explicit FEA solver RADIOSS was used. This section details of the material constitutive and erosion models used and the FEA setup of the impact simulation, and the results obtained.

45.2.1   Material models

Three material models are defined within the impact simulation. Those models represent the materials making up the unreinforced concrete target, the stainless steel (SS) warhead, and the sand used to fill void space in the warhead (simulating explosive filler) during live-fire testing. This subsection details the constitutive models used to describe those three materials in the finite element impact simulation.

For the unreinforced concrete target, the brittle material model of Johnson and Holmquist was used.[7] This model is appropriate for brittle materials subjected to large strains, high strain rates and high pressures, and is also applicable to both Lagrangian and Eulerian formulations.[7] The model determines the equivalent strength (σ^*) of the brittle material as a function of its intact strength (σ_i^*), its fracture strength (σ_f^*), and damage (D). The equivalent strength expression is presented as Equation 45.2.

$$\sigma^* = \sigma_i^* - D(\sigma_i^* - \sigma_f^*) \tag{45.2}$$

In order to fully describe the above parameters, several material constants are required. These constants consist of strength (including Hugoniot elastic limit [HEL] properties) and damage and equation of state (EOS) constants. Parameters describing 7 ksi concrete are presented in Table 45.1.

In Table 45.1, *HEL* represents the Hugoniot elastic limit of the concrete, and *PHEL* represents the pressure in the material at the Hugoniot elastic limit. A, B, N, and C are nondimensional constants, $\dot{\varepsilon}_0$ is the reference strain rate, and $\sigma_{f,max}^*$ is the maximum normalized fractured strength of the material. D1 is the damage constant and D2 is the damage exponent. K1 is the bulk modulus of the material, whereas K2 and K3 are pressure coefficients of the EOS model. Values in Table 45.1 were taken from Tai and Tang, where appropriate.[6]

The elasto-plastic material model of Johnson and Cook (J–C) was used to describe the SS warhead material.[8] This model includes strain rate and temperature effects widely used to model elastoplastic material behavior in explicit FEA codes. In the J–C model, materials behave as linear-elastic when the equivalent stress is below the plastic yield stress. Beyond the plastic yield stress, von-Mises flow stress, σ, is calculated using Equation 45.3.[8]

Table 45.1 Material parameters describing the Johnson–Holmquist constitutive model for 7 ksi concrete

Strength parameters		Damage parameters		EOS parameters	
	2.24×10^{-6}	D1	0.03	K1 (GPa)	17.4
Shear Modulus, G (GPa)	13.567	D2	1	K2 (GPa)	38.8
HEL (GPa)	2.79			K3 (GPa)	29.8
PHEL (GPa)	1.46				
A	0.75				
B	1.65				
n	0.76				
C	0.007				
$\dot{\varepsilon}_0$ *(ms)*	0.001				
$\sigma_{f,max}^*$	0.048				

$$\sigma = (a + b\varepsilon_p^n)\left(1 + c\ln\frac{\dot{\varepsilon}}{\dot{\varepsilon}_0}\right)(1 - (T^*)^m) \qquad (45.3)$$

In Equation 45.3, a is the plastic yield stress (GPa), b is the plastic hardening parameter, n is the plastic hardening exponent, ε_p is the equivalent plastic strain, c is the strain rate coefficient, $\dot{\varepsilon}$ is the strain rate, $\dot{\varepsilon}_0$ is the reference strain rate, T^* is the homologous temperature, and m is the temperature exponent. The argument of the natural logarithm in Equation 45.3, $(\dot{\varepsilon}/\dot{\varepsilon}_0)$ is redefined as ε^*, the dimensionless plastic strain rate. Values for this model are obtained from torsion tests over a range of strain rates and Hopkinson bar tests over a range of temperatures.[8]

The test data gathered by Richards and Liu utilized a warhead fabricated from 15–5 precipitation hardening (PH) SS. J–C parameters for this alloy determined by Mondelin et al.[9] are used in this research, and are presented in Table 45.2. Generic values for precipitation hardening martensitic SS were used where Mondelin et al.[9] did not provide values. Values defined in Equation 45.3 not shown in Table 45.2 are derived by the solver using the additional values in Table 45.2. These values are $\sigma_{max,0}$, the plasticity maximum stress of the material, T_{melt}, the melting temperature of the material, and T_{ref}, the reference temperature or room temperature for the simulation.

For the sand filling the warhead, a linear elastic material law was used. The density of the material is known as 1.6×10^{-6} kg. For this simple material law, only Young's Modulus (E) was needed in addition to density to fully describe the material. According to Berney and Smith, the Young's Modulus of isotropically confined sand varies as a function of the effective mean stress of the sand.[10] As this value is not known *a priori*, and because it will vary throughout the penetration event, an estimate of an appropriate value for this property was made. A reasonable value of E was determined as 1.5 GPa through iterative numeric simulation and comparison with live-fire warhead plastic deformation.

45.2.2 Erosion criteria

For the concrete and steel material models, erosion (or failure) criteria were incorporated in order to allow for penetration of the target, and to allow for the removal of material considered to have failed during the simulation. As noted by Teng et al., erosion criteria (specifically for the target elements) are hypothetical values for the numerical simulation and do not necessarily represent experimental data.[5]

Table 45.2 J–C material model parameters for the steel warhead

J–C parameters		Temperature parameters	
Density, ρ (*mm*$_3$)	7.85×10^{-6}	T_{melt} (K)	1713
Young's Modulus, E (GPa)	212	T_{ref} (K)	298
Poisson's Ratio, v	0.291	m	0.63
a (GPa)	0.855		
b (GPa)	0.448		
n	0.14		
c	0.014		
$\dot{\varepsilon}_0$ (*ms*)	0.001		
$\sigma_{max,0}$ (GPa)	1.24		

For the concrete material model, a simple tensile strain failure criterion is incorporated. This criterion removes any concrete element experiencing a tensile strain of 0.5. When this criterion is met, the element is considered to no longer contribute to the penetration process and is removed from the simulation. Although this process contributes to a small energy imbalance in the simulation where kinetic energy from the warhead is transferred to internal energy of the target element, but is lost when the element is eroded, it is critical to the proper calibration of the impact simulation. The strain value of 0.5 was chosen as it allows the warhead penetration to match live-fire test data.

For the steel material model, the failure criteria of Johnson and Cook was used.[11] In this model, damage, D, is defined by Equation 45.4, where $\Delta\varepsilon$ is a small increment of equivalent plastic strain and ε_f is the equivalent strain to fracture under the material's given conditions. Fracture occurs when $D = 1.0$. The calculation of ε_f is presented as Equation 45.5, where numbered D coefficients are empirically derived constants for the given material.[12]

$$D = \sum \frac{\Delta\varepsilon}{\varepsilon_f} \tag{45.4}$$

$$\varepsilon_f = [D_1 + D_2 \exp(D_3\sigma^*)][1 + D_4 \ln\dot{\varepsilon}^*][1 + D_5 T^*] \tag{45.5}$$

In Equation 45.5, σ^* is the dimensionless pressure ratio defined as the ratio of the average of the three normal stresses and the von-Mises equivalent stress, and $\dot{\varepsilon}^*$ and T^* are the same as previously defined in the J–C elastoplastic material model. The values of constants used in this research are taken from Johnson and Holmquist.[11]

45.2.3 Finite element model

The finite element model used in this research was constructed to represent the live-fire testing of warheads fired by Richards and Liu.[3] In their tests, additively manufactured, ogive-nose stainless steel (SS) projectiles were fired at unreinforced concrete targets at varying velocities and angles of obliquity. Targets consisted of 5 ksi concrete (with an undetermined cure time, likely on the order of years) poured into steel barrels and cut to a length intended to represent a semi-infinite condition. Due to the age of the concrete, data for 7 ksi concrete was used in the concrete constitutive model described previously. Both currently fielded scale models, referred to as the *standard*, penetrator designs and thin-walled optimized designs were fired. All warheads were filled with kiln-dried sand as a surrogate for explosive filler. Due to the complexity of the optimized warhead design, the scaled standard design was used for finite element modeling. As the goal of this research is to develop a warhead that will survive a penetration event, the test considered as the worst-case loading scenario for the warhead was reproduced in the finite element model. Critical aspects of this case were the firing velocity, angle of obliquity, and angle of attack at impact, which all combine to form a dynamic load difficult for a thin-walled warhead to survive.

Preprocessing of the simulation was conducted using Altair Hypermesh. Due to the size and geometry of the model, only a cross section of the original test was reproduced in the model. This approach is similar to Teng et al., and produces results representative of the impact event without requiring an undue amount of computational resources.

Both the warhead and its filler sand were modeled with a Lagrangian mesh, consisting of a combination of 6 and 8 node brick elements. A constraining boundary condition in the z (out of plane) direction was applied in order to enforce a plane strain condition. This type of formulation was chosen over the use of two-dimensional (2D) plane strain elements due to the material to element compatibility available within RADIOSS.[12] When preprocessing the warhead, a finer mesh was applied to the nose in order to maintain high element quality and to properly represent the geometry of the part. Figure 45.2 shows the meshed warhead with sand filler, whereas Figure 45.3 shows a detailed view of the nose of the part.

The target was meshed similarly, using only 8 node brick elements. Again, an out of plane constraint was placed on the mesh. Additionally, the outer edges of the target were clamped in order to simulate the steel barrel surrounding the concrete target in the live-fire testing. Figure 45.4 shows the completed finite element model with the angle of target obliquity and warhead angle of attack incorporated. The size of the target mesh matches the test article in diameter, and is appropriately deep to model the semi-infinite result without requiring large amounts of computational effort. The boundaries of the target

Figure 45.2 Mesh of steel warhead and filler sand.

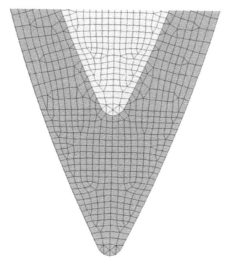

Figure 45.3 Detail of meshed warhead nose.

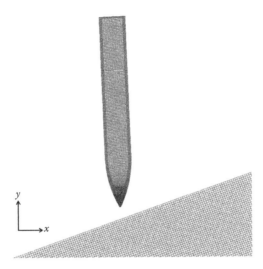

Figure 45.4 The full finite element model, including target angle of obliquity and warhead angle of attack.

are not shown in Figure 45.4 in order to preserve clarity of the mesh. The direction of the imposed velocity for the test was in the y direction.

All elements within the model are formulated with an *Isolid* of 24 as defined by the RADIOSS user's manual.[12] This element formulation is a corotational, underintegrated element with physical hourglass stabilization. Although slightly more computationally intensive than standard reduced-integration elements employing the penalty method for hourglass control, this method is significantly less intensive (by more than a factor of 2) than a full integration scheme and provides good results when used with an appropriately fine mesh.[12]

Contact interfaces between the warhead and the target as well for *self-impact* within the warhead and the target are defined using type 7 penalty method interfaces within RADIOSS. This is a general purpose interface that models contact between a master surface and a group of slave nodes.[12] In the case of the impact interface between the warhead and target, the warhead nodes are defined as slave nodes and all element faces within the target are defined as master surfaces. No contact gap is set, as is often done in impact simulations using shell elements. Deletion of failed elements and nodes connected to failed elements from the contact interface is activated. Coulomb friction between the target and warhead is set to 0.2.

45.2.4 Results of simulation

Results of the standard warhead simulation show good agreement with test data. Overall penetration depth and warhead plastic deformation match observations from live-fire test articles. Figure 45.5 shows the final penetration depth and deformation of the warhead, and the removal of failed target elements. Figure 45.6 shows the transfer of kinetic to internal energy as a function of time throughout the penetration process. Note a loss of 13.9% of the total energy within the model due to erosion of failed elements.

Next relevant loading conditions for application to a topology optimization problem were found using the results of the impact simulation. RADIOSS geometrically distributes

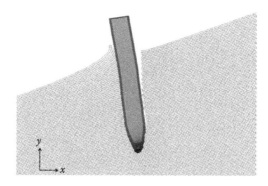

Figure 45.5 The results of the impact simulation at $t = 0.8$ ms. This time represents the warhead's maximum penetration depth.

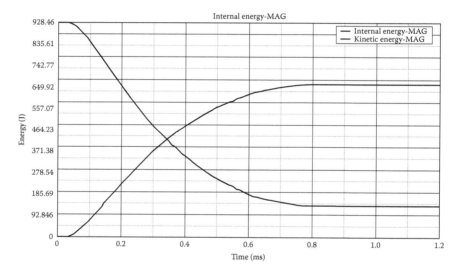

Figure 45.6 Simulation energy balance as a function of time.

the mass of all elements to their attached nodes in its solution process. Additionally, node acceleration (a two-component vector quantity in this case) is available within RADIOSS as an output of the solution process. Consequently, a force distribution for all warhead nodes is found by solving Newton's second law at any moment in time throughout the simulation.

In order to generate useful results for topology optimization, force distributions corresponding to a few critical time periods throughout the penetration were needed. These force distributions would then create the load cases required for topology optimization. To determine which moments in time were most critical to consider, both qualitative and quantitative approaches were taken. First, qualitatively, it is intuitive that the force distribution corresponding to the moment of impact between the warhead and the target is critical to consider. The inertia of the warhead is at its maximum at this time, and the magnitude of the resulting external force is large, even if it only acts over a small

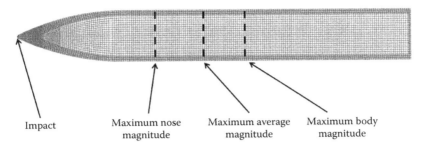

Impact   Maximum nose   Maximum average   Maximum body
                    magnitude   magnitude   magnitude

Figure 45.7 Approximate penetration depths corresponding to critical times for optimization load case generation.

portion of the warhead. Second, more quantitatively, the moment in time corresponding to the largest average magnitude of nodal acceleration was found, and a load case was derived representing that moment. Finally, two further critical times were determined by examination of accelerometer output generated by RADIOSS. Within RADIOSS, the user may attach an *accelerometer* to any node within a model. As the solver runs, it then applies a four-pole Butterworth low-pass filter with a cutoff frequency of 1650 Hz to the raw acceleration output at the node to which the accelerometer is attached. This technique corresponds to class 1000 SAE filtering.[12] At the completion of the solver run, the user is able to view accelerometer output as a function of time throughout the penetration event. Using a distribution of these accelerometers across the length of the warhead, two further critical time frames were chosen. These times correspond to a time where the magnitude of accelerometer output near the nose of the warhead was at its greatest, and to a time where output along the body of the warhead was at its greatest. Figure 45.7 shows the approximate penetration depths corresponding to each one of these four significant moments in time.

45.3   Topology optimization

The ultimate goal of this project is to design a survivable, thin-walled warhead that is supported by a topology optimized internal structure. To achieve this goal, force distributions across a thin-walled warhead were needed for application to the topology optimization problem. As the previously described impact simulation represented a standard warhead penetration, the simulation was rerun with a new, thin-walled warhead. To this end, half of the case mass of the standard warhead design was removed, and the resulting extra void space in the center of the warhead was filled with sand. As the parameters of the impact simulation were properly calibrated to match test data with the standard warhead, it was reasonable to assume that the simulation represents the thin-walled warhead penetration event.

45.3.1   Load step generation

A load step is a combination of forces (referred to as *load cases* within Hyperworks) and constraints applied to a static analysis or optimization. As previously described, four distinct load cases were determined from the impact simulation with the standard warhead. At the completion of the thin-walled warhead solution run, four critical times were again determined according to the procedure described in the previous section. For all

Figure 45.8 Test specimens with center of gravity location noted in red. (From Richards, H. K., *Topology Optimization of Additively Manufactured Penetrating Warheads*, M.S. Thesis, Air Force Institute of Technology, OH, 2015.)

load cases except the one corresponding to the time of impact, four-pole Butterworth filtering with a cut off frequency of 1650 Hz was conducted in order to remove the high-frequency acceleration response of the warhead. This technique essentially replicates the *accelerometer* filtering conducted within RADIOSS, whereas eliminating the need to place an accelerometer at every node in the model. As the accelerations experienced at the time of impact occurred at a frequency beyond 1650 Hz, unfiltered acceleration data was used for this load case. With these four acceleration distributions and the resolved nodal masses output by RADIOSS, the resulting force distributions were resolved using Newton's second law.

In determining how best to constrain the model, test specimens from Richards and Liu were examined.[3] Two of these warheads are shown in Figure 45.8, where the red line indicates the approximate location of the warhead's center of gravity.

Inspection of the test articles shown in Figure 45.8 shows that these warheads either deformed or failed near their respective centers of gravity. This indicates that the inertia carried by the aft end of the warhead generated a large bending moment when combined with the accelerations imparted on the fore of the warhead by contact forces with the target. This bending moment seems to have concentrated stresses at or near the center of gravity of the test warheads. Consequently, it was determined the best means of constraining the warhead for analysis and topology optimization and was to clamp it at the location where the center of gravity was likely to exist in the finished design, slightly forward of the warhead's midpoint. This constraint was applied to all four load cases to define the four load steps used in the optimization of the warhead.

45.3.2   *Optimization setup*

Topology optimization is a concept-level design tool allowing an engineer to determine proper material distribution for a structure in response to given loading conditions.[13] Topology optimizations conducted in the course of this research utilize the Altair products Hypermesh and Optistruct for preprocessing and solving, respectively. Topology optimization within Optistruct utilizes the power-law approach known as the simple isotropic material with penalization (SIMP) method. Within this method, all material properties within a defined design space are held constant (materials are considered as linear-elastic only) with the exception of density, which is variable.[14] When applied in practice, this means the user is responsible for defining a given part's design space, material properties, and loading conditions in order for the solver to define a solution of varying densities between 0 (void space) and 1 (fully dense material) across the design space. Where partially dense material exists, the engineer must interpret the design for their particular application.

45.3.2.1   *Design variable, responses, constraints, and objective*

Every topology optimization problem in Optistruct must define four quantities: the design variable, the response or responses, the optimization constraints, and the objective of the optimization. For this problem, the void space previously occupied by sand was replaced with steel and defined as the design variable. The thin outer case of the warhead was defined as nondesign space. Responses incorporated into the model were a defined volume fraction of design space corresponding to the amount of internal material required for removal, and a weighted compliance of the warhead as a whole, including the design and nondesign space to each of the four load steps. The only constraint placed on the solution was to reach a defined volume fraction of approximately 15.2%. This value corresponds to the fraction of the design space mass equaling the mass of the removed material from the exterior case of the standard warhead. Finally, the objective of the optimization was defined as minimizing the overall weighted compliance of the warhead to the four applied load steps. This last consideration is discussed in further detail in the following subsection.

45.3.2.2   *Member size control*

When defining the design variable in the model, Optistruct allows the user to define three additional quantities valuable in the application of topology optimization. These values are a minimum and maximum member size, and as a minimum gap between members. Properly employing some or all of these parameters allows easy interpretation of the optimization result for manufacturing, as they force the solver to collect partially dense areas into a clearly defined truss or system of trusses. Using these parameters, however, forces the solver to present a solution other than the true mathematically optimized design. The collection of partially dense material into defined trusses can significantly concentrate stresses, and essentially invalidate the optimization solution if not applied correctly.

To determine the proper member size control to apply in this study, a test matrix was carried out. Optimizations using no member size control, minimum member size only, minimum and maximum member size, and minimum and maximum member size with a minimum gap were carried out, and their resulting compliances and maximum stress values were compared. The results of this study showed, as intuitively expected, placing more constraints on member size, increased both compliance and stress. A significant manufacturability benefit was noted with minimal increase in compliance, however, when only a minimum member size was defined. With this boundary condition enforced, the compliance of the warhead increased to an average of 7.58% across all load steps, and the maximum von-Mises stress on the warhead increased to an average of 2.31%. In exchange for this decrease in performance, a greatly improved design in terms of ease of interpretation was achieved. Figure 45.9 shows an example of an optimization solution with no member size control, and the solution for an identical load step with minimum member size control implemented.

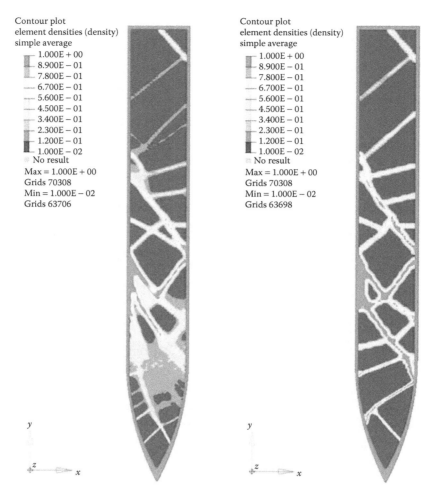

Figure 45.9 The optimization solution on the left uses no member size control, whereas the solution on the right uses a minimum member size constraint. Color contours define material density, with red representing fully dense material and dark blue representing void space.

45.3.2.3 *Weighted compliance considerations*

Weighted compliance is a method used to consider multiple load steps in a topology optimization.[13] When conducting a weighted compliance optimization, the user is asked to define a relative weight between 0 and 1 for each load step considered in the optimization setup. For a given design iteration, the solver determines the design's compliance to each of the given load steps, multiplies the compliance by the respective load step's weight, and then sums the compliance for all load steps. Design iterations are conducted until a minimum sum of weighted compliances is achieved. As a simple example, if the user defines the weight for each of their given load steps as 1, each load step is considered equally in the final optimization result.

Table 45.3 shows pertinent information for the discussion of weighted compliance optimization. The four load steps generated previously are labelled for the ease of discussion, and the compliances resulting from warhead optimizations considering those

Table 45.3 Load step labelling and compliances of single load step optimizations

Load step description	Load step label	Optimized warhead compliance to single load step only (mm/kN)
Time of impact	(a)	0.922
Maximum average acceleration magnitude	(b)	0.0422
Maximum nose acceleration magnitude	(c)	0.0343
Maximum body acceleration magnitude	(d)	0.0261

load steps individually are presented. This information provides a performance baseline against which the weighted compliance solutions are compared.

The effect of a weighted compliance optimization is best understood by visualizing the combination of individual optimization solutions. Figure 45.10 shows how the four optimization solutions representing each individual load step are combined into a weighted compliance solution, when the weight of all load steps is taken as 1.0. Note how the weighted compliance optimization contains characteristics of each of the optimization solutions corresponding to only an individual load step.

Determination of the proper weights for each load step was conducted via an optimization test matrix similar to determining proper member size constraints. Minimum member size control was used for all optimizations in the test matrix. Observing the maximum von-Mises stresses generated by each load step in the minimum member size control study, the load steps were ranked according to the magnitude of the maximum stress they generated. A series of weights was then tested to determine the proper blend of weights for the warhead. The resulting compliances of each weighted compliance designs were then compared to the compliances observed in optimization solutions where only one load step was considered. Essentially, the test was designed to determine the level of performance

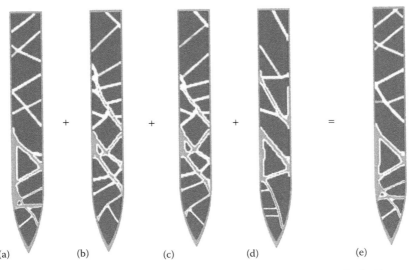

Figure 45.10 The optimization solutions shown from (a) to (d) represent optimizations considering single load steps only (impact, maximum magnitude, maximum nose magnitude, and maximum body magnitude, respectively). Solution (e) is a weighted compliance optimization considering all four load steps equally. Again, Shade of gray contours represent material density.

Table 45.4 Weighted compliance test matrix and results. Note that the compliance increase
columns represent the percentage increase of warhead compliance to a given load step
when the weighted optimized design is compared to the compliance of the optimization
solution considering only the load step in question

Load step	Test 1 weight	Compliance increase (%)	Test 2 weight	Compliance increase (%)	Test 3 weight	Compliance increase (%)	Test 4 weight	Compliance increase (%)
(a)	1	3.41	1	2.71	1	2.55	1	2.12
(b)	1	39.67	.8	45.66	.6	58.77	.5	62.40
(c)	1	54.2	.6	61.57	.4	77.72	.3	81.44
(d)	1	33.48	.4	36.52	.2	43.34	.1	45.31

compromise to each individual load step imposed by considering all load steps simultaneously. Table 45.4 outlines the test matrix and results.

Results presented in Table 45.4 show how much performance the final optimized warhead design looses with respect to designs optimized for each load step individually. It is also important to note that due to the magnitude of the forces on the warhead at impact, the maximum stress experienced as a result of the load step are approximately 1.5–5 times greater than those experienced as a result of other load steps. Consequently, the impact load step is honored to the maximum extent possible, while giving up a minimum of performance in response to all other load steps. Examining Table 45.4, the results of Test 2 show a significant reduction in compliance at impact over the equally weighted optimization, without losing large amounts of stiffness to other load steps. As the weight on other load steps is further reduced in Tests 3 and 4, the response to impact continues to improve, but the responses to other load steps are unacceptably poor. As a result, the weighting shown in Test 2 is determined as the proper weighting for the final warhead design.

45.3.2.4   *Final topology optimization solution*

The final optimization solution considers all previously discussed parameters. Minimum member size control is employed, as is a weighted compliance in the proportion shown as Test 2 in Table 45.4. The resulting topology optimization is shown in Figure 45.11.

Note that the design shown in Figure 45.11 is a response to the oblique impact experienced by the warhead during the impact simulation, and is not symmetric. In a live-fire test, the orientation of the warhead at impact is not known or controlled. Consequently, the design presented here is mirrored on top of itself to give a planar solution capable of withstanding impact with either a positive or negative angle of obliquity. Figure 45.12 shows such an interpretation, modeled in computer-aided design (CAD) software using the final design as presented in Figure 45.11.

Figure 45.11 The final topology optimization solution.

Figure 45.12 The final topology optimization solution, reflected on top of itself and translated in CAD to generate a symmetric solution.

45.4 Conclusion

This work uses techniques from several disciplines in order to achieve a topology optimized warhead design. Explicit FEA, with heavy dependence on high strain rate material constitutive and fracture modeling, is used to simulate a highly dynamic warhead penetration event. Forces are resolved, constraints are applied, and implicit FEA is used to generate a topology optimized design capable of surviving a hard target penetration. Although not discussed here, this work will also rely on AM technology to fabricate the final three-dimensional (3D) warhead design.

45.5 Future work

A great deal of future work remains to bring this project to completion. The authors will translate the 2D solution presented in Figure 45.12 into a computer aided design file, and rotate it to form the internal truss structure of the 3D warhead. The *mass budget* allowable for this operation is limited to only material removed from the exterior case of the standard penetrator design. The rotated structure, therefore, will not remain a solid rotation in its final form, rather it will take on a spoke-like array of trusses designed to support the warhead at any orientation, but without adding any total mass when compared to the standard design. Also, the final warhead's center of gravity must remain in the forward half of the warhead in order to provide aerodynamic stability in flight.

   On completion of the final warhead, test articles will be printed in 15–5 PH stainless steel, heat treated, and live-fire tested. The project will be considered a success if the optimized, thin-walled design survives the penetration event, with at least as much penetration depth as the standard design.

Acknowledgments

First printed in the 57th AIAA/ASCE/AHS/ASC Structures, Structural Dynamics, and Materials Conference, AIAA SciTech, Conference Proceedings in 2016.

References

1. Driels, M. R., *Weaponeering: Conventional Weapon System Effectiveness*. 2nd ed. Reston, VA: American Institute of Aeronautics and Astronautics, 2013.
2. Gurney, R. W., *The Initial Velocities of Fragments from Bombs, Shell and Grenades, BRL-405*. Aberdeen, MD: Ballistic Research Laboratory, 1943.
3. Richards, H. K. *Topology Optimization of Additively Manufactured Penetrating Warheads*, M.S. Thesis, Air Force Institute of Technology, Wright-Patterson AFB, OH, 2015.
4. Richards, H. K. and Liu, D., Topology Optimization of Additively-Manufactured, Lattice-Reinforced Penetrative Warheads. *56th AIAA/ASCE/AHS/ASC Structures, Structural Dynamics, and Materials Conference*, Kissimmee, FL, 2015.

5. Teng, T. L., Chu, Y. A., Chang, F. A. and Chin, H. S., Simulation model of impact on reinforced concrete, *Cement and Concrete Research*, 34 (11), 2004, 2067–2077.

6. Tai, Y. S. and Tang, C. C., Numerical simulation: The dynamic behavior of reinforced concrete plates under normal impact, *Theoretical and Applied Fracture Mechanics*, 45 (2), 2006, 117–127.

7. Johnson, G. R. and Holmquist, T. J., An improved computational constitutive model for Brittle materials, *High-Pressure Science and Technology*, 305 (1), 1994, 981–984.

8. Johnson, G. R. and Cook, W. H., A constitutive model and data for metals subjected to large strains, high strain rates and high temperatures, *Proceedings of the 7th International Symposium on Ballistics*, 21, 1983, 541–547.

9. Mondelin, A., Valiorgue, F., Rech, J., Coret, M. and Feulvarch, E., Hybrid model for the prediction of residual stresses induced by 15-5 steel turning, *International Journal of Mechanical Sciences*, 58 (1), 2012, 69–85.

10. Berney, E. S. and Smith, D. M., Mechanical and Physical Properties of ASTM C33 Sand, ERDC/GSL-TR-06-XX, ERDC, Waterways Experiment Station, Vicksburg, 2006.

11. Johnson, G. R. and Holmquist, T. J., Test Data and Computational Strength and Fracture Model Constants for 23 Materials Subjected to Large Strains, High Strain Rates, and High Temperatures, Los Alamos National Laboratory, Los Alamos, NM. Report No. LA-11463-MS, 1989.

12. Altair Engineering 2015. *The RADIOSS User's Manual Version 13.0.* Troy, MI: Altair Engineering, 2015.

13. Altair Engineering 2015. *The Optistruct User's Manual Version 13.0.* Troy, MI: Altair Engineering, 2015.

14. Sigmund, O., A 99 Line topology optimizaiton code written in MATLAB, *Structural and Multidisciplinary Optimization*, 21, 2001, 120–127.

chapter forty six

Iteration revolution
DMLS production applications

Erin Stone and Chad Cooper

Contents

46.1   Introduction

This chapter explores the role and impact of direct metal laser sintering (DMLS) on rapid prototyping and, more importantly, the recent progression of 3D manufacturing/additive manufacturing (AM) as a disruptive advancement for design and production. First, the history and growth of DMLS is reviewed, followed by a discussion of current challenges facing DMLS technology. Using a leading U.S. DMLS service bureau, i3D MFG™, the chapter then explores how these challenges are being met and surpassed through iteration to raise the bar on DMLS solutions-based production applications.

As a preface to the balance of the chapter, it may be helpful to give a brief explanation as to how DMLS works and what makes it different from traditional manufacturing methods. DMLS, referred to more accurately as direct metal laser welding or laser forging, describes a manufacturing process different than conventional methods in that, instead of milling a part from solid block or casting with molds, DMLS builds up components layer-by-layer using finely powdered metal. DMLS takes 3D CAD models in the form of a .stl file and then converts them into laser build instructions microlayer-by-microlayer using proprietary software. The DMLS machine starts a job by applying a 20–60 micron layer of metal powder to a build platform and then laser welds or forges the powder at exactly the points the CAD file calls out to within ±.001″ –.004″ tolerance of the model. The platform is then lowered slightly and another 20–60 micron layer of powder is deposited across

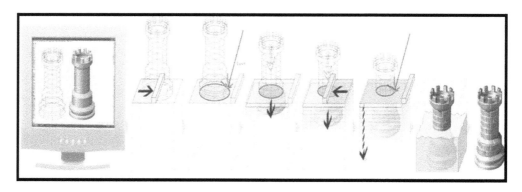

Figure 46.1 DMLS file conversion to DMLS build illustration. (Courtesy of EOS, Krailing, Germany.)

the platform and the metal powder is laser welded/fused to the previous layer at the next predefined model points. Multiple models from multiple customers can be grouped on a single build, enabling mass customization and efficient low volume manufacturing. Figure 46.1 below illustrates the file slicing and DMLS microlayer build process.

46.2   Background and DMLS's role in 3D printing

The early 1990s saw the first 3D plastic machine and the advent of 3D printing for the rapid prototyping of parts. The 3D printing industry focused on meeting design and prepro- duction needs through rapid prototyping and on perfecting various plastic and polymer material combinations using ink-jet based technology. Five years later, laser welding was introduced, enabling powder-based, microlayer forging without binding agents or addi- tives. In 1995, EOS, a German company with exclusive rights to laser-sintering technology, launched the first DMLS machine, the M 250. It was focused solely on rapid tooling to com- pliment the traditional manufacturing and molding sectors. A decade later, EOS, launched the M 270, introducing advancements that enabled direct metal part production with the first fiber laser 3D-printing system. As DMLS enabled mass customization of production parts, dental implants became one of the first segments to adopt DMLS as a means of manufacturing. By 2008, selective laser sintering (SLS) and DMLS moved 3D printing from rapid prototyping squarely into production parts for industrial applications. According to the 2015 Wohler's Service Provider Report, "For 2014, average growth in direct part produc- tion was 20.9%, following growth of 19.9% in 2013 and 16.9% in 2012." (Wohler Associates, *2015 Wohler's Report Service Providers Survey*, July 2015, p. 10.) (Figure 46.2).

The advantages that DMLS offers extend well beyond mass customization and finished parts capabilities. Although other 3D rapid prototyping technologies allow designers and manufacturers the ability to test form, fit, and function, the final production parts still have to be manufactured using traditional methods including computer numerical con- trol (CNC), injection molding, and casting. Typically, this means that expensive tooling or programming is necessary for final part production; requiring designs that fall within the limitations of machining or molding and also require running large production quantities capable of amortizing the cost of tooling. DMLS is changing this entire process equation. With DMLS, proof of concept is production. In other words, once a final pre-production part prints, there is no need to retool, create a new CNC program, nor a complex mold tool. The part, or the mold tooling can be printed as a production part using DMLS as the prototype, and then translating it to 3D metal production efficiently and economically.

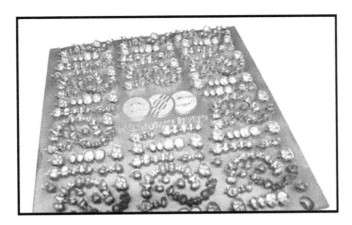

Figure 46.2 DMLS build plate of mass customized dental implants. (Courtesy of EOS, Krailing, Germany.)

With DMLS, prototypes are the pilot parts and can move directly into DMLS production using the same recipe or formula developed during the iterative prototyping phase. Where do DMLS efficiencies or economies make the most sense? The answer is explored in more detail throughout the chapter, however the following are optimal DMLS production parts or projects:

1. Complex parts with an emphasis on organic features that are impossible or incredibly time consuming and expensive to machine because of multiple clamping and tooling change outs with CNC or other traditional methods are the best fit for DMLS.
2. 3D printing facilitates consolidation of multipart assemblies into one printed part, saving assembly time and labor and simplifying the supply chain.
3. Integrating assemblies into single printed parts also creates part performance efficiencies by eliminating potential part failures points such as weld lines, gasket seals, and screw or other attachment junctions.
4. Low to medium production becomes economical and allows for more frequent part updates and greatly reduced R&D time and expense.
5. DMLS is a low waste process that reduces metal scrap costs that often equalizes production pricing for parts made of high cost superalloys and exotic metals.
6. Exotic metals become affordable because of the lack of waste. AM builds material where it is called for whereas traditional subtractive methods carve material away, leaving significant scrap.
7. Conformal cooling lines can greatly increase tool performance and efficiencies by doing two things, reducing cycle times and organically running along parts creating cooling benefits where traditional mold tools could not.

46.3   Past and current DMLS opportunities

Recognizing that DMLS is gaining significant traction as a rapid prototyping and production methodology, it is worth exploring past and current DMLS opportunities for advancement including 3D design practices, technology limitations, surface finish misperceptions, and postprocess requirements. Like all disruptive technologies, 3D printing faces naysayers and entrenched manufacturing establishment interests that point to these challenges as proof that 3D printing is a fad, rather than a game-changing innovation. Certainly computer technology, cell phones, and the Internet could have gone by the wayside; however, the early adopters and technology leaders in these industries converged to catapult these disruptive technologies to mainstream, invaluable technology advances. Although the 3D manufacturing challenges laid out below require solutions to solidify DMLS as the next manufacturing revolution, early DMLS adopters and 3D innovators are committed to pushing the limits toward success, lending convincing evidence that 3D printing, particularly DMLS, is the next disruptive manufacturing technology. DMLS technology limitations and the corresponding solutions or innovations used by i3D MFG™ are examined in this section of the chapter. Common DMLS technology challenges include 3D design gaps, build speed, build envelope size, metal powder development, and surface finish. Certainly, these are not the only challenges or catalysts for DMLS innovations, but they are issues frequently faced within the DMLS industry.

46.3.1   Design for DMLS

As a good 3D design greatly influences the success of the part, the first set of challenges explored are common DMLS design issues and corresponding best practices. To additively manufacture using DMLS, all CAD design files must be converted to .stl files and then, using proprietary OEM software, converted into .sli (slice layer interface) or .cli (common layer interface) files. Layer interface files are a collection of part section cuts, 0.01 mm thick, parallel to the X–Y plane (or build plate). Exposure parameters are then assigned to .sli files; exposure parameters include direct laser power, speed, and distance between laser passes. Starting the process with a high quality, high resolution, and defect free .stl is critical. Defect free is the absence of holes in the surface of the .stl file resulting in a complete and continuous surface geometry. Holes are created by overlapping or missing triangles on the .stl surface that occur during design or during file conversion. Additionally, i3D MFG™ recommends exporting .stl files with a 1.5 degree angle setting and a 3 micron deviation. Resolution is more critical for DMLS parts than for many 3D plastic processes. i3D MFG™ often receives files created for a 3D plastic prototype and must rebuild the files to achieve the necessary resolution required for all part features to print correctly in metal. Although the parent CAD program for modeling is flexible, a parasolid in x_t, .stp, .igs, or .sldprt is best for 3D engineers to smooth and fill in triangles for an effective .stl file clean-up. The flow of the design to DMLS print process is illustrated below (Figure 46.3).

During the initial file review, in addition to adding supports, i3D MFG™ engineers also look for potential build failure points using a patent pending software analysis system. Supports are used to compensate for nonideal conditions in a part design but generally increase stress, add postprocess, increase costs, reduce repeatability, and reduce accuracy in the ability to hold tight tolerances. Past i3D MFG™ projects have benefitted significantly from involving i3D MFG™ DMLS engineers early in the design process to review and assist in optimizing files and reducing support structure. Beyond basic surface clean up and hole repair, supports may need to be added to the design in order for it to grow properly. Typically, supports are added to downward facing surfaces or angles less than 40 degrees. For best results, designs should minimize the need for support structure as this increases build time, requires more postprocess, thermal stress, and increases cost. Part geometry requiring supports are also more likely to experience build failures. Some common i3DTM design strategies used to optimize designs for DMLS include chamfers and fillets. Chamfers provide a transition from one surface to

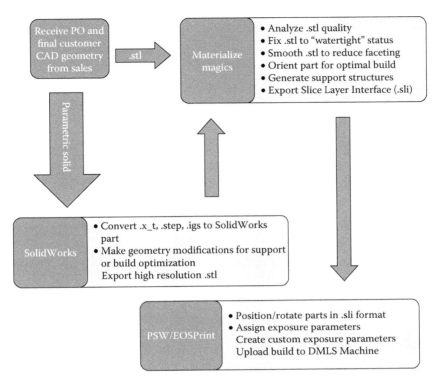

Figure 46.3 DMLS design process to file conversion flow. (Courtesy of Kevin Perry, i3D MFG™ Engineering Technician.)

other using acceptable or ideal downward facing surface angles, eliminating the need for supports. Fillets transition from vertical wall to a horizontal downward facing surface, which does not always eliminate the need for supports. The images below illustrate modeling support structure and an example of parts with significant support structure as well as the finished part (Figure 46.4).

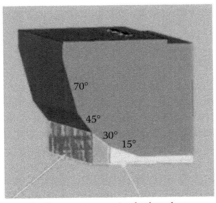

Figure 46.4 DMLS support structure additions to .stl file before printing. (Courtesy of EOS, Krailing, Germany.)

Figure 46.5 Dental prosthesis after manufacturing, with support structures and after completion. (Courtesy of EOS, Krailing, Germany.)

46.3.2   Part orientation

In addition to minimizing support structure, DMLS design optimization also includes minimizing downward facing surfaces, smoothing and creating uniform part transitions while avoiding volume jumps, using buildable angles, and keeping as many of these angles as possible at 40 degrees or more from the X–Y plane, or build plate. Other DMLS design considerations are the total mass of the parts and how well the parts secure themselves to the build plate without creating excessive internal stress created by high cross-sectional areas in the X–Y plane. Balancing mass and part orientation are a DMLS art form, much like a master mold tool maker applies experience and intuition to create tools that create consistent parts. In developing a DMLS manufacturable and reproducible part, i3D MFG™ DMLS engineers and technicians monitor and record part successes and failures and apply this knowledge to new iterations of the part. Highly complex parts require more iteration, whereas components of lower complexity require few, if any, iterations to achieve reliable manufacturing build runs. The combination of material, laser settings, coating, and thermal stresses create a complex engineering challenge that must be overcome by part design, orientation, supports, and exposure strategies (Figure 46.5).

46.3.3   Optimize designs

The bike hub images, Figures 46.6 through 46.10, given below illustrate a traditionally designed 3D part file and an alternate approach that eliminates downward facing surfaces while also smoothing transitions from right angles into more organic curves. DMLS parts are most successful if they start with a sufficient enough flat surface to secure to the build plate and then utilize organic transitions rather than sharp angles. Starting with a contour as a base attached to the plate requires significant support structure that risks being off tolerance as the thermal stresses exert force on the part. However, if the part begins building from an adequate flat surface lasered, or welded, directly to the plate, the part is much more likely to achieve tolerances of .001″–.004″ for the first inch and .0005″–.002″ thereafter. Conversely, once the part starts building from the base, organic part transitions reduce or eliminate downward facing surfaces and angles that also necessitate supports that require postprocessing and add time and cost

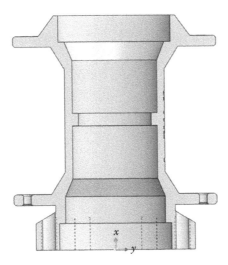

Figure 46.6 Cross section of a traditionally designed bicycle hub. (Courtesy of David Polehn, i3D MFG™ Engineer.)

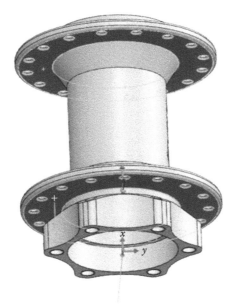

Figure 46.7 Large downward facing surfaces highlighted. (Courtesy of David Polehn, i3D MFG™ Engineer.)

to the build process. A primary goal for effective DMLS design is support reduction or elimination.

46.3.4   Production efficiency

A second concern engineers and production managers raise when considering manufacturing with DMLS is build speed as it relates to production efficiency. Although 3D

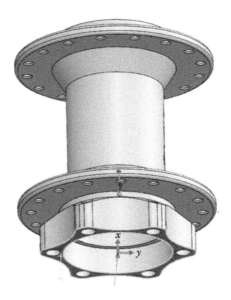

Figure 46.8 Attachment to build plate highlighted. (Courtesy of David Polehn, i3D MFG™ Engineer.)

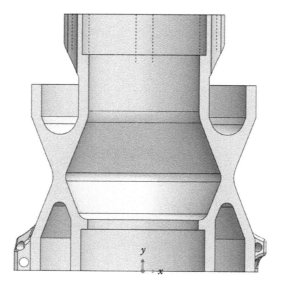

Figure 46.9 i3D MFG™ DMLS designed hub. The model revises the upper spoke features. (Courtesy of David Polehn, i3D MFG™ Engineer.)

printing has proven itself to be a fast and economical method for rapid prototyping and research and development projects, now that DMLS offers the ability to 3D print final parts, is it capable of economical and timely production run quantities? The answer is yes, through the identification of appropriate application and part optimization through itera-tion. However, it is important to recognize that DMLS is not a one-to-one replacement for traditional manufacturing; it is an advanced manufacturing process that revolutionizes the design innovations and corresponding manufacturing capabilities for complex parts

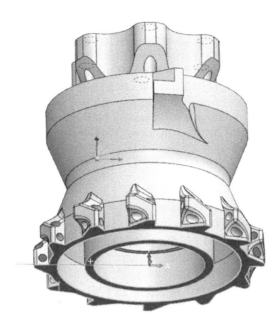

Figure 46.10 All of the downward facing surfaces requiring supports have been removed. The result is good build plate attachment and a part that does not require supports. (Courtesy of David Polehn, i3D MFG™ Engineer.)

and assemblies and low volume production parts. It is not an efficient process for large batch production or routinely stamped, milled, or molded parts. Often, the most highly effective/efficient AM projects are designed, planned and executed synergistically with traditional post machining operations.

Build speed ties directly back to the previous 3D design discussion. The fewer build supports and more optimized the design and part orientation (Z height consideration) the more efficient the part production. A traditional milled manufacturing analogy is square corners. Square corners require many tool changes, slow down the milling process, and require smaller bits, all adding time and cost. Milling operations overcome these issues by using a fillet that matches the required tool, enabling faster part production and reducing costs. i3D MFG™ addresses equivalent DMLS design and machine process obstacles by optimizing DMLS design and adjusting machine build parameters including layer thickness, scan speed, and energy input. Power input is balanced with layer thickness to manage thermal stresses and delicate build areas. Through iteration, i3D MFG™ creates a custom ratio of adjustments until the part meets or exceeds required tolerance and design details and is repeatable. Once the ideal ratio is achieved, the custom build parameters are saved as a final production part file.

Part throughput is another manufacturing efficiency utilized in traditional and AM. For traditional manufacturers, programming and tooling consume time and add cost. Assuming traditional methods can produce a highly complex component, the program time, tool changes, and new tooling frequently create long lead times and expensive piece prices. Complex part efficiency using traditional manufacturing is best realized by large batches with no design changes over years of production; this enables programming and tool cost to be amortized. Similarly, DMLS efficiencies are created by maximizing the number of pieces that can be built in a single build. What distinguishes DMLS is that the

AM process may include multiple parts of an assembly, multiple customer files, or the largest quantity possible of a single file built on the plate. Building full plates of parts spreads the time and cost of powder recoat time across more parts and reduces machine turnover time. Keeping a uniform Z height across the full plate of parts creates additional powder recoat efficiency. Currently, DMLS machines run a single metal during the build process, necessitating that maximized, or nested, builds be grown from one metal. i3D MFG™ furthers its productivity by dedicating a machine to each metal, thereby confining machine turnover to plate removal and job resumption. This saves nearly a day per job in metal changeover times.

46.3.5   Size limitations

Build size is another common customer inquiry. The EOS M 280 and EOS M 290 DMLS build platforms are 250 mm × 250 mm × 325 mm. However, due to laser angle and thermal constraints, the effective EOS DMLS build envelope is 240 mm by 240 mm with a 300 mm Z height. Parts are often angled diagonally to accommodate parts larger than 300 mm but the maximum size is limited to 335 mm. Every DMLS OEM recognizes the opportunity in developing the first reliable build envelope at or beyond 600 mm in each direction. To date, concept laser and EOS have launched new machines with advertised build envelopes of 800 mm × 400 mm × 500 mm and 400 mm × 400 mm × 400 mm respectively. However, the machines currently cost more than double the proven concept laser M2, and EOS M 280, and M 290 machines, and are limited to aluminum and nickel-based alloys for production.

Bigger build envelopes accommodate more parts per build, yet this is only advantageous if the build cycle per part decreases. Larger parts also present build failure risks due to internal stress created by large mass thermal dynamics. Varying expansion and contraction areas produce mechanical stress within the part. Those stresses are from the thermal input of the laser. The larger the part, the larger the thermal stress effect. Extreme stresses result in fractures, warping, part collapses, and potato chipping off the plate. Heat treat, part orientation, and connection to the build plate can mitigate these effects through stress relief. Varying parameters to adjust laser power and speed, printing ghost parts to increase time between layers to give the layers time to cool, and delaying time between exposures can also alleviate thermal stresses. Although the ability to print parts larger than 335 mm is a valuable service, it is worth noting that the DMLS design innovations and efficiencies made possible through 3D manufacturing also create significant opportunity to reduce part size in general, saving material and cost. Because 3D manufacturing is just starting to take root, often, to fully realize this shift in design and engineering mindset takes iteration and partnership between the end-use customer and the 3D service bureau. Once 3D design limitations are erased, build size is generally overcome by design adjustment or optimized assemblies.

46.3.6   Surface finish

A fourth DMLS misperception is surface finish. DMLS parts are nearly 100% dense. Although DMLS stands for direct metal laser sintering, DMLW (direct metal laser welding) or DMLF (direct metal laser forging) would be more accurate. Under a microscope, the melt pool presents itself as a weld bead, suggesting that the part is microwelded or microforged layer-by-layer, producing an extremely dense part. Under EOS standard parameters, this holds true under testing; therefore, any postprocess acts almost identically to a billet machined part and has much less porosity and much greater uniformity than a cast metal part. However, DMLS parts straight out of the machine are more like cast parts in their first surface layer (cookie crumb texture), necessitating a postprocess to achieve fine polished finishes identical to traditional subtractive manufacturing. Different metals have different finishes with variations of vertical and horizontal surfaces that range from 6 microns to 80 microns. Downward facing or shallow angled surfaces are the roughest DMLS finish off the machine. Horizontal flat upward facing surfaces are the second roughest DMLS finish and they range from 6 to 12 microns. Bead blasting and tumbling effectively clean parts up to generally acceptable part smoothness. Through experience, DMLS engineers identify potentially problematic surface areas which prompts the engineers to reorient the part or modify its design to eliminate as many supports as possible. Proper support strategy facilitates clean postprocess EDM cuts, milling, or lathing to reach critical surface call outs. For areas identified for postprocess surface finishing, i3D MFG™ engineers add .005" to .010" material depending on the postprocess precision and desired held tolerance. Again, best results are achieved when projects are designed and planed from the beginning with post process requirements in mind.

The drawback to using secondary postprocesses to achieve desired surface finish is that these processes are limited by the ability of the subtractive process to reach the critical surfaces. If these critical areas are internal channels or difficult to reach, the surface quality must be a function of the DMLS process. To address the needs of customers using DMLS to design conformal cooling channels and highly complex geometries or lattice structures, i3D MFG™ creates application specific parameters that print much finer DMLS only finish. For example, aluminum and titanium contours exposed at significantly higher power than the standard parameters greatly improve surface finish. i3D MFG™'s DMLS innovation enables a fine finish for internal channels not accessible to postprocess finishes and also eliminates the bead blast and tumble time and cost for outside surfaces. The trade-off, thus far, is that the printed part still holds to tolerance but loses detail on some fine features.

46.3.7   *Metal powders*

Finally, the composition on the metal DMLS powder and the atmosphere to which it is exposed has considerable impact on the build and part success and density or porosity. Powder lots, laser power manipulations, and humidity all affect DMLS porosity. Metal powder particle size, flow-ability (spread-ability), and deposition impact the repeatability of part build. Hatch spacing can create porosity on each given layer, whereas energy input fluctuation can create porosity between layers. DMLS metal offerings and development are a huge opportunity within the industry. DMLS service providers are limited by the parameters and corresponding metals determined by the DMLS equipment OEMs. The most commonly offered metals are aluminum, titanium, nickel alloys HX, Inconel® 718 and Inconel® 625, cobalt chrome, maraging (tool) steel, and stainless steels including 15-5, 17-4, 316L, and 304L. The challenge is that the demand within the aerospace and defense industries in particular are growing quickly, and into more and more complex projects that require increasing expensive and exotic metals and a wide variety of density or porosity specifications. The synergy is that DMLS is most cost effective for more expensive and exotic metals because it is a low waste process that drastically reduces the scrap costs associated with traditional subtractive processes. DMLS also expands design options for dense and porous parts because it adds microlayers rather that removing material from a predetermined block at a set density; therefore, density and porosity can be altered by layer within a part. Parameter adjustments enable DMLS to print in additional metals and at varying densities or porosities.

Common nonstandard metal DMLS requests include the following: Monel K 500 Haynes Hastalloy tungsten, copper, copper nickel alloys and titanium alumide. American metal ore and DMLS powder company Additive Metal Alloys offers transparency and domestic traceability to the DMLS material. As they are not tied to a single OEM, and serve multiple industries, they produce uncommon DMLS metal powders in short lead times. Paired with an open parameter set and the physical ability of sintering or welding the metal, material limitations become a lesser concern. Varying particle size distribution (PSD) has a direct effect on the quality of parts produced from selective laser welding AM, referred to in this chapter as DMLS. PSD is the main characteristics in powder flow-ability. Fine particles (0.1–5 microns) tend to clump together and prevent uniform recoating during the manufacturing process. Large particles (60+ microns) minimize layer packing density. A powder's PSD that has an increased amount of fines typically provides higher layer packing density, produces increased density under low laser energy intensity, and generates smoother side surface finishing parts. A powder's PSD that has a decreased amount of fines typically demonstrate better flow-ability. Ultimately a combination of both small and large particles is best suited for DMLS. This allows smaller particles to percolate through the larger particles, filling voids, and aiding in achieving higher density (Figure 46.11).

Each machine does monitor humidity and atmosphere at the machine level but does not directly control it. Builds are generally run in either an argon or a nitrogen environment. Humidity can affect coating and laser power absorption, underexposing random layers and creating unintended porosity. i3D MFG™ observed these effects inadvertently when water entered into the lines of their air compressor and shot spurts of water vapor into its M 280 as the air compressor came on and off. The vapor amounts were slight, but enough to create refraction from the laser, thus changing the energy level into the part. One solution to this particular risk is to run all metals in Argon environments, but this adds additional cost to the material input.

There are challenges to sintering or laser welding various metals that vary depending on the metal itself and its reactions to the thermal dynamics created by the specific part geometry and corresponding parameter settings. For example, fine features or overhanging features in Inconel® are challenging and high volume titanium parts frequently fracture. To compensate, i3D MFG™ engineers alter contouring, design supports that can act as heat sinks, or adjusts the number or orientation of parts on a single build. Temperature

Figure 46.11 Gas atomized metal powder particles. (Courtesy of Additive Metal Alloys.)

gradients affect the laser interaction and require adjustments to insure a consistent density. Each build illuminates another set of considerations that DMLS engineers learn to adjust to reduce failure rates and improve build results. Just as balancing part design with support structure and surface considerations becomes a fluid scenario is greatly influenced by experience, adjusting parameters to compensate for metal performance in the DMLS machine circles back on the earlier analogy to a master tool maker. Similarly to the master mold maker considering how to best interface the tool not only for the machine, but also the material flow and shrinkage for traditional manufacturing, the DMLS engineers and technicians consider the part file as input, and also the powder-metal interactions with the laser speed, recoater arm, and power input much like adjustments for flow and shrinkage.

To summarize this overview of current SLM AM challenges, designing specifically for DMLS and creating or employing 3D metal printing expertise are the keys to successfully manufacturing with DMLS. Challenges including design gaps, build speed and size, surface finish, and metal powder development are being addressed and solved through design improvement, iteration, parameter innovations, and machine advancements. Understanding how to best leverage DMLS for rapid prototyping, research and development and production is greatly improved by relying on the DMLS expertise emerging within DMLS focused contract manufacturers (CM) and partners. The benefits to be gained are an ability to innovate in ways previously impossible due to traditional subtractive manufacturing limitations—including highly complex and organic geometries, latticed internal structures, conformal cooling channels, triple digit design efficiency improvements, low volume production economies of scale, and assembly consolidations.

46.4   Disrupting the disruptive technology

The section 46.3, delved into current challenges and briefly touched on advancements and solutions presently employed to surmount and resolve challenges facing new technology. For DMLS production to truly become a disruptive manufacturing process, innovation must continue to drive improved capabilities. Clayton Christensen, renowned Harvard Business school professor, coined and defined the term disruptive innovation as "a process by which a product of service takes root initially in simple applications at the bottom of a market and then relentlessly moves up market, eventually displacing competitors." (Christensen, Clayton, "Disruptive Innovation," *ClaytonChristensen.* http://www.claytonchristensen.com/key-concepts/.Web. November 8, 2015.) This section reviews some ground-breaking strategies currently being used by i3D MFG™, a DMLS CM recognized for its disruptive, solutions-based approach to metal 3D manufacturing. Three distinctions setting i3D MFG™ apart as a leader in DMLS disruptive innovation are proof-of-concept production, iterative solutions, and application optimization.

46.4.1   Proof-of-concept production

Proof-of-concept differs from prototyping in that proof-of-concept is simply ideating without the physical limits of manufacturing. Past product development models then required several rounds of prototyping and finally, a pilot part to functionally test and use as a template for manufacturing process planning. Using DMLS, i3D shifts this entire premise by approaching the limitless proof-of-concept stage as the initial production stage, effectively eliminating 50% or more of traditional product development process time and outlay. 3D printing was first accepted as a rapid prototyping technology. The ability to additively manufacture ceramic and plastic parts in days without tooling costs, revolutionized research and development in the design and prototype phases of production. Although 3D printed prototypes vastly improve prototype lead times and costs, once a design moves into its final stages, its mechanical and material properties must mirror a production part and pass final part testing to substantiate production. Most additive technologies do not meet these mechanical and material requirements; therefore, time consuming and expensive preproduction tooling is required to traditionally manufacture a proof-of-concept piece. DMLS changes the equation for metal components. The microlayer welding or forging process produces nearly 100% dense parts with almost identical properties to machined parts from raw stock materials. SLS, including DMLS, delivers finished components with comparable traditional milled part material properties while eliminating subtractive manufacturing design constraints; thereby, allowing proof-of-concept to become the prototype and, through iteration, move directly into production. Consider NASA's widely used definition of how rapid prototyping improves design efficiency below (adapted to manufacturing) and then consider i3D's disruptive adaptation that continues to overlap cycles and create design, time, and cost efficiencies. The rapid prototyping model improves on the classic model by reducing steps to gain a more refined product, whereas the i3D MFG™ proof-of-concept as production model improves the product life cycle by reaching the production stage rather than the pilot stage in the same number of steps.

Classic Approach—waterfall cycle	Rapid Protoyping—spiral cycle	Concept as Production—helix cycle
1. Proof-of-concept definition	1. Proof-of-concept definition	1. Proof-of-concept definition
2. Manufacturing requirements definition	2. Prototype additive manufacture	2. Prototype additive manufacture
3. Preliminary design	3. User evaluation and design refinement	3. Functional testing and design refinement
4. Detailed drawing design	4. Reprint of refinement	4. Reprint refinement
5. Prototype implementation	5. User evaluation and design refinement	5. Functional testing and pilot approval
6. Test and accept or revise	6. Reprint of refinement or pilot approval	6. Pilot refinement or production
7. Redesign and retool		

Adapted from http://dsnra.jpl.nasa.gov/prototyp.html#definition. ("Rapid Prototyping," *instructionaldesign.org*. www./models/rapid_prototyping.html, Web, November 8, 2015.) (Courtesy of Matthew Garrett, i3D MFG™ Chief Operations Office.)

The aerospace, medical device, and implant industries are front-runners in taking 3D printed proof-of-concept parts into production. As discussed previously, DMLS production is economical for highly complex parts (some of which can only be built additively), combined assemblies printed as single parts, parts made from exotic or expensive metals, and low volume production parts.

46.4.2   Iteration

Iteration drives the i3D MFG™ proof-of-concept as production model. Iterative solutions to DMLS design, material, parameter, process, and postprocess challenges are aimed at one goal: Refine and document until reaching the optimal, repeatedly manufacturable part with a corresponding DMLS design, parameter, and production recipe. The iterative process is critical to leveraging DMLS as a disruptive manufacturing technology because it furthers understanding between the innovators on both sides of the job. By facilitating a synergy between customer engineering groups and DMLS engineers, ground-breaking design value and the likelihood of project success increase exponentially. As they originated in the rapid prototyping sector, In order for DMLS to have common 3D print service bureau models focus solely on lead time and cost. Longevity as a disruptive manufacturing process close coordination between the engineering teams is critical.

Using 3D models in CAD as the base is a valuable tool that offers clients and i3D MFG™ engineers, the ability to design while taking advantage of many efficiency considerations without traditional manufacturing limitations. Just as any manufacturing process requires adjustments between design and the physical manufacture, DMLS often requires design adjustments. Unlike other conventional manufacturing methods, DMLS is able to validate design iterations by printing small portions of a geometry, multiple iterations in a single build, and then adjust and rerun in days. *Prototype Early and Prototype Often* is the accepted best practice in product development. Design firms are recommending that their customers spend less money up-front validating design through prototyping than making much more expensive tooling and production adjustments later. i3D MFG™ translates this same premise into a parallel philosophy, "Lead and Perfect with Iteration, Did We Make a Better Part Today?" Iteration allows engineers and designers to quickly and efficiently test the limits of 3D manufacturing, often finding solutions that are not yet known to be possible. Using a scope of work to outline the required and desired part specifications creates a framework for the DMLS recipe. i3D MFG™ defines recommended iterations based on the mechanical and finish or surface requirements defined by customer needs. An iteration is considered fully realized when it meets or exceeds the ideal specifications. At that point, i3D MFG™ records the iteration supports (if any), orientation, parameters, and postprocess flow into a unique component recipe to ensure repeatability for production (Figure 46.12).

The first project review is to determine if the file is ready for manufacturing using DMLS. Generally, the geometry is refined to limit or eliminate supports and the build

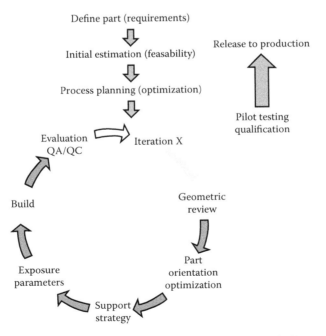

Figure 46.12 i3D MFG™ iterative proof-of-concept as production model. (Courtesy of i3D MFG™.)

parameters are defined to meet mechanical tolerances and come as close as possible to surface finish specifications to reduce postprocess time and cost. In documenting the part recipe or formula, the process is defined to include the setup configuration that optimizes the most efficient DMLS print parameters and the most cost effective component and material throughput. Additionally, the post DMLS process is closely defined in how the part(s) are removed from the build plate, the necessity of stress relief or heat treat, hot isostatic pressing (HIP) call outs, and surface finish add-ons such as anodization, powder coating, or polishing. In some cases, the initial component design reflects the fully realized function and possible design efficiencies, but currently it is more often that design engineers are still transitioning their thought processes toward AM perspectives. 3D manufacturing removes milling pathway confinements, casting and molding flow, and release constraints and stamping, or other subtractive manufacturing restrictions. New DMLS production part possibilities include conformal cooling channels, latticed internal and external features, defined density and porosity variations between or within components, multipart assemblies printed as single parts, and highly precise microgeometries. Using DMLS, reducing overall part size may still achieve equal or improved part efficiencies or surface area.

For example, i3D MFG™ is helping the defense industry optimize DMLS heat sink design. Traditional milling and extrusion manufacturing constraints limit optimal thermal management design strategies. However, AM allows intricate latticed surface structures to be built on heat sink fins; a strategy that other manufacturing methods find physically impossible. This heat sink pictured below (Figures 46.13 and 46.14) is designed to increase

Figure 46.13 Elevation view of looking down a single heat sink fin with a custom surface treatment applied. (Courtesy of David Polehn, i3D MFG™ Engineer.)

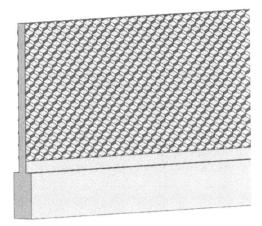

Figure 46.14 Side view showing the custom surface treatment. (Courtesy of David Polehn, i3D MFG™ Engineer.)

the surface area, while maintaining a nominal distance between fins equivalent to that of a standard flat-walled heat sink fin. Through the use of computational fluid dynamics, the end-user can design the part optimized for the DMLS manufacturing process, whereas experiencing efficiency gains in the performance of the heat sink. The real benefit of using DMLS as a manufacturing method for heat sinks, is the ability to manufacture geometries that are not possible in other methods. The end result allows the designer the ability to transfer equivalent heat in a smaller package. All of the physical properties of a heat sink can be manipulated within the constraints of DMLS to allow the designer of the heat sink to make a superior solution as compared to traditional machining techniques.

Finally, i3D MFG™ disrupts the current AM sector in its unique applications-based business model. Standard 3D-printing service bureaus drive the majority of their business online, fulfilling rapid prototyping market demand. There is little customer interface to ensure the design, material, and process-match desired expectations. This model is not well suited toward facilitating design breakthroughs or production parts. Transforming DMLS into a decisively disruptive manufacturing technology requires design and 3D engineering interaction, design and build iteration, and production process planning from the proof-of-concept stage. i3D MFG™'s business model is significantly different than standard 3D service bureaus in its sales-driven, applications-focused approach. Top DMLS equipment manufacturer EOS describes i3D MFG™ as, "very strong in targeting the right markets for layer additive metal and offering unique solutions. They (i3D MFG™) accept difficult and challenging projects that most others would turn away. EOS is working with them as one of our MPA (materials, process, and applications) development partners." By analyzing fits between DMLS and project geometries that complement each other, i3D MFG™ targets industries most likely to benefit from DMLS manufacturing capabilities and assists those industries in creating a competitive advantage through DMLS. Some distinctive approaches include adding inserts during the build process, developing parameters for nonstandard materials, and utilizing application specific substrate platforms to improve or eliminate EDM cut-off time.

During the project analysis, i3D MFG™ accesses its DMLS knowledge-base looking for DMLS best practices to apply to the project in an effort to manufacture the part as efficiently and effectively as possible. The correct fit is a two-way street. On the one hand, the proof-of-concept project may match well or even require DMLS. Conversely, AM may thrust the project into iterative improvements that catapult it into revolutionary advancements because traditional design limitations are lifted. In either case, personal contact between the CM and the client (as opposed to online quick quotes) is essential. Once project design for DMLS is optimized, i3D MFG™ uses proprietary software to simulate layer-by-layer growth and ensure superior part builds. Project iterations may lend additional insight into stress prone areas of the geometry, support structure strategy, improvements in project setup, and efficiencies in postprocess production. To validate repeatability, the part is tested in a controlled environment and all steps are documented. DMLS in-process quality assurance is in development and a critical piece of production validation. The proof of repeatability or manufacturability is the final phase in creating the DMLS recipe or formula.

An important part of the overall project evaluation is, understanding and embracing the integral relationship between AM and traditional manufacturing. DMLS is not a replacement for all conventional metal manufacturing processes, it is a next generation metal manufacturing technology best utilized to create competitive advantages for clients and their manufacturing partners. Because of relative production speed and surface finish limitations, DMLS is highly interdependent on milling and other traditional postprocess services. Identifying parts that are a good fit for DMLS and those that are not is beneficial in building the legitimacy of DMLS as a disruptive manufacturing advancement. Whether in-house or as secondary partnerships, establishing a robust set of subtractive capabilities to complement DMLS production is essential.

46.5   Conclusion

AM technology, and specifically selective laser welding/forging (DMLS or SLS) advanced from a rapid prototyping process into a production parts methodology. Aerospace, defense, and medical industry leaders have been working diligently with OEMs and CMs to leverage this disruptive technology into explosive design innovations. DMLS enables engineers to revolutionize designs by erasing traditional design constraints and facilitating significant weight reduction strategies, material savings, assembly consolidations, and complex internal geometries to create exponential part efficiencies. Although DMLS faces challenges including design gaps, build speed and size constraints, material development needs, and surface finish process improvements, the potential competitive advantages outweigh the drawbacks. Industry users and providers work best in tandem to find innovative solutions and production formulas. i3D MFG™'s customer application focus and iterative production model are explored as the next shift in the 3D-printing sector's business evolution. Through design and part iteration, proof-of-concept becomes production, reducing the product development cycle by as much as 50%. As the AM industry progresses, metals are projected to vastly outpace plastics, particularly as a focal 3D manufacturing technology. i3DTM and its industry partners must continue to push the current AM limitations to create 3D manufacturing solutions.

chapter forty seven

Information storage on additive manufactured parts

Larry Dosser, Kevin Hartke, Ron Jacobson, and Sarah Payne

Contents

One of the primary values of additive manufacturing (AM) is the ability to economically make small numbers of any particular part. This makes the paradigm ideal for making replacement parts on an as-needed basis. To facilitate this, it would be valuable to store information needed for the manufacture of any given part directly with the part. This can facilitate finding full details on the part design, dimensions and fabrication instructions, or possibly eliminate the need to look up for this information. This can be especially important when replacing parts for vehicles or platforms that are intended to remain in service for long periods of time, or are being enabled to continue service through sustainment efforts. In these situations, it is possible that original drawings and specifications may become lost over time with the degradation of records and institutional knowledge. Having key information directly on the part can obviate these concerns.

This chapter addresses two issues: (1) what information is stored with a part and (2) techniques for storing the information. This chapter discusses the information recommended to be stored for the reproduction of a part and suggests different ways to identify each part.

47.1 Types of information to store

The information of interest to store on a part is that which can be used to guide any AM technique in accurate reproduction of the part. This may include dimensions, exact material specification, minimum materials requirements where a variety of materials may be acceptable, directions for specific fabrication techniques (e.g., required equipment or operation settings), directions for finishing the part (e.g., heat treatments, passivation), and directions for required inspection and validation methods. The total amount of this information may be minimal or quite extensive depending on the part.

There are two basic approaches to associating the above information with the part in question. The first is to comprehensively encode all of the required information directly onto the part. The second is to apply only an identifier number that specifies where the complete set of information can be accessed. (Historically, the latter option, in the form of a producer or supplier specific part number, is the only information typically encoded on a part. A modern alternative to this is the globally unique identifier (GUID).

Each of these methods has advantages and drawbacks. The advantage of comprehensive encoding is that no recourse to other resources is required to begin fabrication. This eliminates dependence on third parties for reliable data storage. It also cuts down on time and expense to access information. If one is attempting to fabricate parts in an environment of active military engagement, there is no dependence on communication lines. However, there are also serious limitations to comprehensive encoding. Many parts simply do not have enough space to hold the necessary information. The physical marking of the information on the part may be inordinately time consuming. Also, damage or wear to the part (which is a near certainty given that the part needs to be replaced) is likely to destroy or efface portions of the information on the part. Finally, if details of the part are classified or proprietary, it may be undesirable to include the information on the part itself. In general, comprehensive on-part data storage will only be practical for parts that are simple, nonproprietary, and not too small.

Marking of an ID number on the part reintroduces dependence on an external library of part information, but addresses all the challenges of comprehensive on-part data storage. An ID number takes up relatively little space and is fast to mark. It can be applied with redundancy to increase the chance that a complete and legible number can be discerning after the part suffers damage or wear. Finally, an ID number removes the primary part information to a point where its proprietary nature can be protected. For most parts and situations, ID numbers will remain the best method to store fabrication information on the part, particularly when implementing the GUID concept discussed below.

47.2   Globally unique identifier

GUID is reference number system originally developed to generate unique identifiers in computer software, but the concept is readily adapted for general inventory and database purposes. A GUID is normally represented as a 32-character hexadecimal string (equivalent to a 128-bit binary number).

To use a GUID for part information storage, one must do three things: Generate the GUIDs, apply them permanently to a part, and create and maintain a database that contains the part information associated with each GUID.

Generation of valid GUIDs is trivial. The total number of unique GUIDs ($>10^{38}$) is so large that the probability of random duplication is negligible, even when an enormous number of GUIDs are simultaneously in service. It is literally true that if GUIDs were randomly assigned to every insect on earth[*] (estimated as 10^{19} individuals), the odds that there would be even a single instance of a duplicated GUID is less than 50%. This is so far beyond the number of items to be tracked in any practical database that identifiers can be assigned by any pseudo-random number generator(s) without concern of confusing parts. The GUIDs can be assigned not just to each type of part, but to each and every *individual*

[*] Estimated as 10^{19} individuals by the Smithsonian. http://www.si.edu/Encyclopedia_SI/nmnh/buginfo/bugnos.htm

part. Further, because the numbers can be assigned randomly, they do not need to be assigned by central governing body. Any given fabricator of a part can generate a GUID for each part he makes without any fear of duplicating one already in existence (as long as the generation is done randomly.)

Application of a GUID to a part is straightforward, and methods of doing so are discussed in the following sections.

That leaves the issue of creating and maintaining the GUID database. This is a simple, if potentially large, exercise in information technology and data storage, solvable by many providers with off-the-shelf technology and equipment. The main issues include the following:

- *Determining what data will be stored with each GUID*: With sufficient capacity, CAD files to support AM of each part can be stored in addition to more conventional drawings, specifications, and instructions. Further, as every individual part can have its own GUID, it would be possible to store and update part histories (installation date, last maintenance date, notes taken at last maintenance, hours of cumulative service, and so on).
- *Protocols for accessing the database*: This includes not only the specific technical means of accessing and downloading the information needed to duplicate a part or interest, but also the methods for ensuring the security of the information. In some cases, it may be useful to produce and distribute subset databases that can be stored at a local fabrication facility (perhaps in an area of limited or suspected electronic connectivity).
- *Protocols for adding to the database*: As individual fabricators will be able to generate random GUIDs to cover each part they make, they will need a method for reporting the new GUIDs and part information to the library.

It is beyond the scope of this report to suggest specific methods for setting up such a database.

47.3   Alphanumeric marking

Alphanumeric marking of a part is the most straight forward way of encoding a GUID on a part. It has the advantage of being readily understood by a human operator, but is less well adapted to optical readers. Marking of parts can be accomplished by conventional engraving techniques: dot peening or laser marking. Use of ink generally is not advisable due to likelihood of degradation.

Figure 47.1 shows a typical example of how lettering is applied to metal parts using dot peening. An advantage of dot peening on metal is that it introduces compressive stress, which is generally considered to be safer than engraving with respect to the likelihood of reducing the fatigue life of a part. Dot peening also marks deep enough to be legible after substantial wear. A limitation is that dot peening is only applicable to materials are ductile and will permanently hold a deformation. Therefore it is not appropriate for brittle ceramics and may not retain well in some plastics.

Advantages of laser marking include speed and the ability to address virtually any material. Laser marking requires only line-of-sight to the mark area, and do not access for a physical tool head. Of particular advantage for small parts is the ability to make the font size extremely small. Figure 47.2 shows, for example, lettering marked into the surface of a penny with characters less than 100 microns tall, allowing for redundant or relatively concealed marking.

Figure 47.1 Marking ID numbers via dot peening. Taken from the website of DAPRA, a provider of dot peening equipment. www.dapramarking.com/dot-peen-marking.

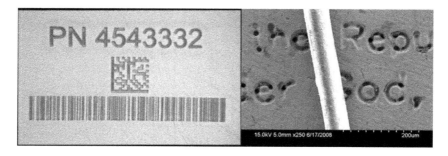

Figure 47.2 Marking ID numbers via laser marking (left) or micromachining (right). Fiber shown in right hand image for scale is ~60 microns wide. Images provided by MLPC, a laser processing company (www.mlpc.com).

47.4 Two-dimensional bar codes

A popular alternative to alphanumerics is barcoding, with 2D (or matrix) barcodes likely to be the most appropriate for most part marking. A 2D barcode encodes information equivalent to alphanumerics as an array of filled and empty cells in a square matrix. Typically, some of the cells are devoted to alignment and registration of the pattern orientation, and the rest are devoted to the actual recorded information. The amount of information that can be stored depends on the size of the matrix. The amount of space that a particular matrix must occupy in is limited primarily by the resolution of the reader technology. A common resolution is 0.33 mm/cell, though better can be achieved with high resolution technologies.

There are a large number of 2D barcode encoding standards, both public and proprietary. An example of a popular format is the quick response (QR) code. The largest QR codes can store 4000+ alpha-numeric characters. They can also be coded with redundancy, up to 30%, by reducing the number of characters. As an example, the QR code shown below encodes this paragraph (Figure 47.3).

Figure 47.3 Sample QR code.

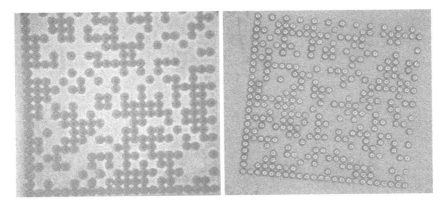

Figure 47.4 Examples of 2D barcodes marked by dot peening (left) and laser micromachining (right). Cells are 0.3 mm wide.

If this QR code were marked at 0.33 mm resolution, it would fit inside a 23 mm (<1″) square. A QR code that contained only a GUID number would fit in a square just 8.25 mm on a side.

The physical marking of barcodes on parts can be accomplished by the same techniques of dot peening and laser marking or micromachining described in section 47.3. Figure 47.4 shows examples of 2D bar codes marked by dot peening (left) and laser micromachining (right). The contrast for laser marking tends to be better, but both have been shown to be compatible with optical readers.

47.5 Radio frequency identification

The possibility of using radio frequency identification (RFID) tags for data storage on parts was investigated but found to be inappropriate. An RFID tag is basically a small antenna with attached integrated circuit chip designed to encode a number and respond to a wireless interrogation device. However, RFID tags cannot be directly produced on a part. They are instead a separate attachment that can become separated from a part. RFID tags typically encode less information than a GUID. Also, they are not very robust. They are more appropriate for inventory and tracking at the warehouse level than for the following individual parts.

47.6   Conclusion

This chapter discussed two issues in information storage for parts: (1) what information should be stored and (2) the current techniques available to store such information. Having key information located on a part can help find full details on and records of a particular part for engineers and technicians. AMs unique capabilities can assist in adding this additional and, quite possibly, vital information.

Acknowledgments

The contents of this chapter were extracted from *Direct Digital Manufacturing*, Report Number 11-S555-0021-05-C1, a research effort sponsored by the Universal Technology Corporation in Dayton, Ohio.

chapter forty eight

Case examples of additive manufacturing initiatives at the air force research lab and air force institute of technology

Adedeji B. Badiru

Contents

A quick review of recent publications and announcements indicate that there are several additive manufacturing activities going on in various defense-related industries around the world. Of particular interest are the leading-edge initiatives going on at the U.S. Air Force Research Laboratory (AFRL) and the U.S. Air Force Institute of Technology (AFIT), Ohio, United States.

AFRL, located within the Wright–Patterson Air Force Base (WPAFB) in Dayton, Ohio is a premier research facility of the U.S. Air Force. It is a global technical enterprise, boasting some of the best and brightest researchers and leaders in the world. The lab prides itself on being revolutionary, relevant, and responsive to the warfighter and the nation's defense. It delivers its mission by unleashing the full power of scientific and technical innovation. This mission includes leading the discovery, development, and integration of affordable warfighting technologies for the nation's air, space, and cyberspace force. Additive manufacturing (AM) features prominently in the new innovation pursuits of AFRL. It is important to note that the city of Dayton is internationally recognized as the birthplace of aviation, thus demonstrating the city's heritage of innovation. The city's prestige of innovation continues today. Adding additive manufacturing to the city's portfolio of innovation fits the theme of this handbook. AFIT, collocated with AFRL at WPAFB, is an internationally-recognized leader for defense-focused technical graduate and continuing education, research, and consultation. The graduate degrees offered by AFIT are predicated on thesis and dissertation research. It is through this research and development avenues that new additive manufacturing initiatives are being pursued at AFIT.

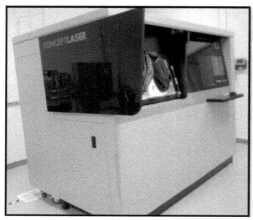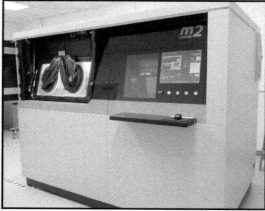

Figure 48.1 AFIT's concept laser cusing M2 metal sintering system, installed in January 2017.

In 2017, AFIT made a significant investment in a new state-of-the-art additive manufacturing equipment by purchasing a Concept Laser M2 Cusing® sintering system, as shown in Figure 48.1. This equipment is globally seen as one of the most-desired high-end powder-bed-based laser metal additive manufacturing systems. Sintering, which is the use of pressure and heat below the melting point to bond metal particles, is the ultimate application of additive manufacturing. It is metal-based printing rather than polymer-based printing of three-dimensional (3D) parts. With this equipment, parts' sizes can range from very tiny to extremely large, thereby creating opportunities to build a variety of parts meeting the needs of the defense industry. In LaserCUSING® machines, application-specific 3D parts with enhanced performance profiles are created in a fully automated digital process. This will facilitate new research and development partnership opportunities between AFIT and collaborators in terms of cost, efficiency, effectiveness, relevance, flexibility, adaptability, modularity, and responsiveness of 3D-printed products. For benchmarking purpose, as of January 2017, Concept Laser's X line 2000R system is the largest metal sintering machine available on the market. AFIT is proud to invest in a Concept-Laser equipment to facilitate research, instruction, and consultation on additive manufacturing.

The leading-edge activities of AFIT in additive manufacturing research and development is evidenced by the fact that Chapters 5, 19, 20, 22, 23, 29, 31, 32, 36, 37, 41, 42, 43, 44, and 45 in this handbook directly convey AFIT's research and instructional activities on the topic. For this reason, this handbook is affectionately and informally referred to as "AFIT Handbook of Additive Manufacturing."

On the AFRL side, a recent article in the *WPAFB Skywrighter* newspaper provides a good account of the latest additive manufacturing research and applications at the lab. The article is reproduced verbatim here, with permission, as a cleared document for public release. Alia-Novobliski (2017) summarizes that its a materials scientist's dream, but as some experts say, also an engineer's nightmare. For scientists and engineers at the Air Force Research Laboratory's Materials and Manufacturing Directorate, additive manufacturing, also known as 3D printing, can be a powerful tool for rapid innovation.

Ultimately, its a new way of looking at manufacturing across the materials spectrum and an area with challenges and opportunities that the Air Force is meticulously exploring. "Additive manufacturing is a huge opportunity for us," said Dr. Jonathan Miller, a materials scientist and the additive manufacturing lead for the directorate. "It allows us to manufacture unique form factors; it provides the opportunity to add functionality and capability to structures that already exist. Essentially, it allows us to redefine manufacturing." Traditional manufacturing methods developed during the times of the Industrial Revolution, when machines began to overtake the human hand for mass production. Many of these processes required material to be molded or milled away from a larger form to produce a specific design. Additive manufacturing, by contrast, is defined by ASTM International as the process of joining materials together, layer-by-layer, based on three-dimensional model data. It increases design possibilities, enhances the speed of innovation, and offers an alternative for creating shapes closer to what an engineer might need, with fewer constraints. "The biggest problem with conventional manufacturing processes is time," said Miller. "Manufacturing is an iterative process, and you never get a part *just right* on the first try. You spend time creating the tools to manufacture a complex part and then spend more time when you realize an initial design needs to be modified. Additive manufacturing offers lower cost tooling and lower lead times. The early mistakes don't hurt you as badly."

48.1   Early days

Though additive manufacturing is receiving a lot of industry interest as of late, it is not new to AFRL. Research into this manufacturing capability for the Air Force started at the same time the concept of rapid prototyping emerged in the industry back in the 1980s. Rapid prototyping was based on the premise that if engineers had an idea and wanted to make a shape, they could visit a shop and *print* the object, usually out of plastic by a printer. "The focus at this time was on creating functional prototypes, or objects that resembled a desired part, but the materials lacked the strength for even minimal use," said Miller. Early additive processing used light to chemically react to specific regions in a volume of gel to build rigid, plastic parts. The technology further evolved to include fused filament modeling, wherein fibers of plastic thread were melted and joined together to form a new object. Additional powder-based processes made use of plastic flakes that were melted by a laser into a shape. In the early 1990s, scientists learned that similar additive manufacturing processes could be used for generating metal objects. However, the technology at the time resulted in crude, large parts with poor surfaces. It was not until the late 2000s that laser technology matured sufficient to truly move forward in this domain. "This spurred the additive revolution pursued today by the entire aerospace industry," said Miller.

48.2   Shift to production parts

Although more affordable lasers and metal powder processes were helping scientists to make better metal products, the *glue gun* route to additive manufacturing of plastics became much cheaper. Small, inexpensive 3D-printing machines began to turn up in garages and schools, to the amateur engineer's delight. "Collectively, these became a new way of thinking about how to make stuff," said Miller. As additive manufacturing thinking evolved from being a way to develop prototypes to a method for actual production, the benefits and applications for the Air Force grew enormously, along with the potential for it to do even more. The manufacturing of customized parts and unique, complex geometric shapes at low production quantities can help to maintain an aging aircraft fleet. Custom tools, engine components, and light-weight parts can enable better maintenance and aircraft longevity. "Additive manufacturing can address a multitude of challenges for us, and there is a big pull to implement these processes from the logistics community," said Miller. "The fleet is aging, and replacement parts for planes built 30 years ago often no longer exist. Rapid production of a small number of hard-to-find parts is extremely valuable." However, the need to develop consistent, quality materials for additive manufacturing still remains a challenge that AFRL researchers are working diligently to address. Engineers need to have full confidence in additive manufactured part alternatives as they implement them as replacements in aging fleets or as system-level enablers in new weapon systems. "There are limits as to how the Air Force can use this technology and for what applications it will work best," said Miller. "That research is the basis of our work here."

48.3   Extension to functional applications

As additive manufacturing has matured over the past few decades, the field has broadened beyond plastic and metal parts. Dr. Dan Berrigan, the additive lead for functional materials at the directorate, is exploring ways to use additive manufacturing processes to embed functionality into structure, such as by adding electronic circuitry or antennas on nontraditional surfaces. As the demand for flexible devices such as activity trackers and performance monitors increases, so does the need to power these sources organically. "Additive processes enable us to deposit electronic devices in arbitrary shapes or in flexible, soft form factors," said Berrigan. "We are looking at different ways to make a circuit that can enable them to bend or adhere to new surfaces or geometries, such as on a dome or patch. Essentially, we are looking at ways to add capabilities to surfaces that already exist." Conventional circuit fabrication requires the lamination of a series of conductive and insulating layers in a patterned fashion, resulting in a rigid circuit board. The electronic properties for these circuits are known and understood, and engineers are able to ensure that the circuit can conduct as intended based on these known concepts. For 3D-printed electronics, a conductive material is divided up into millions of small pieces and suspended in a liquid that is then dispensed from a printer, explained Berrigan. After printing, those individual conductive pieces must maintain contact to enable electrons to move through a circuit and create power. "The demand here is for low-cost, flexible electronic devices, and these direct write, additive processes give the community design capabilities that we cannot achieve otherwise," said Berrigan.

48.4   Additive challenges and future potential

Despite years of development and research into additive manufacturing processes, there are a number of implementation challenges that AFRL researchers need to address in order to enable greater Air Force benefit from the technology, both now and in the future. "Fundamentally, it comes down to a materials processing problem," said Berrigan. The lack of standardized production processes, quality assurance methods, significant material variability, and reduced material performance are just some of the factors AFRL researchers need to overcome. Depending on the applications, material performance can be related to the strength of a part. For example, the electronic proper-ties of an additive manufactured circuit may be worse than those of once traditionally manufactured. "Understanding the safety, reliability, and durability of a part is criti-cal for an aircraft. We know this for parts made through other processes, but we don't know this yet for additive," said Berrigan. Another issue centers on compatibility of basic materials. "There are a lot of different interfaces in additive manufacturing, and ensuring that materials adhere to one another or that a part can support a certain stress or withstand a certain temperature—these are all challenges we need to address," said Miller. The long-term goal, according to Berrigan, is for additive manufacturing to become a well-understood tool in an engineer's toolbox, so that unique components can be design-integrated into a system. It is difficult to go back in a system already built, he said, but additive manufacturing provides the opportunities to build-in greater poten-tial at the start.

"The long-term vision is to have functional and structural additive manufacturing to work more cohesively from the start. Rethinking systems-level design to incorporate func-tionality such as electrical wiring, sensors or antennas is a potential that additive can help us address," he said. "When you build something by layer, why not introduce channels for sensors, cooling, or other functions?" In all, AFRL researchers agree that continued research and time will lead to fuller implementation of additive processes for the Air Force systems of today and future. Innovative technologies are enabling capabilities, and addi-tive technology is the one with limitless opportunities to explore. Figures 48.2 through 48.4 illustrate some of the AFRL activities at AFRL.

Figure 48.2 Dr. Dan Berrigan, the functional additive manufacturing lead for the Air Force Research Laboratory's Materials and Manufacturing Directorate, is exploring new ways to add functionality to existing objects through additive manufacturing. Flexible circuits, embedded antennas, and sen-sors are just a few of the potential manufacturing capabilities provided by additive technologies. (Courtesy of U.S. Air Force/Marisa Alia-Novobilski.)

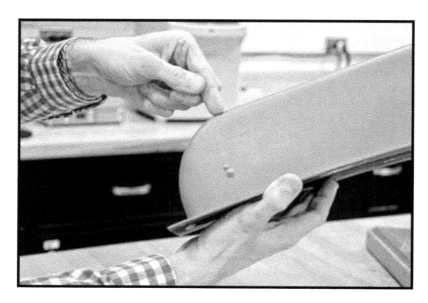

Figure 48.3 Dr. Dan Berrigan points to an embedded antenna on an MQ-9 aircraft part made possible through functional applications of additive manufacturing. Flexible circuits, embedded antennas, and sensors are just a few of the potential manufacturing capabilities his team is exploring using additive technology. (Courtesy of U.S. Air Force/Marisa Alia-Novobilski.)

Figure 48.4 Dr. Mark Benedict, a senior materials engineer and America Makes chief technology adviser at the Air Force Research Laboratory's Materials and Manufacturing Directorate, discusses the potential for additive manufacturing of aircraft components in metal. The complex geometry of the rocket nozzle benefits from the use of additive manufacturing due to its complex, specialized design. (Courtesy of U.S. Air Force/Marisa Alia-Novobilski.)

48.5   Conclusions

Additive manufacturing is revolutionizing manufacturing processes. Although traditional manufacturing has undergone major advancements and new technology developments in recent years, new opportunities are needed to further advancements. Traditional manufacturing relies on tools and techniques developed over several decades of making products. Additive manufacturing, popularly known as 3D printing, brings the efforts to a new level of possibilities. These possibilities are dependent on research and development efforts by organizations such as the Air Force Research Lab and the Air Force Institute of Technology. Collaboration between these two organizations and other, such as the Maker-Movement facilities, will ensure that the much-touted benefits of this new tool continue to be realized far into the future. The premise of this handbook is to empower readers to be ready to participate and or to take advantage of this new landscape of making products.

Reference

Alia-Novobliski, M. (2017). Advances in additive manufacturing create opportunities. *Skywrighter,* 58(4): A1, A10.

Index